CELL CYCLE
CONTROLS

CELL CYCLE CONTROLS

Edited by

George M. Padilla

Department of Physiology and Pharmacology
Duke University
Durham, North Carolina

Ivan L. Cameron

Department of Anatomy
University of Texas
San Antonio, Texas

Arthur Zimmerman

Department of Zoology
University of Toronto
Ontario, Canada
and Ramsay Wright Zoological Labs
Ontario, Canada

ACADEMIC PRESS   New York   San Francisco   London   1974

A Subsidiary of Harcourt Brace Jovanovich, Publishers

ACADEMIC PRESS, INC.
111 Fifth Avenue, New York, New York 10003

United Kingdom Edition published by
ACADEMIC PRESS, INC. (LONDON) LTD.
24/28 Oval Road, London NW1

Library of Congress Cataloging in Publication Data

Padilla, George M.
    Cell cycle controls.

    Includes bibliographical references.
      I.   Cellular control mechanisms.   2.   Cell cycle.
I.   Cameron, Ivan L., joint author.   II.   Zimmerman,
Arthur M., Date   joint author.   III.   Title.
[DNLM: 1.   Cell division.   QH605 P123ca]
QH604.P3   574.8'762   74-10966
ISBN 0–12–543760–9

TABLE OF CONTENTS

Section I. Regulation and Timing of the Cell Cycle

CONTENTS

Section II. Gene Expression in the Cell Cycle

CONTENTS

Section III. Nuclear and Chrosomal Activity in the Cell Cycle

LIST OF CONTRIBUTORS

P. A. Aitchison, Department of Botany, University of Edinburgh, Edinburgh EH9 3JT, Scotland

G. Attardi, Division of Biology, California Institute of Technology, Pasadena, California

Joseph W. Byron, Department of Cell Biology & Pharmacology, School of Medicine, University of Maryland, Baltimore, Maryland

G. B. Calleja, University of the Philippines, Natural Science Research Center, Diliman, Quezon City, The Philippines

Garciela C. Candelas, Department of Cell & Molecular Biology & Dermatology, Medical College of Georgia, Augusta, Georgia

Jen-Fu Chiu, Department of Biochemistry, The University of Texas System Cancer Center, M.D. Anderson Hospital & Tumor Institute, Houston, Texas 77025

R. C. L. S. Collin, Department of Biochemistry, University of Oxford, Oxford OX1 3QU, England

Catherine Craddock, Department of Biochemistry, The University of Texas System Cancer Center, M.D. Anderson Hospital & Tumor Institute, Houston, Texas 77025

James Creanor, Department of Zoology, University of Edinburgh, Edinburgh EH9 3JT, Scotland

Joseph E. Cummins, Department of Plant Sciences, University of Western Ontario, London, Ontario, Canada

Alan W. Day, Department of Plant Sciences, University of Western Ontario, London, Ontario, Canada

J. M. England, Division of Biology, California Institute of Technology, Pasadena, California

Arthur Forer, Biology Department, York University, Downsview, Ontario M3J 1P3 Canada

Ronald S. S. Fraser, Department of Zoology, University of Edinburgh, Edinburgh EH9 3JT, Scotland

Roger Hand, Departments of Microbiology & Medicine, McGill University, Montreal, Quebec and The Rockefeller University, New York, New York

P. Heywood, Department of Human Genetics & Microbiology, Yale University, School of Medicine, New Haven, Connecticut 06510

L. S. Hnilica, Department of Biochemistry, The University of Texas System Cancer Center, M.D. Anderson Hospital and Tumor Institute, Houston, Texas 77025

L. D. Hodge, Departments of Human Genetics & Microbiology, Yale University, School of Medicine, New Haven, Connecticut 06510

Stephen H. Howell, Department of Biology, University of California, San Diego, La Jolla, California 92037

Thomas W. James, Biology Department & Molecular Biology Institute, University of California, Los Angeles, California 90024

Byron F. Johnson, Division of Biological Sciences, National Research Council of Canada, Ottawa, Ontario K1A OR6, Canada

L. Martin, Departments of Hormone Biochemistry & Hormone Physiology (Endocrine Group), Imperial Cancer Research Fund, Lincoln's Inn Fields, London WC2A 3PX, England

D. Mazia, Department of Zoology, University of California, Berkeley, California

M. E. McClure, Department of Cell Biology, Baylor College of Medicine, Houston, Texas

J. M. Mitchison, Department of Zoology, University of Edinburgh, Edinburgh EH9 3JT, Scotland

H. P. Morris, Department of Biochemistry, Howard University College of Medicine, Washington, D.C. 20001

George M. Padilla, Department of Physiology & Pharmacology, Duke University, Durham, North Carolina 27710

C. A. Pasternak, Department of Biochemistry, University of Oxford, Oxford OX1 3QU, England

Livia Pica-Mattoccia, Laboratorio di Biologia Cellulare, C. N. R. via Romagnosi 18 A, Rome, Italy

Robert R. Schmidt, Department of Biochemistry & Nutrition, Virginia Polytechnic Institute & State University, Blacksburg, Virginia 24061

T. Simmons, Departments of Human Genetics & Microbiology, Yale University, School of Medicine, New Haven, Connecticut 06510

J. A. Smith, Departments of Hormone Biochemistry & Hormone Physiology (Endocrine Group), Imperial Cancer Research Fund, Lincoln's Inn Fields, London WC2A 3PX, England

M. C. B. Sumner, Department of Biochemistry, University of Oxford, Oxford OX1 3QU, England

Igor Tamm, Departments of Microbiology & Medicine, McGill University, Montreal, Quebec and The Rockefeller University, New York, New York

S. Taube, Departments of Human Genetics & Microbiology, Yale University, School of Medicine, New Haven, Connecticut 06510

J. Van't Hof, Biology Department, Brookhaven National Laboratory, Upton, New York 11973

K. Wakabayashi, Department of Biochemistry, The University of Texas System Cancer Center, M.D. Anderson Hospital & Tumor Institute, Houston, Texas 77025

D. H. Williamson, National Institute for Medical Research, The Ridgeway, Mill Hill, London NW7 LAA, England

M. M. Yeoman, Department of Botany, University of Edinburgh, Edinburgh EH9 3JT, Scotland

Bong Y. Yoo, Department of Biology, University of New Brunswick, Fredericton, New Brunswick

Erik Zeuthen, The Biological Institute of the Carlsberg Foundation, 16 Tagensvej, DK-2200, Copenhagen N., Denmark

PREFACE

This book reflects the current knowledge of investigators whose chief concern has been to understand cell cycle controls. The material presented stems from studies on many different cell types and cell systems. We have sought to include the most recent studies on the higher eukaryotic cell types as well as many new and exciting insights derived from investigations on lower eukaryotes. We feel that this volume will provide the background for understanding current concepts on cell cycle controls and will stimulate students and researchers to unravel the many exciting questions concerning the cell cycle.

This book should appeal to cellular, molecular, and developmental biologists as well as to many others in the life sciences. It is aimed not just at the specialist in the cell cycle field, but should be of value to all those with a general interest in cellular control mechanisms.

The book is organized into three main sections. Each section begins with a chapter by one of the foremost workers on the cell cycle: Section I includes studies on the regulation and timing of the cell cycle and begins with a chapter by E. Zeuthen of Copenhagen, Denmark; Section II deals with investigations on gene expression in the cell cycle and begins with a chapter by J. M. Mitchison of Edinburgh, Scotland; Section III presents works on the nuclear and chromosomal activity in the cell cycle. It begins with a chapter by D. Mazia of Berkeley, California. Each contributor has thoroughly but concisely reviewed his field and has included new and previously unpublished material to give a critical and up-to-date summary of the subject. Clearly international, the contributors to this book include many of the leading scholars of the cell cycle.

George M. Padilla
Ivan L. Cameron
Arthur M. Zimmerman

A CELLULAR MODEL FOR REPETITIVE AND FREE-RUNNING SYNCHRONY IN *TETRAHYMENA* AND *SCHIZOSACCHAROMYCES*

Erik Zeuthen

The Biological Institute of the Carlsberg Foundation,
16 Tagensvej, DK-2200 Copenhagen N., Denmark

I. Introduction

Some twenty years ago the "induction synchronists" happily brought their results to market. They told the world that synchronous cell divisions could be forced upon populations of cells which thus far had multiplied exponentially with the cells out of phase. This called the "selection synchronists" forth. They selected cells in phase from exponentially multiplying cultures and started synchronous cultures with these. They told us that induction synchrony was no good because the cell cycles studied were distorted, unnatural, not normal. To this no reply was possible. The critics could even be right and in relation to some synchronization work they undoubtedly were.

The heart of the matter is that those who started induction synchrony did this on hope and faith. The hope was that it would be possible to find means to force cells into phase, a situation which should be useful for further work. It would be visibly reflected in periodic displays of synchronized cell divisions. The faith was that normal cell cycle controls and regulations could later be shown to be involved.

So who regulates and sets time in the cell cycle, the cell or the experimenter? The induction synchronist was never quite sure with his cell material, but I believe that on my own behalf, and by this time, I can compromise and say: both in cooperation, and this view shall pervade the following pages.

I want to use this opportunity to report and discuss new results from Copenhagen with a repetitively synchronized *Tetrahymena* system, itself the logical product of careful studies of positive sides, and of weaknesses (58,60), in the earlier multishock system (44). Developments are accounted for in a series of reviews (42,43,52,55,60,62). The fission yeast *Schizosaccharomyces pombe* is now included, and we have started related work on monolayer and suspension cultures of mammalian cells.

II. Tetrahymena

For years our laboratory has worked with an amicronucleate strain of the hymenostome ciliate protozoan *Tetrahymena pyriformis*. The strain used (GL) is Lwoff's original (28): in the United States it has acquired one more capital letter, C - for Copenhagen. In other laboratories a number of different strains - amicronucleate and micronucleate—, and in our own laboratory a different species, (*T. vorax*, strain V_2S) has been synchronized with versions of the multishock procedure. More recent work on the two species shall be surveyed in separate sections of this chapter.

A. *Tetrahymena pyriformis:*
1. The system

First I shall present a rather new *Tetrahymena* system, which we find quite promising by itself and as a model for adventures with other cells (59,60). Fig. 1 shows how repetitive synchrony builds up in exponentially multiplying cells when heat shocks are spaced a full cycle apart. The upper tracing shows the temperature regimen found to be optimal: intervals at 34°C for periods of 30 minutes alternated with intervals at the optimum temperature (28°C) for periods of 150 or (as in Fig. 1) 160 minutes. It is likely to be significant that the time interval between shocks equals the doubling time (158 ± 3 minutes) for cells grown exponentially in the rich medium (2% proteose-peptone, 0.1% liver extract) used here. The next curves shows how the fraction of cells in division and the fraction of cells engaged in DNA-replication fluctuated with time. Each symbol represents a separate experiment arranged around a standard convenient population density. Cell densities increased from 2.5 x 10^3 to 4.8 x 10^5 cells per ml. As reported before (58) the average synchronous division step adds 85% to the population count.

2. Chemistry

a. *DNA, RNA, Protein, Enzymes.* Rather early results (Fig. 2A) indicated that in this system each division step and each replication phase resulted in close to doublings in the number of cells and in the amount of DNA. Enzyme activities followed so far increased discontinuously and mostly in steps. This is true of deoxycytidinemonophosphate deaminase, of α- and β-glucosidase (22). Acid phosophatase (27), and DNA-polymerase (23) show patterns of specific activities resembling those of the glucosidases. On the other hand, RNA and protein increased smoothly (log linearly) through the matching temperature and cell cycles (59).

In Fig. 3, B are shown experiments in which α-and β-glucosidase activities were followed in a system which ran freely at 28°C after synchrony induction with 4 shocks. Two synchronous division are displayed, and each of them is correlated with steps in α- and β-glucosidase activity (22). Thus a stop in the increase in enzyme activity at the end of an enzyme step is not a direct effect of heat. It is an indirect effect the same way as other synchronized phenomena seen in this system.

The chemical data shown in Fig. 2A are fully reflected at the level of accumulation in acid precipitable macromolecules of ^{14}C-thymidine, ^{14}C-leucine, but not of ^{14}C-uridine. Incorporation of the latter came to a stop at every shock, and there were recovery periods after each shock. This way ^{14}C-uridine accumulation tended to lag behind what was seen in normal cells,

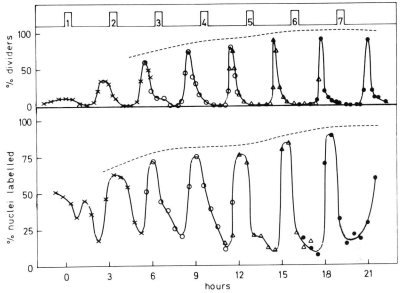

Fig. 1. Temperature induced division synchrony in *Tetrahymena*. The temperature cycle is shown on top and the time is given below. The temperature was changed between 28°C (optimal, held for 160 minutes between shocks) and 34°C (division-blocking, held for 30 minutes in each of the 7 shocks shown). *Upper ordinate:* percentage of cells in division. *Lower ordinate:* percentage of cells in replication. (Indicated as % cells having a labeled macronucleus after a 15 minute pulse with [^{14}C]-thymidine. Points are in the middle of the incorporation intervals). From (59).

and roughly by the duration of the shock, 30 minutes, for every cycle time of 190 minutes.

In terms of label accumulation, repetitively synchronized and free-running synchronized cells were similar in showing log-linear incorporation of ^{14}C-leucine, and stepwise increases in ^{14}C-thymidine incorporation. However, there were signs that synchrony of DNA-synthesis goes wrong after the second free-running division. In free-running cells, ^{14}C-uridine accumulation was log linear through two division cycles, indicating that discontinuities in uptake or incorporation in the repetitive system were indeed direct results of the elevated temperature. Effects on RNA are relevant to discussions about chemical mechanisms in the synchrony induction. I refer the reader to papers by Byfield and Scherbaum (11,12), Christenson (14-17), and to the review by Zeuthen and Rasmussen (62).

b. *Nucleosidetriphosphates.* Nexø (34,35) has followed amounts and concentrations of the normal ribo- and deoxyribonucleoside-triphosphates (r- and dNTP) in *Tetrahymena* cells under normal growth and under repetitive and free running synchrony. The average synchronized cell was slightly larger (59)

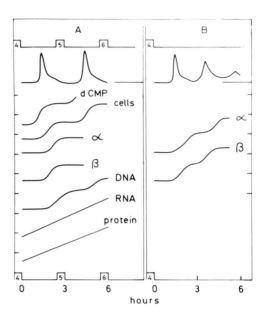

Fig. 2. A. Parameters of repetitively synchronized *Tetrahymena* cells from the 4th to the 6th heat shock. Enzyme activities of deoxycytidinemonophosphate (dCMP); α-glucosidase (α) and β-glucosidase (β) are given in activity/ml of culture. Changes in cell number, DNA, RNA, and protein content are also shown (see text). B. The activities of α- and β-glucosidase in free running system in which synchrony was established with 4 heat shocks. At the top and bottom of both figures are shown the temperature regimens. Just below the top line the changes in the percentage of cells showing division figures are also given (cf. Fig. 1). Enzyme data are from (22), other data from (59).

than the normal cell. This tended to be reflected in the amounts of NTPs per cell. Concentrations hardly differed in normal and synchronized cells. Midway between two divisions 10^6 repetitively synchronized cells (total volume 24 µl) contained 24 nmole ATP. Thus the concentration of ATP was 1 mM. GTP, UTP and CTP range around 200 µM. Concentrations of dTTP, dATP, dCTP and GTP were much lower, around 2 - 8 µM. These values are approximate. There were variations between experiments and through the cell cycle.

There was a fairly smooth accumulation of the four ribonucleotides through the repetitive (from an hour before shock 6 through shock 7) and the free running cycle (through two divisions and 2 S-periods). Concentrations tended to become lower for one half hour after a heat shock. This phenomenon was not fully eliminated if a shock was not applied (free run). Thus, there were faint and probably significant changes in concentration of all four nucleotides through cycles of division and replication at constant temperature. The faint

4

drop in ATP, GTP, UTP and CTP concentrations in free-running cells correlated with the passage of the cells from S into G2. It is relevant here to mention that Hamburger (20) recently found a transitory respiratory decrease mid-way between divisions in normal single cells taken from the exponential population growth phase. The decrease was seen around the time when the cell would be expected to move from S into G2. It would seem that levels of rNTP are controlled by respiration rates, themselves independently controlled, and not by levels of rNDP in the cell as a whole. Cellular compartments might be considered, but this is not the place.

The dNTP data are perhaps more interesting. There was some DNA synthesis between the S-periods (mostly due to imperfections in the synchrony), but synthesis was 10 - 20 x faster during the nuclear S-periods. This situation was reflected on the amounts per cell and on the concentrations of the dNTPs, though not much. They all persisted through the cycle. Fig. 3 shows the *amounts* in cells in a standard volume of culture. The fully drawn curves are for cells which received shocks 6 and 7 and divided (D) and replicated (S) two times. The dotted curves are for cells allowed to run freely after shock 6. First let us disregard the discontinuities in the curves and observe that the overall growth with respect to dNTP paralleled the cell growth. The amounts doubled with each generation. The next observation is that amounts of dTTP, dATP and dCTP, increased during, and are minimal just after both shocks. However, as with rNTPs, these were not clean effects of the temperature increases. Even prior to the time when free-running cells could have "expected" shock 7 - the shock they do not receive - they had passed a point in the S-period when amounts of dNTPs leveled off; in the case of dATP this course lead to a dip in the amounts per cell which was coincident with the beginning of the next S-period, and which repeated what was seen at the initiation of preceding S-periods in the repetitively synchronized system. Thus dATP builds up in amounts during S which stay constant or drop between S-periods. Common for all dNTPs, the amounts and concentrations were maximal somewhere in S.

c. *Experimental Control of Thymidine Supply, Effects on Nucleosidetriphosphates.* The complex medium we used is optimal for cell synchronization but it contains purine and pyrimidine compounds, among them thymidine. At one time I wanted to control DNA replication and cell division independently in the same culture (57,58). I decided to attempt this by controlling the supply of thymidine for new DNA and independently playing with the temperature. These cells can methylate a nucleoside monophosphate and get thymidylate, but they can also take thymidine from the medium. The methylation mentioned is catalyzed by tetrahydrofolic acid and can be inhibited with 0.05 mM methotrexate (M). The flow of exogenous thymidine into new DNA can be checked by addition to the medium of uridine (U) in 20 mM concentration. The two compounds combined (M + U) gave good inhibition of the synthesis of DNA (2,47,56). Together they should cut the cells off from new dTTP, and this occurred (35,36).

Fig. 4 shows the *concentrations* of dTTP and dATP. These fit the curves

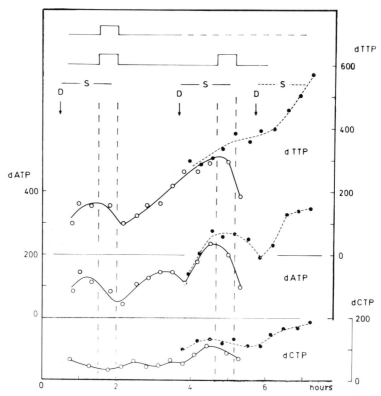

Fig. 3. Changes in deoxynucleosidetriphosphates (dNTP) in *Tetrahymena* cultures with repetitive synchrony around shocks 6 and 7 (open circles), and with synchrony which runs free after 6 shocks (filled circles). Temperature regimens, DNA synthetic periods (S), and times when maximal numbers of cells show division furrows (D) are indicated at the top. Ordinates refer to amounts per unit volume of culture. Calculations were adjusted so that the first sample shows pmole/10^6 initial cells. The abscissa gives the time in hours. From (34,35).

(Fig. 3) for *amounts* per cell nicely. Figure 4 also demonstrates what happened if M + U were added to the system between S-periods (first arrow) and during an S-period (second arrow). When the supply of thymidine was blocked while cells were not in S, a normal increase in dTTP concentration was checked, but there was no decrease. Such a decrease occurred later at the time when an S-period was expected. The S-period came at a normal time and it emptied the cells of available dTTP: thus the cells were stopped early in S, as shown earlier (56). There were also effects on dATP. After a lag, this compound increased to high levels. Such interdependence between dTTP and dATP concentrations is known from other systems (24,25,31,32,41). If the supply of thymidine is blocked during an S-period dTTP siphons out quickly, DNA synthesis stops (2), and dATP goes up. Nexø calculated half-lives for dTTP and dATP to be at around 10 minutes for cells not in S and around 1 minute for cells in S. This is in fair correspondence with the results of Fig. 4, considering that no inhibitor and no synchrony is perfect.

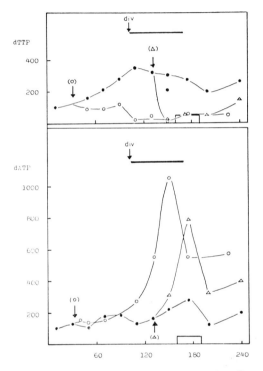

Fig. 4. Amounts (ordinate, filled circles) of dTTP10[6] initial cells and of dATP/10[6] initial cells (cf. Fig. 3), both from the termination of heat shock 6 through shock 7. At arrows, M + U were added to create starvation for thymidine compounds. Open circles show the effects when addition was well ahead of a cell division (div) and the S-period which followed. Open triangles show the stronger immediate effects from addition of M + U when synthesis of DNA synthesis was in progress. The abscissa shows the time in minutes. From (35,36).

d. *Studies on DNA.* Each repetitively synchronized cell division leads to a synchronized S-period exceeding the S-period in normal single cells by 10 - 20 minutes, probably an indication of imperfect synchrony. Studies based on incorporation of BudR into hybrid DNA showed that more than 95% of the cells' DNA replicates in some order in each S-period. The resolution of the method did not permit us to distinguish between replication units smaller than genomic units of which the *Tetrahymena* macronucleus may have many. The replication order in one generation differs completely from the preceding one, which reminds us of the macronucleus of *Euplotes* (itself a ciliate), not of eucaryotes with diploid sets of chromosomes (5). My collaborator, Dr. Helge Andersen and I view the situation as illustrated in Fig. 5. We suggested that the *Tetrahymena* macronucleus possesses units of replication [perhaps Dr. Hytte Nilsson's haploid segregation units (37)] which replicate partially out of phase. Andersen found that an effective block for replication can be established with M + U, two-thirds or three-fourth into the synchronized S-period, and without consequences for the following cell division (2). However, at the time of this division much DNA is extruded. He placed pulses of ^{3}H-thymidine at various segments of the S-period to be interrupted, and found label extrusion to the cytoplasm only of DNA which was pulse labelled immediately before the supply of dTTP was cut off (1). This extruded DNA replicated in the cytoplasm in the next cell generation (4).

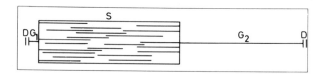

Fig. 5. Time from division, D, to division, D. Between a short G1 and a long G2 the macronucleus replicates its DNA in S. On the basis of the DNA content, the ploidy is 40 - 50. The graph is drawn to suggest that single identical units replicate out of phase. Note that in this illustration G2-functions are initiated before S is yet finished. Drawn in accordance with (5).

We shall see shortly that the point late in S after which DNA replication can be stopped, but not the cell's progression to division, marks a time when structural cytoplasmic changes which normally lead to division are initiated. It thus marks a time when the cell decides that it has enough DNA to go ahead in the cycle. Therefore, the last part of S becomes a period with overlapping S and G2 activities.

Working in our laboratory, Westergaard and Pearlman found that thymine-starvation established with M + U, induced DNA polymerase activity (48) in asynchronous cultures of *Tetrahymena*. Keiding and Andersen (23) now find that this inducibility of DNA polymerase is higher during than between S-periods.

3. Structural Markers of the Cell's Position in G2

The studies of metabolites with fast turnover, which were reported above, gave signs of small enzyme disfunctions, and of their correction, within the first half-hour or less after a shock. In their sum, these and many unknown effects could with some lag, well create a pause after each shock in the cells' progression towards division. Anyway, provided a shock hits at the right time, it separates DNA synthesis from activities later in the cell cycle. Thus it is around the time when S ends and G2 begins that we shall look for immediate and synchronizing effects of a heat shock. Such effects are not likely to be on DNA synthesis because this runs to completion before precursors are used up. They may be on all other levels - from DNA structure and function to biochemical events and cellular functions such as macromolecular assemblies. The first step-activity comes about an hour after the heat shock (dCMP-deaminase step), and this sets an upper limit for the duration of the pause we talk about. However, we have not yet looked at visible structures, i.e. products of macromolecular syntheses assembled into functional structures. Luckily, *Tetrahymena* unlike most cells carries cortical morphological markers of the position of the cell in the G2 phase.

Tetrahymena feeds through a cytopharynx in the anterior end of the slim cell. Division is by constriction of the cell across its short axis. The anterior cell takes the old mouth over, the posterior cell acquires a ready-made mouth which in the normal cell is assembled in the course of the last 84 minutes of the preceding 150 minutes cycle [Frankel (18)]. Thus, a new-born cell has no oral primordium for the first 66 minutes, equal to 44% of its growth-division cycle. After the lapse of this interval the oral primordium emerges in the primordial field as a collection of kinetosomes associated with the inside of the surface membranes (pellicle) of the cell. In multishock-synchronized cells (44) these kinetosomes soon grow ciliary stubs (8) and between them a system of subpellicular fibers develops (50) which seems to be instrumental in the sliding of the kinetosomes on the inside of the pellicle which determines the structural pattern of the finished mouth. Analogies with mitotic movements of centrioles are apparent.

This normal sequence is represented by the steps and curves in Fig. 6 III. We have used Frankel's (24) data to plot the normal morphogenetic sequence over two consecutive cycles, normalized to 160 minutes each, so as to facilitate comparison with our own results. Frankel worked at the time in our laboratory, with the strain we normally use (GL, amicronucleate) and under our normalized conditions. The staging indicated on top of III, also fully comparable, relates to the cycle of cell division and DNA replication (G1, S, G2, D). Data are from the work of Cameron and Nachtwey (13).

Moving one frame up in Fig. 6, to II, we find the repetitively synchronized morphogenetic sequence, recently reported by Dr. Howard E. Buhse, Jr. and myself (10). The cycle time shown is the same as in normal cells, because we have adopted the trick of omitting the two periods at 34°C. Thus in curve II two 30 minutes heat shocks (included in curve I) have been compressed into infinitely short heat pulses, represented by exploding stars. As for frame III, frames II and I are equipped with scales showing the approximate duration of stages in the replication-division cycle. Curve I is a reference with which we

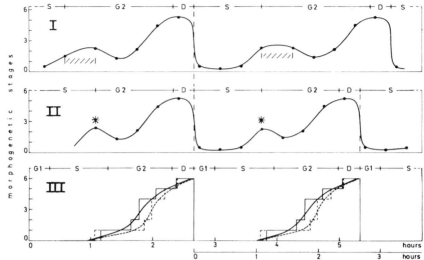

Fig. 6. Sequence of median morphogenetic stages in repetitively synchronized cells (I). Shaded periods indicate temperature shocks numbers 5 and 6. Frame II repeats I except that shock periods are removed (replaced by stars) so that times from division to division equal the generation time of exponentially multiplying cells, those represented in III. Data in I and II are from (10). Morphogenetic data in III are twice repeated from (18). In III fully drawn steps and curves represent exponentially multiplying cells, dashed steps and curves multishock-synchronized cells, (in the latter case time is reckoned from division back in time to the 1 1/2 hours earlier when the final heat shock ends). In III times from division, (D), through S, to D are from (13). Time indications in (18) and (13) are multiplied by 16/15 to permit comparison with our data. Data in I - II are from (10).

shall not be much concerned. Anyhow, difficulties in placing S and G2 precisely (in I) are less pronounced in the case of curve II.

a. *From Division, through Shock, to Division.* We now proceed to a comparison of curves II and I, of the repetitively synchronized and of the normal morphogenetic sequence, and we begin at time zero, as read on the lowest of the two time scales. This is a time when in the three curves a cell division comes to completion.

Many of the newly divided synchronized cells (II) possessed a st. 1 oral primordium (kinetosomes present in the primordial field, but in no order), but there was no further development before a point two-thirds or so into the S-period, at which time morphogenesis was resumed. At around one hour when the heat shock hit (star), the synchronized population had moved on to (median) morphogenetic stage 2 (kinetosomes began to move relative to each other in the plane of the morphogenetic field). For the duration of the shock (I) further morphogenetic progression remained blocked, and after the shock (I, II) there was some structural regression, followed by new progression. The developmental delay in excess of the duration of the shock was 45 - 50

10

minutes, equal to what was observed for the cell division (59,60). This time of 45 - 50 minutes is the recovery time from effects of the shock. However, the greater part of this time was characterized by regression, the smaller by new progression in the cycle, albeit from a more primitive stage than the one prevailing before the shock. So, resumed morphogenetic progression comes some 30 - 40 minutes after the shock. It moves the system on in time, through the point at which it has recovered the situation existing before the shock, and in the end it matures the cells for new division and new replication; the latter runs the cells into the next shock period.

Comparison of curves II and III also indicates that the heat shock acts to fully eliminate the morphogenetic advantage of the newly divided synchronized over the newly divided normal cells. The heat shock-induced morphogenetic pause in the synchronized cells aligns them with the normal cells for the remaining half of the cycle, i.e. through oral morphogenesis and division. However, soon after this next division the synchronized cells are again morphogenetically ahead; furthermore, they skip the G1 phase, demonstrating a readiness to enter S not present in newly divided normal cells. Therefore, synchronized cells reach the point of morphogenetic activiation late in S, a normal G1 time (20 or so minutes) earlier than normal cells, and - as mentioned - in addition they are morphogenetically ahead of normal cells at the same point of the S-period.

In conclusion, the heat shock has acted to revert the system to a structural situation typical of a stage younger in the cycle than the stage hit by the shock, and time is needed after the shock to recover the ability to again perform morphogenesis, i.e. to move forward in the cell cycle. While times can not be given with much precision, it seems to me that repetitively synchronized cells take a normal cell generation's time to display a fairly normal cycle-program. This display begins visibly at 30 - 40 minutes after each shock and it runs 10 - 20 minutes into the next. It covers the time of the intershock-interval but lags somewhat behind. The repetitive cell cycle we have to consider is untraditional. It begins with the activation of normal early G2 events, and it ends with a normal completion of S. It goes from the end of one S-period through the next S-period. We shall be more specific about these matters in the discussion.

b. *From Division to Division, no Shock.* Now is time to compare the repetitive (Fig. 6 II) and the free-running morphogenetic sequences. In Fig. 7 (10) we have followed oral morphogenesis in free-running cells. Events from 0 to 1.5 hours on the upper time scale (time after a shock) represent sequences just reported, so we shall start at time zero on the lower time scale. Unlike repetitively synchronized cells, free-running cells received no heat shock around 1 hour on this scale, and they therefore proceeded smoothly through the stages of mouth formation and they rushed to division in 35 minutes (a time 20 - 25% shorter than normal). A heat shock would have checked precisely this morphogenetic advantage - which, incidentally, came into the system with the very first [Thormar (45)] or the first few heat shocks, those which induced the repetitive synchrony.

11

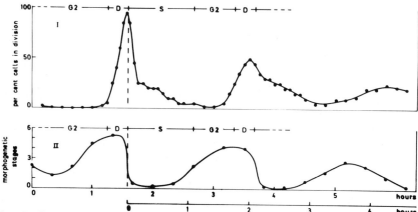

Fig. 7. Free-running synchrony of cell division (I) and of oral morphogenesis (median stages II). From (10).

The situations described by Figs. 6 and 7 may reflect shifts in a balance between production and use of oral precursor macromolecules. After all, overall protein synthesis is not greatly disturbed by heat as is the assembly of proteins into the structures here dealt with. Cells prepared for a series of synchronized divisions by the multi-shock method (44) have been shown to carry excess proteins destined for later incorporation into mouth structures (49). In repetitively synchronized cells the premature appearance of kinetosomes in the oral primordial field indicates either a relative abundance in these cells of partially assembled precursors (kinetosomes) for mouth structures, or the premature appearance of receptor sites for kinetosomes in the primordial field or of the field itself. Anyway, it seems likely that dividing synchronized cells carry structural oral precursors in excess of what is normal. I suggest that they spill over into the next cycle at division and speed developments which lead to the point of morphogenetic activation late in S. The consequences of this phenomenon, in terms of accelerated progression all the way to division are annulled if a next heat shock follows in the repetitive program. On the other hand, events from the *termination* of a heat shock and to division occurred in standard time (80 - 95 minutes) in all systems we have studied. In this period minimal synthesis of new protein was required for structures to be assembled and for division to succeed (39,53).

4. Cell Counting in Batch or Continuous Flow-Cultures, with Asynchronous or Synchronous Mode of Growth

Thanks mostly to the efforts of Dr. Leif Rasmussen of our laboratory, cultures of *Tetrahymena pyriformis* can be continuously monitored by use of automatic electronic cell counting (38,40). One can shift from batch to continuous flow, from asynchronous to synchronous mode of growth. Fig. 8 shows a tracing from a continuous flow experiment in which the temperature of the bath with the growth chamber changed cyclically between half-hour periods at 33.8°C (shocks) and periods of 160 minutes at 28°C. The finest synchrony was obtained when shocks were spaced 150 or 160 minutes apart, which confirms earlier results (59).

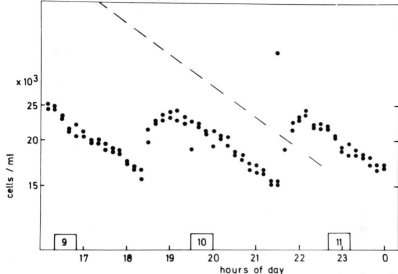

Fig. 8. Continuous flow synchrony. The temperature was regulated as shown on the abscissa; shocks 9 - 11 were applied. Cells were counted automatically. Periods in which the population was gradually diluted alternate with periods with compensatory increases in cell number. The overall course was one of balance between production and loss of cells. The dilution rates observed equal the theoretical (slope of dashed line), assuming that there was no cell division. Compensations reflect expected production periods of new cells by synchronous divisions. Visual inspection of removed samples check with Fig. 1. From (38,40).

We have had 25 ml continuous cultures running for months, with and without repetitive synchrony. We found no signs of change of the latter with time under a fixed temperature program, thus no signs that mutation and selection play significant roles in our experiments.

5. Discussion

Traditionally, the cell cycle goes from the end of one cell division (D) to the end of the next division, with the replication of the nuclear DNA (S) in the middle. However, we are free to consider cycles delimited other ways, and the repetitively synchronized *Tetrahymena* cells insist that we pay attention to a cycle which goes from the end of S to the end of S, with D in the middle. Essentially, this is the sequence these cells display at constant temperature. An S-period runs to its completion during the shock, later events are induced to wait. Thus a new cycle begins after the cells have recovered from effects of the shock. Subsequently the cell first prepares itself for the task of distributing its nuclear DNA and its whole self onto two cellular units, after cell division there is new replication of DNA.

The heat shock separates S from what comes after; it delays not a single but a whole sequence of events, all of them quite complex: morphogenesis, enzyme steps, cell division, DNA replication.

As summarized in Table 1, in this repetitively synchronized *Tetrahymena* system a maximum of cells showed furrows 95 minutes and separated 105 minutes (59) after a temperature shock had been terminated (0 minutes). DNA replication began at 95 minutes and ended some 80 minutes later, i.e. at 175 minutes (5). Blocking of DNA synthesis from 150 minutes had no effect on the subsequent division because at this time a signal had been emitted for the cells to go ahead with preparations for division independent of continued S-activity. The signal was reflected at the level of visible structures: the oral primordium became active (10).

Around 110 minutes there is a time point which relates to later morphogenesis and to cell division: heat shocks prolong the cell generations which they affect, but the relationships between the age of the affected cell in the cycle and the prolongation are such that it appears as if cells of certain ages are all returned to this point when heated after they have passed it; and that after the shock they make new progress towards division starting out from 110 - 115 minutes, the "collection point". By the time of the heat shock the cells spread in biological time past the collection point and, the spread is corrected for by comparison of the population at an apparently younger age. This serves to maintain the synchrony. The collection point is an abstraction. It marks a state of maturation for the later performance of morphogenesis and division; it may have a material basis, but we have no markers. It is placed before the middle of the synchronized S-period, but has no relation to this phase; the morphogenetic reactivation point comes later, two-thirds, perhaps three-fourths, into S, and the transition point (at 170 minutes) comes at or after the end of the synchronized S-period. Neither this transition point nor the division to which it refers are in fact realized in repetitively synchronized cells because the next shock hits before both have occurred (Table 1).

The best conditions for maintenance of synchrony, once induced, exist when a heat shock hits the population prior to a time when significant numbers of cells have undergone transition, i.e. well before 170 minutes after the previous shock has ended. Indeed, experiments show that intervals of 170 minutes between shocks are too long, 160 minute intervals are fine (59,60). On the other hand, a heat shock must not hit the population too early. One would think that all cells should have passed the collection point at 110 - 115 minutes if they all are to be reverted to it by the next shock. Experiments indicate that even 140 minutes between shocks is too short a time, 150 minutes are excellent. Apparently, there are more reasons than have been mentioned why intervals between shocks must not be too short. For example DNA replication might perhaps be disturbed (e.g., by effects on precursors) by heat shocks hitting the system earlier than in the tail of S; or perhaps a heat shock should come after the morphogenetic reactivation so as not to disturb what triggers this event. So by now we have arrived at some preliminary understanding why 150 - 160 minutes between shocks are optimal for this repetitive synchrony.

When shocks are separated by 160 minutes and each shock returns the cells to a situation typical of 110 - 115 minutes, there is a 45 - 50 minute set-back. This was observed (59,60). This time is required after the shocks for the cells to reestablish the morphogenetic situation existing before the shock. After the cells have been returned to 28°C we should therefore observe a sequence of events typical of the interval from the collection point at 110 - 115 minutes through the morphogenetic reactivation around 150 minutes and further on to age 160 minutes. This, I believe, is what we in fact do. For the first 35 - 40 minutes or so after the shock there is no progression, then there is activation of morphogenesis. However, at 35 - 40 minutes the structures have regressed to what is characteristic of cells to be morphogenetically reactivated; and in their progression from the collection point the cells have reached this particular phase just at this particular time. So from 35 - 40 minutes after the shock the cells go normally forward.

Why, then, is there regression for the first 35 - 40 minutes after a heat shock? Perhaps because the cells have been set back in time, perhaps even zeroed, with respect to events - thus to be repeated - which deal with how to make a new mouth. If this is so, the old primordium is left without opportunities for further development, until, with time and regressive development it has found the point at which it is in correspondence with the cell as a whole, as it now pushes towards division after it had been set-back in time by the shock. In conclusion, in the temperature regimen recommended for repetitive synchrony, each new heat shock causes essential removal of an otherwise existing overlap between S and later activities in the cycle. This effect is intimately connected with a synchronizing effect for which there are structural markers and for which the mechanism appears to be one of differential set-back of the cells to a common point in the cycle. These results (even time indications) and views fit with and extend those arrived at in our earlier studies with the multishock system (62).

A few words about "division protein(s)" and the subpellicular fiber system in young oral primordia of *Tetrahymena*. Before multishocked (44) *Tetrahymena* cells can perform synchronous division they must make unspecified new protein(s) (39), termed "division protein(s)" (53). Williams and I proposed that the proteins of the subpellicular oral fibers qualify as "division protein(s)", provided one takes into account fibers not yet seen in the primordial field, either because they are too thin or because they are elsewhere in the cell, perhaps serving other functions. In st. 1 primordia, no fibers can be seen to connect the kinetosomes in the oral primordium. Such fibers appear when morphogenesis becomes active 40 minutes after a heat shock, and they disappear if a temperature shock is applied (50,63). A similar situation might pertain to the repetitively synchronized cells at times around a heat shock. The proposal is the more reasonable because subpellicular oral fibers have features in common with mitotic fibers. The oral fibers contain

15

TABLE 1

TIME POINTS OF INTEREST IN REPETITIVELY SYNCHRONIZED *TETRAHYMENA* CELLS.

Event	At minutes	
	After EH	After div. max.
Heat shock ends	0	
A maximum of cells show furrows, replication begins in new cells	95	0
A maximum of cells liberate daughters	105	10
Collection point	110	15
Morphogenetic reactivation	145-150	50-55
Transition point	170	75
DNA replication runs to completion	175	80
Next heat shock is initiated at	160	65
and ends at	190	95
If no next shock is applied a 2nd div. max. comes at	205	110

microtubules and they interconnect a multitude of "centrioles" (63) in the process of engaging themselves as basal bodies (kinetosomes) for cilia. The kinetosomes are moveable by the connecting fibers in the same way as are the centrioles in mitosis. However, kinetosomes are situated in the cortex of the cell and they do not find chromosomes spilled among their connecting fibers while they move. This mitotic centrioles do.

In mitotic cells, elevated temperature, when precisely controlled, has effects similar to those of colchicine, notable for its capacity to combine with tubulins and thus to interfere with the formation and integrity of mitotic fibers. *Psammechinus* eggs when heated to 26°C [normal is 16°C, (61)] or when immersed in 0.1 mM colchicine (51) go through several cycles of disappearance and reappearance of the nuclear membrane at the locus of the fertilization nucleus. There is no duplication of the nucleus, so chromosome segregation seems inhibited while other aspects of mitosis are preserved.

B. *Tetrahymena vorax:*

Studies on the control mechanism(s) of differentiation are fascinating when the organism under investigation has a choice between several routes of development and when the choice can be experimentally controlled. *Tetrahymena vorax* is such an organism. The strain V_2, subline S, exists during vegetative growth as a small saprozoically feeding cell termed a microstome. Another cell type, called a macrostome, can be induced from the microstome population by either placing microstomes in the presence of prey, *Tetrahymena pyriformis*, or in the presence of a prey factor, stomatin, released from *T. pyriformis*. A macrostome is large and is capable of feeding on smaller species of ciliates (6).

Morphogenetic events leading a microstome to produce a macrostome or a second microstome by division are both accompanied by formation of a new oral primordium. In the former case the oral apparatus of microstome is resorbed and replaced with one of larger dimensions, positioned directly posterior to the degenerating microstome oral apparatus. In the latter case the future oral apparatus of the posterior daughter cell arises in the mid region of the microstome along the stomatogenic meridian.

Working in our laboratory Buhse and Rasmussen (7,8) discovered that cell divisions in populations of *T. vorax* cells could be synchronized (Fig. 9 I) with a version of the multishock procedure for *T. pyriformis* (44). Division occurred around 2.5 hours after the final heat shock. The population to be synchronized should be started with stationary phase cells allowed only one generation before the heat shocks begin. The cells should remain in the complex growth medium. If they were washed into an inorganic medium within 60 minutes after the shocks, they did not perform oral morphogenesis and divide such as *T. pyriformis* would do (21). Instead the old mouth regressed, a new one developed just behind it; and that was a giant mouth enabling the cell to swallow smaller cells of its own species or genus. The transformation was well synchronized; 80% of the population turned into macrostomes between 4.5 and 5.5 hours past the final heat shock (Fig. 13).

III. Other Cells

The differences between a protozoan cell, a yeast cell, a sea urchin egg and a cultured mammalian cell are manifold, and so are the similarities. The cells mentioned have been much used for cell cycle studies. How different, or how similar, are they in their responses to environmental changes which will induce one of them, *Tetrahymena pyriformis*, to divide and replicate in phase?

A. *Schizosaccharomyces pombe:*

To get a first answer to the above question, Dr. Birte Kramhøft and I (26) selected *Schizosaccharomyces pombe* (strain 972 h-) for study and exposed exponentially multiplying populations of this cell to a temperature program like the one which works with *Tetrahymena*.

17

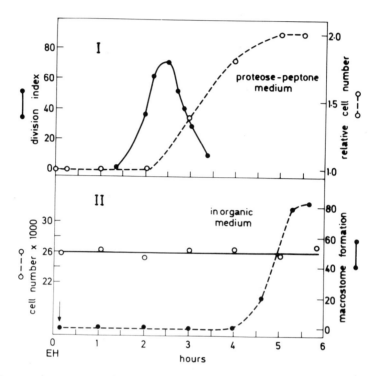

Fig. 9. Synchrony in *Tetrahymena vorax*, strain V₂S. The cells were grown in proteose-peptone, liver, salts, for 6 hours after inoculation. Then they received 6 temperature shocks (29.5°C, 40 minutes) at intervals of 40 minutes at 20.1°C. After the final shock (EH = 0 hours) the temperature was constant at 20°C; the division index and the number of cells/ml (range 2×10^4/ml) followed the courses shown in I. Replacement of the growth medium with saline shortly after EH (arrow in II) suppressed the synchronous division and resulted in a transformation from the microstome to the macrostome type of cell. From (8).

The optimum temperature for exponential growth (32°C), the doubling time (130 minutes) at this temperature and the division-blocking temperature above optimum (41°C) were determined. We now asked if exponentially grown *Schizosaccharomyces* cells could be synchronized by repeated 30 minutes heatings to 41°C with intervals between shocks equal to 130 min. This program was the nearest imitation of the *Tetrahymena* synchronization program which we can propose for *Schizosaccharomyces*.

After a few temperature cycles (41 and 32°C) divisions were limited to short periods between shocks. When shocks were discontinued (free run) two divisions followed each other at 40 minutes closer than normal 130 minute intervals (Fig. 10). This division activity was reflected in fluctuations in the fraction of cells showing a cell plate. We can conclude that, regardless of obvious dissimilarities *Tetrahymena* and *Schizosaccharomyces* react similarly to heat shocks spaced a normal cell generation time apart.

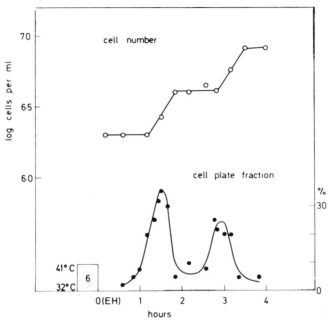

Fig. 10. Free running synchrony in *S. pombe* as induced with 6 heat shocks. The interval between the two divisions is shorter than in normal cells. From (26).

We wanted to check if time and temperature parameters were chosen to be optimal for the synchrony induction. For this we used the synchronized *Schizosaccharomyces* system itself, as obtained with five temperature shocks.

The present system comprises four variable parameters, namely (i) the duration of each heat shock, (ii) the height of the heat shock, (iii) the duration of the intershock interval, and (iv) the intershock temperature. In all experiments the intershock temperature was fixed at the optimal growth temperature, so only the first three parameters were subject to changes. This was done according to the following scheme in which (A), (B) and (C) each represent one series of experiments. The shock that was varied was the sixth one.

(A): Shock height and duration of shock interval were fixed and shock *length* was varied (Fig. 11A).

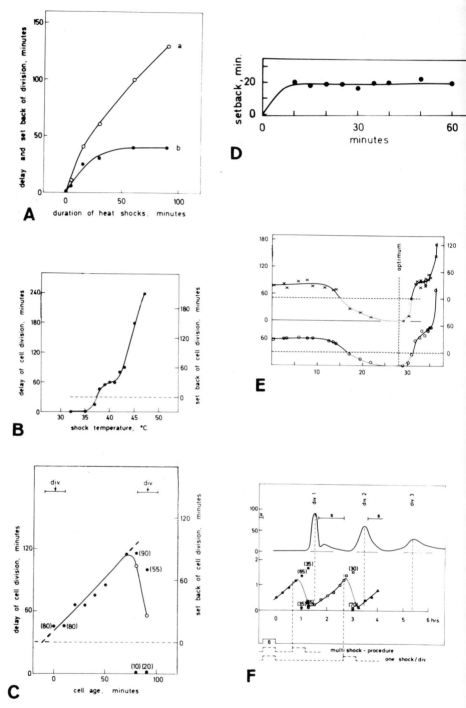

Fig. 11. (For legend see facing page)

(B): Shock length and duration of shock interval were fixed and shock *height* was varied (Fig. 11B).

(C): Shock length and shock height were fixed and duration of the intershock *interval* was varied (Fig. 11C).

(A). Fig. 11A, curve a, shows the delay of what could have been the second free-running division after five shocks, as a function of the duration of the sixth shock. The division was delayed increasingly, from zero to 130 minutes, when the duration of the heat shock increased from zero to 90 minutes. Subtraction of the time for which the cells have been heated from the measured delay gives the *excess delay*. When there was excess delay the cells were farther from the forthcoming division after than before the shock, and this can be viewed as a *"set-back"* of the cells in the cycle. Curve b shows that the set-back was positive for all durations of the heat shock and indicates that the set-back increased for the first hour or so of heating and then assumed a constant value. Apparently the cells can be set back to a standard time point in the cycle, and no farther.

Fig. 11. Responses of *S. pombe (S.p.)* and *T. pyriformis (T.p.)* to a temperature shock. To the left [Figs. A, B, and C, all from (26)] are the results with *S. pombe*, our test organism. To the right [Figs. D, E and F, all from (59)] are comparable results with our model cell, *T. pyriformis*. In Figs. A and D, the duration of the shock was varied. Shown is the delay (open circles) and set-back (filled circles) of the forthcoming cell division by a heat shock which was initiated at the normal time after shock 6 (*S.p.*) or shock 5 (*T.p.*). In Figs. B and E, the temperature in the shock was varied. Shown is the delay (left ordinates) and set-back (right ordinates) by a temperature shock placed at normal time after the preceding heat shock. In both systems we have heated to various temperatures; in *T.p.* we have also studied effects of cold shocks. Durations of shocks are reflected in levels of dashed horizontal lines. In Figs. C and F, the interval between shock 6 (*S.p.*) or 5 (*T.p.*) and the next shock was varied. Shown are delays (in *S.p.* also the set-backs) of a forthcoming cell division by a standard heat shock placed at different times between divisions 1 and 2 (*S.p.*) or after the final heat shock (*T.p.*). In Fig. C, the population was assigned age=0 at the time of the first synchronized division, 96 minutes after shock 5. Set-backs increased linearly up to 90 min duration from age = -10 min ("collection point") to age = 80 min (the transition point relative to the second division indicated in the graph). The transition point cannot be separated from a collection point which relates to a third division not included in this graph. From (26). In Fig. F, the portion of the curve defined by open squares (delimited by 1 and 3 hours) is fully comparable to Fig. C (left ordinate shows delays in hours). Both show delays of division 2 as a function of the time between consecutive transition points. Filled circles and triangles refer to delays of division 1 and division 3, respectively. Transition points occur around 1 and 3 hours referring to divisions 1 and 2, respectively. From (59).

(S. p.) can be synchronized by similar procedures, preparations for a forthcoming cell division must be basically similar in the two cells. Not only does the latter cell in principle obey a synchronization program developed for the former, the two types of cells respond similarly to experimentally arranged variations in the time and temperature parameters which make up the synchronization program. In fact, each of the curves shown here for *S.p.* has its equivalent for *T.p.* (59,60), further illustrated by Figs. 11D, 11E and 11F. While *T.p.* possesses a morphological indicator — the oral primordium — of the position of the cell in the cycle there are no immediately comparable markers in *S.p.* We shall be looking for such, but we shall also keep in mind that our ideas for *T.p.* may need revision as more information becomes available from studies of other cells, e.g., the yeasts.

It has been stated that temperature shocks spaced one normal cell generation apart give a satisfactory synchrony in cultures of *S.p.* Furthermore, we have found that heat shocks in this yeast, like in *T.p.*, cause age dependent set-backs, and that there is transition to insensitivity to heat towards the end of

(B). Also the set-backs are functions of temperature. Fig. 11B shows the delay (left ordinate) and set-back (right ordinate) of the second free-running division as a function of the shock temperature. In this case the duration of the shock was constant at 30 minutes, and this is indicated by the level of the dotted line in the figure. Temperature shocks below 35°C had little or no effect on the forthcoming division. Shocks to higher temperatures delayed the division and the more so the higher temperature. Shocks above 37°C caused set-backs. They tended to plateau in the range 37-42°C and increased greatly above 42°C. This, we believe, separates the effects on the cells' progression in the cycle (plateau) from the effects on cell viability (above plateau).

(C.) The interval between shocks 5 and 6 was varied from 100 to 190 minutes. The first of the two values gives the time from shock 5 to a midpoint in the first forthcoming division step, the second to a midpoint in the second step. The interval between the two free-running divisions was 90 minutes, and it was over this interval that we placed the shock (Fig. 11) with the purpose of measuring delays and set-backs of the second division. Within the time span from - 10 to + 80 minutes (abscissa) set-backs increased linearly from 0 to 90 minutes. Thus, gain in age over a full synchronized cell generation, as measured from a point just prior to one division to a point similarly related to the next division, is fully lost in response to a standard heat shock, regardless of when it is placed on the chosen cellular time scale. Only the second division was affected. We refer to the first time point mentioned as the "collection point"; the second time point is the transition point (39) to heat. The curve tells us that in this system certain temperature-sensitive preparations of a synchronous division (number 2) are continuous, additive and fully reversible over a time corresponding to that of a full synchronized cycle.

IV. Conclusions

Since *Tetrahymena pyriformis (T.p.)* and *Schizosaccharomyces pombe*

the cell cycle. In *S.p.* as well as in *T.p.* the free-running cycle is compressed by a time which corresponds to the set-back a division suffers when its preparations are affected by a next heat shock placed correctly in the regular series used to induce and maintain the repetitive synchrony. In *S.p.* this compression is likely to reflect elimination of normal time from the transition point, through division, to the collection point. Thus the collection point has been moved back in time underneath the time of the division which is triggered at the preceding transition point. The significance of this observation cannot yet be discussed in the terms used with *T.p.* We have yet to place the DNA synthetic interval.

B. Mouse L-Cells

The responses to temperature shocks described for *S. pombe* are closely analogous to the responses long known to exist for *Tetrahymena*, and the possibility now exists that we deal with phenomena which are general for cells and which can be used to bring more cells in synchrony than the two here mentioned. Experiments with sea urchin eggs in early cleavage (61) suggested that the development of mitotic fibers was more sensitive to elevated temperature than were most other components in the mitotic cycle of these non-growing cells. What is the situation with growing mitotic cells?

In collaboration with Dr. Hiroshi Miyamoto we initiated studies of mouse fibroblast cells, line L (29,30). Small clones were cultivated in a nutrient-rich, complex medium in monolayer cultures. Cell multiplication was followed by time-lapse photography or video recording. In both cases exposures were made every hour, a frequency which is sufficient to keep track of the original cell and its progeny from the one-cell to the 128-cell stage, or for about a week. Cell lines were followed, "family trees" were constructed and individual and average generation times were calculated.

The average durations of the early generations were rather constant. Division synchrony deteriorated with time but could still be perceived at the end of 7 generations. As would be expected from these studies the coefficient of variation of generation times within one single clone in our experimental system was low enough (about 9%) to permit us to calculate prolongations of generation times which result from chillings or heatings to defined temperature for defined time, usually one hour. Prolongation is the difference in duration of a treated generation, as measured from division to division of single cells, minus the duration of the generation of the mother cell. Fig. 12 gives results and interpretations of our experiments as shown in the lower and upper frames respectively. With exclusion of some observations (circles, squares), curves were fitted to the points by eye and thin lines were drawn on either side of each curve at such a distance that about two-thirds of the points fall on the band between the lines.

Chillings to 10°C elicited no other responses than a delay of 1 hour, regardless of cell age. The width of the band fitted to these points is an indication of normal spreads in generation time of untreated cells. Generally,

23

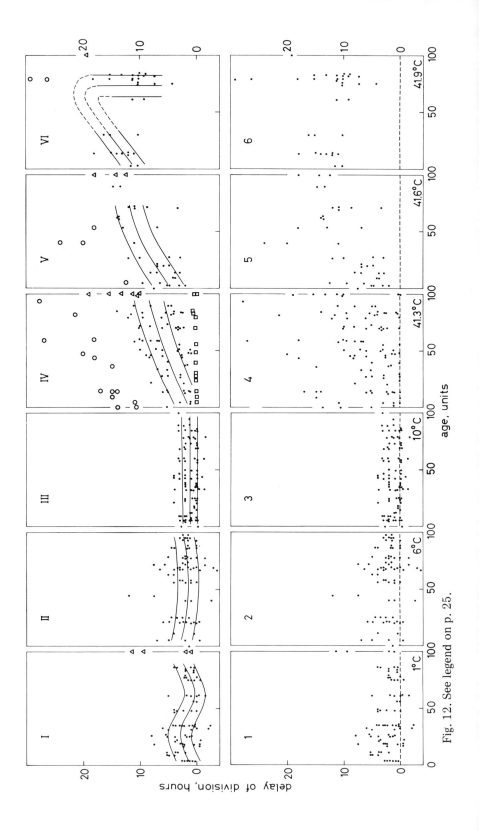

Fig. 12. See legend on p. 25.

one hour chillings of these cells [though not of all cell types (33)] .yielded results of limited interest: 1 hour exposures to temperatures between 41ºC and 42ºC are of interest. Heatings to either 41.3, 41.6 or 41.9ºC tended to induce very considerable recovery times, as illustrated in frames 4, 5 and 6. The results scatter greatly, especially after exposure to the lowest of the three temperatures: In frame 4, a sizeable fraction of the points represent cells with very long generation times, or cells not delayed at all (open circles and squares in IV). Thus, for these two fractions we can speak of an all-or-none reaction to heat shocks. However, most cells showed intermediate delays, and they tended to be the more delayed by the heat shock the more they had progressed in the cycle at the time heated. The curve and band in frame IV represent these cells. Characteristically, heating to only slightly higher temperature, 41.6º or 41.9ºC, delayed *all* cells with respect to the forthcoming division, and the more so, the higher is the shock temperature. Cells in mitosis (triangles) showed moderate delays and were *not* delayed more by the higher temperatures than by 41.3ºC. Taking an overall view at the points and interpretations we see some similarity with *Tetrahymena* and *S. pombe*. As newly divided cells grow older they develop increasing sensitivity to heat shocks, at least maximal sensitivity is in the latter half of the cycle. Furthermore, (frame VI) we believe we see a late transition point with respect to sensitivity to heat shocks, another similarity between the ciliates and the yeast. Thus, in our hands, responses of L-cells are far more heterogenous than those of the 2 organisms mentioned above. This is not the best basis for synchrony-induction. Nonetheless we have made attempts at this which we now describe.

From the above discussion is clear that asynchronous monolayer cultures of L-cells can be expected to react to heat pulses by showing an early long pause in cell multiplication. A period of perhaps faster than normal growth then should follow. We find the pause, but not the faster growth (Fig. 13 hollow circles, filled squares). However, a second shock at the time when multiplication would just have been resumed created this effect (filled triangles). Consequently, shocks collected more cells in one phase than did one shock, which is reminiscent of *Tetrahymena* exposed to heat shocks at short intervals such as in the earlier procedure for synchrony induction (44). We had hoped for better results with three shocks, but were disappointed (filled circles). In these experiments the shocks tended to block cell division for the duration of the sequence of shocks and induced a division-burst after a lag which exceeded the duration of the intershock interval. There must be

---

Fig. 12. L-cells: Relations between cell age (abscissa, unity scales from 0 to 100) and delays of the forthcoming cell division (ordinate, h) in response to chillings or heatings for 1 hour. Raw data are shown in frames 1-6. These are fitted with bands of interpretation in frames I - VI. Frames 1, 2 and 3: chillings to 1, 6 and 10ºC, respectively. Frames 4, 5 and 6: heatings to 41.3, 41.6 and 41.9ºC, respectively. From (29).

preparations for division between each 2 shocks, and these are reversed by each new shock — as in *Tetrahymena*. In conclusion we feel that the detailed analysis we have initiated are logical first steps in attempts to synchronize mammalian cell populations by use of temperature shocks. At present, even with our best observations, we are no farther along in this respect than were our predecessors. At the moment we have, at the best, parasynchrony.

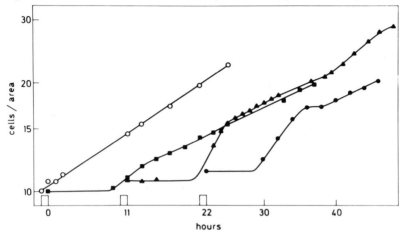

Fig. 13. Cell multiplication in 4 replicate cultures of L-cells. O, control culture maintained at constant temperature: ■ , ▲ and O represent multiplication after 1, 2 or 3 exposures to heat shocks (41.6°C for 1 hour), respectively. From (29).

IV. Concluding Remarks

In *Tetrahymena* temperature shocks — to the cold and to the warm side of the optimum — cause disturbance of equilibrium reactions which the cell shifts in one direction on its way from division to division, and which a temperature shifts in the opposite direction. The oral primordium is a sensitive structural indicator of such changes in equilibria. I suggest that the cells go from division to division by assembling structures having the ultimate coordinated functions of dividing the nucleus and the cell into two; and that some, if not all, structures go through heat sensitive stages.

Clearly, between divisions the cells must do more than make division-relevant structures, and the case of the repetitively synchronized *Tetrahymena* cells suggests that for defined intervals in the cycle other activities can be given first priority. Oral morphogenesis is initiated in newly dividied cells, but it is soon repressed and not again derepressed before two-thirds into S. We believe that this concept of morphogenetic repression and derepression giving leeway to G_1 (whenever it is there — and whatever it is there for) and DNA synthesis is new, at least for *Tetrahymena*.

If, turning to mitotic cells, one refers to the development of the mitotic apparatus and to the placement and deepening of the division furrow as morphogenesis, and if one accepts that centrioles behave as early markers of stages in morphogenesis — then one sees interesting analogies between ciliate oral morphogenesis followed by cell division, and ordinary mitosis. It has often been said that the behaviour of the centrioles in the mitotic cycle indicates that mitosis is initiated at telophase, suspended through interphase and resumed at the time when the cell as a whole is ready for its next division. Thus mitotic interphase may be a morphogenetic pause comparable to the pause found in *Tetrahymena*. Now we attempted synchronization of L cells with heat shocks, and the question may become important if also in mammalian cells there are signs that morphogenetic derepression is coupled to a late stage in the S period. We have no opinion, but cannot dismiss as insignificant that late replicating DNA is surrounded by mystery in mammalian cells, the same as in *Tetrahymena*.

For good repetitive synchrony to be obtained in *Tetrahymena*, each new heat shock must be placed so as to split the cycle open, separating overlapping S and G_2 activities, letting the former run to completion and reverting the latter. This leaves us with the cycle the system presents us with; the cycle in which preparations for the cell's physiological reproduction, and this event itself, precede the cell's genetic reproduction — to use phrases coined by Dr. Daniel Mazia. In this sequence events follow each other in a natural way and at constant temperature. The state of affairs is quite reasonable if, as we consider, our synchrony induction is directed at morphogenesis and cell division. Synchronous DNA replication follows synchronous division because natural mechanisms trigger newly divided cells into S. If we affect the relationships between division and replication unfavourably by spacing heat shocks differently than described in this chapter, we may arrange beautiful division synchrony and rather miserable DNA replication synchrony. At least in *Tetrahymena* the two events are surprisingly dissociable. This we should keep in mind now when we are taking steps to apply results from *Tetrahymena* to other cells. Will a population of yeast cells or of mammalian cells fall into synchrony if we treat them as if they were *Tetrahymenas*. The yeast, *Schizosaccharomyces pombe*, will, mouse cells are less willing, even though they evidently share features with the two other cells. Both can be excessively delayed by heat shocks, in both division delays increase with age of the cells in the cycle, in both heat shocks can collect many cells in a common phase so that after the shocks they divide together.

The yeast system is now in a state which permits a multitude of cell cycle studies, the L cells are not yet.

Acknowledgement

I am deeply indepted to Drs. H. E. Bushse Jr., H. Miyamoto, L. Rasmussen, H. A. Andersen, Kirsten Hamburger and Birte Kramhoft for contributions reported in this chapter. Also I drect my thanks to Mrs. Helga Langelo

Sorensen and Eugenie Palludan for daily assistance in the laboratory and in the secretariat.

References

1. H. A. Andersen, *Exptl. Cell Res. 74*, 610 (1972).
2. H. A. Andersen, *Exptl. Cell Res. 75*, 89 (1972).
3. H. A. Andersen, Abstract from 4th International Congress on Protozoology, Clermont-Ferrand, 2-9 September, 1973.
4. H. A. Andersen, *J. Cell Sci. 14*, In press, (1974).
5. H. A. Andersen and E. Zeuthen, *Exptl. Cell Res. 68*, 309 (1971).
6. H. E. Bushe, Jr., *J. Protozool. 13*, 429 (1966).
7. H. E. Bushe, Jr. and L. Rasmussen, *Norwegian J. Zool. 21*, 185 (1973).
8. H. E. Bushe, Jr. and L. Rasmussen, *Compt. Rend. Trav. Lab. Carlsberg*, In press, (1974).
9. H. E. Buhse, Jr., S. J. Stamler and J. O. Corliss, *Trans. Amer. Micros. Soc. 92*, 95 (1973).
10. H. E. Buhse, Jr. and E. Zeuthen, *Compt. Rend. Trav. Lab. Carlsberg*, In press (1974).
11. J. E. Byfield and C. L. Young, *J. Protozool. 17*, 445 (1970).
12. J. E. Byfield and O. H. Scherbaum, *Proc. Nat. Acad. Sci. 57*, 602 (1967).
13. I. L. Cameron and D. S. Nachtwey, *Exptl. Cell Res. 46*, 385 (1967).
14. E. G. Christensson, *Zeitschrift fur Biologie 116*, 143 (1968).
15. E. G. Christensson, *Zeitschrift fur Biologie 6*, 441 (1971).
16. E. G. Christensson, *J. Protozool. 17*, 496 (1970).
17. E. G. Christensson, *(Thesis) Studentlitteratur, Lund*, 1, (1971).
18. J. Frankel, *Compt. Rend. Trav. Lab. Carlsberg 33*, 1 (1962).
19. J. Frankel and N. E. Williams, In: *Biology of Tetrahymena*, A. M. Elliott (ed.), Dowden, Hutchison and Ross, Inc., New York, In press, (1974).
20. K. Hamburger, *Compt. Rend. Trav. Lab. Carlsberg*, To be published (1974).
21. K. Hamburger and E. Zeuthen, *Exptl. Cell Res. 13*, 443, (1957).
22. V. Hansen, (Thesis) The Biological Institute of the Carlsberg Foundation, Copenhagen (1970).
23. J. Keiding and H. A. Andersen, Personal Communication.
24. H. Klenow, *Biochim. Biophys. Acta 35*, 412 (1959).
25. H. Klenow, *Biochim. Biophys. Acta 61*, 885 (1962).
26. B. Kramhoft and E. Zeuthen, *Compt. Rend. Trav. Lab. Carlsberg 38*, 351, (1971).
27. M. Lasman, *J. Protozool. 17*, Suppl., 28, (1970).
28. A. Lowff, *Compt. Rend. Acad. Sci., Paris 176*, 928 (1923).
29. H. Miyamoto, L. Rasmussen and E. Zeuthen, *J. Cell Sci. 13*, 889 (1973).
30. H. Miyamoto, E. Zeuthen and L. Rasmussen, *J. Cell Sci. 13*, 879 (1973).
31. J. Neuhard, *Biochim. Biophys. Acta 129*, 104 (1966).
32. J. Neuhard and A. Munch-Petersen, *Biochim. Biophys. Acta. 114*, 61 (1966).

33. A. A. Newton, In: *Synchrony in Cell Division and Growth*, E. Zeuthen (ed.), Interscience, New York, pp. 441, (1964).
34. B. A. Nexø, (Thesis) The Biological Institute of the Carlsberg Foundation (1973).
35. B. A. Nexø and E. Zeuthen, Abstract from the 6th Meeting of the European Study Group for Cell Proliferation, Moscow, 25-29 September, 1973.
36. B. A. Nexø and E. Zeuthen, To be Published (1974).
37. J. R. Nilsson, *J. Protozool. 17*, 539 (1970).
38. L. Rasmussen, *Veröffentlichungen der Universitat Innsbruch 77*, 10 (1973).
39. L. Rasmussen and E. Zeuthen, *Compt. Rend. Trav. Lab. Carlsberg 32*, 333 (1962).
40. L. Rasmussen and E. Zeuthen, *Norwegian J. Zol. 21*, 192 (1973).
41. P. Reichard, *Ciba Lectures in Microbial Biochemistry*, John Wiley & Sons Ltd., New York, pp. 1-77, (1967).
42. O. H. Scherbaum, *Ann. Rev. Microbiol. 14*, 283 (1960).
43. O. H. Scherbaum and J. B. Loefer, In: *Biochemistry and Physiology of Protozoa 3*, S. H. Hutner (ed.), Academic Press, New York, pp. 9-59 (1964).
44. O. Scherbaum and E. Zeuthen, *Exptl. Cell Res. 6*, 221 (1954).
45. H. Thormar, *Compt. Rend. Trav. Lab. Carlsberg 31*, 207 (1959).
46. H. Thormar, *Exptl. Cell Res. 28*, 269 (1962).
47. I. Villadsen and E. Zeuthen, *J. Protozool. 16*, Suppl., 31 (1969).
48. O. Westergaard and R. E. Pearlman, *Exptl. Cell Res. 54*, 309 (1969).
49. N. E. Williams, O. Michelsen and E. Zeuthen, *J. Cell Sci. 5*, 143 (1969).
50. N. E. Williams and E. Zeuthen, *Compt. Rend. Trav. Lab. Carlsberg 35*, 101 (1966).
51. E. Zeuthen, *Pubbl. Staz. Zool. Napoli 23*, Suppl., 47 (1951).
52. E. Zeuthen, *Adv. Biol. Med. Physics 6*, 37 (1958).
53. E. Zeuthen, In: *Biological Structure and Function 11*, T.W. Goodwin and O. Lindberg (eds.), Academic Press, London, pp. 537-548 (1961).
54. E. Zeuthen, Abstract from the Proceedings of the First International Conference on Protozoology, Prague, 22-31 August, 1961, *Progress in Protozoology*, 218 (1963).
55. E. Zeuthen, In: *Synchrony in Cell Division and Growth*, E. Zeuthen (ed.), Interscience Publishers, New York, pp. 99-158, (1964).
56. E. Zeuthen, *Exptl. Cell Res. 50*, 37 (1968).
57. E. Zeuthen, *Excerpta Medica International Congress Series 68*, 88 (1968).
58. E. Zeuthen, *Exptl. Cell Res. 61*, 311 (1970).
59. E. Zeuthen, *Exptl. Cell Res. 68*, 49 (1971).
60. E. Zeuthen, In: *Advances in Cell Biology 2*, D. M. Prescott, L. Goldstein and E. McConkey (eds.), Appleton-Century-Crofts, New York, pp. 111-152 (1971).
61. E. Zeuthen, *Exptl. Cell Res. 72*, 337 (1972).

62. E. Zeuthen and L. Rasmussen, In: *Research in Protozoology 4*, T. T. Chen (ed.), Pergamon Press, New York, pp. 9-145, (1972).
63. E. Zeuthen and N. E. Williams, In: *Nucleic Acid Metabolism Cell Differentiation and Cancer Growth*, E. V. Cowdry and S. Seno, (eds.), Pergamon Press, Oxford, pp. 203-217, (1969).

A CELL CYCLE NOMOGRAM:
GRAPHS OF MATHEMATICAL RELATIONSHIPS ALLOW ESTIMATES OF THE TIMING OF CYCLE EVENTS IN EUCARYOTE CELLS

Thomas W. James

Biology Department and Molecular Biology Institute
University of California, Los Angeles, California 90024

I. Introduction

The large number of techniques by which events in the cell cycle can be examined and assigned a pattern of change appear to be predicated on the idea that knowing the temporal order of events will lead to understanding the dependence or independence of cellular processes on the cyclical duplication of constituents. Although such thinking may be frought with the possibility of logical fallacies and it should be recognized that *effects* cannot precede *causes*, there is heuristic value in establishing the temporal order of events. However, in the case of the cell cycle, there is the added difficulty that cycles do not have beginnings or ends and *causes* and *effects* are often confused. Fortunately, investigators have not been immobilized by such dilemmas and following the course of least effort have used the most obvious cycle markers. The most desirable is cell age, measured from fission to fission since the act of fission can be used as a marker for zero age. The order of events has been determined by autoradiographic techniques (1), synchronization of cell populations by induction or selection methods (2), isolation of age-related cell size classes (3), time-lapse cinemicrography (4), and the use of conditional cell cycle mutants (5). Another alternative which is not generally recognized is to use the skewed relationship in the age distribution of a logarithmically growing culture of cells to determine the timing of events in the cell cycle. This procedure can be employed to supplement other approaches.

In several of these procedures, the precision of determining the order of events is plagued by common problems, namely, the inherent variability of cell parameters such as generation time, cell volume, the nonuniformity of stages in the cell cycle and unbalanced growth under the conditions of expansion of the culture. These are exclusive of the cases in which the change measured is not uniquely coupled to the cell cycles. Variability is easily overlooked when one is examining a small number of cells, but when observations are extended to properties that require the use of a population, they cannot be ignored. Unfortunately, there is no simple statistic that can be applied. It is therefore necessary to consider a formal derivation of the ideal age distribution as our point of departure (6, 7).

The purpose of this report is to present a simple and easily applied method for estimating the timing of events in a cycle from a minimum of information. For the reasons given it is founded on a model which does not take into account the statistical variances. Since it is applicable to an ideal culture and offers a ready means of obtaining independent estimates of timing, it may prove a bonus since information can be obtained from data which are often available from the techniques mentioned above. Alternately, it may offer a test of how close to ideal an experimental condition may be.

This method uses several pieces of commonly available data, i.e., the average amount of a constituent or an activity per cell obtained on a logarithmic culture, the average per daughter cell from the same culture, and either the duration or the time of the beginning or end of the synthetic period. It is an extension of previous analyses (6, 7, 8, 9) and offers a set of graphs and a nomogram for quick estimations. Derivations are given so that the limits of its applicability may be better understood.

II. Ideal and Real Age Distributions in Cell Populations

Before deriving the method, a brief look at some limitations must be presented. To avoid confusion about the nature of ideal and real populations, it is important to recognize that the unique structure of the age distribution in an actively growing population of cells is generated by logarithmic expansion, balanced growth and a uniform generation time. The basic distinction is that an ideal age distribution has an invariant generation time, while a real age distribution has the same mean generation time but that of individual cells is variably distributed. The difference is illustrated in Fig. 1 in which the solid line graph is the ideal case, while the dashed line portion shows its departure to

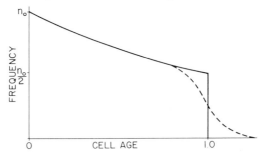

Fig. 1. A comparison of an ideal and a real cell age distribution. The departure of the real from the ideal is indicated by the dashed line at the right end of the distribution. The source of this difference is the variability of the generation time of ± 0.10.

a real distribution. The real case is constructed from photomicrographic time-lapse data on a growing colony of the yeast *Schizosaccharomyces pombe*. Twenty-six cell cycles were examined. In using these data, the generation time distribution rather than the inverse function (10) was used and converted to a sigmoidal cumulative distribution. In this form it expresses the number of cells that remain undivided as a function of time (dashed line).

The ideal age distribution shown as the solid line, gives the frequency of cells at each age from zero to 1 and is constructed from the well-known distribution function,

$$f(t) = kn_o e^{-kt} \qquad\qquad I$$

where t is normalized to a fraction of the generation time, k is the growth constant and n_o is the number at zero age. As a simplification $f(t)$ can be expressed using the base 2 and it becomes

$$f(t) = n_o 2^{-t} \qquad\qquad II$$

Since t takes on values from zero to 1, the base 2 makes the growth constant equal to the reciprocal of the generation time, i.e., one. This was used previously. However, other convenient bases can be employed, i.e., 4, where quadrupling of cell number occurs each cell cycle (5). Bases greater than 2 will generate age distributions which show a much larger proportion of young cells to old cells and therefore allow greater reliance on the estimation of the timing of events. A base less than 2, i.e., an incomplete doubling, presumably may also be of value. However, care must be taken in such cases to justify their use through a special derivation if they are to be applied to conditions showing unbalanced growth. These are special cases and will not be covered here.

From the function in Equation II, it can be readily seen that for the ideal case the frequency of cells at t_o equal to zero will be n_o, while at age t_g equal to one the frequency will be $n_o/2$. The intermediate frequencies also will be determined by the function and will follow a logarithmic decrease between n_o and $n_o/2$ as shown by the dark lines of Fig. 1. In the real case for *S. pombe*, the generation time distribution approximates normality and the standard deviation of the generation time is ± 0.10. When this generation time distribution is expressed as a cumulative distribution with its mean set equal to 1 and made to coincide with the terminal age of the ideal age distribution, the real distribution of Fig. 1 results (solid and dashed lines).

III. The Effects of Age and Generation Time Variability

It is apparent that if there is an abrupt doubling of a cycle-dependent constituent in a single cell, it would not be reflected as such in the real population since the statistical variables mentioned above would come into play. The same would be true for an abrupt cessation of a cycle-dependent function leading to an inhibition of cell division which would generate a predictable decay in the culture growth pattern (5). Some may argue that if an event is truly coupled to cell age it would not have a dispersion greater than

33

that of the generation time. However, the effect introduced by the variability of cell age can only be considered if we have some measure of it or if we assume it to be some value irrespective of whether it is greater than, equal to, or less than the generation time dispersion. Unfortunately, generation time dispersions are generally unknown so the calculations that follow have been confined to ideal distributions. There is no reason why they could not be extended to real distributions. We will consider the ideal case presented here as a first step. Such statistical complications are the fundamental limitation in the use of this method. It must be recognized that raw data from a population are always a result of the addition of the actual patterns of change in the individual cell and the dispersion contributed by the variability of age-related events in the cell cycle. In most instances statistical errors will lead to an overestimate of the duration of an event.

IV. The Unit Culture

We have invented a general scheme for the analysis of cycle events. The foundations of it can best be understood by referring to Fig. 2. It consists of a three dimensional plot of cell frequency (nominal Z axis), amount per cell (Y

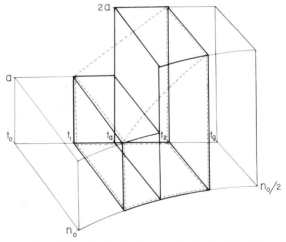

Fig. 2. A volume integral to illustrate the method of calculating the averages and the timing of synthesis over the cell cycle.

axis) and cell age (X axis). We define the volume of this figure as a *unit culture* which is the sum of all age-related products. Each age interval is the product of the relative cell number at a specific age and the property or activity per cell at that age, and the sum of these is the volume. If the figure's volume integral is divided by the cell age-cell frequency area, the average per cell is obtained for a logarithmically growing culture. We designate this average by A, which is the common experimentally obtained culture average.

Fig. 2 is a generalized volume. Limiting cases for linear synthesis may be

34

represented by (i) a plane of abrupt synthesis at a right angle to the age axis at any point designated at t_a and (ii) an inclined plane of linear doubling commonly referred to as S, which in the limit goes from a at t_o to 2a at t_g (not shown). The inclined plane shown is for a general case. The continuous synthesis of a constituent over the whole cycle is a simple case and was presented by Cook and James in 1962 (7). It is generally recognized that DNA synthesis in many procaryote cells approximates this simple pattern. The general case shown by the dotted line is for a synthetic period that starts at t_1 which is greater than t_o and ends at t_2 which is less than t_g. To a first approximation the DNA change in eucaryote cells follows this pattern: first, a plateau (G_1), second, an idealized linear doubling (S period) and, third, a plateau (G_2) and the mitotic period (M). This model applies not only to DNA but to any constituent that doubles.

A simple estimate of the timing of the S period in this eucaryote pattern requires more data than the procaryote case and is more difficult to make. A solution has been presented by Collins and Bostock in an appendix to a report by Bostock (9). They used the average amount of DNA per cell in a logarithmic culture, A, and the average amount per daughter cell, a, analogous to the continuous case (7). They extended their model to cells with a G_1, S, G_2, and M pattern. The solution of their equations (9) requires a series expansion for each case which becomes tedious. Of course, to specify either t_1 or t_2, one or the other must be known. We have avoided the need for a series expansion by doing an inverse systematic exploration of the relationship between three variables, t_1, t_2, and t_a, the latter being a function which can be uniquely specified and which shows a definable relationship to the center of the S period.

To initiate our study, one important geometric condition was recognized, i.e., the volume integral (Fig. 2) has specific boundary conditions which change for various S periods and the logarithmic age distribution sets these limits. To cover these conditions, we first developed the relationship and a graph relating a series of A/a values to a single function, the time of abrupt doubling, t_a. It covers the range of abrupt synthetic patterns that are possible for any given ratio, A/a.

Explicitly, this procedure starts with an estimate of the time of abrupt synthesis, t_a, that uniquely fits a given ratio. These are obtained as follows. The volume integral of the unit culture is represented as V and in this case can be considered as the sum of two integrals shown in Fig. 2 as covering the range between t_o to t_a and t_a to t_g. This sum is presented in Equation III. To obtain an average, this volume is divided by the area, N, of the age distribution given as Equation IV.

$$V = n_o a \int_{t_o}^{t_a} 2^{-t} dt \; + \; n_o 2a \int_{t_a}^{t_g} 2^{-t} dt \qquad\qquad III$$

$$N = n_o \int_{t_o}^{t_g} 2^{-t} dt \qquad\qquad IV$$

V is the volume of the unit culture, n_0 is the number representing the frequency of cells at zero age, a is the amount per daughter cell, t_0 is zero age, t_g is the average generation time, t_a is the time of abrupt doubling of the constituent and N is the area of the age distribution. Obviously, V/N is the average amount or activity per cell, i.e., A, as determined on cells from a logarithmic culture. The a is obtained by analysis of daughter cells from a synchronized culture or small cells in an age-related cell size isolation (3). Therefore, if A and a have been determined experimentally, the quotient V/N of the integrals will equal A and the equations can be solved in terms of t_a and a. Since a given ratio uniquely determines the time at which an abrupt doubling, t_a, could occur, it can be presented graphically or calculated. Solution of the combined equations given above allows t_a to be expressed in terms of the ratio A/a, namely,

$$2^{t_a} = 2a/A \text{ or } t_a = \ln(2a/A)/\ln 2 \qquad\qquad V$$

The values of A/a and t_a are plotted in Fig. 3.

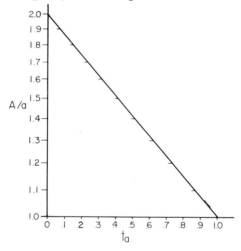

Fig. 3. A plot of the ratio of a logarithmic culture per cell average, A, to a daughter cell average, a, against the time of an instantaneous doubling, t_a.

V. Synthetic Intervals in the Unit Culture

Determining t_a is an intermediate step for (i) the estimation of the beginning or end of a linear S period (t_1 to t_2) and (ii) a determination of the middle of the S period. The second procedure will be given in the section "The Center of the Synthetic Period."

The object of introducing t_a is its unique relationship to A/a. Furthermore, by inspecting Fig. 2, it can be seen that a limited geometrical relationship is forced onto any set of values of t_1 and t_2 for any given value of t_a. It is therefore possible to solve for a series of paired values of t_1 and t_2 if t_a is known. The time, t_a, represents an instantaneous doubling, but as cases that are not instantaneous are considered, the S plane will cross t_a as it rotates and shifts. (See Fig. 2.) The level at which the synthetic curve crosses t_a and shifts will vary with its slope. The t_a will not be midway between t_1 and t_2 on the age axis because of the logarithmic character of the age distribution. By reviewing the effects of changing the angle of the synthetic plane and changing the values of t_1 and t_2, a parametric condition can be found which will preserve the volume of the unit culture and continue to yield the same A/a ratio and therefore apply to the same t_a. This condition will be satisfied by one in which the position and the slope of the synthetic plane yield a prismatic subvolume which can be equated to a pair of subvolumes on either side of the time, t_a. In Fig. 2, the subvolumes to be equated are represented by the volume enclosed by the dotted lines and the volumes enclosed by the dark lines respectively. It should be noted that the portion of the volumes below a on the amount axis must be included in order to maintain the equalities of the total volume. Therefore, a family of synthetic lines between t_1 and t_2 can be found for a specific t_a which will satisfy the condition that the total volume of any specified volume unique to a given t_a will not be altered as the synthetic pattern goes from an abrupt condition to one which occurs over an extended period of time. Fig. 2 is one such case. An equation relating t_1, t_2, and t_a can be set up and solved by equating the integrals of the two subvolumes, i.e., the sum of the two rectangular volumes (the dark lines) is set equal to the triangular prism (the dotted lines). This is the condition in which there will be no change in the figure volume. These integrals can be more clearly appreciated from a consideration of Equation VI.

$$n_o a [\int_{t_1}^{t_2} 2^{-t} dt + 1/(t_2 - t_1) \int_{t_1}^{t_2} t 2^{-t} dt - t_1 (t_2 - t_1) \int_{t_1}^{t_2} 2^{-t} dt] =$$

$$n_o a [\int_{t_1}^{t_a} 2^{-t} dt + 2 \int_{t_a}^{t_2} 2^{-t} dt]. \qquad\qquad VI$$

The left-hand member of this equation has some similarity to that used by Collins and Bostock (9) except that the base 2 is used, different limits are assigned, and it is solved for t_a rather than A/a. When solved for the relationship between t_2, t_1, and t_a, it yields

$$2^{-t_a} = [2^{-t_1} - 2^{-t_2}] / [\ln 2 (t_2 - t_1)]. \qquad\qquad VII$$

Since t_a can be known from Equation V if either t_1 or t_2 are known or restricted, the corresponding time can be estimated. t_1 or t_2 are restricted in instances where either t_1 equals t_0 or t_2 equals t_g or when other events in the cell cycle limit their time of occurrence. In these cases the slope of the synthetic plane becomes uniquely determined. The nomogram (Fig. 4) has been constructed by programming a Hewlett-Packard 9100B calculator and

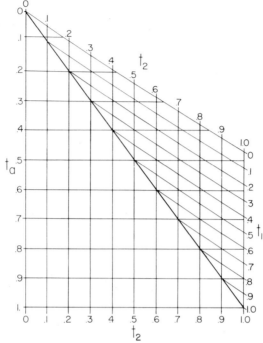

Fig. 4. A nomogram for estimating the range of the beginning t_1 and end t_2 of a synthetic doubling period for given values of the time of abrupt doubling t_a.

9125A plotter to solve and plot Equation VII. Solutions were obtained by fixing values of t_1 and varying t_2 such that t_a could be iterated and plotted. Interpolation can be used to obtain intermediate values or more exact values can be determined with the equation.

VI. The Center of the Synthetic Period

The previous work (7) on this problem yielded averages for linear synthesis which continued over the entire cell cycle and answered the questions of what age cell was best represented by the logarithmic culture cell average. A was found to be equal to 1.44 times the amount per daughter cell, i.e., A/a was 1.44. This value used in Equation V or Fig. 3 gives a t_a equal to 0.47 which is the time at which an abrupt synthesis would give the above ratio.

If a value of A/a is found equal to 1.44, it can be recognized that in one case t_1 and t_2 may equal zero and one respectively, and the midpoint of synthesis would then be 0.50, distinguishing it from the value obtained for abrupt synthesis, i.e., 0.47. This difference, although small, is real and is a result of the weighting that accompanies the disproportionate number of young cells found in the age distribution. It is obvious to ask how a t_a value is related to the center of the synthetic period. An empirical function can be found which will correct for this weighting effect such that limits can be placed on the value for the center of the synthetic period, t_r if t_a is known. The limits are calculated by considering a series of synthesis curves beginning at t_o and ending each at one of a series of times in the cell cycle out to t_g. In effect this consists of calculating the midpoint of synthesis, i.e., $(t_2 - t_1)/2$, on a single cell synthesis pattern and calling it t_r. These same values of t_1 and t_2 will also yield a t_a from Equation VII or the nomogram and the two can be compared. Plotting values of t_a against the corresponding value of t_r gives distinct limiting equations. These were obtained by the use of the Hewlett-Packard calculator and plotter. The two equations are developed as follows. The first is for a series of values of t_2 when t_1 is set equal to t_o, and the second is for a set of values of t_1 when t_2 is set equal to t_g. By this means the t_r relation to t_a is approached from two directions. For t_a between 0 and 0.47 they are represented by the equation

$$t_r = .00035 + .99434\, t_a + .13606\, t_a^2$$

and for t_a between 0.47 to 1 represented by

$$t_r = .10064 + .80221\, t_a + .09659\, t_a^2$$

These coefficients illustrate the degree to which the limiting curves are gentle parabolas and they are seen plotted as the solid lines of Fig. 5 with a break at t_a equal to 0.47. Given any one t_a value, t_r may have a value anywhere between the solid and the dashed line depending on the difference between the

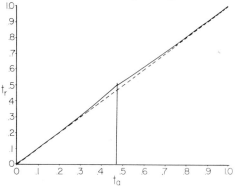

Fig. 5. A curve for determining a corrected center of a linear synthetic period. t_r is the corrected value and t_a is the time of abrupt synthesis. Points between the solid and dashed curve along a line parallel to the t_r axis will represent the the range of corrections that are necessary for various combinations of t_1 and t_2 values.

t_1 and t_2 values for which it applies. The difference between t_a and t_r for a set of t_1 and t_2 values will vary between the maximum difference where the synthetic period is longest to no difference where $t_1 = t_2 = t_a = t_r$. The greatest range is found where t_a is 0.47. The separation between the solid and dashed lines crossed by the vertical line at 0.47 is approximately 0.03 in Fg. 5. This is the case where t_1 is zero and t_2 is one.

VII. Use of the Nomogram

The curves and nomogram can be used to establish either the beginning or end of a synthetic period if one or the other is known, or its midpoint, or its beginning and end if its duration is known, by the following procedure.

First, a value for A/a must be obtained using appropriate experimental data.

Second, a value for t_a is determined either by the use of the relationship in Equation V or from Fig. 3.

Third, the t_a value is located on the nomogram and any set of conjugate values for t_1 and t_2 are obtained as projections. The t_2 is obtained from the point of intersection of the t_a line and the t_1 line. In this instance, the t_1 must be established from separate experimental evidence. Since t_1 is either equal to or less than t_2, the data points are restricted to the triple line section in the upper right-hand portion of Fig. 4. In a case where the ratio is 1.50, a value of t_a equal to .415 is found from Equation V, and if t_1 is estimated to be 0.1 by independent experiments, the two values will intersect to give a t_2 of 0.75; while if t_1 is zero the same ratio will yield a t_2 of 0.86.

Fourth, to obtain a corrected center of the synthetic period, Fig. 5 is used to locate the range of projections of the t_a value between the curved and the dashed lines yielding a t_r value on the ordinate. In the case of t_a equal to 0.47, the value of t_r will be found to be 0.50 on the solid line if synthesis extends over the entire cell cycle. However, if the period of synthesis is less than the entire cycle with t_a equal to 0.47, the value of t_r will fall somewhere between the solid curve and the straight dashed line along the 0.47 vertical line. In these cases in which t_1 and t_2 are not at the extremes, they can be calculated by using midpoints on a single cell synthesis curve. By using the single cell estimates of t_r, i.e., the average $(t_2 - t_1)/2$, to generate an appropriate set of t_r values and by establishing corresponding t_a values using the nomogram, a good fit can be obtained. It should again be noted that t_a is never different from t_r by more than 0.03 parts at the 0.50 position. Although the statistical limitation mentioned in the first portion of this report will tend to increase the apparent differences between t_1 and t_2, it will not have a major effect on estimating the center of the period of synthesis. Thus the center of the synthetic period can be determined with acceptable accuracy by the estimation of t_a.

Finally, if one knows the duration of the synthetic period, i.e., $(t_2 - t_1)$ and the ratio A/a, and hence the t_a, an estimate of both t_1 and t_2 can be obtained by selecting the point on the t_a line such that the t_1 and t_2 values corresponding to it will satisfy the condition that $t_2 - t_1$ equals the duration of the synthesis. The estimate will be as reliable as the values used to determine the A/a ratio and the duration.

40

VIII. Examples

A typical example can be taken from the data on *S. pombe* mentioned earlier and from the literature (2). Growth curves of the volume of this organism approximate a linear increase for the first three quarters of the cell cycle. From the time-lapse photographs mentioned, we estimate the volume of a daughter cell to be $60\mu^3$. The average cell volume in a logarithmitcally growing culture obtained with a Coulter counter is 94 μ^3. These data yield an A/a ratio of 1.57 and a t_a of 0.35. On the basis that the increase in volume starts at zero age, $t_1 = 0$, t_2 is found to be equal 0.75 which is in good agreement with other finds (2), including ours.

In the case of the DNA S period of *S. pombe*, the value of A/a is 1.17 for a ratio of 0.0353/0.0302 (9) and 0.035/0.030 (3). These will yield a t^a of 0.77 which in turn is a t_r of 0.78. If it is assumed that the S period is 0.05 of the cycle, t_1 becomes 0.75 and t_2 is 0.80. If the duration is twice this value, i.e., 0.10, t_1 will be 0.73 and t_2 0.83. If, on the other hand, t_2 is equal to 1.00, t_1 would have to start at 0.58, all of which can be estimated from the nomogram or by application of the equations. Obviously, the graphs and the nomogram do not allow estimates to three significant figures.

These methods are applicable to all cell constituents where a doubling and only a doubling occurs in the cell cycle. In other words, they apply to cultures which are in "balanced growth" defined in a very special way. As an ideal, we define a "balanced system" as one in which the constituents per cell double each generation time and the average generation time is invariant over an indefinite period of time. Obviously, such conditions are never realized but other qualified definitions are difficult to be agreed upon or to apply universally.

IX. Summary

It can be shown that the ideal age distribution of cells in a logarithmically growing population provides a basis for deriving the midpoint of doubling of a cellular constituent or activity from a knowledge of the logarithmic per cell average and the daughter cell average. A consideration of the boundary conditions set by the age distribution and specific synthesis curves yields a set of graphs and nomogram by which the beginning and the end of the synthetic period in a cell cycle can be estimated from a minimum of data. The method of using these graphs and the means by which they were derived is given.

Acknowledgements

This work was supported by an NSF grant GB-18500 to T. W. James.

I wish to thank Dr. Collins for his detailed correspondence on the nature of his equations.

References

1. H. Quastler and F. G. Sherman, *Exptl. Cell Res. 17*, 420 (1959).
2. J. M. Mitchison, *The Biology of the Cell Cycle*, Cambridge University Press, London, 25, (1971).
3. J. M. Mitchison and W. S. Vincent, *Nature 205*, 987 (1965); J. Sebastian, B. L. Carter and H. O. Halvorson, *J. Bact. 108*, 1045 (1971); J. R. Wells and T. W. James, *Exptl. Cell Res. 75*, 465 (1972).
4. J. E. Sisken, *Meth Cell Physiol.*, *1*, 387 (1964).
5. L. H. Hartwell, J. Culotti and B. Reid, *Proc. Nat. Acad. Sci. U.S. 66*, 352 (1970); S. H. Howell and J. A. Naliboff, *J. Cell Biol. 57*, 760 (1973).
6. T. W. James, *Ann. N.Y. Acad. Sci. 90*, 550 (1960).
7. J. R. Cook and T. W. James, *In Synchrony in Cell Division and Growth*, (E. Zeuthen, Ed.) Interscience Publishers, Inc., New York, 485 (1964).
8. E. O. Powell, *J. Gen. Microbiol. 15*, 492 (1956).
9. C. J. Bostock, *Exptl. Cell Res. 60*, 16 (1970).
10. H. E. Kubitschek, *Exptl. Cell Res. 26*, 439 (1962); H.E. Kubitschek, *Nature 209*, 1039 (1966).

REGULATION OF CELL PROLIFERATION

J. A. Smith and L. Martin

Departments of Hormone Biochemistry and Hormone Physiology
(Endocrine Group), Imperial Cancer Research Fund, Lincoln's
Inn Fields, London WC2A 3PX, England.

I. Introduction

Subdivision of the intermitotic period into four "phases", G_1, the period between mitosis and the onset of DNA synthesis; S, the period of DNA synthesis; G_2, the period between completion of DNA synthesis and mitosis; and M, mitosis itself (1) provided a strong stimulus to attempts at analysing the processes controlling cell proliferation To understand which of these phases was important for control, it was first necessary to have estimates of the duration of each phase under different conditions. Methods of analysis quickly reached a high level of sophistication and have been extensively reviewed (2,3). A tendency for different mammalian cell types to have similar S, G_2 and M durations, but widely varying G_1 durations has repeatedly been adduced as evidence that controlling events occur in G_1 (4,5,6). In the majority of mammalian cells S phase lasts about 6-10 hr, G_2 3-5 hr and mitosis about 1 hr (7). By contrast G_1 can be apparently absent, as in Chinese hamster lung cells in culture (8), rabbit pre-enamel cells (9) or Ehrlich ascites cells in the peritoneal cavity of male mice (10), while, at present, no upper limit can be put to its possible duration.

More significant perhaps, is the fact that experimentally induced changes in population growth rate are also due mainly to changes in the duration of G_1 (4). For example Tobey *et al* (11) found that the population doubling time of Chinese hamster ovary cells in culture varied widely when different batches of serum were used in the medium; the duration of $S + G_2 + M$ was not appreciably altered, the effect being due to differences in G_1. It is also clear that very slowly proliferating cells, such as normal hepatocytes, uterine epithelium of ovariectomised mice, serum starved or "contact inhibited" cells in culture are mostly in G_1, since when stimulated to rapid proliferation, the cells go through a period of DNA synthesis before dividing. Although in some tissues some cells may be "held up" in G_2 (12,13) the weight of evidence favours the conclusion that the *major* control mechanisms operate in G_1.

Two general approaches to the elucidation of these mechanisms have been used. The first is to dissect G_1 in proliferating cells that have been synchronised in some way, for example by physical selection of mitotic cells (14,15,16) in the hope that the presumed orderly sequence of events leading up to initiation of DNA synthesis may be revealed and possible control points discovered. It can be argued that this approach can only tell us what a cell has to do to get from mitosis to DNA synthesis, and does not tell us much about the controlling processes. Despite much work along these lines it remains true that the first unambiguous indication that a cell has been "committed" to division is the onset of DNA synthesis. An alternative approach has been to study changes following administration of mitogenic stimuli to cells which are not proliferating (17,18,19,20,21,22,23). Such studies seem to show that non-proliferating cells are "arrested" in G_1, whether the cells are experimentally prevented from proliferating or are in a natural state of proliferative rest. It has been thought that the stages in G_1 at which cells become arrested could be identified as important control points. However, in may so-called quiescent populations, a proportion of the cells, albeit small, is still entering S, and this creates difficulties for the idea of "arrest" in G_1. A different interpretation, based largely on *in vivo* systems, is that if, for some reason, cells cannot proceed through G_1, they go into a distinct state of proliferative rest, designated G_0 (24). Under given conditions some of the cells may avoid this fate, but once a cell has gone into G_0 it requires a stimulus to reenter the proliferative pathway. In that case the processes involved in retrieving cells from G_0 need bear no relation to the normal processes of unimpeded progress through G_1 (24). If this is true studies of induced proliferation may be irrelevant to the problem of regulating continuously proliferating populations. It is therefore surprising that little attention has been given to the experimental regulation of proliferation rates, other than the "on/off" systems. However, there is a considerable body of evidence from *in vivo* systems with variable proliferation rates, including tumours, which suggest that varying rates of entering or leaving G_0 could play an important part in regulation (5,25). It is frequently observed that mean generation times estimated directly from fraction of labelled mitosis curves (see below) are considerably shorter than those estimated indirectly from thymidine labelling indices (25,27). To reconcile these discrepant estimates it has been suggested that, within apparently homogeneous populations, only a proportion of cells are engaged in proliferation, the rest being in G_0 but capable of entering the proliferative pool after stimulation (28). It is argued that after mitosis a fraction of the cells always leaves the cycle and enters G_0. The overall proliferation rate is then very much dependent on this fraction. Burns and Tannock (29) reversed the argument and suggested that the same data could be explained by assuming that *all* cells in these populations enter G_0, but then always have a certain probability of re-entering the proliferative cycle without specific external stimulation. This view has not yet been generally accepted. However, consideration of the behaviour of androgen responsive Shionogi mammary carcinoma cells in culture (30) led us to a similar conclusion.

These cells, (S115), grow in culture without androgens in the medium, but the presence of androgens induces faster population growth. A peculiarity of the cells is that the growth curve is linear, not exponential, over a wide range of population densities. The rate of cell production in the presence of androgens is approximately double that of controls.

The linear population growth was not due to continuously increasing cell death, because the thymidine labelling index was proportional to the growth rate/cell. It might have resulted if half the cells entered G_0 after mitosis and stayed there (decycling probability of 0.5). In that case the effect of the hormone would be to halve the generation time of the cycling cells without affecting the decycling probability, otherwise the population would no longer increase arithmetically.

The cells grew as monolayers in which, even at low overall density, islands of confluent cells predominated. Within the confluent areas the labelling index was inversely proportional to the local population density, and at any given density was approximately doubled in the presence of androgen (Fig. 1). These relationships were difficult to reconcile with a decycling probability of 0.5. It

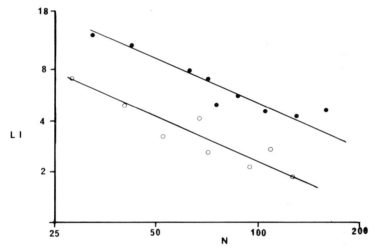

Fig. 1. The relationship between local thymidine labelling index and local population density in S115 cells without hormone (o) and after treatment for 24 hr with 10^{-8} didhydrotestosterone (●). Cultures were incubated in the presence of ^3H thymidine for 1 hr and random fields scored for the number of cells, total and labelled. In both treated and untreated cultures the labelling index was halved for each doubling of population density, indicating a linear relationship between proliferation rate and local population density. At all densities the labelling index was approximately doubled in the presence of hormone.

seems that all the cells proliferate, but the rate of cell production per cell (Kp) is inversely proportional to cell number. The density related labelling indices varied from about 15% to 1% or less, which implies an extraordinarily wide range of generation times. Since both the thymidine labelling index and Kp were inversely related to population density and both were doubled by androgen, changes in labelling index were not due to change in duration of S.

It seems reasonable to suppose that the mean duration of G_1 was the main variable.

How could the hormone change the mean duration of G_1? It could not be due to an increase in the rate at which cells pass through G_1, because the effect of doubling the G_1 rate would take one half of the unstimulated mean G_1 time to become maximally effective: yet even when the mean local generation time exceeded 250 hr, the local labelling index was doubled within 24 hr of adding androgens (Fig. 1). In trying to resolve this puzzle we were led to suppose that cells must be reversibly "blocked" in G_1, but that they always had a certain chance of becoming "unblocked." The probability of a cell overcoming the block was influenced by population density, but at all densities androgen treatment doubled that probability, thereby allowing a rapid attainment of a new maximum rate of entry into S. If one equates the "blocked" state with G_0, this is the same as Burns' and Tannock's proposal mentioned earlier. However, since G_0 is not normally regarded in this way, and because we wish to apply the hypothesis to rapidly proliferating cells, which are not usually regarded as having a G_0 phase, we prefer the non-commital term "A-state."

In the following pages, we present further evidence that the intermitotic period is composed of two fundamentally different parts. One is the orderly sequence of events necessary for division and includes S phase, G_2, M and *part* of G_1. It is convenient to refer to these phases collectively as the B-phase. Some time after mitosis cells enter the A-state, in which they do not progress towards division. A cell may remain in the A-state for *any* length of time, but always has a certain chance of re-entering the B-phase. This "transition probability" (P) does not change, provided environmental conditions remain constant. It may be supposed to be a characteristic of the cell type, but capable of modification by environmental or developmental factors. We suggest that modification of transition probability provides a major means of controlling cell proliferation.

II. Evidence for the Existence of an A-state and Transition Probability

The introduction of a probabilistic A-state into G_1 predicts that this phase should be of very variable duration between cells of the same population. This is well known to be the case. In fact, in 1962, Cattaneo, Quastler and Sherman (31) noticed that the variability of G_1 in growing hair follicles was such as to suggest that cells in G_1 have a constant probability, independent of time, of leaving this phase.

A population of cells in the A-state with constant P is analogous to a population of radioactive elements in that it will decline with a constant half-life. The frequency distribution of times spent in the A-state will therefore be exponential. If the duration of B-phase (T_B) were identical for all cells in a population the frequency distribution of intermitotic times (T_I) would also be exponential.

In practice, T_B is demonstrably not invariate; nevertheless, provided the variability of T_B is relatively small, the exponential distribution of A-state durations should be detectable. Some time after mitosis all cells enter the A-state and immediately begin to leave it at a rate determined by transition probability. Some time after leaving the A-state a cell will divide, and if the B-phase were of constant duration, the proportion of cells remaining in interphase (α) should also decline exponentially with age, beginning at time T_B after mitosis. Therefore, a semilogarithmic plot of "α" against age after mitosis

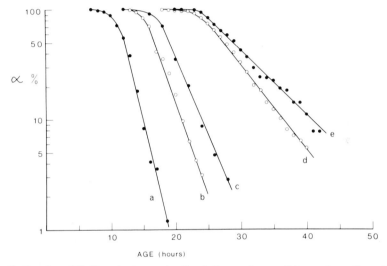

Fig. 2. Graphs of "α" against age at mitosis for various cell types in culture (32).

should show a linear decay after a lag period equal to T_B. In fact Fig. 2 shows that there is an initial downward curvature before linearity is attained. This is expected if T_B is variable. The minimum T_B is the time for the first cells to divide, and the maximum T_B the time when the curve becomes linear.

In a previous paper (32) we have shown the hypothesis to be qualitatively and quantitatively consistent with a wide range of experimental data, and that the actual distribution of intermitotic or G_1 times is close enough to the exponential for this to be used without serious error.

A more rigorous demonstration has been suggested by Minor (33). Ex hypothesi the times spent in A-state by a pair of sister cells can be expected to vary just as much as those of a pair of unrelated cells with the same transition probability. Yet intermitotic times are often found to vary less between sister cells than between unrelated cells (34,35,36,37,38), i.e., both sisters tend to have shorter or longer T_I than the average. Analysis of two sets of experimental data (original measurements on BHK21 C13 cells, and published data for Euglena gracilis (34)) both of which exhibit a correlation between the T_I of sister cells, has shown that the differences between sibling cells are precisely as expected on the basis of transition probability, with a virtual identity of T_B in

47

sister cells, but not in the population as a whole. If (but only if) both members of a sibling pair have the same T_B then differences in T_I would represent differences in the time spent in the A-state. If two cells are in the A-state with the same transition probability P, and one leaves, the other still has a probability P of leaving in the next hour. Conversely, it has a probability of 1-P of *remaining* in the A-state for at least one hour more. Therefore, if the transition probability is known the proportion (β) of sibling pairs with differences in T_I greater than any specified time t, can be calculated. A semilog plot of β against t would be a straight line originating at β = 1, t = o whose slope, determined by P, is the same as that observed in the linear "α curve" for the population as a whole. This was clearly seen in the results for *Euglena*, shown in Fig. 3, and a similar result was obtained with BHK21 C13

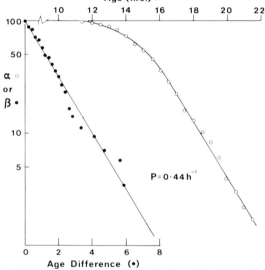

Fig. 3. Graphs of "α" against age at mitosis for cells of *Euglena gracilis* and of "β" against age difference at mitosis for sibling cells. "β" is the proportion of sibling pairs with age differences greater than that specified on the abscissa (see text for explanation). Calculated from data published by Cook and Cook (34). The slope of the "β" line is the same as that of the linear part of the "α" curve.

cells. It seems therefore that the intrapopulation variation in intermitotic times is due to a combination of an exponential distribution generated by a random process during the intermitotic period (our A − B transition) and some other distribution which is largely determined by variations in the state of cells at the time of mitosis.

III. The Role of Transition Probability in the Regulation of Cell Proliferation

We have argued that the regulation of cell proliferation in S115 cells by population density and by hormones is due to variation in the probability of overcoming the A-B transition block. In principle proliferation rate could be regulated by changes in P, T_B or some other parameter such as cell loss or differentiation. However, it seems likely that variation in P is a mjor means of regulation. This would imply that *variability* in G_1 increases as proliferation rate decreases. Consideration of fraction of labelled mitosis curves (FLM's) tends to support this. The FLM technique is widely used to estimate mean cycle times. It consists in labelling cells in S with a "pulse" of ^3H thymidine and discovering the time course with which they subsequently pass through mitosis (39). At a time equal to G_2, cells which were at the end of S when labelled enter mitosis; thereafter, for a time equal to T_S, only labelled cells enter mitosis, so the fraction of labelled mitoses (FLM) = 100%. Once cells labelled early in S have passed through mitosis, the FLM falls to zero and remains so until cells labelled late in S again enter mitosis. If the cell cycle were invariant, there would follow a peak of labelled cells identical with the first, and so on *ad infinitum*. In practice, no such thing happens. The first peak usually fits well, but the second seldom, if ever, reaches 100% and is much more widely spread. Such "damping" is always observed and may be of any degree. It is attributed to variability in cell cycle time, primarily in G_1. In some cases, especially with slowly proliferating tissue, no clear second peak occurs at the expected time (40,41,42). Nevertheless it is maintained that the technique yields values for the mean cycle time, measured as the time between maxima of first and second peaks. There are, however, many instances in which estimates of cycle times based on FLM curves give values considerably shorter than those estimated from thymidine labelling indices. These discrepancies have been interpreted to mean that only a fraction of cells proliferates, and variation in this fraction is postulated to account for differences in proliferation rate of populations with the same "cycle time." On the hypothesis proposed here, however, the discrepancy is expected because the time between peaks of an FLM is not a valid measure of mean cycle time. Although the first peak is interpreted in the usual way, the second is not. Since T_B is the minimum T_I it is only after this that labelled mitotic cells have a chance of re-entering mitosis. These cells contribute to the pool of labelled mitoses at a rate determined by P. Their contribution rises abruptly after T_B, and then declines exponentially. Successive groups behave in the same way; thus the total of labelled mitoses in the second peak is the sum of a staggered set of exponential decay curves. The second peak starts T_B after the beginning of the first, and rises for a period equal to T_S, whatever value P may take. Variation in P alone simply alters the height of the second peak *without* changing its position. Thus the distance between peaks is not equal to, nor even a function of "mean cycle time" but is close to T_B, the minimum cycle time. In general this will be lower than the actual mean, although the discrepancy will be small when P is high. It is therefore unnecessary to postulate *distinct* subpopulations of "resting" and "cycling" cells. "Proliferative pools" could simply reflect, instead, different proportions of cells in B-phase. This proportion is a function of P, being 1

49

when P = 1 and 0 when P = O.

If this interpretation is correct, FLM curves of cells with the same T_B but different P should have second peaks occurring at the same time but rising to different maxima. Such is the case, for example, in mouse seminal vesicle (43) and mouse uterus (44) under continuous androgen or oestrogen stimulation respectively. In both the thymidine labelling index first increases and then decreases, over a period of several days. FLM curves constructed in periods when the labelling index is high and when it is low have been compared. Only minor changes were seen in G_2, S and the distance between peaks. However the height of the second peak was drastically reduced when labelling index was low, indicating that the lowered LI was accompanied by a great increase in the variability of G_1, as expected.

Other evidence comes from the effects of "mitogenic" stimuli on quiescent cells. If these act by increasing transition probability from a low level, they should cause cells to enter S phase at a new rate, rather than cause the stimulated cells to enter S simultaneously. Although "synchronous entry of cells into S" is often alluded to, this is loose terminology, arising from the fact that "pulse" labelling with thymidine at various times after stimulation gives a fairly narrow peak of "cells in S" (45,46,47). However, this peak always rises to its maximum over a period of several hours. Considering the pulse labelling index alone, it may appear that different *numbers* of cells are induced to enter S by different amounts of stimulation (45,48), but continuous labelling experiments show that it is the *rate* of entry into S which is dose dependent and reveal a prolonged period after the peak during which cells continue to enter S. It is therefore somewhat misleading to speak of mitogenic stimuli as if they were "triggering" arrested cells back into cycle; they actually regulate the rate at which cells enter S. This is precisely what is expected if they act on transition probability. Ideally, if the stimulus induced an abrupt change in P, the rate at which cells enter S at any time would be proportional to the

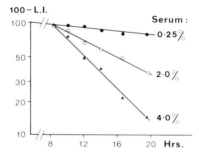

Fig. 4. Quiescent (serum starved) BHK21 C13 cells were stimulated with different concentrations of calf serum and 25 µM adenosine. ^3H Thymidine was present in the medium throughout the experiment. The proportion of unlabelled cells declined exponentially after a lag, at rates determined by the serum concentration (Brooks (49)).

number of cells remaining in the A-state. In a "quiescent" culture, nearly all the cells are initially in A-state, and the number of cells remaining unlabelled with thymidine after stimulation should decline exponentially. On this basis, theoretical curves for pulse labelling index are readily calculated, and are in qualitative agreement with observation (32). There is no reason why P should change *abruptly* after stimulation. Nevertheless, Brooks (49) has found that under certain conditions this is the case. He studied the kinetics of entry into S in BHK cells after stimulation with serum, by the technique of continuous thymidine labelling. With serum alone the rate at which the unlabelled population declined indicated a continuously increasing transition probability over the period of the experiments. In the presence of 25 μM adenosine as well as serum the unlabelled population declined *exponentially* from the time the first cells entered S, whatever concentration of serum was used, but the "half-life" was a function of serum concentration (Fig. 4). He concludes that serum alone is effective by changing transition probability slowly, and that adenosine increased the rate of change of P, which becomes extremely rapid. Whatever the final interpretation of these results may be, they present a clear case of cells entering S with a constant probability per unit time, and a demonstration that this probability can be experimentally manipulated.

IV. What is the A-B Transition?

The interpretation of transition probability is that there is *one* event, which occurs at random and "commits" a cell to division. It is therefore important to discover what this event may be, and how its randomness is generated. So far, all we have said is that the A-state, and therefore the critical $A — B$ transition, occurs in G_1. However, G_1 and the A-state are not synonymous, because there is usually some minimum period which cells spend in G_1. This is shown in cell culture by the lag between plating out mitotic cells and the first appearance of cells in S-phase. This is not an artefact due to the experimental manipulation, because a similar minimum G_1 can be detected in asynchronous cultures. Sisken and Morasca, (50) for example, made time lapse films of human amnion cells in culture and, at the end of filming, added ^3H thymidine to the medium. In this way they were able to show by autoradiography that the first cells entered S-phase 4-5 hr after mitosis. Since, by definition, the minimum duration of A-state is zero, the minimum G_1 must be in B-phase, and may be designated G_1B. The question arises whether G_1B represents a period of post-mitotic reorganisation which must be accomplished before cells enter the A-state, or a period following the A-B transition in which the cells prepare for DNA synthesis, or both. A period of post-mitotic reorganisation seems probable, as the chromosomes presumably need to unwind and nucleoli reform, before the cell can resume its normal activities. Equally, a "run-up" to initiation of DNA synthesis has seemed likely, and the controlling events have been placed conceptually some time before actual initiation. In support of this, stimulated populations of cells always exhibit a lag period before the onset of

DNA synthesis. On the present hypothesis this lag could equally well be interpreted as the time taken for the stimulus to effect a change in P. We have been unable to find any evidence demanding the existence of a "pre-replicative" G_1 following the A-B transition. On the other hand there is some positive evidence suggesting that this transition is equivalent to initiation of S-phase.

There are some cells for which no part of G_1 is essential for entry into S-phase. These are cells with no detectable minimum G_1 referred to before. In addition Burk has shown that cells stimulated to divide show the usual lag before the first round of cell division, but may enter a second S-phase with no detectable G_1 (51). Thus these cells must be capable of entering S-phase without any time-consuming preparation. Further circumstantial evidence may be seen in the behaviour of a hepatoma cell line in culture (52). In these cells treatment with dexamethasone induces tyrosine aminotransferase synthesis but has no effect on growth rate. Enzyme induction is demonstrable in contact inhibited cultures, i.e. when nearly all the cells are in A-state, but, in synchronous cultures, there is a period of 5-6 hr after mitosis during which induction is not observed. This suggests that there is indeed a period of reorganisation of considerable duration following mitosis. The data indicate that the recovery of inducibility coincides with recovery of the ability to enter S-phase. If recovery of inducibility could be taken to signal the attainment of A-state, any pre-replicative phase would have to be very short.

More direct evidence comes from experiments employing metabolic inhibitors of DNA synthesis like hydroxyurea, FUDR, and high levels of thymidine. These materials have been used to produce synchronous cultures of cells, on the assumption that if cells cannot enter S-phase they should accumulate at the G_1/S boundary. Using a double block to prevent complications arising from blocking cells when they are already in S, the technique does result in synchrony. At first sight this suggests that cells can indeed be accumulated at the G_1/S interface. However, after such a double block, there is still a considerable period during which cells enter S. Data for Chinese hamster ovary cells synchronised by various regimens of this type (53) show that, after release, the number of cells remaining in G_1 in fact declines more or less exponentially, at a rate compatible with the idea that the cells were accumulated in the A-state. If this is true it implies that cells prevented from commencing DNA synthesis remain in the A-state.

Further evidence is provided by Pfeiffer and Tolmach (54) who synchronised Hela cells by mitotic selection. Some cultures were incubated in the presence of 2.5 mM hydroxyurea for the first 16.7 hr, during which virtually all of the untreated cells entered S-phase. If the treated cells had accumulated at the G_1/S boundary, this should have produced a dramatic improvement in synchrony. It did not. The rate at which cell number increased following removal of the hydroxyurea was not significantly different from that of untreated controls but its onset merely delayed. This again suggests that cells unable to enter S remain in the A-state.

We have preliminary evidence indicating the same thing. Quiescent (low serum) BHK cells were treated with 2% or 5% serum and 25μM adenosine, to produce abrupt changes in transition probabilty (see above). Just before cells began to enter S, 3 mM hydroxyurea was added to some cultures for 6 hr. All the cultures contained ^{3}H thymidine throughout the experiment, so the rate of entry of cells into S could be measured autoradiographically. We found no

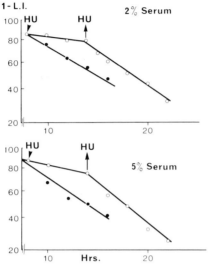

Fig. 5. Quiescent serum starved BHK cells were stimulated with 2% or 5% serum and 25 µM adenosine. ^{3}H Thymidine was present in the medium throughout the experiment. At the time when cells began to enter S-phase at an increased rate (8 hr) some of the cultures were treated with 3 mM hydroxyurea. This was washed out after 6 hr (see text for explanation).

evidence of accumulation of cells in a pre-replicative phase in the presence of hydroxyurea. Immediately after the block was removed, the cells began to leave the unlabelled compartment exponentially at rates determined by the serum concentration (Fig. 5).

The only obvious interpretation of these results is that the A-B transition coincides with initiation of the S-phase, and that neither can occur in the absence of ability to make DNA.

V. Comparison with Other Possible Interpretations

We have presented arguments and evidence in favour of the view that the initiation of S-phase is a random event. During G_1 conditions are produced which make initiation possible, but having reached that state (the A-state) actual initiation remains only probable.

In presenting the arguments we have ignored alternative interpretations, principally because there are no well-established theories purporting to cover the data as a whole. In this section we shall review the major points in the light of other possible explanations.

1. In the past it has been tacitly assumed that in continuously proliferating populations cells progress through a regular sequence of events leading from one mitosis to the next, and that variability in T_I is due to variation in the rate of progress through this sequence. In fact if one assumes a normal distribution of rates the frequency distribution of T_I plotted as "α" against time can be very similar to that observed, provided the coefficient of variation of the rates is about 15%, i.e. it exhibits an initial curvature followed by a tail which cannot be distinguished experimentally from a straight line. Our dissatisfaction with this hypothesis was aroused by the behaviour of S115 cells, which could not have had the required age distribution. There are three other pieces of evidence which cannot readily be accommodated in this scheme.

a) The distribution of differences in T_I between sibling cells is exponential throughout its range and has the same parameter (P) as the exponential part of the α curve. It is not obvious what distribution of differences would be expected if variability were due to differences in rate. The most likely consequence is a distribution of differences in rates between siblings, with a smaller variance than the population as a whole. This would not give the observed result. Other, more elaborate conditions might be found to give the "right" answer, but we have so far been unable to discover them.

b) BHK cells stimulated to proliferate by addition of serum and adenosine leave the quiescent state exponentially. The cells in the quiescent population were proliferating, albeit at a very low rate, as judged by the thymidine labelling index. They would therefore be expected to be distributed throughout the cycle, with a greater proportion in the early part than the later. Therefore, an increase in the overall rate of tranversing G_1 would necessarily give a continuously increasing rate of entry into S with time. This is clearly not the case. It is possible that the age-distribution effect could be balanced by time dependent changes in the rate of traversing G_1 so as to give exponential curves, but that requires a considerable degree of ingenuity.

c) Temporarily preventing cells from entering S-phase does not result in accumulation of cells at the G_1/S interface, as required.

2. The variability in T_I could result from cumulative errors in the timing of different steps in the cycle, giving rise to a log normal distribution. This can also be used to describe the observed distribution of T_I adequately (55). This is to imply that G_1 is composed of a number of separate, but connected, processes each of which requires some time for its completion. Control of proliferation could be due to variation in any one of these steps. However, changing the duration of any of these steps is, in effect, to change the overall rate of progress through G_1, so the difficulties encountered by the previous postulate arise again, as far as the sister cell data and DNA block are concerned. It does avoid the problem of induced entry into S, because it can be argued that the quiescent cells are predominantly blocked at one stage of G_1. It would

54

then be necessary to say that the stimulus to proliferate imparted a certain probability of overcoming that block, in order to account for exponential release.

3. The gamma distribution, resulting from the existence of a number of transitions each with a different transition probability, is sometimes used to describe the frequency distribution of T_I (56,57). This is obviously close to the hypothesis we advocate, except that we claim there is only one important transition; if there are others they must have very high transition probabilities. This follows from the sister cell data, for the distribution of differences between siblings could not be exponential if the differences were the result of more than one transition probability, it would have to be a gamma. Similarly, if there were more than one transition probability direct inhibition of entry into S should result in subsequently increased rate of entry into S and mitosis.

4. G_O with "decycling probability" is a special theory devised primarily to account for the behaviour of slowly proliferating tissues. It says that after mitosis a cell makes a choice between undergoing a further division cycle or entering a resting phase, G_O (25,28). In order to maintain a steady proliferative pool it is necessary to postulate a "recycling" probability or an elaborate feedback system acting on the decycling probability. We argue that the data on which this interpretation is based can equally well be explained by transition probability and insofar as transition probability can be established for rapidly growing systems this explanation should be preferred. It remains to be seen whether all such data can in fact be accounted for on the present theory. That being so it may be unwise at this time to equate our "A-state" with the "G_O" of others.

Although we claim that changes in transition probability are of major importance in growth regulation, we do not deny that other mechanisms exist. For example the duration of some component of B-phase could undergo changes more important than changes in transition probability. This seems to be the case in mouse epidermis (58). In this tissue FLM curves have shown two well defined peaks of labelled mitoses separated by about 100 hr. Induced proliferation (by plucking the hair) caused the time between peaks to be reduced to about 50 hr. This must have been due to a change in the duration of the post-mitotic G_1B, and not to a change in P.

VI. A Possible Basis for Transition Probability

Our argument so far leads to the conclusion that initiation of S-phase depends on the random occurrence of some critical state. We have suggested that this must be acted upon as soon as it arises, otherwise the chance is lost i.e. the critical state is a transient one. However, if the opportunity is missed (e.g. through a DNA synthesis inhibitor) it must be able to arise again, otherwise cells would be unable to divide after a DNA block. One can imagine that, if DNA synthesis were blocked, this critical state would occur with a random frequency, so that the probability of its occurring in any time interval would remain constant. This probability would be the transition probability, and it can be regarded as the result of random fluctuations in a "state of readiness." One can imagine a random fluctuation in the amount of some "S-phase initiator" which has to exceed a certain threshold in order to activate the process, or else a fluctuation in the "receptivity" of the chromatin for such an initiator. In either case transition probability could be a function of mean concentration of initiator. One can ask whether the random element is generated in the cytoplasm or nucleus. The most direct evidence bearing on this problem comes from studies on multinucleate cells. In a recent review of nuclear synchrony in DNA synthesis Johnson and Rao (59) concluded that in general the nuclei of multinucleate cells, whether naturally occurring or experimentally induced, are synchronous with respect to DNA synthesis. This could only occur if initiation of S-phase were due to fluctuations of something in the cytoplasm, for if it were due to fluctuating states in the nucleus they would act independently of each other.

We are left, therefore, with one old problem and two new ones: what conditions are necessary for initiation of S-phase, how is its randomness generated, and how is its frequency of occurrence controlled?

First of all, it is necessary to distinguish between initiation of DNA synthesis and the initiation of S-phase, for S-phase is not simply DNA replication. During S-phase a specific pattern of protein synthesis occurs involving histones and many of the enzymes actually involved in DNA synthesis. On the other hand it seems that all of the enzymes and co-factors necessary for DNA synthesis are present in adequate amounts at all stages of the cell cycle, so it is unlikely that these are involved in controlling the onset of S (61).

The replication of chromosomal DNA follows a heritable sequential pattern through S, involving the initiation of DNA synthesis in many replication units at specified times (60). Furthermore, initiation of chromosomal DNA synthesis can occur without involving replication of the whole genome, for example in the amplification of the DNA content of certain loci in larval insects (62,63), and the rRNA cistrons during oogenesis (64).

In view of the orderly nature of the processes involved in chromosome replication, it seems probable that some master initiator is necessary to set them going. Such a substance might be expected to act by combining with DNA. Several investigators have tried to find DNA-binding proteins in the cytoplasm which could fulfill this function (65,66,67,68). One of these, called P8, seems a good candidate (68): it has been found in all of the many types of mammalian cell so far tested. It is synthesised more slowly in physiologically quiescent cells, which could indicate that its concentration is related to transition probability. It increases in amount when entry into S is blocked by high concentrations of thymidine, which could be because this results in an accumulation of cells in the A-state. It is apparently not synthesised in G_1B (66), nor is its synthesis detectable during S-phase. As yet however there is no direct evidence that it initiates S-phase.

Supposing some such protein is involved, how could it act at random? It would either have to be synthesised sporadically, or its concentration be subject to temporal fluctuation. If it were regulated so that its mean level was below a certain critical threshold, then departures from the mean, due to imperfect regulation, or merely to biochemical "noise," could occasionally result in that threshold being exceeded. The frequency with which this happened would depend upon the mean level: if that were controlled close to the threshold the chance of its exceeding threshold would be high; if far, low. A rather sharp threshold is easily imagined in terms of co-operative binding i.e. the binding to a sufficient number of sites causes a sharp increase in the binding affinity of other sites [cf. "gene 32 protein" (69)].

It is possible that the mean level of "S-phase initiator," whether a protein or not, is related to the overall protein synthetic capabilities of the cell. Thus, conditions permitting a rapid rate of growth would also permit a rapid rate of proliferation. Such a correlation between protein synthesis and proliferation can be demonstrated in S115 cells, in which both can be limited in a rather unspecific way by environmental conditions (cell crowding), but can be specifically increased by hormones.

There is a wide variety of slowing proliferating cell populations which proliferate rapidly when exposed to some specific stimulus. These have been extensively reviewed (24) from the point of view that stimuli set in train some sequence of events leading to initiation of DNA synthesis. Despite the heterogeneity of the systems used, there are some striking similarities among the changes observed in response to proliferative stimuli. The first is the invariable occurrence of a delay between application of the stimulus and the increased rate of entry of cells into S-phase. This lag varies in duration from about 5 hr in the oestrogen stimulated uterine epithelium (70) to a day or more in many other systems. Equally variable is the rate of entry into S-phase following stimulation. Clearly none of the stimuli directly induces initiation of DNA synthesis. As we have argued above, it seems likely that proliferation rate is increased by raising transition probability. But if we are right in saying B-phase begins with initiation of S-phase, the lag period must represent the time taken to produce a detectable change in transition probability. In cases where there is no background proliferation it is possible that the cells have first to be rendered capable of DNA synthesis before transition probability can be expressed.

The simple and direct way to raise transition probability would be to stimulate, specifically, the synthesis of a hypothetical S-phase initiator. The idea that induced DNA synthesis in general involves specific "gene activation" in some way, has progressively gained favour, but it must be admitted that there is no conclusive evidence, only suggestive indications that it may be so in some cases (71-75).

In reality the responses are more complex and in nearly all cases involve increased synthesis of all types of RNA and most of not all species of protein. So far none of the changes following stimulation can be unequivocally identified as being of paramount importance in initiation of DNA synthesis. Since, in our view, the increased rate of proliferation may be due to a change in concentration of a single unidentified substance this is hardly surprising. It may be that the S-phase initiator simply increases as a consequence of the overall changes in protein synthetic capacity. The lag period would then represent the time taken to increase rate of synthesis of the initiator and for this to be reflected in detectable changes in its concentration. It is possible, therefore, to regard "induced DNA synthesis" as a consequence of increased protein synthesis.

References

1. A. Howard & S. R. Pelc, *Expt. Cell Res. 2*, 178, (1951).
2. J. D. Thrasher, *Methods Cell Physiol. 2*, 323, (1966).
3. P. K. Lala, *Methods Cancer Res. 6*, 3, (1971).
4. D. M. Prescott, *Cancer Res. 28*, 1815, (1968).
5. O. I. Epifanova, & V. V. Tershikh, *Cell Tissue Kinet. 2*, 75, (1969).
6. R. Baserga, *Cancer Res. 25*, 581, (1965).
7. J. M. Mitchison, *The Biology of the Cell Cycle.* Cambridge University Press, Cambridge, (1971).
8. E. Robbins & M. D. Scharff, *J. Cell Biol. 34*, 684, (1967).
9. W. E. Starkey, *J. Brit. Dental Ass. 115*, 143 (1963).
10. R. Baserga, *Arch. Pathol. 75*, 156, (1963).
11. R. A. Tobey, E. C. Anderson & D. F. Peterson, *J. Cell Biol. 35*, 53, (1967).
12. S. Gelfant, *Methods Cell Physiol. 2*, 359, (1966).
13. J. Van't Hof, in this volume.
14. T. Terasima & L. J. Tolmach, *Expt. Cell Res. 30*, 344, (1963).
15. G. S. Stein & T. W. Borun, *J. Cell Biol. 52*, 292, (1972).
16. M.D. Enger & R. A. Robey, *J. Cell Biol. 42*, 308, (1969).
17. H. L. Cooper & A. D. Rubin, *Blood, 25*, 1014, (1965).
18. G. J. Todaro, G. K. Lazar & H. Green, *J. Cell. Comp. Physiol. 66*, 325, (1965).
19. G. Sauer & V. Defendi, *Proc. Nat. Acad. Sci. 56*, 452, (1966).
20. J. W. Grishman, *Cancer Res. 22*, 842, (1962).
21. T. Barka, *Expt. Cell Res. 39*, 355, (1965).
22. C. A. Perrotta, *Am. J. Anat. 111*, 195, (1962).
23. G. Hodgson, *Proc. Soc. Expt. Biol.* (N.Y.), *124*, 1045, (1967).
24. R. Baserga, *Cell Tissue Kinet. 1*, 167, (1968).
25. F. Bresciani, *Europ. J. Cancer, 4*, 343, (1968).
26. M. Mendelsohn, *J. Nat. Cancer Inst. 28*, 1015, (1962).
27. I. F. Tannock, *Brit. J. Cancer, 22*, 258, (1968).
28. L. G. Lajtha, R. Oliver & C. W. Gurney, *Brit. J. Haematol. 8*, 442, (1962).
29. F. J. Burns & I. F. Tannock, *Cell Tissue Kinet. 3*, 321, (1970).
30. J. A. Smith & R. J. B. King, *Expt. Cell Res. 73*, 351, (1972).
31. S. M. Cattaneo, H. Quastler, & F. G. Sherman, *Nature, 190*, 923, (1961).

32. J. A. Smith & L. Martin, *Proc. Nat. Acad. Sci. 70*, 1263, (1973).
33. P. D. Minor & J. A. Smith, *Nature*, (In Press).
34. J. R. Cook & B. Cook, *Expt. Cell Res. 28*, 524, (1962).
35. J. E. Sisken & R. Kinosita, *Expt. Cell Res. 22*, 521, (1961).
36. W. T. McQuilkin & W. R. Earle, *J. Nat. Cancer Inst. 28*, 763, (1962).
37. K. B. Dawson, H. Madoc-Jones & E. O. Field, *Expt. Cell Res. 38*, 75, (1965).
38. D. Killander & A. Zetterberg, *Expt. Cell Res. 38*, 272, (1965).
39. H. Quastler & F. G. Sherman, *Expt. Cell Res. 17*, 420, (1959).
40. S. J. Gibbs, & G. W. Casarett, *Radiat. Res. 40*, 588, (1969).
41. O. I. Epifanova, *Expt. Cell Res. 42*, 562, (1966).
42. G. G. Steel, *Cell Tissue Kinet. 5*, 87, (1972).
43. A. R. Morley, N. A. Wright & D. Appleton, *Cell Tissue Kinet. 6*, 239, (1973).
44. A. E. Lee, L. A. Rogers & G. Trinder, *J. Endocr.* (In Press).
45. H. M. Temin, *J. Cell Physiol. 78*, 161, (1970).
46. R. Baserga & S. Heffler, *Expt. Cell Res. 46*, 571, (1967).
47. L. Martin, C. A. Finn & G. Trinder, *J. Endocr. 56*, 303, (1973).
48. G. D. Clarke, M. G. P. Stoker, A. Ludlow & M. Thornton, *Nature, 227*, 798, (1970).
49. R. F. Brooks, Ph.D. Thesis, University of London, (1973).
50. J. E. Sisken & L. Morasca, *J. Cell. Biol. 25*, 179, (1965).
51. R. R. Burk, *Expt. Cell Res. 63*, 309, (1970).
52. D. Martin, G. M. Tomkins & D. Granner, *Proc. Nat. Acad. Sci. 62,*, 248, (1969).
53. R. A. Tobey & H. A. Crissman, *Expt. Cell Res. 75*, 460, (1972).
54. S. E. Pfeiffer & L. J. Tolmach, *J. Cell Physiol. 71*, 77, (1968).
55. D. S. Nachtwey & I. L. Cameron, *Methods Cell Physiol. 3*, 213, (1968).
56. M. Takahashi, *J. Theoret. Biol. 13*, 202, (1966).
57. B. Jansson, *F.O.A.P. Report* C8299-M6, (1971).
58. M. A. H. Hegazy & J. F. Fowler, *Cell Tissue Kinet. 6*, 17, (1973).
59. R. T. Johnson & P. N. Rao, *Biool. Rev. 46*, 97, (1971).
60. D. M. Prescott, *Adv. Cell Biol., 1*, 57, (1970).
61. J. D. Watson, *Adv. Cell Biol. 12*, 1, (1971).
62. H. G. Keyl, *Experientia, 21*, 191, (1966).
63. C. Pavan & A. B. da Cunha, *An. Rev. Genet. 3*, 425, (1969).
64. H. Wallace & M. L. Birnstiel, *Biochim. Biophys. Acta, 114*, 296, (1966).
65. B. M. Alberts & G. Herrick, *Methods Enzymol 21*, 198, (1971).
66. T. O. Fox & A. B. Pardee, *J. Biol. Chem. 246*, 6159, (1971).
67. Y. Hotta & H. Stern, *Nature, New Biol. 234*, 83, (1971).
68. R. L. Tsai & H. Green, *J. Mol. Biol. 73*, 307, (1973).
69. B. M. Alberts & L. Frey, *Nature, 227*, 1313, (1970).
70. L. Martin, C. A. Finn & G. Trinder, *J. Endocr. 56*, 133, (1973).
71. I. Lieberman, R. Abrams & P. Ove, *J. Biol. Chem. 238*, 2141, (1963).
72. M. Fujioka, M. Koga & I. Lieberman, *J. Biol. Chem. 238*, 3401, (1963).

73. R. Baserga, R. D. Estensen & R. O. Peterson, *Proc. Nat. Acad. Sci. 54,* 745, (1965).
74. R. B. Church & B. J. McCarthey, *J. Mol. Biol. 23,* 459, (1967).
75. R. B. Church & B. J. McCarthey, *Biochim. Biophys. Acta, 199,* 103, (1970).

ANALYSIS OF INDUCED MITOTIC ACTIVITY IN MOUSE EAR EPIDERMIS UNDER *IN VITRO* CONDITIONS

Graciela C. Candelas

Department of Cell and Molecular Biology and Dermatology
Medical College of Georgia Augusta, Georgia

I. Introduction

In attempting to elucidate the mechanisms controlling a biological process, it is a vital requirement to consider the particular characteristics of the experimental system which may play a significant role and may even become of a limiting nature in the cellular process under scrutiny.

It is, therefore, necessary to meet this requirement prior to embarking on a study of the control of mitosis.

Upon exposure to a common set of experimentally imposed variables, it should neither be unreasonable to expect nor surprising to obtain results which show: differences in behavior patterns between eukaryotic and prokaryotic cells; differences in response between rapidly and slowly proliferating tissues; a display of different sensitivities from cells in tissue culture with reference to cells from whole tissues; a variance between responses originating from undifferentiated cells when compared to those arising from the differentiated counter parts; different answers from cells from young tissues in contrast to those comprising older tissues.

In summary, to achieve a high degree of significance, in studies on the control of a cellular process, they should be carried out in an experimental system where the cellular process has been fairly well characterized. In this respect, mouse epidermis qualifies for studies on the control of mitosis.

II. Experimental System

Mouse epidermis, particularly the ear, has been the subject of a considerable number of studies which have yielded a vast amount of information regarding parameters of fundamental significance to the study of the control of mitosis.

Baseline values have been obtained for a number of measurable activities and, therefore, it is an attractive model for *in vitro* studies.

The characteristics of mouse epidermis may be summarized as follows:

(1) Mitotic activity is low *in vivo* and it has been categorized as repressed in this system (1-4).

(2) The cell cycle has a rather long G_1, as much as several days. The S phase is longer than the average values for other cells 18 hours for the ear, somewhat less for body skin. The G_2 phase is relatively short, (4 hours)

and M is normal (5-7).

(3) The basal layer of epidermis is heterogeneous in the sense that it is composed of discrete cell populations with respect to their patterns of movement through the cell cycle. Two distinct populations have been detected experimentally: a cycling one and a considerably smaller non-cycling population which is blocked at the G_2 phase of the cell cycle (8-11).

(4) Mitotic activity may be induced with the proper stimuli: by hair plucking and wounding in the body skin, and by heat and wounding in the ear (12,9,10).

(5) Upon stimulation, cells from both populations may be elicited to enter into mitosis (8-10). The point to be emphasized here is that the non-cycling cells which have been blocked at G_2 retain their potential to divide under the proper conditions. This presents the system with a supply of readily available and fast responding cells for mitosis.

(6) There is a diurnal cycle of the mitotic activity detected in this tissue (6,13).

(7) Epidermal cells produce within themselves some inhibitory substance which helps to limit the mitotic rate and is epidermis-specific (2), also reported in human skin (14). This will be referred to in this chapter as the "chalone concept." The regulatory mechanism seems to function as a negative feedback system (4). This regulatory substance has been partially purified from skin extracts (15) and seems to be resolvable into two active components under the proper experimental procedures: one with a G_1 inhibiting activity and a second fraction containing G_2 inhibiting activity residing in separate cell layers (16-19). For an exhaustive and comprehensive treatment of the epidermal chalone concept, see articles by Houck, Bullough, Rothberg and Arp, Marks, Elgjo, Voorhees et al., Laurence, and Simmit and Fisher (20).

The *in vitro* mouse ear epidermis system was originally devised by Bullough and Johnson (21), and has been extensively used by both his group and by Gelfant (c.f. Gelfant references). Thus, it has been fully described and has withstood long and rigorous testing. Simple histological preparations permit light microscopic identification and scoring of colcemid metaphase-arrested mitosis. Highlighted should be the fact, that, within the proper experimental design the cells pertaining to the non-cycling G_2 blocked population can be detected from the bulk of cycling cells by using labelled precursors and appropriate autoradiographic techniques (8,11).

Within the framework of this available information, a series of experiments were designed and performed under *in vitro* conditions with the aim to analyze the mitotic response elicited in mouse ear epidermis. The general features of the experimental approach which predominate throughout the experiments consisted of removing uninjured ears from healthy male C57 BL/6J mice, cutting these into strips and incubating them under standard conditions and in a defined medium for a period of five hours as described by Gelfant (12). Thus,

62

mitotic activity was elicited in the tissue by virtue of the cut. This evoked response was submitted to analysis and dissection with respect to the synthetic metabolic requirements for its elaboration by using specific inhibitors as analytical tools. The inhibitors recruited for this analysis do not fall within the category of cell cycle specific agents. Their currently accepted mode of action has been reviewed and defined by Cameron (22).

Comparative quantitation of the mitotic response, obtained as a result of the experimental conditions, was used in interpreting the synthetic processes that may or may not be required for the elaboration of the mitoses. Within the current interpretation of the regulation of macromolecular synthesis in eukaryotic cells (23), the magnitude of the response in the presence and/or absence of the inhibitors has been used in attempting to elucidate and pinpoint the possible sites of control and the nature of the regulatory mechanism.

The experimental design took into account the heterogeneous nature of the tissue (previously described) and allowed for the resolution of the two populations. It is not unreasonable to expect that the two distinct cell populations display different behavior patterns in their response to the various insults to which the tissues were subjected. Temporal location of events within the cell cycle is feasible to some extent.

III. Experimental Work

Typical experiments have been selected from the bulk of the data to illustrate the lines of approach and the sequence of our studies. Some of this work has been presented previously (24,25,7,26,27).

The experiments were designed to take into consideration the features of the experimental system as outlined in the introduction. Thus, factual knowledge pertaining to other types of systems and which are not considered relevant to our system, has been disregarded in the experimental design and will not be reviewed in the discussion.

The initial approach for the analysis of the mitotic response elicited by cutting was to enquire into the synthetic processes which might be required, or not required, for the elaboration of the induced event: mitosis. Three levels of synthetic processes were analyzed. This was accomplished through the use of specific inhibitors which serve as tools for the dissection of each level. For DNA synthesis, hydroxyurea served the purpose. The level of transcription was analyzed by using actinomycin D and ethidium bromide in high concentrations. Three translational inhibitors of protein synthesis in eukaryotes served as analytical tools for this level: puromycin, chloramphenicol and cycloheximide.

Table I summarizes the average mitotic counts in the sections of tissue strips which were exposed to the inhibitors while in the process of entering into mitosis elicited by the cut. From the counts obtained, it may be concluded that the stimulus-responding cells entered into mitosis without having to engage in DNA synthesis, nor into the process of transcription. There was a

TABLE I

Effect of Inhibitors on Epidermal Mitosis *In Vitro*

Inhibitor	Concentration	Average Mitoses/cm*	Control**
Cycloheximide	5.0 µg/ml	3.2 ± 0.9	10.2 ± 1.2
	10.0	0.8 ± 0.05	10.2 ± 1.2
Puromycin	1	8.2 ± 1.1	9.2 ± 0.8
	5	0.4 ± 0.1	9.2 ± 0.8
Chloramphenicol	10	6.2 ± 0.8	9.2 ± 0.8
	1000	1.0 ± 0.7	9.2 ± 0.8
Actinomycin D	0.1	3.8 ± 0.5	3.4 ± 0.7
	1	12.6 ± 0.7	3.4 ± 0.7
	10	13.0 ± 0.9	3.4 ± 0.7
Ethidium Bromide	50	11.0 ± 1.2	9.5 ± 1.1
	100	13.0 ± 1.3	9.5 ± 1.1
	200	pycnotic nuclei	9.5 ± 1.1
Hydroxyurea	200	8.6 ± 1.3	7.3 ± 1.2

The following holds true for all the tables:

*Each number represents the average counts obtained from 5 cm from 5 different animals.

**Controls display the mitoses elicited by the cut, mitosis is extremely low in the uncut ear.

demand, however, at the level of translation. This is supported by the fact that although mitosis was elicited by the cut (see controls), it could not be elaborated in the presence of the translational level, protein synthesis inhibitors.

That the mitotic response can bypass the DNA synthesis was expected since the response was obtained within five hours, and therefore the responding cells had to be in the G_2 phase of the cell cycle in these tissues (8-10). One might infer from the data that transcription was not a requirement for the synthesis of the macromolecule (s) leading to mitosis, and that these were, therefore, translated from a stable or long-lived messenger.

The paradoxical effects of high concentrations of actinomycin D captured our attention and were considered worthy of pursuit. Repeated trials with various sources and concentrations of actinomycin D indicated that the enhancing effect was reproducible, but was not concentration dependent (Table II). Other features of this response were that the mitoses were not randomly distributed along the ear section. They were located predominantly

TABLE II

Effect of Actinomycin D on Mitotic Activity in Mouse Ear Epidermis *In Vitro*

	No. Mitoses/cm length of epidermis				
		Actinomycin D (μg/ml)			
Expt.	Control	1	5	10	20
1	3 + 0.9	-----	16 + 1.6	21 + 1.8	-----
2	7 + 0.7	23 + 1.2	16 + 1.8	16 + 1.7	-----
3	9 ± 1.7	-----	29 + 3.3	35 ± 3.4	30 + 3.3
4	7 ± 0.8	24 + 2.0	21 ± 1.0	19 ± 0.8	-----
5	3 + 1.0	3 ± 0.8	15 + 0.8	18 ± 1.6	-----
6	4 ± 0.6	-----	40 ± 2.6	42 ± 2.9	-----

along the mid-portion of the longitudinal section. Mitoses were rarely seen in the tip or at the base of the ear. Another observation that bears mentioning is the fact that the mitoses had a tendency to show up in foci (see Fig. 1).

TABLE III

Effect of Inhibitors on Epidermal DNA Synthesis *In Vitro*

Inhibitor + [3]HTdr	No. labeled interphase nuclei per cm epidermis
None	58 ± 1.1
Actinomycin D (1 μg/ml)	53 ± 1.9
Actinomycin D (10 μg/ml)	52 ± 1.3
Ethidium Bromide (50 μg/ml)	51 ± 1.2
Ethidium Bromide (100 μg/ml)	42 ± 1.5
Hydroxyurea (20 μg/ml)	0

Since reports on paradoxical effects of actinomycin D had not yet come to our attention, we directed our efforts towards posing questions as to how actinomycin D might be maneuvering the enhancement of mitosis. Thus, we questioned the possibility that it would be activating DNA synthesis within the five hour period of incubation. The strips were, thus, incubated in a medium containing various inhibitors concomitantly with ^{3}H-Thymidine ([3]HTdr). This experiment is summarized in Table III. Inasmuch as the autoradiographic count of labelled interphase nuclei remained virtually the same in either the presence or absence of actinomycin D, we feel that it is justifiable to discard DNA synthesis as the target site of the effects of actinomycin D. Hydroxyurea performed as it is expected by totally inhibiting DNA synthesis.

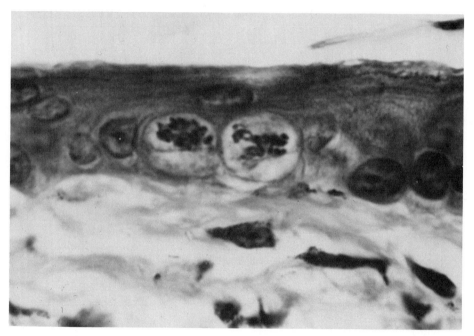

Fig. 1. Epidermal mitoses obtained in the presence of actinomycin D 10 μg/ml) in the incubating medium.

The next series of experiments led to the conclusion that actinomycin D was exerting its effect on cells which were in the G_2 compartment of the cell cycle and that it displayed, as accounted by the experimental data, a preferential effect on the non-cycling G_2 blocked cells. In order to probe into the system and thus obtain this information, the following experimental procedure was followed: mice were exposed to continuous labelling with ^{3}HTdr for 48 hours prior to the removal of the ears and the stimulation by the cut. The strips were subsequently incubated for five hours in the presence of the inhibitors and ^{3}HTdr. Table IV summarizes both the mitotic counts and the breakdown of the mitoses into labelled and unlabelled. The rationale for

TABLE IV

The Effect of Actinomycin D on the G_2 Period of the Cell Cycle

Controls				Actinomycin D 10 μg/ml			
Mitoses/cm 3.1 ± 0.9				Mitoses/cm 18 ± 1.2			
Labelled	%	Unlabelled	%	Labelled	%	Unlabelled	%
2.3	75	0.7	25	4.5	25	13.5	75

this analysis is that an unlabelled metaphase-arrested figure pertains to a cell which entered into mitosis post DNA synthesis (S), i.e. from the G_2 phase; and since the [3]HTdr had been made available for 48 hours prior to the application of the stimulus and the *in vitro* incubation, unlabelled mitoses reflect cells which were in G_2 during this period.

An alternative effect of actinomycin which would result in an increased mitotic yield was that it might be acting as a metaphase arrester. If this were the case, then its effect might be additive to those produced by the colcemid. This question was answered when the stimulated tissues were incubated in the presence of actinomycin D and in the presence and absence of colcemid (Table V). The number of mitoses collected in the absence of colcemid was equivalent to those collected in its presence. Thus the effects of actinomycin D did not result from its action as a metaphase arrester.

TABLE V

The Effect of Actinomycin D on Epidermal Induced Mitosis
in the Absence of Colcemid

Expt. Medium	Mitoses/cm (10 strips from different animals)
Control + Colcemid	4.8 ± 0.2
Control − Colcemid	1.6 ± 0.8
Actinomycin D (5 μg/ml) + Colcemid	18.7 ± 1.6
Actinomycin D − Colcemid	2.0 ± 1.1

The next point of enquiry concerned the relationship between the mitotic activity induced by the cut and that elicited by actinomycin D. Was the former mechanism a prerequisite for the latter? Both strips and whole ears were exposed to actinomycin D, during an *in vitro* incubation. The results obtained suggested that the cut was an obligatory precedent to the action of actinomycin D (Table VI), the two displaying a synergistic behavior. Epidermis is permeable to actinomycin D, not only under *in vitro* conditions (Candelas unpublished), but when applied topically *in vivo* (28).

The data discussed so far are summarized diagramatically in Fig. 1.

TABLE VI

The Effect of Actinomycin D in the Absence of Cutting and in the
Presence of DNA Synthesis Inhibitor

	Mitoses/cm length of epidermis	
	Ear Fragments	Whole Ear
CONTROLS	4.5 ± 0.4	0.1 ± 0.03
Actinomycin D 10 µg/ml	17.2 ± 1.3	0.1 ± 0.04
Hydroxyurea 200 µg/ml	5.8 ± 0.6	0.3 ± 0.02
Actinomycin D 10 µg/ml and hydroxyurea 200 µg/ml	16.5 ± 1.2	0.1 ± 0.03

IV. Discussion

A. The Effects of Actinomycin D on the Eukaryotic Cell Cycle.

The only generalization feasible under this heading is that the effects are quite variable, and, although a review of this subject does not fall within the scope of this chapter, we wish, nevertheless, to touch on it lightly to stress what we consider to be a fundamental point. The literature in this field is quite vast. However, a few selected samples will suffice to substantiate our point.

In mammalian culture cells, Kishimoto and Lieberman (29), found an actinomycin D sensitivity in late S or early G_2. Baserga et al. (30), showed that low doses of actinomycin D had no effect on the flow of cells from G_2 to M in Erhlich ascites tumor cells. Chinese hamster ovary cells in tissue culture were found to be sensitive to the drug early in G_2 (31). Kim and collaborators offered relatively low concentrations of actinomycin D (1 µg/ml) to HeLa cells in cultures at G_1, and prevented DNA synthesis. Also in HeLa cells, Stein and Matthews (32), treated them with actinomycin D during mitosis and found

MOUSE EAR EPIDERMAL MITOSIS

A Slow Renewal Tissue

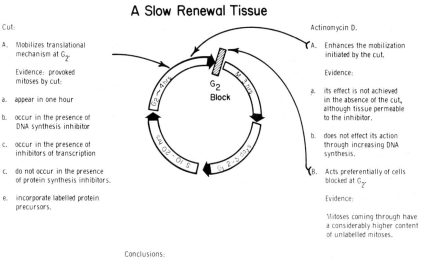

Cut:

A. Mobilizes translational mechanism at G_2.

Evidence: provoked mitoses by cut:

a. appear in one hour

b. occur in the presence of DNA synthesis inhibitor

c. occur in the presence of inhibitors of transcription

c. do not occur in the presence of protein synthesis inhibitors.

e. incorporate labelled protein precursors.

Actinomycin D.

A. Enhances the mobilization initiated by the cut.

Evidence:

a. its effect is not achieved in the absence of the cut, although tissue permeable to the inhibitor.

b. does not effect its action through increasing DNA synthesis.

B. Acts preferentially of cells blocked at G_2.

Evidence:

Mitoses coming through have a considerably higher content of unlabelled mitoses.

Conclusions:

1. The cut mobilizes the translation of a stable messenger (s) into protein (s) leading to mitosis.

2. Actinomycin D enhances the effect at a control level beyond transcription. Opens up G_2 block.

3. The syneristic effects of actinomycin D are not due to metaphase arresting.

Fig. 2. Summary of experimental data.

that the cells entered G_1 and synthesized non-histone chromosomal proteins during that period in the presence of the inhibitor. Cultured mouse leukemia cells displayed a greater sensitivity to actinomycin D during G_1 (33), and Mittermayer and co-workers (34) found that actinomycin D blocked the passage from G_1 to DNA synthesis in cultured mouse L-cell fibroblasts.

In liver regeneration Fujioka et al. (35), found that actinomycin D affected RNA synthesis after partial hepactectomy, but not prior to it, thus indicating a variation of sensitivity concommitant with proliferation. Schwartz et al. (36) demonstrated that actinomycin D was toxic to proliferative hepatocytes only. Melvin (37) also in liver regeneration experiments, reported an increase in mitotic counts evoked by actinomycin D attributable to a prophase lag rather than to an increase in mitotic rate. Wheeler and Simpson-Herren (38), based on several studies, concluded that the cytotoxicity of cells to drugs is different for proliferating and non proliferating cells. This was also observed in the intestine where Schwartz, Steinberg and Phillips (39) reported that actinomycin D destroyed the regenerating regions of the intestinal crypts without damaging non-dividing cells.

Wessels (40) has shown that as cells differentiate they lose their sensitivity to actinomycin D and Gross (41) demonstrated that early cleavage cells in sea urchin embryo were actinomycin D insensitive, but become sensitive later in embryogenesis. In whole organism, Samuels (42) in his review, states that the toxicity of actinomycin D first manifests itself in tissues with a rapid turnover, i.e., gut epithelium, bone marrow, and lymphoid tissues. Revel and co-workers (43) detected an actinomycin D sensitivity in liver cell homogenates that was not present in whole cells.

The point we wish to stress makes itself evident with these few cases. It would be redundant to summarize and discuss it any further.

For further references and reviews on the subject see (42,44,45), and for effects on hepatic cells and their regeneration see (46).

B. The Effects of Actinomycin D on Epidermis

Possibly the earliest studies with this tissue were those conducted by Flamm and co-workers (47). Topical application of actinomycin D to mouse epidermis resulted in the inhibition of ribosomal RNA synthesis and no detectable inhibition of mRNA synthesis. Low dosages of actinomycin applied topically did not significantly alter the incorporation of ^{3}H-leucine into protein, and high concentrations of the drug resulted in an increase in the rate of leucine incorporation. The paper presents a limited study of the effects of the topical applications of actinomycin D on epidermal mitoses. Examination of the data reveals that the application of high dosages augmented mitotic activity in epidermis when the tissues were analyzed for mitoses eight days after the topical treatment.

Elgjo and his group (18) treated mouse skin with $100\,\mu g$ of actinomycin D. They subsequently sacrificed the animals and made water extracts of the treated area. Whereas the water extracts obtained from the skin of the un-treated animals inhibited epidermal mitoses in a test system, the actinomycin D treated skin extracts had lost their mitotic inhibiting activity, and specifically, the G_2 inhibiting capacity. This is presented and interpreted within the framework of the current chalone concept (16-18).

Those two sets of investigations are in agreement with our findings and support our data on the enhancement of mitosis by actinomycin D. The work of Elgjo and his group (16-18) supports the fact that actinomycin D exerts its effects in the G_2 phase of the cell cycle.

C. The Effects of Actinomycin D on Other Eukaryotic Cellular Processes

Here again, we are confronted with a considerable bulk of publications, from which, we have selected those we consider to be pertinent to our work and our results.

70

D. Superinduction Phenomenon

The phenomenon of superinduction is often observed in eukaryotic cells. It is characterized by the increase of an inducible biological process (synthesis of a protein) as a response to the inhibition of RNA synthesis, commonly brought about by actinomycin D.

Tomkins and collaborators (48) have reviewed this subject and they have compiled a rather impressive list of biological processes in a diverse number of systems which display superinduction. A ubiquitous feature of this phenomenon is that the response to actinomycin D or other superinducers is restricted to one or a limited number of proteins within a cell type.

The most extensive studies in superinduction have been conducted by Tomkins and his group (48-51) primarily with the enzyme tyrosine amino transferase (TAT). Based on their data, they have drawn a hypothetical scheme in which induced TAT is synthesized from a structural mRNA of some stability whose expression is regulated by a postulated labile repressor. Actinomycin D supposedly achieves superinduction subsequent to the hormonal induction of TAT by inhibiting the synthesis of the hypothetical repressor's mRNA, thus allowing the structural gene's expression (50,51).

Since the regulatory role in this scheme was assigned to a hypothetical labile repressor molecule whose identity has not been demonstrated experimentally, the scheme necessitates further testing. Regardless of the mechanism, the synthesis of TAT has been experimentally shown to be superinduced by actinomycin D, and regulated at a level beyond transcription (48).

Inasmuch as our results, thus far, had a striking parallelism with those of the superinduction of TAT, we originally presented a scheme assigning a regulatory role to a hypothetical labile repressor. This repressor represented the target of actinomycin D inhibition. Thus, in the presence of actinomycin D, the structural mRNA's for the mitotic protein(s) were freed from repression, achieved translation and enhanced mitotic activity over that induced by the cut. The cut was visualized as mobilizing the posttranscriptional mechanisms leading to the elaboration of mitosis. Our scheme also remains to be tested (7).

However, as the synthesis of TAT, epidermal mitosis is a superinducible cellular process and its regulation lies in a level beyond that of transcription. Since it is a eukaryotic system, the levels of posttranscription and translational modulation (23) are two of the possibilities.

Palmiter and Schinke (52) accounted for the selective superinduction elicited by actinomycin D in chick magnum in quite a different manner. The fact that ovalbumin was superinduced in the presence of actinomycin D and in the absence of a conomitant increase in total ovalbumin mRNA activity led them to discard the regulatory role of a repressor molecule in their system. They offer an alternate scheme based on their data and which relies on two

71

cellular features: 1) the differential stability of cellular mRNA's and 2) the competition between the cellular mRNA's for the rate limiting factors common to their translation. Their viewpoint is that in the presence of actinomycin D the shorter-lived messenger RNA molecules ceased to be synthesized, thus, the long-lived messengers increased proportionally in the cell and succeeded in their translation by the diminished competition for the rate limiting factors for translation. This scheme provides a most attractive framework for the interpretation of our data, especially in view of the fact that actinomycin D had a selective effect on the G_2 blocked non-cycling cell population. This population certainly had time to accumulate messengers (53), of some stability. These messengers are excellent candidates for the mechanism proposed by Palmiter-Schimke (52), especially since our data support the fact that the mRNA(s) leading to mitosis are of a stable nature.

Open to question, of course, is the possibility that the mode of action of actinomycin D in superinducing mitosis is not executed via RNA synthesis inhibition. Another alternative worthy of consideration is that actinomycin D might be superinducing mitosis by a differential stabilization of the cells mRNA's since several cases of this nature have been reported (54,55).

Singer and Penman (56) have shown, not only that actinomycin D lengthens the half-life of HeLa cell mRNA's, but that it also has an additional effect of slowing the initiation of the attachment of ribosomes onto mRNA. Actinomycin has even been speculatively assigned a role in facilitating permeability across the mitochondrial mebrane (57).

E. The Chalone Concept

Insofar as our data are compatible with the epidermal chalone concept and inasmuch as chalones seem to bear the responsibility for the control of mitosis in epidermis and other tissues we would like to present an interpretation of our results within the framework of the chalone concept. (For an exhaustive treatment of the epidermal chalone concept see 19, 58-60).

The intracellular concentration of chalone (effective antimitotic activity) in an *in vitro* system is determined by the rate of synthesis of the chalone minus the rate of loss from the tissue to the environment (16). Upon exposure to actinomycin D in the medium, no mitotic activity was observed if the whole ear was incubated. If, however, tissue strips were exposed to these conditions, actinomycin D enhanced the mitotic activity elicited by the cut. A testable hypothesis originating from these experiments would be that in the whole ear, actinomycin D interfered with the synthesis of the chalone, and therefore the intracellular concentrations would decrease. However, the remaining endogenous concentration in the whole structure would suffice to maintain the inhibition during the five hour incubation period. In the strips, the intracellular concentrations would decrease. However, the remaining endogenous concentration in the whole structure would suffice to maintain the inhibition during the five hour incubation period. In the strips, the intracellular concentration of chalone would be altered by an increase in the rate of loss. Thus the inhibition of mitosis would be partially relieved. In the presence of actinomycin D, the synthesis of chalone would be shut off, its intracellular concentration would be lowered additively, and the inhibition of mitosis would be further alleviated and as a result an enhacement of mitosis would be expressed in the presence of the inhibitor.

F. Cyclic AMP

In a series of recent papers, Voorhees and his collaborators (61-66) have correlated epidermal mitoses with intracellular cyclic AMP concentrations. They have presented evidence to satisfy Sutherland's criteria for an event to qualify as being cyclic AMP mediated (67). The role of cyclic AMP in the regulation of mammalian cell division has also been reviewed by Abell and Monohan (68).

Iversen (69) has presented some evidence that the adenyl cyclase-cyclic AMP system is involved with G_2 inhibition, and Marks (58) and Voorhees and co-workers (64) have demonstrated that in epidermis cyclic AMP and theophylline are potent G_2 inhibitors. On a strictly theoretical approach one could raise the possibility that actinomycin D could be affecting the intracellular cyclic AMP level by producing alterations in either the synthesis or activity of the cyclase, phosphodieterase or any protein moiety related to cyclic AMP. Thus far, no relationship has been demonstrated between cyclic AMP and the epidermal chalones except at a speculative level (61,66). The latter paper presents an up to date evaluation of the present state of the role of cyclic AMP in epidermal mitosis.

V. Concluding Remarks

Independently of what the mechanism(s) of actinomycin D superinduction turn out to be in this system, the highlight of the present work is the fact that the drug exerts its effects at the G_2 phase of the cell cycle and that it acts preferentially on the non-cycling G_2 blocked cell population. A control mechanism operating at G_2 has temporal and other advantages over those which operate at G_1 or S. Cells stimulated to enter into mitosis from the G_2 compartment can do so more readily and at faster rate than those from the other compartments. G_2 blocked cells have had ample time to accrue the requirements for mitosis. The G_2 cells equip a tissue with a supply of readily available cells for a mitotic demand. Since these non-cycling cells differ from cycling cells, at least in this respect and also their DNA content, it should not be surprising that they have functional differences and that they are under different control mechanisms.

Acknowledgements

I wish to thank Dr. Fernando Renaud from the Biology Department, University of Puerto Rico, Rio Piedras, Puerto Rico for his critical reading of this manuscript.

References

1. W. S. Bullough, *Biol. Rev. 37*, 307 (1962).
2. W. S. Bullough, *In Cellular Control Mechanisms and Cancer* (P. Emmelot and Muhlbock, eds), Elsevier Publishing Company, Amsterdam (1964).
3. W. S. Bullough, *Cancer Res. 25*, 1683, (1965).
4. W. S. Bullough, *Br. J. Derm. 87*, 187 (1972).
5. F. G. Sherman, H. Quastler and D. R. Wimber, *Exp. Cell Res. 25*, 114 (1961).
6. C. Pilgrim, W. Erb and W. Maurer, *Nature 199*, 863 (1963).
7. S. Gelfant and G. Candelas, *J. Invest. Dermatol. 59*, 7 (1972).
8. S. Gelfant, *Exp. Cell Res. 16*, 527 (1959).
9. S. Gelfant, *Exp. Cell Res. 32*, 521 (1963a).
10. S. Gelfant, *Int. Rev. Cytol. 14*, 1 (1963).
11. S. Gelfant, *Meth. Cell Physiol. 2*, 359, (1966).
12. S. Gelfant, *Exp. Cell Res. 16*, 527 (1959).
13. W. S. Bullough and E. B. Laurence, *Proc. Roy. Soc. B. 154*, 540, (1961).
14. D. P. Chopra, R. J. Yu and B. A. Flaxman, *J. Invest. Dermatol. 59*, 207 (1972).
15. W. H. Boldingh and E. B. Lawrence, *Europ. J. Biochem. 5*, 191 (1968).
16. K. Elgjo, H. Hennings and W. Edgehill, *Virchows Arch. Abt. B Zell. Path. 7*, 342 (1971).
17. K. Elgjo, O. D. Laerum and W. Edgehill, *Virchows Arch. Abt. B. Zell. Path. 8*, 277 (1971).
18. K. Elgjo, O. D. Laerum and W. Edgehill, *Virchows Arch. Abt. B. Zell Path. 10*, 229 (1972).
19. K. Elgjo, *Nat. Can. Ist. Monog. 38*, 71 (1973).
20. J. C. Houck, Nat. *Can. Inst. Monog. 38*, 1 (1973).
21. W. S. Bullough and M. Johnson, *Exp. Cell Res. 2*, 445 (1951).
22. I. L. Cameron, *In Drugs and the Cell Cycle* (A. M. Zimmerman, G. M. Padilla and I. L. Cameron, eds.), Academic Press, New York (1973).
23. J. E. Darnell, W. R. Jelinek and G. R. Molloy, *Science 181*, 1215 (1973).
24. G. C. Candelas and S. Gelfant, *J. Cell Biol. 47*, 30a (1970).
25. G. C. Candelas and S. Gelfant, *Abs.* 11th Ann. Meeting Am. Soc. *Cell Biol.* New Orleans (1971).
26. G. C. Candelas, Presented at 5th International Cell Cycle Conference, Toronto (1973).
27. G. C. Candelas and S. Gelfant, 9th International Congress of Biochemistry, Stockholm, p. 152, (1973).
28. K. Elgjo and H. Hennings, *Virchows Arch. Abt. B. Zell. Path. 7*, 342 (1971).
29. S. Kishimoto and I. Lieberman, *Exp. Cell Res. 36*, 92 (1964).

30. R. Baserga, R. D. Estensen and R. O. Petersen, *Proc. Nat. Acad. Sci. USA* *54*, 745 (1965).
31. R. A. Tobey, D. F. Petersen, E. C. Anderson and T. T. Puck, *Biophys. J. 6*, 567 (1966).
32. G. S. Stein and D. Matthews, *Science 181*, 71, (1973).
33. Y. Doida and S. Okada, *Cell Tissue Kinetics 5*, 15 (1972).
34. G. Mittermayer, P. Kaden, U. Trommershaeuser and W. Sandretter, *Histochemie 14*, 113 (1968).
35. M. Fujioka, M. Koga and I. Lieberman, *J. Biol. Chem. 238*, 3401, (1963).
36. H. S. Schwartz, J. E. Sodergren, M. Garofalo and S. S. Sternberg, *Cancer Res. 25*, 1867 (1965).
37. J. B. Melvin, *Exp. Cell Res. 45*, 559, (1967).
38. G. Wheeler and L. Simpson-Herren, *In Drugs and the Cell Cycle* (A. Zimmerman, G. M. Padilla and I. L. Cameron, eds), Academic Press, N.Y., pp. 250-293 (1972).
39. H. S. Schwartz, S. S. Sternberg and F. S. Phillips, *Cancer Res. 23*, 1125, (1963).
40. N. K. Wessels, *Develop. Biol. 9*, 92 (1964).
41. P. R. Gross, *Exp. Cell Res. 33*, 368 (1964).
42. L. D. Samuels, *New Eng. J. Med. 271*, 1252 (1964).
43. M. Revel, H. H. Hiatt and J. P. Revel, *Science 146*, 1311, (1964).
44. R. Baserga, *Cell Tissue Kinetics 1*, 167 (1968).
45. J. Hoffman and J. Post, *In Drugs and the Cell Cycle* (A. M. Zimmerman, G. M. Padilla and I. L. Cameron, eds), Academic Press, New York (1972) pp. 219-242).
46. J. W. Grisham, *In Drugs and the Cell Cycle* (A. M. Zimmerman, G. M. Padilla and I. L. Cameron, eds), Academic Press, New York, (1972) pp. 95-127.
47. W. G. Flamm, M. R. Bannerjee and W. B. Counts, *Cancer Res. 26*, 1349 (1966).
48. G. M. Tomkins, B. Levinson, J. D. Baxter and L. Dethlefsen, *Nature New Biol. 239*, 9 (1972).
49. L. D. Garren, R. R. Howell, G. M. Tompkins and R. M. Crocco, *Proc. Nat. Acad. Sci. USA 52*, 1121 (1964).
50. G. M. Tomkins, T. D. Gelehrter, D. K. Granner, B. Peterkofsky and E. B. Thomson, *In Exploitable Molecular Mechanisms and Neoplasia* Williams and Wilkins, Baltimore (1969).
51. G. M. Tomkins, T. D. Gelehrter, D. K. Granner, D. W. Martin, Jr., H. H. Samuels and E. P. Thompson, *Science 166*, 1474 (1969).
52. R. D. Palmiter and R. T. Shinke, *J. Biol. Chem. 248*, 1502 (1973).
53. R. A. Flickinger, D. M. Kohl, M. R. Lauth and P. J. Stambrook, *Biochim. Biophys. Acta 209*, 260 (1970).
54. Y. Endo, H. Seno, H. Tominaga and Y. Natori, *Biochim. Biophys. Acta 299*, 114, (1973).
55. M. Izawa and S. Ichii, *Endocrinol Japonica 19*, 225 (1972).

56. R. H. Singer and S. Penman, *Nature 240*, 100 (1972).
57. M. K. Mostafapour and T. T. Tchen, *Biochem. Biophys. Res. Comm. 48*, 491 (1972).
58. R. Marks, *Br. J. Derm. 86*, 543 (1972).
59. W. S. Bullough, *Nat. Can. Inst. Monog. 38*, 5 (1973).
60. F. Marks, *Nat. Can. Inst. Monog. 38*, 79 (1973).
61. J. J. Voorhees and E. A. Duell, *Arch. Dermatol. 104*, 352 (1971).
62. J. A. Powell, E. A. Duell and J. J. Voorhees, *Arch. Dermatol. 104*, 359 (1971).
63. E. A. Duell, J. J. Voorhees, W. H. Kelsey and E. Hayes, *Arch. Dermatol. 105*, 601 (1971).
64. J. J. Voorhees, E. A. Duell and W. H. Kelsey, *Arch. Dermatol. 105*, 384, (1972).
65. J. J. Voorhees, E. A. Duell, J. J. Bass, J. A. Powell and E. R. Harrell, *Arch. Dermatol. 105*, 695 (1972).
66. J. J. Voorhees, E. A. Duell, J. J. Bass and E. R. Harrell, *Nat. Can. Inst. Monog. 38*, 47 (1973).
67. G. A. Robinson, R. W. Butcher and E. W. Sutherland, (eds) *Cyclic AMP*, Academic Press, N.Y. (1971).
68. C. W. Abell and T. M. Monahan, *J. Cell Biol. 59*, 549 (1973).
69. O. H. Iversen, *In Ciba Symposium, Homeostatic Regulators* (A. E. W. Wolstenholm and J. Knight, Eds), London and Churchill, 29 (1969).

CONTROL OF THE CELL CYCLE IN HIGHER PLANTS*

J. Van't Hof

Biology Department, Brookhaven National Laboratory, Upton, New York 11973

I. Introduction

In higher plants mitotic activity occurs primarily within the confines of a complex tissue called a meristem. The proliferative cells of a tissue are constrained not only by their neighbors but also by nutritional and hormonal substances that may originate in another rather distant organ. Also, it is essential to the survival of the individual that most dividing cells of growing organs eventually differentiate and carry on the specialized functions of mature cells. The central theme of this paper is applicable to embryonic, growing, and maturing cells of higher plants. It postulates that the cell cycle is controlled at two positions, one located in G1 and another in G2. These controls determine whether cells will proceed from G1 to S and from G2 to M. They are called principal controls because subsequent functions such as DNA synthesis and mitosis, which in themselves are highly regulated, do not take place until the requirements of the principal controls are met. In other words, a cell arrested in G1 need not synthesize molecules required for S unless it is metabolically committed to progress to S; likewise, a cell in G2 need not prepare for mitosis unless it is going to divide. At the moment it is unknown if the postulated principal controls function in a manner that involves sequential enzyme syntheses or activation of the sort proposed by Mitchison (1,2). Most of the evidence that supports the hypothesis is biological and is derived from cytokinetic and microspectrophotometric analyses. The evidence, however, is not confined to higher plant cells and the reader is referred to the review articles of Gelfant (3) and Epifanova and Terskikh (4) for information about mamalian cells.

II. Cells of Embryonic Root Tissue

The progenitor of the primary root tip is the radicle, a distinctive organ of the embryo in the seed. The potentially profliferative cells of the radicle are

---

*Research carried out at Brookhaven National Laboratory under the auspices of the U.S. Atomic Energy Commission. Abbreviations of cell cycle periods are as follows: M, mitosis; G1, presynthetic period; S, period of DNA synthesis; G2, postsynthetic, premitotic period. Cells in G1 have a relative nuclear DNA content of 2C, those in G2 have a 4C DNA content, cells in S label with ^{3}H-thymidine.

arrested nonrandomly in the cell cycle during dormancy. In some species all the cells have a nuclear DNA content of 2C, which is characteristic of G1, and in other species the radicle is composed of cells some with a 2C and others with a 4C amount of DNA (Table 1). The cells with 4C nuclei are in G2. The

Table 1. The percentage of cells with either a 2C or 4C nucleus in the radicle of dormant seed of several plant species. Cells in G1 have nuclei with a 2C amount of DNA, those in G2 have a 4C amount.

Species	% 2C	% 4C	Reference
Aegilops longissima	98	2	5
Allium cepa var. Bianci di Maggio	98	2	6
A. cepa var. Rossa Fiorentino	98	2	6
A. cepa var. Evergreen Bunching	100	0	7
Helianthus annuus	100	0	8
Lactuca sativa	100	0	9
L. sativa var. Grand Rapids	100	0	10
Pinus pinea	100	0	9
Pisum sativum var. Alaska	65	35	11
Triticum aestivum var. Indus	100	0	11
T. diccoccum	97	3	5
T. durum var. Aziah	94	6	12
T. durum var. Cappelli	75	25	13
Vicia faba var. Major	83	17	14
V. faba var. Minor	71	29	14

absence of cells in S or in M in dormant embryonic tissue indicates that potentially proliferative cells naturally arrest in the G1 and in the G2 periods of the cycle.

III. Dividing Cells of the Growing Root Tip

When seed germinate cell division and DNA synthesis commence ultimately forming a primary root tipped at the apex with a meristem. The dividing cells of the meristem constitute a quasi-steady state proliferative system and represent growth fractions (15), which are species specific, that range between

40 and 90% (16-18). The age distribution of the cells in the cycle of a root meristem is not known but current experiments with *Helianthus, Pisum, Triticum* and *Vicia* indicate that neither an exponential nor an uniform distribution, as proposed by Sinclair for mammalian cells (19), are satisfactory (Evans and Van't Hof, unpublished results).

IV. Initiation of DNA Synthesis and Mitosis Controlled Experimentally

Normally the meristematic cells of a growing root divide asynchronously. Such root tips are defined as having proliferative phase meristems and are in contrast to those with cells neither in S nor in M which are said to be in a stationary phase (S.P.). A stationary phase meristem is established by detachment of the primary root tip and starving it of carbohydrate in White's medium (20) for 48 hours (21-24). During starvation the asynchronously

Table 2. The percentage of cells arrested in G1 and in G2 of carbohydrate starved primary root meristems of eight plant species determined by microspectrophotometry of Feulgen stained nuclei and ^{3}H-thymidine autoradiography (from ref. 25).

Species	% G1	% S	% G2	% M
Crepis capillaris	74	0	26	0
Glycine max	50	0	50	0
Helianthus annuus	77	2	21	0
Lycopersicon esculentum	50	0	50	0
Pisum sativum	44	0	56	0
Triticum aestivum	64	0	36	0
Vicia faba	29	0	71	0
Zea mays	53	0	47	0

dividing cells arrest in G1 and in G2. The meristems have no cells in M and very few if any, in S as signified by ^{3}H-thymidine autoradiography and microspectrophotometry (Table 2). Stationary phase meristems have been established in every plant tested to date but the number of cells arrested in G1 and G2 differs from one species to the next as shown in Table 2. For example, *Helianthus* consistently has 75% or more cells in G1 and *Vicia* with equal regularity has as few as 29%. Conversely, in G2, *Vicia* has as many as 71% while *Helianthus* has as little as 21%.

When cells of S.P. meristems are replenished with carboyhdrate, they enter

S and M after a delay of 2 to 14 hours depending on the duration of starvation (26). The delay before the onset of DNA synthesis is noted as T_{ds} and that which precedes mitosis is T_{dm}. In most species cell entry into S is linear occurring at a rate R, which is determined to a large degree by the number of cells in G1 in the S.P. meristem (Table 3). Included in Table 3 are the values of

Table 3. The percentage of cells arrested for 72 hours in G1 and in G2 in stationary phase root meristems of five plant species and cell population kinetic parameters after carbohydrate replenishment. R, rate of increase of ^{3}H-thymidine labeled interphase cells; T_{ds}, delay of entry into S from G1; T_{dm}, delay of entry into M from G2.

| | % cells | | R | T_{ds} | T_{dm} | |
Species	G1	G2	(%/hr)	(hr)	(hr)	Reference
Helianthus annuus	77	21	5-8	2-4	2-3	23, 27
Lycopersicon esculentum	50	50	2	-	-	
Pisum sativum	44	56	2	10-11	10-11	22, 26
Triticum aestivum	64	36	3	10	6	
Vicia faba	29	71	0.6-1	7-8	7	24

T_{ds} and T_{dm} of four species. In each case the duration of starvation was 72 hours and the tissue experienced a single nutritional shift, i.e., after excision the roots were transferred once from medium without sucrose to medium with sucrose and ^{3}H-thymidine. While T_{ds} and T_{dm} may be as short as 2 hours in *Helianthus* and as long as 11 hours in *Pisum*, each species utilizes this time for the synthesis of proteins needed for the transition to the next cycle period. Treatment with inhibitors of protein synthesis in conjunction with carbohydrate replenishment prevents cells in G1 from progression to S and those in G2 from division in *Helianthus*, *Pisum* and *Vicia* (22,24,27). In *Pisum* the syntheses that occur during T_{ds} and T_{dm} are energy dependent and require oxygen in addition to carbohydrate (28). It is assumed that cells of other species have requirements that are not different from those of *Pisum*. The change from a proliferative to a S.P. meristem is ascribed to a deficiency of factors, presumably proteins, associated with the postulated principal controls in G1 and in G2. Only when a sufficient amount of these factors is synthesized can cell progression in the cycle be initiated again. The same or similar requirements are shared by cells of the radicle of dormant *Vicia* seed as certain proteins, and not RNA, must be synthesized before the onset of cell division during germination (29,30). Thus, *Vicia* S.P. meristems, whether established naturally during dormancy or by artificial nutritional starvation in culture, are mediated via mechanisms that involve the loss of proteins needed for the G1 to S and the G2 to M transitions.

V. The Arrest of Cell Division and Tissue Maturity

The root tip meristem represents a quasi-steady state system of proliferative cells that gradually diminishes in cell number with age (31). After mitosis a little more than half of the newly formed cells are determined to differentiate and the remainder return to the proliferative pool (32). Concomitant with differentation and subsequent maturation is a low level of proliferation which is accompanied by cell arrest in the cycle in a nonrandom manner. Tissue 2 cm above the root tip meristem is generally considered to be mature but not old. Microspectrophotometry of the nuclei of cells of mature root tissue of four plant species indicated that cell arrest accompanying maturity was not uniform. In *Vicia* and *Pisum* most cells have nuclei with a 4C amount of DNA while in *Helianthus* and *Triticum* the majority are 2C (Table 4). Thus, cell

Table 4. The percentage of cells with a 2C, 4C, and more than 4C amount of nuclear DNA in mature primary root tissue 2 cm from the tip of four plant species.*

Species	% 2C	% 4C	% >4C	% in S or M	Age of seedling[+] (days)
Helianthus annuus	91	2	0	7	4
Pisum sativum	15	65	16	4	4
Triticum aestivum	65	24	7	4	4
Vicia faba	22	63	7	8	7

*From ref. 33.

[+]Day zero equals time of planting.

arrest during maturation is preferential favoring 4C nuclei in some species and 2C in others. Moreover, if meristematic cells of a certain species favor arrest in G1, then the mature cells will preferentially arrest at the 2C level. The reverse is also true, i.e., if carbohydrate starved cells of the meristems arrest mostly in G2, then most of the mature cells will be 4C. (For verification of these observations the data of Tables 2 and 4 should be consulted.) Because meristematic and mature cells of a given species prefer to arrest in the cycle at same nuclear DNA level, and because the preferred DNA level differs among species, the common physiological functions of the root and the cell cycle itself may not be major factors in the determination of where in the cycle cell populations arrest. The cell cycle of each species mentioned in this paper adheres to the model of Howard and Pelc (34) yet their patterns of arrest are different. Is it possible that cell populations within a species are genetically determined to arrest in certain cycle periods? There is no unqualified answer to this question but there is a strong possibility that species within the same phylogenetic family control proliferating or potentially proliferative cells in a similar manner. For example, dormant cells of the radicle of *Helianthus annuus* are all 2C (8) and the dormant cells of a different organ of a different but related species, the tuber of *Helianthus tuberosum* are also 2C (35,36).

VI. Subpopulations of Meristematic Cells and a G2 Factor

The concept of control of the cell cycle in terms of a decision to proliferate or not to proliferate can be traced to Gelfant (37,38). Working with stimulated mouse epidermis he found that many cells remain in G2 for an extended length of time. He postulated that the epidermis was composed of two distinct cell populations, one which arrested in G1 and another that stopped in G2. When stimulated the cells in G1 initiate DNA synthesis and eventually divide and those in G2 begin to enter mitosis. The notion of cell arrest in G1 and in G2 was later incorporated into a cell cycle model by Epifanova and Terskikh (4). They proposed the alternation of "rest" periods with those of active proliferation. The "rest" periods were positioned after DNA replication in G2 and after mitosis in G1. Epifanova and Terskikh did not imply that their model considered separate, distinctive cell populations but Gelfant's concept does. Our experiments with root tips of different species indicate that both concepts are probably correct.

Cytokinetic experiments with *Pisum* showed that cells previously arrested in G1 also stopped in G2, if given the opportunity by manipulation of the culture medium (26). It was postulated that such cells have two principal controls each. Until recently the full implication of this observation was overlooked. However, current experiments by Evans and Van't Hof (39) demonstrate the existence of a substance that originates in the cotyledons and promotes cell arrest in G2. This substance, called the G2 factor, affects approximately 20 to 30% of the root meristem cells of *Pisum*. If present during carbohydrate starvation responsive cells arrest in G2; if absent the cells arrest in G1. The number of cells affected by the G2 factor represent about one-half of the growth fraction; therefore, those remaining must arrest preferentially in G1 or in G2. These cytokinetic tabulations suggest that the root meristem of *Pisum* has three cell populations, one prescribed to stop in G1, one prescribed to stop in G2, and a third that arrests in either G1 or G2 depending on the presence or absence of the G2 factor.

It is undetermined if *Helianthus* has a G2 factor equivalent to that of *Pisum* but it is likely that *Helianthus* root meristems have more than one cell population. During carbohydrate starvation the behavior of a small group of cells indicates that they have only one principal control located in G2 (23,27). When starved, 75 to 90% of the cells arrest in G1, 5 to 20% stop in G2, and 0.5 to 5% slowly leave G1, transit S, and arrest in G2. Members of the latter small population are referred to as "leaky" cells, a term that describes their behavior while other cells are arrested. Because "leaky" cells are in S when all the other cells are in either G1 or G2, they are easily marked by ^{3}H-thymidine autoradiography. The number of "leaky" cells is small and with prolonged starvation they ultimately accumulate in G2. "Leaky" cells synthesize DNA in the presence of inhibitors of protein synthesis and after exposure to relatively large (up to 7200 R) doses of gamma rays. Rost and Van't Hof (27) originally concluded that "leaky" cells were randomly distributed throughout the meristem but other investigators (P.L. Webster, personal communication) show them to be localized near the distal tip, a postion that suggests the possibility that they are initial or stem cells.

VII. The Effect of Ionizing Radiation on the G1 to S and the G2 to M Transitions

The meristematic cells of *Helianthus* that arrest in G1 are sensitive to ionizing radiation and inhibitors of protein synthesis. Either treatment produces impairment or prevention of the G1 to S transition. However, all cells, including the "leaky" population, when in G2 exhibit mitotic delay after an exposure of gamma rays. These cells are also prevented from mitosis when treated with inhibitors of protein synthesis.

The effect of ionizing radiation on cell progression in the cycle shows a curious coincidence with the pattern of arrest of a particular cell population. For example, "leaky" cells stop only in G2 and here, not in G1, they are detained by gamma rays. Also, "leaky" cells appear to be refractory to inhibitors of protein synthesis except in G2. Earlier experiments with *Pisum* suggested that sensitivity to inhibitors of protein synthesis and radiation-induced mitotic delay may be related to the cytochemistry of the treated cells (40-42). Cells of S.P. meristems that are arrested in G1 lack the proteins needed to complete G1 and to initiate DNA synthesis. Such cells when exposed to 300 R of gamma rays and then supplied carbohydrate are delayed in G1 for an additional 3 to 5 hours. They then traverse S and G2 and divide without further delays. The effect of the gamma rays on cell progression is restricted to G1. However, when cells are synchronized at the G1/S boundary and irradiated, a different response is noted. During the process of synchronization these cells make the proteins of G1 and those needed to initiate DNA synthesis and they require only carbohydrate to progress in the cycle. When supplied with sucrose and treated with inhibitors of protein synthesis, these cells advance from G1 to S unimpaired. An exposure to 300 R likewise neither impairs the G1 to S transition of synchronized cells nor extends the duration of S but the cells are delayed in G2 for 3 to 5 hours before they divide. Thus, cells are delayed in G1, if starved, or delayed in G2, if synchronized. One interpretation of these experiments is that many cells of *Pisum* have two principal controls and that radiation-induced delay reflects the time required to remove damage before cells prepare for the G1 to S transition. Starved cells though detained in G1 are not again delayed in G2 because recovery of the two controls is simultaneous. Synchronized cells, on the other hand, are prepared to enter S and damage to the G1 control has no immediate effect. Damage to the control in G2 remains, however, and it must be removed before mitosis occurs. The delay in G2 is proposed to be the time during which this removal takes place.

VIII. Conclusions

1. The dormant seed and in carbohydrate starved root meristems potentially proliferative cells arrest in G1 and in G2 in proportions that differ between plant species. Cell cycle arrest in mature tissue also occurs in G1 and in G2 in proportions that differ between species.

2. The lack of proteins needed to complete G1 and G2 and to initiate the G1 to S and G2 to M transitions is responsible for cell cycle arrest in dormant and starved cells alike.

3. Certain meristems may be composed of three cell populations. One that preferentially arrests in G1, another that prefers G2, and a third that will arrest either in G1 or in G2.

4. Cell cycle arrest and cell division in the primary root meristem, in the embryo, and in mature or differentiating tissue is governed by principal controls that operate in G1 and in G2.

5. Ionizing radiation produces damage that interferes with the operation of the two cell cycle controls.

Acknowledgements

To Dr. L. S. Evans for use of preliminary data, and to Mrs. S. Yagi for technical assistance, I express my gratitude.

References

1. J. M. Mitchison, *Science 165*, 657 (1969).
2. J. M. Mitchison, *The Biology of the Cell Cycle*, Cambridge University Press, London, pp. 159-180 (1971).
3. S. Gelfant, *Methods Cell Physiol. 2*, 359 (1966).
4. O. I. Epifanova and V. V. Terskikh, *Cell Tissue Kinet. 2*, 75 (1969).
5. G. Ancora, A. Brunori and G. Martini, *Caryologia 25*, 373 (1972).
6. A. Brunori and G. Ancora, *Caryologia 21*, 261 (1968).
7. T. R. Bryant, *Caryologia 22*, 140 (1969).
8. A. Brunori, J. Georgieva and F. D'Amato, *Mutation Res. 9*, 481 (1970).
9. A. Brunori and F. D'Amato, *Caryologia 20*, 153 (1967).

10. A. H. Haber, D.E. Foard and S. W. Perdue, *Plant Physiol. 44*, 463 (1969).
11. L. S. Evans and J. Van't Hof, unpublished data.
12. S. Avanzi, A. Brunori and G. Giorgi, *Mutation Res. 3*, 426 (1966).
13. S. Avanzi, A. Brunori and F. D'Amato, *Develop. Biol. 20*, 368 (1969).
14. A. Brunori, S. Avanzi and F. D'Amato, *Mutation Res. 3*, 305 (1966).
15. M. L. Mendelsohn, *Science 132*, 1496 (1960).
16. J. Van't Hof and H.-K. Ying, *Cytologia 29*, 399 (1964).
17. P. L. Webster and D. Davidson, *Cell Biol. 39*, 332 (1968).
18. F. A. L. Clowes, *Ann. Bot. 35*, 249 (1971).
19. W. K. Sinclair, in *Radiation Research*, G. Silini, Editor, North-Holland Publishing Co., Amsterdam, pp. 607-631 (1967).
20. P. R. White, *A Handbook of Plant Tissue Culture*, Cattell and Co., Lancaster, Pa., (1943).
21. J. Van't Hof, *Amer. J. Bot. 53*, 970 (1966).
22. P. L. Webster and J. Van't Hof, *Amer. J. Bot. 57*, 130 (1970).
23. J. Van't Hof and T.R. Rost, *Amer. J. Bot. 59*, 769 (1972).
24. J. Van't Hof, D. P. Hoppin and S. Yagi, *Amer. J. Bot. 60*, 889 (1973).
25. J. Van't Hof, *Brookhaven Symp. Biol. 25*, in press.
26. J. Van't Hof, in *Proceedings IV International Congress of Radiation Research*, H. Duplan, Editor, Gordon and Breach, London, in press.
27. T. R. Rost and J. Van't Hof, *Exptl. Cell Res. 55*, 88 (1969).
28. P. L. Webster and J. Van't Hof, *Exptl. Cell Res. 55*, 88 (1969).
29. K. M. Jakob and F. Bovey, *Exptl. Cell Res. 54*, 118 (1969).
30. K. M. Jakob, *Exptl. Cell Res. 72*, 270 (1972).
31. R. Brown, *J. Exptl. Bot. 2*, 96 (1951).
32. J. Van't Hof and H.-K. Ying, *Nature 202*, 981 (1964).
33. L. S. Evans and J. Van't Hof, *Amer. J. Bot.*, submitted.
34. A. Howard and S. R. Pelc, *Heredity 6* (Suppl.) 261 (1953).
35. D. Adamson, *Can. J. Bot. 40*, 719 (1962).
36. J. P. Mitchell, *Ann. Bot. 31*, 427 (1967).
37. S. Gelfant, *Exptl. Cell Res. 26*, 395 (1962).
38. S. Gelfant, *Exptl. Cell Res. 32*, 521 (1963).
39. L. S. Evans and J. Van't Hof, *Exptl. Cell Res.*, in press.
40. J. Van't Hof and C. J. Kovacs, *Radiation Res. 44*, 700 (1970).
41. C. J. Kovacs and J. Van't Hof, *Radiation Res. 48*, 95 (1971).
42. C. J. Kovacs and J. Van't Hof, *Radiation Res. 49*, 530 (1972).

CELL MECHANISMS INFLUENCING THE TRANSITION OF HEMOPOIETIC STEM CELLS FROM G_0 INTO S.

Joseph W. Byron

Department of Cell Biology and Pharmacology, School of Medicine,
University of Maryland, Baltimore, Maryland.

I. Introduction

Mouse bone marrow contains a primitive pluripotent cell (stem cell) which following transplantation produces macroscopic colonies in the spleen of heavily irradicated mice (1); each colony arises from a single cell (2,3). These cells are called spleen colony forming units or CFU-S (4).

With respect to the cell cycle, CFU-S of normal adult bone marrow have generally been considered to be in a resting stage (G_0) or to have long cell cycles with a relatively small fraction occupied by DNA synthesis. Accordingly, they are insensitive to the cytocidal actions of high specific activity ^{3}H-thymidine(5) and of inhibitors of DNA synthesis; a more accurate description of the cell cycle state of resting VFU-S may be that they are held up in late G_1; i.e., close to the G_1-S boundary (6). Under conditions requiring increased numbers of differentiated bone marrow cells, CFU-S may enter into cell cycle; hence their sensitivity to ^{3}H-thymidine and to inhibitors of DNA synthesis will increase.

The hemopoietic stem cell possesses the property of self-renewal, and also gives rise to hemopoietic cells committed to differentiate toward a particular bone marrow end-cell. Accordingly, e.g., the erythropoietin sensitive cell (ESC) which gives rise to erythrocytes and the *in vitro* colony forming cell of bone marrow (CFU-C) are thought to be derived from the hemopoietic stem cell. The latter progenitor cell, CFU-C, gives rise to both macrophages and granulocytes. The committed progenitor cells are thought to have a finite number of cell divisions and hence a finite life span. Day-to-day growth and differentiation of bone marrow cells is maintained by these cells derived from CFU-S. In contrast to the stem cell in adult mouse bone marrow, the committed progenitor cells are in a cell cycle, actively synthesise DNA and are, therefore, susceptible to killing by high specific activity ^{3}H-thymidine or by other agents which kill cells in cycle. Increased demand for differentiated bone marrow cells can alter the cell cycle state of CFU-S such that these cells enter into an active cell cycle and now become sensitive to ^{3}H-thymidine killing.

Mechanisms regulating growth and differentiation processes are of basic importance to an understanding of cell kinetics in the bone marrow, and for understanding pathophysiological conditions connected with bone marrow.

Molecular events influencing the hemopoietic system are far from understood. The studies to be described were directed towards the uncovering of cellular mechanisms that influence or possibly serve to regulate the cell-cycle state of the hemopoietic stem cell. More precisely, systems were investigated that were with the triggering of CFU-S from G_0 into DNA synthesis (Fig. 1).

A variety of agents have been shown to increase the sensitivity of CFU-S to the cytocidal action of high-specific activity ^3H-Thymidine (^3H-TdR); i.e., they trigger CFU-S from G_0 into DNA synthesis. (Fig. 1)

AGENTS on CELL CYCLE of CFU-S

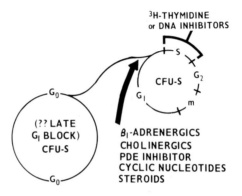

Fig. 1. Some compounds that trigger the hemopoietic stem cell from G_0 into DNA synthesis (PDE=phosphodiesterase). Upon entering S, stem cells become sensitive to the cytocidal actions of either ^3H-TdR or inhibitors of DNA synthesis, e.g., hydroxyurea. These cytocidal agents will inhibit the ability of CFU-S to form colonies in the spleens of whole body irradiated mice. The compounds used to trigger CFU-S into DNA synthesis help in defining certain properties of the hemopoietic stem cell.

The studies suggest some mechanisms involving drug receptor sites and cyclase systems that serve to initiate DNA synthesis in the stem cell. The mechanisms may involve β_1 — adrenergic receptor sites and the adenyl cyclase system and cholinergic receptor sites and the guanyl cyclase system. Steroid-induced DNA synthesis appears to be independent of cyclic nucleotide mediation.

II. Experimental Methods

The technique (7) [in its more recent modifications, (8,9)] used to study the action of drugs on CFU-S of mouse bone marrow has been described in detail elsewhere. Briefly, bone marrow cells are exposed *in vitro* to drugs for 2.5 hours. Cultures are then treated for an additional 0.5 hours with 200 μCi/ml of high specific activity ^{3}H-TdR. Control cultures receive a comparable amount of unlabeled thymidine. After exposure to thymidine the cultures are diluted. The diluted suspension is injected intravenously (i.v.) into irradiated recipients, and spleen colonies are counted 9 days after transplantation. A comparison is made between numbers of colonies per spleen after transplanting control (^{1}H-TdR) and ^{3}H-TdR treated cultures. The per cent loss of CFU-S is then calculated. For agents switching-on DNA synthesis the per cent loss of splenn colonies due to ^{3}H-TdR is expected to increase.

As an additional test as to whether cells are synthesising DNA, 10^{-3}M hydroxyurea is substituted for high specific activity ^{3}H-TdR. Hydroxyurea (HU), however, is added at the start of cultures and is generally present for 3 hours.

Where possible, agents were selected for study having known mechanisms of action, at least in other tissues. Hence, such compounds served as analytical tools for defining properties of the hemopoietic stem cell.

The compounds used were:
    (a) agents known to stimulate particular drug receptor sites,
    (b) compounds known to block drug receptor sites,
    (c) dibutyryl derivatives of cyclic nucleotides,
    (d) an inhibitor of cyclic nucleotide phosphodiesterase,
    (e) a stimulator of cyclic nucleotide phosphodiesterase, and
    (f) an inhibitor of protein synthesis.

As this system is free from serum or plasma, the experimental technique may allow for a more realistic assessment of the concentration of free drug molecules needed to act on cells since extracellular factors that might consume compounds are not present.

III. Evidence for the β_1-Adrenergic Mechanism

When exposed to low concentrations of the β-adrenergic receptor stimulating agent, isoproterenol, CFU-S greatly increased their sensitivity to the cytocidal actions of ^{3}H-TdR and of hydroxyurea (8). At low concentrations of isoproterenol this effect was blocked by the β-adrenergic receptor blocking agent, propranolol (Figure 2). These data suggest that a β-adrenergic receptor site is associated with the stem cell, and that its stimulation results in the initiation of DNA synthesis. Isoproterenol acts on both β_1 - and β_2 adrenergic receptors. Further analysis suggested that the receptor was β_1 in character. This was inferred from studies using agents such as orciprenaline and salbutamol, which acts primarily on β_2 receptors. These agents initiated DNA synthesis in CFU-S. However, relative to isoproterenol, higher concentrations were required and the plateau responses to both of these agents were lower than that seen after isoproterenol treatment.(8) Table 1 shows that blocking the β-receptor after the initiation of DNA synthesis by

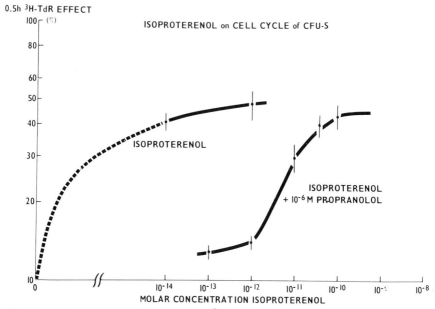

Figure 2. Per cent loss CFU-S due to ^{3}H-thymidine ($\pm$ S.E.M.) in bone marrow cultures treated with isoproterenol alone or isoproterenol in the presence of 10^{-6} M propanolol. Each point is made up from a minimum of 24 - 32 spleens. The point at 10^{-14}M isoproterenol, however, is from a total of 81 spleens (8).

TABLE 1

β-ADRENERGIC BLOCKADE ON DL-ISOPROTERENOL-INDUCED DNA SYNTHESIS

Treatment	CFU-S/Spleen ± S.E.M.		Loss at 3 hr. due to ^{3}H-Thymidine (%)
	^{1}H-Thymidine	^{3}H-Thymidine	
Isoproterenol (10^{-13} M) Alone	12.9 ± 0.75 (13)	7.9 ± 0.57 (14)	38.8
Propanolol (10^{-6} M) 5 Minutes before Isoproterenol (10^{-13} M)	11.7 ± 0.7 (13)	10.2 ± 0.7 (12)	12.8
Propanolol (10^{-6} M) 1.5 hr. after	13.05 ± 0.55 (14)	8.35 ± 0.6 (13)	36.0

(   ) = number of spleens counted

isoproterenol does not switch off DNA synthesis in CFU-S.

CFU-S respond to the dibutyryl derivative of adenosine 3':5'-cyclic monophosphate (cyclic-AMP) with increased sensitivity to ^{3}H-TdR. Adenosine-5'-monophosphate (5'-AMP) is not active on CFU-S. The concentration-response curve to dibutyryl cyclic-AMP is bell-shaped giving a peak effect at 10^{-8} M but no effect at 10^{-6} M. Figure 3 compares the response of CFU-S to dibutyryl cyclic-AMP and 5'-AMP with their response to an

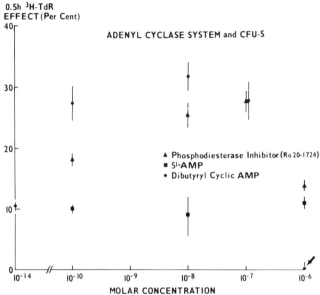

Figure 3. Per cent loss CFU-S due to ^{3}H-thymidine ($\pm$ S.E.M.) in bone marrow cultures treated with dibutyryl cyclic-AMP, 5'-AMP or Ro 20-1724. Each point is made up from a minimum of 24 - 32 spleens.

inhibitor of cyclic nucleotide phosphodiesterase, 4-(3-butoxy-4-methoxybenzyl)-2-imidazolidinone; Ro 20-1724 (10). Given alone, Ro 20-1724 increased sensitivity to ^{3}H-TdR. Activation of phosphodiesterase with 4 mM imidazole suppressed the actions of both isoproterenol and of dibutyryl c-AMP on CFU-S. (8)

The studies with isoproterenol and with chemically related β-adrenergic stimulating compounds suggest that a β_1-adrenergic receptor site is associate with CFU-S, and that chemical excitation of this receptor initiates DNA synthesis. The adenyl cyclase system probably mediates the action of isoproterneol. This may be deduced from the findings that (a) CFU-S respond to dibutyryl cyclic-AMP, (b) phosphodiesterase activation suppresses the triggering actions of both β-adrenergic stimulation and dibutyryl cyclic-AMP and (c) inhibition of phosophodiesterase initiates DNA synthesis. The latter finding also suggests that CFU-S synthesize cyclic nucleotides, and that phosphodiesterase may act to maintain concentrations at sub-triggering levels in G_O cells; hence it may be suggested that the enzyme phosphodiesterase

91

may act to regulate the cell cycle state of the hemopoietic stem cell.

IV. Evidence for a Cholinergic Mechanism

Following chemical excitation of bone marrow with carbamylcholine or with acetylcholine in the presence of an inhibitor of cholinesterase, i.e. neostigmine, ^{3}H-TdR decreased the colony forming ability of suspensions of mouse bone marrow (Fig. 4). Low concentrations (10^{-14} to 10^{-12}M) of the

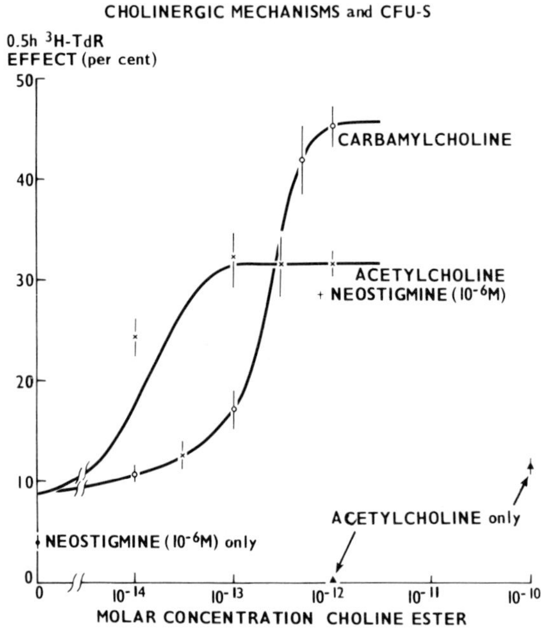

Fig. 4. Per cent loss CFU-S due to ^{3}H-thymidine ($\pm$ S.E.M.) in bone marrow cultures treated with carbamylcholine, acetylcholine, neostigmine, or with both neostigmine and acetylcholine. Each point is made up from a minimum of 24 - 32 spleens. The point for 10^{-12}M carbamylcholine alone is from four experiments giving a total of 98 spleens.

cholinergic agents were effective. Alone, neostigmine did not stimulate DNA synthesis. If added 5 minutes prior to carbamylcholine, the anticholinergic agent d-tubocurarine (10^{-6} M), inhibited the effect of low concentrations of carbamylcholine. This inhibition could be overcome by increasing the concentration of carbamylcholine (11). These studies suggest that a cholinergic receptor site may be part of a mechanism initiating DNA synthesis in CFU-S. Initiation through this site may be independent of the β_1-adrenergic receptor mechanism (Table 2), since β-adrenergic blockade does not antagonise the effect of cholinergic stimulation and cholinergic blockade does not antagonise -adrenergic stimulation.

TABLE 2

INTERCHANGE OF RECEPTOR BLOCKING AGENTS

	CFU-S per spleen ± S.E.M.		Loss at 3 hr. due to
	[1]H-Thymidine	^{3}H-Thymidine	^{3}H-Thymidine (%)
1-Isoproterenol bitartrate $(10^{-13}$ M) (B-adrenergic stimulation).	13.2 ± 0.9 (14)	9.6 ± 0.8 (13)	27.3
1-Isoproterenol bitartrate $(10^{-13}$ M) + d-tubo curarine chloride $(10^{-6}$ M)	11.8 ± 0.7 (13)	8.7 ± 0.6 (13)	26.3
Carbamylcholine chloride $(10^{-12}$ M)	12.5 ± 0.6 (13)	7.8 ± 0.7 (13)	37.6
Carbamylcholine chloride $(10^{-12}$ M) + propanolol $(10^{-6}$ M).	12.1 ± 0.7 (14)	7.3 ± 0.5 (13)	39.7

(   ) = number of spleens counted.

The question arises as to what mechanisms may mediate in the cell cycle action of cholinergic compounds. Elevated levels of guanosine 3':5'-cyclic monophosphate (cyclic-GMP) have been reported in the myocardium of the rat following perfusion of the isolated heart with acetylcholine (12) and also in the mouse brain tissue following treatment with oxotremorine, a cholinergic agent. (13).In both reports, the effect of the cholinergic agents on cyclic GMP levels could be prevented by pre-treatment with the anti-cholinergic agent, atropine. Cholinergic stimulation, however, does not lead directly to the activation of adenyl cyclase (13,14), and hormones that elevate cyclic-AMP levels in tissues do not raise the level of cyclic-GMP (15,16).

The dibutyryl derivative of cyclic-GMP increased sensitivity of CFU-S to ^{3}H-TdR, and activation of phosphodiesterase with 4 mM imidazole suppressed the actions of both carbamylcholine and of dibutyryl cyclic-GMP (11).Table 3 shows the effect of cycloheximide, an inhibitor of protein synthesis, on dibutyryl cyclic-GMP induced DNA synthesis. Cycloheximide (4 x 10^{-6} M) inhibited the action of this cyclic nucleotide.

The studies with carbamylcholine, and with acetylcholine in the presence of cholinesterase inhibition, suggest that a cholinergic receptor site may be part of

TABLE 3

CYCLOHEXIMIDE ON DIBUTRYRL CYCLIC-GMP INDUCED DNA SYNTHESIS

Treatment	Exp. No.	CFU-S/Spleen ± S.E.M.		Loss at 3 hr. due to ^{3}H-Thymidine (%)
		^{1}H-Thymidine	^{3}H-Thymidine	
Dibutyryl Cyclic GMP (10^{-8} M)	1	12.0 ± 0.89 (14)	3.79 ± 0.38 (14)	68.4
	2	9.97 ± 0.45 (14)	5.29 ± 0.69 (14)	46.9
Cycloheximide (4×10^{-6} M)	1	12.04 ± 0.83 (14)	11.4 ± 0.86 (14)	5.3
+				
Dibutyryl Cyclic GMP (10^{-8} M)	2	12.18 ± 0.93 (13)	9.07 ± 0.54 (14)	25.5

( ) = number of spleens counted

a mechanism initiating DNA synthesis in CFU-S. Initiation through this site appears to be independent of the β_1-adrenergic mechanism previously discussed. Cyclic-GMP may mediate in the cell cycle action of carbamylcholine and presumably acetylcholine as (a) dibutyryl cyclic-GMP induces DNA synthesis in CFU-S, and (b) both the effects of carbamylcholine and of dibutyryl cyclic-GMP are suppressed by phosphodiesterase activation. (c) The triggering action of the phosphodiesterase inhibitor, Ro 20-1724, could lead to elevation of cyclic-GMP levels. Hence there is the possibility that the triggering of CFU-S into DNA synthesis, whether through β_1-adrenergic receptors or via a cholinergic receptor, may be mediated by a 3':5'-cyclic mononucleotide. Cholinesterase is associated with hemopoietic cells. The function of this enzyme in cells from bone marrow has not been clarified. The acetylcholinesterase, e.g., in the circulating erythrocyte, may be the remains of a cholinergic mechanism associated with DNA synthesis in the hemopoietic stem cell. The inhibition of dibutyryl cyclic-GMP induced DNA synthesis by cycloheximide suggests that protein synthesis is associated with the actions of dibutyryl cyclic-GMP. Attempts thus far to block DNA synthesis through the use of actinomycin D have failed. These results, if only tentative, are compatible with the suggestion(6) that the information for DNA synthesis is already present in CFU-S. Initiation by the agents studied may therefore be a rapid unmasking of this already available information.

V. Evidence for a Steroid Mechanism

Given *in vivo* (17) or *in vitro* (7), testosterone triggers CFU-S from G_O into S. The mechanism of testosterone's effect is not clear. In contrast to the triggering actions of isoproterenol, carbamylcholine and cyclic nucleotides, however, pre-treatment of cultures with 4 mM imidazole does not suppress the initiation of DNA synthesis by testosterone hydrogen succinate (8). This suggests that testosterone does not act through cyclic nucleotides. Stucture-activity studies indicate that steroids, biologically potent as androgens, are not required for triggering CFU-S into DNA synthesis. The stem cell action of testosterone appears to be independent of erythropoietin-mediated effects of androgens on hemopoiesis (7,17). The physiological action of many steroids is preceded by a binding of steroids to cytoplasmic receptors followed by a migration of a steroid-complex to the nucleus and binding to nuclear receptors. In the case of androgens, steroids reduced at position 5, eg., 5α-dihydrotestosterone, participate in the reaction. Whether such a mechanism is involved in the cell cycle action of testosterone on CFU-S is a matter for speculation. Using the Tfm mutant mouse (18), studies are in progress to determine whether steroid-binding receptors mediate in the action of testosterone on bone marrow progenitor cells. The Tfm mutant mouse has lost the ability to bind reduced testosterone to cytoplasmic receptors (19). Hence, this mouse does not respond to androgen. It would be of interest to know whether the stem cell of the Tfm mutant could be triggered into DNA synthesis by testosterone or by any of its metabolites.

The studies thus far presented used the ^{3}H-TdR killing technique as a measure of DNA synthesis in CFU-S. Table 4 shows that following treatment with carbamylcholine, DL-isoproterenol, dibutyryl cyclic-AMP or testosterone hydrogen succinate, the sensitivity of CFU-S to the cytocidal action of

TABLE 4

HYDROXYUREA (HU) ON DRUG-INDUCED DNA SYNTHESIS

Treatment	CFU-S/Spleen ± S.E.M.		% Loss due to HU
	Without HU	HU	
None	11.2 ± 0.7 (13)	10.4 ± 0.9 (13)	7.1
Carbamylcholine (10^{-12}M)	10.5 ± 0.51 (16)	5.2 ± 0.59 (16)	50.5
DL-Isoproterenol (10^{-14}M)	12.9 ± 1.6 (12)	8.2 ± 0.80 (10)	36.4
Dibutyryl Cyclic-AMP (10^{-8} M)	11.1 ± 0.5 (13)	7.5 ± 0.5 (13)	32.4
Testosterone Hydrogen Succinate (2.4×10^{-8} M)	13.4 ± 1.35 (11)	8.9 ± 0.7 (10)	33.6

( ) = number of spleens counted.

hydroxyurea increases. These findings help to substantiate the results obtained using high-specific activity ^{3}H-TdR. Deoxynucleosides can prevent the inhibitory action of hydroxyurea on DNA synthesis. Addition of either the 2' deoxyriboside of thymine or cytosine prevents the inhibitory effect of hydroxyurea observed in cultures treated with testosterone hydrogen succinate (6,20). A similar effect of the 2' deoxyriboside of thymine was also observed in dibutyryl cyclic-AMP treated cultures (unpublished observations).

VI. Discussion

The three mechanisms described above influence the cell cycle state of the hemopoietic stem cell. A question of importance is which, if any, of these mechanisms might serve, at the *physiological level*, to regulate the proliferation of the hemopoietic stem cell. The data presented only allow for speculations. It is unlikely that the testosterone mechanism should be given consideration. The remaining mechanisms must attract attention on the basis of the low concentrations of β-adrenergic and cholinergic agents needed to initiate DNA

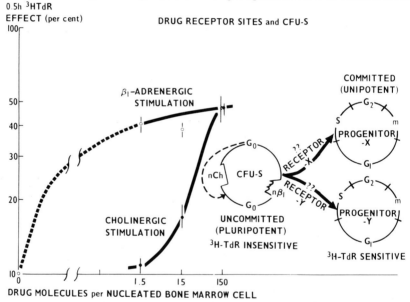

Fig. 5. The molar concentrations of isoproterenol (β_1-adrenergic stimulation) and carbamylcholine (cholinergic stimulation) are translated into molecules of drug per nucleated bone-marrow cell. For reference, 10^{-14}M is equal to 1.5 molecules per nucleated bone-marrow cell. CFU-S are pictured as having cholinergic receptors (nCh) and β_1-adrenergic receptors ($n\beta_1$) on their surfaces. The different shapes of the receptor sites represent their independence with respect to triggering CFU-S into S, and also serve to raise the question as to whether a particular receptor might determine the direction in which CFU-S differentiates.

96

synthesis. Fig. 5 shows the number of molecules of drug per nucleated bone marrow cell needed to trigger CFU-S into S. At 10^{-14} M, only a few molecules per cell are required. It may be, however, the CFU-S have an unusually strong attraction for isoproterenol or for cholinergic molecules and can concentrate molecules at these receptors.

The receptor sites, or possibly enzymes along the chain of reaction between receptors and initiation of DNA synthesis, may be points for the physiological triggering of DNA synthesis. As previously mentioned, the enzyme phosphodiesterase could act as a regulator of cell cycle state.

An endogenous source of agents capable of stimulating or blocking receptor sites must be located in order to consider the receptor sites as part of a regulatory mechanism. Nerve fibers, both myelinated and unmyelinated pass throughout the bone marrow and terminate in close proximity to bone marrow cells (21). This is no proof that neurotransmitters from these nerve endings influence or regulate CFU-S. It would, however, seem of interest to determine whether the stimulation of neurons leading to bone marrow leads to the transition of CFU-S from G_O into S. Future studies might also consider whether cells, other than nerve fibers, might serve as a source of compounds capable of acting on either adrenergic or cholinergic receptor sites; i.e., cell-cell communication via drug receptors. Given *in vitro*, the cholinesterase inhibitor neostigmine did not trigger CFU-S into DNA synthesis (Fig. 4). This experiment would indicate that bone marrow cells, per se, may not serve as a source of agents capable of stimulating cholinergic receptors. Preliminary studies in our laboratory have shown that neostigmine given *in vivo* to mice very markedly increases the sensitivity of CFU-S to the cytocidal action of hydroxyurea. This effect has been demonstrated in two F_1 hybrids (B6AF$_1$/J; C3D2F$_1$/J). These findings suggest that endogenous acetylcholine may act, in some way, to eventually trigger CFU-S into DNA synthesis.

Receptor sites studied thus far serve to switch-on DNA synthesis in the hemopoietic stem cell. The possibility exists that receptor sites may be present which serve to switch-off DNA synthesis; i.e. receptors which return CFU-S to G_O. Further studies should, therefore, involve a search for such receptor sites. Preliminary studies to determine whether a-adrenergic receptor sites might serve to switch-off DNA synthesis in CFU-S have been negative.

To implicate the enzyme phosphodiesterase as a regulator of cell cycle state, requires the uncovering of endogenous mechanisms that could alter the activity of this enzyme. Endogenous inhibitors of phosphodiesterase have not been reported. This, however, does not exclude their existence. The phosphodiesterase enzymes vary from tissue to tissue. This variation may provide the very basis for tissue specific inhibition or stimulation of cell proliferation. In time, when (and if) the much talked about tissue specific 'chalones' have been adequately characterized, it may be fruitful to study their effects on the activity of an enzyme such as phosphodiesterase.

CFU-S may differentiate into the various committed bone marrow progenitor cells, e.g., CFU-C and ESC. The mechanism by which this occurs is not clear. Fig. 5 also raises the question for future study as to whether receptor sites might influence the direction in which the stem cell differentiates.

Many of the drugs used in clinical treatment interact with drug receptor sites now known to be associated with bone marrow cells. The biochemical basis for drug-induced hypoplasia or aplasia of the bone marrow is not clearly understood. It might be reasonable to suggest that some adverse effects, in part, may be due to an unusual blocking by therapeutic agents of receptor-mediated mechanisms that might underly proliferation and differentiation in the hemopoietic system. In accordance with this suggestion, it is of interest to note that large doses of the β-adrenergic blocking agent, propanolol, have resulted in the development of aplastic anemia (22). Careful analysis of the effects on hemopoietic precursor cells of drugs known to be toxic to bone marrow, may give fruitful results that may lead to a more logical treatment of drug-induced bone marrow aplasia.

On the assumption that drug receptor sites play a role in regulating the proliferation and differentiation of the hemopoietic stem cell, the question arises as to what role, if any, these receptors might play in the leukemogenic process. Accordingly, it would be of interest to determine whether the sensitivity of drug receptor sites to agents that initiate DNA synthesis alters prior to, and during the development of leukemia. Preliminary studies in our laboratory have demonstrated that, *in vitro*, the potential carcinogen, methylazoxymethanol acetate triggers the hemopoietic stem cell from G_0 into DNA synthesis. The carcinogenic activity of this molecule depends upon the formation of methylazoxymethanol following deacetylation of the parent molecule by cholinesterase. The triggering action of the acetate derivative in bone marrow is prevented if cholinesterase activity is inhibited. Hence, it would be of added interest to know whether the interaction of potential carcinogens with receptor sites mediates in chemically-induced malignancy; e.g., chemically-induced leukemia.

Acknowledgements

A major portion of the experimental findings reported in this study were obtained when the author was a staff member at the Paterson Laboratories, Christie Hospital and Holt Radium Institute, Manchester, U.K.

I wish to thank Joyce K. Punchard for technical assistance during these experiments. Propranolol was a gift from Imperial Chemical Industries, Ltd., and testosterone hydrogen succinate was a gift from the Upjohn Co. Ro20-1724 was a donation from the Hoffman — La Roche Co.

References

1. J. E. Till and E. A. McCulloch, *Rad. Res.*, *14*, 213, (1961).
2. A. J. Becker, E. A. McCulloch and J. E. Till, *Nature*, *197*, 452, (1963).
3. J. H. Fowler, A. M. Wu, J. E. Till, E.A. McCulloch and L. Siminovitch, *J. Cell. Physiol.*, *69*, 65, (1967).
4. D. J. A. Sutherland, J. E. Till and E. A. McCulloch, *J. Cell. Physiol.*, *75*, 267, (1970).
5. A. J. Becker, E. A. McCulloch, L. Siminovitch and J. E. Till, *Blood*, *26*, 296, (1965).
6. J. W. Byron, *Blood*, *40*, 2, (1972).
7. J. W. Byron, *Nature*, *234*, 39, (1971).
8. J. W. Byron, *Exp. Cell. Res.*, *71*, 228, (1972).
9. J. W. Byron and J. K. Punchard, Second International Workshop on Hemopoiesis in culture. Warrenton, Virginia, May 23-26, 1973. *In press*.
10. H. Sheppard and G. Wiggan, *Biochem Pharmacol.*, *20*, 2128, (1971).
11. J. W. Byron, *Nature (New Biol.)*, *241*, 152, (1973).
12. W. J. George, J. B. Polson, A. G. O'Toole and N. D. Goldberg. *Proc. Nat. Acad. Sci.* (U.S.A.), *66*, 398, (1970).
13. J. A. Ferrendelli, A. L. Steiner, D. B. McDougal and D. M. Kipnis, *Biochem Biophys. Res. Commun.*, *41*, 1061, (1970).
14. H. McIlwain, *Effects of Drugs on Cellular Control Mechanisms*, (edit. by B. R. Rabin and R. B. Freedman, Macmillan, London), *281*, *(1971)*.
15. A. L. Steiner, A. S. Pagliara, L. R. Chase and D. M. Kipnis, *J. Biol.Chem.*, *247*, 1114 (1972).
16. G. Schultz, K. H. Jakobs, E. Bohme and K. Schultz, *Eur. J. Biochem.* *24*, 520, (1972).
17. J. W. Byron, *Nature*, *228*, 1204, (1970).
18. M. F. Lyon and S. G. Hawkes, *Nature*, *227*, 1217, (1970).
19. U. Gehring, G. M. Tomkins and S. Ohno, *Nature (New Biol.)*, *232*, 106, (1971).
20. J. W. Byron, Second International Workshop on Hemopoiesis in culture. Warrenton, Virginia, May 23-26, 1973, *in press*.
21. W. Calvo, *Amer. J. Anat.*, *123*, 315, (1968).
22. J. Duttera, *personal communication*.

EXPRESSION OF THE MITOCHONDRIAL GENOME
DURING THE CELL CYCLE IN HELA CELLS

James M. England, Livia Pica-Mattoccia * *and Giuseppe Attardi*
Division of Biology, California Institute of Technology
Pasadena, California 91109

I. Introduction

Mitochondria of all eukaryotic cells contain a unique DNA and a distinctive protein synthesizing system, with specific ribosomes, tRNA and amino acyl-tRNA synthetases, initiation and elongation factors (1-3). The expression of the mitochondrial genome appears to occur mainly, if not exclusively, through this protein synthesizing system. In particular, the mitochondrial genome specifies most, if not all, RNA components of the mitochondrial protein synthesizing system and probably also several polypeptides which are synthesized within the organelles and which are essential for mitochondrial function, e.g. components of cytochrome c oxidase (4-6) and oligomycin-sensitive ATPase (7-8).

Several lines of evidence suggest that the mitochondrial genome is functionally coupled with the nuclear genetic system and that its expression is integrated with the overall processes of cellular growth and division. In dividing cells, the average amount of mitochondrial DNA (mit-DNA) and the average mitochondrial functional capacity per cell remain constant from one generation to the next. This implies some form of control on the part of extra-mitochondrial compartment on mit-DNA replication and transcription. Furthermore, the assembly of the mitochondrial protein synthesizing apparatus and of functionally active mitochondria is under the control of both the nuclear and the mitochondrial genetic systems, pointing to a coordination and interdependence between the two systems. Useful information concerning the nature and the mechanisms of this coupling could come from an analysis of the expression of the mitochondrial genome during the cell cycle.

In this paper we report the results of experiments aimed at investigating the temporal patterns of mitochondrial DNA, RNA and protein synthesis during the cell cycle of HeLa cells synchronized by the selective detachment technique (9).

II. Mit-DNA Replication during the Cell Cycle

---

*Present address: Laboratorio di Biologia Cellulare, C.N.R. via Romagnosi 18A, Rome, Italy.

Mit-DNA synthesis has been previously reported to proceed throughout the cell cycle, with a possible acceleration in correspondence to nuclear DNA synthesis, in *Tetrahymena pyriformis* (10-12) and *Physarum polycephalum* (13,14). In yeast mit-DNA, replication has been shown to occur synchronously, or almost so, after nuclear DNA synthesis, 20-30 minutes before bud separation (15,16).

In the work reported here, synchronous cultures of HeLa cells (S3 line) were obtained by selective detachment of mitotic cells (9) and labeled for 1 hour with [methyl-^{3}H] thymidine at different stages of the cell cycle. The DNA was extracted from the mitochondrial fraction and analyzed in a CsCl-ethidium bromide density gradient (17). From the amount of radioactivity associated with closed circular mit-DNA, after correcting for differences in the labeling of intra-mitochondrial pools of thymidine and its phosphorylated derivatives, the relative rate of mit-DNA synthesis per cell during the cell cycle was estimated.

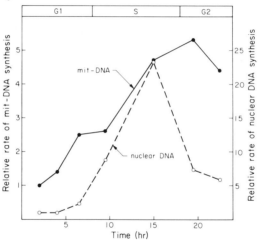

Fig. 1.   Relative rate of mit-DNA synthesis during the cell cycle in HeLa cells. ●—●—● mit-DNA; o---o---o nuclear DNA. (Drawn from 18).

Figure 1 summarizes the data thus obtained in several of these experiments. (18). One can see that there is relatively little replication of mit-DNA in G_1 cell populations, and that the rate of mit-DNA synthesis per cell starts accelerating in "early S" populations and reaches a maximum in "late S" and in "G_2" populations: this maximum corresponds to more than 5 times the rate in "G_1" populations. Therefore, mit-DNA replication in HeLa cells appears to occur predominantly, if not exclusively, in a restricted portion of the cell cycle covering the S and G_2 phases. It is not certain, on the basis of these results, whether any mit-DNA replication occurs in cells in the G_1 phase. In these experiments, 70 to 80% of the selectively detached cell population consisted of cells in mitosis. Assuming that in an unsynchronized HeLa cell population there are about 50% cells in G_1 phase, about 30% in S phase and 15% in G_2

phase (19), each stage having the rate of mit-DNA synthesis observed in the present work, then the 20 to 30% of unsynchronized cells present in the original population would account for at least 50% of the level of mit-DNA synthesis detected in "G_1" populations. It could indeed account for all or almost all of it if, as is likely, the rate of mit-DNA synthesis in cells in G_2 phase is considerably higher than that observed in "G_2" populations, because of the substantial contamination of the latter by cells in S phase, mitotic cells, and cells in G_1 phase.

It should be pointed out that, in the interpretation of the results given here, the assumption has been made that all, or a constant proportion, of mit-DNA is in the form of covalently closed circles at all times during the cell cycle.

III. Mit-DNA Transcription during the Cell Cycle

Mit-DNA is very actively transcribed in HeLa cells. The transcription products consist of discrete species with sedimentation constants of 16 S, 12 S, and 4 S, and of heterogenously sedimenting species. (20,21). The 16 S and 12 S RNA species represent the structural RNA components of the large and small subunits, respectively, of mitochondria specific 60 S ribosomes, which are involved in mitochondrial protein synthesis (22); at least a part of the 4 S RNA is represented by mitochondria-specific transfer RNA species, as shown for rat liver (23) and HeLa cells (24).

In order to measure the rate of mitochondrial RNA synthesis during the cell cycle, synchronous cultures of HeLa cells (S3) (obtained by selective mitotic detachment) were pulse-labeled at different times with [5-^{3}H] uridine (25). After addition of a constant excess of unlabelled cells, the mitochondrial fraction was isolated by differential centrifugation, washed with EDTA (26), and the mitochondria-associated RNA extracted and analyzed in sucrose sodium dodecyl SO_4 gradients.

Figure 2 (a), (b) and (c) show the sedimentation patterns of RNA extracted from the mitochondrial fraction of synchronized HeLa cells exposed to a 20-minute [5-^{3}H] uridine pulse at 2.5 hours ("G_1"), 12.5 hours ("S") and 17 hours ("G_2"), respectively after resuspension in fresh medium (see insert of nuclear DNA labeling). In the OD_{260} profile one can recognize two peaks at 28 S and 18 S, corresponding to the RNA components of ribosomes of the contaminating rough endoplasmic reticulum not removed by the EDTA washing (27). The mitochondrial 16 S and 12 S RNA components are not resolved from the 18 S RNA peak in these short centrifugal runs. The radioactivity profile shows a broad peak centered around 16 S, and heavier heterogeneous components with sedimentation coefficients up to 50 S and more. In longer sedimentation runs which fully display the components sedimenting slower than 28 S, the initial labeling of the discrete 16, 12 and 4 S RNA species can be clearly seen after this labeling time (26). A peak at about 33 S is reproducibly recognizable in the radioactivity pattern emerging over the background of heterogeneous RNA. Previous work from this laboratory (21) has shown that the rapidly labeled heterogeneous RNA associated with the mitochondrial fraction from HeLa cells hybridizes with great efficiency with

103

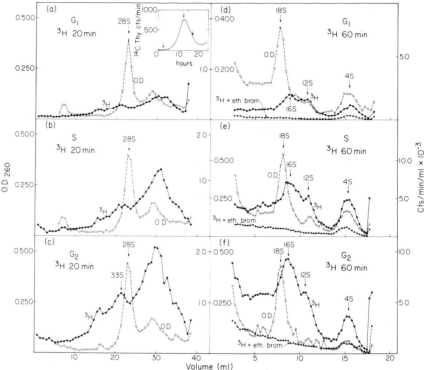

Fig. 2. Labeling of mitochondria-associated RNA from HeLa cells pulsed with [5-^{3}H] uridine for 20 minutes [(a), (b), and (c)], or 60 minutes [(d), (e), and (f)], at different stages of the cell cycle.

The insert in (a) shows the rate of [methyl-^{14}C] thymidine incorporation into total cell DNA in the 20 minute pulse labeling experiment. Arrows point to times at which samples were removed from the synchronized population for mitochondrial RNA labeling.

In panels [(d), (e), and (f)], three different OD_{260} scales have been used in order to make the 18 S RNA peaks (pertaining mostly to the unsynchronized unlabeled carrier cells) of similar height, and the radioactivity scales have been made in proportion, so that the labeling profiles are directly comparable. (In different experiments, the amount of ribosomal RNA pertaining to ribosomes of the rough endoplasmic reticulum which contaminates the mitochondrial fraction has been found to be proportional to the amount of mitochondria, as estimated from the cytochrome oxidase activity, thus providing a convenient internal standard to correct for differences in cell breakage.)

--o--, OD_{260}; ⬤—⬤ ^{3}H radioactivity; -- -- -- ^{3}H + ethidium bromide (25). △—△

mit-DNA. Furthermore, ethidium bromide, an intercalating phenanthridine dye which selectively inhibits, at suitable concentrations, the synthesis of mitochondrial RNA(28), suppresses almost completely the labeling of this heterogeneous RNA at a concentration of 1 ug/ml(21); at this dose, in agreement with the data reported by Zylber et al (28), the drug has been found not to have any detectable effects for at least 4 hours on the labeling of total cell RNA, to which nuclear RNA synthesis contributes by more than 95% (26 and unpublished results). These observations strongly suggest that the mitochondrial fraction, as prepared here, is substantially free of RNA leaked from the nuclei or from nuclear fragments.

One can see from Fig. 2 that on a per cell basis, the labeling, during a 20-minute [5-^{3}H] uridine pulse, of different size components of the mitochondrial heterogeneous RNA in "G_2" populations is four to five times higher than that in "G_1" populations, and almost twice as high as that in "S" populations. Differences in the specific activity of the precursor pools in different phases of the cell cycle could conceivably account for the above mentioned incorporation results. In order to test this possibility, the specific activity of the intra-mitochondrial UTP pool was determined as a function of the cell cycle following a 20-minute [5-^{3}H] uridine pulse. It was found that the labeling of the mitochondrial UTP pool was very similar in "S" and "G_2" populations and only 30% lower in "G_1" populations. It is clear, therefore, that the observed differences in the labeling of mit-DNA-coded heterogeneous RNA during the cell cycle cannot be accounted for by pool labeling effects and must reflect differences in the rate of synthesis of this RNA.

Figure 2 (d), (e) and (f) show the sedimentation patterns of the RNA components sedimenting slower than 28 S from the mitochondrial fraction of synchronized HeLa cells exposed to a 60-minute [5-^{3}H] uridine pulse [carried out in the presence of 0.04 ug/ml actinomycin D to inhibit selectively the synthesis of cytoplasmic ribosomal RNA (29-31)] at 2.5 hours ("G_1"), 11.75 hours ("S") and 16 hours ("G_2"), respectively, after resuspension in fresh medium. One can see that the labeling of the discrete 16, 12 and 4 S RNA components in cells at different stages of the cell cycle shows the same behavior observed for the 20-minute pulse labeling of the heterogeneous mitochondrial RNA. It appears also from Fig. 2 (d), (e) and (f) that the labeling of the 16 S and 12 S RNA, and of most of the 4 S RNA and of heavier heterogeneous components, is inhibited by ethidium bromide at 1 µg/ml, in confirmation of previous observations (28, 21). The relatively small amount of ethidium bromide-insensitive heterogeneous RNA presumably represents, in part at least, RNA of nuclear origin associated with rough endoplasmic reticulum. The increase in the labeling of the ethidium bromide-insensitive heterogeneous RNA in "S" and "G_2" populations, as compared to that in "G_1" populations, is in agreement with previously reported observations indcating a two- to threefold increase in the rate of incorporation of uridine into heterogeneous RNA of nuclear origin across interphase (32-34).

Table 1 summarizes the results of the labeling of mitochondrial RNA after a 20-minute pulse and a 60-minute pulse (in the latter case only the ethidium bromide-sensitive incorporation has been considered) in cells at different stages

105

TABLE 1

Incorporation of [5-^{3}H] uridine into Mitochondrial RNA after a 20-minute or a 60-minute Pulse into HeLa Cells in Different Phases of the Cell Cycle

Cell population	20-min pulse			60-min pulse	
	Total cts/min incorporated	Relative incorporation of [5-^{3}H] uridine	Relative rate of RNA synthesis	Ethidium bromide-sensitive cts/min incorporated	Relative incorporation of [5-^{3}H] uridine
"G$_1$"	7,390	1.0	1.0	12,600	1.0
"S"	18,100	2.5	1.9	33,820	2.7
"G$_2$"	32,250	4.4	3.4	64,690	5.1

The data of total [5-^{3}H] uridine incorporation after a 20-min pulse and of ethidium bromide-sensitive incorporation after a 60-min pulse in the experiments illustrated in Fig. 2 have been normalized for differences in recovery of ribosomal RNA. The relative rates of synthesis of mitochondrial RNA during a 20-min [5-^{3}H] uridine pulse have been calculated by correcting the incorporation data for differences in specific activity of the intramitochondrial UTP pool (From 18).

of the cell cycle. The data have been normalized for differences in recovery of mitochondria, due to variations in cell breakage, as estimated from the amount of ribosomal RNA of the contaminating rough endoplasmic reticulum (see legend of Fig. 2). The 20-minute pulse incorporation data have also been converted to relative rates of RNA synthesis by using the values of specific activity of the intra-mitochondrial UTP pool.

In the present experiments, the original selectively detached population contained 15 to 20% of unsynchronized cells. This fraction of cells could presumably account for only 20 to 30% of the relatively low level of labeling of mitochondrial RNA observed in the present work in "G_1" populations after a 20-minute or a 60-minute [5-^{3}H] uridine pulse. This suggests that mitochondrial RNA synthesis occurs, although at a reduced rate, also in the G_1 phase. Taking into consideration the relatively small fraction of unsynchronized cells in the selectively detached population and their doubling within 20 to 24 hours, it appears that the increase in the rate of synthesis of mitochondrial RNA per cell observed in "S" and especially in "G_2" populations is considerably greater than can be accounted for by the doubling of mit-DNA templates occurring in the synchronized cells during the S and G_2 phases (Section II). It should be further noticed that the substantial contamination of the "G_2" populations by originally synchronized cells which have lagged behind the S phase or progressed to the G_1 phase, would mask to an appreciable extent the increase in the rate of mitochondrial RNA synthesis per cell in G_2 cells. Therefore, it is reasonable to conclude from the present observations that a considerable acceleration of mitochondrial RNA synthesis, independent of the expected increase in templates, occurs in HeLa cells in the S and especially in the G_2 phase.

IV. Mitochondrial Protein Synthesis during the Cell Cycle

A "minor" protein synthesizing system, which is both structurally and functionally distinct from the main cytoplasmic protein synthesizing system, is located in the mitochondrial compartment of eukaryotic cells (for reviews, 1-3). It seems likely that the effects of factors which control mitochondrial gene expression manifest themselves through changes in the rate of overall or specific mitochondrial protein synthesis.

In order to measure the rate of mitochondrial protein synthesis during the cell cycle, synchronous cultures of HeLa cells (F-315 line) (obtained by selective mitotic detachment) were exposed, at different times, for 1 hour to [4-5-^{3}H] leucine in the presence of 100 µg/ml emetine, an inhibitor of extra-mitochondrial protein synthesis (35). Parallel cultures labeled in the presence of emetine and 200 µg/ml chloramphenicol (CAM), an effective inhibitor of mitochondrial protein synthesis (36-38), were used to determine any emetine resistant non-mitochondrial incorporation. In this work, about 70% of the emetine-resistant leucine incorporation was CAM-sensitive; the products of emetine-resistant CAM-sensitive leucine incorporation were found to be associated in their great majority with mitochondria and presumably result from mitochondrial protein synthesis(39).

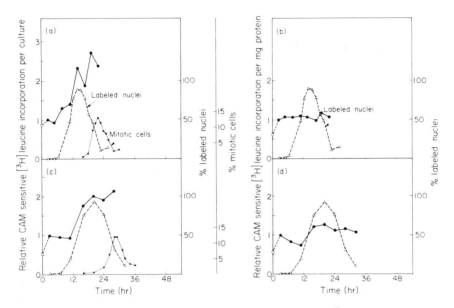

Fig. 3. Chloramphenicol-sensitive incorporation of [4,5-^{3}H] L-leucine into protein in HeLa cells at various stages of the cell cycle. Equal samples of physiologically synchronized HeLa cells were grown in Petri dishes at 37°C (a and b) or 34.5°C (c and d), and at various times during the cell cycle exposed either to [4,5-^{3}H] L-leucine (1 hour at 37°C) for protein labeling, or to [methyl-^{3}H] thymidine (30 minutes at 37°C) for determination of the percentage of labeled nuclei by authoradiography. The percentage of mitotic cells was determined in parallel cultures stained with 1% acetic orcein (39).

As appears in Fig. 3(a), there is approximately a twofold increase during interphase in the average rate of CAM-sensitive [^{3}H] leucine incorporation into protein per cell. Figure 3(b) shows that, when normalized for the increase in cell mass, the CAM-sensitivie [^{3}H] leucine incorporation into protein proceeds at a constant rate throughout the cell cycle, with the exception of the mitotic period. In this period, the rate of [^{3}H] leucine incorporation into protein, on a per unit mass basis, appears to be somewhat depressed with respect to the G_1 phase.

In the experiment illustrated in Fig. 3(a), the short duration of the G_1 phase did not allow a precise investigation of the behavior of mitochondrial protein synthesis in the early portion of the cycle. It was felt, therefore, to be desirable to try to extend the G_1 phase by lengthening the cell generation time with a reduced incubation temperature.

Figures 3(c) and (d) show the rates of CAM-sensitive [^{3}H] leucine

108

incorporation into protein in synchronized cells grown at 34.5°C. As compared with the growth at 37°C, the generation time at 34.5°C was lengthened to 29 hours and the peak of nuclear DNA synthesis was shifted to 20 hours. In this experiment, it can be seen that the relative rate of CAM-sensitive [^{3}H] leucine incorporation into protein is lower in mitotic cells as compared to "G_1" populations, both on a per cell and a per unit mass basis [Fig. 3(c) and (d)]. The rate of CAM-sensitive [^{3}H] leucine incorporation into protein per culture remains fairly constant during G_1, increases rapidly in the first half of the S phase and then stabilizes or increases slowly in the late S-G_2 phases [Fig. 3(c)]. On a per unit mass basis, the rate of incorporation appears to decrease slightly during G_1, increases at the beginning of the S phase to a level somewhat higher than that of early G_1, and then stabilizes in the rest of the cell cycle [Fig. 3(d)].

In the experiments described above it was conceivable that cell cycle dependent variations in the uptake of [^{3}H] leucine and/or the size of the mitochondrial leucine pool could mask the real behavior of mitochondrial protein synthesis. However, an analysis of the labeling of the extra- and intra-mitochondrial leucine pools after [^{3}H] leucine pulses revealed no significant variation during the cell cycle (39). Therefore, it seems reasonable to conclude that the [^{3}H] leucine incorporation rates measured in different stages of the cell cycle reflect fairly closely the rate of mitochondrial protein synthesis.

A relatively constant rate of total mitochondrial protein synthesis during the cell cycle would be compatible with either a different rate of synthesis of various protein species in different portions of the cycle or a constant rate synthesis of all individual species. In order to distinguish between these two alternatives, synchronized cell cultures maintained at 37°C, at various stages of the cell cycle, were labeled for 1 hr with [^{3}H]-leucine in the presence of emetine or emetine and CAM, and mixed with a constant excess of unsynchronized unlabeled cells. A sodium dodecyl SO_4 lysate of the mitochondrial fraction from these cells was analyzed by sodium dodecyl SO_4-polyacrylamide gel electrophoresis. Such a method of fractionation is known to separate proteins on the basis of their mol. wt. Previous work (40) has shown that relatively simple electrophoretic profiles are obtained for proteins *in vivo* or *in vitro* by HeLa cell mitochondria. Fig. 4(a) shows the electrophoretic distribution of proteins labeled *in vivo* in mitochondria of unsynchronized HeLa F-315 cells during a 1-hour L-[4,5-^{3}H]-leucine pulse. The profile exhibits a main group of not well resolved components migrating in the region corresponding to 30,000 to 50,000 mol. wt. and another group, less abundant, in the range 12,000 to 25,000 in good agreement with the electrophoretic pattern observed for the mitochondrially synthesized proteins from HeLa S3 cells (40). In the presence of 200 μg/ml CAM, the labeling of the

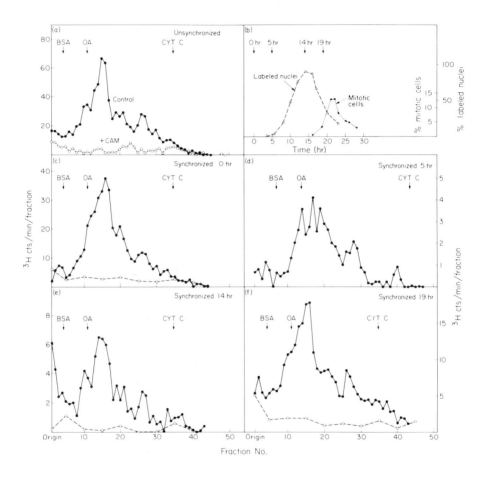

Fig. 4. Polyacrylamide gel electrophoretic analysis of proteins of a crude mitochondrial fraction from HeLa cells labeled in the presence of emetine or emetine + CAM at various stages during the cell cycle. (a): Lysate of the mitochondrial fraction from unsynchronized HeLa cells pulse-labeled for 1 hour with [^{3}H] leucine in presence of emetine (control) or emetine + CAM. (b): Time pattern of nuclear DNA synthesis and mitotic activity in the synchronized cells utilized in this experiment. (c), (d), (e) and (f): Lysates of the mitochondrial fraction from synchronized cells which were labeled as in (a) at 0 hour, 5 hour, 14 hour, and 19 hour, respectively, after mitotic selection (39).

discrete components is eliminated, and the low amount of residual incorporation is distributed fairly uniformly throughout the gel. Figures 4(c), (d), (e), and (f) show that the electrophoretic profiles of proteins newly synthesized in mitochondria of cells in mitosis (0 hour), "G_1" cells (5 hour), "S" cells (14 hour), and "G_2" cells (19 hour) are very similar to each other and to the profile obtained from unsynchronized cells. This result indicates that, within the limits of resolution of the analysis applied here, proteins of the various size classes are synthesized in HeLa cell mitochondria at the same relative rate during the cell cycle.

V. Discussion

The evidence presented here indicates that mit-DNA is replicated and transcribed discontinuously during the cell cycle in HeLa cells. The observation that the synthesis of both mitochondrial DNA and RNA reaches a peak during late S and G_2 phases in physiologically synchronized cultures suggests that there might be a link between the two processes. Furthermore, these results indicate that extramitochondrial factors related to the cell cycle, possibly nuclear signals, influence the replicative and transcriptive activity of mit-DNA in HeLa cells. In agreement with this possibility, Storrie and Attardi (41) reported that inhibition of cytoplasmic protein synthesis by cycloheximide caused a progressive, fairly rapid decline in the rate of synthesis of mitochondrial RNA and DNA. Although several factors may be responsible for this decline, an interesting possibility is that it is due to the progressive disappearance, by turnover or utilization, of some specific protein which needs to be continually produced in the cytoplasm to maintain the normal level of mitochondrial RNA and DNA synthesis during the "active" period of the cell cycle. One may speculate that this protein is a signal involved in the cell cycle dependent regulation of mitochondrial nucleic acid synthesis.

The situation found in HeLa cells as concerns the temporal pattern of mit-DNA replication during the cell cycle appears to be analogous to that described in yeast, where the mit-DNA replicates synchronously, or almost so, during the period following nuclear DNA synthesis and before bud separation (15,16). In both cases, extra-mitochondrial signals appear to determine the onset of mit-DNA synthesis. The existence of labile nuclear signals produced once per cell cycle would ensure a coordination of cell division and mitochondrial reproduction.

The detection, in "G_1" populations, of a level of mitochondrial RNA synthesis which cannot be accounted for completely by the contaminating unsynchronized cells, suggests that a significant, though relatively low rate of mit-DNA transcription occurs during the G_1 phase of the cell cycle. The most plausible interpretation is that the mitochondrial RNA synthesis occurring in the G_1 phase represents the continuation, at a decreasing rate, of the burst of mit-DNA transcriptive activity initiated in the late S and G_2 phases. In agreement with this interpretation is the observation that HeLa cells arrested in metaphase by colcemide after a double thymidine block synthesize mitochondrial RNA (42). In contrast to mitochondrial RNA synthesis, very little, if any, mit-DNA synthesis occurs in the G_1 phase. This suggests that the cycle of mit-DNA transcriptive activity initiated in the late S and G_2 phases extends beyond mit-DNA replication, although both appear to begin at approximately the same time.

One can only speculate at present on whether mit-DNA synthesis in the individual mitochondria of each "active" cell proceeds continuously throughout the replicative phase of the cell cycle, or whether, on the contrary, synthesis is discontinuous at the level of the individual mitochondria and asynchronous in the mitochondria of the same cell. If the average rate of mit-DNA replication in HeLa cells is similar to that estimated for mouse L cells (0.1 μ/min) on the basis of the frequency of replicating molecules (5 to 6%)(43-44), one can estimate that the replication of each mit-DNA molecule would require about 50 minutes. Assuming a content of 2 to 6 mit-DNA molecules per mitochondrion in HeLa cells, as in L cells (45), the replication of the entire DNA complement of a mitochondrion would require about 50 minutes if the mit-DNA molecules in the same organelle replicate simultaneously, 100 to 300 minutes if they replicate successively without any time interval. These times are shorter than the S and G_2 phases of the cell cycle. The calculations above would, therefore, tend to favor the idea of asynchronous mit-DNA synthesis in the mitochondria of the same cell, in agreement with what has been described for mit-DNA synthesis in *Tetrahymena* on the basis of autoradiographic experiments (11).

At present, there is no direct evidence regarding the time pattern of mitochondrial RNA synthesis in the mitochondria of the same cell and the temporal relationship between replication and transcription in the individual organelles. If the period of mit-DNA transcription in each mitochondrion extends over a longer portion of the cell cycle than the period of mit-DNA replication, one would expect a less pronounced asynchrony of RNA synthesis, as compared to DNA synthesis, in mitochondria of the same cell.

In contrast to mit-DNA replication and transcription, whose rate per cell is greatly accelerated in the S phase and reaches a peak in the G_2 phase corresponding to at least 4 or 5 times the early G_1 level, the rate of mitochondrial protein synthesis in HeLa cell mitochondria increases only about two-fold, on a per cell basis, during interphase. This increase does not appear to be uniform throughout the cell cycle. In fact, the observations made on cultures incubated at 34.5°C, which exhibit a more expanded G_1 phase that is better resolved from the S phase than at 37°C, indicate that the rate remains fairly constant during G_1 phase, and accelerates at the onset of the S phase, in coincidence with the acceleration of mitochondrial RNA synthesis.

On a per unit mass basis, the rate of mitochondrial protein synthesis at 37°C appears to be constant throughout the cell cycle, except during mitosis. On the contrary, at 34.5°C, as expected from the incorporation per cell data, there appears to be a slight decline in the rate of mitochondrial protein synthesis per unit mass during the G_1 phase, which is compensated by an increase in the early S phase. This observation suggests that the apparent constancy observed at 37°C is the result of poor resolution between the G_1 and S phases at this temperature.

As concerns the time during the cell cycle when new units of functional mitochondrial protein synthesizing apparatus are assembled, the evidence obtained here of a decline during G_1 phase in the rate of organelle specific protein synthesis per unit mass suggests the possibility that the assembly of new active protein synthesizing machinery does not occur throughout the cell cycle in parallel with the increase in cell mass. Instead, the data would be more compatible with a process of assembly occurring mainly in the S and G_2 phases, in correspondence with the synthesis of new mitochondria-specific ribosomal and transfer RNA species (section III); this process of assembly would add new protein synthesizing units to the pre-existing ones in proportion to the overall increase in cell mass, or even in excess if the protein synthesizing apparatus is somewhat unstable. There is indeed evidence for a certain metabolic instability of mitochondrial ribosomal RNA in cultured animal cells (26,46). If the above interpretation is correct, the protein synthesizing activity measured in the G_1 phase must be due to the pre-existing units; this activity would be expected to decrease on a per unit mass basis during G_1, as indeed appears to be the case. Further investigations are needed to establish this point conclusively.

The observation that the electrophoretic pattern of the products of mitochondrial protein synthesis remains essentially unchanged in cells in different stages of the cell cycle suggests that there is little or no cell cycle dependent differential control, at the tarnscription or translation level, of their synthesis. These products, which have been shown to be components of the inner mitochondrial membrane (see review by Beattie, 47) that are essential for the respiratory functions of mitochondria(41); see also review by Ashwell and Work (1), are apparently synthesized at the same relative rate during the cell cycle. Whether they are continuously utilized for the assembly of an active inner mitochondrial membrane remains to be established. As mentioned above, there are in mitochondria functional complexes such as mitochondrial ribosomes (for summary, see 3), cytochrome c oxidase (4-6) and oligomycin-sensitive ATPase(7,8) which require the contribution of both the mitochondrial and the extra-mitochondrial genetic and protein synthesizing systems. An analysis of the timing during the cell cycle of the synthesis of the components of these complexes which are under the control of the nucleo-cytoplasmic system, and of their assembly with the mitochondrially synthesized components, will be very informative concerning the nature and mechanism of coupling between these two systems.

As far as has been established, the major cell cycle dependent control of the expression of the mitochondrial genome appears to occur at the level of transcription(25). This transcriptional control, however, while essential for increasing the overall mitochondrial protein synthesizing capacity of the cell in proportion to its mass, does not appear to play any significant role in the regulation of synthesis of different size classes, nor possibly of individual species, of proteins. The present observations imply that most of the RNA species which serve as messenger for mitochondrial protein synthesis, independent of their origin, either are relatively stable or are synthesized throughout the cell cycle.

While the results reported here suggest the lack of cell cycle-dependent control of mitochondrial protein synthesis in exponentially growing cells, nothing can be said about the possible existence of regulation in response to physiological stimuli. The modulation of the mitochondrial function which has been described in response to thyroid hormones (48) oxygen tension (49) or changes in respiratory demands (50), may indeed involve a translation or transcriptional control of the synthesis within the organelles of some proteins having an essential role in oxidative phosphorylation.

Acknowledgements

This work was supported by grant GM-11726 from the National Institutes of Health and a Dernham Fellowship of the American Cancer Society, California Division (D-196) to J.M.E. It is a pleasure to acknowledge the valuable assistance of Mrs. Benneta Keeley, Mr. James Posakony and Mrs. LaVerne Wenzel.

References

1. M. Ashwell and T. S. Work, *Ann. Rev. Biochem. 39*, 251 (1970).
2. P. Borst and L. A. Grievell, *FEBS Letters 13*, 73 (1971).
3. G. Attardi, P. Costantino, J. England, M. Lederman, D. Ojala and B. Storrie, *Sixth Karolinska Symposium on Research Methods in Reproductive Endocrinology* 263 (1973).
4. A. M. Kroon and H. DeVries, In *Symposium on Autonomy and Biogenesis of Mitochondria and Chloroplasts.* North Holland, Amsterdam, 318 (1971).
5. B. Kadenbach, In *Symposium on Autonomy and Biogenesis of Mitochondria and Chloroplasts.* North Holland, Amsterdam, 360 (1971).
6. G. Schatz, G.S. P. Grott, T. Mason, W. Rouslin, D. C. Wharton and J. Saltzgaber, *Fed. Proc. 31*, 21 (1972).
7. A. Tzagoloff and P. Meagher, *J. Biol. Chem. 247*, 6511 (1972).
8. A. Tzagoloff, A. Akai and M. F. Sierra, *J. Biol. Chem. 247*, 6511 (1972).
9. T. Terasima and L. J. Tolmach, *Exp. Cell Res. 30*, 344 (1963).
10. J. A. Parsons, *J. Cell Biol. 25*, 641 (1965).
11. J. A. Parsons and R. C. Rustad, *J. Cell Biol. 37*, 683 (1968).
12. R. Charret and J. Andre, *J. Cell Biol. 39*, 369 (1968).
13. E. W. Guttes, P. C. Hanawalt, and S. Guttes, *Biochim. biophys. Acta 142*,

114

181 (1967).

14. T. E. Evans, *Biochim. biophys. Acta 182*, 511 (1969).
15. D. Smith, P. Tauro, E. Schweizer and H. O. Halvorson, *Proc. Nat. Acad. Sci. (U.S.A.) 60*, 936 (1968).
16. S. T. Cottrell and C. J. Avers, *Biochem. Biophys. Res. Comm. 38*, 973 (1970).
17. B. Hudson and J. Vinograd, *Nature 216*, 647 (1967).
18. L. Pica-Mattoccia and G. Attardi, *J. Mol. Biol. 64*, 465 (1972).
19. P. N. Rao and J. Engelberg, In *Cell Synchrony: Studies in Biosynthetic Regulation* (eds. I. L. Cameron and G. M. Padilla) Academic Press, New York, 332 (1966).
20. B. Attardi and G. Attardi, *Nature 224*, 1079 (1969).
21. G. Attardi, Y. Aloni, B. Attardi, D. Ojala, L. Pica-Mattoccia, D. L. Robberson and B. Storrie, *Cold Spr. Harb. Symp. Quant. Biol. 35*, 599 (1970).
22. G. Attardi and D. Ojala, *Nature New Biology 229*, 133 (1971).
23. C. A. Buck and M. M. K. Nass, *J. Mol. Biol. 41*, 67 (1969).
24. J. B. Galper and J. E. Darnell, *Biochem. Biophys. Res. Comm. 34*, 205 (1969).
25. L. Pica-Mattoccia and G. Attardi, *J. Mol. Biol. 57*, 615 (1971).
26. B. Attardi and G. Attardi, *J. Mol. Biol. 55*, 231 (1971).
27. B. Attardi, B. Cravioto and G. Attardi, *J. Mol. Biol. 44*, 47 (1969).
28. E. Zylber, C. Vesco and S. Penman, *J. Mol. Biol. 44*, 195 (1969).
29. R. P. Perry, *Nat. Cancer Inst. Monograph 14*, 73 (1964).
30. D. T. Dubin, *Biochem. Biophys. Res. Comm. 29*, 655 (1967).
31. S. Penman, C. Vesco and M. Penman, *J. Mol. Biol. 34*, 49 (1968).
32. L. J. Bello, *Biochim. biophys. Acta 157*, 8 (1968).
33. S. E. Pfeiffer, *J. Cell Physiol. 71*, 95 (1968).
34. G. N. Pagoulatos and J. E. Darnell, *J. Cell Biol. 34*, 49 (1968).
35. S. Perlman and S. Penman, *Biochem. Biophys. Res. Comm. 40*, 941 (1970).
36. A. M. Kroon, *Biochim. biophys. Acta 108*, 275 (1965).
37. A. Linnane, In *Biochemical Aspects of the Biogenesis of Mitochondria* (eds. Stater *et al.*) Adriatica Editrice, Bari, *333* (1968).
38. M. Lederman and G. Attardi, *Biochem. Biophys. Res. Comm. 40*, 1492 (1970).
39. J. M. England and G. Attardi, submitted for publication (1973).
40. M. Lederman and G. Attardi, *J. Mol. Biol. 78*, 275 (1973).
41. B. Storrie and G. Attardi, *J. Mol. Biol. 71*, 177 (1972).
42. H. Fan and S. Penman, *Science 168*, 135, (1970).
43. D. L. Robberson, H. Katsamatsu and J. Vinograd, *Proc. Nat. Acad. Sci. (U.S.A.) 69*, 737 (1972).
44. D. L. Robberson and D. A. Clayton, *Proc. Nat. Acad. Sci. (U.S.A.) 69*, 3810 (1972).
45. M. M. K. Nass, *J. Mol. Biol. 42*, 521 (1969).

46. D. T. Dubin, *J. Biol. Chem. 247*, 2662 (1972).
47. D. S. Beattie, *Subcell. Biochem. 1*, 1 (1971).
48. D. B. Roodyn, K. B. Freeman and J. R. Tata, *Biochem. J. 94*, 628 (1965).
49. D. A. Pious, *Proc. Nat. Acad. Sci. (U.S.A.) 65*, 1001 (1970).
50. A. Hamberger, N. Gregson and A. L. Lehninger, *Biochim. biophys. Acta 186*, 373 (1969).

SURFACE CHANGES DURING THE CELL CYCLE

C. A. Pasternak, M.C.B. Sumner and R.C.L.S. Collin,

Department of Biochemistry,
University of Oxford,
Oxford OX1 3QU.

I. Introduction

There is much evidence that external effectors of cellular growth act at the cell surface (1,2). Since surface architecture itself alters during the cell cycle, it is possible that internal controls, such as the initiation of S, may also be mediated via the cell surface (3). It is therefore important to delineate any surface changes that occur, and to determine in what way they are superimposed upon the doubling of surface components that takes place in every cell cycle.

For these studies we used cells (mouse P815Y) that grow in suspension and that remain spherical throughout the cell cycle. This has the advantage of avoiding side effects due to morphological changes (rounding-up and flattening) and to alterations in contact with substratum and neighbouring cells. Spherical cells generally double in volume between G_1 and G_2 (e.g. ref 4). This means that surface area increases by $2^{2/3}$, or 1.6-fold, between G_1 and G_2; cytokinesis presumably provides the extra 0.4-fold increase in surface area necessary to produce two daughter cells having the original (G_1) amount of surface area. If surface components are inserted so that the 'density' (mass/surface area) of the plasma membrane remains constant, therefore, insertion must continue throughout the cell cycle. If insertion occurs only during interphase or only during mitosis, plasma membrane 'density' will fluctuate (5). Such changes in 'density' may reflect alterations either in thickness or in ruffling of the membrane.

The object of the present experiments was two fold; (i) to determine at what stage of the cell cycle the major components of the plasma membrane are inserted and (ii) to evaluate the extent to which the expression of surface sites is modified during the cell cycle.

II. Insertion of Major Components

Cells were separated according to their position in the cell cycle by zonal centrifugation (4,5). Fractions corresponding to particular periods in the G_1, S and G_2 phase were pooled and the washed cells disrupted by nitrogen cavitation (6). Plasma membrane was isolated by high-speed zonal centrifugation through sucrose (6). The yield of plasma membrane proteins (and protein in endoplasmic reticulum and mitochondria) was relatively constant throughout the cell cycle. Since total protein doubled gradually between G_1 and G_2 this must be true of protein in plasma membrane, endoplasmic reticulum and mitochondria also. Our earlier observation on microsomal and mitochondrial protein assembly (7) are therefore confirmed.

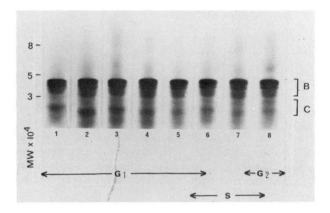

Fig. 1. Polyarcrylamide gel electrophoresis of plasma membrane proteins. Plasma membranes, prepared (6) from cells at various stages of the cell cycle (5), were disolved in 1% sodium dodecyl sulphate and subjected to electrophoresis (2mA/tube), with a running buffer of tris/HCl at pH 8.9. After fixation, proteins were stained with 1% Coomassie Brilliant Blue. Apparent molecular weights are based on the mobility of standard proteins run simultaneously. Fractions were allocated to particular phases of the cell cycle on the basis of (a) cell size, (b) [^{3}H] thymidine incorporation and (c) content of DNA. Though not apparent from the photograph, the relative intensity of individual bands showed little variation between G_1 and G_2.

No major differences in pattern of plasma membrane proteins separated by polyarcylamide gel electrophoresis in sodium dodecyl sulphate were detected (Fig. 1). Minor variations, however, do occur. When surface proteins are labelled with [^{125}I], a gradual increase in [^{125}I] binding is observed between G_1 and G_2 (Fig. 4). During this time the distribution of [^{125}I] fluctuates; in G_1, for example, components of molecular weight between 25,000 and 50,000 are the major labelled species, whereas later on in the cell cycle there is a relative increase in components of higher molecular weight. Whether this is indicative of net changes in amount, or of alteration in accessibility to iodination, is being investigated. A further indication that different proteins are synthesised at discrete times is revealed by studies on [^{14}C] glucosamine incorporation (Fig. 3).

The phospholipid:protein ratio of isolated plasma membrane varies little

(130 - 150 μg phospholipid/mg protein) between G_1 and G_2; hence phospholipid, like protein, is inserted gradually throughout interphase.

Thus the most abundant components (protein and phospholipid) of the plasma membrane roughly double in amount between G_1 and G_2. Hence the 'density' of the plasma membrane gradually increases during interphase; it is restored by some kind of stretching process during cytokinesis (5). Such a mechanism of plasma membrane assembly is compatible with other studies (8) which suggest that cytokinesis is predominantly a physical process in animal cells. It is also in agreement with results of freeze-fracture experiments which show that during mitosis and early G_1, intra-membrane particles become diluted-out, to be restored again in S and late G_2 (9). Whether the components of the plasma membrane are added as small vesicles (10) or as individual molecules (e.g. ref. 11) has now to be determined. Assay of plasma membrane enzymes does not reveal any 'precursor pool', suggesting that synthesis is rapidly followed by insertion (5). Growth of plasma membrane during interphase obviously conflicts with the assertion that total phospholipid is synthesised only during G_2 and mitosis (12); the suggestion that such synthesis might reflect the creation of new plasma membrane (12), seems to ignore the fact that only a fraction of cellular phospholipid is in the plasma membrane.

In the case of carbohydrates, a somewhat different result is obtained. Individual sugars (measured by gas liquid chromatography of plasma membrane fractions) fluctuate during interphase (Fig. 2), as a result of which total carbohydrate increases predominantly in G_1. In this instance the contributing species (glycoproteins, gangliosides, neutral glycolipids and surface mucopolysaccharides) do not alter in concert.

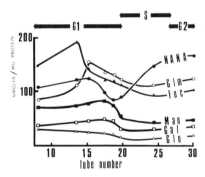

Fig. 2. Content of sugars in plasma membranes. Portions of the plasma membrane fractions of Fig. 1 were dried, converted to the trimethyl silyl sugar derivates and assayed by gas-liquid chromatography (31). Sugars are expressed per mg of plasma membrane protein.

NANA, N-acetylneuraminic acid; Glm, N-acetylglucosamine;
Fuc, Fucose; Man, Mannose; Gal, galactose; Glu, glucose.

The fluctuations in carbohydrate content (Fig. 2) agree with similar changes detected in KB cells (13). Increases during G_1 of particular carbohydrate-containing species have recently been reported for surface components in 3T3 (14) and CHO (15) cells and seems to be true of certain glycolipids in NIL cells also (16); histocompatability antigens, which are glycoproteins, are likewise synthesised predominantly in G_1 (see below). In fact synthesis of glycoproteins in general, as measured by incorporation of $[^{14}C]$ glucosamine, occurs predominantly in G_1. The compounds that are labelled most are species of apparent molecular weight 35,000-50,000 (bands B of Fig. 1); the rate of synthesis falls off during S (Fig. 3). Other components (bands C Fig. 1, of apparent molecular weight 25,000-30,000) are also synthesised in G_1 and then remain constant. Mitotic cells are not obtained as a discrete fraction by the zonal centrifugation procedure, and it is possible that synthesis is commenced already in M. The extent to which decreases in carbohydrate content after G_1 reflect net turnover or secretion (15) remains to be seen.

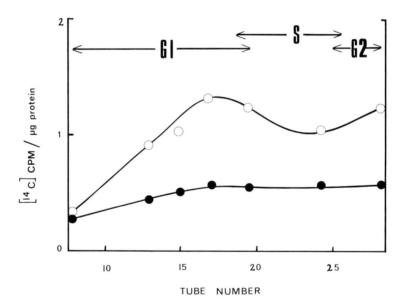

Fig. 3. Incorporation of $[^{14}C]$ glucosamine into plasma membrane glycoproteins. Plasma membrane fractions (Fig. 1) from cells exposed (at 5 x 10^7/ml) to $[^{14}C]$ glucosamine (0.2 µc/ml; 4µM) for 75 mins. prior to zonal centrifugation (5), were separated by polyacrylamide gel electrophoresis as described in Fig. 1. Sections from the gels were cut out, dissolved in Soluene 100 and radioactivity measured. Components corresponding to region B (Fig. 1), o—o; components corresponding to region C, •—•.

III. Expression of Surface Sites

It has recently become clear that *attachment* to, and *expression* of, surface sites are two distinct processes. For example normal and transformed cells bind the same amount of lectin such as concanavalin A, but only in transformed cells does this lead to agglutination. We decided to investigate the relation between binding and expression of H-2 histocompatability antigens during the cell cycle. Expression, measured by cytolytic release of intracellularly accumulated ^{51}Cr by anti H-2 serum and complement, proved to be maximal in M and early G_1, falling to a minimum in S (17). Binding, however, measured by direct titration with $[^{125}I]$ antibody prepared from H-2^d antiserum kindly

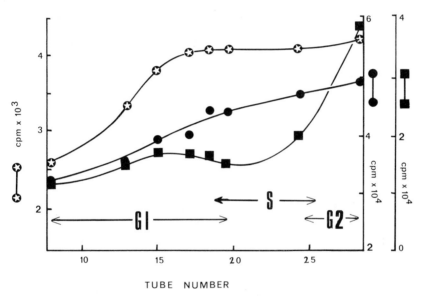

TUBE NUMBER

Fig. 4. Accessibility of surface components. H-2 histocompatibility antigens on 5×10^6 cells at different stages of the cell cycle (Fig. 1) were assayed (18) by treatment with $[^{125}I]$ labelled anti-H-2^{d}Ig G (5×10^6 cpm; total of 1 g) for 20' at 4° in 1 ml; radioactivity on 5×10^6 washed cells, ☆——☆. Con A binding sites were assayed by incubating 5×10^6 cells with $[^{125}I]$ labelled con A (5×10^7 cpm; total of 1 µg) for 30' at 4° in 1 ml; radioactivity on 5×10^6 washed cells, ■——■. Control values (binding in the presence of 1M α-methyl glucoside) have been subtracted. Surface proteins were iodinated by incubating 5×10^6 cells with Na^{125}I (50µc; 0.25 µmole) in the presence of lactoperoxidase (24µg) and hydrogen peroxide (0.24 µmole) for 15' at 37° in 0.4 ml, precipitating in 5% trichloroacetic acid and assaying radioactivity in the pellet corresponding to 5×10^6 cells, ●——● .

121

donated by Dr. D.A.L. Davies) or by 'back-titration' with [^{51}Cr] labelled cells, follows a different course (18). That is, surface antigen appears to double in G_1 and then remains accessible to antibody for the rest of the cell cycle (Fig. 4). Since complement attachment, like antibody binding, does not alter (19) in the way that expression of H-2 does (17) either, other reasons for the latter fluctuation must be sought.

One possibility is that cytolysis in general is facilitated in M and some G_1 cells and this has proved to be the case. Thus non-immune lysis caused by detergents, hypotonicity of freeze-thawing, shows exactly the same changes as immune lysis (18). In other words, mitotic cells are more fragile than interphase ones; the extent to which late G_2 and early G_1 cells behave like M cells is difficult to assess with the resolution of the present separation method. While this result explains the fluctuation in immune cytolysis measured by [^{51}Cr] release (17), it does not explain the increase in H-2 measured by binding of fluorescent antibodies (20). However visual revelation of fluorescence may depend on factors other than just binding (e.g. ref. 21).

In the case of lectins, binding of isotopically-labelled agglutinins (22) *does* change in the same way as fluorescent-labelled material (3,21), and suggests a considerable 'unmasking' at mitosis. It therefore seemed worthwhile to examine the binding of [^{125}I] concanavalin A in the present system. Unlike binding of [^{125}I] anti-H-2, binding of [^{125}I] concanavalin A shows only a slight rise in G_1, followed by a larger increase in late S and G_2, perhaps reaching a maximum in mitosis (Fig. 4); experiments with synchronously-growing cells are being carried out to establish the latter point. In other words, the results confirm those of Noonan *et al* (22) and show that lectin receptors, unlike H-2, are subject to topographical variations; alternatively their synthesis is predominantly a late interphase or mitotic event.

Another approach to investigating the disposition of surface components is to label proteins *in situ* with [^{125}I] iodide (23). When this was applied in the present system, a result intermediate between that for the binding of [^{125}I] anti-H-2 and [^{125}I] concanavalin A was observed; that is a gradual increase in accessibility of surface proteins between G_1 and G_2 (Fig. 4). Since surface and plasma membrane proteins both increase throughout interphase, this is the expected result. Mild trypsin (0.025% for 15 mins.) removes more than 90% of bound ^{125}I at all times; hence the label is exterior to the cell. Cultured cells are capable of binding large amounts of serum components to their surface (2,24), and the nature of the [^{125}I] labelled proteins was therefore investigated by polyacrylamide gel electrophoresis. The pattern obtained was distinct from that of [^{125}I] labelled serum. Moreover the distribution fluctuates between G_1 and G_2. During G_1, for example, most of the ^{125}I is in components of molecular weight 25,000-50,000; later relatively more is in components of higher molecular weight. These do resemble serum components somewhat. The possibility that some of the proteins on the surface of P815Y cells are derived from serum, cannot therefore be eliminated as effectively as has been done in the case of NIL cells (25).

A greater fragility in the surface membrane of mitotic cells has been correlated (5) with a greater susceptibility to viral fusion (26). The initial event in fusion is probably some sort of membrane destablisation (27), which is reflected in a leakage of intracellular metabolites (28). Leakage can be dissociated from subsequent events by carrying out the assay under conditions of non-agglutination, or with P815Y cells which show leakage but do not fuse (29). When leakage was measured with P815Y cells at different stages of the cell cycle, it was found that mitotic or early G_1 cells were *not* more susceptible than cells at later periods. In fact no difference in susceptibility to leakage between G_1 and G_2 was observed (29). It is therefore likely (a) that the results of Stadler and Adelberg (26) are due to fluctuation in some parameter subsequent to the initial destablisation and (b) that fragility to lysis (measured by release of phosphorylated intermediates) reflect different aspects of surface topography. Just what these are, is under investigation.

We are also examining transport processes (30) during the cell cycle of P815Y (5). For alterations in the concentration of intra cellular metabolites may be the means by which the cell surface controls the initiation of particular phases of the cell cycle. On the other hand the link between surface and nuclear events (3) may not be as tight as originally thought (22).

IV. Summary

1. The bulk components (phospholipid and protein) of the plasma membrane are inserted throught interphase. Cytokinesis is therefore a physical, stretching process, rather than one involving *de novo* membrane growth.

2. Carbohydrate-containing compounds show fluctuations depending on the molecular species; many components are synthesised predominantly in G_1.

3. The synthesis of H-2 histocompatability antigens occurs mainly in G_1; concanavalin A receptors, on the other hand, appear to be synthesised (or exposed) towards the end of the cell cycle. Total surface proteins, as measured by accessibility to iodination, increase throughout interphase.

4. Mitotic cells are more susceptible to immune and non-immune lysis than cells in S. Susceptibility to virally-mediated leakage, however, is constant between G_1 and G_2.

5. It is concluded that generalisations concerning the topographical distribution of surface components should be made with caution.

Acknowledgements

We are indebted to Dr. J. M. Graham for help in isolating plasma membrane, and gratefully acknowledge financial support and a scholarship (M.C.B.S.) from the Medical Research Council, financial support and the opportunity to attend the Cell Cycle Controls conference from the Cancer Research Campaign and a grant (R.C.L.S.C.) from Tenovus.

References

1. A. B. Pardee, *In vitro*, 7, 95 (1971).
2. P. Knox and C. A. Pasternak, *In: Mammalian Cell Membranes* (G. A. Jamieson and D. M. Robinson, eds.) Vol. 5, Butterworths & Co., London (1974).
3. T. O. Fox, J. R. Sheppard and M. M. Burger, *Proc. Nat. Acad. Sci.*, *68*, 244 (1971).
4. A. M. H. Warmsley and C. A. Pasternak, *Biochem J.*, *119*, 493 (1970).
5. J. M. Graham, M.C.B. Sumner, D. H. Curtis and C. A. Pasternak, *Nature*, *246*, 291 (1973).
6. J. M. Graham, *Biochem J.*, *130*, 113 (1972).
7. A. M. H. Warmsley, B. Phillips and C. A. Pasternak, *Biochem. J.*, *120*, 683 (1970).
8. D. Marsland, *In: High Pressure Effects on Cellular Processes* (A.M. Zimmerman, ed.) p. 259, Academic Press Inc., New York, (1970).
9. R. E. Scott, R. L. Carter and W. R. Kidwell, *Nature NB*, *233*, 219 (1971).
10. H. Hirano, B. Parkhouse, G. L. Nicolson, E. S. Lennox and S. J. Singer, *Proc. Nat. Acad. Sci.*, *69*, 2945 (1972).
11. W. C. McMurray and W. L. Magee, *Ann. Rev. Biochem.*, *41*, 129 (1972).
12. H. B. Bosmann and R. A. Winston, *J. Cell Biol.*, *45*, 23 (1970).
13. M. C. Glick, E. W. Gerner and L. Warren, *J. Cell. Physiol.*, 77, 1 (1971).
14. K. Onodera and R. Sheinin, *J. Cell Sci.*, 7, 337 (1970).
15. P. M. Kraemer and R. A. Tobey, *J. Cell Biol.*, *55*, 713 (1972).
16. B. A. Wolf and P. W. Robbins, In press.
17. C. A. Pasternak, A.M.H. Warmsley and D. B. Thomas, *J. Cell Biol.*, *50*, 562 (1971).
18. M. C. B. Sumner, R.C.L.S. Collin and C. A. Pasternak, *Tissue Antigens*, 3, 477 (1973).
19. R. A. Lerner, M.B.A. Oldstone and N.R. Cooper, *Proc. Nat. Acad. Sci.*, *68*, 2584 (1971).
20. M. Cikes and S. Friberg, *Proc. Nat. Acad. Sci.*, *68*, 566 (1971).
21. J. Shoham and L. Sachs, *Proc. Nat. Acad. Sci.*, *69*, 2479 (1972).
22. K. D. Noonan, A. J. Levine and M. M. Burger, *J. Cell Biol.*, *58*, 491 (1973).
23. J. J. Marchanlonis, R.E. Cone and V. Santer, *Biochem. J.*, *124*, 921 (1971).
24. P. Knox and C. A. Pasternak, in preparation.
25. R. O. Hynes, *Proc. Nat. Acad. Sci.*, *70*, 3170 (1973).
26 J. K. Stadler and E. A. Adelberg, *Proc. Nat. Acad. Sci.*, *69*, 1929 (1972).
27. G. Poste, *Int. Rev. Cytol.*, *33*, 157 (1972).
28. C. A. Pasternak and K. J. Micklem, *J. Membrane Biol.*, in press (1973).
29. C. A. Pasternak and K. J. Micklem, in preparation.
30. G. Sander and A. B. Pardee, *J. Cell Physiol.*, *80*, 267, (1972).
31. J. R. Clamp, T. Bhatti and R. E. Chambers, *Methods Biochem. Anal.*, *19*, 229 (1971).

SEQUENCES, PATHWAYS AND TIMERS IN THE CELL CYCLE

J. M. Mitchison

Department of Zoology, University of Edinburgh,
Edinburgh EH9 3JT, Scotland

I. Introduction

One way of looking at the cell cycle is to regard it as a series of events which mark the passage of a growing cell from its inception as a daughter cell to its ultimate fate in division. There is nothing very new in this prospect of a sequence of periodic events, stemming as it does from the discovery more than twenty years ago that DNA synthesis is a periodic event in interphase in nearly all cells. What is novel is the increasing interest in the analysis of this sequence and, in particular, how far an observed sequence can be dissected into a series of parallel sequences with independent causation or timing controls. My main purpose in this short review is to display a few of these recent dissections, without any pretension to cover all the ideas and results in a rapidly growing field. It may be appropriate to start with a series of simplified sequences. These are versions, dusted-off and renamed, of models that I described in an earlier volume in this series (1). They were originally produced *in vacuo*, but there is now a good body of experimental evidence which can give them real substance.

Fig. 1 shows a number of possible relationships between four observable events (A, B, C and D) in a temporal sequence through a cell cycle. There can be a direct causal connection between successive events such that each one cannot occur unless the preceding one has occurred [Fig. 1(a)]. I called this originally a "causal fixed sequence," but I suggest that a "dependent sequence" is better. It is more euphonious, and it is in accord with the use of "dependence" by others (2,3). I think that "sequence" is more suitable than the alternative "pathway," because "pathway" carries a flavour of direct biochemical pathways which may not be appropriate for morphological events like mitosis.

In Fig. 1(b) the events are independent and do not have direct causal connections, but there is a master timing mechanism which initiates them in sequence. This can be called an "independent, single timer sequence" or "IST sequence." "Clock" is an attractive substitute for timer, but it has been pre-empted in the field of circadian rhythms for timing mechanisms that are independent of temperature. Although there are some cycle sequences that are independent of growth rate (such as chromosome replication in fast-growing *Escherichia coli*), none of them appear to be independent of temperature. An example of an IST sequence is the "linear reading" theory for the control of

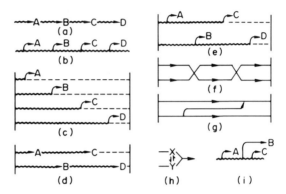

Fig. 1. Models for sequential events during the cell cycle. The wavy lines indicate the progress of timers. (a) Dependent sequence. (b) Independent, single timer (IST) sequence. (c) Independent, multiple timer (IMT) sequence. (d) Two parallel dependent sequences. (e) Two parallel IST sequences. (f) Two sequences with two check points. The events of each sequence before a check point have to be completed if there is to be progress beyond the check point. (g) Two sequences in which an early event in the lower sequence has to have been completed before a late event in the upper sequence can occur. (h) "Interdependency" of two events. (i) Delay in the full expression of an event B in an IST sequence may alter its order.

periodic enzyme synthesis (4). In the simplest case of bacteria, the hypothesis is that an RNA polymerase moves along the chromsome once per cycle and acts as a timer for the successive synthesis of enzymes.

The events can be independent, and in addition they can each have a separate timing mechanism. Such an "independent, multiple timer sequence" or "IMT sequence" is shown in Fig. 1(c). Howell (3) calls this the "race track" theory and contrasts it to the "domino" theory of a dependent sequence. An example of an IMT sequence is the "oscillatory repression" theory of the control of periodic enzyme synthesis (4). The synthesis of each enzyme would be timed independently by the oscillations of its end-product acting as a repressor.

Two or more sequences can run in parallel, and this is an important concept in recent work on the cell cycle. Fig. 1(d,e) shows two dependent sequences and two IST sequences. They are shown as starting from one vertical line and finishing at another. This raises a major problem about the relations between parallel sequences. The timers could run independently without any cross-connections, or alternatively they could all start at one time and finish at another. Similarly, several sequences might have to be completed before some major event like cell division could take place. The cell cycle might consist of sequences which converge at a series of check points which would ensure that all previous events had been completed [Fig. 1(f)]. The checks also need not be at the same time in two sequences. Situations could arise where an early event in one sequence had to be completed before a later one in a second sequence could occur [Fig. 1(g)]. The dependence relations in branched pathways can be complex, and Howell (3) discusses them later in this volume. Hereford and Hartwell (5) distinguish the "interdependence" of two events (as well as their "dependence" and "independence"). Here two events have to be completed before a sequence can proceed but *neither* can happen unless the other one takes place. These would normally be simultaneous events before a convergence [Fig. 1(h)]. Another dependency relation is that the initiation of two events may follow an IST sequence but their completion shows dependency. For example, the initiation of DNA synthesis and of nuclear division might be started at different times by a single timer but the completion of nuclear division would not occur unless DNA synthesis had taken place. A last point that is worth bearing in mind is that there may not be a clear distinction between an event and a timer. DNA synthesis, for example, can be regarded as a single event, or two events (start and stop of synthesis) or multiple events (initiation of replicons). But it also may have importance as a timing mechanism in its own right. The process of DNA synthesis acts as timer between initiation and completion (see 39).

Despite the complexities of real life within the cell, it has been profitable to think in terms of these very simple models, and it will continue to be so until they prove inadequate. Models are only valuable if they can be tested, and the obvious way of testing the dependency of sequences is by perturbing them (1,3,6). Blocking an event in a dependent sequence should prevent all subsequent events, and perturbing the sequence might alter the timing of the events but not their order. In an IMT sequence, blocking one event would not necessarily affect the other events, and perturbations could affect the order as well as the timing. In principle, the IST sequence is intermediate. Blocking an event should not affect the later events, but the order should be invariant. In practice, however, the effect of perturbation might be to increase the delay between the initiation of one event by the timer and the full expression. In this case, the order of the events would be changed [Fig. 1(i)]. The ways of testing other models are not difficult to devise, in theory at any rate. If, for example, there are two dependent sequences running in parallel, one of them should continue even though the other was blocked. Hereford and Hartwell (5) give an elegant description of how to separate dependency, interdependency and independency.

Three main techniques have been used in the last decade for cycle disturbance: specific inhibitors, especially those that stop DNA synthesis; changes in physiological conditions, for example the use of different nutrients for *E.coli* or heat shocks for *Tetrahymena*: and, most recently, temperature-sensitive mutants which affect the cycle. I will now describe some of the results that come from using these methods.

II. The DNA-Division Cycle and the Growth Cycle in Fission Yeast

My colleagues and I have been developing for the last few years the concept of two parallel sequences of events during the cell cycle of the fission yeast *Schizosaccharomyces pombe*. This work has been described in several recent publications (4,7,8,9) so only a summary will be given here. The concept is illustrated in Fig. 2 in a general form to cover a wide variety of cell types, and there is no direct evidence that some of the events occur in yeast. The first sequence, the "DNA-division cycle" or "DD cycle", has as its main events, DNA synthesis (the S period), nuclear division and cell division. It would also probably include histone synthesis and various transition points. A transition point defines operationally the effect of a treatment on cells at different stages of the cycle with respect to a cycle event. The event is usually, but not necessarily, cell division. If the treatment is with an inhibitor of protein synthesis, cells at stages later in the cycle than the transition point will continue through the cycle and divide, whereas those which are earlier in the cycle will not divide. It is assumed that the last protein needed for division has been synthesised by the time of the transition point. The second sequence, the "growth cycle", includes the main processes of protein and RNA synthesis which cause cell growth. Many of these processes may be continuous but there is good evidence that a number of enzymes are synthesised periodically, in "step" or "peak" patterns, and these provide discrete events which can act as markers of the growth cycle.

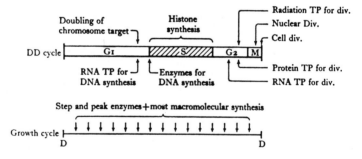

Fig. 2. The "DNA-division cycle" and the "growth cycle". TP, transition point; D, division. From Mitchison (4).

Our reason for suggesting that there are two sequences is that we have evidence that the growth cycle can continue when the DD cycle is blocked. This comes from two experimental situations. In one of them, synchronous cultures made by selection after gradient centrifugation were treated with agents which inhibit events in the DD cycle (8). These cultures were assayed for three step enzymes, aspartate transcarbamylase, ornithine transcarbamylase and alcohol dehydrogenase. When treated with hydroxyurea (an inhibitor of DNA synthesis) towards the end of one cycle, the synthesis of DNA and the cell division of the next cycle was considerably delayed. But the enzyme steps were not delayed and occurred at the same time as those in untreated control cultures. Similar results have been found after DNA blocks in *Bacillus subtilis* (10) and in mammalian cells (11,12,13). There was the same absence of an effect on the enzyme steps with another inhibitor, mitomycin C. This is not a direct inhibitor of DNA synthesis in *S. pombe* but it does delay nuclear division, possibly by causing cross-links in the DNA which can in time be repaired. When nuclear division was delayed so also were the subsequent cell division and DNA synthesis, which suggests that these main events of the DD cycle are in a dependent sequence.

Our second experimental situation was a somewhat more complicated case' involving induction synchrony (7,9). An asynchronous culture was pulsed for three hours with deoxyadenosine, an inhibitor of DNA synthesis (Fig. 3). DNA synthesis was blocked and so after time was cell division (the delay here being due to the long G2 in this yeast). DNA synthesis started shortly before the end of the block and was followed by two synchronous divisions marked by peaks in the cell plate index. This type of induction synchrony, which has a long history in mammalian cell cultures, relies on the DD cycle being a dependent sequence. Cells accumulate at the G1/S boundary during the block and, when released, proceed synchronously through DNA synthesis and division. Although the sequence of the DD cycle events was unaltered during this induction synchronisation of yeast, there were two interesting differences in timing. The normal G2 gap between DNA synthesis and division was considerably reduced before the first synchronous division, and the cycle time between the first and second synchronous division was unusually short (60% of normal).

Although the events of the DD cycle were blocked and thereafter synchronised, growth continued. This is shown by the RNA and optical density curves in Fig. 3. As a result, the cells at the first synchronous division were not only 70-80% larger than normal dividing cells but they were also more variable (Fig. 4). They recovered rapidly, however, and by the second synchronous division the dividing cells had the same length and variability as those in normal asynchronous cultures or after selection synchrony. These results suggest that the growth cycle had not been synchronised by the induction treatment, but a more specific test comes from assaying step enzymes which are the markers or events in the growth cycle. Fig. 5 shows that there were no doubling steps in aspartate transcarbamylase and ornithine transcarbamylase after induction synchrony and little difference from an unsynchronised control. Similar results were also obtained with a third enzyme, alcohol dehydrogenase. Presumably each cell after induction synchrony is running through the events of the two cycles but the relative timing between cycles varies from cell to cell.

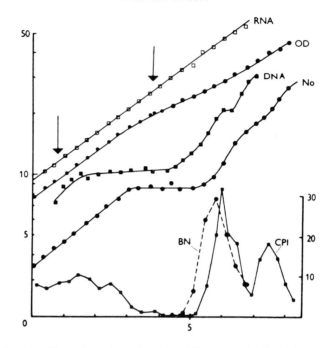

Fig. 3. *Abscissa*: time (hours); *ordinate*: arbitrary units. Induction synchrony in *Schizosaccharomyces pombe*. Effect of a 3 h treatment with deoxyadenosine (2 mM) on various cell parameters of an exponential phase culture. The treatment commences at the first arrow and finishes at the second arrow. RNA, 1 arbitrary unit on left ordinate AU=1 μg/ml. OD, optical density at 595 nm, 1 AU=0.0255 absorbance units. DNA, 1 AU=0.01 μg/ml. No, cell number, 1 AU=0.715 X 10^6 cells/ml. CPI, cell plate index as % on right ordinate. BN, proportion of binucleate cells without cell plates as 1.5 X % on right ordinate. From Mitchison and Creanor (7).

These results support the concept of two sequences which can be dissociated and can run independently. The growth cycle continues when the DD cycle is blocked. Can the reverse happen? We have no clear evidence on this, but there is a suggestion that it can occur from some earlier experiments (14). If *S. pombe* is deprived of a nitrogen source, growth is much reduced but the cells divide once or twice and produce abnormally small progeny. This situation needs to be re-examined.

The relations between the cycles is interesting but obscure. Our results, particularly those with mitomycin, indicate that the growth cycle can recommence without the preceding nuclear division or cell division. There is not therefore a necessary "start" signal for the growth cycle which is dependent on the completion of the previous DD cycle. On the other hand, there is the rapid recovery of normal cell size and variation by the second

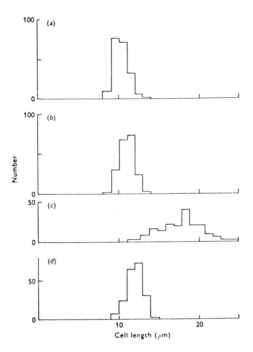

Fig. 4. Histograms of length of cells with cell plates (dividing cells) in cultures of *Schizosaccharomyces pombe*. (a) From an asynchronous culture; (b) from the first synchronous division after selection synchrony by gradient centrifugation; (c) from the first synchronous division after induction synchrony with deoxyadenosine; (d) from the second synchronous division after induction synchrony. From Mitchison and Creanor (7).

division after induction synchrony. The diminution of cell size can be explained by the abnormally short inter-division time, but the controls may be more subtle than this because the variation is also reduced. It might be that long cells divide quicker or grow slower than short cells, and this needs to be looked into. Our impression is that the relations between the cycles involve interlock controls but that they are loose ones that can be uncoupled.

Finally, there is the question of the dependency relations within these cycles. Our present evidence suggests that the three main events of the DD cycle - DNA synthesis, nuclear division, and cell division lie in a dependent sequence. The results with mitomycin also indicates that DNA synthesis in one cycle is dependent on the completion of nuclear division in the preceding cycle. This is in accord with Hartwell's model considered later. We have no direct evidence about the dependency and timing in the growth cycle, though it seems most unlikely, on general grounds, that it is a single dependent sequence.

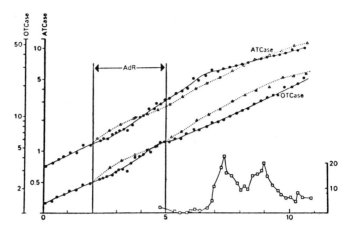

Fig. 5. *Abscissa*: time (hours); *ordinate*: (*left*) 10^3 X enzyme units/ml; (*right*) cell plate index (%). Activities of ATCase and OTCase during induction synchrony. ●—● , experimental culture with AdR (to 2 mM) added between 2 and 5 h; △---△ , control culture with no AdR added. CPI, cell plate index. Initial cell conc. 2.0 X 10^6 cells/ml. From Sissons, Mitchison and Creanor (9).

III. Cell Cycle Mutants in Budding Yeast

One of the most interesting aspects of research on the cell cycle in the last few years has been the work of Hartwell and his colleagues on temperature-sensitive mutants (*cdc* mutants) which affect the cell division cycle in *Saccharomyces cerevisiae*. It is particularly important because it demonstrates the power of a relatively new tool in dissecting the events of the cycle in eukaryotes. Cells which are shifted from the permissive temperature (23°C) to the restrictive temperature (36°C) before a particular point in the cycle will not proceed through the cycle and divide, but if they are shifted after this point they will divide. This has been called the "execution point" and it is operationally the same as a transition point. The *concept* here is that this is the point at which the thermolabile gene product completes its function at the permissive temperature. Further analysis can show the "initial defect" as the first cell cycle event (among those observed, e.g., initiation of DNA synthesis) which fails to take place at the restrictive temperature. In addition, most of the mutants show a characteristic morphology at the "termination point" when left for some hours at the restrictive temperature. Both Howell (3) and Hartwell *et al* (2) point out the resemblances between inhibitors and temperature-sensitive (TS) mutants as means of blocking and analysing the cycle. The mode of action of inhibitors is often better known than in TS mutants. On the other hand, the mutants "permit more detailed conclusions, both because of the greater number of specific cell cycle blocks in a single organism and because of the greater assurance that a single gene defect directly affects one and only one event in the cell cycle" (2).

This work has been well reviewed recently (2) and all I wish to do here is to emphasise some points and make a few comparison. The main conclusions on the dependence of cycle events in budding yeast are shown in Figs. 6 and 7. From "Start" there are two dependent sequences. One of them runs through DNA synthesis, nuclear division to cell division. The other runs through bud emergence and nuclear migration into the bud and then converges with the first sequence on cell division. The reason for allocating separate sequences are that there are mutants which permit bud emergence but are blocked in DNA synthesis and *vice versa*. The reason for the convergence is partly that cell division (cytokinesis and cell separation) does not in fact occur in a mutant which permits bud emergence but are blocked in DNA synthesis and *vice versa*. The reason for the convergence is partly that cell division (cytokinesis and cell separation) does not in fact occur in a mutant which completes nuclear division without bud emergence, and partly because it would be difficult to conceive of a situation with budding yeast in which this could possibly happen. The fork after the end of nuclear division is because there are mutants in which cell division is blocked but which show multiple rounds of DNA synthesis, nuclear division and bud emergence.

Hartwell *et al* (2) conclude that two dependent sequences are sufficient to explain these causal relations, and they reject an "independent pathways" model which is the same as the IST sequence in Fig. 1(b). There is, however, one interesting situation in a mutant (*cdc* 4) which, at the restrictive temperature, goes through cycles of forming buds (up to five) from a single mononucleate cell. They suggest that there is a timer which initiates bud emergence (or "start" followed by bud emergence) and functions once per cycle. This timer would not coordinate the later events of the two dependent sequences but it might provide the link between these sequences and cell growth.

When comparing these results with those from other eukaryotes, one should remember that bud emergence and nuclear migration are peculiarities of budding yeasts (2). If this sequence is removed, what is left is the DD cycle, with the addition of a "start" signal for this cycle. Our results from the use of inhibitors on *S. pombe* are in complete agreement with this concept of a dependent sequence, though we have less information. It is also clear that it has a wide application to other eukaryotic cells (4). The fork after nuclear division which implies that the sequence can recommence without cell division is illustrated by the many cases in which multinucleate cells occur naturally (e.g. liver, *Physarum* and many fungi) or can be induced by chemical agents (e.g. caffeine or cytochalasin). In some cases, there can be a fork at an earlier stage between nuclear division and DNA synthesis since there are situations, both natural and artificial, where a new cycle of DNA synthesis can start without an intervening nuclear (or cell) division (4, p. 249).* There are only a very few exceptions to the two main rules of dependency in the DD cycle; that cell

---

*An elegant example of this in sea urchins is discussed by Mazia in this volume.

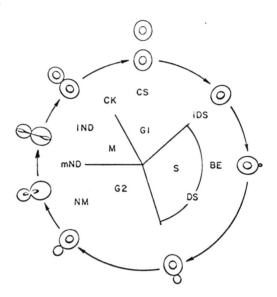

Fig. 6. The sequence of events in the cell division cycle of yeast; iDS, initiation of DNA synthesis; BE, bud emergence; DS, DNA synthesis; NM, nuclear migration; mND, medical nuclear division; lND, late nuclear division; CK, cytokinesis; CS, cell separation. Other abbreviations: G1, time interval between previous cytokinesis and initiation of DNA synthesis; S, period of DNA synthesis; G2, time between DNA synthesis and onset of mitosis; and M, the period of mitosis. From Hartwell, *et al* (2) with permission of the Publishers.

division is dependent on nuclear division, and that nuclear division is dependent on DNA synthesis. In rare cases, mostly with polyploid cells, division can take place without prior DNA synthesis (4, p. 249). Cell division without nuclear division will produce enucleate cells. This is well known in bacterial mutants but very uncommon in eukaryotes. The best example is the activated enucleate sea urchin egg which goes through several divisions without any nuclei present (15), but there is a recent report of enucleate yeast cells being formed under conditions of very fast growth (16).

The evidence from fission yeast is that growth continues when the DD cycle is blocked. This also seems to be the normal situation with the cycle mutants in budding yeast. Except in three out of the 32 *cdc* genes, most of the mutants continue to enlarge at the restrictive temperature after the termination of development round the DD cycle (17). It would be interesting to follow under these conditions the patterns of synthesis of step enzymes, of which many are known in budding yeast. Even more interesting would be to isolate TS mutants of these enzymes and apply to the growth cycle the techniques that have been so useful in dissecting the DD cycle.

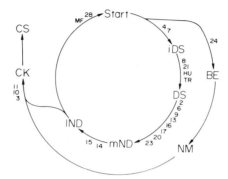

Fig. 7. The circuitry of the yeast cell cycle. Events connected by an arrow are proposed to be related such that the distal event is dependent for its occurrence upon the prior completion of the proximal event. The abbreviations are the same as in Fig. 6. Numbers refer to *cdc* genes that are required for progress from one event to the next; HU and TR refer to the DNA synthesis inhibitors hydroxyurea and trenimon, respectively; MF refers to the mating factor, α factor. From Hartwell, *et al* (2) with permission of the Publishers.

IV. Cell Cycle Mutants in *Chlamydomonas*

Another important application of TS cell cycle mutants has been by Howell and his colleagues on *Chlamydomonas reinhardi*. This work is summarised by Howell in this volume (3) and there is, therefore, even less necessity than with the yeast work to describe it here. Very briefly, this organism can be synchronised either by induction or by selection (18), but asynchronous cultures were used for most of the mutant analysis. The position of the "block point"[1] for division for each mutant was determined from the amount of residual cell division in the culture after a switch from the permissive to the

---

[1]It is a pity that there are now three names in the literature, "transition point," "execution point" and "block point," for a point in the cycle which is defined operationally in exactly the same way, even though there may be differences in the underlying concepts. Howell (3) discusses this and suggests that "block point" is reserved for the genetic approach and "transition point" for the physiological approach. My view is that this ought not to be a real distinction and that the use of two names will only encourage confusion. Each name has certain merits, but I think that "transition point" should be generally adopted because it has been in use for much longer.

restrictive temperature. If the block point is early in the cycle, there is more residual cell division than with a late block point. This technique has been used earlier for analysing the effects of inhibitors and radiation (e.g. 19). It can be applied to a wide variety of systems, particularly those difficult to synchronise, but it demands accurate growth curves or a good differential measure of progress through the cycle (e.g. mitotic index). It can also be used to measure the block points for other events in the cycle, e.g. DNA synthesis.

A number of TS mutants have been isolated and they have been shown to have block points distributed throughout the cycle, though with a tendency to cluster in some regions. They are now being used to analyse the dependence of various events in the cycle and the arguments and results are set out by Howell (3). The point I wish to emphasise is how well the preliminary results on dependency fit in with those from the work on yeast. In the mutant *ts 10009*, both DNA synthesis and cell division are dependent on the expression of the mutant gene. But the expression of two enzymes, glutamate dehydrogenase and aspartate transcarbamylase, are independent and their synthesis continues when division has been blocked at the restrictive temperature. This result is very similar to the situation in fission yeast when the synthesis of three steps enzymes (one of which was aspartate transcarbamylase) continued after blocking DNA synthesis and cell division. It will be interesting to see how far further investigations support the concepts of the DD cycle as a dependent sequence and the existence of an independent growth cycle.

V. Sequences in the Bacterial Cell Cycle

There is a tendency for work on the prokaryotic and the eukaryotic cell cycles to proceed independently with a mutual lack of attention. This is unfortunate because it is quite likely that there are real similarities between them. Theories of the control of enzyme synthesis, for example, draw on evidence from both systems. It is, therefore, worth considering one of the most recent models for the control of cell division in *E. coli* put forward by Donachie *et al.* in a recent review (20). This model is shown in Fig. 8, together with a descriptive legend. The essential elements are a timer, varying with growth rate; and two sequences, probably dependent, which do not vary with growth rate(within limits). These sequences are independent of each other but both have to be completed before cell division can occur.

There are two lines of evidence which suggest independence between the sequences. There is normally a delay of 20 minutes between the end of a round of replication and the succeeding division. But if DNA synthesis is blocked for a time and the block is then released, this delay does not occur and the cells divide almost immediately after the completion of the round of replication. The suggestion is that the division protein sequence has continued to run when the DNA sequence has been stopped, so the division protein is complete by the end of the replication round. Notice here the similarity to the situation in *S. pombe* described earlier where the normal G2 gap between DNA synthesis and

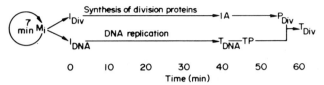

Fig. 8. Model of the cell cycle in *Escherichia coli*. Doubling of the initiation mass (M_i) takes place every mass doubling time (τ min). At each doubling two processes are initiated approximately simultaneously. These are the initiation of DNA replication (I_{DNA}) and the initiation of the sequence of events leading to division (I_{Div}). Termination of chromsome replication (T_{DNA}) at 40 min induces the synthesis of termination protein (TP). The first 40 min of the division sequence involves protein synthesis which is then followed by the initiation of assembly (IA). (N.B. Assembly could actually start earlier). After 15-20 min more the cell has reached a stage (P_{Div}) where interaction between some septum 'primordium' and termination protein leads to cell division (T_{Div}).

Our picture of the timing of the main events in the cycle is therefore of a periodic event which occurs at intervals equal to the mass doubling time of the cell and at multiples (2^n) of a constant cell mass (Mi). This event triggers two parallel but separate sequences of events which take *constant* period of time to complete, largely independent of the rate of cell growth. One of these processes is chromsome replication and the synthesis of the termination protein, and requires 40-45 min to complete. The other is a sequence of protein synthesis, followed by another process which may be assembly of some septum precursor. This sequence requires nearly 60 min and, at the end of it, there is an interaction between the septum precursor(s) and the termination protein to give the final septum and cell division. This last event takes only a few minutes. From Donachie, Jones and Teather (20) with permission of the Publishers.

division is considerably reduced after the temporary inhibition of DNA synthesis in induction synchrony. A marked shortening of G2 has also been found in some mammalian cells after double thymidine blocks to DNA synthesis (21). It may be, therefore, that the model with two sequences converging on division applies to many eukaryotic cells.

A second piece of evidence for independent sequences comes from mutants of *E. coli* in which division can continue in the absence of DNA replication, producing "enucleate" cells (e.g. 22). At first sight, this would suggest that cell division was not dependent on DNA synthesis, but the answer may be that it *is* dependent in normal cells but not in the mutant. A mutation can stop the synthesis of a protein but it can also alter its control, as in constitutive mutants. If so, it may alter the dependence of events in sequences. This is a point which must be borne in mind in the use of cell cycle mutants.

VI. Timers

Cell cycle sequences involve timers. In dependent sequences, the timers operate in what may be the relatively short intervals between the events; but in independent sequences, the timers may have to run for the length of the cycle or longer. So far, two main kinds of timer have been suggested. One of them involves linear transcription along the genome with the timing being dependent on the rate of movement of the RNA polymerase molecules. This is the basis of the linear reading theory of the control of periodic enzyme synthesis, which has been mentioned earlier. Watson (23) has also suggested linear transcription as a timer both for cell cycles and over the longer time span of embryological differentiation.

The other, more popular, model assumes the build-up of a substance (or a structure) which eventually triggers an event such as division. Alternatively, an inhibitor can be produced at some point and diluted by growth until it ceases to be effective. These models are usually defined in rather general terms, but more precise ones have been formulated. Kauffman (24), for example, discusses a "mitotic oscillator" and draws on sophisticated techniques which have been developed for the analysis of relaxation oscillators (which generate saw-tooth patterns) in circadian rhythms (25). These techniques are particularly valuable in predicting what should happen when the iscillator is perturbed.

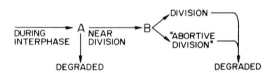

Fig. 9. Model of a mitotic timer. Substance A (perhaps a protein) accumulates during interphase. It can be degraded by various treatments, e.g. heat shocks. Normally, however, it reaches a critical *concentration* near the end of the cycle. This triggers its conversion into a structure B which is essential for division. After division, it is degraded. If there is an insufficient *quantity* of A to allow completion of B, division is abortive and does not occur.

I wish to try my hand at building a model of a mitotic timer. It is not expressed in quantitative terms nor is it particularly novel, but it does attempt to reconcile three different sets of experiments. Let us assume that substance A in Fig. 9 increases in *concentration* in the cytoplasm during interphase. It could be synthesised or it could be accumulated from the medium, and these processes do not necessarily have to start at the beginning of the cycle. It could be synthesised at the same rate as the general rate of increase of cell mass, but, if so, it would have to start at a low concentration so as to allow this to increase through the cycle. The simplest situation is to postulate zero concentration initially. When A reaches a critical concentration level near the time of mitosis, the next stage of the process is triggered and A is converted into B. It is best to think of B as a structure of finite size, for a reason that is given later. When B is complete, division proceeds. After division, B breaks down and the components are degraded. The accumulation of A then starts again and the new cycle is initiated. With the appropriate control connections between A and B, this system is a simple relaxation oscillator.

The first set of experiments against which this model should be tested are the fusion experiments which have been carried out with *Physarum* (26,27,28) and with mammalian cells (29,30). If *Physarum* plasmodia from two parent cultures at different stages of the cycle are fused, all the nuclei divide synchronously at a time about half-way between the next mitoses in the parent cultures. Some nuclei are therefore delayed beyond their normal time of mitosis and others are accelerated. This is easy to interpret in terms of a cytoplasmic initiator of mitosis, such as A, whose concentration rises through the cycle, and this is in fact the control mechanism suggested by many of the workers in this field. The fused plasmodium would have a concentration of A intermediate between those of the two parent cultures. This would ensure synchrony and produce the appropriate advances and retardations of nuclear divisions.

The second set of experiments are those which support the "division protein" hypothesis of Zeuthen and his colleagues (31). This concept was originally developed for *Tetrahymena*, but it may well have wider application (4,32). Division protein is built up during the cycle, and after a transition point is stabilised as a structure which is essential for division. Before the transition point, however, various treatments such as heat shocks or temporary blocks to protein synthesis will destroy the existing stocks of division protein. This has to be remade, so the cells are delayed in their progress towards division, and the later they are in the cycle (before the transition point) the greater the delay. The model in Fig. 9 fits in with these experiments, providing the arrow is put in under A to indicate its destruction by heat shocks, etc. Indeed the model is very nearly the same one suggested some time ago by Rasmussen and Zeuthen (33). The transition point is when A is converted into B. One problem, however, should be pointed out. The concentration of A is supposed to trigger its conversion to B, so the simplest way to consider A is a population of single molecules. On the other hand, there is the concept that a temporary inhibition of protein synthesis not only stops the production of division protein but also causes the existing stock to be broken down. It is easier, but not essential, to explain this in terms of A being a protein structure, for which there is an analogy in the oral apparatus in ciliates which is resorbed after a temporary block to protein synthesis (34).

The third set of experiments, which are less well known, are those on amputation of *Amoeba*, first done by Hartmann (35) and repeated and amplified by Prescott (36). They showed that removal by microsurgery of a substantial portion of the cytoplasm delayed the forthcoming division. If this process was repeated every day, an *Amoeba* could be kept from dividing for four months, during which time a control clone would go through 65 divisions. Removing cytoplasm will not alter the *concentration* of a cytoplasmic initiator such as A, but it will alter the total *quantity*. It would be difficult for a cell to measure the total quantity of a substance if it was distributed uniformly through the cytoplasm. But it would be possible if the substance was used to build one or more structures which had to be completed. What might therefore be happening in the *Amoeba* experiments is indicated by the "abortive division" pathway in Fig. 9. The amputated *Amoeba* would reach a critical concentration of A and the conversion to B would proceed. There would not, however, be sufficient A to complete the structure of B and division would not happen. Since the existing stock of A was used and then destroyed, it should take a full cell cycle before division could be triggered again. This is a prediction which can be tested.

The amputation experiments reduce the ratio of cytoplasm to nucleus. What would be the effect of increasing it? The model would predict that the quantity of A would be more than sufficient to complete B and there should be a surplus left over. This would be carried into the next cycle and would shorten it since the critical concentration of A would be reached earlier. A cell larger than normal at division would produce daughters with shorter cycles than normal and therefore smaller size at division. This could be a normal homeostatic mechanism for regulating cell size.

This explanation might also account for the situation after induction synchrony in *S. pombe* which has been mentioned earlier. The large cells at the first synchronous division could have accumulated a surplus of A which would account for the marked shortening of the following cycle. Evidence against this, however, comes from the effect of heat shocks applied before this first division (37). These shocks, which would be expected to degrade A, do delay the first division but they do *not* increase the time between the first and second division. Some other explanation seems needed, at least for this special case. It could be that the rate of accumulation of A in the large cells after the first division is considerably higher than usual so that the critical concentration is reached in a shorter time even though the cell volume is larger.

Before leaving this model, it is reasonable to ask whether B corresponds to any known structure. One possibility is that it might be the mitotic apparatus. A structure of microtubules might be neither stable nor effective until it was complete. Zeuthen and Rasmassen (31) have also pointed out some of the similarities between division protein and microtubules.

Theories about single mitotic timers, like the model above, may seems somewhat remote from the detailed analysis of sequential events within the cycle. Even though timers are required in theoretical schemes or in Hartwell's analysis, it seems unlikely that the increase of substance A could be responsible for initiating a whole sequence of different events at successive levels of concentration. I would suggest that the mitotic timer is one of two or more parallel dependent sequences, all of which have normally to be completed before division can occur. This is very much like the bacterial model in Fig. 8, where the same phrase of "division protein" is used. Cases have been quoted in that section in which the division timer seems to proceed when DNA replication is blocked. There are also situations where the reverse can occur, and DNA replication is reinitiated though nuclear division has been blocked. This happens, for example, during multiple heat shocks in *Tetrahymena* (31) or after a single heat shock applied late in the cycle in *Physarum* (38). But the exact relations between these sequences and between other sequences remain to be explored and they will certain be a major theme of future research on the cell cycle.

Acknowledgements

It is a pleasure to acknowledge the help of my colleagues in criticising and refining these models.

References

1. J. M. Mitchison, *In* "The Cell Cycle. Gene-enzyme Interactions" (G. M. Padilla, G. L. Whitson and I. L. Cameron, eds) pp. 361-372 Academic Press, New York (1969).
2. L. H. Hartwell, J. Culotti, J. R. Pringle and B. J. Reid, *Science 183*, 46 (1974).
3. S. H. Howell, This volume (1974).
4. J. M. Mitchison, "The Biology of the Cell Cycle," Cambridge Univ. Press, London (1971).
5. L. M. Hereford and L. H. Hartwell, *J. Mol. Biol.* In press (1974).
6. R. R. Schmidt, This volume (1974).
7. J. M. Mitchison and J. Creanor, *Exp. Cell Res. 67*, 368 (1971).
8. J. M. Mitchison, *Symp. Soc. Gen. Microbiol. 23*, 189 (1973).
9. C. H. Sissons, J. M. Mitchison and J. Creanor, *Exp. Cell Res. 82*, 63 (1973).
10. M. Masters and W. D. Donachie, *Nature, 209*, 476 (1966).
11. J. R. Churchill and G. P. Studzinski, *J. Cell Physiol. 75*, 297 (1970).
12. A. S. Gelbard, J. H. Kim and A. G. Perez, *Biochim. Biophys. Acta 182*, 564 (1969).
13. R. R. Klevecz, *J. Cell Biol. 43*, 207 (1969).
14. Mazia, *In* "The Cell" (J. Brachet and A. E. Mirsky, eds) Vol. 3 p. 101. Academic Press, New York (1961).
15. E. B. Harvey, *Ann. N. Y. Acad. Sci. 51*, 1336 (1951).
16. D. Vraná, J. Lieblová and K. Beran, *Proc. 3rd. Internat. Specialized Symp. on Yeasts* (H. Suomalainen and C. Waller, eds). *2*, 285. Helsinki (1971).
17. L. H. Hartwell, R. K. Mortimer, J. Culotti and M. Culotti, *Genetics 74*, 267 (1973).
18. G. Knutsen, T. Lien, Ø. Schreiner and R. Vaage, *Exp. Cell Res. 81*, 26 (1973).

141

19. Y. Doida and S. Okada, *Radiat. Res. 38*, 513 (1969).
20. W. D. Donachie, N. C. Jones and R. Teather, *Symp. Soc. Gen. Microbiol.* 23, 9 (1973).
21. G. Galavazi and D. Bootsma, *Exp. Cell Res. 41*, 438 (1966).
22. Y. Hirota, A. Ryter and F. Jacob, *Cold Spring Harbor Symp. Quant. Biol. 33*, 677 (1968).
23. J. D. Watson, "Molecular Biology of the Gene" 2nd ed. p. 528. Benjamin, New York (1970).
24. S. Kauffman, *Bull. Math. Biophys.* In press (1974).
25. A. T. Winfree, *J. Theor. Biol. 28*, 327 (1970).
26. H. P. Rusch, W. Sachsenmaier, K. Behrens and V. Gruter, *J. Cell Biol. 31*, 204 (1966).
27. W. Sachsenmaier, U. Remy and R. Plattner-Schobel, *Exp. Cell Res. 73*, 41 (1972).
28. B. Chin, P. D. Friedrich and I. A. Bernstein, *J. Gen. Microbiol. 71*, 93 (1972).
29. P. N. Rao and R. T. Johnson, *Nature. 225*, 159 (1970).
30. R. T. Johnson and P. N. Rao, *Biol. Rev. 46*, 97 (1971).
31. E. Zeuthen and L. Rasmussen, *In* "Research in Protozoology" (T-T. Chen, ed.) Vol. 4. pp. 11-145. Pergamon Press, Oxford (1972).
32. H. Miyamoto, L. Rasmussen and E. Zeuthen, *J. Cell Sci. 13*, 889 (1973).
33. L. Rasmussen and E. Zeuthen, *C. R. Trav. Lab. Carlsberg. 32*, 333 (1962).
34. J. Frankel, *J. Cell Physiol. 74*, 135 (1969).
35. M. Hartmann, *Zool. Jb. 45*, 973 (1928).
36. D. M. Prescott, *Exp. Cell Res. 11*, 94 (1956).
37. M. M. Polanshek, Ph.D. Thesis. Univ. of Edinburgh (1973).
38. E. N. Brewer and H. P. Rusch, *Exp. Cell Res. 49*, 79 (1968).
39. M. E. McClure, This volume (1974).

PROTEIN SYNTHESIS AND THE CONTROL OF NUCLEAR DNA REPLICATION IN YEAST AND OTHER EUKARYOTES

D. H. Williamson

National Institute for Medical Research,
The Ridgeway, Mill Hill, London NW7 1AA.

It is not an entirely novel thought that man and all his works are merely devices developed by DNA to ensure its own replication and survival. This may be too fanciful for some, but few would seriously protest the view that the dividing cell is just such a device — or at least that the proper replication of chromosomal DNA is a key event in the cell division cycle. Nor would many claim that we are closer than the fringe of understanding the mechanisms by which this intricate process is co-ordinated and controlled. This problem is of course encountered in all types of cells, but is at its most complex in the eukaryote with its multiple large chromosomes, and it is in this type of cell that we might expect the most elaborate controls. There are many ways of approaching this problem. Where methods are available, direct assay of enzymes such as the replicase or ligase may be of some value, though usually this can do little more than show that a particular enzyme is available in the cell at the time it is required, and in any case this approach is restricted to assayable enzymes.

The genetic approach, exemplified in the case of yeast by Hartwell and his colleagues (1) is likely to be one of the most fruitful, for it can identify particular genes and may tell us not only the activities of particular gene products at the molecular level, but also the time at which these products act in the cell cycle.

Time is of course the essence of the cell cycle, and temporal aspects of regulation of all cell cycle processes, but particularly of DNA replication, deserve our careful scrutiny. A rather crude approach that relates to the question of time of synthesis of proteins involved in controlling DNA replication, rather than their time of action, is the simple procedure of inhibiting protein synthesis at different stages in the cell cycle and measuring the effect on entry into, or passage through, the S period.

In the case of bacteria, this simple stratagem, using either amino-acid starvation or inhibitors such as chloramphenicol (2, 3, 4) has revealed a seemingly uncomplicated situation. This is that once replication of the bacterial chromosome is under way, it is essentially unaffected by inhibition of protein synthesis. In fact it has recently been shown that this is not quite true; although in a gross sense replication of the chromosome proceeds to completion in the absence of protein synthesis, there is a tiny amount of replication, probably near the terminus, that is not undertaken unless protein synthesis is allowed (5). However, the DNA concerned amounts to no more than 0.5% of the genome, and it is a fair conclusion that most of the bacterial chromosome can be replicated without the need for concomitant synthesis of proteins.

On the other hand, initiation of replication, as distinct from propagation, cannot take place in the inhibited bacterial cell, though the nature of the proteins that have to be made, and the time they are normally synthesised in relation to the actual initiation event, is unresolved.

Applied to eukaryotes, this same trick of using inhibitors of protein synthesis, commonly cycloheximide (CHI) or puromycin (PM), has yielded altogether different and more complex results. In the first place we should exclude mitochondrial DNA (mitDNA) from consideration, for this DNA is evidently controlled separately from its nuclear counterpart. In yeast, incorporation into mitDNA can apparently continue for some time in the absence of cytoplasmic protein synthesis (6) though in mammalian cells, replication of mitDNA is affected only a few hours after exposing cells to CHI (7).

Replication of the nuclear DNA of eukaryotes, however, is markedly sensitive to inhibition of cytoplasmic protein synthesis. Moreover, in striking contrast to the prokaryotes, it is not just the initiation of replication that is affected; chain growth already in progress is also sensitive. Until recently it seemed that in almost all eukaryotes examined, progress through S ceased soon after inhibition of protein synthesis, even in individuals already well embarked on this stage of the cell cycle.

There are numerous examples to draw on. Littlefield & Jacobs (8) found that PM added to partially synchronised mouse fibroblasts during DNA synthesis markedly inhibited the latter process. Similarly, Schneiderman, Dewey & Highfield (9) found CHI to block DNA synthesis in Chinese hamster cells in S, though according to Highfield and Dewey (10) S-phase Chinese hamster cells in culture continued to synthesise DNA at a greatly reduced rate after addition of either CHI or PM. The plant cell is no exception to the general rule; Bloch *et al* (11) found that continuation of DNA synthesis in onion root meristem required concurrent protein synthesis, for addition of PM to cells in S caused immediate inhibition of replication. However, inhibition is not always instantaneous. Thus in *Physarum* (12) and mammalian cells (13, 14) addition of inhibitors to S-phase cells allowed replication to proceed for a short while before ceasing altogether. An extension of this apparent anomaly is provided by the sand dollar. According to Young, Hendler and Karnofsky (15), completion of the first S period following pronuclear fusion can occur in the presence of CHI provided the inhibitor is not added too much before the onset of replication. Subsequent S periods however obey the general eukaryotic rule, showing marked sensitivity to the addition of the inhibitor.

These few examples, by no means comprehensive, suffice to illustrate the main trend, namely that passage through S is almost invariably slowed down or halted by inhibition of protein synthesis. On the other hand the detailed response to such inhibition does vary somewhat between systems. This may reflect the variety of techniques used by different experimenters. However it is just as likely to result from differences in the way various cell types control their passage through the S period. One particularly instructive example deserves a closer examination. This is the work carried out on the eukaryotic slime mould *Physarum polycephalum*. Many authors have contributed to the realisation that synthesis of this organism's nuclear DNA, which takes place in the multinucleate plasmoidal syncytium in perfect natural synchrony, requires concomitant protein synthesis for its completion, and that particular pieces of the genome are replicated in a sequence which is adhered to in successive S periods (12, 16, 17). Building on the work of Cummins and Rusch (12), Muldoon *et al.* (18), using CHI, showed that the sequence of DNA replication can be defined operationally in terms of about 10 so-called "replicative units" (RUs). The initiation of each RU is sensitive to CHI, but propagation of replication of a unit already in progress when the inhibitor is added can go to completion. The RUs are replicated sequentially, so that the gross effect of adding CHI to the plasmodium is typically eukaryotic in that synthesis of DNA continues for a short while after adding the drug, but then completely ceases. However, at the molecular level, the individual RU behaves in a manner formally analogous to the bacterial chromosome; initiation requires protein synthesis but propagation does not.

The nature of the RU is something of a mystery. *Physarum* has about 50 chromosomes (19), so an average RU must contain five chromosomes or pieces of many more. This makes it unlikely that we can equate the CHI Replicative Unit of *Physarum* with the "replication unit" defined by Huberman and Riggs (20) in mammalian cells by DNA fibre autoradiography. Tandem assays of units of the latter type, usually nowadays referred to as "replicons," have been seen in various eukaryotic cells (including, recently, yeast (T.D. Petes, personal communication)). The origins of these units must be the centres at which chain growth, definable by autoradiography, is initiated. There seem invariably to be many replicons per DNA fibre, and it will be surprising if the chromosomes of *Physarum* turn out to be an exception. Even if it were, there must be at least one replicon per chromosome, i.e. 50 per nucleus, while there are only 10 CHI RUs. It is possible that each RU comprises a collection of replicons sharing a common initiator protein, perhaps specific for that RU, which is made only when replication of the previous RU terminates (Fig. 1). Other models are of course possible and we shall have to wait for further developments before this situation is understood. In the meantime it is instructive to consider the possibility that all eukaryotic DNA replicating systems may be organised around such a framework of RUs. It might be that the apparently diverse responses of the DNA replication machinery of different cell types to inhibition of protein synthesis reflect differences in the numbers, sizes, and temporal interrelationships of their RUs.

With this background, it seemed desirable to investigate this general

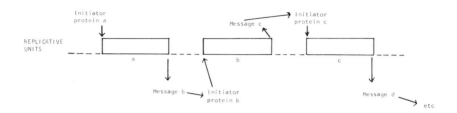

Fig. 1. Possible model for replication of the *Physarum* genome, illustrating the ideas discussed by Cummins & Rusch (12) and Muldoon *et al.* (18). Each bar represents one of ten Replicative Units. Replication of each is initiated by a protein synthesised in response to a signal generated on completion of the preceding unit.

situation in the yeast *Saccharomyces cerevisiae*. Hitherto, observations on the effect of protein synthesis on eukaryotic DNA replication have been made on relatively highly evolved eukaryotes with large and more or less complex chromosomes. It has been pointed out previously that there are good reasons for regarding the yeast chromosome as being a relatively simple, possibly primitive structure (21). It does not seem to undergo the intense condensation process required for mitosis in the more highly evolved cells, and it is continuously available for transcription throughout nuclear division. More recently it has been clearly shown that the chromosomes [about 17 (22)] are unineme, i.e. each contains only a single DNA duplex (23, 24). It can be calculated that the average DNA content of a yeast chromosome is only about 3% that of most mammalian chromosomes (25). If packaging of the yeast chromosomes for mitosis can be managed without the intensive condensation and coiling up undergone by the HeLa chromosome for the same purpose, it seemed possible that the replication process of the yeast genome might also be organised in a comparatively simple manner.

As it happened this somewhat naive line of reasoning turned out to be justified. In experiments reported in detail elsewhere (26) it was shown that addition of 100 g/ml CHI to a synchronously dividing yeast culture in mid-S had no effect on the rate of incorporation of exogenous adenine into nuclear DNA in a 20 min pulse immediately following the CHI (Fig. 2). On the other

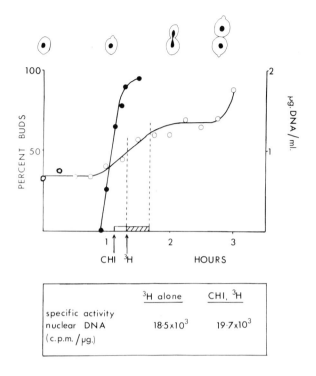

Fig. 2. Effect of CHI (100μg/ml) on incorporation of ³H-adenine into a synchronous culture in mid—S. ●——● , percent young buds (i.e., percent cells initiated S); o——o , total DNA/ml culture. The essential morphological changes in this type of culture are illustrated diagrammatically. The interval between inoculation and the initial round of budding is referred to as a lag period (effectively G_o), which does not recur. The first cell cycle proper starts with S as the first buds appear, and ends around 3 hours when cell division occurs and the next S period starts. The bars indicate the periods for which separate portions of the culture were exposed to CHI followed by ³H-adenine, or ³H-adenine alone. The figure is from Williamson (26).

hand, addition of this same amount of CHI to the culture at times up to a critical point about 10 minutes before the normal start of the S period, completely prevented entry into S, i.e. no synthesis of nuclear DNA occurred at all. However, colorimetric analyses showed that exposure to the inhibitor at times after this critical point allowed limited amounts of DNA synthesis to take place in the culture as a whole (Fig. 3). To investigate the distribution of

147

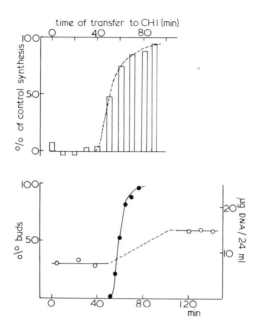

Fig. 3. The effect on DNA synthesis of exposing synchronous cells to CHI (199µg/ml) at different times during the initial lag phase and first S period. ●——●, percent buds (initiations) in the master culture; o——o , total DNA per 24 ml culture (the broken line indicates the usual time course of DNA synthesis in this system). The histogram bars show the amounts of DNA synthesised in separate portions of the culture transferred to CHI at the indicated times and incubated to 140 min. The ordinate in this case is the amount of DNA synthesised as a percentage of the amount synthesised in the master culture in the course of the S period. The figure is from Williamson (26).

this synthesis between individual cells in the population, experiments involving single cell autoradiography of DNA-labelled cells from synchronised cultures were performed. The protocol of the experiment and the resulting frequency distributions of grain counts are shown in Fig. 4. Label (^{3}H-adenine) was added to the synchronised population just before the start of S, and incubation continued until well after the time when the control normally finished S. This gave the symmetrical frequency distribution of grain counts seen in Fig. 4a. If CHI was added with the label, essentially all the incorporation was into mitDNA, since under these conditions entry into S was prohibited (Fig. 4b). If on the other hand the inhibitor was added early in S, at a time when some cells had initiated but essentially none had completed their round of replication (Fig. 4c), and incubation continued as before, the resulting frequency

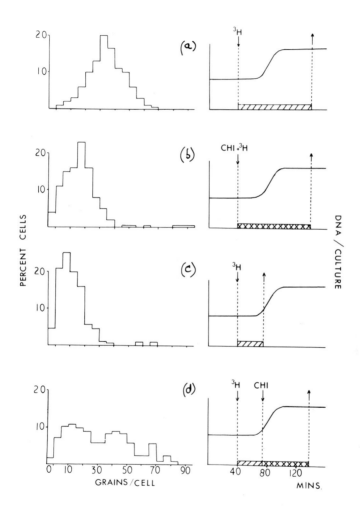

Fig. 4. Frequency distributions of DNA grain counts over individual cells following the various labelling routines illustrated diagrammatically on the right. Approximately 200 cells were counted in each preparation. The cells were prepared for DNA autoradiography by the method of Williamson (29). The figure is from Williamson (26).

distribution of grain counts was bi-modal (Fig. 4d). The more heavily labelled cells had a mean grain count approximately that of the control cells labelled for a whole generation (Fig. 4a). The less radioactive cells on the other hand, had grain counts comparable with those of the fully CHI-blocked cells (Fig. 4b). Taking into account the other results mentioned above, these data were interpreted as evidence that in this organism there is a 'transition point' (27) for the effect of CHI on nuclear DNA replication. Prior to this point, assembly of proteins essential for entry into and completion of S is taking place. By the time the transition point is reached all such proteins are ready and the S period can start and go to completion even when further protein synthesis is prevented. Strictly speaking, these experiments do not rule out the possibility that replication in the inhibited cells stops just short of completion, as described above for bacteria. However, Hereford and Hartwell (28), in some ingenious experiments using CHI in combination with temperature-sensitive DNA replication mutants, provided good evidence that the DNA synthesised in CHI blocked cells is both normal and complete, since mutant cells incubated for a period in CHI at the permissive temperature acquired the capacity to divide once when the inhibitor was removed and the temperature shifted to the non-permissive level. In other respects the results obtained by these authors provided further support for the view that all the proteins required for one complete round of replication of the yeast genome are made by a critical point in time just before initiation of replication takes place.

It thus appears that the yeast nuclear genome behaves in a manner analogous to that of the bacterial chromosome, and might be said to comprise a single 'Replicative unit' in the sense that term was used by Muldoon *et al.* (18). The significance of this for our overall understanding of the control of DNA replication in eukaryotes is not clear. It suggests some element of simplicity in the functional organisation of the yeast nucleus as compared with more highly evolved eukaryotes, and it will be of interest to see if this situation holds good for other simple eukaryotes with uncomplicated chromosomes. However, it would be unwise to press this line of thought too far. Already there is good evidence from electron microscopy (23) and DNA fibre autoradiography (Petes, personal communication) that yeast chromosomes exhibit numerous tandemly arranged replicons, and in this respect therefore the yeast cell reveals a typically eukaryotic degree of complexity. The only difference between yeast and other eukaryotes may be the time at which proteins required for intiating individual replicons are synthesised. The timing of their activities, and the mechanism which governs such timing, has yet to be determined, and this mechanism could turn out to be a good deal more complicated than we presently anticipate.

Perhaps this is really the moral of this story. It is not enough to define cell cycle controls merely in terms of *what* particular proteins do. To fully comprehend the situation, we will have to learn more, at the molecular level, about the mechanisms which fix not only the time of synthesis of gene products, but also the time at which they act.

References

1. L. H. Hartwell, R. K. Mortimer, J. Culotti and M. Culotti, *Genetics 74*, 267 (1973).
2. O. Maaloe and P. C. Hanawalt, *J. Mol. Biol. 3*, 144 (1961).
3. K. G. Lark, R. Repko and E. J. Hoffman, *Biochim. Biophys. Acta 76*, 9 (1963).
4. H. Yoshikawa, *Proc. Nat. Acad. Sci. Wash. 53*, 1476 (1965).
5. T. Marunouchi and W. Messer, *J. Mol. Biol. 78*, 229 (1973).
6. L. I. Grossman, E. S. Goldring and J. Marmur, *J. Mol. Biol. 46*, 367 (1969).
7. B. Storrie and G. Attardi, *J. Mol. Biol. 71*, 177 (1972).
8. J. W. Littlefield and P. S. Jacobs, *Biochim. Biophys. Acta 108*, 652 (1965).
9. M. H. Schneiderman, W. C. Dewey and D. P. Highfield, *Exptl. Cell Res. 67*, 147 (1971).
10. D. P. Highfield and W. C. Dewey, *Exptl. Cell Res. 75*, 314 (1972).
11. D. P. Bloch, R. A. MacQuigg, S. D. Brack and J-R Wu, *J. Cell Biol. 33*, 451 (1967).
12. J. E. Cummins and H. P. Rusch, *J. Cell Biol. 31*, 577 (1966).
13. G. C. Mueller, K. Kajiwara, E. Stubblefield and R. R. Rueckert, *Cancer Res. 22*, 1084 (1962).
14. E. W. Taylor, *Exptl. Cell Res. 40*, 316 (1965).
15. C. W. Young, F. J. Hendler and D. A. Karnofsky *Exptl. Cell Res. 58*, 15 (1969).
16. R. Braun, C. Mittermayer and H. P. Rusch, *Proc. Natl. Acad. Sci. Wash. 53*, 924 (1965).
17. R. Braun and H. Wili, *Biochim. Biophys. Acta 174*, 246 (1969).
18. J. J. Muldoon, T. E. Evans, O. F. Nygaard and H. H. Evans, *Biochim. Biophys. Acta 247*, 310 (1971).
19. I. K. Ross, *Amer. J. Bot. 53*, 712 (1966).
20. J. A. Huberman and A. D. Riggs, *J. Mol. Biol. 32*, 327 (1968).
21. D. H. Williamson, *in Cell Synchrony* (I. C. Cameron and G. M. Padilla, eds) p. 81, Academic Press, New York, (1966).
22. R. K. Mortimer and D. C. Hawthorne, *Genetics 74*, 33 (1973).
23. T. D. Petes, C. S. Newlon, B. Byers and W. L. Fangman, *Cold Spring Harb. Symp. Quant. Biol.* (1973) *in press.*
24. J. Blamire, D. R. Cryer, D. B. Finkelstein and J. Marmur, *J. Mol. Biol. 67*, 11 (1972).
25. R. Holliday, *in Organization and Control in Prokaryotic and Eukaryotic Cells* (H.P. Charles and B.C.J.G. Knight, eds) p. 359, University Press, Cambridge (1970).
26. D. H. Williamson, *Biochem. Biophys. Res. Commun. 52*, 731 (1973).
27. J. M. Mitchison, *The Biology of the Cell Cycle*, University Press, Cambridge (1971).

28. L. M. Hereford and L. H. Hartwell *Nature New. Biol.* *244*, 129 (1973).
29. D. H. Williamson, *J. Cell Biol.* *25*, 517 (1965).

CELL DIVISION IN YEASTS. II. TEMPLATE CONTROL OF CELL PLATE BIOGENESIS IN *SCHIZOSACCHAROMYCES POMBE**

Byron F. Johnson, Bong Y. Yoo***, and G. B. Calleja**

Division of Biological Sciences, National Research Council
of Canada, Ottawa, Ontario K1A OR6
and
***Department of Biology, University of New Brunswick,
Fredericton, New Brunswick

I. Introduction

About twenty years ago, template notions were abundant. They were bandied about to rationalize any and every unknown biological assembly process. One of the least probable of those old template speculations, that protein might be formed on nucleic acid templates, has borne fruit so plentifully that the notion seems almost naive today—surely the finest of fates for a speculation. Other apparently probable template notions, for instance, antigen acting directly as a template for antibody formation, were soon seen to have little if any heuristic value.

It is curious that template models are less frequently proposed today, even though the nature of most biological assembly processes still merits speculation. Green (11), in a recent erudite book review on assembly processes, does mention templates but envisages unnecessary restrictions upon them.

We have recently described electron microscope observations of cell plate biogenesis and fission in the fission yeast, *Schizosaccharomyces pombe* (16). It is our intention here to expand that description by direct comparison with fluorescence mocrographs. The use of fluorochromes for cytological purposes derives from the studies of von Provazek in 1914 (22), but their useful employment for yeast cytology is recent (25,26). Fluorescent brighteners were first used on biological objects about 10 years ago (6,7) and are just now finding use as cytological fluorochromes (4, 12, 24 and personal communication from Dr. D. Bush).

Micrographs based upon primuline fluorescence have been available to describe fission (26), but not cell plate biogenesis. We have been using a fluorescent brightener (CFW) in order to examine cell plate biogenesis. All of our results suggest that cell plate biogenesis of the fission yeast involves a polysaccharide template whose own construction and destruction take place

*Issued as National Research Council of Canada Publication No. 13761
**Current Address: University of The Philippines, Natural Science Research Center, Diliman, Quezon City, The Philippines.

153

within closely defined spatial and temporal limits and which functions to guide the deposition of the final, permanent wall. Because the fluorescence observations can be directly linked with the electron microscope observations, it now becomes possible to examine control mechanisms operating in cell wall and cell plate biogenesis of *S. pombe* as a function of its cell cycle.

II. Experimental

A. Methods and Considerations

Cultures of *S. pombe* (N.C.Y.C. 132, A.T.C.C. 26192) were grown as described before (16). Samples of cells from either mid-log or late log phase cultures were suspended in a 0.1% solution (in 0.1 M phosphate buffer, pH = 8.0 [12]) of CFW (Calcofluor White M2R New, a laundry brightener) and immediately centrifuged. The pellet of cells was resuspended in a small volume of the supernatant and directly examined, or was mixed with an equal volume of 3% agar (Difco Purified, molten and maintained at about 43°C). The cell-agar suspension was quickly placed upon a microscope slide, covered and immediately inverted so that some cells might settle to touch the coverglass before the agar cooled. As an additional precaution against motion in all of these preparations, the corners of the coverglasses were anchored with a lacquer. All fluorescence studies were carried out with a Reichert Zetopan microscope using light from a high pressure mercury arc. Photomicrographs were taken using a Wild MKa4 automatic camera and Kodak Tri X 35 mm film.

All structures made secondarily fluorescent in living cells of *S. pombe* by CFW are either wall or wall-associated, because CFW is excluded by the cell membrane (unpublished observations). Cytoplasmic fluorescence occurs only when cells have been photolysed by the near-UV exposure or have been damaged in other ways. In this respect, CFW is wholly comparable with another fluorochrome, primuline, whose value as a viability indicator, based upon dye exclusion by the membrane, has been carefully examined with yeast (10).

Technics for electron microscopy were detailed earlier (16).

B. Observations

Although the fission yeast cells were ordinarily made entirely visible by dim fluorescence of various cell wall structures, the cell plate region of dividing cells was their only region marked by bright fluorescence. The bright cell plate fluorescence appeared first as a fine bright ring around the surface of the cell, more or less midway between the ends of the cell (Figs. 1A, 2A). The ring of brightness grew centripetally like a closing iris diaphragm (Figs. 1B, C, D, and 2B, C, D, J, K), and eventually constituted a bright disk which bisected the cell (Fig. 2E). Then, as fission took place, the diameter of the bright disk became smaller (Figs. 1E, F, G, and 2F, G) until finally, the bright fluorescence disappeared coincident with separation of the two daughter cells (Fig. 1H; Fig.

2H still bears the last traces of brightness.)

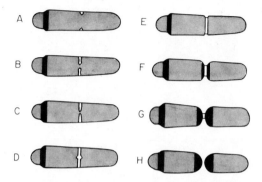

Fig. 1. Cell division of *S. pombe* as viewed by CFW fluorescence (diagrammatic). A, Initiation of fine bright fluorescent ring; dark ring near end is old fission scar. B - D, Centripetal growth of bright ring (showing intensely fluorescent leading edge in B and D) becoming bright disk in D. E, Beginning of fission marked by diminished diameter of bright disk. F - G, Fission progresses; cells swell; new fission scars apparent as new dark rings. H, Fission complete; bright material gone.

The new cell ends generated by fission of the cell plate are characterized by their very dim fluorescence (noted also in [26]). These dark scars remain easily identified after many generations (see Fig. 2G, a cell with five old scars). Coincident with the division process is a generalized swelling of the cell (16) which is most easily marked by examination of cells with two or more old fission scars. The region between two scars always has greater cylindrical diameter than regions between the ends of the cell and fission scars (Figs. 2A, B, C, D, E, F, G, H; Fig. 2G shows a cell which has divided and swollen so often that it is becoming spheroidal as a consequence of its present cell division).

The leading edge of the centripetally growing bright structure was often intensely bright. This intense brightness could often be dispersed by prolonged exposure to the exciting beam, but the brightness described in the preceding paragraph resisted photolytic dispersion. This intense bright leading edge is illustrated in some of the diagrams (Fig. 1B, D) and can be seen in some of the micrographs. Thus, in Fig. 2B, it is apparent in the cell with the brighter ring of the two. It is discernable in Fig. 2C as brightness medial to the old cell wall at the right side of the cell. It is clearly marked in Fig. 2D where it also shows that the circular leading edge is displaced toward the bottom of the picture. This evidence for displacement toward one end of the living cell is useful, for many electron micrographs of dividing fission yeasts show displacement of the cell plates (e.g. Figs. 5A, C, D, E, F, 6A, and [17, his Fig. 2]), and this always

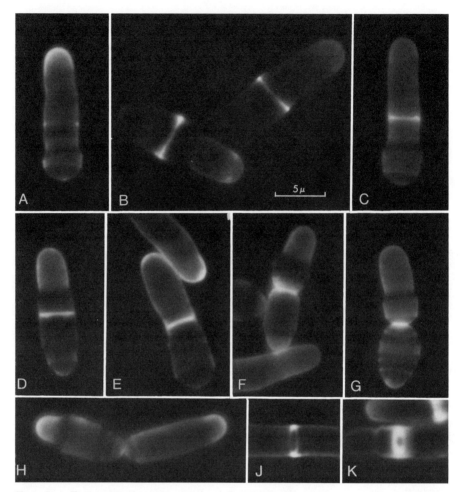

Fig. 2. Fluorescence micrographs of cell division in *S. pombe*; original magnification = 900x for all. A, Initiation of bright ring of fluorescence (in focus at edges of cell, out of focus over longitudinal axis); narrow end still fluorescent from previous extensile growth; two dark fission scars at large end. B, Two cells with bright rings after some centripetal growth; left cell with three dark fission scars, right cell with two. C, Bright ring with intensely fluorescent leading edge, seen as broader brightness near right edge of ring; two dark fission scars. D, Bright ring with intensely bright leading edge displaced toward lower end of micrograph, three dark fission scars; narrow end still fluorescent from previous extensile growth. E, One cell with bright fluorescent disk; three dark fission scars. Other cell (cut off) with bright fluorescent extensile end. F, Fission in progress, with diminished diameter of bright disk; three old dark

156

raised concern about possible fixation artefacts. Finally, it is interesting to note that regions of extensile growth of *S. pombe* (13,19) often fluoresce brightly (Fig. 2A, D, E, H).

Observations made by electron microscopy have been described in detail elsewhere (16), but the micrographs are included here (Figs. 3, 4, 5, 6, 7) for ease of comparison.

III. Discussion

A. *Comparison of Observations*

The presented pattern of change of bright fluorescence in the cell plate region during cell division of *S. pombe* is quite striking (Figs. 1, 2): i) A bright ring appears around the cell about midway between the ends. ii) The bright rings grows centripetally toward the middle, eventually closing to approximate a bright disk which bisects the cell. iii) The bright disk becomes smaller through loss of fluorescence at its periphery in synchrony with the fission process. iv) The bright fluorescence disappears with the completion of fission.

Careful examination of the electron micrographs of dividing fission yeast cells (Figs. 3, 4, 5, 6, 7) shows only one structural element with similar behaviour — the electron transparent AR (annular rudiment). The precisely parallel behaviour of the bright fluorescence and the electron transparent AR strongly suggest that the one identifies the other. Henceforth we discuss them as identified, for convenience.

The intense fluorescence found at the leading centripetal edge of the bright band in some cells seems to correlate with no structural element which we can identify in the electron micrographs. It might correspond with the most recently synthesized leading edge of the AR, or with the very thin periplasmic layer which lies between the centripetal edge of the AR and the cytoplasm. The fact that the intense brightness can be photolysed suggests that it is not a function of special dye-binding to an organized structure like the brightly fluorescent AR. However, this is not so strong a suggestion as to direct all our attention to the possibility that it could be evidence for a differentiated region within the periplasmic dark layer. For reasons discussed above, we doubt that the intense fluorescence has a cytoplasmic locus.

Similar ambiguity exists with respect to bright fluorescence occasionally seen at the growing ends of the cells. Whether the CFW dye is bound to

---

fission scars, new fission scars still somewhat bright. Lower cell photolysed, with cytoplasmic fluorescence. G, New dark fission scars obvious near shrinking bright disk; five old fission scars obvious. H, Fission nearly complete; two new and two old fission scars; both old ends bright from fluorescence associated with newly initiated extensile growth. J, Cell suspended in agar and oriented obliquely to plane of focus; bright fluorescent ring well formed and closing. K, as J, but centripetal growth of bright ring much further advanced.

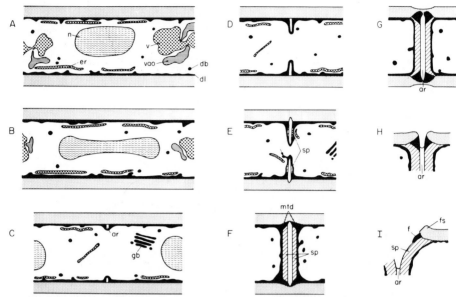

Fig. 3. Fission of *S. pombe*. A - C, Mitosis and initiation of annular rudiment (AR). D, Establishment of dense fillets at base of AR. E, Initiation of scar plug layer (SP[26]); fusion of dense body (DB) with dark layer (DL, arrow). F, Migration of dense fillet into wall to become material triangulaire dense (MTD); closure of pore. G, Growth of SP layer to make contact with old cell wall; initiation of fission by erosion of old cell wall. H, Outward movement of MTD to become fuscannel (F); fission scar (FS) ridge becomes apparent; AR material is lost at edge of layer. I, Fission almost complete; F at final disposition; SP layer new end of daughter cell; AR material nearly all lost.

recently synthesized cell wall or to regions of the periplasmic dark layer which are differentiated according to the rates of synthesis of cell wall materials is a question which cannot be answered with available data. Unfortunately, these staining ambiguities are not resolved by knowledge of the staining affinities of the dyestuff.

B. Specificity of CFW Binding

High specificity of binding has often been claimed for fluorochromes. For instance, alkaline aniline blue is supposed to bind with great specificity to the β $(1\rightarrow3)$ glucans, characteristic of callose, in sieve plates of higher plants (9). However, careful examination suggests that binding of at least some of the fluorochromes might be rather non-specific. Specificity, in terms of dyeing substantivity for textiles, of a variety of fluorescent brighteners has been discussed in some detail (28). An interesting approach has been that of Maeda and Ishida who examined the reaction of CFW with ten polysaccharides and a

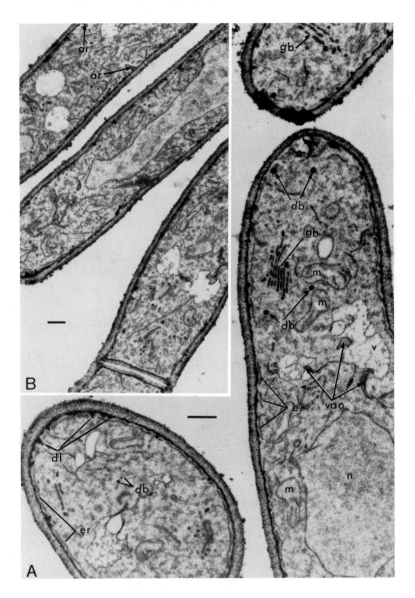

Fig. 4. A, Normal premitotic cytoplasm; all magnification bars equal 0.5 μm. B, first signs of annular rudiment (ar) in post-mitotic cell, shown just above a mitotic cell. Abbreviations: ar, annular rudiment; db, dense body; dl, dark layer; er, endoplasmic reticulum; gb, Golgi body; m, mitochondrion; n, nucleus; v, vacuole; vao, vacuole-associated organelle.

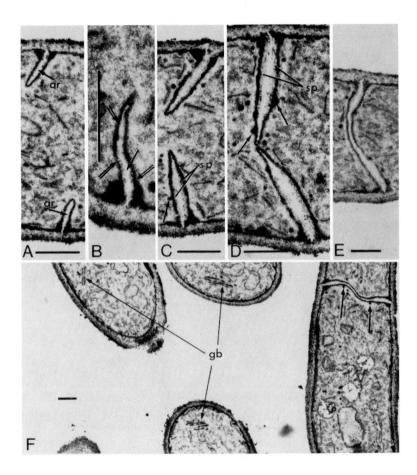

Fig. 5. A, Centripetal growth of annular rudiment (ar) layer with thickening of dark layer to form fillets at base of ar layer; all magnification bars equal 0.5 μm. B, Highly magnified view of one-half cell plate, showing ar layer (single-shafted arrow), dark layer, and fillet covered entirely by cytoplasmic membrane (double-shafted arrow). C, Initiation of scar plug (sp) layer, and fusion of dense body with dark layer (arrow). D, Closing the pore; fusion of dense body with dark layer (arrows). E, F, Closure of cell plate; note Golgi bodies in neighboring cells; fusion of dense body with dark layer (arrows).

variety of microorganisms (18). Of the ten polysaccharides, three were α-linked, and none of those reacted. The other seven were β-linked hexapyranosyl polysaccharides, one β (1→3) linked and the other β (1→4) linked, and all reacted with the fluorochrome. The infrared spectra of three of

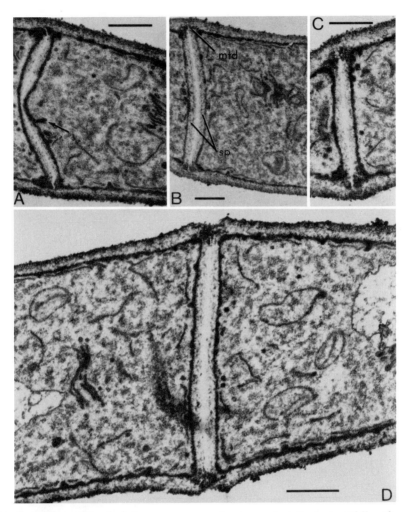

Fig. 6. A, Scar plug (sp) layers now nearly grown together at middle of cell plate; fusion of dense body with dark layer continues (arrow); all magnification bars equal 0.5 µm. B, Material of fillet has migrated outwardly to become material triangulaire dense (mtd); sp layers now appear complete centrally and begin attachment to old cell wall behind mtd. C, Progressive outward movement of mtd. D, Erosion of outer surface of old cell wall and first appearance of mtd at outside (now fuscannel).

the binding polysaccharides were modified in the presence of the dye, and "electrodialysis of the cellulose-brightener complex (for) over 70 hours in

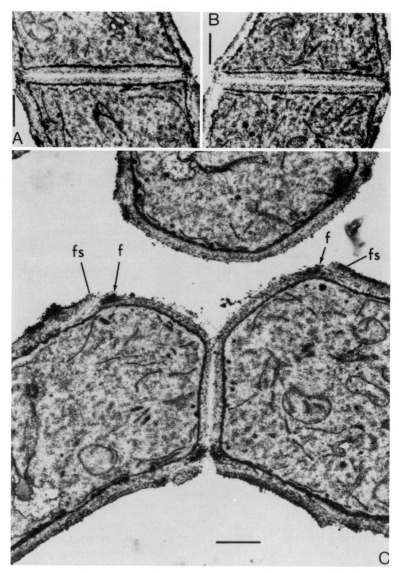

Fig. 7. A, B, Progressive contact of scar plug layer with old cell wall; fuscannel (f) now external; all magnification bars equal 0.5 μ m. C, Eccentrically dividing cell showing new fission scar (fs) and f at top, and slightly earlier stages at bottom; new ends of cell round as old annular rudiment material disappears.

162

M/100 NaCl at 200 V did not cause a noticeable decrease of the fluorescence". However, strength of binding is not specificity, and the now fair sized list of microorganisms found to be stainable (7,18) with CFW suggests that caution should be exercised. Curiously, lysozyme treatment caused *Candida* to lose some CFW-induced secondary fluorescence (18), and consideration of the known specificity of lysozyme implies that CFW was binding to chitin in the wall of that yeast (23 [p. 234]). Others have found that CFW binding to bud scars of *Saccharomyces* cells could be reduced but not eliminated by extraction with concentrated HCl, which should have removed chitin (24). It would be easy to draw a parallel between the chitin-ring in the bud scar of *Saccharomyces* (1,5) which is electron-transparent and binds CFW, and the electron-transparent, fluorochrome-binding AR of *S. pombe*. However the bright staining during cell plate biogenesis seen above should not now be ascribed to specific binding of CFW to chitin, if for no other reason than that chitin has been sought but not yet found in the walls of *S. pombe* (4,8,21,23). Finally, another reason for reserving judgment about the specificity of staining with CFW is the fact that photolysed cells absorb the dye strongly and generally in their cytoplams (c.f. one cell in Fig. 2F).

C. The Fluorochrome-Binding AR as a Template

The scar plug (SP [26]) layers of the cell plate become the new ends of the divided progeny. It is readily apparent (Fig. 3F, G, and Fig. 5C, D) that the SP layers are laid down upon the previously formed AR and indeed, their site of initiation on the AR is not proximal to the old wall. It is also obvious that the brightly fluorescing AR material is lost at fission (Fig. 1E, F, G, H; Fig. 2G, H; Fig. 3H, I; and Fig. 7C). Thus the AR acts as a mold or template, which the cell "uses" temporarily while establishing the permanent SP layers, and then discards.

During biogenesis of the AR, which is presumed to be polysaccharide from its reaction to fixation and staining for electron microscopy (and might even be chitin in spite of the earlier discussion), enzymes active in its elaboration are working (activated?) outside the cytoplasmic membrane (Fig. 5B). After the AR has been generated, there is a brief period during which biogenesis of the SP layers is completed. The enzymes responsible for elaboration of the SP layers also act outside the cytoplasmic membrane. By this time the daughter cells are physiologically separate units, but must act in coordination lest the remaining complex activities fail to produce complete physical separation from one another.

The intricacies of attachment of the SP layers to the old cell wall during fission, and of movement of periplasmic dark material to the outside of the fission scars to form the fuscannel have received attention previously (16). The template functions of the AR are finished, and presumably it has no other function. At any rate it seems firmly bonded to the SP layers for even after erosion of the old cell wall at initiation of fission, the SP layers of what are now the daughter cells are held flat against it (Fig. 1E; Fig. 2F; Fig. 3H; and

Fig. 7A, B). Only after lytic removal of the AR can turgor pressure cause the new ends of the cells to round up (Fig. 2F, G, H, and Fig. 7C). The firmness of binding of the new cell ends (the SP layers) to the AR template and the legendary strength of yeast cell walls combine to suggest that covalent bonds play a major role in the assembly process contrary to the template speculations of Green (11) which were based primarily upon considerations of membrane biogenesis. All of these enzymes which have been responsible for erosion of the old cell wall at fission and for the complete disappearance of the AR material have been active at particular sites and working at particular times, the coordination of synthetic and lytic activities being a function of two cells rather than the original one.

D. Cellular Extension, Cell Plate Elaboration and Cell Cycle Controls

Extensile growth of fission yeasts has received much attention (13,15,19,26,27). A model for extensile growth has been proposed (15) involving coordinated lytic activity and synthetic activity. Thus the glucan layer (assumed to be responsible for form and strength of the yeast cell wall) grows by enzymatic insertion of glucose or oligosaccharides into cuts made by hydrolytic enzymes in preexisting glucan molecules. The basic premises of the model have recently been expanded in detail and applied to extensile growth of other fungi (2,3).

It seems improbable that such a mechanism can be applied to cell plate biogenesis. Both the AR and the SP layers seem to appear *de novo*. If this be true, the basic premise of growth by molecular intussusception into preexisting polysaccharides is inapplicable. However, lytic enzymes should have a role as discussed above, but are conceived of as acting *after* biogenesis rather than as a preliminary step. Thus, after the AR is completely formed and has functioned as a template, it is lysed away. After the SP layers have bonded to the insides of the old cell wall, there is lysis of the old cell wall as the initiation of fission.

Viewed as a subject for study of cell cycle controls, extension seems very difficult. The rate of wall extension is not constant in the cell cycle, but increases as a function of cell length (13). Hence extension involves ever increasing activities of both lytic and synthetic enzymes until extension abruptly stops at about the third quarter of the cell cycle (19). Coordination must be very precise because lysis of yeast cells, which is the penalty for faulty coordination, is a rare phenomenon in normal growth conditions (14,20).

The inversed coordination patterns at template (AR) formation, of synthetic activity followed by lytic activity, with the activities separated by space and time, with the activities constrained to a material which is present at a unique phase of the cell cycle, and with the activity occurring in only a short portion of the cell cycle indicate a complex of controlling elements which now seems experimentally separable.

IV. Summary

Fluorescence micrographs of fission yeast cells in cell division show a bright ring which grows centripetally to finally become a bright disk bisecting the cell. The behaviour of the brightly flourescent structure corresponds in detail with that of an electron transparent structure previously described in the cell plate. The electron transparent structure acts as a template, upon which the new ends of the daughter cells are formed. At fission the brightly fluorescent template is removed concomitant with release of the two daughter cells. The stepwise sequence of generation and removal should allow closer scrutiny of cell division controls in yeasts than has heretofore been possible.

Acknowledgements

We thank Robert Whitehead for assistance with the photomicrography and for preparation of the final plates; Donna Kelly for general technical assistance; F. G. Villaume of American Cyanamid Company for a generous gift of the fluorochrome Calcofluor White M2R New; The American Society for Microbiology for permission to use copy-righted material. B. F. J. thanks Dr. E. Cabib for stimulating discussion of cell division in yeasts. B. Y. Y. was supported by National Research Council of Canada grant A3651. G. B. C. was a National Research Council of Canada Postdoctoral Fellow and currently holds a Canadian International Development Agency Research Associateship.

References

1. J. S. D. Bacon, E. D. Davidson, D. Jones and I. F. Taylor, *Biochem. J. 101*, 36C (1966).
2. S. Bartnicki - Garcia, 23rd *Symp. Soc. Gen. Microbiol.* 245 (1973).
3. S. Bartnicki - Garcia and E. Lippman, *J. Gen. Microbiol. 73*, 487 (1972).
4. D. Bush, M. Horisberger and P. Würsch, *Experientia 29*, 769 (1973).
5. E. Cabib and B. Bowers, *J. Biol. Chem. 246*, 152 (1971).
6. M. A. Darken, *Science 133*, 1704 (1961).
7. M. A. Darken, *Appl. Microbiol. 10*, 387 (1962).
8. J. Deshusses, S. Berthoud and Th. Posternak, *Biochim. Biophys. Acta 176*, 803 (1969).
9. W. Eschrich and H. B. Currier, *Stain Technol. 39*, 303 (1964).
10. R. K. Graham, *J. Inst. Brew. 76*, 16 (1970).
11. D. E. Green, *ASM News 39*, 677 (1973).
12. K. Gull, P. M. Moore and A. P. J. Trinci, *Trans. Brit. Mycol. Soc. 59*, 79 (1972).
13. B. F. Johnson, *Exp. Cell Res. 39*, 613 (1965).
14. B. F. Johnson, *J. Bacteriol. 94*, 192 (1967).
15. B. F. Johnson, *J. Bacteriol. 95*, 1169 (1968).
16. B. F. Johnson, B. Y. Yoo and G. B. Calleja, *J. Bacteriol. 115*, 358 (1973).
17. N. MacLean, *J. Bacteriol. 88*, 1459 (1964).
18. H. Maeda and N. Ishida, *J. Biochem. Tokyo 62*, 276 (1967).
19. J. M. Mitchison, *Exp. Cell Res. 13*, 244 (1957).
20. O. Necas, *Nature 177*, 898 (1956).
21. H. J. Phaff in The Yeasts, Vol. 2, p. 135, A. H. Rose and J. S. Harrison,

eds., Academic Press, London, 1971.

22. S. von Provazek, *Kleinvelt 6*, 30 (1914) (Referred to in Ref. No. 7).
23. H. J. Rogers and H. R. Perkins, Cell Walls and Membranes, E. and F. N. Spon Ltd., London, 1968.
24. O. Seichertová, K. Beran, Z. Holan and V. Pokorný, *Folia Microbiol. 18*, 207 (1973).
25. E. Streiblová and K. Beran, *Folia Microbiol. 8*, 221 (1963).
26. E. Streiblová, I. Malek and K. Beran, *J. Bacteriol. 91*, 428 (1966).
27. E. Streiblová and A. Wolf, *Z. Allg. Mikrobiol. 12*, 673 (1972).
28. F. G. Villaume, *J. Amer. Oil Chem. Soc. 35*, 558 (1958).

EARLY EVENTS IN THE GERMINATION OF *S. POMBE* ASCOSPORES

George M. Padilla*, James Creanor and Ronald S. S. Fraser

Department of Zoology, University of Edinburgh, Edinburgh, Scotland

I. Introduction

In the context of embryogenesis and organogenesis, cellular differentiation is an extension of subcellular specialization.

Multicellularity probably evolved with the advent of the compartmentalization of the cytosol into discrete organelles. In fact, the physiological and metabolic diversity of differentiated cells is dependent not only on the maintenance of subcellular organelles but on the fidelity of their replication from cell division to cell division. It is at this level that genetic mechanisms must operate in directing the synthesis or renewal of enzymes, macromolecules and structures specific to individual subcellular organelles. In a diseased state such as oncogenesis, the loss of this controlling elements leads in no small measure to de-differentiation. This is one of the reasons why much attention is presently being directed towards achieving a greater understanding of the relationships between the repetitive events which comprise the cell cycle and those transitions which lead from one differentiated state to another. Such interactions are seen in developmental systems where differentiation and the cell cycle impinge upon one another. They are also found in the spore-cell transition in bacteria (1).

A similar sequence of events is seen in conjugation, sporogenesis and germination in yeast (2). This is a system in which subcellular differentiation and genetic control of the cell cycle come into direct interplay. The H90 sportulating haploid strain of *Schizosaccharomyces pombe* is particularly suited for these studies. It yields populations of spores in excess of 80%, which, as will be discussed in this communication, can be isolated in pure form and induced to germinate in a controlled manner. Unlike the budding yeasts, *S. pombe* possesses other advantages worth mentioning. It is haploid during most of its growth and cell cycle (3-5) and assumes a diploid level of DNA only as a zygote which is formed following conjugation between two vegetative cells. Two meiotic divisions ensue usually resulting (6,7) in the formation of four ascospores per ascus. The spores are spontaneously liberated and may be stored indefinitely in a freezer. When placed in a nutrient medium, they readily germinate and proceed into vegetative cell growth and division.

We thus have a well delineated sequence of events and complex biological

---

*Present Address = Department of Physiology and Pharmacology, Duke University, Durham, North Carolina, U.S.A.

processes which are separated in time and may be subjected to experimental manipulation. The synchronization of any one of these phases would be of considerable practical significance. It would allow one to study the underlying controlling mechanisms (e.g., sequential enzyme synthesis or induction, genomic readout, etc.) during the transition from one state to another.

In the present report, we have focused our attention on the early events of germination and will describe the experimental approaches used to study the initial macromolecular events underlying that process.

II. Experimental

Vegetative cells of *S. pombe* (H90 strain, NCYC 132) were grown in the chemically defined EMM 2 medium or malt extract broth (4). Sporulation was induced by plating cells from a stationary or late-log culture onto the surface of 2% agar (w/v) containing 1.5% malt extract broth (Oxoid), and incubating at 32º for 7-14 days. The spores and cells were harvested by washing the agar surface with distilled water. They were freed of debris by two centrifugations (5 min, 1500 RPM) in a bench top centrifuge. The pelleted cells and spores were stored frozen at -20º until needed.

Although this strain will routinely yield large quantities of spores (80±5%), they cannot be reproducibly freed from the residual vegetative cells by a single centrifugation through linear sucrose density gradients using conventional low speed swinging bucket centrifuges (3). Such a method of separation was developed by Mitchison and coworkers to select highly synchronous populations of vegetative *S. pombe* cells (8,9). In addition, it was observed that the individual spores were highly heterogenous with respect to size and onset of germination (Mitchison, personal communication and 10). We thus investigated alternative methods of spore isolation, the most successful approach being the use of the analogue, 2-deoxyglucose, to produce not only "weak walled" spores, but to differentially induce lysis of the residual vegatative cells as shown by previous investigators (11,12).

The time course of germination was followed by time lapse photomicroscopy of spores that were resuspended in EMM 2 and spread over small agar pads placed in a wax-sealed coverslip (13). The coverslip preparations were incubated at 32º. It was thus possible to follow the development of spores whose dimensions were determined from enlarged photographs or projections. Parallel sets of spores were also resuspended in liquid media and placed in a water bath set to reach 32º by a delayed time device (4). Techniques for isotopic labelling, chemical analyses, etc., used with *S. pombe* have been summarized by Mitchison (4).

A. Size Distribution of Ascospores

In a recent study, Yoo et al (14) found that the onset of ascosporogenesis in a homothallic haploid strain of *S. pombe* was asynchronous, even for spores within the same ascus. Spores reached different levels of "maturity" even though they presumably arose from the same zygote. In addition, spores were formed by having unequal quantities of cytoplasm enclosed by the enveloping "spore membranes." Thus spores of different sizes were seen within the same ascus. If conjugation itself is asynchronous, there would be little reason to expect that the ensuing germinative events would be synchronous, unless spores could be selected into narrow size or age classes. The biological significance of asynchrony of ascosporogenesis is unknown. Asynchrony may simply reflect the randomness usually associated with growing populations of cells, or it may be part of a mechanism designed to prevent self-mating between spores released from the same ascus. In any event, it became necessary to determine the extent of the heterogeneity in the H90 spores and to see how it would affect the timing of germination.

Newly formed spores were harvested from the agar surface as previously described. Aliquots were diluted with formalin saline (0.9% NaCl, 1.0% formalin) to a density of approximately 10^5 cells/ml to minimize coincident loss of electronic counting. The spores were dispersed by sonication (3,4) and counted with a Coulter counter (Model B, 70 μm aperture). Parallel counts were taken with a hemocytometer.

The counter was calibrated by measuring the size distribution of latex spheres of known diameters (7.6±2.3 μm, Dow Chemical Corp., Diagnostic Products, Indianapolis, Ind.) and spores that were isolated into distinct size classes by zonal density centrifugation (10). The diameters of the spores were measured from enlarged photographs taken under dark-field phase microscopy (4,13). (The Coulter counter had a resolving power of from 1-2.5 μm^3/threshold unit, depending on the settings used.)

The size distribution of a mixture of spores and cells is shown in Fig. 1. The size distribution of spores found in the presence of 2-deoxyglucose is also shown. (This part of the experiment will be discussed later.) The median volume of the control cells is 37 μm^3. It is clear that the size distribution is bimodal. The distribution to the left is presumably generated by the spores. The range of volumes is from about 20 to 90 m^3. This is in good agreement with the range of spore sizes determined photographically (17.8-93.9 μm^3, 10). The smaller distribution to the right is probably generated by the residual vegetative cells that failed to conjugate and sporulate. As shown by Mitchison (15), vegetative cells of *S. pombe* had an average volume of 140±22 μm^3 at the time of cell plate formation. Considering that this value was derived from cells in exponential growth and under different culture conditions than used in this study, the agreement with our determination is good. It falls well within the volume range of ca. 100-220 μm^3 shown in Figure 1. Measurement of the relative proportions of these two distributions by planimetry showed that the smaller distribution is 14.7% of the total distribution. This is the same as the proportion of residual vegetative cells (see Table 1).

The lack of overlap between the size distributions of the two populations suggested that they could be easily separated by density gradient centrifugation. Indeed, zonal density centrifugation in the A-XII rotor system

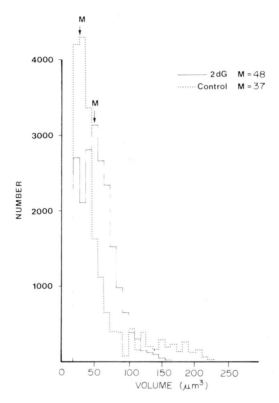

Fig. 1. Size distribution of *S. pombe* ascospores formed in the presence and absence of 2-deoxyglucose. Sizing was done with a Coulter counter as described in the text. The median volumes are indicated by the arrows.

(16, 17) was subsequently employed to separate the spores into distinct size classes (10). As mentioned before, we used an alternate approach to eliminate the vegetative cells. We added 2-deoxyglucose to malt extract agar to induce lysis of the residual vegatative cells.

This effect is illustrated by Fig. 1 and Table 1. *S. pombe* cells taken from liquid cultures in late log or early stationary phase of growth were plated onto malt extract broth agar (slants containing varying concentrations of 2-deoxyglucose added just before autoclaving). After 7-15 days, the cells were harvested and analyzed microscopically. They were also sized with the Coulter counter as previously described. It is clear that 2-DG had at least two concentration dependent effects: (a) At concentrations between 20 and 50 μg/ml, it reduced the percentage of residual vegetative cells to one-fifth of that of the control cells (Table 1). Fig. 1 shows this effect for cells exposed to 40 μg/ml. In terms of size distribution, there were very few cells with volume

170

TABLE 1

Effect of 2-Deoxyglucose on Sporogenesis of S. *pombe*

Concentration (μg/ml)	Spores	Asci (Percent)	Vegetative Cells
0	78.8	1.3	19.9
20	92.2	0.6	7.2
40	95.9	0.8	3.3
50	94.3	-	5.7
77	68.0	16.0	16.0
80	20.8	7.1	72.1
154	16.0	25.0	59.0

greater than 100 μm^3. A secondary effect was also evident, the median volume increased to 48 μm^3. The spores reared in the presence of 2-deoxyglucose are larger than control spores, but as will be discussed later, will germinate later. (b) At concentrations above 50 μg/ml, 2-deoxyglucose interfered with the process of ascosporogenesis. The spore yield dropped considerably, the proportion of asci, many of which contained poorly matured spores, rose and the number of residual vegetative cells increased sharply. They were enlarged, misshapen, and many of them had multiple cell plates. In addition, they were easily lysed during the washing procedure in harvesting. Because of this, in all subsequent work we used concentrations of 2-DG of 40 μg/ml.

B. Kinetics of Germination

The results described above raised the question as to the effect of 2-deoxyglucose on the kinetics of germination. An experiment was performed in which S. *pombe* cells were allowed to conjugate and sporulate in the presence of 2-deoxyglucose (40 μg/ml). The spores were harvested and analyzed by time-lapse photomicrography as previously described. Care was taken, at harvest time, to assure that the spores were freed of residual medium. They were resuspended in EMM 2 and allowed to germinate on small agar pads (2% agar in EMM 2) using the coverslip preparation (13). They thus germinated in the *absence* of 2-deoxyglucose. Table 2 summarizes the data from this experiment. It shows the relationship between the initial volume (calculated from diameters measured on enlarged photographs) and time at which the onset of outgrowth and cell plate formation occurred. Outgrowth is defined as the time when spores become markedly elliptical ("pear shaped"). Photographs were taken at 30 minute intervals.

Although the number of spores analyzed in this experiment was small (2-DG, n=39; control, n=37), the data show that larger spores germinated earlier. This is in agreement with a prallel study in which a larger sample was analyzed (10). 2-DG introduced approximately a two hour delay in the onset of outgrowth (i.e., for spores of the same size). The time interval between the

171

TABLE 2

Kinetics of Germination of *S. pombe* spores as a Function of Volume
and Sporogenesis in the Presence of 2-Deoxyglucose (2DG)

Sporulation Conditions	Volume (μm^3)	Outgrowth (hrs $\pm$ SD)	Division (hrs $\pm$ SD)
	40	6.8 + 0.8	11.3 + 1.0
CONTROL	63	6.4 + 0.6	10.4 + 0.6
	94	5.2 + 0.6	9.3 + 0.9
	40	8.2 + 0.5	12.7 + 0.2
2-DG (40 µg/ml)	50	7.6 + 0.9	11.9 + 1.5
	63	7.0 + 1.0	11.3 + 1.0

onset of outgrowth and the first cell division (i.e., cell plate formation) remained constant in both situations. It is about 4.5 hours long.

The kinetics of germination were further analyzed on a larger sample of ascospores separated into a narrower size class by centrifugation through a 15-40% linear sucrose density gradient in an A-XII zonal rotor as described elsewhere (16,17). The spores were harvested as described above and allowed to germinate on EMM 2 agar pads. (Fig. 2) The time course of germination is expressed as the germination index (No outgrown/total number x 100). It was determined on at least 150 spores per point. The mean and standard deviation of the volume of the 2-DG spores was 33.0 ± 0.02 μm^3 and that of the control was 31.5 ± 0.02 μm^3. It is clear that the control spores germinated much more rapidly than the 2-DG reared ones, even though the latter began to outgrow earlier. The time taken for the germination index to rise from 10 to 90% was 5 hours for the 2-DG spores and 3 hours for the control ones. Although the onset of the first vegetative division was not measured, it presumably would follow similar kinetics as shown in Table 2.

Notwithstanding the fact that control spores germinate more "synchronously," they were not as "fragile" as the 2-DG spores. For example, preliminary experiments showed that in order to achieve a breakage in excess of 75%, the control spores had to be subjected to repeated passes through an Eaton Press (which is a simplified version of the French Press). On the other hand, over 90% of the 2-DG spores were broken within 3 minutes with a Vibromixer (Shandon Scientific, London). This instrument vibrates a fluted stainless steel probe dipping into a slurry consisting of 5 ml containing 9.6×10^8 cells and 4 gms of glass beads (0.2 mm in diameter). This provides a rapid and convenient method of breaking up spores and was the method used in all subsequent experiments for the extraction of the RNA. All the experiments to be described below were performed on spores which were formed in the presence of 40 µg/ml 2-DG for this reason.

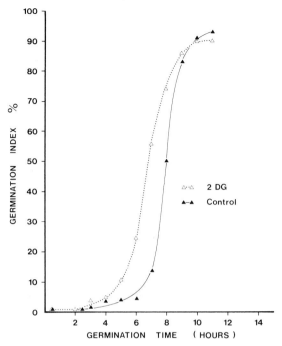

Fig. 2. Kinetics of germination of *S. pombe* ascospores that had sporulated in the presence and absence of 2-deoxyglucose. Spores were separated by zonal density centrifugation. Germination was at 32⁰ on EMM 2 agar pads as described in the text.

C. Early Macromolecular Events During Germination

1. Protein and RNA Synthesis. Synthesis of RNA and protein, measured by the incorporation of labeled precursors into TCA-precipitable material, began within the first hour of germination (Fig. 3). In this experiment, spores were resuspended in EMM 2 to which the appropriate precursors were added. The kinetics of germination of these spores are shown in the insert. It is clear that even though that outgrowth did not begin until the 3rd hour and a 50% index was not reached until the 8th hour, incorporation of both labels began within the first 30 minutes of germination. In a parallel study, we found that the relative *rate* of uptake of ^{3}H-adenine increased at a linear rate with doubling in rate by the 5th hour of germination.

In order to determine which species of RNA were being synthesized during this interval, we isolated the RNA of spores that had been exposed to a 20-minute pulse of ^{3}H-adenine at 30, 180 and 300 minutes after being resuspended in EMM 2. The pelleted spores were then broken by the

173

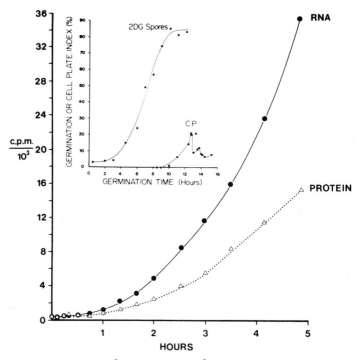

Fig. 3. Incorporation of ³H-leucine and ³H-adenine into trichloroacetic acid precipitable material of *S. pombe* spores continuously exposed to the label in EMM 2. The kinetics of germination are shown in the insert. *Insert.* Kinetics of germination of spores formed in the presence of 2-deoxyglucose and separated by centrifugation through a sucrose density gradient. The germination and cell plate indices (ordinate) were determined photographically as described in the text.

Vibromixer and nucleic acids were extracted by a detergents-phenol procedure (21). The DNA was not removed. The RNA was fractionated on 3.0% polyacrylamide gels (22) for 3 hours at 8 V/cm; 5 mA/gel. The gels were scanned for ultraviolet absorption at 265 nm with a Joyce-Loebl Gel Scanner. The gels were frozen in solid CO^2 and sliced transversely at 0.8 mm intervals with a Mickle Gel Slicer, incubated with NH_4OH, mixed with scintillation fluid and counted on a Packard spectrometer (18). Fig. 4 shows the results of these analyses. The specific radioactivity is expressed as counts per minute/unit of ribosomal RNA. The quantity of rRNA on the gel was calculated from the peak areas of 18 and 26S rRNAs on the ultraviolet absorbance scan (23). Note the change in the scale in each panel.

Thirty minutes after resuspension in EMM 2, the spores had begun to synthesize RNA of all sizes. The 26 S and 18 S rRNA species had specific

174

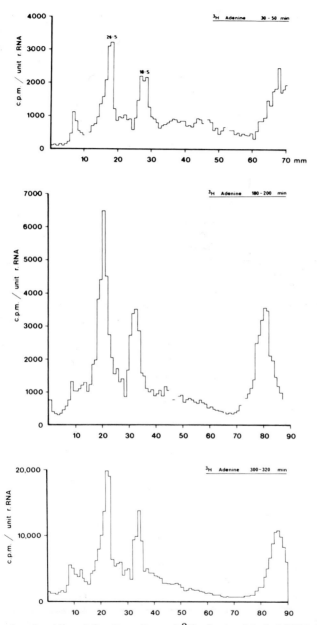

Fig. 4. Acrylamide gel fractionation of ³H-adenine labeled RNA isolated from *S. pombe* spores. Spores were resuspended at t=0 in EMM 2 and exposed to 100 μCi, ³H-adenine and 1μ cold adenine/ml for 20 minutes. Top panel t=30-50 min; middle panel, t=180-200 min; bottom panel, t=300-320 min. The ordinate gives the specific activity as counts per minute/unit ribosomal RNA.

activities in excess of 3×10^3 and 2×10^3 cpm/unit rRNA, respectively. At 180 minutes, the specific activities had doubled and by 300 minutes of germination they had essentially tripled. Note that the spores were synthesizing proportionately equal molar quantities of these two species of RNA, suggesting that the synthesis is coordinated and that complete ribosomes were being made. In addition, polydisperse and high molecular weight RNAs were being actively synthesized.

2. *Messenger RNA Synthesis.* The labeling of polydisperse RNA suggested that messenger RNAs were being synthesized soon after the onset of germination. It has been reported that in budding yeast, some messenger RNAs contain a polyadenylic acid [poly(A)] sequence (24) and we have found that up to half of the polydisperse RNA of vegetative cells of *S. pombe* contains a poly(A) sequence (Fraser and Creanor, unpublished). We, therefore, looked for poly(A) sequences in the RNA of germinating spores. We used the technique developed by Fraser and Loening (18) to assay for the presence of messenger RNAs which contain poly(A) sequences. This technique involves hybridization of ^{3}H-polyuridylic acid [poly(U)] to the poly(A) sequences. The hybrids may then be fractionated, and freed from excess poly(U) by sedimentation on sucrose gradients, or the excess poly(U) may be removed by RNAs, the hybrid co-precipitated with carrier RNA and counted on glass fiber filters. The method avoids the need to label mRNA directly, and can be used to assay the total poly(A)-containing messenger RNA content of the cell, including pre-existing, stable messengers.

Approximately 2×10^8 spores, sporulated in the presence of 2-DG, were allowed to germinate in 15 ml of EMM 2 at 32°. At various times, 5 ml of spores were collected and the RNA was extracted as described by Fraser *et al* (21). A portion of the RNA was fractionated through polyacrylamide gels to quantitate the amount of RNA present. The remainder of the RNA was mixed with an excess of tritium labeled polyuridylic acid (0.01 µg/ l, 0.2 µCi/µg, Miles Laboratories, Inc.) dissolved in 0.5% sodium dodecylsulfate, 150 mM NaCl; 1 mM EDTA; 50 mM Tris-HCl, pH 7.6. The SDS was removed by precipitation with 50µl M KCl (18). The supernate was digested with 10µg RNase/ml for 10 minutes at room temperature. Exactly 100µg of carrier RNA were then added and the mixture was treated with 3 volumes of 95% ethanol at -20° for 2 hours to precipitate the hybrids. They were collected on 20 mm glass filters (Whatman, GFA) and washed 3 times with 95% ethanol. The filters were dried and placed in scintillation fluid and counted with a Packard Tri-Carb scintillation spectrometer (18).

Figure 5 shows the results of this experiment. The ordinate gives the relative quantities of polyadenylic acid sequences per unit ribosomal RNA in each sample. It is clear that there was a three-fold increase in the relative amount of adenylate-rich RNA species within 30 minutes of germination. It is also apparent that spores that had not yet germinated had measurable quantities of these RNAs (t=0). When RNA was fractionated on a sucrose gradient, and each fraction was challenged with ^{3}H-poly (U) and treated with RNase as above, the

176

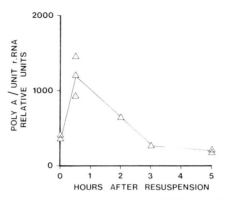

Fig. 5. Changes in the relative concentration of RNA containing polyadenylic residues as a function of germination. Spores were resuspended in EMM 2 at t=0, the RNA was extracted and hybridized with ^{3}H-labeled polyuridylic acid. See text for details.

sizes of those RNAs containing poly(A) sequences was shown by the distribution of radioactivity on the gradient. We found that poly(A) was broadly distributed from about 5 S to 30 S. This means that poly(U) hybridization was not measuring free poly(A), which is much smaller (24), but probably poly(A) sequences in messengers RNAs. In particular, poly(A) present at zero time was not a pool of free poly(A), but was in high molecular weight RNAs. Since the quantities of the mRNAs were determined in terms of the rRNA present, and since the latter was being synthesized rapidly at this time, there was a drop in the ratio of mRNA to rRNA between 30 minutes and 3 hours of germination. We interpret the results to show that the rate of rRNA synthesis was high at the start of germination. The rate of rRNA accumulation then increased faster than the rate of increase of mRNA content. Thus one of the first events in the germination process is the active synthesis of mRNA. (Please note that the method of poly(U) doesn't itself measure mRNA synthesis only cell content of mRNA. So the decline in the poly(U)/rNA data does not necessarily mean that the rate of mRNA synthesis is declining, only that the rate of rRNA accumulation is probably increasing).

3. DNA Synthesis. As shown by Bostock (3), the quantity of DNA in the spores of *S. pombe* is at the haploid level (ca 0.0146 picograms per spore). The time course of DNA synthesis during germination is not known. We tried to measure the bulk quantity of DNA but were unable to obtain a consistent value, largely because of the technical difficulties in the quantitative breakage of the spores. We measured, instead, the kinetics of incorporation of ^{32}P into the DNA at various times during germination.

A one milliliter aliquot (containing 2×10^8 spores) of the same 2-DG spores used above were resuspended in 20 ml of EMM 2. At different times, 2 ml were removed and pulsed with 100 µCi of ^{32}PO$_4$ (Radiochemical Centre,

177

Amersham, U.K., specific activity 1-9 mCi/ml) for 30 minutes. The tubes were rapidly chilled, centrifuged and the supernatant discarded. The spores were then broken with the Vibromixer and the nucleic acids were extracted as described above. The samples were treated with RNase (40 g/ml, 20 minutes at room temperature) and fractionated through 3% acrylamide gels. Slices of gels containing ^{32}P were dried and counted in a Geiger tube as previously described (18). Total ^{32}P incorporation into DNA was determined by addition of counts in slices making up the DNA radioactivity peak. The results are shown in Fig. 6.

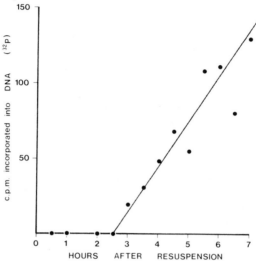

Fig. 6. Incorporation of ^{32}P-orthophosphate into the DNA of *S. pombe* spores following resuspension in EMM 2. See text for details.

There was no incorporation of ^{32}P into the DNA in the first 2.5 hours of germination. From that time the incorporation was linear, although the scatter in the points makes it impossible to state this with certainty. In any event, it is clear that at the outset, the spores were not engaged in any measurable synthesis of DNA. This would suggest that the onset of DNA synthesis must await the synthesis of a considerable quantity of RNA, and possibly also enzymes required for DNA synthesis. This is at best only a guess at this time.

III. Summary and Concluding Remarks

Germination of *S. pombe* spores can be separated into distinct macromolecular and cytological phases of activity. Upon resuspension in a nutrient medium, protein and RNA synthesis are initiated almost immediately and continue for several hours with a constantly increasing rate. Although we cannot detail the diversity of the proteins being synthesized, we do know a little more about the types of RNAs that are made. It seems that ribosomal RNA synthesis is coordinated so that both species of rRNA (26 S and 18 S) are made concurrently. Secondly, although the spores have measurable quantities of putative messenger RNAs (i.e., RNA species rich in adenylic acid residues), there is a rapid increase in synthesis of this kind of RNA within the first 30 minutes of germination. The biological implications are obvious, the spore has a store of information sufficient for a "start in life."

As for the DNA, it would appear that there is a presynthetic period of approximately 2.5 hours duration, as indicated by the lack of incorporation of ^{32}P into the DNA. DNA synthesis then continues in a linear fashion up to the time of the first vegetative division. Thus, there is probably no G_2, spores are "G_1 types."

Cytologically the onset of outgrowth seems to be a function of initial spore size, while the interval between outgrowth and the first vegetative division is constant. This would imply that spores must achieve a minimal level of cellular growth before they can proceed towards cell division. What this implies in terms of subcellular replication is unknown, since we do not know exactly to what extent organelles such as mitochondria, Golgi, etc. were disassembled within the spore.

We have also shown that 2-deoxyglucose can be used to achieve the separation of spores from the residual vegetative cells. A dividend is the formation of spores that are much more easily ruptured. There is, however, an indication that a delay is introduced into the timing of the onset of outgrowth by 2-DG and possibly also a loss of synchronization. This secondary effect, may however, be artifactual in that the separation of spores on the basis of size alone may not be sufficiently precise to yield a high degree of synchrony of germination.

The conjugation-sporulation-germination-division sequence is of biological significance in itself. It offers us an opportunity not merely to follow complex changes in state, but to manipulate the reproductive events and thus alter the subsequent expression of the genetic complement. With the advent of specific mutants in this strain of yeast, it will be possible to use crossbreeding techniques in conjunction with macromolecular methods to pin point the controlling mechanisms in the expression of a spore's genetic complement.

Acknowledgements

This work was supported by the Medical Research Council of the U.K. and the Food and Drug Administration of the U.S.A. (Grant no. 5 RO1 FD-00120 to GMP). The senior author was a recipient of a Postdoctoral Research Fellowship from the Medical Research Council while on a sabbatical leave of absence in Professor Mitchison's laboratory. We also wish to thank Dr. B. L. A. Carter for his assitance in the zonal centrifugation and Miss Anne Baker for excellent technical assistance.

References

1. A. Keynan. The transformation of bacterial endospores into vegetative

cells. *In:* Microbial Differentiation (Ed. J.M. Ashworth and J. E. Smith) Cambridge Univ. Press, Cambridge, pp. 85-123 (1973).

2. A. S. Sussman and H. O. Halvorson. Spores, Their Dormancy and Germination. Harper and Row, New York 354 pp (1966).
3. C. J. Bostock. *Exptl. Cell Res. 60:*16 (1970).
4. J. M. Mitchison. *Meth. Cell Physiol. 4*:131 (1970).
5. J. M. Mitchison and J. Creanor. *J. Cell Sci. 5:*373 (1969).
6. R. R. Fowell. Sporulation and Hybridization of Yeasts. *In:* The Yeasts Vol. 1 (Ed. A. H. Rose and J. S. Harrison) Academic Press, New York, pp. 303-383 (1969).
7. H. Gutz. *Science 158:*796 (1967).
8. J. M. Mitchison and W. S. Vincent. *Nature 205:*987 (1965).
9. J. M. Mitchison. The Biology of the Cell Cycle. Cambridge Univ. Press, Cambridge, 313 pp (1971).
10. G. M. Padilla, J. M. Mitchison and B. L. A. Carter. In preparation.
11. B. F. Johnson and C. M. Rupert. *Exptl. Cell Res. 48:*618 (1967).
12. H. C. Birboim. *J. Bact. 107:*659 (1971).
13. M. M. Swann. *Nature 193:*1222 (1962).
14. B. Y. Yoo, G. B. Calleja and B. F. Johnson. *Arch. Mikrobiol. 91:*1 (1973).
15. J. M. Mitchison. *Exptl. Cell Res. 30:*521 (1963).
16. N. G. Anderson.*Nat. Cancer Inst. Mon. 21:*9 (1966).
17. H. O. Halvorson, B. L. A. Carter and P. Tauro, *Methods Enzymol. 21,:* 462 (1971).
18. R. S. S. Fraser and U. E. Loening. *Eur. J. Biochem. 34:*153 (1973).
19. J. H. Parish and K. S. Kirby. *Biochim. Biophys. Acta 129:*554 (1966).
20. U. E. Loening, K. Jones and M. L. Birnstiel. *J. Mol. Biol. 45:*353 (1969).
21. R. S. S. Fraser, J. Creanor and J. M. Mitchison, *Nature 244,* 222 (1973).
22. U. E. Loening, *Biochem. J. 102,* 251 (1967).
23. R. S. S. Fraser, *Virology 45,* 804 (1971).
24. C. S. McLaughlin, J. R. Warner, M. Edmonds, H. Nakazato and H. Vaughn. *J. Biol. Chem. 248,* 1466 (1973).

THE CELL CYCLE REGULATION OF SEXUAL MORPHOGENESIS IN A BASIDIOMYCETE, *USTILAGO VIOLACEA*

Joseph E. Cummins and Alan W. Day

Department of Plant Sciences, University of Western Ontario
London, Ontario, Canada

I. Introduction

Sexual morphogenesis in the anther smut fungus, *Ustilago violacea*, consists of the cooperative construction of a cylindrical fusion tube between a pair of uninucleate, yeast-like cells of opposite mating type. Following copulation the haploid nuclei remain separate and a dicaryotic mycelium is established. This dicaryon is an obligate parasite of a host plant (family Caryophyllaceae). Nuclear fusion (karyogamy) is normally restricted to the hyphae inhabiting the anthers of a host plant and nuclear fusion leads to the formation of a diploid brandspore. When the brandspore germinates its nucleus divides meiotically to re-establish the haploid sporidial phase. In this basidiomycete copulation and nuclear fusion are temporally separated and the mechanisms governing each can be analysed separately (1).

This chapter is primarily concerned with the genetic and molecular aspects of the synthesis and assembly of the copulatory organelle. The morphogenetic induction of this organelle is providing revealing insights into the regulation of gene action during the cell cycle and the mechanism restricting cellular morphogenesis to a particular phase of the cell cycle. Furthermore, the relatively small genome size of this smut (2) and the ease with which genetic analyses can be made (1,3) provide a powerful genetic tool for the study of cellular morphogenesis and its regulation.

II. The Morphology and Kinetics of Mating

A. Developmental Stages and Kinetics

The development of the copulatory organelle between compatible cells is a relatively complex morphogenetic event involving several stages (4). Approximately 1-1.5 hours after wild type cells are mixed and plated on mating medium (2% water agar) they form loosely bound couplets, in which the two cells are connected and move together even though they may be separated by as much as the width of a cell (about 3 μm). No strands connecting them have been seen with the light microscope, but electron microscopic preparations of thin-sectioned and freeze-etched cells show that the glycocalyx is 'hairy' (4 and unpublished data) (Fig. 1a). At about 2

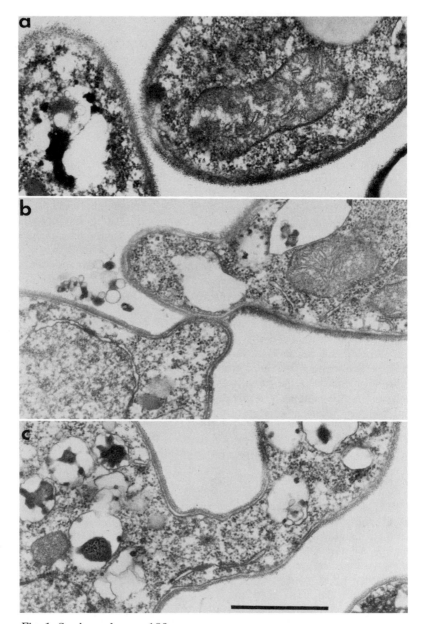

Fig. 1. See legend on p. 183.

hours after the initiation of mating, the cells appear to be aggregated into clumps, which later break up into pairs of closely attached cells. Electron micrographs of this stage show that the cell walls have fused but that the plasma membranes are still intact (Fig. 1a). The copulatory organelle is assembled within the next 45-60 minutes. First small bumps appear on each cell at the site of contact, then each bump elongates into a peg, (Figure 1b) and finally the fused cell walls and plasma membranes at the point of contact of each peg break down to form a connecting tube between the cells (Figure 1c). Further elongation of this tube takes place over the next 4-8 hours until it is approximately 5 µm long, although sometimes it may reach a length of 15-20 µm.

Auxotrophic strains designated 1.C2 (histidine requiring, his_1 and a_1 mating type) and 2.716 (lysine requiring, lys_1 and a_2 mating-type) develop conjugation tubes at a similar rate to wild type cells but take about 1.5-2 hours longer to initiate mating. Thus intimately paired couplets are not seen until at least 3.5 hours after mating is initiated between these auxotrophic strains. The stages in copulation together with their approximate timings are shown in Fig. 6.

In about 2-5% of copulations involving haploid cells, several cells are involved and a *multiple fusion* results. Such multiple fusions become much commoner in crosses involving a diploid, triploid or tetraploid strain with a haploid partner (9). Thus multiple fusions in an $a_1a_2a_2$ x a_1 mixture account for over 50% of the conjugations. In some cases a single triploid cell may be involved in conjugation with as many as 6 haploid partners.

B. Budding and Conjugation

In view of experiments reported later in this chapter on the effect of the cell cycle on conjugation, we have investigated the correlation between budding and conjugation in *U. violacea*. Conjugations of a_1 and a_2 cells are initiated either between two unbudded cells or between an unbudded and a budded cell, but not between two budded cells. A previous treatment with a high level of

Fig. 1. Thin-section electron micrographs of conjugation in *Ustilago violacea*.
a. Cells at an early stage in mating showing intimate contact and the beginnings of fusion between the cells walls.
b. An intermediate stage in the development of the conjugation tube, showing the production of 'pegs' from each partner. The cell walls of the pegs are completely fused at the point of contact, but the plasma membranes are still intact.
c. A pair of cells after assembly of the copulatory organelle (tube) but before elongation. The tube is completed following breakdown of the plasma membranes at the point of contact of opposing pegs (as in 1b).
The bar represents 1 µm. See Poon *et al.*, (4) for further details, techniques etc.

dextrose can be used to 'mark' the cells of one mating type as such cells accumulate large numbers of lipid bodies. Use of this technique shows that a_1 cells mate only before budding, while a_2 cells initiate mating in either an unbudded or a budded condition (4). As unbudded cells have been demonstrated to be in G_1 and to enter S phase when they initiate a bud (unpublished data), these results are in complete agreement with our more precise experiments on the mating ability of the two alleles at different stages of the cell cycle, reported in the next section.

Further observations show that a_1 cells do not initiate a bud later in mating during assembly or elongation of the copulatory organelle. On the other hand a_2 cells may complete bud formation during mating or may initiate a bud within the first 1-2 hours after mating. New buds are not initiated after this time, however, although control cultures of either mating type continue to bud slowly for up to 10 hours on the non-nutritive mating-medium.

C. The Fate of Conjugated Cells

We have described the fate of conjugated cells in natural conditions as well as on artificial media in a previous paper (1). In natural conditions the nuclei associate but do not fuse, and dicaryotic hyphae are produced which are obligately parasitic and grow only in the cells of the host plant (a member of the *Caryophyllaceae*). Such dicaryotic hyphae are systemic and constitute the longest phase of the life cycle. Karyogamy occurs many cell generations later in the anthers of the host plant, and the thick walled diploid brandspores replace the pollen. The haploid sporidial phase is re-established following a meiotic division in the short promycelium produced from the brandspore.

No medium has been found which will support growth of the dicaryon. When conjugated cells are removed from the non-nutritive mating medium to a nutritive medium, the majority of them revert to haploid budding. The cell cycle which is arrested in G_1 on the non-nutritive mating medium is activated again and each mating type begins to bud independently of the other. In a very few pairs (about 3 in 10^4) the nuclei fuse and a diploid budding phase is established. Such diploids can be selected for by the use of complementary, compatible auxotrophs (1). These diploids form stable cultures but can be induced to undergo mitotic crossing over or haploidization (3). Homozygous a_1a_1 or a_2a_2 diploids formed after UV induced mitotic crossing-over in heterozygous diploids, have been selected and used to form triploid and tetraploid strains (2,5). These stocks make possible the examination of dosage effects of alleles at the mating type locus. The frequency of karyogamy can be increased up to 10,000 fold by treatment with ultra-violet light (254 nm) (6; Day and Day, in preparation) or 100 fold by using stationary phase cells rather than log phase cells before conjugation (unpublished data). In view of this latter effect we are analysing the effect of the cell cycle on the frequency of karyogamy in conjugated cells.

III. The Cell Cycle Control of Mating in Haploid Cells

In this section we discuss the relationship between the cell cycle and formation of the copulatory organelle in haploid strains. We show that mating is dependent upon the stage in the cell cycle of conjugating sporidia. In *U. violacea* conjugation consists of a synchronous series of biochemical events that is initiated during courtship and terminates with the assembly and elongation of the fusion tube. As indicated earlier distinct morphological changes are not observed during courtship but it is during this period that the genes governing mating are transcribed and translated.

An analysis of the influence of the cell cycle on the mating activity of a mixed culture of the two mating types, using sucrose gradients to synchronize the sporidia, showed that conjugation occurs primarily in the G_1 phase of a synchronous mixed population and that it is inhibited during the S phase. The observation that conjugation is inhibited during S phase was substantiated by using hydroxyurea to block DNA synthesis in synchronous cultures (7). A population in which DNA synthesis was prevented maintained a high level of conjugation (70%) even though budding continued. As the inhibited cells passed through the block in DNA replication the conjugation level dropped to less than 10%. Thus a blockage in DNA replication has little influence on budding yet maintains the mixed culture in a state of readiness for mating. When the cells escape the block and enter the S phase they lose their ability to mate (7).

To test whether or not the two mating types were under similar control during the cell cycle we first devised an "out of phase" experiment. In this experiment sporidia of the two mating types were separately synchronized using sucrose gradient centrifugation. Small unbudded (G_1) cells were recovered from the sucrose gradients and these cells were either allowed to progress around a cell cycle or they were maintained as "testers" blocked in the G_1 phase by holding them in water at a reduced temperature. Cells from cultures passing through a cell cycle were then mated at intervals with cells of the opposite mating type blocked in the G_1 phase. The results of an out of phase experiment are shown in Figure 2. The experiment indicates that the cell cycle effect observed in synchronous mixed cultures (in phase cultures i.e. Xa_1 x Xa_2) is due solely to the refractoriness of mating type a_1 during phases of the cell cycle beyond G_1 as shown by the curve Xa_1 and G_1a_2. Mating type a_2 remains sexually active throughout the cell cycle (curve G_1a_1 x Xa_2). The curve G_1a_1 x G_1a_2 is a control showing that low temperature starvation does not adversely influence the mating ability of cells bearing either allele.

The results of the out of phase experiment were confirmed and extended by using a re-orienting zonal rotor to synchronize cells of one mating type. Fractions containing highly synchronous cells in different phases of the cell cycle were mated with asynchronous cells of the opposite mating type. The results of a typical experiment are shown in Fig. 3. The results of the experiment confirm the fact that cells bearing mating-type a_1 are restricted in

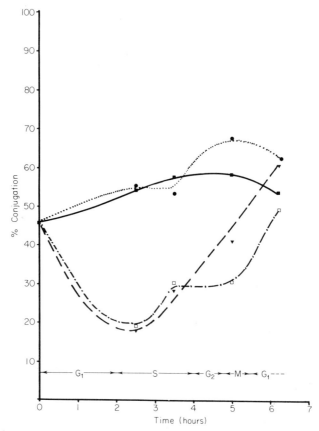

Fig. 2. *"Out of phase" experiment.* Top fractions (G_1 cells) of a_1 and a_2 cells in sucrose gradients were collected separately and each was divided into two samples. One sample was washed and stored in a water suspension at 4°C for the duration of the experiment (7 hrs) during which time the cells remain in an unbudded G_1 condition. The other sample was washed and incubated in complete medium at 22°C for the 7 hours. Such cells passed through a complete cell cycle in about 5½ hours as indicated on the figure. At intervals samples were removed from each of the four cultures, washed, mixed in various combinations and allowed to mate on water agar. The combinations used were i) $G_1.a_1$ x $G_1.a_2$ (■) a control showing that mating remains high even in these cold, starved cultures; ii) $X.a_1$ x $X.a_2$ (□) where X is the population of cells passing through the cell cycle in complete medium (the exact phase in the cell cycle at any sampling time is indicated on the figure, and was determined by correlation of the proportion and size of buds with a determination of the state of DNA replication, Cummins & Day, in preparation); iii) $X.a_1$ x $G_1.a_2$ (▼); iv) $G_1.a_1$ x $X.a_2$ (●).

186

their sexual activity to the G_1 phase of the cycle while cells bearing mating type a_2 are active throughout the cell cycle. We have introduced the terms *stringent* and *relaxed* to describe the behaviour of these mating type alleles during the cell cycle. The regulation of mating type a_1 is termed a *stringent* cell cycle control because the activity of the allele a_1 is restricted to the G_1 phase while the regulation of allele a_2 is termed a *relaxed* control because the allele functions at any stage of the cell cycle.

The experiments that we have described and the experiments described in a following section of transcription and translation of the genetic information for morphogenesis are consistent with an hypothesis that there is a sex message which is a messenger RNA molecule or molecules that is transcribed from the gene or genes governing sexual morphogenesis. The sex message is translated into proteins that organize and assemble the organelle for copulation (the conjugation tube).

We are attempting to establish whether or not the stringent or relaxed behaviour of the mating type alleles is intrinsic to these alleles or whether it is governed by a separate locus. The results presently available are not extensive enough to provide an unequivocal answer but the failure to observe a mixture of stringent a_2 and relaxed a_1 cells in single cell isolates of the meiotic products of numerous crosses argues that a locus for cell cycle stringency would have to be tightly linked to the mating type locus.

Furthermore, the stringent restriction of mating activity in allele a_1 is comparable to periodic enzyme systhesis observed during the cell cycle of many different yeast strains (8). Recently, however, Carter and Halvorson (9) have found that the cell cycle periodicity of enzyme formation of at least two enzymes was not greatly influenced by conditions of induction, repression or even by the constitutiveness of the operon governing enzyme formation. Carter and Halvorson (9) suggest that the temporal regulation of enzyme synthesis is based on sequential transcription of the genome. Our results on the cell cycle regulation of morphogenesis suggest that either sequential transcription does not govern the morphogenetic activity of allele a_2 (the relaxed form) or there is a positive control that acts to maintain the a_2 locus in an induceable configuration throughout the cell cycle.

The requirement that mating type a_1 undergo sexual morphogenesis during the G_1 phase of the cell cycle is reminiscent of the morphogenetic behaviour of most higher eukaryotes whose cells differentiate almost exclusively during the G_1 phase of the cell cycle. The mating allele a_1 has, like the chromosomes of higher organisms, a stringent cell cycle regulation over the morphogenetic activity that it affects. We believe these observations of stringent and relaxed control of natural alleles of a locus that regulates a major morphological event to be unparalleled. A fuller elucidation of the molecular basis of the control difference between alleles a_1 and a_2 could enhance an understanding of the mechanisms that limit the differentiation of higher cells to the G_1 phase.

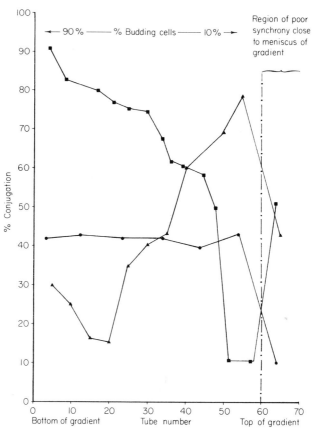

Fig. 3. *The mating ability of haploid and diploid cells during the cell cycle.* In separate experiments cells of a_1 (▲), a_2 (●) and a_1a_2 (■) strains were synchronized in a 15-30% sucrose gradient in a Sorvall re-orienting zonal rotor. Cells were collected in fractions starting at the bottom of the gradient (tube 1-90% large, budded G_2 and M cells, tubes 55-60 - 10% small budded cells, 90% unbudded G_1 cells). The cells were washed and mixed with an equal quantity of the appropriate mating type (a_1 tested with a_2 cells, a_2 and a_1a_2 tested with a_1 cells). The difference in % conjugation reached in the synchronized a_2 haploid (about 40%) and G_2 and S populations of the a_1a_2 diploid (about 80%) is because log phase tester a_1 cells were used for the former and stationary phase a_1 cells for the latter. The synchronized a_1 cells were tested with log phase a_2 cells. The top of gradient (tubes 60+) consisted of a random population of cells, including dead ones associated with proximity to the meniscus. Hence the % conjugations obtained here are ignored in the interpretation of these results. (See also 7,23).

188

IV. The Cell Cycle and Dominance

Several intriguing questions are raised by the discovery that different cell cycle controls operate on each of the two mating type alleles in haploid strains. How do these controls interact in diploid and polyploid systems? Does each allele retain its own control system or is one control system dominant? Does the concept of specific cell cycle controls for different alleles affect our understanding of the underlying mechanisms of genetic dominance? To attempt to answer these questions we review here firstly observations of dominance at this locus in polyploid strains of *U. violacea*, and secondly analyses of the cell cycle control system operating in heterozygous diploid strains. Finally, we compare these results with those from other yeast like organisms, and discuss genetic dominance as it appears in the light of these observations.

A. Dominance at the Mating-type Locus

Asynchronous log phase populations of the a_1a_2 diploid of *U. violacea* will conjugate normally with a_1 or a_1a_1 cells, but never with a_2 or a_2a_2 cells (5). Similarly, heterozygous triploids, ($a_1a_2a_2$ or $a_1a_1a_2$) and tetraploids ($a_1a_1a_2a_2$) also mate only with a_1 haploids or a_1a_1 diploids. In all these pairings involving log phase populations, at least 50% conjugation was observed, a comparable figure to that obtained for log phase haploid x haploid pairings. Thus allele a_2 appears to behave as a fully dominant allele in the classical sense and its dominance is complete even in the ($a_1a_1a_2$) triploid.

However, a closer examination of the behaviour of heterozygous cells reveals that dominance is modified by the age of the cells. While homozygous (a_1a_1 or a_2a_2) or hemizygous (a_1 or a_2) cells conjugate in either log phase or stationary phase, we found that heterozygous cells (a_1a_2, $a_1a_1a_2$, $a_1a_2a_2$, or $a_1a_1a_2a_2$) mate only in log phase, during which allele a_2 is dominant.

As soon as these heterozygous cells enter stationary phase they lose the ability to conjugate and become *neutral* (5) (Fig. 4a). Consequently we term the dominance of allele a_2 *age-related*. This age-related dominance of the a_2 allele suggests that dominance is indeed affected by the cell cycle. The next experiments provide clear evidence of this.

B. The Cell Cycle Control of Mating in a_1a_2 Diploids

As the mating ability of a_1a_2 cells is lost when the cells enter stationary phase, it was clearly important to determine at what phase in the cell cycle such cells were arrested. A chemical determination of the DNA content of haploid and diploid cells showed that the stationary phase value was lower than the value in log phase. It is clear that stationary phase cells are arrested in G_1 phase just as in most other species. In the results described in the previous section, we demonstrated that allele $\underline{a_1}$ is under stringent cell cycle control and is active only in G_1 phase, while allele a_2 has a relaxed cell cycle control and is active at all phases of the cell cycle. In considering the possible interaction of these alleles in the a_1a_2 diploid a plausible model to explain age-related dominance became apparent (Fig. 4b).

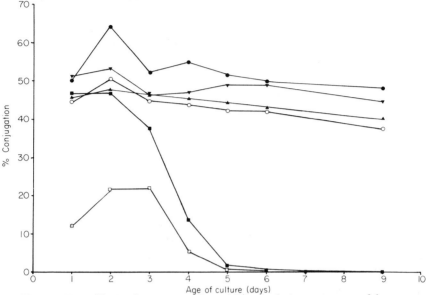

Fig. 4a. The effect of age on mating ability in heterozygous and homozygous strains. Strains were inoculated at low concentration to liquid complete medium and incubated at 22°C on a shaker for 9 days. Samples were removed every 24 hours mixed in various combinations and the % conjugation determined by plating on water agar (see Day & Cummins, 1974). The combinations used were i) a_1 x a_2 (●); ii) a_1a_1 x a_2 (▼); iii) a_1 x a_2a_2 (▲); iv) a_1a_1 x a_2a_2 (o); v) a_1 x a_1a_2 (■); vi) a_1a_1 x a_1a_2 (□). All cultures grew at approximately the same rate and entered stationary phase after 3.5-4.5 days.

We postulated that i) the cell cycle controls of each allele were retained in the diploid, and ii) that at any point in the cell cycle when both alleles were active, they would effectively 'cancel each other out' and produce a sexually neutral cell, (10). This model would predict therefore that during G_1 when both alleles are transcribing sex message, the a_1a_2 cell would be neutral, while during the rest of the cell cycle, allele a_2 would be the only allele transcribed and the cells would have a_2 mating ability.

The neutral behaviour of stationary phase (G_1) cells is in agreement with this model, and further evidence was obtained from an analysis of the different cell cycle stages in a log phase population of a_1a_2 cells. Fig. 3 shows clearly that the predictions of the model are upheld in that G_1 cells are neutral while other phase cells show a_2 mating activity (the converse of the behaviour of a_1 cells). Similar results have also been obtained from experiments in which we

190

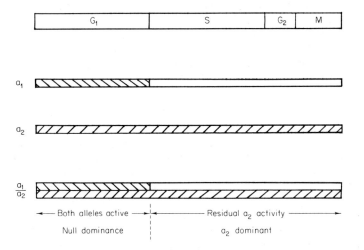

Fig. 4b. A model of dominance determined by temporal allelic interaction. Haploid a_1 cells mate only in G_1 (shaded area) while a_2 haploids mate at all phases of the cell cycle. The diploid shows a_2 mating activity in S and G_2/M phases. This behaviour of the diploid is explained by assuming i) the cell cycle controls observed in each haploid strain operate independently in the a_1a_2 diploid, ii) during G_1 both alleles are active and neutralize each other, so that the cell is neutral (null dominance). The dominance of allele a_2 during most of the cell cycle is thus attributed to its longer period of transcription than allele a_1 - i.e., a temporal cause rather than a functional cause.

have followed the mating ability of an a_1a_2 population as it escapes from stationary phase during growth on fresh medium (10). In contrast, homozygous diploids show the same cell cycle controls as the corresponding haploids.

The basic features of the model described above are shown in Fig. 4b. The dominance of allele a_2 observed at the phenotypic level is a function of its longer period of activity (probably transcription) than allele a_1. We have termed this new kind of dominance, deriving from cell cycle control systems, an example of *temporal allelic interaction* (10). In the particular genetic situation reported here, we have postulated that alleles a_1 and a_2 interact negatively in that they effectively neutralize each other in G_1 when they are both active in the same nucleus. It is perhaps most probable that interaction occurs after translation of the primary gene product, e.g., the two specific allele products may bind together to produce a dimer or polymer in which the active sites are unavailable to initiate the development of the copulatory organelle. Such dimer or polymer formation would not occur when the two

191

alleles were in different cells e.g., (a_1 x a_2), and full mating activity would be retained. However, it is also possible to envisage neutralization occuring during the regulation of transcription, or by an interaction of mRNA's, or by complex interactions in the metabolic pathways initiated during mating. To test these and other possibilities, we are comparing transcription and translation of sex message in a_1a_2 cells at various stages in the cell cycle. Another approach that we are using is to select mutants of either mating type with altered cell cycle controls and to determine their effects on dominance.

C. The Cell Cycle and Mating in Ustilago, Saccharomyces, and Hansenula

A considerable mass of data has been accumulated recently on mating in the yeasts *Saccharomyces* and *Hansenula*. Although the work on these species has so far not concentrated on the cell cycle aspects a review of the data accumulated indicates that cell cycle control systems do operate on mating ability in these yeasts, and that there are some interesting differences from the situation outlined above for the basidiomycete, *U. violacea*.

Several workers have shown that mating occurs only between unbudded a and α cells of *Saccharomyces cerevisiae* (11-13). These unbudded cells are in G_1 since just as in *Ustilago* the onset of budding coincides quite closely with the initiation of DNA synthesis (12). In contrast to *Ustilago*, however, each mating type of *Saccharomyces* appears to repress DNA synthesis in the opposite mating type (11). Thus in yeast the cell cycle is altered by mating activity and cells of opposite mating type are *actively synchronized* in the G_1 phase. In *Ustilago*, however, the cell cycle is unaffected at least initially by mating activity and the mating cells are not synchronized. It is probable that this difference is associated with the very different fate of the conjugated cells in each species. It may be essential for the immediate karyogamy that occurs in yeast that the two cell cycles be synchronized in the G_1 phase. In *Ustilago*,, where karyogamy is normally delayed for many cell generations the lack of synchrony may favour the establishment of the dicaryon. The rare somatic nuclear fusions that do occur following conjugation may then derive from infrequent conjugations in which the nuclear cycles were in synchrony and in the correct phase. It is clear that further comparative investigations of the cell cycle control of plasmogamy and karyogamy in organisms such as smuts and yeasts may contribute significantly to our knowledge of how organisms regulate their life cycles.

Heterozygous diploid, triploid and tetraploid yeast cells differ from those of *U. violacea*, in that they are always neutral (14, 15). Duntze *et al.* (16) have shown that heterozygous diploid or triploid cells of *Saccharomyces* resemble a or aa forms in that they do not produce an oligopeptide mating stimulant characteristic of α or ααforms. Thus allele a may appear to be dominant in heterozygous cells, but as these cells are neutral in mating ability it is perhaps more appropriate to consider that both alleles are repressed or their products neutralized and that there is therefore neither dominance or repressiveness. We

192

describe this situation as one of 'null dominance' (see below). In this connection it is clear that if both alleles were to retain their normal 'haploid' functions in the diploid (co-dominance) the diploid might not be able to replicate its DNA.

The relationship of mating to the cell cycle has not been determined for *Hansenula wingei*. However, Crandall and Brock (17) have much interesting information on the chemical basis of the agglutination that normally precedes conjugation. An intriguing and potentially important aspect of their finding is that the heterozygous diploid obtained from two haploid agglutinative strains (5 and 21) which is normally non-agglutinative and sexually neutral can be induced to synthesize either product. The diploid synthesizes 5 factor and agglutinates with strain 21 when the cells enter late stationary phase in the presence of certain trace metals, while it synthesizes 21 factor and agglutinates with strain 5 when log phase cells are grown in the presence of EDTA (18). The cells in both cases remain heterozygous.

Thus the dominance of the two alleles affecting agglutination in *Hansenula* varies according to the presence or absence of trace metals and the age of the cell. This latter effect is of particular interest here, and indicates that a cell cycle analysis of mating in the haploid and diploid strains would be most informative. We might speculate on the basis of our results with *Ustilago* that this represents another kind of temporal allelic interaction with the diploid synthesizing 21 factor mainly in S and/or G_2 phases (i.e., many of the log phase cells) and 5 factor mainly in G_1 cells (including the majority of stationary phase cells). Even if this is not the correct explanation, this cell cycle effect on allelic expression and dominance remains a fascinating observation of great significance to the anlaysis of cell cycle controls.

D. The Cell Cycle and Dominance

The behaviour of the mating type alleles in diploid cells of *Ustilago*, *Saccharomyces*, and *Hansenula* represents a kind of interaction that is not easy to describe with current genetic terminology. Indeed these temporal allelic interactions represent a hitherto undetected kind of dominance (10). In these situations, both alleles produce functional products, but one allele is dominant because it is transcribed during a phase of the cell cycle when the other allele is repressed or 'silent'. The use of the word 'dominant' in this case is fully justified as the term is used to describe a *phenotypic* observation.

The better known, but perhaps not necessarily more frequent, kind of dominance derives from the inability of one allele (the recessive) to produce a functional product. The heterozygous cell therefore has the ability to produce at least one 'dose' of gene product and if this is not limiting then there is said to be complete dominance.

As little is known of the genetic structure and function of the mating-type locus, the use of the term "allele" for the mating type determinants is perhaps questionable. However, until more precise information is available on these points, we retain use of the term noting that no recombinants have yet been obtained and that the overall function (mating control) is the same for both forms. In most fungi that have been studied the mating type locus has proved to be large and subdivided into cistrons (see 19). It is certainly possible that the base sequences differ markedly in one or more of the cistrons of each form, as in other sex determining loci or chromosomes.

Current teminology used in defining types of dominance (e.g., *complete*, *partial*, *no* dominance, *co*-dominance and *over*-dominance) is inadequate to describe situations in which both alleles produce products that may be very different from each other, yet may be complementary (+ and -) and may neutralize each other so that the heterozygote shows absence of either function (0). Thus the term 'no dominance' is better reserved for cases where the heterozygote is exactly intermediate (++) between two homozygotes with phenotypes differing in quantity of expression (+++ and +). We prefer the term "null-dominance" to describe situations in which the two alleles produce functional products which interact negatively ('cancel out' or 'neutralize') so that the cell does not carry out the function.

In cases where the two alleles produce functional products which act independently the term co-dominance is normally used (e.g., the ABO blood group system). Similarly, when the two alleles give products which interact additively this would be defined as "over-dominance". We have suggested that temporal allelic interactions between alleles having additive effects may account for some heterosis or hybrid vigour (10). For instance if transcription is a limiting factor in a particular morphogenetic step, a heterozygote carrying alleles which have different time periods for transcription may have a longer total transcription period than either homozygote.

Experimental evidence is so far lacking for heterosis caused by such temporal allelic interactions in systems where the two alleles have an additive effect on the phenotype. However, the data reviewed here for mating type loci in fungi show that similar kinds of temporal interactions *do* occur in systems where the two alleles interact negatively and give null dominance. Thus a_1a_2 cells of *U. violacea* show a cyclical dominance pattern alternating between phases of null dominance (G_1 cells) and phases showing complete dominance of the a_2 allele (S and G_2 cells) (Fig. 4b). *Saccharomyces cerevisiae* appears to show a non-cyclical null dominance, i.e., the two alleles have the same transcription period in the cell cycle (during G_1). In *Hansenula wingei*, the cell cycle controls are not yet clear but one possibility is that the transcription periods of the two alleles are so different that there is a cyclical pattern of dominance involving first complete dominance of one allele during one stage of the cell cycle, and then complete dominance of the other allele at a later stage of the cycle. There may also be a phase or phases of null dominance. The exact nature of the cell cycle controls operating on this species clearly need to be investigated.

The results discussed in this section point out pitfalls in the analysis of cell cycle controls in diploid organisms. In some cases, heterozygous strains may have cell cycle behaviour which is the result of an interaction between two very different control systems. Careful comparison with haploid or homozygous forms is necessary in interpreting such results.

V. Transcription and Translation of the Sex Message

As indicated earlier conjugation in *Ustilago violacea*, involves a period of courtship followed by the cooperative assembly of a conjugation tube between paired sporidia of opposite mating type. In this section we show that information is exchanged during courtship which directs transcription and translation of a "sex message". Once translation is completed the copulatory organelle is assembled. The evidence supporting this formulation of the sexual process is derived from studies on the induction, transcription and translation of the 'sex message' of each allele. The studies include an analysis of the influence of ultra-violet (UV) light on production of the sex message, and the use of UV-sensitive mutants to investigate the roles of each mating type allele. Further studies include experiments using i) inhibitors of transcription and translation and ii) radioactive tracers, to determine when the sex message is synthesized and decoded and the copulatory organelle assembled.

A. *The Effects of Ultraviolet Light on Conjugation*

As indicated earlier, auxotrophic cells of opposite mating type assemble a conjugation tube after about 3-4 hours of mating on nutrient free medium. Low doses of ultra-violet light (100-300 ergs/mm^2) delay but do not prevent conjugation if given in the first 2-5 hours of mating. However, irradiation later than this period has little effect and conjugation proceeds normally. The UV effect is photoreactivable and we conclude that UV induces dimers which affect transcription of specific messenger RNA molecules needed for conjugation ("sex message"). Our evidence suggests that the dimers may cause mistranscription rather than the complete prevention of transcription (20).

Thus total RNA synthesis was not significantly affected by such low doses and we infer that the accuracy of transcription of the sex message is probably diminished by the presence of UV-induced dimers in the gene or genes concerned with conjugation. It is possible that the mRNA synthesized had altered bases opposite the dimers, or more likely perhaps that it is prematurely terminated, or has a 'frame-shift'.

The effect of UV on conjugation in reciprocal crosses of UV-sensitive and wild-type strains indicates clearly that both partners must complete transcription of sex messages in order to conjugate. Inactivation of either partner before transcription prevents conjugation, but conjugation proceeds when either cell is inactivated after transcription of the sex messages has occurred. These results suggest that a mutual and reciprocal exchange of information between the two mating-types occurs prior to the assembly of the conjugation tube. The information exchange precedes assembly at a time when the membranes of the individual cells are intact and the sole cell to cell contact is by means of the hairs on their glycocalyxes. It is not yet clear whether each cell, once induced by the other, continues to transcribe sex messages until the message is complete even when its partner is inactivated, or whether there is a mutual and sequential induction of genes which is halted if one partner is inactivated. What our experiments do show, however, is that *both* partners must complete the transcription of the sex message to complete conjugation (20).

B. The Effects of Inhibitors of Replication, Transcription and Translation on Conjugation

Hydroxyurea and fluorodeoxyuridine do not affect assembly of the organelle in mating populations even though these inhibitors block DNA replication in actively growing cell populations at the levels employed in this study (21, 10, and unpublished data). Preliminary experiments indicated that mating in *Ustilago violacea* was insensitive to the inhibitors of transcription, actinomycin C and D and rifampin. A combination, however, of the transcription inhibitors and a fungistatic agent, amphotericin B, that reduces the permeability barrier of the cytoplasmic membrances of fungi (22) was found to inhibit both cell growth and copulation. The cogent feature of these experiments was that transcription inhibitors prevented conjugation if provided during the period of courtship.

Furthermore, cycloheximide, an inhibitor of translation in the large 80S eukaryote ribosome effectively prevented conjugation provided that treatment was initiated prior to assembly of the copulatory organelle, while the inhibitor of the small 70S mitochondrial ribosomes, chloramphenicol, had no inhibitory influence on mating.

To quantitate any increase in conjugation following inhibitor treatment it was necessary to combine an analysis of the kinetics of conjugation with inhibitor treatments during the course of mating. Figure 5 shows the kinetics of conjugation of auxotrophic strains and the effect of the inhibitors, cycloheximide and rifampin. Inhibitor treatments given during the period of active assembly did not immediately prevent all further conjugation but allowed a limited increase. This increase corresponded to between 45 and 60 minutes of uninhibited copulation in the cycloheximide treated cultures and to between 60 and 90 minutes in the rifampin treated cultures. Poon, *et. al.*, (4) estimate that assembly of the conjugation organelle requires roughly 45 minutes. Microscopic examination of the inhibited cultures showed single cells or conjugated cells but no intermediate pairing stages. From these data it is reasonable to assume that the last proteins essential for assembly are completed about 15 minutes prior to assembly of the conjugation tube (23).

Cells that have completed translation of the specific conjugation proteins thus go on to complete conjugation even in the presence of inhibitors. The absence of cells blocked in intermediate stages suggests that irreversible mating changes do not occur until all the sex proteins are manufactured.

Amino acid incorporation studies showed that even though the cellular

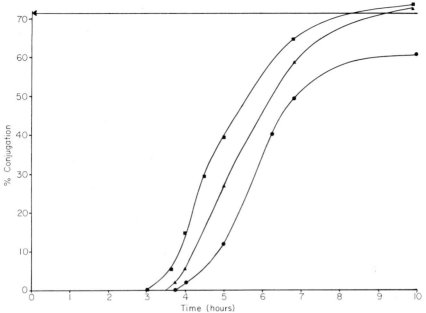

Fig. 5. *The effect of cycloheximide and rifampin on conjugation.* Auxotrophic a_1 and a_2 cells were mixed on water agar. The kinetics of normal conjugation were determined (●). Samples of the mating cells were treated at various times during conjugation with the inhibitors cycloheximide 100 µg/ml (▲) or rifampin 100 µg/ml + amphoterecin, 100 µg/ml (■) and left in contact with these inhibitors until the % conjugation was determined at 24 hours. The time axis for the inhibitors thus represents the time of addition of the inhibitor.

protein levels remain constant during courtship there was extensive amino acid turnover as sporidia adjusted to mating conditions. An enhancement of amino acid turnover by mating, as opposed to unmixed sporidia, was not detectable. Both proto- and auxotrophic strains showed extensive RNA turnover during courtship as evidenced by radioactive precursor incorporation but a specific enhancement of turnover due to mating was not detectable from the labeling pattern alone. A round of DNA replication was completed during courtship but mating activity did not influence the gross pattern of DNA replication in either of the mating types.

The attempt to distinguish alterations in the gross patterns of macromolecule synthesis during the phase of courtship prior to assembly of the copulatory organelle of the anther smut was not successful. Differences between mating and unmixed sporidia were not evident in the patterns of tracer incorporation and macromolecule stability. Unlike the active synchronization of nuclear replication reported in courtship of yeast *Saccharomyces cerevisae* (11, 12) the fundamental metabolic machinery of smut appeared to carry on "business as usual" during courtship even though one of the mating types (a_1) is restricted in its activity to the G_1 phase of the cell cycle (7). In other words, courting yeast actively synchronize the cell cycles of both mating partners while smut passively grasp the chance liaison between any of the sporidia of mating type a_2 and a sexually oriented cell of mating type a_1.

In spite of the observed absence of gross alterations in the patterns of nucleic acid and protein synthesis, inhibitor studies showed that nucleic acid synthesis and protein synthesis are both essential prior to assembly of the copulatory organelle. This specific macromolecule synthesis is very likely masked in the labelling pattern by the change-over in cellular metabolism brought about by the transition from rapid logarithmic growth in complete medium to survival in a medium lacking an exogenous energy source (23).

Earlier we discussed the influence of ultraviolet light on assembly of the copulatory organelle, and showed that during courtship conjugation is prevented if either partner is inactivated with UV light. The experiments indicated that there was a mutual exchange of information between the mating cells, prior to the major morphological changes. The information exchange precedes assembly at a time when the sole cell to cell contact is by means of the hairs on their glycocalyxes. In other words, both of the mating partners must complete transcription of the sex message for both to complete conjugation. Since the period of maximum sensitivity to ultraviolet light corresponds to the period of transcription of sex message it appears likely that the information transferred through glycocalyx contact acts at the level of the genome. As we have indicated, the ultraviolet damage is photoreversible and probably due to dimers in the genes concerned with transcription of the sex message. The nature of the substance or substances transmitting regulatory information to the genome is presently unclear but current evidence suggests that the transfer depends upon cell to cell contact or close cellular proximity (4).

The observation that transcription and translation of the sex message precedes assembly of the copulatory organelle of mating smut indicates that this assembly involves rearrangements of previously synthesized structural units in a manner comparable to the differentiation of water mold zoospores into a vegetative germling (24) or the assembly of the mitotic apparatus and reconstructed nucleus in the slime mold *Physarum* (25).

Resolution of the questions raised in this chapter about transcription and translation of the sex message will depend upon the isolation and characterization of the RNA and protein molecules concerned with the sexual process in smut. These analyses are presently being undertaken.

VI. Summary

The cell cycle control of sexual morphogenesis in the smut fungus, *Ustilago violacea* has been considered in this chapter. The morphogenetic aspects of mating were separated into five stages ranging from a period of "courtship" to the final elongation of the conjugatory organelle. Both the morphogenetic alterations and the molecular alterations leading to them appear synchronously in a mating population.

Cell cycle studies of mating haploid cells showed that the two alleles (a_1 and a_2) governing the induction of the fusion tube are differently regulated during the cell cycle. Allele a_1 is restricted in its activity to the G_1 phase of the cell cycle (a *stringent* cell cycle control) while allele a_2 is active during all phases of the cell cycle (a *relaxed* cell cycle control) (7).

This observation of a pair of alleles, differently restricted during the cell cycle in their regulation of a morphological change, when fully clarified at the molecular level, should enhance an understanding of the mechanism restricting cellular differentiation of higher eukaryotes to the G_1 phase.

An analysis of the mating activity of a diploid heterozygous for the mating alleles indicated that one of the alleles (a_2) was dominant. Furthermore the dominance was age related as only young diploid cultures showed full sexual vigour and aging cultures lost their ability to mate. Homozygous diploids do not have an age related loss in sexual vigour (9). Cell cycle analysis of the mating activity of the heterozygous diploid showed that conjugation was limited to the S and G_2 phases of the cell cycle most probably because both mating alleles were active during G_1 and neutralized each other. The age related loss in sexual activity was thus related to the accumulation of stationary cells in the G_1 phase. This cell cycle analysis uncovered a hitherto undescribed mechanism of dominance, one termed *temporal allelic interaction* (10).

Finally, analyses of the transcription and the translation of the "sex message" in synchronously mating cultures showed that the synthesis of gene product governing sexual morphogenesis was completed during "courtship" prior to the assembly of the copulatory organelle (23). Experiments on the influence of ultraviolet light on assembly using UV sensitive and resistant strains showed that the mating cells mutually exchange information that regulate "read out" of the genes governing morphogenetic induction. This information exchange occurs during "courtship" and prior to cell fusion or assembly of the conjugation tube (20).

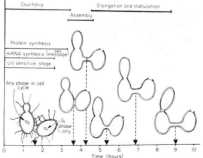

Fig. 6.

Developmental stages and molecular events during conjugation of haploid cells in *Ustilago violacea*. From (23).

199

The main thrust of our future work lies in the isolation of mutants of the cell cycle regulation system, conditional mutants of morphogenesis and in the isolation and characterization of the sex message and the proteins into which it is translated.

Acknowledgements

We would like to thank the following for their enthusiastic help with this work, Mr. J. Hodges, Mrs. V. Marnot, Mr. J. Martin, Miss M. Morden, Miss K. Neilsen. We are also grateful to the National Research Council of Canada for research support (Grant No. A5062).

References

1. A. W. Day, and J. K. Jones, *Genet. Res. 11*, 63 (1968).
2. A. W. Day, *Can. J. Genet Cytol. 14*, 925 (1972).
3. A. W. Day, and J. K. Jones, *Genet. Res. 14*, 195 (1969).
4. N. Poon, J. Martin, and A. W. Day, *Canad. Jour. Microbiol.* (in press).
5. A. W. Day, *Nature New Biology 237*, 282 (1972).
6. L. L. Clements, A. W. Day, and J. K. Jones, *Nature 223*, 961 (1969).
7. J. E. Cummins, and A. W. Day, *Nature, Lond. 245*, 259 (1973).
8. P. Tauro, and H. Halvorson, *J. Bact. 92*, 652 (1966).
9. B. L. Carter, and H. Halvorson, *Exptl Cell Res. 76*, 152 (1973).
10. A. W. Day, and J. E. Cummins, *Nature, London. 245*, 260 (1973).
11. E. Bucking-Throm, W. Duntze, L. Hartwell, and T. Manney, *Exptl Cell Res. 76*, 99, (1973).
12. L. Hartwell, *Exptl. Cell Res. 76*, 111 (1973).
13. E. Sena, D. Radin and S. Fogel, *Proc. Nat. Acad. Sci. (USA) 70*, 1373 (1973).
14. S. Pomper, K. Daniel and D. W. McKee, *Genetics 39*, 343 (1954).
15. N. Gunge and Y. Nakatomi, *Genetics 70*, 41 (1972).
16. W. Duntze, V. MacKay and T. Manney, *Science, 168*, 1472 (1970).
17. M. Crandall and T. Brock, *Bacteriol. Rev. 32*, 139 (1968).
18. M. Crandall and J. Caulton, *Exptl. Cell Res.* (in press).
19. J. Fincham and P. Day, *Fungal Genetics 3rd Ed.*, Blackwell, Oxford and Edinburgh, (1971).
20. A. W. Day and J. E. Cummins, *J. Cell Science*, (in press.)
21. R. Esposito and R. Holliday, *Genetics 50*, 1009 (1964).
22. G. Medoff, G. Kobayashi, G. Kwan, D. Schlessinger and P. Venkou, *Proc. Nat. Acad. Sci (USA) 69*, 196 (1972).
23. J. E. Cummins and A. W. Day, (Submitted for publication.)
24. D. Soll and D. Sonneborn, *Proc. Nat. Acad. Sci. (USA) 68*, 459 (1971).
25. J. Cummins, E. Brewer and H. Rusch, *J. Cell Biol. 27*, 337 (1965).

TRANSCRIPTIONAL AND POST-TRANSCRIPTIONAL CONTROL
OF ENZYME LEVELS IN EUCARYOTIC MICROORGANISMS

Robert R. Schmidt

Department of Biochemistry and Nutrition
Virginia Polytechnic Institute and State University
Blacksburg, Virginia 24061

I. Introduction

Current concepts regarding the regulation of gene transcription and enzyme levels have been largely derived from research with procaryotes. Although research with procaryotes has and will continue to contribute to our knowledge of cellular control mechanisms, these organisms are devoid of the subcellular organelles typically found in eucaryotes and therefore they might lack certain regulatory mechanisms required to coordinate the timing of expression of nuclear and extra-nuclear (*i.e.*, organelle) genes and to control the levels of the proteins or other macromolecules for which they code.

The research in this laboratory has had as its objective to use synchronous cultures of the highly compartmentalized eucaryotic microorganism, *Chlorella*, as a model system to study regulation of gene expression and enzyme levels during the cell cycle of eucaryotic cells. Answers to a number of basic questions are being sought. Is the timing of expression of genes, located within the same or different organelles but which code for catabolic and biosynthetic enzymes on the same or different metabolic pathways, regulated by similar or different transcriptional or post-transcriptional mechanisms in the same eucaryotic cell? Are the levels of enzymes, associated with the same or different organelles or cytosol or with different metabolic pathways, regulated by different post-transcriptional mechanisms (*e.g.*, active degradation, endogenous stabilizers, *etc.*) in the same eucaryotic cell?

A problem arises as to which and how many genes and their enzymes should be studied during the cell cycle of a eucaryotic cell. Should the research be restricted to an in depth study of one or two genes and their enzymes or initially should a reasonably large number be studied in lesser detail to obtain a better idea as to whether different transcriptional or post-transcriptional mechanisms are operative in regulating gene expression for different classes of enzymes? The experimental approach taken in this laboratory reflects the latter viewpoint, *i.e.*, first demonstrate whether or not different classes of enzymes are regulated differently and then select examples of each to be studied in detail.

The rationale of this approach stems in part from the reports (1) that catabolic and biosynthetic enzymes are regulated differently at the transcriptional level (*i.e.*, cAMP dependent *vs.* non-dependent transcription, respectively) in procaryotes. Moreover, research from a number of laboratories (2,3) has revealed that many of the physical and biochemical characteristics of the DNA and the protein synthesizing systems in mitochondria and chloroplasts are very similar to those in procaryotes. In fact, it has been proposed that these organelles once may have been bacteria which parasitized a primitive eucaryotic cell. Therefore, it seems reasonable to predict that genes located in organelles might be regulated as in procaryotes (*i.e.*, oscillatory repression) and might be continuously available for transcription during the eucaryotic cell cycle, whereas transcription of at least some nuclear genes might be restricted to certain periods of the cell cycle. For example, the availability of some nuclear genes in eucaryotes might be unavailable for transcription when the chromosomes are condensed during mitosis, an event not shared with procaryotes. In view of these possibilities, caution should be exercised before making generalizations about the nature of the mechanism(s) regulating gene transcription and enzyme levels in either higher or lower eucaryotes until a large number of genes and enzymes are studied in detail in the same eucaryote.

In addition to summarizing the research results obtained from cell cycle studies with *Chlorella* in this laboratory, an attempt has been made in this chapter to discuss and to propose concepts, experimental approaches, and interpretations of data from studies with other eucaryotic microorganisms as well.

II. Concepts and Experimental Approaches for Studies on Regulation of Inducible Enzymes during the Eucaryotic Cell Cycle

Two experimental approaches have been used to study the availability of inducible genes for transcription at different times during the cell cycle of eucaryotes. The first approach consists of culturing synchronous cells in the absence of inducer, harvesting aliquots of cells from the parent culture at periodic intervals, and then challenging the harvested cells to synthesize inducible enzyme(s) by addition of inducer(s) to their culture medium (4,5). The second approach involves continuously challenging preinduced synchronous cells to synthesize inducible enzyme(s) throughout the cell cycle by culturing them in the presence of inducer(s) for the entire cell cycle (6,7).

Interpretation of data obtained with the first experimental approach is facilitated by the absence of or low basal level of the inducible enzyme at the time of inducer addition throughout the cell cycle. The continuous or periodic inducibility of both stable and unstable enzymes can be readily demonstrated by the occurrence or absence of induced enzyme-accumulation at different times during the cell cycle. However, before induced enzyme-accumulation can be said to reflect the availability of a structural gene for transcription, inducibility must be shown to be dependent upon both *de novo* RNA and protein synthesis.

Attempts have been made to relate induction kinetics to the gene dosage of

202

the cells (4,5,8,9). Ideally, if a structural gene is fully induced, the *rate* of enzyme *synthesis* should be proportional to the gene dosage. However, the *in vivo* stability characteristics of the enzyme will dictate the kinetics of its *accumulation* after inducer is added to previously uninduced cells. Whereas a stable enzyme will continue to accumulate in a linear manner (per ml of culture) when its synthesis becomes constant, an unstable enzyme will cease to accumulate after its rate of synthesis becomes constant (Fig. 1). The level of an unstable enzyme will increase to a plateau at which the rates of synthesis and degradation are equal. The time required for an unstable enzyme to reach a steady-state level once its rate of induced synthesis becomes constant will depend upon the magnitude of its rate constant for decay (k_d). As k_d increases, the time required to reach a steady still will shorten. Thus, the induced rate of accumulation of an unstable enzyme will accelerate at the beginning of the induction period and decelerate as its induced steady-state level is approached. Even the basal (uninduced) level of an unstable enzyme represents a steady state between synthesis and breakdown. Thus, to obtain a reliable relationship between the rate of induced-accumulation of an unstable enzyme and the dosage of its structural gene, the rate of enzyme accumulation should be measured exactly between the basal and plateau levels. Moreover, the new plateau level reached under fully-induced conditions should also be proportional to the gene dosage.

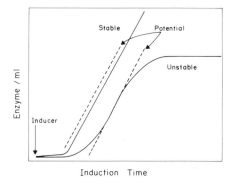

Fig. 1. Kinetics of induction of stable and unstable enzymes and the relationship between the rate of induced enzyme accumulation and enzyme potential in previously uninduced cells.

The interchangeable use (10,11) of the terms inducibility and potential has led to some confusion (12) in the interpretation of induction data from cell cycle studies. Inducibility (or derepressibility) should be used only to indicate whether or not the synthesis of an enzyme can be induced (or derepressed) at a particular time during the cell cycle. In the strictest sense, this term should be used only in a positive or negative manner and should not be used to describe the rate of induced (or derepressed) enzyme synthesis. Keumpel *et al.* (8) used the term *potential* in an attempt to describe the rate of enzyme synthesis under fully-induced (or derepressed) conditions. They describe potential as "the maximum ability for enzyme synthesis such as should be obtained by complete derepression." In practice, however, these workers measured the *rate* of enzyme *accumulation (i.e.,* slope of accumulation curve) after addition of saturating levels of a derepressing agent or inducer to previously repressed or uninduced cells. Since the rate of accumulation of an unstable enzyme is not necessarily equivalent to its rate of synthesis, either the term potential cannot be used to describe the rate of induced (or derepressed) accumulation of unstable enzymes or the definition of potential needs to be modified. We prefer the definition to be modified to read: *Potential is the maximum rate of enzyme accumulation under conditions for complete induction or derepression.*

In certain bacteria, the change in potential for a stable inducible enzyme has been shown to increase in proportion to the increase in gene dosage at the time of gene replication (9). The same relationship could presumably hold for an unstable enzyme provided the k_d for the enzyme remained constant during the cell cycle and provided potential was measured at the point of inflection between the initial and induced steady-state levels of the enzyme as discussed earlier (Fig. 1).

The importance in distinguishing between potential and inducibility becomes apparent when one considers that synthesis of an enzyme can be partially induced (or derepressed) without maximal rates being achieved. For example, with suboptimal levels of inducer, it would be possible to demonstrate continuous inducibility without achieving maximal induction rates. Alternately, the inducer could be added at saturating levels to cells harvested at different times in the cell cycle, but the cultural conditions could be changing (*i.e.,* energy level, pH, nutrient concentration, *etc.*) during the cycle so as to limit the rate of total protein synthesis with the resultant decrease in the rate of induced enzyme synthesis as the cycle proceeds. In other words, the conditions to demonstrate inducibility are much less stringent than those for measurement of potential. Thus, to determine if a direct proportionality exists between the change in potential and the change in gene dosage, it is essential to add a saturating level of inducer or derepressing agent and to maintain very constant culture conditions throughout the cell cycle.

When interpreting data from experiments in which enzyme potential is measured in eucaryotes, certain concepts and methods of plotting data must be considered. First, one must consider the possible relationship between the rise in potential and gene replication in the eucaryotic cell with multiple nuclear chromosomes. Nuclear chromosomes in eucaryotes appear to replicate in segments called replicons (13,14). Different replicons within the same and different chromosomes appear to replicate at distinct times within the S-phase for total DNA replication (14-19 See also, Williamson, this monograph). Thus, even in a synchronous culture, in which some degree of randomness still exists, the earliest and latest periods of the S-phase should be enriched in different replicating genes, *i.e.,* the time of replication of different genes will not be symmetrically distributed around the mean time of the S-phase. Since a eucaryotic chromosome is not replicated in a sequential manner as in bacteria

(20,21), the individual replicated segments might be isolated for a period of time before being linked by a ligase to adjacent later-replicating segments. If the genes encompassed by a given replicon can be transcribed soon after replication, even though adjacent DNA segments have not replicated, the rise in potential for different inducible enzymes would be expected to be initiated at different times within the S-phase (Fig. 2A). In this case, since the time required for replication of a single gene would be much shorter than the duration of the entire S-phase, the rate of increase in potential for each inducible enzyme should be greater than that for the rate of total DNA accumulation (Fig. 2A). However, if genes in newly replicated segments within a chromosome cannot be transcribed until all segments are replicated and linked, the rise in potential for all inducible genes within that chromosome would be initiated at the same time in the S-phase. In this latter case, the time at which the replication of each chromosome is completed in S-phase would determine when potential would rise for a group of encoded enzymes. Although little is known as to the order of completion of replication of individual chromosomes in the S-phase, it is known that DNA which is replicated early in S-phase has a higher buoyant density (*i.e.*, higher G-C content) than DNA replicated in mid- or late-S-phase (19,22). Moreover, some evidence (14) indicates that heterochromatic segments and genetically inactive regions are late replicating, suggesting that active, unique DNA might be replicated early while repetitious DNA replicated late in the S-phase.

Second, some of the problems encountered in analyzing the relationship between enzyme potential and the S-phase can be illustrated by plotting hypotehtical data in different ways (Fig. 2A, B). In Figure 2A, the structural genes for five different stable inducible enzymes are assumed to be inducible at all times during the cell cycle. A direct relationship is assumed to exist between enzyme potential (P_1, P_2, P_3, P_4, P_5) and the gene dosage for the five enzymes whose structural genes replicate at different times during the S-phase in a synchronous culture of cells. The cells are assumed to be incompletely synchronized so that the change in gene dosage and the accompanying increase in potential will take a discrete fraction of time of the S-period. The genes are assumed to be transcribable shortly after their replication. The S-phase is assumed to be initiated and terminated at times S_o and S_t, respectively; whereas S_m is designated as the midpoint of the S-phase.

There are at least three different ways in which data of this type can be plotted for interpretation: (a) comparison of the timing and kinetics of increase in potential and total DNA plotted as cell cycle patterns (Fig. 2A); (b) cell cycle time-map in which the timing of the midpoints of enzyme potential and the S-phase are compared (Fig. 2B-1); and (c) cell cycle time-map in which the timing of the midpoints in enzyme potential is compared to the interval of the S-phase (Fig. 2B-2). Because the first method reveals the fold-increases and the relative rates of increase in enzyme potential and in DNA, in addition to indicating the time at which the potential of a given enzyme rises within the S-phase, this method of plotting is obviously the method of choice for

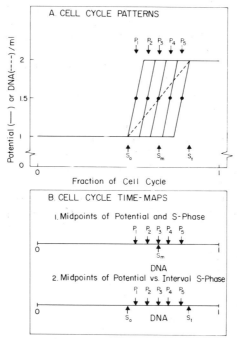

Fig. 2. A, B. Relationship between the time of gene replication and the change in enzyme potential during the cell cycle of synchronous cells in which the rate of induced enzyme synthesis is directly proportional to the gene dosage. P_1, P_2, P_3, P_4 and P_5 represent the midpoint times for the step-increases in potential for five different inducible enzymes whose genes replicate during the same time interval as the rise in potential for each enzyme during the S-phase (i.e., period of DNA replication). S_o, S_m, and S_t represent the onset, midpoint, and termination of the S-phase (--------), respectively.

comparing data from within a single cell cycle experiment.

Although the cell cycle time-map procedure (Fig. 2B) can be used to compare data obtained within a given experiment, this procedure was primarily developed to compare enzyme potential and DNA data from different experiments. The necessity developed because synchrony procedures employed with some organisms do not yield enough synchronous cells to allow DNA and different enzyme potential analyses to be performed in the same cell cycle experiment. Therefore, the midpoints for the step-increases in potential for different enzymes and in DNA have been compared (4) by the midpoint cell cycle time procedure shown in Fig. 2B-1. Moreover, to establish more accurately the midpoint times for a given enzyme potential or DNA, the midpoint times from different experiments have been averaged. Several problems exist with the midpoint procedure (Fig. 2B-1). By examining only the midpoint cell cycle time-map (Fig. 2B-1), it is not possible to ascertain whether or not the midpoints for the different enzyme potentials fall within the S-phase, *i.e.*, only when the midpoints for enzyme potential and the S-phase are coincident (*e.g.*, P_3 and S_m, Fig. 2B-1) can one be confident that enzyme potential increases within the S-phase. As stated earlier, the period of DNA replication cannot be considered to be a homogeneous cellular event such as the period of synthesis of a single enzyme. Different genes apparently replicate at specific times within the S-phase of synchronous eucaryotic cells. Therefore, since the timing of replication of different genes is not symmetrically distributed around the midpoint of the S-phase, the midpoints for the increases in potential are more properly related to the interval (Fig. 2B-2) rather than to the midpoint (Fig. 2B-1) of the S-phase. If these concepts are not considered, it is possible that one could falsely conclude from a plot, such as Figure 2B-1, that the rise in potential for most enzymes is displaced in time from the S-phase. In fact, if by chance the enzyme potential of only late replicating genes was measured, it would appear solely from the midpoint plot (Fig. 2B-1) that a delay might exist in the transcription of newly replicated genes. It is recommended, therefore, that DNA and enzyme potential measurements *always* be made within the *same* cell cycle experiment, and that comparison of these two parameters should be made by the procedures shown in either Fig. 2A or 2B-2.

Other factors also can obscure the possible relationship between the timing of gene replication and the timing of the increase in enzyme potential. For example, because of the length of the lag period after addition of inducer and also the length of the period of induced enzyme accumulation required to define a constant slope for estimation of enzyme potential for any given harvest period in the cell cycle, the problem arises as to where on the *cell cycle time scale* to plot the experimentally determined values for potential. This plotting problem is complicated by the cells usually continuing to grow and, if in the S-phase, to replicate DNA during the period in which enzyme potential is measured. The potential can be plotted at three possible times on the cell cycle time scale: (a) the time corresponding to the time of cell harvest from the parent synchronous culture and the addition of inducer, (b) the time corresponding to the end of the lag period or the beginning of the period of constant induced-enzyme synthesis, or (c) the time corresponding to the midpoint of the period of induced enzyme accumulation.

To decide which of the plotting times would give the closest correlation between the time of replication of a specific gene and the time of increase (or midpoint of the step increase) in potential of its encoded enzyme, a series of hypothetical cell cycle experiments was graphically constructed to simulate potential measurements. The simulated data included different: (a) frequencies of sampling times from the hypothetical parent culture prior to enzyme induction, (b) lag periods (*i.e.*, no lag to long lag periods) after addition of inducer, (c) length periods of induced enzyme accumulation after the lag period (d) times within the S-phase for specific gene replication, and (e) length S-phases.

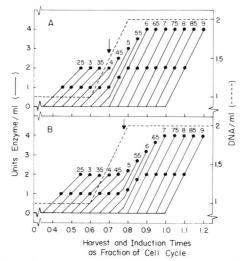

Fig. 3. A, B. Hypothetical induction data to show the effect of the length of the induction period on the time at which the rate of induced enzyme accumulation (i.e., potential) appears to increase relative to the time of gene replication in cells in which potential is directly proportional to the gene dosage. The inducer is assumed to be added to cells at the time of harvest at 0.10 cell cycle intervals during the cell cycle. The induction lag-period is assumed to be 0.10 cell cycles long. The dark circles mark the end of 0.10 and 0.20 cell cycle intervals of induced enzyme accumulation beginning at the end of each lag-period. The numbers at the end of each induction plot indicate the time of cell harvest for that induction period. A and B differ only in the time of gene replication (indicated by arrows) within the S-phase (------).

Typical examples of these simulated data are shown in Figures 3A, B and 4A,B,C in which enzyme potential is assumed to be directly related to the gene dosage. Because enzyme induction periods must be long enough to establish accurately the rate of induced enzyme accumulation, potential measurements will overlap periods in which genes are and are not replicating (Fig. 3A,B). In these simulations a sharp change in rate is shown within an enzyme accumulation period when it overlaps into a period of gene replication. However, in practice, because of the spacing of the sampling times and the intrinsic variation in enzyme analyses, etc., this sharp transition will not be seen; instead one will obtain essentially a "straight line" with an average slope representing the whole induction period. Thus, when the values for enzyme potential are plotted on the cell cycle time scale at either the time of harvest or at the end of the induction lag, this overlap will result in the displacement of the midpoint of step-increase in enzyme potential to a time prior to that for gene replication (Fig. 4A,B,C). This displacement occurs for these two plotting procedures even if there is no induction lag. However, as the period of induced enzyme accumulation is shortened, the midpoint of the rise in potential moves closer to the time of gene replication (Fig. 4A *vs.* 4B). Moreover, as the period of induced enzyme accmumulation is shortened, the enzyme potential appears to increase more sharply (Fig. 4A *vs.* 4B).

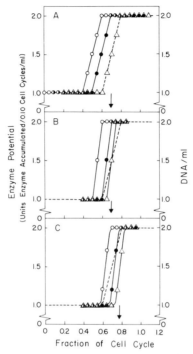

Fig. 4. A,B.C. Effect of the different ways of plotting enzyme potential on the apparent relationship between the midpoint for the step-increase in potential and the time of gene replication in the S-phase in cells in which the potential is assumed to be directly proportional to the gene dosage. Enzyme potential is plotted at three possible times on the cell cycle time-scale: O, the time corresponding to the time of cell harvest;● , the time corresponding to the end of the lag-period;△ , the time corresponding to the midpoint of the period of induced enzyme accumulation. The values for potential and the timing of gene replication in A and B were from Fig. 3A, and those in C from Fig. 3B. The potential values in A and B were calculated from periods of induced enzyme accumulation of 0.10 and 0.20 cell cycles in length. In C, the period was 0.10 cell cycles. The arrows indicate the time of gene replication in the S-phase.

From these simulations, it is apparent that the procedure of displacing the potential on the cell cycle time scale by the time corresponding to the midpoint of the period of induced enzyme accumulation gives the closest correlation between the midpoint of the step-increase in potential and the time of specific gene replication (Fig. 4A,B,C). This plotting procedure has been wisely recommended by Mitchison and coworkers (4,23) for experiments in

209

which the period of enzyme induction is long relative to the cell cycle or S-phase. With this procedure the rates of increase in potential are also affected by the length of the period of induced enzyme accumulation (Fig. 4A vs. 4B).

Although other relationships between enzyme potential and gene replication could be discussed, the foregoing discussion seems to illustrate the problems in interpreting enzyme potential data. If enzyme potential and gene dosage are directly related, the addition of an inhibitor of DNA replication to cells immediately prior to or at the time of inducer addition should eliminate most of the aforementioned plotting problems. If such an inhibitor is not available or it is only slowly permeable, the plotting problems can be minimized by harvesting cells with high frequency from the uninduced parent culture, keeping the period of induced enzyme accumulation very short relative to the length of the S-phase, and plotting enzyme potential data by the method recommended by Mitchison and coworkers (4,23).

In the second experimental approach used to study the availability of inducible genes for transcription during the cell cycle, it cannot be ascertained solely from the patterns of induced enzyme accumulation whether or not an enzyme is continuously or periodically inducible during the cell cycle. The *in vivo* stability characteristics of an inducible enzyme must be known for the complete interpretation of enzyme patterns obtained from preinduced synchronous cells growing in the continuous presence of inducer. For example, step patterns of enzyme accumulation can be derived from genes which are either periodically or continuously transcribed, depending upon whether the enzymes for which they code are stable or unstable, respectively. Figure 5A,B,C,D, and the discussion in the next three paragraphs summarize some of the different types of induced enzyme patterns predicted for synchronous cells growing in the continuous presence of inducer. Some of the concepts related to patterns of inducible and repressible enzymes under either basal or fully-induced (or derepressed) conditions in procaryotes have been discussed in an excellent review by Donachie and Masters (11).

If an inducible enzyme is unstable *in vivo* and its structural gene is transcribed continuously under either basal or fully-induced conditions, and if post-transcriptional factors are not differentially rate-limiting during the cell cycle, the enzyme is predicted to increase from one steady-state level to another in a single step during the period of DNA replication (Fig. 5A). The step-increase should be proportional to the increase in gene dosage. The number of steps of the same enzyme activity, within a single cell cycle, will depend upon the number of similarly regulated (*i.e.*, respond to same inducer) non-allelic structural genes coding for isozymes. Moreover, a mutation leading to constitutive synthesis of an unstable inducible enzyme is predicted to give the same step pattern for the enzyme as that for the wild-type in which transcription is fully-induced and is continuous during the cell cycle. Although the basal (uninduced) and fully-induced levels of an unstable enzyme will differ in magnitude, the timing and pattern of the step-increase should be identical for each. In contrast, if the structural gene of a stable inducible enzyme is

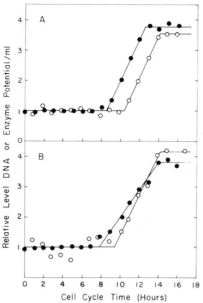

Fig. 5. A,B. The potentials of isocitrate lyase and NADPH-specific glutamate dehydrogenase and their relationship to DNA during the cell cycle of *Chlorella pyrenoidosa* (strain 7-11-05). A. Isocitrate lyase; B. NADPH-specific glutamate dehydrogenase; ●, DNA; O, enzyme potential.

continuously transcribed, the enzyme is predicted to accumulate in a linear manner with a change in rate of linear accumulation occurring at the time of gene replication (Fig. 5B).

If in the continuous presence of inducer the structural genes of inducible enzymes are periodically rather than continuously available for transcription, two types of induced enzyme patterns are predictable during the cell cycle. The level of an unstable enzyme will increase sharply during the period of transcription of its structural gene and then will decrease exponentially thereafter (Fig. 5C). In contrast, a stable enzyme will exhibit a step pattern (Fig. 5D). These periodic or step-increases in enzyme level can theoretically occur at any time during the cell cycle and would not necessarily be correlated with the time of replication of the structural genes. The amount of enzyme synthesized would be proportional to the length of the period of transcription and the half-life of the mRNA.

In conclusion, it should be apparent that when cells are cultured in the continuous presence of inducer, DNA and *in vivo* enzyme turnover measurements (performed under the same cell cycle conditions) are essential for interpretation of step patterns of inducible enzymes. Thus, since many enzymes are known to exhibit rapid turnover *in vivo* in higher eucaryotes (25-27), the first experimental approach would appear to be the most direct one for determining the availability of inducible genes for transcription during the eucaryotic cell cycle.

III. Studies on Enzyme-inducibility (or derepressibility) and -potential during the Eucaryotic Cell Cycle

In our laboratory, continuous inducibility (or derepressibility) has been shown for isocitrate lyase (24) and a NADPH-specific glutamate dehydrogenase (5) throughout the cell cycle of the eucaryote *Chlorella*. Cycloheximide and actinomycin D separately inhibit the induced (or derepressed) synthesis of these enzymes, strongly suggesting that the structural genes of these enzymes are continuously available for transcription throughout the cell cycle.

The potential for each enzyme begins to rise at a different time within the S-phase and the fold-increases in enzyme potential and DNA are essentially equal (Fig. 5). These data are consistent with the inference that the rise in enzyme potential and gene replication are directly related as in procaryotes. In earlier reports (5,24) in which the time of increase in potential of these enzymes was compared to the DNA pattern, the values for enzyme potential were plotted on the cell cycle time scale at a time corresponding to the time of cell harvest. In view of the findings revealed by the simulated plots in Figures 3 and 4, the values for enzyme potential are now displaced in Figure 5A,B by a time corresponding to the midpoint of the period of induced enzyme accumulation (*i.e.*, 50 minutes for isocitrate lyase and 57 minutes for the NADPH-specific glutamate dehydrogenase). With this time displacement, the midpoint for the step-increase in isocitrate lyase potential occurs near the end of the S-phase, whereas the midpoint for the NADPH-specific glutamate dehydrogenase falls close to the middle of the S-phase. Since in these cell cycle experiments two rounds of DNA replication occur within the S-phase, a given gene will replicate during each of the DNA doublings. If the rise in potential is directly related to gene replication in *Chlorella*, the midpoint for the step-increase in potential should fall between the two times in the S-phase that the gene actually replicates. Thus, it can be inferred from the data in Figure 5A that a short delay might exist between the replication and transcription of isocitrate lyase genes in *Chlorella*.

Additional evidence (unpublished) supports the concept that a direct relationship exists between the increase in potential and gene replication in this eucaryote. Under certain cultural conditions, the normally periodic DNA pattern can be converted to a semi-continuous pattern (28). The period of DNA accumulation is initiated at the usual time near the end of the cell cycle; however, instead of DNA accumulating in a periodic manner, it accumulates continuously into the next cell cycle. The increase in isocitrate lyase potential parallels this continuous increase in DNA. In future experiments, inhibitors of DNA replication will be used to establish whether or not a direct relationship exists between the potential of these two enzymes and gene replication.

At present it is not known whether isocitrate lyase synthesis is inducible or derepressible in the strain (7-11-05) of *C. pyrenoidosa* used in our laboratory.

212

Since isocitrate lyase synthesis can be initated and its constant rate of synthesis sustained solely by placing the cells into the dark, it is possible that synthesis of the enzyme is derepressible rather than inducible in this organism. Synthesis of the enzyme might be repressed by a compound (*i.e.*, corepressor) which accumulates in the light and is utilized in the dark. Alternately, an endogenous inducer might only be synthesized in the dark. Because another strain (211-8p) of *C. pyrenoidosa* was reported (29) to require acetate in addition to darkness to "induce" the synthesis of this enzyme, 0.2% acetate was included in the culture medium during the dark incubation of cells of strain 7-11-05 in our earlier studies (24). However, in strain 7-11-05, acetate has been found to have no effect on the initiation or rate of synthesis of the enzyme until it exceeds a concentration of 0.5% and becomes inhibitory to enzyme synthesis.

Because synthesis of the enzyme can be induced (or derepressed) by darkness alone in strain 7-11-05, the cell cycle experiments (Fig. 5A) were performed with synchronous daughter cells selected directly from continuously lighted cultures by use of the isopycnic density gradient procedure developed in this laboratory (30,31). The cells were maintained at approximately 3° during the harvest and centrifugation procedures to prevent synthesis of the enzyme from occurring. Moreover, because the cells increase fourfold in volume during synchronous growth, a constant turbidity of the parent culture was maintained by continuous dilution with fresh culture medium to prevent the increase in culture turbidity from lowering the effective light intensity per cell to a level at which synthesis of isocitrate lyase might be partially induced or derepressed. Enzyme synthesis was measured in the dark with cells harvested from this synchronously growing parent culture.

Although our laboratory has shown (24) that isocitrate lyase is inducible (or derepressible) at all times during the cell cycle of strain 7-11-05, McCullough and John (32) have presented excellent evidence that this enzyme is not inducible (or derepressible) for a period during the last 0.5 of the cell cycle of strain 211-8p. As mentioned earlier, this latter strain requires both darkness and acetate for the synthesis of the enzyme. The requirement for acetate does not necessarily mean that enzyme synthesis is inducible rather than derepressible in this organism. In fact, synthesis of the enzyme might also be repressed by a compound (*e.g.*, corepressor) which accumulates in the light; however, utilization of this compound in the dark might require exogenous acetate. The inability to induce or derepress the synthesis of the enzyme at a certain period of the cell cycle might be related to the cell's inability to synthesize an endogenous inducer or to lower the concentration of corepressor rather than due to a physical inaccessibility or unavailability of the structural gene of this enzyme for transcription. This interesting phenomenon of restricted inducibility might yield important information about post-transcriptional rather than transcriptional controls in this eucaryotic microorganism.

In contrast to isocitrate lyase synthesis which occurs only in the dark in strain 7-11-05, the ammonium-induced synthesis of the NADPH-specific glutamate dehydrogenase (Fig. 5B) occurs only in the light in this organism. Because in *Chlorella* the rate of total protein synthesis and the rate of induced synthesis of the NADPH-specific glutamate dehydrogenase were shown (5) to be proportional to the effective light intensity per cell at a given gene dosage, the culture turbidity of the parent synchronous culture from which cells were harvested for the hourly inductions was held essientially constant by continuous dilution during synchronous growth. Moreover, after harvest from the parent culture, the cells were rapidly centrifuged from their original medium, resuspended in the same volume of ammonium-containing medium, placed in "mini-chambers" with the same length light-path (*i.e.*, internal thickness) as the large parent culture chamber, and then placed under what seemed to be identical cultural conditions as the parent culture. In spite of these rigidly controlled cultural conditions, considerable variation existed between the potential measurements even during the period prior to DNA replication (Fig. 5B). This variation has been traced to differences in the bubbling rate of the CO_2-air mixture through the medium of the individual culture chambers at the different harvest times. Although the rate of flow of the CO_2-air mixture into the large parent culture was precisely controlled by a flowmeter, the flowrate (or bubbling rate) in the mini-chambers was estimated by eye. The bubbling rate of the cultures was found to affect the rate of total protein synthesis and hence the rate of induced enzyme synthesis. This problem has been described in detail to illustrate how variations in the culture environment can possibly obscure relationships between enzyme potential and gene dosage during the eucaryotic cell cycle.

Knutsen (12) has used synchronized cultures of another strain (211-8b) of *C. pyrenoidosa* to measure the inducibility and potential of nitrite reductase during the cell cycle. Although the enzyme was clearly inducible throughout the cell cycle, the potential of the enzyme decreased dramatically after increasing during the S-phase. Since the inductions of nitrite reductase were performed with a constant cell number rather than with a constant turbidity of cells harvested throughout the cell cycle, the rapid increase (exponential) in culture turbidity near the end of the cell cycle undoubtedly reduced the effective light intensity per cell to a level where it restricted the rate of induced enzyme synthesis and the measurement of true potential.

Continuous inducibility (or derepressibility) also appears to be a characteristic of genes of a number of enzymes during the cell cycle of both fission and budding yeasts. Mitchison and Creanor (4) observed that surcrase synthesis can be derepressed throughout the cell cycle of the fission yeast, *Schizosaccharomyces pombe*. However, whereas in bacteria (11) both the increase in potential and the changes in rate of linear accumulation of stable enzymes under basal (fully-repressed) conditions occur within the period of DNA replication, Mitchison and Creanor (4) have interpreted their data as indicating that both of these events occur well after the period of DNA replication in this fission yeast. The step-increase in surcrase potential was interpreted as being delayed approximately one-third of a cell cycle after the

214

period of DNA replication. Moreover, they also concluded that timing of the changes in the linear accumulations of sucrase, acid phosphatase, and alkaline phosphatase measured under basal conditions was also displaced beyond the period of DNA replication. They have suggested that this delay might represent "a substantial delay between chemical replication of the genome and what can be called functional replication." The time at which the newly replicated genes are proposed to become available for transcription was termed the "critical point."

Because these workers compared the timing of the midpoint for the step-increase in surcrase potential and the timing of the rate changes in linear enzyme accumulation under basal conditions with the timing of the midpoint of the S-phase in a cell cycle time-map plot as in Figure 2B-1, it seems possible that this plotting procedure (4,33) might be deceptively indicating a longer delay between gene replication and transcription than actually exists. As mentioned in an earlier section in this chapter, a delay in transcription of newly replicated genes could presumably occur if the newly replicated DNA segments (i.e., replicons) could not be transcribed until linked by a ligase at the end of the S-phase.

In the aforementioned studies by Mitchison's group, synchronous cells were selected from asynchronous cultures by use of sucrose density gradients. However, recently they have achieved induction synchrony with deoxyadenosine (23,34). The use of this compound apparently results in the synchronization of DNA replication but not in synchronization of those processes not directly coupled to DNA replication. Whereas autoregulated enzymes increase in a periodic fashion after selection synchrony, they increase continuously in cells after induction synchrony. In contrast, after either selection or induction synchrony, enzyme potential increases in a stepwise manner. Sucrase and maltase show step-increases in potential within the S-phase after induction synchrony, supporting the concept that the increase in enzyme potential is directly related to DNA replication in this eucaryote. However, it is difficult to rationalize the observation that, after induction synchrony, enzyme potential for sucrase rises early during S-phase, whereas after selection synchrony, sucrase potential is displaced until late in the S-phase or delayed for a period after this event. Does this apparent difference in timing of potential reflect a difference in the timing of transcription of newly replicated genes in cells prepared by the two procedures, or does it reflect some variation due to differences in frequency or length of the enzyme induction relative to the length of the S-phase in the two synchrony procedures (see discussion associated with Figs. 2, 3 and 4)? For whichever reason, this important finding deserves further study in S. pombe and in other eucaryotes.

The early studies by Halvorson and coworkers (6,7) on synchronous cultures of different species and of interspecific hybrids of the budding yeast, Saccharomyces, are often cited as providing experimental evidence for the concept that transcription of specific structural genes of enzymes is restricted to discrete periods of the cell cycle in these eucaryotic microorganisms.

215

However, if one reexamines these studies with respect to concepts for interpreting patterns for stable and unstable enzymes for sunchronous cells growing in the continuous presence of inducer (Fig. 6), it is readily apparent that insufficient evidence exists in these early experiments to justify the restricted gene transcription concept.

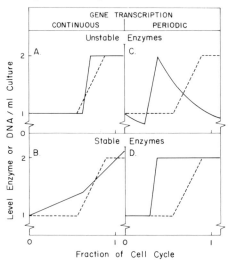

Fig. 6. A,B,C.D. Predicted patterns for stable and unstable inducible enzymes, when their genes are transcribed continuously or periodically, in synchronous cells growing in the continuous presence of inducer. —————— and ------, enzyme and DNA levels, respectively. mRNA assumed to be labile and post-transcriptional factors non-limiting throughout the cell cycle.

In their original model, Halvorson and coworkers (7) predicted a linear enzyme pattern for continuous transcription and a step enzyme pattern for periodic transcription in synchronous cells growing in the continuous presence of inducer. Because of the close correspondence between the timing of the *step-patterns* of α-glucosidase in the basal and induced cultures, these step-patterns were interpreted as supporting a restricted gene transcription model. If α-glucosidase is stable *in vivo* in synchronous cells growing in the continuous presence of inducer, this interpretation is correct. However, since the *in vivo* stability characteristics of this enzyme were (are) not known for these cultural conditions and the different species of *Saccharomyces* employed, these early data are also consistent with a model in which the enzyme is unstable *in vivo* and its structural gene transcribed continuously (see Fig. 6A and related discussion). Because DNA measurements were not made during the same cell cycles in which the basal or induced enzyme patterns were measured, it is still rather unclear whether or not the step-increases in -glucosidase occur within or outside of the S-phase. Since the synchrony procedure employed in these early studies has been improved, it is not accurate to compare DNA data

216

of current vintage to the earlier enzyme patterns. For example, it is unclear whether the reported (35, 36) differences in the duration (0.25 to 0.70 cell cycles) of the S-phase of the different species of *Saccharomyces* reflect intrinsic differences among these organisms or a difference in degree of synchrony in the early and latter work from Halvorson's laboratory.

Consistent with continuous rather than restricted gene availability during the cell cycle of *Saccharomyces* is the recent report by Sebastian *et al.* (37) that, when inducer is added to synchronous cells growing in the absence of inducer, ornithine transaminase and β-galactosidase are inducible at all times during the cell cycle. In apparent contradiction, however, these same enzymes showed step patterns in cells cultured in either the continuous presence or absence of inducer. Moreover, Carter and Halvorson (38) observed step patterns for wild-type and constitutive synthesis of UDP-galactose-4-epimerase. If these enzymes are unsable *in vivo* under basal and fully-induced conditions, all of these seemingly paradoxical results are indeed compatible with the model of continuous gene availability (see Fig. 6A and related discussion). The major argument against this possibility appears to be the observed (37) apparent *in vivo* stability of ornithine transaminase in previously uninduced cells to which both cycloheximide and inducer have been added. However, because all of these experimental results can be rationalized so easily, if these enzymes are continuously synthesized and unstable *in vivo*, it might prove rewarding to use another method (*i.e.*, radioimmunoprecipitin procedure) to determine whether or not enzyme turnover occurs during periods of constant enzyme activity under basal and fully-induced conditions in synchronous *Saccharomyces.* Although enzyme turnover *in vivo* may not be responsible for these seemingly paradoxical results, it should be throughly tested before being eliminated as a possibility.

Not many studies have been concerned with the availability of inducible genes for transcription during the cell cycle of higher eucaryotes. The most definitive studies have been those by Tomkins and coworkers (39) on the hormone-induced synthesis of tyrosine transaminase in mammalian cells. This enzyme is inducible at all phases in the cell cycle with the exception of G2, M, and early G1. Although their model of regulation proposes that this restriction in inducibility during G2, M, and early G1 is at the transcriptional level, a post-transcriptional mechanism limiting inducibility during these periods has not been entirely ruled out.

In conclusion, continuous inducibility (or derepressibility) during the cell cycle appears to be a general characteristic of many inducible enzymes in eucaryotic microorganisms, provided enzyme induction (or derepression) is performed on previously uninduced cells. A notable exception to continuous inducibility (derepressibility) in eucaryotic organisms is observed for isocitrate lyase in one strain of *C. pyrenoidosa.* Because of the limited number of cell cycle studies on inducible enzymes in higher eucaryotes, it is currently unclear as to whether or not restricted inducibility of tyrosine aminotransferase in G2, M, and early G1 will be a general characteristic of other inducible or derepressible enzymes in mammalian cells or in other higher eucaryotes. Moreover, the question of whether or not inducible genes are continuously available for transcription in synchronous budding yeast cultured in the continuous presence of inducer still remains an open question. Since enzyme turnover is such a general phenomenon in higher eucaryotes, one must be careful, in working with lower eucaryotes, not to assume that a constant enzyme level reflects the absence of enzyme synthesis.

The concept of continuous inducibility for enzymes involved in the utilization of exogenous substrates makes more biological sense than one of restricted inducibility. In nature, a eucaryotic microorganism would be at a decided competitive disadvantage if it had to grow to a particular stage of its cell cycle before an enzyme could be induced to allow effective utilization of an exogenous substrate. When restricted inducibility is observed, it is more likely that this phenomenon will occur over a short period rather than the major part of the cell cycle of eucaryotic microorganisms.

Most of the aforementioned studies have been concerned with the availability of genes which code for enzymes that are inducible by exogenous substrates. However, there is another class of inducible enzymes in eucaryotes whose regulation during the cell cycle has yet to be studied. These are enzymes which are inducible by endogenous compounds (*i.e.*, pathway intermediates). In *Neurospora* and *Saccharomyces*, for example, synthesis of the first enzyme on the pyrimidine pathway is regulated by repression while a number of subsequent enzymes on this pathway are sequentially induced by the product of the preceding enzyme (40-42). Are the structural genes of these inducible enzymes available for transcription throughout the cell cycle? Questions of this type indicate the continued need for research on inducible enzymes in higher and lower eucaryotes.

IV. Concepts and Experimental Approaches for Studies on Regulation of Biosynthetic Enzymes during the Eucaryotic Cell Cycle

Although considerable experimental evidence supports the model that synthesis of biosynthetic enzymes is regulated by oscillatory repression (11,43) during the cell cycle of procaryotes, there is a scarcity of evidence to support the oscillatory repression model or any other model for cell cycle regulation of synthesis of biosynthetic enzymes in eucaryotic cells. The lack of understanding of cell cycle controls of biosynthetic enzyme synthesis in eucaryotes stems from the greater number of possible options that must be considered in interpreting enzyme patterns and other cell cycle data from eucaryotes.

In higher eucaryotic plant (27) and animal (25,26) cells, and in at least some eucaryotic microorganisms (28), enzyme turnover plays an important role in regulating cellular enzyme levels. Because many enzymes are stable in procaryotes, a constant enzyme level within these cells is usually assumed to reflect the absence of enzyme synthesis. In eucaryotes, however, a constant enzyme level can often reflect a steady state between synthesis and breakdown of an unstable enzyme. Thus, periodic enzyme *accumulation* cannot be equated with periodic enzyme *synthesis* in eucaryotes.

In procaryotes, enzyme synthesis is regulated primarily at the transcriptional level; therefore, periodic enzyme synthesis during the cell cycle usually reflects periodic gene transcription. However, in eucaryotes, a number of post-transcriptional mechanisms can possibly obscure the timing relationship between gene transcription and enzyme synthesis or accumulation during the cell cycle. These post-transcriptional mechanisms include: (a) the conversion or processing of precursor mRNA (*i.e.*, heterogeneous nuclear RNA, HnRNA) into mRNA (44), (b) the transport of mRNA with poly-A segments and associated proteins from nucleus to cytoplasm (45), and (c) the factors regulating the half-life of the mRNA and its encoded enzyme(s). For example, a condition can be visualized in which gene transcription is continuous and one of the post-transcriptional processes (*e.g.*, HnRNA processing) is discontinuous during the eucaryotic cell cycle. A periodic enzyme pattern generated under this condition would be indistinguishable from one generated by periodic transcription with a continuous post-transcriptional process. A complete understanding of enzyme patterns from eucaryotes will require the use of procedures for measuring the rates of synthesis and processing of specific HnRNAs, transport and turnover of specific mRNAs, and the synthesis and breakdown of their encoded inducible enzymes.

Another degree of complexity added to the interpretation of biosynthetic enzyme patterns in eucaryotes is that these organisms can have both repressible and inducible enzymes on the same biosynthetic pathway (40-42). By contrast, in procaryotes inducible enzymes and repressible enzymes are usually on different pathways, catabolic and biosynthetic, respectively. On the pyrimidine biosynthetic pathway in a number of eucaryotes (40-42), the first enzyme is regulated by end-product repression and a number of successive enzymes are sequentially induced by the product of the previous enzyme in the pathway. In this case, although the pattern of accumulation and *in vivo activity* (*i.e.*, end-product inhibition must be considered) of the first enzyme would program the *synthesis* of the second enzyme, the pattern of *accumulation* of the second enzyme and successive enzymes would depend upon their respective *in vivo* stabilities. Thus, the study of cell cycle regulation of the synthesis of an enzyme within a eucaryotic biosynthetic pathway requires the simultaneous study of the levels and activities of the previous enzymes on the pathway.

Cellular compartmentalization in eucaryotes creates major problems in the design and interpretation of experiments in which exogenous compounds are used to attempt the repression, derepression, inhibition, *etc.*, of enzyme synthesis during the cell cycle. If the compound of interest is absorbed by the cell but does not have its desired effect, how does one know that the absorbed compound reached the desired intracellular site? Evidence is steadily accumulating which indicates that compounds synthesized within a eucaryote exist in separate pools from those which are absorbed from exogenous sources. Related to this problem is the one of acquiring the knowledge of what intracellular levels of possible corepressors, enzyme stabilizers, *etc.* exist at the site of the gene or the enzyme in a highly compartmentalized cell. Analysis of total cellular levels of compounds yields little useful information to support or refute a particular model of regulation, such as oscillatory repression, in a eucaryotic cell.

After considering some of the major problems associated with the design and interpretation of cell cycle experiments with eucaryotes, some of the different models which predict the nature of the control mechanisms which regulate the timing of accumulation of biosynthetic enzymes can be considered.

A restricted gene-transcription model (6,7) predicts that differences in the timing of accumulation of different biosynthetic enzymes results from a sequential transcription of the genome during the cell cycle. In this model specific genes are visualized as being transcribed only at given times in the cell cycle. Alternately, in view of recent findings, this model could be modified by replacing periodic transcription with periodic or sequential processing of different precursor mRNAs (*i.e.*, different HnRNAs) or periodic transport of specific mRNAs from the nucleus during the cell cycle.

The oscillatory repression model (11,43) predicts that genes are continuously available for transcription throughout the cell cycle. Periodic enzyme accumulation is visualized as resulting from the operation of feedback repression circuits in which the enzyme and corepressor (*e.g.*, pathway end product) form part of a closed feedback system in which the concentration of one determines the rate of synthesis of the other. As a consequence of these interactions, the cellular concentration of both enzyme and corepressor oscillate. Since inducible biosynthetic enzymes must also be considered in eucaryotes, the model might be more appropriately called the oscillatory repression-induction model.

What types of enzyme patterns might be anticipated for biosynthetic enzymes regulated by either of these proposed mechanisms? The restricted gene-transcription mechanism predicts that only during periods in which a gene is accessible for transcription would the typical Jacob-Monod type of regulation of gene activity be operative. If this is the situation, step patterns (Fig. 6D) are predicted for stable enzymes, during the cell cycle. Moreover, linear enzyme patterns would not be observed. A mutation resulting in constitutive enzyme synthesis would affect only the amount of enzyme synthesized during the period of gene accessibility and would not presumably result in enzyme synthesis outside of this period. Although it can be argued that a mechanism of this type can order the sequence of biochemical events required to program the cell's development through the cell cycle, it can also be argued that a cell's biosynthetic system cannot readily respond to changes in the external environment under such a mechanism of control.

If genes are continuously accessible for transcription, as proposed in the oscillatory repression-induction model, a large number of patterns of accumulation of biosynthetic enzymes are possible during the cell cycle. If the oscillatory repression plays a major role as in procaryotes, then changes in the growth rate of the cells, with probable changes in corepressor levels (relative to the gene dosage), would likely result in significant changes in the timing or pattern of specific enzyme accumulation. Inhibitors of transcription, post-transcription steps, and translation must be used along with the altered-timing approach to reveal at which biochemical level enzyme accumulation is controlled.

In conclusion, a number of altered-timing procedures can be used to study the availability of structural genes of biosynthetic enzymes for transcription at different times in the eucaryotic cell cycle: (a) attempting to derepress the synthesis of enzymes at different times in the cell cycle by addition of compounds which will inhibit the synthesis of cellular intermediates suspected to be corepressors; (b) attempting to derepress the synthesis of enzymes at different times in the cell cycle by removal, from the culture medium, of a required nutrient which is either the suspected corepressor or which leads to formation of the corepressor in a limited number of biosynthetic steps (auxotrophic mutants are particularly useful); (c) attempting to change the timing of synthesis of an enzyme to different times in the cell cycle by attempting to perturb the normal pattern of change in intracellular corepressor level by periodic addition and removal from the culture medium of compounds which are not normal nutrients but are corepressors or which lead to formation of corepressors; (d) attempting to change the timing of synthesis of enzymes to different periods of the cell cycle by changing the cultural conditions (temperature, energy or carbon source, light intensity, etc.) or synchrony procedure in an attempt to alter the normal fluctuation in corepressor levels during the cell cycle; and (e) attempting to isolate constitutive mutants whose structural genes are no longer controlled by fluctuating corepressor levels. Although this experimental approach is simple, it does offer a means for obtaining an initial insight into the mechanisms which regulate gene expression and enzyme levels during the cell cycle of eucaryotes.

V. Studies on the Regulation of Levels of Biosynthetic Enzymes during the Eucaryotic Cell Cycle

Biosynthetic enzymes are observed to accumulate in step patterns during the cell cycle of most eucaryotic microorganisms (10,46). The strain (7-11-05) of *C. pyrenoidosa* used in this laboratory appears to be an exception in that linear and exponential patterns are seen in addition to step patterns during the cell cycle of this organism (Fig. 7). Furthermore, step and linear patterns are exhibited by enzymes which are stable *in vivo* and step and exponential patterns are observed for unstable enzymes during the cell cycle.

The genes for both stable and unstable enzymes appear to be continuously available for transcription during the cell cycle of this strain of *Chlorella*. There is no evidence so far to indicate that transcription of genes of biosynthetic enzymes is restricted to certain phases of the cell cycle. By changing the culture conditions during synchronous growth, it has been possible to change

level of repression is constant (*i.e.*, corepressor or inducer not oscillating), a stable enzyme should accumulate in a linear manner with a change in linear accumulation occurring at the time of gene replication, whereas an unstable enzyme should accumulate in a step pattern with the step-increase also occurring in the S-phase. When an unstable enzyme is not accumulating under these conditions, a steady state will exist between synthesis and breakdown of the enzyme. Constitutive synthesis of stable and unstable biosynthetic enzymes will generate patterns identical to those predicted for conditions in which the level of repression is constant. When the level of repression (or induction) is oscillating, the patterns of both stable and unstable enzymes will approach step patterns, and these steps can occur at any time in the cell cycle.

In the oscillatory repression-induction model, the binding of corepressor (or inducer) with the aporepressor (or repressor) is assumed to be reversible as is the repressor-operator interaction. Therefore, the rate of transcription and thereby the rate of enzyme synthesis will approach zero asymptotically but will never become zero. Thus, when a corepressor is oscillating in a type of sine-wave manner, and the gene dosage remains constant, a stable enzyme will accumulate in a linear manner (at a low rate) between the major bursts of enzyme accumulation. However, the level of an unstable enzyme will remain constant between major bursts of enzyme accumulation. This relationship is seen for these enzymes because the rate of transcription and the rate of enzyme synthesis becomes essentially proportional to the gene dosage at high (or low) corepressor levels.

Although the "step" pattern is usually envisoned as the general type of enzyme pattern predicted by the oscillatory repression-induction model, other types of patterns are possible under this type of control mechanism. One must consider that, in addition to the activity of the repressible enzyme, the rate of increase in cell and organelle volume and the rate of utilization of corepressor for anabolic and catabolic processes also affect the cellular levels of corepressor and thereby the rate of enzyme synthesis. Consider the situation in which the rate of utilization of corepressor continued to exceed the rate at which its repressible enzyme was repressed at the beginning of the cell cycle, a continued exponential rate of decrease in the size of the corepressor pool could result in the exponential synthesis and the continuous accumulation of the enzyme.

What types of experimental approaches can be used to distinguish between restricted gene transcription and oscillatory repression-induction as mechanisms of control of biosynthetic enzymes during the eucaryotic cell cycle? One approach involves attempting to change the timing of synthesis or accumulation of biosynthetic enzymes to periods of the cell cycle in which they do not "normally" accumulate. For example, if periodic enzyme accumulation is regulated by a mechanism which restricts transcription of specific genes to certain periods of the cell cycle, then the timing of specific enzyme synthesis should not be altered significantly by changes in the growth rate of the cells, *i.e.*, specific enzyme synthesis should be initiated at the same fractional time of the cell cycle whether the cycle is short or long. However, if

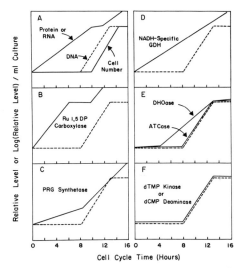

Fig. 7. A,B,C,D,E,F. Patterns for total protein, RNA, DNA, cell number and seven different biosynthetic enzymes during the cell cycle of *Chlorella pyrenoidosa* (strain 7-11-05) growing synchronously under constant environmental conditions. A. Log (relative level total protein, RNA, DNA, and cell number); B. Relative level ribulose 1,5-diphosphate carboxylase; C. Relative level phosphoribosylglycinamide synthetase; D. Log (relative level NADH-specific glutamate dehydrogenase); E. Relative level aspartate transcarbamylase and log (relative level dihydroorotase); F. Relative level dTMP kinase or dCMP deaminase; ---------, DNA level.

Corrected in Proof: The semi-step pattern described for the ATCase in the text has been corrected to the step pattern shown in Fig. 7E. (See Ref. 60 for related discussion.)

step patterns to continuous patterns and *vice versa*, indicating that gene transcription is responsive to the cell's environment. The genes of a number of biosynthetic enzymes appear to be either free from repression (*i.e.*, expressed constitutively) or under a constant level of repression so that the *in vivo* stability characteristics of the enzyme, the gene dosage, and fluctuations in the levels of endogenous stabilizer(s) regulate the pattern of accumulation of these enzymes.

Enzyme turnover appears to play an important role in regulating the levels of a number of biosynthetic enzymes in *Chlorella*. Eucaryotic microorganisms with long cell cycles are confronted with similar problems as the eucaryotic cell in tissues of higher organisms in which cell division occurs infrequently. Both cell types must be able to reduce the levels of certain enzymes in response to environmental change in the absence of frequent cell division. Therefore, it is likely that long-cycled eucaryotes, such as *Chlorella*, have evolved mechanisms

223

for turnover of many enzymes while short-cycled eucaryotic microorganisms, such as the budding and fission yeasts, have not been under selective pressures to evolve mechanisms other than cell division to reduce the levels of unwanted enzymes. However, although many enzymes in yeast are probably stable in exponentially growing cells, there are probably mechanisms for controlling the turnover of key regulatory enzymes during the normal cell cycle, and also mechanisms for degrading many proteins for cell maintenance during the stationary phase of non-growing cells.

The cell cycle of strain 7-11-05 of *C. pyrenoidosa* is characterized (47) by a long G1 phase, an S-phase near the end of the cell cycle, and a short G 2 phase (Fig. 7A). Total protein and RNA increase exponentially with the same exponential coefficient during the cell cycle (Fig. 7A). These macromolecular fractions cease to increase in level for a short time interval following the onset of DNA replication. The significance of this period of reduced protein and RNA accumulation is currently unknown and should be the focus of future research. This period of reduced accumulation is usually seen in synchronous cells selected by equilibrium centrifugation (30,31) from asynchronous cultures growing in continuous light; however, it is often not apparent in light-dark synchronized cells. Since many enzymes are synthesized and accumulate during this period, does this period reflect a reduced rate of synthesis of ribosomal RNA and structural proteins?

The first evidence that a structural gene of a biosynthetic enzyme is continuously available for transcription during the cell cycle of this organism came from studies (48,49) with ribulose-1,5-diphosphate carboxylase. When synchronous cells were cultured under a constant light intensity of 550 footcandles and the culture turbidity was held constant by continuous dilution, this chloroplast enzyme accumulated in a step pattern (Fig. 7B). However, under a constant light intensity of 1,100 footcandles, the carboxylase accumulated in a continuous manner closely paralleling the exponential increase in total cellular protein. When added at any time during the cell cycle, inhibitors of protein (*i.e.*, cycloheximide) and RNA (*i.e.*, actinomycin D or azaserine) synthesis blocked carboxylase accumulation and the enzyme activity decayed in a first order manner with a half-life of 4.4. hours. In synchronous cultures with a division number of 4 (Fig. 7B), inhibitor studies indicate that a steady state exists between carboxylase synthesis and breakdown during the period of constant activity between the sixth and ninth hours of the cell cycle. Thus, the dependence of both RNA and protein synthesis for the increase or maintenance of carboxylase levels throughout the cell cycle indicates that the carboxylase structural gene is transcribed continuously and its encoded enzyme synthesized continuously throughout the cell cycle. Rather than an on-off mechanism of control of gene transcription, the rate of transcription would appear to be regulated.

The different patterns of carboxylase accumulation are explainable by a model in which oscillatory feedback repression, by a photosynthetic product acting as a corepressor, regulates synthesis of the carboxylase during the cell cycle (48). The close similarity in the patterns of accumulation of total portein and the carboxylase at the high growth rate at 1,100 footcandles, suggests that the concentration of the corepressor does not reach repressive levels because its utilization for growth approaches the photosynthetic capacity of the cells. At the lower growth rate at 550 footcandles, the cessation in enzyme accumulation with continued accumulation in total cellular protein suggests that enough enzyme had accumulated by approximately 0.5 of the cell cycle to catalyze sufficient CO_2 fixation to support exponential growth and to repress net synthesis of the carboxylase for almost the remainder of the cell cycle. A reduction in the pools of photosynthetic products during nuclear and cellular division could account for the high photosynthetic rate observed (50) after cell division and in the complete derepression of carboxylase synthesis as indicated by the identical rates of accumulation of the enzyme during the first 0.5 of the cell cycle in cells growing at both low and high growth rates.

The initial step-increase in the carboxylase occurs outside the period of DNA replication. As stated in the previous section of this chapter, the step-increase in an unstable enzyme can occur outside the S-phase if the level of corepressor is oscillating and reaches minimum levels when DNA is not replicating.

Phosphoribosylglycinamide synthetase, the second enzyme on the purine biosynthetic pathway, was observed (51) to increase in a linear manner with a change in rate of linear accumulation occurring at the onset of the period of DNA replication (Fig. 7C). The increase in enzyme activity was dependent upon total protein and RNA synthesis. The activity of this enzyme remained constant during a 6-hour period in which total protein and RNA synthesis were inhibited in separate cell cycle experiments. These data indicate that the enzyme is stable *in vivo* under cell cycle conditions. Thus, the linear pattern of accumulation of this stable enzyme and its change in rate of linear accumulation early in the period of DNA replication provides evidence that the structural gene of this enzyme is transcribed continuously and is either free from repression or is under nearly a constant level of repression during the cell cycle. Moreover, this structural gene appears to be early replicating and to be transcribed shortly after its replication.

Since there is evidence that the structural genes of a number of enzymes within the pyrimidine biosynthetic pathway might be inducible rather than repressible, the pattern of this purine biosynthetic enzyme might reflect that of a fully-induced gene throughout the cell cycle.

Based on previously observed enzyme patterns in eucaryotic microorganisms, Mitchison (10) posed the question as to whether there is a tendency for biosynthetic enzymes and catabolic enzymes to follow periodic and continuous (linear) patterns, respectively, during the cell cycle. The observed linear pattern of accumulation of phosphoribosylglycinamide synthetase during the cell cycle of this organism indicates that biosynthetic enzymes can also exhibit continuous patterns in eucaryotes.

Recent research by Dunn and Schmidt (52) with dihydroorotase, the second

225

enzyme on the pyrimidine biosynthetic pathway, has revealed that this enzyme also is stable *in vivo* but accumulates in essentially a "step pattern" rather than in a linear pattern as observed for the stable phosphoribosylglycinamide synthetase. Cycloheximide inhibits the step-increase in activity observed at the fourth hour of the cell cycle. Moreover, the activity of the enzyme does not decay if this inhibitor of protein synthesis is added at either one or four hours into the cell cycle, indicating the *in vivo* stability of the enzyme. The enzyme accumulates slowly in a linear manner before and after the major step-increase in activity. This pattern is predicted for a stable enzyme whose synthesis is being controlled by oscillatory-repression or -induction (*i.e.*, endogenous inducer). When the endogenous corepressor level is high (or endogenous inducer level is low), a stable enzyme should accumulate at a low rate. A high rate of enzyme accumulation should be seen as the endogenous corepressor level decreases or the endogenous inducer increases. Since accumulation of dihydroorotase occurs during the period of accumulation of aspartate transcarbamylase, the product (carbamylaspartate) of aspartate transcarbamylase could be inducing the synthesis of dihydroorotase. The pattern of dihydroorotase synthesis should follow the increase in the *in vivo* activity of the first enzyme, not its level as measured *in vitro*. Aspartate transcarbamylase activity *in vivo* is regulated by feedback inhibition. Thus, the patterns of accumulation of the first and second enzymes need not be kinetically similar even if the first is regulating the synthesis of the second.

The maximal activity of *Chlorella* dihydroorotase is dependent upon an endogenous low molecular weight compound. Removal of this compound from crude extracts by dialysis, *etc.*, results in a 70-75% loss in enzyme activity. Activity can be restored by addition of NADPH or dithiothreitol. The identification and possible regulatory role of this compound is currently being studied.

In nitrate cultured cells, there is only a single form of glutamate dehydrogenase in *Chlorella* (5). The enzyme is five times more active on NADH than NADPH; therefore, it is referred to as being an NADH-specific glutamate dehydrogenase. When *Chlorella* cells are synchronized by intermittent illumination and then placed into continuous light, the activity of this enzyme remains constant for 4 hours (following the dark period) and then increases exponentially into the second cell cycle. The measurement of glutamate dehydrogenase activity, after various dilutions of homogenates from different times in the cell cycle, did not reveal the presence of inhibitors or activators of enzyme activity. The possibility that the change in concentration of modulators of enzyme activity are responsible for differences in the enzyme patterns at the beginning of the first and second cycles is thereby ruled out. The observed increase in enzyme activity at the beginning of the second cycle but its failure to increase during the first 4 hours following a dark period is more consistent with a model in which the structural gene of this enzyme is continuously available for transcription and is controlled by oscillatory repression rather than with model in which transcription of this gene is restricted to a discrete period of the cell cycle. Repression of glutamate dehydrogenase synthesis could presumably result from accumulation of corepressor during the dark period.

Because the enzyme accumulates exponentially throughout the second cycle, the pattern under constant environmental conditions is assumed to be a continuous exponential pattern (Fig. 7D). If interpretation of the second cycle pattern is correct, the pattern in the first cycle also should be continuous and exponential in synchronous cells selected by equilibrium centrifugation from continuously illuminated asynchronous cultures.

The pattern of accumulation of aspartate transcarbamylase, the first enzyme on the pyrimidine biosynthetic pathway, was observed (28) to be very similar to the step pattern for DNA except for a period of gradual enzyme accumulation immediately prior to the period of DNA accumulation (Fig. 7E). As stated earlier, if the structural gene of an unstable enzyme is transcribed continuously and is either free from repression or under a constant level of repression, and the unstable enzyme is synthesized continuously, the pattern of accumulation of the enzyme is predicted to be very similar to the DNA pattern.

If aspartate transcarbamylase is unstable *in vivo* and its structural gene regulated as stated above, the following relationships should exist: (a) a modified DNA pattern should give rise to a similarly modified aspartate transcarbamylase pattern, and (b) a steady state between enzyme synthesis and degradation should exist during periods of constant aspartate transcarbamylase activity. The step DNA pattern can be modified to a semi-continuous pattern by increasing the light intensity continuously during the cell cycle (28). When DNA accumulation was changed to a semi-continuous pattern, the enzyme accumulated in a similar manner, suggesting that a direct relationship might exist between gene replication and accumulation of the enzyme. Furthermore, when either cycloheximide or actinomycin D was added to synchronously growing cells during the first 4 hours of the cell cycle, a rapid decay in aspartate transcarbamylase activity was observed. This observation supports the model that a steady-state level of the enzyme exists during periods of constant activity.

The gradual increase in enzyme level prior to the step increase in DNA appears to be caused by the accumulation of an endogenous stabilizer which reduces the rate of enzyme degradation during this period of the cell cycle. The endogenous stabilizer is organic, low molecular weight (1000), heat stable 100°), acid soluble, Norit A absorbable, and non-cationic. The endogenous stabilizer rapidly increases in concentration between the fourth and tenth hours of the cell cycle and decreases thereafter. When cycloheximide is added to growing synchronous cells at the tenth hour of the cell cycle, the increase in enzyme level is blocked but there is no decay in enzyme activity as seen early in the cycle. Furthermore, the enzyme is stable in homogenates prepared from cells harvested at the tenth hour of the cell cycle. Removal of the endogenous stabilizer from these homogenates by either ultrafiltration, g e l filtration, ammonium sulfate precipitation, or sucrose density-gradient centrifugation results in the rapid first order decay of enzyme activity even at 0-3º. The apparent feedback inhibitor, UMP, and also uridine (a less effective inhibitor) are the only compounds, with properties consistent with those of the endogenous stabilizer, which are effective as stabilizers of the enzyme *in vitro*. UMP and uridine completely stabilize the enzyme (freed of endogenous stabilizer) at concentrations of 0.2 mM and 0.4 mM, respectively (53). The activity of the enzyme is not inhibited *in vitro* until the UMP concentration reaches 1 mM, and maximal inhibition is not reached until 10 mM. Because an endogenous inhibitor of aspartate transcarbamylase activity and total nucleotide-P increase at the same time in the cell cycle as the endogenous stabilizer, UMP with both its stabilizer and inhibitor properties is likely to be the endogenous stabilizer.

If UMP is the endogenous stabilizer, these data provide the first evidence that a feedback inhibitor might play an important role in regulating both the level and the activity of the first enzyme on a biosynthetic pathway during the eucaryotic cell cycle. Since *Chlorella* aspartate transcarbamylase can be stabilized by low concentrations of UMP which are not inhibitory, this nucleotide can act potentially as both a positive and negative regulator of pyrimidine nucleotide synthesis during the cell cycle. Low concentrations of UMP could lead to stabilization and accumulation of the unstable enzyme and accelerated UMP synthesis, whereas high concentrations of UMP could lead to feedback inhibition of enzyme activity and reduced UMP synthesis. One novel aspect of this regulatory mechanism is that an unstable enzyme is conserved, when its activity is inhibited, so that it can readily respond to the cell's needs once inhibition is relieved.

Recent studies have been aimed at determining the state of repression of the aspartate transcarbamylase gene. Attempts to derepress the synthesis of the enzyme with 6-azauracil during periods of constant enzyme activity during the cell cycle have not been successful. The compound blocks conversion of OMP to UMP, and leads to shrinkage of the pyrimidine nucleotide pool. In bacteria, low concentrations of 6-azauracil have been successful in derepressing the synthesis of aspartate transcarbamylase (54). However, when added to *Chlorella* cells during the period of constant aspartate transcarbamylase activity, this compound at low concentrations (1-3 μg/ml of culture) had no effect on the enzyme level whereas at higher concentrations (5-30 μg/ml of culture) it resulted in the decrease of the enzyme to a lower level which paralleled the enzyme pattern in the untreated culture. The enzyme decreased to a new level which was approximately proportional to the per cent inhibition of total RNA synthesis. These data are consistent with a model in which the synthesis of the enzyme is already fully-depressed (or free from repression) and any reduction in the rate of mRNA synthesis slows the rate of synthesis of the enzyme, resulting in its decay to a new steady-state level.

Attempts to repress the synthesis of *Chlorella* aspartate transcarbamylase also have been unsuccessful. Although at high concentrations radioactive-uracil

(29 mM) and -uridine (40 mM) were absorbed and incorporated into RNA, these compounds, in the culture medium for an entire cell cycle, did not repress the synthesis of the enzyme (52).

Tentatively, these derepression and repression studies are consistent with a model in which the enzyme is synthesized constitutively or is fully-derepressed and cannot be repressed by exogenous uracil or uridine.

All of these findings have been summarized into the following model (28) for the regulation of aspartate transcarbamylase levels during the cell cycle of *Chlorella:* the structural gene of the enzyme is transcribed continuously and is either expressed constitutively or is free from repression throughout the cell cycle. The enzyme is synthesized continuously and its rate of synthesis is proportional to its structural gene dosage. The enzyme is unstable, and decays in a first order manner in the absence of an endogenous stabilizer, presumably a nucleotide such as UMP. During the first 4 hours of the cell cycle, the stabilizer concentration does not reach levels sufficient to stabilize the enzyme. Thus, the constant rate of synthesis of the unstable enzyme results in a steady state between enzyme synthesis and breakdown and maintenance of a constant enzyme level. Between the fifth and eighth hours, the stabilizer reaches levels which retard enzyme breakdown and a slow net accumulation of the enzyme occurs. Because the rate of enzyme synthesis is proportional to the gene dosage, an increase in the rate of enzyme accumulation occurs during the period of DNA replication (8th to 12th or 13th hours). The stabilizer level decreases near the end of the cell cycle (10th hour), and a new steady-state level of the enzyme is reached which is proportional to the increase in gene dosage. If this model is correct, then the cellular concentration of aspartate transcarbamylase is controlled at both the transcriptional and post-transcriptional *(i.e.,* post-translational) levels during the cell cycle of *Chlorella.*

In early publications (55,56) from this laboratory, the patterns of accumulation of dCMP deaminase and dTMP kinase, two enzymes on the dTTP biosynthetic pathway, were compared to the DNA pattern during the cell cycle of light-dark synchronized cells of *Chlorella.* The levels of these enzymes and of the total DNA were expressed on a specific activity basis (*i.e.,* percentage of total cellular-P or cell volume). The patterns of these enzymes were essentially identical near the end of the cell cycle. Although their patterns were also similar to the DNA pattern, the peak in specific activity of DNA was displaced about 1 to 1.5 hours later in the cell cycle from the peak in specific activity of the enzymes. The DNA pattern reported in these early studies (57) has recently been shown (47) to be incorrect. Certain non-DNA substances which increased near the end of the cell cycle contributed chromophores in the procedure (57) used to analyze for DNA in earlier studies. When the patterns of these two enzymes were expressed on a per ml of culture basis and compared to a corrected DNA pattern expressed on the same basis, the enzyme and DNA patterns were essentially coincident during the S-phase (Fig. 7F). There was a slight increase in activity of these enzymes prior to the onset of DNA replication. Because the synchrony of these cells was not improved by equilibrium centrifugation after the light-dark synchronization procedure, the gradual increase in enzyme activity prior to the period of DNA replication is interpreted as being due to incomplete synchrony. Wanka and Poels (58) observed the levels of dTMP kinase and uridine kinase to remain constant and then to increase in parallel fashion in a step pattern during the period of DNA replication of another strain (211-8b) of *C. pyrenoidosa.*

Because of the relationship between the timing of the increase in dTMP kinase or dCMP deaminase and the increase in DNA reported earlier (55,56), it was suggested earlier that either of these two enzymes might be rate-limiting the synthesis of DNA in strain 7-11-05 of *C. pyrenoidosa.* However, since the patterns of accumulation of these enzymes are essentially the same as the corrected DNA pattern, it is more likely that DNA replication is regulating the levels of these enzymes. The close similarity in the enzyme and DNA patterns predicts that these enzymes might be unstable *in vivo* with their structural genes transcribed continuously (and are either free from repression or under a constant level of repression) and the enzymes synthesized continuously so that they increase from one steady-state level to another during the period of gene replication. This inference is supported by the observed (unpublished data) decrease in the levels of these enzymes when the rate of protein synthesis is slowed by placing the cells of strain 7-11-05 in the dark. Wanka and Poels (58) also have observed the level of dTMP kinase to decay in cells placed into darkness. If these three nucleotide biosynthetic enzymes are indeed unstable enzymes *in vivo* in both strains of *Chlorella* as tentative evidence suggests, then apparently their accumulation is regulated solely by the increase in gene dosage.

In conclusion, controlled changes in culture conditions which result in changes in the growth rate or in DNA patterns have proven to be an excellent tool for gaining insight into the mechanisms regulating gene expression and enzyme levels during the cell cycle of strain 7-11-05 of *C. pyrenoidosa.* The observed changes in the patterns of enzyme accumulation in response to environmental changes is not compatible with a model in which transcription of specific genes or synthesis of specific enzymes is restricted to discrete periods of the cell cycle. These changeable enzyme patterns are more consistent with a model in which synthesis of biosynthetic enzymes is regulated by oscillatory-repression or -induction, or by changes in gene dosage with post-transcriptional factors (*i.e.*, enzyme turnover, etc.) affecting the accumulation of enzymes not controlled by repression.

Because different synchrony procedures have not resulted in altered enzyme patterns in synchronous or synchronized budding yeasts (6) and a number of enzyme patterns are the same in the first and second cell cycles in continuous light following the light-dark synchronization of strain 211-8p of *C. pyrenoidosa* (59), there has been a tendency to discount oscillatory repression as a mechanism for control of biosynthetic enzymes in these organisms. However, it should be noted that the patterns of *Chlorella* aspartate transcarbamylase accumulation are identical in light-dark synchronized cells and in synchronous cells selected from asynchronous cultures growing in continuous light. The enzyme patterns are also identical in cells growing at low or high growth rates under a constant light itensity of 550 or 1,100 footcandles, respectively. However, when the light intensity is gradually increased throughout the cell cycle from either 550 or 1,100 footcandles, an altered aspartate transcarbamylase pattern is obtained. Different synchrony procedures do not alter the pattern of ribulose 1,5-diphosphate carboxylase; however, the pattern can be altered by culturing cells at different growth rates at a constant light intensity. These observations serve to illustrate that the regulatory systems of different enzymes are sensitive to different environmental changes. Thus, a negative response to a given environmental change does not necessarily mean that synthesis of an enzyme is restricted to a given period of the cell cycle. Thus, until step patterns are exhaustively analyzed under a wide variety of culture conditions and subjected to detailed biochemical and genetic analyses, it is premature to consider that either the expression of genes or the synthesis of biosynthetic enzymes is restricted to discrete periods of the cell cycle of eucaryotic microorganisms.

Acknowledgements

The author extends his gratitude to the former and present members of his laboratory group whose creative and productive research efforts made this chapter possible. Thanks are also extended to Drs. Daniel Israel, Herbert Jervis, Joe Saunders, Mrs. Judy Wilkins, and Mr. Larry Hull for critically reviewing the manuscript, and to Mrs. Margaret Walrath for her excellent job of typing and editing the manuscript.

The research discussed in this chapter has been supported by grants from the National Science Foundation (GB 17305) and the National Institutes of Health (GM 19871-01).

References

1. R. L. Perlman and I. Pastan, in *Current Topics in Cellular Regulation* (Horecker, B. L., and E. R. Stadtman, eds.), Vol. 5, Academic Press, New York, 1971, p. 117.
2. S. S. Cohen, *Am. Scientist 58*, 281 (1970).
3. S. S. Cohen, *Am. Scientist 61*, 437 (1973).
4. J. M. Mitchison and J. Creanor, *J. Cell Sci. 5*, 373 (1969).
5. D. J. Talley, L. H. White and R. R. Schmidt, *J. Biol. Chem. 247*, 7927 (1972).
6. P. Tauro and H. O. Halvorson, *J. Bacteriol. 92*, 652 (1966).
7. H. O. Halvorson, R. M. Bock, P. Tauro, R. Epstein and M. La Berge, in *Cell Synchrony — Studies in Biosynthetic Regulation* (Cameron, I. L. and G. M. Padilla, eds.), Academic Press, New York, 1966, p. 102.
8. P. Keumpel, M. Masters and A. B. Pardee, *Biochem. Biophys. Res.*

Commun. 18, 858 (1965).

9. M. Masters and A. B. Pardee, *Proc. Natl. Acad. Sci. 54*, 64 (1965).
10. J. M. Mitchison, *Science 165*, 657 (1969).
11. W. D. Donachie and M. Masters, in *The Cell Cycle: Gene-Enzyme Interactions* (G. M. Padilla, G. L. Whitson and I. L. Cameron, eds.), Academic Press, New York, 1969, p. 37.
12. G. Knutsen, *Biochim. Biophys. Acta 103*, 495 (1965).
13. J. H. Taylor, *J. Cell Physiol. 62* (Suppl. I), 73 (1963).
14. J. H. Taylor, in *International Review Cytology* (Bourne, G. H. and J. F. Danielli, eds.), in press.
15. J. H. Taylor, *J. Biophys. Biochem. Cytol. 7*, 455 (1960).
16. R. Braun, C. Mittemayer and H. P. Rusch, *Proc. Natl. Acad. Sci. U.S. 53*, 924 (1965).
17. G. C. Mueller and K. Kajawara, *Biochim. Biophys. Acta 114*, 108 (1966).
18. J. H. Taylor, T. L. Myers and H. L. Cunningham, *In Vitro 6*, 309 (1971).
19. C. Bostock and D. M. Prescott, *Exptl. Cell Res. 64*, 481 (1971).
20. J. Cairns, *J. Mol. Biol. 6*, 208 (1963).
21. R. Bird and K. G. Lark, *Quant. Biol. 33*, 799 (1968).
22. A. M. Tobia, C. L. Schildkraut and J. J. Malo, *J. Mol. Biol. 54*, 499 (1970).
23. C. H. Sissons, J. M. Mitchison and J. Creanor, *Exptl. Cell Res.*, in press.
24. F. S. Baechtel, H. A. Hopkins and R. R. Schmidt, *Biochim. Biophys. Acta 217*, 216 (1970).
25. R. T. Schmike, in *Current Topics in Cell Regulation* (Horecker, B. C. and B. R. Stadtman, eds.), Vol. I, Academic Press, New York, 1969, p. 77.
26. R. T. Schimke and D. Doyle, *Annu. Rev. Biochem. 39*, 939 (1970).
27. H. R. Zielke and P. Filner, *J. Biol. Chem. 246*, 1772 (1971).
28. A. A. Vassef, J. B. Flora, J. G. Weeks, B. S. Bibbs and R. R. Schmidt, *J. Biol. Chem. 248*, 1976 (1973).
29. P. J. Syrett, *J. Exp. Bot. 17*, 641 (1966).
30. T. O. Sitz, A. B. Kent, H. A Hopkins, and R. R. Schmidt, *Science 168*, 1231 (1970).
31. H. A. Hopkins, T. O. Sitz and R. R. Schmidt, *J. Cell Physiol. 76*, 231 (1970).
32. W. McCullough and P. C. L. John, *Biochim. Biophys. Acta 269*, 287 (1972).
33. J. M. Mitchison and J. Creanor, *Exptl. Cell Res. 69*, 244 (1971).
34. J. M. Mitchison and J. Creanor, *Exptl. Cell Res. 67*, 368 (1971).
35. H. O. Halvorson, J. Gorman, P. Tauro, M. LaBerge and R. Epstein, *Federation Proc. 23*, 1002 (1964).
36. P. Tauro, E. Schweizer, R. Epstein and H. O. Halvorson, in *The Cell Cycle: Gene-Enzyme Interactions* (Padilla, G. M., Whitson, G. L. and I. L. Cameron, eds.), Academic Press, New York, 1969, p. 101.
37. J. Sebastian, B. L. A. Carter and H. O. Halvorson, *Eur. J. Biochem. 37*, 516 (1973).
38. B. L. A. Carter and H. O. Halvorson, *Exptl. Cell Res. 76*, 152 (1973).

39. G. M. Tomkins, T. D. Gelehrter, D. Granner, D. Martin, Jr., H. H. Samuels, and E. B. Thompson, *Science 166*, 1474 (1969).
40. F. Lacroute, *J. Bacteriol. 95*, 824 (1968).
41. D. F. Caroline and R. H. Davis, *J. Bacteriol. 100*, 1378 (1969).
42. R. Jund and F. Lacroute, *J. Bacteriol. 109*, 196 (1972).
43. B. C. Goodwin, *Nature 209*, 479 (1966).
44. J. E. Darnell, W. R. Jelinek and G. R. Molloy, *Science 181*, 1215 (1973).
45. S-W. Kwan and G. Brawerman, *Proc. Natl. Acad. Sci. U.S. 69*, 3247 1972).
46. H. O. Halvorson, B. L. A. Carter, and P. Tauro, *Adv. Microb. Physiol. 6*, 47 (1970).
47. H. A. Hopkins, J. B. Flora and R. R. Schmidt, *Arch. Biochem. Biophys. 153*, 845 (1972).
48. G. R. Molloy and R. R. Schmidt, *Biochem. Biophys. Res. Commun. 40*, 1125 (1970).
49. T. O. Sitz, G. R. Molloy and R. R. Schmidt, *Biochim. Biophys. Acta 319*, 103 (1973).
50. C. Sorokin, *Physiol. Plantarum 10*, 659 (1957).
51. G. R. Molloy, T. O. Sitz and R. R. Schmidt, *J. Biol. Chem. 248*, 1970 (1973).
52. J. H. Dunn and R. R. Schmidt, *J. Biol. Chem.*, in preparation.
53. J. Wilkins and R. R. Schmidt, *J. Biol. Chem.*, in preparation.
54. M. Masters, P. L. Kuempel and A. B. Pardee, *Biochem. Biophys. Res. Commun. 15*, 38 (1964).
55. S. R-C. Shen and R. R. Schmidt, *Arch. Biochem. Biophys. 115*, 13 (1966).
56. R. Johnson and R. R. Schmidt, *Biochim. Biophys. Acta 129*, 140 (1966).
57. E. C. Herrmann and R. R. Schmidt, *Biochim. Biophys. Acta 95*, 63 (1965).
58. F. Wanka and C. L. M. Poels, *Eur. J. Biochem. 9*, 478 (1969).
59. P. C. L. John, W. McCullough, A. W. Atkinson, Jr., B. G. Forde, and B. E. S. Gunning, in *The Cell Cycle in Development and Differentiation* (Balls, M. and F. S. Billet, eds.), Cambridge University Press, 1973, p. 61.
60. Schmidt, R. R., in *Regulation and Significance of Protein Turnover* (Schimke, R. T., and Katunuma, N., eds.). Academic Press, New York, in press.

233

AN ANALYSIS OF CELL CYCLE CONTROLS IN TEMPERATURE SENSITIVE MUTANTS OF *CHLAMYDOMONAS REINHARDI*

Stephen H. Howell

Department of Biology, University of California San Diego
La Jolla, California 92037

I. Interference with the Normal Cell Cycle

It is difficult, if not impossible, to study the control of the cell cycle without means to upset that control. To demonstrate convincingly any mechanism of cell cycle control, one must be able to perturb the controlling mechanism(s) and show a subsequent effect on cell cycle events. Information of this type has been obtained by using selective inhibitory agents which interfere with the normal cell cycle. The kinds of inhibitory agents which have been used include antimetabolites, antibiotics, analogs of compounds required for growth, a variety of general chemical agents and physical perturbations (e.g., heat, radiation or pressure). Another technique for interfering agents which have been used include antimetabolites, antibiotics, analogs of compounds required for growth, a variety of general chemical agents and physical perturbations (e.g., heat, radiation or pressure). Another technique for interfering with the cell cycle involves starvation for a single growth requiring substance, such as an essential amino acid (1). An important alternative to the use of external agents or starvation is the use of temperature sensitive (*ts*) mutants. The advantage of this approach is that defects in *ts* mutants are products of lesions in gene controlled functions. Hence, one can study features of the genetic control of the cell cycle by using such mutants.

What information can be gained from the use of selective agents, starvation or conditional mutations to block the cell cycle? First, it is clear that any perturbation may have a ramifying effect on the cell cycle. The finding that the entire cell cycle can be interrupted by blocking a single cellular process or "cell cycle event" (i.e., process which occurs during a discrete interval of the cell cycle) indicates a strong interdependence of cell cycle events. But this is not always the case. Therefore, to understand the relationships of cell cycle events to each other, one can determine what consequences to subsequent cell cycle events result from blocking a prior one. Second, by interfering with the cell cycle one can learn a) what process, sensitive to the agent or altered by mutation, is required for cell cycle progress and b) when the sensitive process occurs in the cell cycle.

The effects of agents which interfere with the cell cycle are ordinarily studied in synchronous cell cultures. If the action of an agent is reversible or recoverable, then it is administered for short intervals ("pulsed") at different

235

cell cycle stages, and the subsequent mitotic delay is measured. An example of a recoverable action is the effect of low dose radiation resulting in radiation induced delay (e.g., 2). An agent with irreversible action can be studied in nearly the same way, except that its effect is often to block, and not simply delay, the subsequent division. A consistent finding in most synchronous cell systems is that there is a *transition point* before which an agent will affect division in the ongoing cell cycle and beyond which the agent will not, but will affect division in the subsequent cycle (see 3, for definition and review).

Transition points for agents which inhibit general metabolic processes (ATP generation, protein synthesis, etc.) are mostly grouped near the end of the cell cycle (see review, 4), because many of the processes sensitive to these agents are required nearly until the time of division. Toward the end of the cycle many general metabolic processes, including respiration and protein synthesis, decline. Because of the decline, this period of time in the cell cycle was considered to be singularly significant and was referred to as the *critical point* (5). However, as the time scale of cell cycle events became refined, specific *transition points* for various agents could be individually resolved (see, for example, 2). An important concept gained from recent studies of *transition points* in numerous cell systems is that various processes required for division terminate at different cell cycle stages and, therefore, these processes prepare the cell for subsequent division during different intervals of the cycle.

A counterpart to the use of selective inhibitors for interfering with the cell cycle has been the use of *ts* mutants. Hartwell, Culotti and Reid (6), working with budding yeast, introduced the notion that cell cycle events were gene-mediated functions, and the genes controlling such functions could be mutated. They isolated a series of *ts* mutants (called cell division cycle or *cds* mutants) that were defective in functions indispensable to the operation of the cell cycle. To describe the cell cycle stage when the defective gene product in a *cdc* mutant would normally function, they coined the term *execution point*. Operationally, it specified the cell cycle stage in a *cdc* mutant after which a shift to restrictive conditions would permit cells to progress through an additional cell cycle. The *execution point* was identified photomicroscopically by following the fate of individual mutant cells from an asynchronous culture after shift to restrictive temperature. The initial and final cell cycle stages of any budding yeast cell could be conveniently monitored by comparing the relative size of bud to parent cell. Hartwell's laboratory has currently isolated numerous *cdc* mutants belonging to separate complementation groups and has identified their execution points.

II. Cell Cycle Analysis in *Chlamydomonas reinhardi*

Chlamydomonas reinhardi is an ideal organism for studying the genetic control of the cell cycle. It is haploid during vegetative growth, and, therefore, experimentally induced mutations are immediately expressed. Such mutants are readily identified and isolated by standard replica-plating techniques.

236

C. reinhardi can be synchronized by "entrainment" to alternating light-dark cycles. Entrainment synchrony lies somewhere between selection and induction synchrony in terms of the degree to which a synchronization procedure disturbs a culture (see 3, for description of selection and induction synchrony). In illumination entrainment synchrony, light-dark cycles are adjusted to about a normal generation time in asynchronous culture, and cells are apparently "gated" at some cell cycle stage(s) to phasing in the light-dark program (7). After several illumination cycles, the cell cycle position of any cell within the culture requires minimal adjustment. Hence, the synchronization procedure should not be strictly categorized as induction synchrony, where by use of a blocking agent, cells are accumulated at a single cell cycle stage and then released.

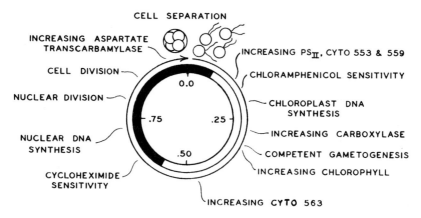

Fig. 1. Cell cycle clock for *C. reinhardi*. Cultures are synchronized by 12 hr light-12 hr dark illumination cycle as indicated by inner light-dark bar. Cell cycle progresses from zero time to one, where zero is time of cell separation, usually late in the dark period. Other cell cycle event markers are indicated. Marker events have been compiled from the studies of Armstrong *et al* (12), Chiang and Sueoka, (13), Howell and Walker (14), Kates and Jones (15), Mihara and Hase (9), Schmeisser *et al* (16) and Schor *et al* (17).

Synchronous *C. reinhardi* cultures clearly show that many biochemical and cytological events occur during discrete intervals of the cell cycle. Fig. 1 illustrates the positions of several marker events (only the estimated midpoints for these processes are shown). It should be pointed out that these marker events are well distributed throughout the cycle. Of course, there are many processes which are more or less continuous, such as cytoplasmic synthesis which has been reported to be nearly continuous throughout the cycle (8), with the possible exception of the cell division stage. Nonetheless, it is interesting that there is only a short cell cycle interval when protein synthesis inhibition causes mitotic delay (9). Therefore, for at least one apparently continuous process there may be only a short period during an ongoing cycle when this process is required for a subsequent division.

A. Cycle-blocked (cb) Mutants

To study the genetic control of the cell cycle in *C. reinhardi* we have isolated a series of *ts* mutants with defects which interfere with the cell cycle. These mutants grow normally at permissive temperature but are unable to complete a full cell cycle at restrictive (higher) temperature. Because these mutants are defective in functions required for normal cell cycle progress, they are called cycle-blocked (*cb*) mutants.

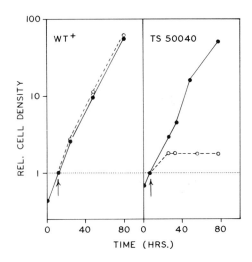

Fig. 2. Growth of non-mutant (wt⁺, strain 137 c mating type +) and *cb* mutant *ts* 50040 at 21⁰ C ●—● and after shift to 33⁰ C o---o. Cultures were shifted at the time indicated by the arrow (↑).

Cb mutants were selected for the inability of single cells to produce more than four daughter cells at restrictive temperature. (*C. reinhardi* normally produces four daughter cells per generation under our experimental conditions). That feature can be seen in the growth curves in Fig. 2. In asynchronous culture, nomutant (wt⁺) cells grow at the same rate at either permissive or restrictive temperature. Mutant cells (*ts* 50040 is shown as a typical example) at permissive temperature grow much the same as nonmutants. However, when shifted to restrictive temperature, mutant cells soon lose their ability to divide. After a shift to restrictive temperature, a mutant culture will continue to increase in cell number for a short period of time and then will stop growing. In no case will a *cb* mutant culture more than quadruple in cell density at restrictive temperature.

238

B. *The* Block Point

Operationally, the *block point** in a temperature sensitive *cb* mutant is defined as the last point in the cell cycle when a shift to restrictive temperature will *block* the subsequent cell division. It is interpreted to be the cell cycle stage when the defective (heat-labile) gene product would normally terminate its function. This interpretation assumes that in the *cb* mutant there is a specific cell cycle interval during which the defective gene product would normally function at permissive temperature, and, accordingly, this interval terminates at some unique point in the cell cycle. If a *cb* mutant is shifted to restrictive temperature during that interval, then the heat-labile gene product would be denatured and would not complete the function ultimately required for the next cell division. Such a cell would arrest in the ongoing cell cycle. On the other hand, if a *cb* mutant is shifted to restrictive temperature *after* that interval, the heat-labile gene product would still be denatured (if it is present), but its activity would no longer be required for the next cell division.

Ordinarily, one could determine the *block point* for a *cb* mutant by periodically shifting synchronous mutant cultures from permissive to restrictive temperature and measuring subsequent cell division. This procedure is similar to the techniques used to measure *transition points* in synchronous cell populations.

However, to simplify and extend the applicability of the *block point* measurement for other cell systems (including those which are difficult to synchronize), we developed a technique to determine the *block point* in asynchronous cultures. The technique was suggested by the observation that for any single *cb* mutant the amount of residual cell division in an asynchronous culture after shift to restrictive temperature was constant. Also,

---

*It is suggested that the term *transition point* be applied when inhibiting agents and deprivation of a substance required for growth are used to interfere with the cell cycle and *block point* be applied when conditionally defective gene products are used to interfere with the cycle. Either term describes the last point in the cell cycle when a subsequent cell cycle event (particularly cell division) can be prevented. *Transition point* denotes a physiological approach and *block point* a genetic approach in determining this point.

Execution point, coined by Hartwell, Culotti and Reid (6), originally described the cell cycle stage when the defective gene product in a *cdc* mutant would normally function in the nonmutant cell. This explanation of the term has recently been corrected by Hartwell (10) to be the cell cycle stage when the defective gene product in a *cdc* mutant would normally *complete*, not *perform*, its function. Because *execution point* now refers to an end point in the interval during which a defective gene product would normally function, the term is no longer suitable. *Block point* is recommended in lieu of *execution point*.

the amount of residual division in a culture differed among various *cb* mutants.

We used this easily measured parameter, residual cell division after temperature shift, to determine the *block points* for *cb* mutants. It was reasoned if the *block point* for a *cb* mutant was early in the cell cycle (Fig. 3A), then when an asynchronous culture is shifted to restrictive temperature, most of the cells (subpopulation I_2) would lie beyond the *block point*. These cells would undergo a single cell division before encountering the *block point*.

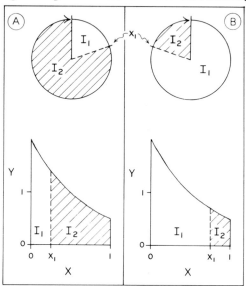

Fig. 3. Effects of early,A, and late,B, *block points* on the extent of cell division after temperature shift. Upper figures represent cell cycle clocks and lower ones are age distributions of *C. reinhardi* cells in the cell cycle. Age distribution profiles plot the relative number of cells, Y, at cell cycle stage, X (where X varies from 0 to 1). *Block points* (indicated at x_1) are shown both in the cell cycle clock and in their equivalent positions in the age distribution. When the *block point* (x_1) is early in the cell cycle, most cells (subpopulation I_2) are beyond the *block point* but before the time of cell division. When x_1 is late, few cells (I_2) are beyond the *block point*.

On the other hand, if the *block point* was at a late cell cycle stage (Fig. 3B), then most of the cells (subpopulation I_1) would lie before the *block point* when the culture is shifted to restrictive temperature. These cells would be prevented from undergoing division. Hence, one can see that the amount of residual cell division in an asynchronous *cb* mutant culture is related to the position of the *block point* in the cell cycle.

The relationship between *block point* position and residual cell division is somewhat complicated by the fact that cells in an asynchronous exponential culture are not distributed uniformly around the cell cycle (see lower graphs in Fig. 3). The age distribution is skewed such that there are more younger (newly divided) cells than older cells in a culture. Intuitively, this skewing arises because a single older cell gives rise to four daughter cells (in the case of *C. reinhardi*) at each division. In the age distribution profiles shown in Fig. 3, if we assume that x_1 is the position of the *block point* in the cell cycle (where x, varying from 0 to 1, is cell cycle stage), then the subpopulation (I_1) before the *block point* will not divide after shift to restrictive temperature. The subpopulation (I_2) beyond the *block point* will divide once. Therefore, when the culture reaches growth cessation the final cell population, measured as the relative increase in cell density after temperature shift (N/No) will be

$$N/No = I_1 + \alpha I_2 \tag{1}$$

where $I_1 + I_2 = 1$ and where α is the number of daughter cells produced during each cell cycle. ($\alpha = 4$ for *C. reinhardi*.) Since $I_1 = 1 - I_2$, then

$$N/No = 1 - I_2 (\alpha - 1), \tag{2}$$

$$\text{where } I_2 = \frac{\ln \alpha}{\alpha - 1} \int_{x_1}^{1} \alpha^{(1-x)} \, dx$$

The relationship of N/No to the cell cycle stage (x_1) of the *block point* is, therefore,

$$N/No = \chi^{(1-x_1)}. \tag{3}$$

Solving for x_1,

$$x_1 = 1 - \frac{\ln (N/No)}{\ln \alpha} \tag{4}$$

With equation 4 one can easily determine the *block point* for any asynchronous *cb* mutant culture. It might be added that this equation can also be used to measure *transition points* in asynchronous cultures for inhibitory agents which block the cell cycle.

While this technique obviates the need for synchronous cultures, such cultures can be used to corroborate calculated *block points*. Howell and Naliboff (11) have compared the calculated *block points* determined from asynchronous culture with those obtained by periodically shifting the temperature of cells in synchronous culture. In general, there is close correspondence between the two determinations. It was significant that in synchronous culture studies there is a sharp transition in the temperature sensitivity of the subsequent division at the *block point* on all *cb* mutants studied. This observation reinforced the supposition that the *block point* in any *cb* mutant is a discrete termination point for a specific cell cycle function. An alternate, but unsupported view is that the *block point* may represent some average cell cycle stage in a mutant population in which cells experience a deficiency in a slowly depleting substance required for cell division. The sharp transition at the *block point* in synchronous cultures argues against that view.

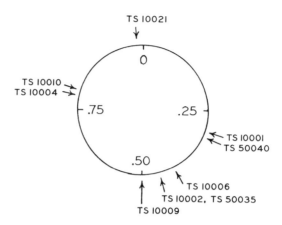

Fig. 4. Positions of *block points* for a sampling of *cb* mutants.

Block points for a sampling of *cb* mutants are shown in Fig. 4. It can be seen that *block points* are located throughout the cycle. However, here (and among other *cb* mutants studied so far) there is a pronounced clustering of *block points* in the second and fourth quarters of the cycle. This observation prompts speculation that there may be, at least, two major groups of processes required for cell division which terminate at about the same time. What these groups of processes may be is suggested by a similar clustering of identified cell cycle events and *transition points* as determined from inhibitor studies. Several processes relating to the synthesis of chloroplast components are located in the first half of the cycle, while those relating to general nuclear or cytoplasmic processes are found in the second half (Fig. 1.). Furthermore, *transition points* for inhibitors of organellar macromolecular synthesis (especially RNA and protein synthesis) lie in the second quarter of the cycle while those for inhibitors of general metabolic functions (ATP formation, cytoplasmic protein synthesis, nuclear DNA synthesis, etc.) are found in the fourth quarter (Howell and Drew, in preparation). Hence, one can speculate that in *C. reinhardi* many organellar processes utltimately required for cell division terminate in the second quarter while other general nuclear and cytoplasmic activities complete their functions in the fourth quarter. In cultures synchronized by illumination cycles, this interesting scheduling would mean that chloroplast replication activities would occupy the light phase while other activities are completed in the dark phase.

III. Interdependence of Cell Cycle Events

While the temperature-sensitive defects in *cb* mutants are described in terms of their consequence to cell division, their immediate effects may not necessarily be associated with division. The lesion in a *cb* mutant may affect a cell cycle event that is quite unrelated to the division process. The defect manifests itself in cell division, however, because the cell is arrested in the cycle prior to the divisional event. Since all cb mutants have been selected for an inability to perform cell division at restrictive temperature, it follows that in b mutants cell division is conditionally related to, though perhaps many steps removed from, the defective process. Therefore, cb mutants are a special class of mutants — a class in which the defective cell cycle event is related to the successful execution of cell division in the ongoing cell cycle.

When considering the effects of cb mutant lesions on other cell cycle events, there are two extreme views which can be taken which account for the interrelationships of cell cycle events with each other and with cell division. These views are metaphorically classified as the "domino" and "race-track" theories. In the "domino" theory, all cell cycle events are conditionally related to each other such that any single cell cycle event is dependent upon the prior expression of another event. The nonoccurrence of any cell cycle event would prevent the expression of any succeeding event. The "race-track" theory represents the opposite point of view. In this theory, cell cycle events are not conditionally related to each other — meaning that one cell cycle event does not depend on the prior expression of another. In the "race-track" theory, however, all events would mark time from (and depend upon) a common starting event. The blocking of one cell cycle event would have little effect on the others.

Probably neither of these theories gives an accurate picture of the encompassing control of all cell cycle events. Nonetheless, one of these views may correctly describe the interactions between a limited number of events. Certainly by a pairwise consideration of cell cycle events, it can be easily determined whether one event depends on the prior completion of another.

Consider three cell cycle events A,B and C which occur in sequence. Let A represent the expression of event A and $\sim$A its nonexpression. A convenient designation to describe the pairwise relationship between two events is B $\supset$ A, "if B then A," or B is *dependent* upon the prior expression of A. Alternatively, B $\supset (A_V \sim A)$ means "if B then A or not A," or B is *independent* of the prior expression of A. Assume in this instance that A is a temperature sensitive cell cycle event in a cb mutant. B and C are measurable cell cycle events such as chlorophyll and nuclear DNA synthesis. If one prevents the normal expression (or completion) of A by shifting the cb mutant culture from permissive to restrictive temperature, then one can measure the subsequent expression of B or C. If event B is expressed, then B is independent of the prior expression of A, i.e., B $\supset (A_V \sim A)$.

With two pairs of dependency measurements, B on A and C on A one can effectively describe a "conditional relationship pathway" (CRP) between the three events (see Table 1). For any two pair of conditional relationships there exist four general pathways. For example, if B $\supset$ A and C $\supset (A_V \sim A)$, then the CRP is → A →B. This means that A and B are on a single pathway where B

→C

depends directly or indirectly on prior expression of event A. C is on a separate pathway, immediately independent of A or B. Two general CRPs are nonunique (Table 1) in the sense that there exist two alternatives to describe them. The alternatives are denoted by dotted lines. For example if B ⇁A and C ⊃A, then →A→B→C. This means that C is either on a direct pathway with B, →A→ B→ C, or on a branched pathway →A–[B C. The alternatives can only be

distinguished by measuring the intermediate dependencies, i.e., C on B (as shown in Table 1). A unique CRP is possible in that case only if B can be specifically blocked while C is measured. By continuing this procedure one could ultimately construct an entire cell cycle map of CRPs. There are obviously preferred techniques for building such a map. Measurement of the dependence of many events on such a single one, e.g., D, E, F, G.......on A, leads to the construction of nonunique pathways. Obviously, it is advantageous to build small unique pathways and join them by common cell cycle events.

TABLE 1

Conditional Relationship Pathways of Cell Cycle Events

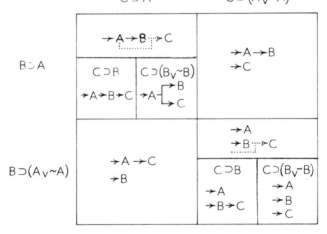

An extensive mapping procedure would require one to measure the expression of numerous cell cycle events in various *cb* mutants after shift to restrictive temperature. This procedure could be quite demanding, if it was necessary to shift various synchronous *cb* mutants cultures prior to their *block points* and periodically measure the expression of subsequent cell cycle events. The procedure can be simplified, as in the case of the block point measurement, by use of asynchronous cultures. The logic for this procedure is as follows: Cosider that B is the stepwise synthesis of a "cell cycle substance," an enzyme or subcellular component normally involved in the replication of the cell. Assume, also, that B is primarily regulated by the position of the cell

244

in the division cycle. In this case, the amount of "cell cycle substance" normally synthesized in a single cycle is proportional to the relative increase in daughter cells ( ) during any cell cycle. (This observation can be confirmed by measuring the amount of the cell cycle substance produced in synchronous cultures. For those events which have been previously studied, e.g., DNA synthesis, step enzymes, components of the photosynthetic apparatus, etc., this relationship is generally true.) Two points in the cell cycle of a *cb* mutant are significant (Fig. 5). Here, we will consider only the case where the first point, x_1, is the block point of the temperature sensitive *cb* mutant (event A) which is determined as previously described. The second point, x_2, is the cell cycle stage of B, which must be independently determined in synchronous culture. (For simplicity, the cell cycle stage of an event such as the synthesis of an enzyme or the replication of an organellar component will be considered to be the midpoint of its period of synthesis).

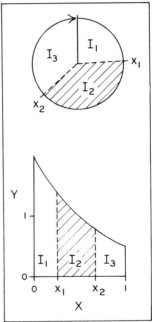

Fig. 5. Diagrammatic representation of the cell cycle clock (above) and age distribution of cells in the cell cycle (below) as described in Fig. 3. Positions of the *block point* for a *cb* mutant (x_1) and midpoint of cell cycle event B (x_2) are shown in both the cell cycle clock and age distribution.

If in asynchronous culture one measures B (the amount of cell cycle substance produced) after shift to restrictive temperature, then one of two situations should follow. First if B depends upon the prior expression of event A, (B ⊃A) in the same way as cell division depends upon A, then those cells in subpopulation I_2 (Fig. 5) will produce the cell cycle substance at x_2. Since subpopulations I_1 and I_3 are before x_1, the block point for the *cb* mutant, but beyond x_2, they will be prevented from expressing event B. In this case, the block point for cell division is the same as that for B. The final concentration of cell cycle substance produced in asynchronous culture, C, relative to the concentration at temperature shift, C_0, will be

$$C/C_o = I_1 + \alpha I_2 + I_3. \tag{5}$$

Since $I_1 + I_2 + I_3 = 1$, then

$$C/C_o = I_2 (\alpha - 1) + 1 \tag{6}$$

where $I_2 = \dfrac{\ln \alpha}{\alpha - 1} \displaystyle\int_{x_1}^{x_2} \alpha^{(1-x)} dx = \dfrac{1}{\alpha - 1}(\alpha^{(1-x_1)} - \alpha^{(1-x_2)})$

If, however, B is independent of A, $B \supset (A_v \sim A)$, then the value of C/C_o will be considerably higher. How much higher? It is impossible to predict. But let us assume, for example, the case where B is independent of A but ultimately dependent upon cell division. In this case subpopulation I_1 (Fig. 5) will produce the cell cycle substance at x_2, but because it is before the block point, it will not undergo division in the ongoing cell cycle. I_2 will also synthesize the cell cycle substance at x_2, but will undergo division and carry out another round of synthesis at x_2 again. Since subpopulation I_3 is beyond x_2, it will carry out division and produce the cell cycle substance during its only encounter with x_2. Hence the total increase in the cell cycle substance will be

$$C/C_o = \alpha I_1 + \alpha^2 I_2 + \alpha I_3 = \alpha[I_2 (\alpha - 1) + 1]. \tag{7}$$

As might be expected, it is rare when calculated and experimentally determined C/C_o values would exactly agree. There are several reasons for this. One is that cell cycle functions are not instantaneous events. They are, indeed, limited to a period in the cell cycle, but it is a simplification in this analysis to use the midpoint of an event as determined from synchronous cultures. Second, the cell cycle stage of a *block point* as it relates to cell division may be different as it relates to another event. This might be the case when only a portion of the period during which the gene product normally functions is required for the subsequent cell division. A different portion of that same period might be required for the subsequent expression of a different cell cycle event. Third, C/C_o values can be adversely by general degenerative effects of the high temperature treatment. Most of these effects, including cell lysis, etc., are easily distinguished by morphological observation or by careful scrutiny of the kinetics of C/C_o value increases.

An example of use of these measurements is found in the analysis of one of the *cb* mutants (ts 10009) as shown in Table 2. In these cells we investigated the effect of the mutant defect on four subsequent cell cycle events. The events were DNA synthesis (DNA), the stepwise synthesis of two autoregulated enzymes, aspartate transcarbamylase (ATC) and glutamate dehydrogenase (GD), and cell separation or division (CS). Table 2 shows that experimentally determined values for DNA and CS are very close to the calculated values for dependency, while those for ATC and GD are considerably higher. Therefore, DNA and CS depend on the prior expression of A, while ATC and GD are independent.

TABLE 2

Summary of C/C_0 Values for *cb* Mutants ts 10009

Cell Cycle Event	Cell Cycle Stage (x)	Calculated C/C_0 for Dependency on A	Experimental C/C_0	Conditional Relationship
A. Temperature sensitive cell cycle event (ts 10009)	0.50 (block point)	----	-----	-------
B. Nuclear DNA synthesis	0.75[1]	1.59	1.58	$B \supset A$
C. Aspartate transcarbamylase stepwise synthesis (ATC)	0.87[2]	1.80	8.60	$C \supset (A \lor \sim A)$
D. Glutamate dehydrogenase stepwise synthesis (GD)	0.92[2]	1.88	11.00	$D \supset (A \lor \sim A)$
E. Cell separation (CS)	1.00	2.00	1.94	$E \supset A$

[1]Howell and Walker (14)
[2]Kates and Jones (15)

247

Using the notation in Table 1 the CRPs between the five events can be written as follows

defective event
$$\longrightarrow \text{ts } 10009 \longrightarrow \text{DNA} \dashrightarrow \text{CS}$$

$$\longrightarrow \text{ATC} \dashrightarrow \text{GD}$$

Similar studies with other mutants and selective inhibitors are currently in progress in an effort to construct a cell cycle map of conditionally related events.

IV. Conclusions

It has been the intent of this paper to show that rigorous analysis of cell cycle control can be carried out in normal asynchronous cultures using mutants (or inhibitory agents) to interfere with the cell cycle. These experiments demonstrate that cycle blocked *cb* mutants are especially powerful tools to study such control. While the specific defects in *cb* mutants are yet unidentified, these mutants can still be used as means to interfere with or block the cell cycle. The blocking action allows one to determine the time when various defective gene controlled functions would normally operate in the cell cycle. It also permits one to examine what chain of events (CRP) is linked to the defective cell cycle event in the mutant cell.

Although the number of mutants and cell cycle events reported here is limited, the effort has been made to introduce a new technique for rapidly identifying *block points* (or transition points) and conditional relationship pathways among cell cycle events. General application of this technique to cell cultures other than *Chlamydomonas* will be invaluable in understanding cell cycle controls in a wide variety of cell systems.

Acknowledgement

This work was supported by a grant from the National Science Foundation.

References

1. R. A. Tobey and K. D. Ley, *J. Cell Biol. 46*, 151 (1970).
2. Y. Doida and S. Okada, *Radiat. Res. 38*, 513 (1969).
3. J. M. Mitchison "The Biology of the Cell Cycle" pp. 216-233. Cambridge Univ. Press, Cambridge (1971).
4. R. A. Tobey, D. F. Petersen and E. C. Anderson, in "The Cell Cycle and Cancer" (R. Baserga, ed.), pp. 309-353. Marcel Dekker, New York (1971).
5. M. M. Swann *Cancer Res. 18*, 727 (1958).
6. L. H. Hartwell, J. Culotti and B. Reid, *Proc. Natl. Acad. Sci. U.S. 66*, 352 (1970).
7. L. N. Edmunds and R. R. Funch, *Science 165*, 500 (1969).

8. R. F. Jones, J. R. Kates and S. J. Keller, *Biochim. Biophys. Acta 157*, 589 (1968).
9. S. Mihara and E. Hase, *Plant & Cell Physiol. 12*, 237 (1971).
10. L. H. Hartwell, *J. Mol. Biol. 59*, 183 (1971).
11. S. H. Howell and J. A. Naliboff, *J. Cell Biol. 57*, 760 (1973).
12. J. J. Armstrong, S. J. Surzycki, B. Moll and R. P. Levine, *Biochem. 10*, 692 (1971).
13. K.-S. Chiang and N. Sueoka, *Proc. Natl. Acad. Sci. U.S. 57*, 1506 (1967).
14. S. H. Howell and L. L. Walker, *Proc. Natl. Acad. Sci. U.S. 69*, 490 (1972).
15. J. R. Kates and R. F. Jones, *Biochim. Biophys. Acta 145*, 153 (1967).
16. E. T. Schmeisser, D. M. Baumgartel and S. H. Howell, *Dev. Biol. 31*, 31 (1973).
17. S. Schor, P. Siekevitz and G. E. Parade, *Proc. Natl. Acad. Sci. U.S. 66*, 174 (1970).

CONTROL OF PERIODIC ENZYME SYNTHESIS IN DIVIDING PLANT CELLS

P. A. Aitchison and M. M. Yeoman

Department of Botany, University of Edinburgh

Edinburgh, Scotland

I. Introduction

In a uniform, non-differentiating, exponentially growing culture in which the daughter cells start each cycle, after division, in the same condition as the parent cell did in the previous generation, there must be a doubling of all material in the course of one complete cycle. In such a situation the accumulation of new material may be continuous, for example linear or logarithmic with time through the cycle (1). This is frequently the case with RNA and total protein (2,3,4) and the synthesis of certain enzymes (5). Alternatively, production may be discontinuous, so that when the quality of a particular product is plotted against time in the cell cycle, a stepped or peaked pattern is revealed. Such periodic syntheses are a common feature of dividing cells (6) and present the investigator with a suitable system for the study of biosynthetic regulation. Unfortunately the investigation of such control systems in higher plants has so far proved difficult, because highly synchronous cell populations with the ideal characteristics described above are not available. Indeed, examples of synchronous populations of higher plant cells are very few, and such as do exist are far from ideal. Street and co-workers (7) have developed suspension cultures of sycamore cells which can show a sequence of synchronous divisions but these cultures have not yet been exploited to investigate the control of discontinuous syntheses. The Jerusalem artichoke system (8) shows several synchronous divisions but suffers from a number of disadvantages: a) some of the early events which occur in culture seem to be associated with a wound response and are not related to a normal cell cycle (9); b) the first cycle has an extended G_1-phase; c) after the first cycle, the average cell size decreases with successive divisions and therefore all the cellular material does not double from cycle to cycle (10). On the other hand, the system is an integrated multicellular aggregate which must bear a closer relationship to the organisational situation in the whole plant than do isolated cells in suspension culture, and is therefore more likely to be subject to the same control systems as operate in the whole plant. The system does, moreover, display discontinuous syntheses of several enzymes and, bearing in mind its disadvantages, it can be used effectively to determine the factors controlling the levels of activity of such enzymes during the cell cycle.

251

II. Control of Enzyme Activities During the Artichoke Cell Cycle

The best known example of discontinuous synthesis in a cell cycle is the period of DNA synthesis and it might be expected that the synthesis of enzymes necessary to complete DNA synthesis would precede the onset of the S period. The general pattern of change in activity of several of these enzymes is shown in Fig. 1, in relation to the onset of DNA synthesis as measured by incorporation of labelled thymidine. As can be seen, these enzymes do not show any marked increase in activity before the start of the S-phase. The start of the increase in their activities is coincident with the onset of DNA replication and most of the increase occurs during the S-period. Indeed, when enzyme activity is plotted against the increase in DNA a very close correlation is observed (11). Moreover, when DNA synthesis is blocked by fluorodeoxyuridine (FUdR), an inhibitor which blocks DNA synthesis selectively in this tissue, with very little effect on RNA or protein synthesis, the increase in activity of DNA polymerase is also prevented (Fig. 2). Data for thymidine kinase (TdR kinase) and thymidine monophosphate kinase (dTMP kinase) also show that FUdR prevents the increase in these enzymes (11). If it is assumed that the increased activity reflects an increased synthesis, then the increased rate of synthesis of these enzymes may be dependent on the new DNA.

This group of enzymes shows a similar pattern of synthesis at the same time as DNA synthesis in many other cells in culture (12), whereas the synthesis of most other stepped enzymes does not show such a correlation with the amount of DNA. The fact that the rate of synthesis (or more correctly the ratio of the rate of synthesis to the rate of degradation) only changes when the amount of DNA present changes, suggests that the rate of synthesis is directly controlled by the number of gene copies available, i.e., that the rate of transcription is being limited by availability of DNA and that subsequent steps are not rate-limiting and therefore do not constitute an effective control system. This is obviously something of a simplification, as an explanation in terms of a simple gene dosage effect would predict a doubling of levels of enzymes as cells progress from a 2C to a 4C state (i.e. doubling DNA content). In fact the observed increases, especially for DNA polymerase, do not represent a strict doubling in levels. It is nevertheless reasonable to suggest that regulation of activity by numbers of gene copies acts as a coarse control system, with transcriptional, translational or other mechanisms acting as a fine control.

It might be possible to test this hypothesis by determining the levels of such enzymes in cells of common origin of different ploidy and correlating activity with mean C-value. As a corollary, if the rate limiting control system is operated at the DNA level it might be predicted that the rate of synthesis could not be increased by altering other conditions in the cell. This would of course necessitate measuring the actual rate of synthesis of enzyme protein to ensure any effects were not due to a change in the rate of degradation. The majority of enzymes, however, do not show periods of change in levels that can be

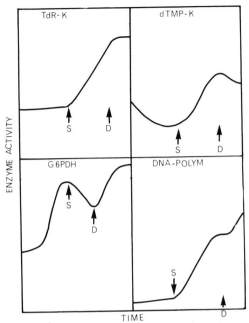

Fig. 1. The pattern of change of activities of Thymidine Kinase (TdR-K), Thymidine Monophosphate Kinase (dTMP-K), Glucose-6-Phosphate Dehydrogenase (G6PDH) and DNA Polymerase (DNA-Polym) during synchronous growth of artichoke explants. The times of onset of DNA synthesis (S) and cell division (D) are indicated.

correlated with changes in DNA. In the artichoke for example the stepwise increases in Glucose-6-Phosphate Dehydrogenase (G6PDH) and ATP-Glucokinase (ATPGK) occur during the G_1-phase (9). These increases, as in the case of TdR kinase and dTMP kinase, represent somewhat more than a doubling of activities. Therefore it is obvious that the control mechanism suggested above for 'DNA enzymes' cannot operate in determining the actual (as opposed to potential) levels of such G_1 enzymes.

It is possible to envisage several ways of controlling the effective levels of G_1 enzymes, (measured by assaying, enzyme activities), (13). These include three broad categories, a) activation and inhibition; b) control at translation; c) control at transcription. Examples of each of these may be found in the literature but usually involve differentiated tissues, and few studies have been directed to control systems operating during the cell cycle, especially in higher plants.

In *Helianthus tuberosus* tissue cultures, both G6PDH and ATPGK showed no significant increase in activity when cultures were grown in the presence of cycloheximide (CH) added at the outset. This was taken as evidence that the increase in levels was due to increased levels of enzyme protein (9). However, CH is rather potent in the artichoke system. It not only prevents amino-acid incorporation into proteins immediately after it is added, but also inhibits

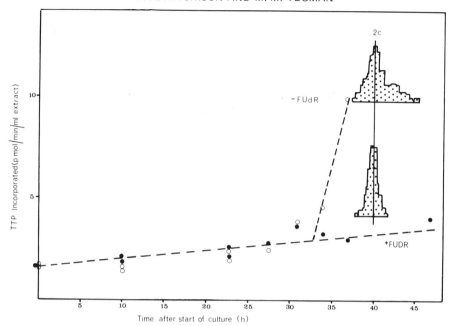

Fig. 2. Effect of FUdR on DNA polymerase activity. Histograms show the spread of DNA (Feulgen) per nucleus in populations of cells after 37h in culture with and without FUdR. From Harland, Jackson & Yeoman (11).

nucleoside incorporation into RNA (Fig. 3). It might be argued that its effect on levels of activity of these enzymes is not direct but through changes in amounts of activators or inhibitors. However, extracts prepared at different times during the cycle, when the apparent activities of these enzymes differed markedly, showed quantitative addition of activities when mixed after extraction or when tissues were homogenised together. Nor did extracts alter the activity of an added sample of purified G6PDH (14).

It cannot be excluded that the actual levels of enzyme are partly determined by changing rates of degradation, but given that enzyme synthesis is at least involved in regulation of the levels of activity through the cycle, how is it possible to differentiate between control at the level of transcription (with the translation element not limiting the rate of synthesis) or at the level of translation (with free availability of the necessary RNAs)? Clearly an agent was required that would selectively block RNA synthesis without affecting protein synthesis, and this proved somewhat difficult to obtain. Actinomycin D does not penetrate artichoke cells very efficiently, and 5-fluorouracil, (5FU), while inhibiting RNA synthesis did not do so completely. Eventually 6-methyl purine (6-MP) was chosen. 6-MP efficiently and rapidly blocks nucleoside incorporation into RNA (Fig. 3). It has no direct and immediate effect on amino-acid incorporation into protein although inhibition can be detected after a period of 1h. This effect, which is cumulative and reaches 95% by 16h, is probably largely an indirect one due to a progressive reduction in availability of all classes of RNA. This means that considerable care must be taken in interpreting the effects of 6-MP in terms of the process being affected.

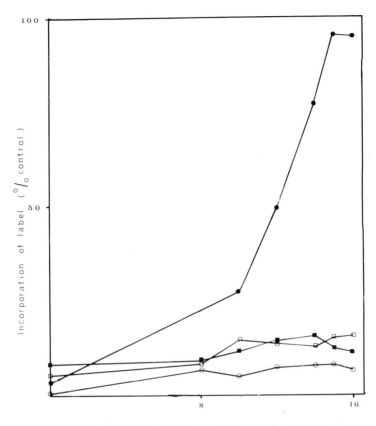

Fig. 3. The effect of CH and 6-MP on amino-acid incorporation into protein and uridine incorporation into RNA. 6-MP (2×10^{-5} M) or CH (2×10^{-5} M) was added at different times to growing cultures of artichoke explants. 16h after the start of culture, a sample of either [^{14}C]-protein hydrolysate or [^{3}H]-uridine was added. Incorporation in duplicate samples into protein or RNA was determined after a further hour. Incorporation of amino-acids in presence of CH (O) or 6-MP (●); incorporation of uridine in presence of CH (□) or 6-MP (■).

However, it was shown that addition of 6-MP to growing cultures had an almost immediate effect on the levels of G6PDH. If added during the period in which G6PDH activity was increasing, 6-MP had the effect of freezing the activity at that level (or only slightly higher). The effect was observed considerably before 6-MP could have had any effect on protein synthesis and it is therefore probable that the increase in activity is dependent on RNA synthesis. This suggests that RNA synthesis is necessary for regulation of the levels of G6PDH during 'G_1', though no conclusions can be drawn as to the class of RNA involved.

255

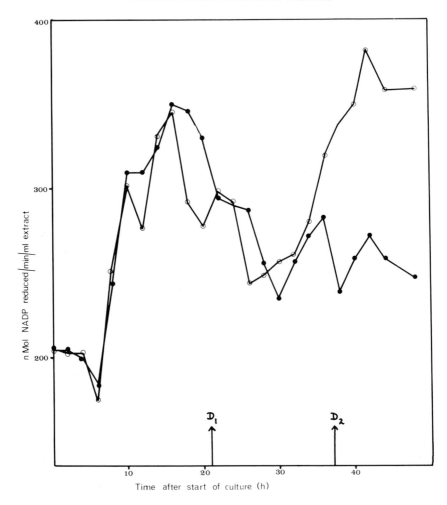

Fig. 4 The effect of FUdR on G6PDH activity. G6PDH was extracted and assayed as previously described (14) in cultures with (●) or without (○) FUdR. The times of onset of cell division are shown (D).

The suggestion that the synthesis of DNA does not affect the timing of increase of G6PDH (and possibly 'G₁' enzymes generally) was supported by the observation that FUdR added at the start of culture did not affect the increase in this enzyme in the first cycle (9). FUdR did, however, prevent any further increase in this enzyme in the second cycle (Fig. 4). This contrasts with the effect of FUdR on stepped enzyme increases in some other cells, for example in *B. subtilis* where successive increases in ornithine transcarbamylase continue after DNA synthesis is blocked by FUdR, and it was argued that the

256

TABLE 1

The effect of FUdR on increase of total RNA in growing
cultures of artichoke explants.

Time	RNA (units/explant)			
	control		+ FUdR	
	exp. 1	exp. 2	exp. 1	exp. 2
0	2.64	2.02	2.64	2.02
6		2.32		2.42
8	2.54		2.38	
12	2.98	2.63	2.90	2.57
16		2.91		2.79
20		3.21		3.02
24	3.96	3.70	3.66	3.20
28		4.35		3.50
29	5.60		4.20	

RNA was determined in perchloric acid extracts as described previously (25) in
batches of 5 explants from cultures with or without FUdR. In experiment 1
DNA synthesis started at 10½h and cell division at 21h; in the second at 15 and
27h respectively.

level of the enzyme was determined by a feedback system causing periodic
changes independent of the process of DNA replication (15). In the artichoke
it seems that some other factor is operating to restrict the level of G6PDH after
the initial increase in the presence of FUdR. Further synthesis may be
dependent on RNA normally synthesised in 'G$_2$' only after a round of DNA
synthesis and there is some preliminary evidence that while RNA synthesis
occurring in the early part of growth is not significantly reduced in the
presence of FUdR, that occurring in the period covering 'G$_2$' (in control cells)
is reduced in FUdR treated cells (Table 1). FUdR seems to act by fixing these
cells in a 'G$_1$' condition and makes them unable to proceed to the second
cycle, not only in respect of DNA synthesis and cell division, but also in
respect of RNA synthesis of 'G$_2$,' and enzyme synthesis normally associated
with the next 'G$_1$'.

III. Periodicity of Synthesis in Relation to the Cell Cycle

The periodicity of such discontinuous syntheses seems to be strictly
controlled in that the point in the cell cycle at which synthesis occurs is a
characteristic of the material used and is the same in succeeding generations of
a well-synchronized cell line. In some cultures for instance there is evidence
that the relative order of increase in activities (or extent of inducibility) of
different enzymes is always the same (16). However, the significance of

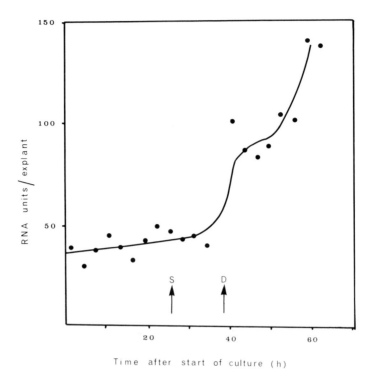

Fig. 5 Changes in transfer RNA during culture of artichoke explants. Explants were homogenised in tris buffer (30 mM, pH 7.6) containing sucrose (0.3 M), magnesium acetate (0.1 mM), KC1 (50 mM), dithiothreitol (5 mM), sodium napthalene 1,5-disulphonate (0.5%). Sodium tri-isopropylnapthalenesulphonate and 4-aminosulphonate were added to final concentrations of 1 and 4% respectively. The extract was deproteinized in phenol/cresol and the phenol layer after centrifugation (3000 r.p.m., 15 min) was discarded. The aqueous phase and interfacial precipitate were re-extracted and NaCl added to the aqueous phase (to 3%), which was again extracted with phenol/cresol. Nucleic acids were precipitated by mixing the aqueous phase with 2 volumes of ethanol and storing at -20° for 12h. RNA was fractionated on 2.2% polyacrylamide gels scanned in a Joyce-Loebl chromoscan. tRNA was estimated from gel peak areas. The time of onset of DNA synthesis (S) and cell division (D) are indicated. From Fraser (18).

periodic synthesis of particular cell constituents in the context of the whole cell cycle is not very well understood. Is it a question of ordered transcription correlated with position on the genome? The siting of a particular gene in a chromosome set is known to have marked phenotypic effects in some cases, but what significance this could have in a non-differentiated population of dividing cells is not known. It has been suggested that the ordered sequence of syntheses described above reflects an ordered progression of the cell through a series of metabolic states, each chronologically dependent on the preceding states and each enabling the cell to progress to the next, culminating in cell division, whereupon the whole sequence and pattern is repeated. In this view, which is easy to acquire when considering dedifferentiating material, cell division is seen as an event which deprogrammes the cell and allows it to start afresh (17).

However, this may not be the best way to regard these events, some of which occur very late in the cycle and it is difficult to see how they can be concerned with the cycle in which they occur. For instance, according to Fraser (18), there is a slow steady synthesis of tRNA during the initial stages of growth of cells of the Jerusalem artichoke tuber cultured with the hormone 2,4-D in the medium, but the greatest increase in tRNA occurs in the G_2-phase (Fig. 5). It is just possible that part of this synthesis represents the production of tRNA species coding for amino-acids particularly well-represented in mitotic (e.g. spindle) proteins. This group of proteins contains a much higher proportion of basic amino-acids than most cell proteins. If a large proportion of the protein synthesising machinery was concerned with production of a limited range of proteins for a restricted period, which may be the case for mitotic proteins, it is reasonable to expect there to be a necessary enrichment of particular tRNAs at the onset of this period. Resolution of this point needs detailed information on the turnover rates of tRNA species and whether the spectrum does change in such a manner preceding mitosis. However, Fraser (18) has shown that RNA synthesis occurring during the later part of the cell cycle does not seem to be concerned with the first division. When 5-FU (plus thymidine to eliminate a direct inhibition of DNA synthesis) was added during early culture, up to the beginning of the S-phase, no cell division occurred. When added at the start of the S phase, cell number increase reached the same level as in controls, but was delayed. If added in the middle of the S-phase or later, 5-FU + thymidine affected neither the extent nor timing of the first division, suggesting that cells are already competent, as far as RNA synthesis is concerned, to complete the first division by the middle of the S-phase. This suggests the possibility that the RNA synthesis observed after mid S-phase is not relevant to the first cycle, but to the next cell cycle. This is supported by the observation that 5-FU + thymidine added in the late S-phase, strongly inhibited the second division, more so indeed than when added at the end of the first cycle (Fig. 6). The dependence of increased synthesis of certain enzymes on new DNA, suggests that initiation of DNA synthesis does not depend on a previous increase in levels of these enzymes during 'G_1', and that the previously quiescent cells are already competent to initiate DNA synthesis. In turn, this suggests that the increased activities are more relevant to the daughter cells making them in their turn competent to initiate DNA synthesis without further increases in levels of these enzymes in 'G$_1$'; that is, the observed increases constitute a preparation for the second division.

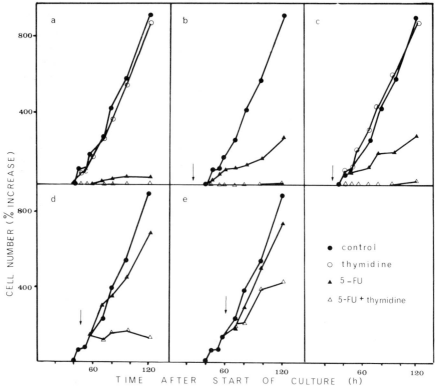

Fig. 6 Effect of 5-FU on cell division. From Fraser (18). Thymidine (final concn. 0.33 mM) 5-FU (0.1 mM) or both were added to cultures at the following times: a,Oh; b, 23h; c, 31h; d, 45h; e, 57h. Cell numbers were estimated using a modification of the method of Brown and Rickless (25).

One interesting question which may be posed is whether the newly synthesised DNA is immediately available for transcription or whether it only becomes active when separated from parental molecules, in daughter cells. One possible but unlikely interpretation of the effect of FUdR on the DNA related enzymes is that these are only synthesised on newly synthesised DNA strands. There is some preliminary evidence (Table 1) that FUdR present from the start of culture does cause an inhibition of RNA synthesis in the later stages of the first cycle, which could be interpreted as showing that this synthesis is partly dependent on the new DNA. It could also arise, however, if the progression through the stage of DNA synthesis was necessary for RNA synthesis normally associated with the post-S-phase and affected by a control system which limited its synthesis to this period.

It is not known how widespread this apparent direct dependence on DNA level may be in determining levels of enzyme activity. It is certainly curious that in so many tissues there is a close temporal relationship between increases in levels of this group of enzymes and DNA synthesis, though it is difficult to be sure that the same cause-effect relationship holds in all these cases as seems to operate in the artichoke. Nevertheless it seems that determination of levels of activity of enzymes, or quantities of any synthetic product, mainly by the C-level or by events consequent upon DNA synthesis, offers perhaps the simplest regulatory mechanism of ensuring that daughter cells in their turn are equipped to progress part of the way through the cell cycle. In this case the increased rate of synthesis of enzymes during the S-period seems to be irrelevant to the first cycle but is concerned with rendering the progeny capable of the initiation of DNA synthesis.

An alternative view can be put forward to explain this particular localisation of synthesis in some cells. There are many reports that synthesis of protein is necessary for subsequent DNA synthesis. Observations that inhibitors of protein synthesis have a greater effect when added at the onset, or just before the S-phase than when added later in the S-phase have led to the suggestion that certain proteins may be needed to initiate DNA replication, which then proceeds independently of further protein synthesis (19). In practice, initiation of DNA replication is not an unique event, but different parts of the total DNA complement seem to initiate replication at different times during the S-phase. Cummins and Rusch (20) have suggested that the S-period in *Physarum* may be divided into several sequentially dependent rounds of limited DNA replication and that protein synthesis is necessary for initiation of each new round but not for its continuation. This is based on the observation that about 25% of total DNA synthesis still occurs when CH is present at the start of the S-phase. It might seem at first that the enzymes associated with DNA synthesis could constitute a fraction of the protein necessary for initiation of DNA synthesis, in which case the restriction of increases of activity of these enzymes to the period of DNA synthesis would reflect the continuing need for such proteins for new initiations through the S-phase. In the case of unstable proteins it is necessary that they should be produced either continuously (which is less economical) or that the period of their synthesis should coincide with the immediate period of their use, and such a situation may exist in HeLa cells where the levels of TdR kinase and dTMP kinase are very low in 'G_1' and reach appreciable levels only during the S-phase (21), and in mouse L-cells where TdR kinase and deoxycytidine monophosphate kinase have short half-lives (2h) (22). However, where the enzymes are relatively stable, there is a less compelling reason for restriction of their synthesis to any particular point in the cell cycle. In the artichoke, for instance, levels of TdR kinase and dTMP kinase at least are detectable from the start of culture and do not drop markedly during 'G$_1$', suggesting that they are relatively stable, and this explanation of restricted synthesis becomes less plausible.

Similarly the effect of FUdR on G6PDH suggests that while cells are capable of undertaking an increased rate of synthesis during the first 'G$_1$' in culture, this potential is limited, and some event probably occurring in the first S-phase or 'G$_2$' is necessary before any further increase can occur in the next 'G$_1$'.

Another impression that a large proportion of the synthetic processes observed in a cell culture are concerned with providing material for a subsequent cycle, rather than for the immediate one, is gained from a comparison of the protein complement of cells which are going to divide with cells remaining in a 'resting' condition. Quiescent tuber parenchyma cells of the Jerusalem artichoke do not divide when cultured in a mineral salts and sucrose medium but remain in a 2C, 'G$_1$' state. If cultures are also supplied with a hormone, such as 2,4-D, the metabolic pattern changes, cells synthesize DNA and embark on a series of synchronous divisions. However, when the complement of soluble proteins is compared using polyacrylamide gel electrophoresis profiles, there are only minor differences to be seen up to the time at which cultures supplied with 2,4-D started to synthesize DNA (23). By this time such cells already seem committed to division, and it seems plausible that the subsequent differences reflect syntheses more relevant to the next cell cycle. It is suggested that cells in the non-dividing culture do not lack most of the synthetic machinery necessary to proceed with a division cycle, they merely lack the trigger. Such differences in protein synthesis at least as do occur in the early part of culture are not of sufficient magnitude to be detected by the present gel techniques.

Some cells seem to be able to 'plan' even further ahead. In cultured mouse cells, events which occur in one cycle may not have an effect until the next, or even the one after that (24). Certain cytostatic agents which normally cause arrest in 'G$_1$', if presented to synchronous cells in 'G$_1$' cause arrest not in that 'G$_1$', but in the next. If presented at any later stage, arrest occurs only after 2 periods of cell division, that is, cells once reaching G$_1$/S are committed not just to one, but to two cell divisions.

It would seem to us to be misleading always to consider the cell cycle from division to division. Much of the synthetic capacity of a cell is already available in G$_1$ without need for further specific syntheses. The events of each cycle combine not only to produce daughter cells at cell division, they ensure that the daughter cells start their lives with a considerable legacy.

Acknowledgements

The authors gratefully acknowledge a grant from the SRC in support of this research.

References

1. J. M. Mitchison, *J. Cell. Comp. Physiol. 62*, Suppl. 1, 1 (1963).

2. A. Zetterberg and D. Killander, *Exp. Cell Res. 39*, 22 (1965).
3. M. D. Enger and R. A. Tobey, *J. Cell Biol. 42*, 308 (1969).
4. D. M. Prescott, *J. Cell Biol. 31*, 1 (1966).
5. J. M. Mitchison, *Science, N.Y. 165*, 657 (1969).
6. J. M. Mitchison, The Biology of the Cell Cycle, Camb. Univ. Press. (1971).
7. P. J. King and H. E. Street, *In Plant Tissue and Cell Culture*, Ed. H.E. Street, Blackwell's Botanical Monographs, 269 (1973).
8. M. M. Yeoman and P. K. Evans, *Ann. Bot. 31*, 323, (1967).
9. M. M. Yeoman and P. A. Aitchison, *In* Cell Cycle in Development and Differentiation, *Symp. Br. Soc. Devl. Biol.* (1973).
10. M. M. Yeoman. *Int. Rev. Cytol. 29*, 383 (1970).
11. J. Harland, J. F. Jackson and M. M. Yeoman, *J. Cell Sci. 13*, 121 (1973).
12. P. Eker, *J. Biol. Chem. 240*, 2607 (1965).
13. J. E. Varner. In Control Mechanisms of Growth and Differentiation, *Symp. Soc. Exp. Biol. 25*, 197 (1971).
14. P. A. Aitchison and M. M. Yeoman. *J. Exp. Bot. 24* (In press) (1973).
15. M. Masters and W. D. Donachie, *Nature, Lond. 209*, 476 (1966).
16. M. Masters and A. B. Pardee, *Proc. Nat. Acad. Sci. U.S.A. 54*, 64 (1965).
17. M. M. Yeoman and P. A. Aitchison, *In Plant Tissue and Cell Culture*, Ed. H. N. Street. Blackwells Botanical Monographs, 240 (1973).
18. R. S. S. Fraser, Ph.D. Thesis. Univ. of Edinburgh, (1868).
19. J. H. Kim, A. S. Gelbhard and A. G. Perez, *Expl. Cell Res. 53*, 478 (1968).
20. J. E. Cummins and H. P. Rusch, *J. Cell Biol. 31*, 577 (1966).
21. T. P. Brent, J. A. V. Butler and A. R. Crathorn, *Nature, Lond. 207*, 176 (1965).
22. C. Mittermayer, R. Bosselmann and V. Bremerskov. *Eur. J. Biochem. 4*, 487 (1968).
23. P. G. Sealey. Unpublished observations.
24. B. Thomas. Personal communication.
25. M. M. Yeoman and J. P. Mitchell, *Ann. Bot. 34*, 799 (1970).
26. R. Brown and P. A. Rickless, *Proc. R. Soc. B. 136*, 110 (1949).

THE CHROMOSOME CYCLE IN THE CELL CYCLE

Daniel Mazia *

Department of Zoology
University of California, Berkeley, California 94720

I. Introduction

Before the term "Cell Cycle" came into general use, some of us liked to think of the idea as a "Life History of the Cell" (1). Very early on, only the distinction between the interphase and the mitotic period was evident. The mitotic period attracted the greatest interest—and still does for many—because it contains the dramatic climax of the history, visible to the microscopist. The interphase could once be described (with apologies) as a "resting phase," a time of peaceful growth between the more exciting chapters of the history. We began to appreciate that different things were happening at different times in interphase when we began to explore growth curves, but were only compelled to divide it into the phases we now use when it was first discovered that the replication of the chromosomes, certainly a crucial event of the mitotic cycle itself, takes place during interphase and that the timetable of the S period is quite regular for a given kind of cell.

We have been writing the life history of the cell in terms of historical phases that have served us well so far, but will not be surprised if further research leads us to modify our chronology. For one thing, we have long had intimations of a "decision" (2) that takes place some time after the end of mitosis, which determines whether the cell will enter a proper G1 that leads to S or will become a non-cycling cell in a state that many like to call G_0. For another, the techniques of genetics are now being applied to the study of the cell cycle and may well locate important turning-points that we do not perceive in our present analyses of the phases. Other authors in this book deal with both of these considerations (3,4).

In dealing with the chromosome cycle in the cell cycle, I have less to say about the chromosomes and mitotic apparatus themselves than about the events which switch the cycle from one phase to the next. We can think conservatively of "transitions," then ask whether these transitions involve distinct events of their own, switches or signals. We do not have to assume the existence of such switches or signals; the transition could be a smooth and automatic passage from one phase to the next without any special event which could be identified as the switch for the next phase. A transition we identify

*Work of the author's laboratory described in this paper was supported by Program-Project Grant GM-13882 from the National Institutes of Health.

could be a signpost, not a toll gate. But evidence for the existence of crucial decisive events has been accumulating.

Two epochal events in the chromosome cycle will be considered: the turning on of DNA replication which commits the cell to division and the initiation of chromosome condensation which is the sign that the cell is entering mitosis. In presenting my own recent work, I acknowledge the fact that it has been done on an atypical kind of cell cycle, that of the sea urchin egg. This cell offers us a stripped-down cycle; growth is negligible, the chromosome cycle proceeds rapidly and if there are major biosyntheses other than those of chromosomal constituents they are likely to be related directly to mitosis.

II. Turn-on of Chromosome Replication

We have had every reason to think of the turning-on of chromosome replication as a decisive — literally — event in the reproduction of the cell; a commitment to division long before division. It would not, logically, have to be a distinct event in the sense that there is an identifiable switch or signal; logically, there is nothing wrong with the idea that ingredients of DNA replication are building up during the G1 period and replication begins when the cell is equipped to proceed. On the other hand, we could have thought of the onset of S as a change-of-state of the whole cell from the fact that replication usually begins synchronously when there are several nuclei in the same cytoplasm but not when the nuclei are in separate cells (5). The most striking evidence for the existence of a distinct event came only with the technique of cell fusion. We now have the demonstration by Rao and Johnson (6) that a nucleus in G1 begins DNA replication when that cell is fused with a cell that is in S. Moreover, work on hybrids made between cell types having different cell-cycle schedules shows that the replication of DNA from different species in the hybrid *begins* simultaneously (7). After synthesis begins, the chromosomes replicate and terminate the replication according to their own programs. It is argued, then, that the switch for replication is something that pervades the whole cell and is not specific. It is a switch in the sense that it only *starts* the replication which, once started, follows the normal program for each set of chromosomes. We find in the literature many speculations on the events of the transition from G1 into S. The following discussion explores evidence that DNA synthesis is switched on by changes in the internal environment of the cell. This possibility, which takes into account the fact that the switch pervades the cell and is not specific for nuclei of different species, is quite in line with the resurgent interest in the regulatory role of ions, small molecules and activities of membranes. It is no more remarkable to propose that changes of this kind are switches in the cell cycle than to acknowledge that the regulation of free Ca ions controls the contraction and relaxation of muscle.

Let me speak of some of my own recent work which bears on the question. The material is the sea urchin egg, an extravagant example of the turning on of chromosome replication and cell division in a cell in which growth and biosynthesis of molecules other than chromosomal components has seemed to be a minor consideration. Everything seems to be there; the cell cycle is turned on by fertilization of the rather inert unfertilized egg. Our entry into the problem of immediate interest was through a bioelectric study of fertilization. That is interesting in itself, especially the evidence of an action potential at the moment of contact of egg with sperm (8), but I will not be speaking of fertilization as such. The problem of membrane biology which we attacked was the changes immediately after insemination, the development of a resting potential which we could show was the development of K-conductance. Studying that problem (9), we discovered that the polarization of the fertilized egg following its first interaction with a sperm cell could be induced in an unfertilized egg by exposing it to NH_4OH — the technique was merely to put the eggs in sea water brought to pH 9 with NH_4OH. We interpreted this action as a change in internal pH; it was not obtained, or only very slowly, in control experiments with NaOH, a strong base which does not penetrate the membrane as does NH_4OH. Let me stress that the egg remains unfertilized by the usual criteria, that this is not exactly a parthenogenetic activation. There is no cortical reaction and, as we will see, the egg can be fertilized later.

We have induced in an unfertilized egg some membrane changes which ordinarily follow fertilization. Chromosome replication normally begins shortly after fertilization. Could we have turned on chromosome replication in unfertilized eggs?

The answer (10) is that unfertilized eggs treated with NH_4OH do begin to make DNA. The continuous presence of NH_4OH is not necessary; in fact, it is better to transfer the eggs back to normal sea water after a period of exposure to NH_4OH-sea water; then we can see successive cycles of thymidine incorporation. The thymidine incorporation in these eggs is a true DNA replication which can be seen later as the doubling of the chromosomes.

A chromosome cycle started by fertilization is not exactly a typical cycle. In a growing cell, the limiting factors in the initiation of DNA synthesis could be any of the components of the synthesis itself which are performed in the sea urchin egg but have to be made by an ordinary cell. The present experiment is a demonstration that, given the ingredients for the synthesis, the switch can be a change in the intracellular environment. Here the effect of NH_4OH is interpreted as a change of intracellular pH which could itself be the switch or could flip a switch such as the release of another ion into the cell water from the outer environment or from intracellular lakes.

No one will mistake the simplicity of the agent (OH ion or, say, Ca ion) for a simplicity of the switching process. It is easy for me to dump some ammonia into the medium but an explanation of how the living cell will change its internal pH can not be simple and only brings to the foreground our ignorance of the intracellular environment and the complexities of membrane biology.

III. Initiation of Chromosome Condensation

I turn now to the other large event of the chromosome cycle in the cell cycle; the transition from interphase into the mitotic phases. Here is a major transformation of the whole cell, seen conspicuously as the condensation of the chromosomes, the disappearance of the nuclear membrane in the typical case and the formation of a mitotic apparatus. Shall we think of it as a smooth transition which follows directly from the completion of interphase events or shall we think of a distinct event, a switch or signal? Again, I think, the use of cell fusion has directed us to look for a distinct event. We have the remarkable experiments of Johnson and Rao (11), which tell us that chromosomes of a cell in any stage of interphase will be forced into condensation when the cell is fused with a cell in which the chromosomes are condensed. Again, the factor or condition which forces the condensation is something that pervades the cell and is not specific for a kind of cell.

I can add a little to that remarkable story by following up what I have said about the turn-on of DNA synthesis in the unfertilized sea urchin egg with NH_4OH. In these experiments, the cells are returned to sea water after the exposure to the NH_4OH-sea water.

The thymidine incorporation I have spoken of is not only DNA synthesis but chromosome replication. After a round of synthesis, the chromosomes condense and go through an interesting cycle (details are described in [12]). Remember that these are unfertilized eggs; they should not have active centrioles, and they do not form anything like a normal mitotic apparatus nor do they divide. A complete chromosome cycle is seen, but all the phases of chromosome condensation and decondensation are executed and some are even better displayed than in normal mitosis. Much observation remains to be done, but my present view is that the fully condensed chromosomes move to the periphery of the nuclear zone—the nuclear membrane does dissolve—then clearly split, move inward as they decondense, from nuclear vesicles and finally an interphase nucleus. The splitting of the chromosomes, the evidence that they have indeed replicated, is especially well-displayed. Moreover, the cycle is repeated and one ends up with single eggs with a great number of chromosomes.

This is an interesting story in itself, but it might seem to contradict what I said about a distinct event of chromosome condensation. The chromosome replicated and condensation followed; why postulate a signal for condensation? The system lends itself to a test of the hypothesis that a cell with condensed chromosomes will force other chromosomes within it to condense. I said before that these eggs in which the chromosome cycle was turned on by NH_4OH are unfertilized eggs; they can be fertilized later. Thus one can start the paternal chromosomes into the cycle some time behind the maternal chromosomes.

In the normal fertilized egg, both maternal and paternal chromosomes condense synchronously, about 90 minutes after fertilization. In unfertilized eggs whose DNA replication is turned on by NH_4OH, the maternal chromosomes condense in 90 minutes. If now, I start unfertilized eggs into DNA replication with NH_4OH and then — say, an hour later — I fertilize them, I have put paternal chromosomes into the cycle an hour behind the maternal chromosomes. Yet at 90 minutes both condense, the maternal chromosomes on time, the paternal chromosomes an hour prematurely. The prematurely condensed paternal chromosomes look thin and stringy and I do not yet know whether they accomplished their replication in the mere 30 minutes they had before they were forced to condense. Thus, in a quite different system, the observations of Johnson and Rao on the existence of a pervasive switch for chromosome condensation are confirmed. The experiments with the sea urchin egg can be done simply and on a large scale and the material may be helpful in the search for the factor which induces chromosome condensation.

IV. Initiation of the Formation of the Mitotic Apparatus

The experiment also permits us to deal with another question about the passage of the cell into mitosis. In the normal cycle, the condensation of the chromosomes is closely coordinated with the formation of the mitotic spindle. It is appealing to think that the same event evokes both of these changes. I have just described the premature condensation of male chromosomes when eggs are fertilized some time after the chromosome cycle has been started by exposure to NH_3. Now I add the observation that a mitotic apparatus is formed, prematurely with respect to the normal time required between fertilization and mitosis but at the right time if it is formed under the same conditions as determine chromosome condensation. (The fact that the mitotic apparatus forms at all, when none was formed as a result of the NH_4OH treatment, is easy to explain by the classical view that centrioles are introduced by the spermatozoon and are needed to form a mitotic apparatus. I confess that I have not yet studied the details of this mixed-up mitosis.)

It is worthwhile to speculate on the possibility that the initiation of chromosome condensation and the formation of the mitotic apparatus have a common cause. For one thing, the problem of the coordination of the chromosome cycle and the formation of the mitotic apparatus is a fundamental problem of mitosis; here would be a simple solution; that they are switched on by the same event. Moreover, while we know nothing at all about the conditions for chromosome condensation we begin to know something about the conditions for the formation of the mitotic apparatus. Recent work on the assembly of microtubules, which would account for spindle fibers if not for the rest of the mitotic apparatus, turns our attention again to the intracellular environment, in this case to the free Ca ions (13). Given a pool of molecular units (tubulin) which is indeed found in many cells and certainly in the sea urchin egg, one condition for the assembly of these molecules into microtubules is a lowering of the free Ca concentration. One can well imagine that the event leading to the formation of spindle fibers (and perhaps to

269

simultaneous chromosome condensation) is a sequestering of free Ca from the intracellular environment.

It so happens that just about the time these exciting discoveries about the assembly of microtubules were being made, our laboratory was studying a Ca-activated ATPase in relation to the mitotic cycle. We found the enzyme in the mitotic apparatus (14). A series of studies by Petzelt (15,16) describes a cycle of the activity of the enzyme that is closely related to the mitotic cycle, with a peak at the time of formation of the mitotic apparatus around metaphase and a minimum of activity around telophase when the mitotic apparatus is breaking down. In a great variety of experiments, Petzelt confirmed the close coordination between the visible mitotic cycle and the cycle of enzyme activity; I stress activity, there is as yet no basis for thinking of synthesis.

In thinking of the biochemistry of mitosis, we have been in the habit of borrowing ideas from the students of muscle contraction; indeed, that was the basis of our looking to Ca-activated ATPase in the first place. In this connection, such an activity can be related to myosin, but has another interpretation: that we are observing the enzymatic aspect of an internal Ca pump which controls the transport of Ca into internal membrane compartments. (Just as measurements on Na-K ATPase of our membranes tells us about Na and K transport.) If that were so, we would predict that the pump would be most active at the time of assembly of spindle fibers — the assembly depends on the lowering of Ca by sequestration of some kind. Conversely, the pump would be least active when the MTs were breaking down. Clearly the data conform to prediction, but even if we are wrong about the enzyme cycle — which is interesting in itself — we would think that the formation of the mitotic apparatus corresponded to a low free Ca level in the cytoplasm and that the absence of an MA in interphase represented a higher free Ca level.

Thus we are led to the hypothesis that a lowered level of free Ca is a condition for the condensation of chromosomes.

V. Conclusion

Let me summarize in a speculative way. The two main events of the chromosome cycle in the cell cycle are (1) the turning on of chromosome replication and (2) the transformation of an interphase cell into a mitotic cell, conspicuously the condensation of chromosomes and the development and action of the mitotic apparatus. There is evidence that both of these are initiated by distinct events; signals or switches. These events are changes that pervade the whole cell. They are not specific. One thinks, then, of a change in the internal environment of the cell; diffusible factors found in all cells; ions, small molecules, agencies that can be influenced by the action of membranes. The agents may be simple but the cellular mechanisms are correspondingly more complex; it is harder to explain in detail the change of an ionic concentration within a living cell than to uncover the synthesis of a hypothetical enzyme, for example. This deceptively simple interpretation of the problem is attractive mainly because it links the cell cycle to the external experience of the cell; to membranes; the action of environmental agents; cell contact; organismal factors, even the action of the nervous system in cases where it influences cell proliferation.

More specifically, the work I have discussed leads to the hypothesis that the chromosome cycle reflects a cycle of free Ca in the internal environment. We have an abundance of evidence of the special significance of external Ca for the growth of cells, at least mammalian cells (17). The exploration of internal Ca — or the whole ionic picture — in the cell cycle will be more difficult but will be made possible by new tools such as the Ca-ionophores and, it is to be hoped — Ca-sensitive electrodes. The unexpected proposition is that chromosomes condense under low-Ca conditions and decondense when the free Ca is higher. Experience with isolated chromatin shows that divalent cations cause clumping of chromatin and we have thought that they might cause condensation — but orderly condensation and clumping are not the same thing. One would not have supposed that chromosomes condensed under low-Ca conditions before it was discovered that the simltaneous formation of microtubules require low-Ca conditions.

If I am suggesting a new direction in the study of the cell cycle — a search for switches and signals as changes in the ions and small molecules of the internal environment of the cell — I am only speaking of something that is already happening, in the increasing volume of studies of environmental factors other than nutrients, the growing body of information obtained by intracellular electrodes and the general interest in membrane changes in the cell cycle.

References

1. D. Mazia, *American Scientist 44*,. 1 (1966).
2. H. Quastler and F. Sherman, *Exptl. Cell Res. 17*, 420 (1959).
3. J. A. Smith and L. Martin, *Proc. Nat. Acad. Sci. USA 70*, 1263, (1973); J. A. Smith and L. Martin, Chapter in this volume.
4. S. H. Howell, Chapter in this volume.
5. R. T. Johnson, *Biol. Rev. 46*, 97 (1971).
6. P. N. Rao and R. T. Johnson, *Nature 225*, 159, (1970).
7. J. Graves, *Exptl. Cell Res. 73*, 81 (1972).
8. R. A. Steinhardt, L. Lundin, and D. Mazia, *Proc. Nat. Acad. Sci. USA 68*, 2426 (1971).
9. R. A. Steinhardt and D. Mazia, *Nature 241*, 400 (1973).
10. D. Mazia and A. Ruby, *Exptl. Cell Res.*, In press (1974).
11. R. T. Johnson and P. N. Rao, *Nature 226*, 717 (1970).
12. D. Mazia, *Proc. Nat. Acad. Sci. USA*, In press (1974).
13. R. C. Weisenberg, *Science 177*, 1104 (1972).
14. J. F. Whitfield, J. P. McManus, and D. J. Gillan, *J. Cell. Physiol. 81*, 241

(1973).

15. D. Mazia, C. Petzelt, R. O. Williams, and I. Meza, *Exptl. Cell Res. 70*, 325 (1972).
16. C. Petzelt, *Exptl. Cell Res. 70*, 333 (1972).
17. C. Petzelt, *Exptl. Cell Res. 74*, 156 (1972).

DNA REPLICATION: INITIATION AND RATE OF CHAIN GROWTH IN MAMMALIAN CELLS

Roger Hand and Igor Tamm

Departments of Microbiology and Medicine
McGill University, Montreal, Que. and
The Rockefeller University, New York, N.Y.

I. Introduction

The original observations of Cairns (1) and Huberman and Riggs (2,3) showed that the long fibers of which eukaryotic chromosomal DNA is composed are made up of many tandemly joined sections in each of which DNA is replicated at a fork-like growing point. Huberman and Riggs (3) also showed that replication proceeded in opposite directions at adjacent growing points. This has been confirmed by several investigators (4-7). Huberman and Riggs (3) offered the definition of a replication section as a stretch of DNA replicated by a single growing point and proposed the term "replication unit" to mean a basic unit of control in the initiation of replication, presumably an adjacent pair of diverging replication sections. The size of such replication units was of the order of 60 μm or less and the rate of progression of individual replication forks on replication sections was of the order of 0.5 to 1.2 μm per minute. They also suggested that initiation on individual replication units occurred throughout the S phase of DNA replication.

The rate of DNA replication in the S phase mammalian cell is determined by two factors: frequency of initiation of synthesis of the new DNA chains and the rate of chain growth. Control of the processes of initiation and chain propagation is not well understood, although it seems clear that concurrent cellular protein synthesis is necessary to maintain the normal rate of DNA replication in both prokaryotic and eukaryotic cells (8,9).

The study of the rate of DNA initiation and chain propagation in mammalian cells can be approached in 2 ways: indirectly, by the use of equilibrium or velocity sedimentation methods (10-14) or directly, by use of DNA autoradiography (1-4,6,7,15-23). In this chapter, we will present some of our studies over the past several years on DNA replication in mammalian cells in tissue culture using the technique of DNA autoradiography, and consider some of the recent advances in the knowledge of DNA replication in mammalian cells.

II. Radioisotope Labelling Procedures

For most of our studies, monolayers of L-929 cells (a mouse fibroblast line) in logarithmic growth were labelled according to one of the other of two protocols. We added 2×10^{-6} M (4×10^{-7} M in some earlier experiments) 5-fluoro-2'-deoxyuridine (FdUrd) to cells 30 minutes before the onset of labelling with radioactive isotope in both protocols to exhaust the endogenous thymidine (dThd) pool. As a result, sharp high-density labelling was achieved as soon as high specific activity ^{3}HdThd (50 Ci/mmol, 5×10^{-6}M) was added for 30 minutes in protocol A or 10 minutes in protocol B (the hot pulse). The FdUrd was left in the medium during the labelling period to increase incorporation of exogenous ^{3}H-dThd. In earlier experiments, dThd solutions were prepared using ^{3}H-dThd at 50 µCi per ml and 22-26 Ci per mmole. In later experiments ^{3}H-dThd solutions at 250 µCi per ml were prepared using ^{3}H-dThd of specific activity 40-60 Ci per mmole. The hot pulse was followed by a warm pulse (5 Ci/mmol, 5.5×10^{-5} M) for 30 minutes (protocol A) or 2 hours (protocol B). At the conclusion of the labelling period the cells were washed with phosphate-buffered saline without Ca^{++} or Mg^{++} (PBS-def) and trypsinized to suspend them. We then processed the cells for DNA fibre autoradiography.

III. The Technique of DNA Fibre Autoradiography

DNA fibre autoradiography was first developed by Cairns (15,16). In these studies he demonstrated the circularity of the *Escherichia coli* chromosome and the fork-line nature of the growing points. Cairns (1) and Huberman and Riggs (2,3) applied the technique to the study of DNA replication in mammalian cells. In these studies, cells were pulse-labelled with ^{3}H-dThd. Suspensions prepared from these cells were diluted to a concentration of 500 cells per ml in a hypertonic buffer containing 1.0 M sucrose, 0.05 M NaCl, and 0.01 M ethylenediaminetetraacetate (EDTA). One ml of this cell suspension was placed in a short cylindrical glass chamber, the ends of which were formed by membrane filters. The cells were lysed by dialyzing the contents of the chamber against a buffer containing the same concentrations of sucrose, salt and EDTA plus 1% sodium dodecyl sulphate. The sucrose was then removed from the cell suspension by further dialysis against a buffer containing 0.05 M NaCl and 0.005 M EDTA.

After the chambers were removed from the dialysis bath, one filter on each chamber was punctured with a 26 guage needle and the fluid was allowed to drain slowly from it. The DNA fibres in solution were caught on the filters during drainage. The filters were dried in the air, cut from the glass chamber, mounted on slides and covered with AR10 stripping film. These autoradiographs were exposed for 2 to 6 months and, after development, the stripping film was removed and examined by light microscopy. We (21) and Callan (4) used this technique in some studies. Although it produced satisfactory results the technique had several drawbacks. It was difficult to find a waterproof adhesive that would keep the membrane filters attached to the glass chambers during prolonged dialysis. The glass chambers were not available commercially and had to be specially constructed. Some batches of membrane filters darkened the photographic emulsion during the exposure period, raising the background level of silver grains to a degree that the autoradiographs could not be interpreted. Lark *et al.* (18) introduced a significant modification of the technique that speeded up the production of the autoradiographic slides and eliminated the drawbacks discussed above. We have used this technique for all our DNA fibre autoradiographic experiments since 1971. At the completion of the labelling period with ^{3}H-dThd the cells are detached from the Petri dishes by trypsinization. They are collected by centrifugation, washed and suspended in PBS-def at a concentration of 40,000-60,000 cells per ml. Cells grown in suspension culture may be used equally well as cells grown in monolayer culture. A drop containing 2,000-3,000 cells is placed on a subbed glass slide near a drop of PBS-def containing 1% sodium dodecyl sulphate and 0.01 M EDTA. A clean glass rod is then touched simultaneously to both drops and the rod drawn gently over the surface of the slide. The material from the two drops is mixed by this maneuver and the DNA fibres from the lysing cells are spread and extended over the slide. Spreading may also be accomplished by touching the 2 drops with the edge of a second glass slide, allowing the drops to spread by surface tension along the edge of the slide and the moving the second slide down the surface of the first slide (24) in a manner used to prepare haematologic smear. Slides prepared in either manner are allowed to dry in air, washed 4 times in 5% trichloroacetic acid and dehydrated in graded alcohols. They are then coated with Kodak NTB-2 emulsion and exposed at -20°C for periods of 3 to 6 months. At the end of the exposure period the slides are warmed slowly to room teperature and developed in Kodak D-19 for 2 minutes, rinsed, fixed for 4 minutes and then placed in hypo clearing agent (Kodak) for 2 minutes before prolonged washing. We originally exposed our slides at a temperature of 4°C. After experiencing some difficulties with latent image fading (25) we changed our exposure temperature to -20°C. Latent image fading has been less of a problem at the lower temperature. Our present method still has the drawback of requiring a long exposure period. We have experimented with several of the methods designed to shorten autoradiographic exposure time (26,27). In our hands these techniques have not increased the density of autoradiographic grains (Hand and Tamm, unpublished data) during short term exposure over background.

IV. Scoring of Autoradiographs

When cells are labelled by a hot pulse followed by a warm pulse, we observed two types of autoradiograms. Those sites where replication begins after the beginning of the hot pulse are represented by linear arrays of heavy grain density, flanked on either sides by arrays of lighter density (post-pulse initiation figures); see Fig. 3c, d. When replication begins in a unit before the beginning of the hot pulse we observe a small clear area indicating DNA replicated before the start of the pulse, flanked by arrays of heavy grain density proceeding to lighter grain density arrays (pre-pulse initiation figures);

see. Fig. 3e, f. The grains represent two nascent daughter chains lying side by side.

For determinations of rate of DNA chain growth, pre-pulse initiation figures are scored, since these provide definite markers for the beginning and end of the hot pulse. The length of the autoradiogram produced during the hot pulse is measured and the rate of DNA chain growth calculated by dividing the length by the time of the hot pulse. The distance between initiation points in the mammalian cell lines that we use is large enough that hot pulses of 30 minutes are of short enough duration to allow measurement of rate of chain growth in individual replication sections.

The distance between the centres of adjacent replication units is measured to determine the interval between initiation sites. With DNA fibres labelled according to protocol A, we can only assume that two neighbouring replication units are on the same fibre, and the further apart two such units are, the less likely that they are adjacent. With protocol B, many replication sections show direct continuity with their neighbours as a result of extensive fusion of newly replicated chains during the two-hour warm pulse.

In all experiments, only internal autoradiograms (figures that are clearly within labelled DNA chains) are scored. DNA autoradiograms at the end of the long DNA chains could represent breaks in the DNA fibre.

V. Intervals Between Initiation Sites in Mammalian Cells

We determined the distances between DNA initiation sites in mouse L-929 cells in monolayers after labelling with ^{3}H-dThd according to either protocol A or B.

Figure 1 shows the patterns of silver grains obtained. With either protocol, densely labelled initiation sites are distributed at intervals along DNA fibres up to 500 μm in length. The pattern obtained with protocol, A, 30-minute hot and 30-minute warm pulses, shows clearly defined replication units with heavily labelled centre sections flanked by lightly labelled sections (Fig. 1 a). The pattern obtained with protocol B, i.e. a 10-minute hot followed by a 2-hour warm pulse, shows multiple densely labelled segments (Fig. 1b) occurring at intervals along stretches of lightly labelled DNA. Cells were labelled with warm ^{3}H-dThd after the hot pulse rather than before so that cell cycle kinetics would be altered minimally by exposure to large amounts of radioactivity before the beginning of the hot pulse.

Figure 2 compares the frequency distribution of the distances between initiation sites using the two labelling protocols. The modal interval of 40-50 μm is the same with both protocols. If the regions labelled during the 30-minute hot pulse (protocol A) are composed of multiple initiation sites that fused during the pulse, then the modal interval should be smaller with the 10-minute hot pulse (protocol B). Thus, it does not appear likely that initiation at the level of the replication unit occurs frequently at intervals shorter than those seen in Fig. 2. The distribution of distances derived from protocol B is broad and is skewed to the right with no evident periodicity. If initiation sites

276

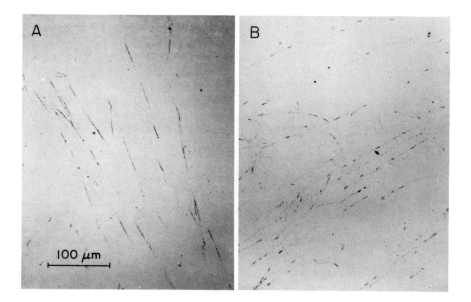

Fig. 1. Pattern of DNA replication in L cells. Cells were labelled with ³H-dThd according to (a) protocol A or (b) protocol B. From Hand and Tamm (23), with the permission of Academic Press, London.

occur at regular intervals along chromosomes, we would expect to see a sharper peak and periodicity at multiples of the modal distance. The lack of skew in histogram A is due to the difficulty in knowing whether widely spaced labelled regions represent adjacent replication units.

These observations show that functioning initiation sites occur in clusters (Fig. 1) and that most sites are spaced at irregular intervals of 20-70 μm (Fig. 2). Hori and Lark (19) have also suggested this based on similar observations in Chinese hamster cells using autoradiographic techniques.

Non-regular size of replication units has also been suggested by Callan (4) based on his interpretation of the data of Huberman and Riggs (3), and Callan's observations that the number of initiation sites on replication units may vary within eukaryotic cells. In *Triturus*, the premeiotic S phase in spermatocytes is much longer than the S phase in somatic cells, and the latter is longer than the S phase in embryonic cells. Callan suggests that the changes in the length of S phase in the cells are caused by variation in size of the replication units, and that initiation need not be at fixed regular intervals along the chromosome.

The mean distance between initiation points in protocol A is 45.4 μm, and with protocol B, 62.8 μm. Since the mean approaches the mode in protocol A,

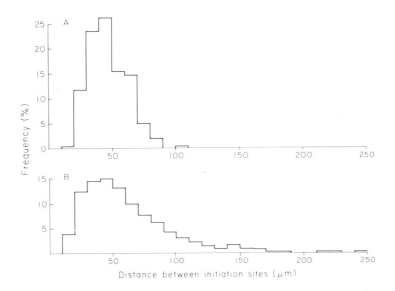

Fig. 2. Distances between initiation sites in L cells. (a) Cells labelled according to protocol A. (b) Cells labelled according to protocol B. From Hand and Tamm (23) with the permission of Academic Press, London.

we used it to compare the distance between initiation sites in several different types of mammalian cells (Table 1). There is more than a two-fold variation in the mean distance between initiation sites in different cell lines.

VI. Intervals Between Initiation Sites in Cells Treated with FdUrd

We wished to determine whether FdUrd had any effect on initiation in L cells under the conditions of our experiments. Cells were labelled according to protocol B with or without FdUrd. FdUrd increases the incorporation of ^{3}H-dThd into DNA and the FdUrd-treated autoradiograms have a heavier grain density. We exposed some of the autoradiograms without FdUrd for 6 months and measured the initiation site intervals and compared these with the intervals in cells that were treated with FdUrd. There is no significant difference in the distance between initiation sites in cells treated or not treated with FdUrd. Since heavier labelling is obtained after using FdUrd, we have used this compound in all experiments.

TABLE 1[+]

INTERVALS BETWEEN INITIATION SITES

Cells	Interval, μm* ($\bar{x} \pm$ s.e.)
L-929, Mouse	45. 4 ± 1.19
BHK, Hamster	30.1 ± 1.03
MDBK, Bovine	17.3 ± 0.77
CV-1, Monkey	42.2 ± 4.61
HEK, Human	22.7 ± 0.60

*The mean distance between the centers of adjacent internal autoradiograms was determined. The cells were labelled according to protocol A. More than 100 intervals were measured for each cell type (except for CV-1 cells in which, 73 intervals were measured) and the measurements taken from experiments performed on at least 2 different occasions.
+Reproduced from Hand and Tamm (23) with the permission of Academic Press, London.

VII. Synchrony of Initiation Events

On each slide, we observed some DNA autoradiograms where cell lysis and DNA spreading was incomplete (Fig. 3b). We also observed fan-shaped arrangements of the DNA fibres (Fig. 3a). These observations suggest that the DNA over a localized area of a slide may come from topographically contiguous areas of DNA. Autoradiograms from individual replication units are either pre-pulse initiation or post-pulse initiation figures. We observed in any one microscopic field (an area approximately 0.16 mm^2) with well-spread DNA fibres, the autoradiographic figures are usually pre-pulse or post-pulse, but uncommonly a mixture of the two (Fig. 3c-f). This was quantitated by determining the ratio of pre-pulse to post-pulse initiation figures in individual microscopic fields in cells labelled according to protocol A. Most fields (71.5%) showed a preponderance of one figure or the other (ratios of pre-pulse to post-pulse figures of greater than 7:3 or less than 3:7). This suggests that initiation in localized regions of DNA occurs in a synchronous manner. Initiation does not occur in absolute synchrony since the size of the gap in pre-pulse initiation figures and the size of the heavily labelled segments in post-pulse initiation figures varies. In addition, occasionally a replication unit with a pre-pulse initiation figure is adjacent to one with a post-pulse initiation figure. However, on the basis of our observations of a partial synchrony of initiation in DNA visualized within single microscopic fields and the clustering

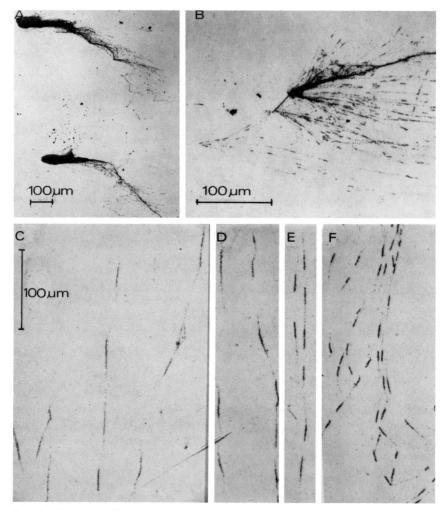

Fig. 3. Representative autoradiograms suggesting the topographical contiguity of neighbouring DNA strands and partial synchrony of initiation events. The mangification is the same in c-f. (a) Labelled DNA originating from two discrete nuclei where spreading of the DNA from the lysed nuclei was incomplete. (b) Labelled DNA fibres showing fan-shaped spreads suggesting that neighbouring DNA fibres were closely associated spatially *in vivo*. (c,d) Examples of localized areas of a slide in which all initiation figures were of post-pulse variety. Cells labelled according to protocol A. (e,f) Examples of autoradiograms from localized areas of a slide in which all initiation figures were of the pre-pulse variety. Cells labelled according to protocol A. From Hand and Tamm (23) with the permission of Academic Press, London.

of initiation events within segments of DNA, it seems reasonable to suggest that initiation events occur in bursts over topographically localized areas of template DNA within the nucleus. It is also reasonable to postulate that these areas of DNA might be derived from limited regions of single chromosomes. Autoradiographic analyses of pulse labelled metaphase cells have shown that DNA replication occurs at different times in different regions of individual chromosomes, with each chromosome showing a characteristic labelling pattern (28-32). These findings established that at any time in S phase, particular chromosomal regions are active in DNA replication and they may correlate with our findings of clustering and synchrony of initiation events.

Remington and Klevecz (33) and Klevecz and Kapp (34) have suggested that S phase DNA synthesis occurs in large bursts based on their data showing several peaks of dThd incorporation throughout the S phase of synchronized cells.

If initiation sites occur at intervals shown in Fig. 2 (mean interval 62.8 μm), then there are more than 16,000 such sites in a mammalian cell with a DNA fibre length of approximately 1 metre. This estimate is similar to previous estimates obtained by sedimentation techniques (11). Estimates based only on the rate of chain growth are much lower than this (1). Thus, frequency of initiation rather than rate of chain growth seems to be the main factor that determines the rate of DNA replication and the duration of S phase. This type of control mechanism is compatible with chromosomal initiation events occurring in bursts as discussed above.

VIII. Rate of DNA Chain Growth in Mammalian Tissue Culture Cells

Cells were labelled according to protocol A but with ^{3}H-dThd levels of 50 μCi per ml (see pulse labelling techniques). We measured the length of hot pulse autoradiograms in 30-minute pre-pulse initiation figures (Table 2). The rate of chain growth varied from 0.6 to 0.8 μm per minute in several different experiments. Two such determinations are shown in Table 2. The most reproducible estimates of rate of chain growth are obtained from pre-pulse autoradiograms. A frequency distribution of the rates in various replication sections is shown in Fig. 4. The lengths of DNA produced during the shorter pulse (10 min.) were not proportionally reduced and estimates of rate are therefore 25% higher (Table 2, Fig. 4). These shorter autoradiograms are more difficult to measure accurately, and this may account for the difference, although we cannot rule out a slowing of the rate during the later part of the hot pulse as an explanation.

By measuring the lengths of segments produced during the warm pulse, we were also able to estimate the rate of DNA chain growth using post-pulse autoradiograms. In these particular experiments, the length of the warm pulse was varied from 10 to 20 minutes. Estimates of rate are 1.1 μm per minute from autoradiograms produced by warm pulses of 10 minutes, and 0.87 μm per minute from autoradiograms produced by warm pulses of 20 minutes (Table 2). These estimates of rate are probably somewhat high because of the

TABLE 2

RATE OF DNA CHAIN ELONGATION IN L-929 CELLS

Autoradiogram measured	Pulse time (min.)	Rate of chain elongation (μm/min, $\bar{x} \pm$ s.e.)
"hot"	30	0.66 ± 0.012
"hot"	30	0.80 ± 0.045
"hot"	10	0.90 ± 0.040
"warm"	20	0.87 ± 0.068
"warm"	10	1.10 ± 0.075

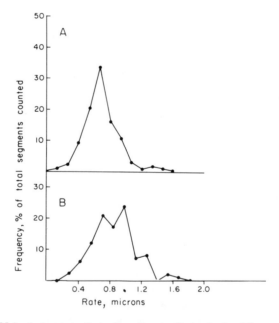

Fig. 4. Rate of DNA chain growth in L cells. A Data derived from cells that underwent a 30-minute hot pulse. B Data derived from cells that underwent a 10-min. hot pulse. Rate, plotted on the abscissa, is calculated by dividing the length in μm of the hot pulse autoradiogram by the time in minutes of the hot pulse. From Hand and Tamm (22), with the permission of Academic Press, New York.

difficulty in measuring autoradiograms of low grain density.

Our estimates of the rate of DNA chain growth in mammalian cells is in the range found by other investigators using either this technique (3,18) or sedimentation techniques (10-12). Painter and Schaeffer (35) have shown the rate of chain growth to vary during S phase in HeLa cells, with early S phase chain growth proceeding at a rate of 0.3 to 0.4 μm per minute and mid-S phase cells at a rate of 1.0 to 1.1 μm per minute. This variation may be reflected in the distribution seen in unsynchronized cells (Fig. 4).

IX. DNA Chain Growth During Inhibition of Synthesis of DNA Precursors

We have studied the effect of the inhibition of synthesis of DNA precursors on the rate of chain elongation. In these studies we have employed two compounds, hydroxyurea and FdUrd.

We could not use a direct method to measure rate of chain growth with FdUrd. The block induced by FdUrd at the level of thymidylate synthetase is bypassed by added dThd (reviewed by Heidelberger, 36). ^{3}H-cytidine (^{3}H-Cyd) was used as a label, and although it produces readable autoradiograms, the labelling density is lighter than with ^{3}H-dThd and changes in grain density cannot be interpreted with confidence. Cells were exposed to 3.3 x 10^{-6} M ^{3}H-Cyd (25 Ci/mmol, 80μCi/ml), and after 1,7, or 15 minutes, FdUrd, 2 x 10^{-6} M was added. The experiment was terminated after 30 minutes. Controls received no FdUrd. If FdUrd blocks chain growth then the mean lengths of simple pulse autoradiograms of DNA labelled with ^{3}H-Cyd would be reduced. Table 3 shows that control autoradiograms produced by 30-minute labelling with ^{3}H-Cyd have a mean length similar to that obtained after labelling with ^{3}H-dThd (21). FdUrd, present in the cultures for 23 or 29 of the 30 minutes of the ^{3}H-Cyd pulse, reduces the length of DNA segments. FdUrd present for 15 of the 30 minutes has little effect. The results can be explained by postulating that FdUrd inhibits chain elongation but requires 15-23 minutes in order to exert this effect in L cells. This may represent the time required to deplete the cellular pool of dThd, and it agrees well with estimates by alternative methods (Holford, 1965, as quoted in Cleaver, 37).

In experiments with hydroxyurea, a compound that inhibits ribonucleotide reductase (28,29), we were able to employ dThd as the label without reversing the inhibitory effect of the hydroxyurea. Cells were labelled according to protocol A. The 30-minute hot pulse was begun at various times after the addition of hydroxyurea (2 x 10^{-3} M). All cells had been exposed to 2 x 10^{-6} M FdUrd for at least 30 minutes before the start of the pulse with 5 x 10^{-6} M ^{3}H-dThd, and FdUrd remained in the medium during the pulse. Thus the cells were exposed to both inhibitory compounds during the pulse. However, we assumed that the high concentration of ^{3}H-dThd reversed the effect of the FdUrd. The pattern of DNA replication is altered by treatment with hydroxyurea for 15 minutes. The grains are arranged in multiple short segments below the minimal length permitting accurate measurement. Linearity is preserved however, and this suggests that the autoradiograms

TABLE 3*

EFFECT OF FLUORODEOXYURIDINE ON DNA CHAIN GROWTH

Time of addition of FdUrd after beginning of ^{3}H-Cyd pulse (min)	Length of time FdUrd was present during ^{3}H-Cyd pulse (min)	Length of DNA segments in m ($\bar{x} \pm$ s.e.)
—	0	20.9 ± 0.84
15	15	18.9 ± 0.80
7	23	13.1 ± 0.72
1	29	12.1 ± 0.58

*Reproduced from Hand and Tamm (7) with the permission of The Rockefeller University Press.

represent small nascent chains of DNA or intact template strands. The most marked effect of hydroxyurea is a slowing of the rate of chain growth. A defect in ligation is also possible. On each slide there are a few autoradiograms that show the longer tandem arrays characteristic of the inhibited controls. These segments tend to be clustered together, suggesting that they are derived from contiguous areas of template. In these few segments, the rate of chain growth was normal. In cells treated for 30 minutes and 1 hour with hydroxyurea, these long arrays are rarely seen, and after treatment for 2 hours none are seen. The reason for the resistance to hydroxyurea of these few segments is unknown, however it might be related to deoxynucleoside pool-size differences in certain sub-poupulations of cells.

X. DNA Chain Growth During Inhibition of Protein Synthesis

We determined the effect of inhibition of cellular protein synthesis on DNA chain growth using puromycin at two different concentrations. Cells were treated with puromycin for 2 hours and rate of chain growth then determined. Puromycin, 200 µg per ml, reduces the rate of DNA chain growth (Fig. 5). However, at a dose of 20 µg per ml there is little effect on the rate of DNA chain propagation, even though a significant inhibition of cellular protein synthesis and ^{3}H-dThd uptake is produced. Our results at the dose of 20 µg per ml are similar to those obtained by Hori and Lark (19).

Cycloheximide is an inhibitor of protein synthesis that acts differently from puromycin. We compared the kinetics of inhibition of chain growth during puromycin or cycloheximide treatment.

Cells were treated with puromycin 200 µg/ml and the rate of DNA chain growth determined. Rate of DNA chain growth is unaffected for 30 minutes but is reduced more than 50% by 120 minutes (Table 4). Control cells maintain

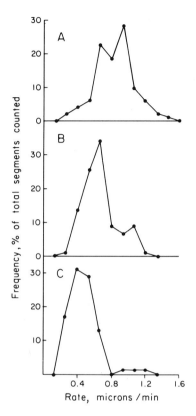

Fig. 5. Rate of DNA chain elongation in puromycin-treated cells. (A) Untreated control. (B) Cells treated with 20 μg/ml for 2 hr. ^{3}H-dThd incorporation was 27.8% of that in the untreated controls. (C) Cells treated with 200 μg/ml for 2 hrs. H-dTh incorporation was 13.0% of that in untreated controls. Cells exposed to hot pulse for 30 min. From Hand and Tamm (22) with the permission of Academic Press, New York.

a rate of chain elongation of 0.6-0.7 μm/minute and they maintain this rate throughout the experiment. The reduction in the rate of chain growth is always less than the degree of inhibition of ^{3}H-dThd uptake as determined by pulse labelling in replicate puromycin-treated cultures that had not been treated with FdUrd or large amounts of ^{3}H-dThd (Table 4).

Similar experiments were performed using cycloheximide at 50 μg/ml. In contrast to the results with puromycin, cycloheximide causes an immediate reduction in the rate of DNA chain elongation which is similar in magnitude to the inhibition of incorporation of ^{3}H-dThd (Table 4). Control cells maintain a rate of chain growth of 0.6-0.7 μm/minute for the duration of the experiment.

285

TABLE 4[+]

COMPARISON OF THE DEGREE OF INHIBITION OF ^{3}H-dThd
INCORPORATION AND RATE OF DNA CHAIN GROWTH CAUSED BY
CYCLOHEXIMIDE OR PUROMYCIN

Drug	Time after addition of inhibitor (min)	^{3}H-dThd incorporation* (% of control)	Rate of DNA chain growth (% of control)
Puromycin,	15.	48.2	83.9
200 µg/ml	30	33.6	87.0
	60	25.9	63.7
	120	23.9	40.3
Cycloheximide,	15	49.2	47.6
50 µg/ml	30	43.6	54.2
	60	40.9	51.6
	120	44.8	41.4

*Control or inhibitor treated cells were pulse-labelled with ^{3}H-dThd (1.0 µCi/ml) for 15 min. The incorporation of radioactivity into trichloroacetic acid-insoluble material was determined by liquid scintillation counting.

+Reproduced from Hand and Tamm (7) with the permission of The Rockefeller University Press.

Weintraub and Holtzer (9) and Gautschi and Kern (40) have also measured the rate of DNA chain growth during cycloheximide treatment using methods based on a shift in density of nascent DNA strands labelled with bromodeoxyuridine. Their results are very similar to ours, with a 50% reduction of chain growth occurring almost immediately after introduction of the compound and good correspondence between the reduction in rate of chain growth and dThd incorporation.

The effects of puromycin and cycloheximide on DNA chain growth clearly differ. The rapid onset of the cycloheximide effect might be due to direct action on DNA replication. This has been suggested (41), however there is also evidence that cycloheximide does not inhibit DNA synthesis except through its effect on protein synthesis (9). The modes of action of puromycin and cycloheximide on ribosomal protein synthesis differ and perhaps this results in a differential effect on the synthesis of the proteins required in DNA chain propagation. Puromycin causes premature release of unfinished polypeptides (42). Some prematurely released peptides near completion at the time of puromycin addition may retain physiologic activity. It has been suggested (Huberman and Weintraub, personal communication) that such peptides perhaps maintain DNA chain growth for a longer period of time after inhibition of protein synthesis by puromycin as opposed to inhibition by cycloheximide where the polypeptide chain is not released but is merely slowed down on the ribosome (43).

XI. Summary

We have discussed several aspects of DNA initiation and chain growth in this review. The main points may be summarized:
1. Initiation of DNA replication occurs multifocally as clusters of initiation events on topographically contiguous areas of template.
2. Initiation events occur in bursts and there appears to be a degree of synchrony among individual events.
3. The rate of DNA chain growth in tissue culture cells is slightly less than 1 μm/min.
4. Inhibition of DNA synthesis by hydroxyurea and FdUrd slows the rate of chain growth.
5. Inhibition of cellular protein synthesis by cycloheximide or puromycin slows the rate of chain growth, although the onset of action of cycloheximide is more rapid than puromycin.

Acknowledgements

The investigations reviewed here were supported by U.S. Public Health Service Grant A1-03445. Much of the work was performed while R.H. was The Katherine Rhodes Blackwell Fellow and Senior Investigator of The New York Heart Association.

References

1. J. Cairns, *J. Mol. Biol.* 15, 372-373 (1966).
2. J. A. Huberman and A. D. Riggs, *Proc. Nat. Acad. Sci. U.S. 55*, 599 (1966).
3. J. A. Huberman and A. D. Riggs, *J. Mol. Biol. 32*, 327-341 (1968).
4. H. G. Callan, *Proc. R. Soc. Lond B. Biol. Sci. 181*, 19 (1972).
5. H. Weintraub, *Nature New. Biol. 236*, 195 (1972).
6. F. Amaldi, F. Carnevali, L. Leoni and D. Mariotti, *Exp. Cell Res. 74*, 367 (1972).
7. R. Hand and I. Tamm, *J. Cell Biol. 58*, 410 (1973).
8. K. G. Lark, *Ann. Rev. Biochem. 38*, 569 (1969).
9. H. Weintraub and H. Holtzer, *J. Mol. Biol. 66*, 13 (1972).
10. J. H. Taylor, *J. Mol. Biol. 31*, 579 (1968).
11. R. B. Painter and A. W. Schaeffer, *J. Mol. Biol. 45*, 467 (1969).
12. A. R. Lehmann and M. G. Ormerod, *Biochim. Biophys. Acta 204*, 128 (1970).
13. Y. Fujiwara, *Cancer Res. 32*, 2089 (1972).

14. W. P. Cheevers, J. Kowalski and K. K.-Y. Yu, *J. Mol. Biol.* 65, 347 (1972).
15. J. Cairns, *J. Mol. Biol.* 4, 407 (1962).
16. J. Cairns, *Cold Spring Harbor Symp. Quant. Biol.* 28, 43 (1963).
17. J. A. Huberman and A. Tsai, *J. Mol. Biol.* 75, 5 (1973).
18. K. G. Lark, R. Consigli and A. Toliver, *J. Mol. Biol.* 58, 873 (1971).
19. T. A.Hori and K. G. Lark, *J. Mol. Biol.* 77, 391 (1973).
20. F. Amaldi, M. Buongiorno-Nardelli, F. Carnevali, L. Leoni, D. Mariotti and M. Pomponi, *Exp. Cell Res.* 80, 79 (1973).
21. R. Hand, W. D. Ensminger and I. Tamm, *Virology* 44, 527 (1971).
22. R. Hand and I. Tamm, *Virology* 47, 331 (1972).
23. R. Hand and I. Tamm, *J. Mol. Biol.* In Press (1974).
24. D. M. Prescott and P. E. Keumpel, *Proc. Nat. Acad. Sci. U.S.* 69, 2842 (1972).
25. R. Baserga and D. Malamud, *Autoradiography; techniques and applications,* Harper, New York, p. 54 (1969).
26. A. T. Wilson, *Biochim. Biophys. Acta 40,* 522 (1960).
27. H. A. Fischer, H. Korr, H. Thiele and G. Werner, *Naturwissenschaften 58,* 101 (1971).
28. J. H. Taylor, *J. Biophys. Biochem. Cytol.* 7, 455 (1960).
29. J. German, *J. Cell Biol.* 20, 37 (1964).
30. T. C. Hsu, *J. Cell Biol.* 23, 53 (1964).
31. E. J. Stubblefield, *J. Cell Biol.* 25, 137 (1965).
32. F. Gavosto, L. Pegoraro, P. Masera and G. Rovera, *Exp. Cell Res.* 49, 340 (1968).
33. J. A. Remington and R. R. Klevecz, *Exp. Cell Res.* 76, 410 (1973).
34. R. R. Klevecz and L. N. Kapp, *J. Cell Biol.* 58, 564 (1973).
35. R. B. Painter and A. W. Schaeffer, *J. Mol. Biol.* 58, 289 (1971).
36. C. Heidelberger, *Prog. Nucleic Acid Res. Mol. Biol. 4,* 1 (1965).
37. J. E. Cleaver, *Thymidine metabolism and cell kinetics.* North Holland Publ. Amsterdam, pp. 70-103 (1967).
38. R. L. P. Adams and J. G. Lindsay, *J. Biol. Chem. 242,* 1314 (1967).
39. C. W. Young, G. Schochetman, and D. A. Karnovsky, *Cancer Res.* 27, 526 (1967).
40. J. R. Gautschi and R. M. Kern, *Exp. Cell Res. 80,* 15 (1973).
41. E. F. Gale, E. Cundcliffe, R. E. Reynolds, M. H. Richmond and M. J. Waring, *The molecular basis of antibiotic action.* John Wiley & Sons Ltd., London, p. 361 (1972).
42. D. Nathans, *Antibiotics.* I. *Mechanism of action.* (Ed. Gottlieb, D. and Shaw, P.D.) Springer-Verlag, New York, pp. 259-277 (1967).
43. H. D. Sisler and M. R. Siegel, *Antibiotics.* I. *Mechanism of action* (Ed. Gottlieb, D. and Shaw, P.D.) Springer-Verlag, New York, pp. 283-307 (1967).

APPROACHES TO THE STUDY OF THE REGULATION OF NUCLEAR RNA SYNTHESIS IN SYNCHRONIZED MAMMALIAN CELLS

T. Simmons, P. Heywood, S. Taube and L. D. Hodge

Departments of Human Genetics and Microbiology,
Yale University, School of Medicine, New Haven, Connecticut 06510

I. Introduction

Logarithmically growing mammalian cells in culture are a convenient experimental system for the study of nuclear RNA metabolism. This has led to rapid advances in the knowledge of the nucleolar synthesis, processing, and partial assembly of ribosomal particles (1). A role for heterogeneous nuclear RNA (hn-RNA) as a precursor of cytoplasmic messenger RNA (mRNA) has been implied in studies of its structural characteristics and post-transcriptional modifications (2). Functions associated with the synthesis, processing and transport to the cytoplasm of nuclear RNA appear to occur continuously in randomly growing cultures. However, studies with synchronous cultures have established that biochemical and morphological changes in individual cells occur as ordered sequences of events (3-6). If regulatory events occur in nuclear RNA metabolism, it is reasonable to assume that they can be defined and investigated more precisely in synchronous cultures. The cell cycle in mammalian cells consists of a period of mitosis (M) followed by an interval (G_1) before the onset of replicative DNA synthesis; a phase in which DNA is replicated (S) followed by an interval (G_2) before the next mitosis. To effectively explore the possibilities of cell cycle related regulation in RNA metabolism, it will be necessary to define precise events and to identify and analyze specific RNA sequences.

The synthesis of nucleolar pre-ribosomal RNA and total heterogeneous nuclear RNA occurs continuously throughout interphase (7-10). Although there are conflicting reports concerning their relative rates of synthesis during interphase (9,10), it is well established that synthesis of these species ceases at metaphase (11-14). Thus, the entry into or the exit of cells from mitosis represent times in the cell cycle in which "coarse" controls in the synthesis of major nuclear RNA species may be investigated. Previous attempts to demonstrate "fine" control in the metabolism of RNA species, the functions of which are not required continuously throughout interphase, have been inconclusive (15-17). For example, competitive hybridization experiments between unfractionated preparations of hn-RNA and/or cytoplasmic mRNA lacked the sensitivity necessary to establish cell cycle related differences in RNA composition. For this type of analysis to succeed, specific RNA sequences must be defined and compared. Cell lines transformed by

DNA-containing viruses appear to offer systems with which to investigate the "fine" control of mRNA metabolism since virus-specific RNA sequences can be identified and isolated.

We have begun to correlate cell cycle related events to nuclear RNA metabolism in two systems. In the first, events in the transition of mitotic cells into G_1 have been examined to correlate the ultrastructural and biochemical changes that accompany the postmitotic restoration of RNA synthesis. In the second, RNA metabolism in an adenovirus transformed mammalian cell line is being probed for cell cycle related phenomena in the regulation of the synthesis, processing, and cytoplasmic utilization of specific viral transcripts.

II. Cells and Methods of Synchronization

Mitotic HeLa S_3 cells were obtained by the method of selective detachment (18-20). With this technique, populations consisting of 85-95% metaphase cells were collected which had not been exposed to physical or chemical inhibitors during the time of mitosis. Continued incubation at 37°C in suspension culture, allowed these cells to pass through the remaining stages of mitosis into G_1. To obtain the large number (4×10^7) of cells necessary for these experiments, mitotic cells were prepared by combining the techniques of double thymidine blockade (21,22) and selective detachment. The cells were incubated for 15 hrs in the presence of 2.0 mM thymidine followed by 9 hrs in the asbsence of inhibitor and were then incubated for an additional 14 hrs in the presence of thymidine. At the end of the second blockade, the inhibition was reversed by resuspension in medium lacking thymidine and the resulting S phase cells were transferred to monolayer culture. Mitotic cells were then collected 9 to 11 hrs later.

Rat embryo cells transformed *in vitro* by adenovirus type 2 (Ad2-T) were grown in suspension as spinner cultures. The cells were cloned according to the soft agar technique of Puck *et al.* (23) and the experiments reported below were performed with the progeny of one clone. This derivative demonstrated levels of T antigen production and virus-specific RNA synthesis at least equivalent to those of the uncloned population (24). In the majority of experiments, cells were synchronized by double thymidine blockade as described above except that 2.5 mM thymidine was employed. After release these cells were used directly as S phase populations. After the second removal of thymidine, incorporation of radioactive thymidine into DNA began immediately and reached a maximum at about 3 hrs. Autoradiography using [^{3}H]-thymidine revealed that 85% of the cells synthesized DNA during the period 1-5 hrs following release, while only 22% synthesized DNA at 7-11 hrs. At least 70-80% of the cells divide during a 2-3 hr interval beginning at about 6 hrs after release from thymidine. This division was indicated both by the increase in cell number and by the mitotic index monitored in cultures arrested in metaphase with colcemid.

To study cells in metaphase further synchronization was imposed after thymidine blockade. Colcemid (0.0025 μg/ml) was added to cells at the time of the second release which resulted in the accumulation in metaphase of approximately 60% of the cells after 7 hrs incubation at 37°C.[1] Metaphase-arrested cells were collected by means of discontinuous albumin gradients. The cells were resuspended in 30% bovine albumin. Gradients consisted of 30% (containing unfractionated populations), 23%, 20%, 17%, and 10% albumin in spinner salts and were centrifuged in a Spinco SW65 rotor at 17,500 rpm for 45 min at 5°C. The band of cells which rose to the 17-10% interface routinely consisted of 95-97% metaphase-arrested cells.

III. Restitution of Nuclear RNA Synthesis at the Mitosis to G_1 Transition

During mitosis the ultrastructural events include formation and dissolution of the mitotic apparatus, condensation and decondensation of chromatin, and dispersal and reformation of both the nucleolus and the nuclear envelope (25). Mitchison (26) has suggested that the condensation of chromosomes at metaphase could be responsible for the depression of RNA synthesis. More recently the hypothesis has been advanced that the sites of DNA previously available for transcription are masked by proteins added to the chromosomes during condensation (27,28). Reconstruction experiments between metaphase chromosomal proteins and either non-mitotic or mitotic DNA have shown that template combined with metaphase proteins is a poor substrate for microbial RNA polymerases. Regardless of the etiology, the marked reduction in nuclear RNA synthesis in metaphase cells enabled us to study the ultrastructural events that accompanied the resumption of RNA synthesis as such cells proceeded into G_1. The technique of electron microscopic autoradiography was used to study the resumption of RNA synthesis in individual cells exposed to [^{3}H]-uridine at the time of synchronization in metaphase. Incubation of these cells at 37° for 40 min yielded a population that consisted of cells in late anaphase; early, mid and late telophase, and early G_1. This spread of late mitotic stages represents 20 to 25 min of the cell cycle. With this experimental approach, it was possible to correlate the appearance of silver grains with the successive stages of nuclear reformation in late mitosis.

A. *Ultrastructural Events and RNA Synthesis*

Examination of cells in early telophase indicated a pronounced increase in the degree of nuclear labeling over that present in anaphase cells (Compare Fig. 1c with Fig. 1a and 1b). The appearance of grains in these cells coincided with the reformation of the nuclear envelope. Furthermore, the density of grains was higher over the periphery of the chromosome mass, at or near the site of the reforming nuclear envelope. This is apparent both from Fig. 1c and Fig. 2, as well as from quantification based on cells in various stages of telophase and early G_1 (Table 1). The peripheral distribution of grains appeared before

---

[1]It appears that an additional 20 to 30 per cent of the population was arrested in prophase.

decondensation of the chromosomes and before reformation of the nucleoli began to appear. This change was reflected by the marked increase in the density of silver grains over the non-nucleolar nucleoplasm. In spite of this, the relative grain density over the nuclear envelope was still appreciably higher than over the remainder of the nucleoplasm (Table 1). This differential distribution became even less distinct further into G_1. Therefore, it would appear that the resumption of nuclear RNA synthesis in HeLa cells occurs at or near the reforming nuclear envelope and represents extranucleolar RNA synthesis. Histograms, constructued according to the half-distance method of Salpeter et al. (29) of the grain distribution over telophase and early G_1 nuclei substantiate these conclusions (30).

It is not known whether nuclear envelope-associated RNA synthesis continues throughout interphase. This question may be unanswerable in mammalian cells using the techniques of electron microscope autoradiography. Once cells have entered G_1, the grains over the nucleoplasmic areas become increasingly heavier and more generalized. Also, the reformation of nucleoli results in yet additional sites of RNA synthesis which become heavily associated with silver grains. Unlike mitotic cells, interphase cells are continuously synthesizing and processing nuclear RNA for transport into the cytoplasm. Even under conditions of brief exposure to radioactivity, transient association of labeled RNA with the nuclear envelope during transport may further complicate the interpretation of the autoradiographs. These processes would expalin the lack of any preferential association of newly-synthesized RNA with the nuclear envelope in randomly growing exponential cultures (31,32). An alternative explanation is that membrane-associated RNA synthesis is unique to the restoration occurring in telophase cells and cannot be generalized to synthesis during interphase.

B. Ultrastructural Events during Mitosis in other Eukaryotic Organisms

The fact that changes in chromosomes, nucleoli, and nuclear envelope are so tightly coupled during mitosis in mammalian cells makes it difficult to determine which of these structural changes, either alone or in combination, is

Fig. 1. Mitotic HeLa cell nuclei. (a) An oblique serial section through a late anaphase cell. x 5,600. (b) A continuous serial section of the large chromosome group of the cell in panel a. Few silver grains were found over anaphase cells. Three of seven of these grains observed over panels a and b are near sites of completed nuclear envelope: of these grains, two occur in similar locations on the serial sections (arrowheads). x 5,600. (c) An early telophase nucleus. At this stage the chromatin remains condensed although individual chromosomes are not distinguishable. Reformation of the nuclear envelope is nearly complete. Note the random, peripheral, pattern of grains (arrowheads) at or near the nuclear envelope. x 6,400. (d-f) Three contiguous serial sections through a mid to late telophase nucleus in which areas of chromosome decondensation are present. Grains occur at the nuclear envelope and over the chromatin. x 6,400. (Reprinted with permission of the Rockefeller Press.)

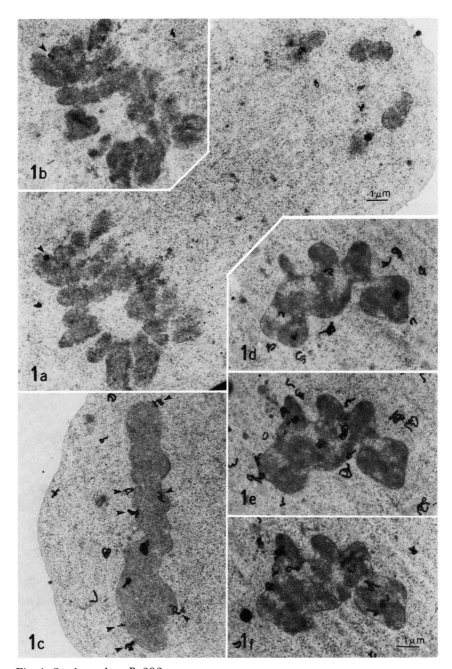

Fig. 1. See legend on P. 292.

associated with the resumption of RNA synthesis. Mitosis in other eukaryotic organisms may provide additional systems with which to explore the correlations between ultrastructural changes and RNA synthesis during the mitosis to G_1 transition. In the fungus *Saccharomyces* (33) and two groups of algae, the Dinophyceae (34) and the Euglenophyceae (35), the nuclear envelope and nucleolus remain intact during mitosis. How does this relate to a restriction of RNA synthesis, if indeed such a restriction occurs? The slime mold *Physarum flavicomum* provides an example of an organism in which two types of nuclear envelope behavior occur during mitosis (36). In the haploid myxamebae stage the nuclear envelope is dispersed in late prophase, whereas in the diploid plasmodial stage the nuclear envelope is intact except for limited polar dissolution late in mitosis. In the filamentous green alga *Spirogyra* nucleolar material is not dispersed until telophase: it persists through the earlier stages of mitosis as a layer on the surface of the chromosomes (37). The pattern of restriction and restitution of RNA synthesis in *Physarum* and in *Spirogyra* deserves careful investigation and should be correlated with the unusual morphological events during mitosis in these organisms. Even the possibility that chromosome condensation is responsible for the reduction of RNA synthesis in metaphase could be tested in organisms such as yeast which apparently lack a cycle of chromosome condensation.

C. Inhibitor Studies on the Restitution of RNA Synthesis

The electron microscope autoradiographic studies previously discussed provide evidence for the resumption of nuclear RNA synthesis in early telophase at or near the site of the newly reformed nuclear envelope. We have also approached the investigation of this synthesis biochemically by both *in vivo* and *in vitro* systems. For example, to explore the kinetics of the reappearance of nuclear RNA synthesis, metaphase cells were collected in the presence or absence of inhibitors of RNA or protein synthesis. The incorporation of $[^{14}C]$-uridine into acid precipitable radioactivity for successive 15 min intervals for the first 15 to 90 min after synchronization was determined. During this time period, the mitotic index fell from 90% or greater to 10% or less; the cell number approximately doubled; and the interphase cellular morphology was re-established. The initial rate of increase in uridine incorporation in mitotic cells collected in the presence of inhibitors of protein synthesis, was similar to that observed in control cells for 60 to 75 min post-synchronization (Fig. 3A). Subsequently, the rate of incorporation declined. In cycloheximide-treated cultures, the rate of uridine incorporation declined until a plateau level of approximately 25% of the maximum was reached at 120 to 135 min after synchronization (30). However, the effect of the inhibitor was reversible. After 90 min of incubation with cycloheximide, cells resuspended in inhibitor-free medium approached interphase levels of

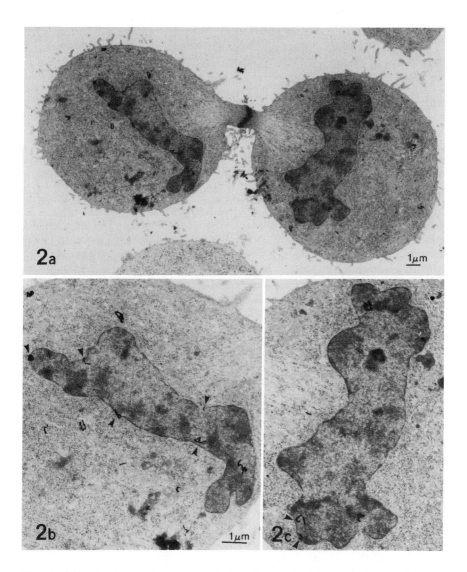

Fig. 2. Mid to late telophase cell. (a) Progeny cells are connected by a relatively short intercellular bridge. x 3,500. (b and c) The progeny nuclei have similar morphology, degree of chromosome decondensation, and peripheral location of silver grains (arrowheads). x 6,400. (Reprinted by permission of the Rockefeller Press).

295

STAGE IN CELL CYCLE	COMPARTMENT	NUMBER OF GRAINS	RELATIVE AREA (%)	RELATIVE GRAIN DENSITY
Telophase	Nuclear Envelope	101	2.3	1.00
	Nuclear	103	13.8	.17
	Cytoplasmic	440	83.9	.12
Early G-1	Nuclear Envelope	85	2.9	1.00
	Nuclear	522	28.5	.62
	Cytoplasmic	185	68.6	.09

Table 1. Relative grain density in telophase and early G_1 cells. Cellular area and grain counts were determined from autoradiographs of 32 telophase and 36 G_1 cells. x 14,060. The relative area of each cellular compartment was estimated with a uniformly dispersed grid of points by dividing the points per compartment by the total number of points. The relative grain density represents the grains per area for each compartment normalized to that of the nuclear envelope.

uridine incorporation at 45 to 90 minutes following resuspension. The concentration of inhibitors employed were sufficient to prevent the restoration of $[^{14}C]$-mixed amino acid incorporation to interphase levels and reformation of polyribosomes (30). These results indicate that the initiation but not the maintenance of RNA synthesis during the mitosis to G_1 transition occurs independently of renewed protein synthesis.

In analogous studies, mitotic cells were collected in the presence or absence of low (0.04 µg/ml) and high (1-4 µg/ml) doses of actinomycin D (Fig. 3B). In randomly growing cells, low doses of the drug preferentially inhibit nucleolar ribosomal precursor RNA synthesis, while high doses inhibit both nucleolar and extra-nucleolar RNA synthesis (38). This differential effect of low and high doses of actinomycin D on uridine incorporation into RNA was also observed for synchronous populations entering G_1. In the presence of low doses of the drug, the rate of incorporation of $[^{14}C]$-uridine became constant at 45-50% of the untreated control rate by 45 to 60 min post-synchronization (Fig. 3B). Presumably, this residual level of incorporation represents extranucleolar RNA synthesis. It appears that almost equal amounts of nucleolar and extranucleolar RNAs are made during the mitosis to G_1 transition.

The results of these studies raise several questions regarding the classes of nuclear RNA involved in the restitution of synthesis and affected during the

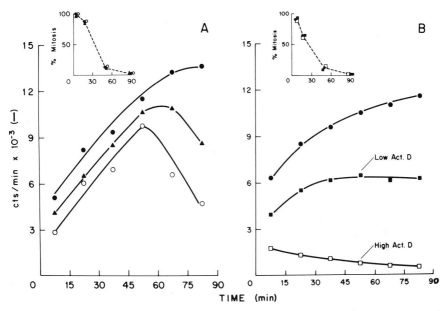

Fig. 3. Effect of inhibitors of macromolecular syntheses on the restitution of RNA synthesis. Preparations of mitotic cells were collected in the absence or presence of inhibitors of macromolecular synthesis and were maintained at 37°C as described (see Cells and Methods of Synchronization). At the indicated times 4.0 ml aliquots of each culture were incubated at 37°C for 15 min with 1.0 μCi [^{14}c]-uridine and total acid-precipitable radioactivity was determined. The data are plotted at the midpoint of each labeling period and have been normalized for an initial cell count of 43 x 10^4 cells per ml. Inserts: The mitotic index of control and inhibitor-treated cells at various times after synchronization ascertained by phase microscopy. (A) Control cells (●——●); cells collected with 60 μg per ml cycloheximide (▲——▲); cells collected with 25 μg per ml puromycin (O————O). (B) Control cells (●——●); cells collected with 0.04 μg per ml actinomycin D (■——■); cells collected with 2.0 μg per ml actinomycin D (□——□). (Reprinted with permission of the Rockefeller Press.)

inhibition of protein and RNA synthesis. For example, does cycloheximide treatment affect the synthesis of ribosomal precursor RNA and heterogeneous nuclear RNA coordinately? To analyze the molecular species of RNA made, mitotic cells collected in the presence or absence of inhibitors were exposed to radioactive uridine for a 15 min period at 60-75 min after synchronization. The nucleolar precursor RNA and heterogeneous nuclear RNA repsectively were prepared (39) and were subjected to sedimentation analysis in linear 15-30% SDS-sucrose gradients (FIg. 4). In comparison to untreated control cells, the

297

synthesis of both major classes of nuclear RNA was reduced in cycloheximide-treated cells. This was found for extracts made from cells at a time when the decreasing rate of RNA synthesis was first noted (Fig. 3A). Significant restoration of only heterogeneous nuclear RNA synthesis occurred in cultures exposed to the low dose of actinomycin D. Although there is no obvious 45S ribosomal precursor RNA or 32-28S pre-ribosomal RNA in the nucleolar fraction from these cells, the usual contamination of this preparation by heterogeneous RNA is apparent (40).

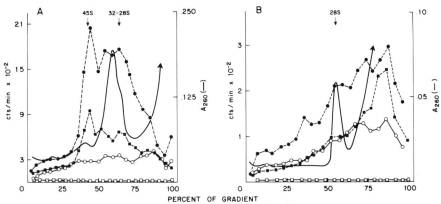

Fig. 4. Synthesis of nuclear heterogeneous RNA and pre-ribosomal RNA in the presence of inhibitors. Mitotic cells were collected in metaphase as described (see Cells and Methods of Synchronization). Cultures were incubated at 37°C in the presence or absence of inhibitors of macromolecular synthesis. Cultures were exposed to [14C]-uridine for 15 min at 60 min after synchronization when the proportion of mitotic cells was less than 20%. Heterogeneous nuclear RNA and nucleolar pre-ribosomal RNA were subjected to sedimentation analysis in 15-30% linear SDS-sucrose gradients by centrifugation for 16 hr at 20,000 rpm in an SW27 rotor. The data were normalized for cell concentration and were corrected for recovery after phenol extraction. (A) Nucleolar pre-ribosomal RNA: control cells (●——●); cells incubated with 100 µg per ml cycloheximide (■——■); cells incubated with 0.04 µg per ml actinomycin D (O————O); and cells incubated with 4.0 µg per ml actinomycin D (□——□). (B) Heterogeneous nuclear RNA: control cells (●——●); cells with cycloheximide (■——■); cells with low actinomycin D (O————————O); and cells with high actinomycin D (□——□). (Reprinted with permission of the Rockefeller Press.)

Several conclusions may be drawn from the experiments using inhibitors of macromolecular synthesis. The differential inhibition of RNA synthesis by actinomycin demonstrates that the restoration of heterogeneous nuclear RNA synthesis occurred independently from that of ribosomal precursor RNA synthesis. This may be interpreted to mean that the mechanisms regulating the restitution of nucleolar and extranucleolar synthesis in the mitosis to G_1 transition are distinct. The differential effect of the inhibitors of protein synthesis on the initiation and maintenance of RNA synthesis has several implications. The initial increase in the rate of RNA synthesis implies that the activity of pre-existing RNA polymerase(s) persists through mitosis. The failure of maintenance of the interphase rate of RNA synthesis in cycloheximide-treated cells, which affects both major classes of nuclear RNA, implies that the synthesis of new enzyme(s) and/or factor(s) in both synthetic pathways is necessary in early G_1. Taken together, these results support the conclusions that at least pre-existing DNA dependent RNA polymerase I and II (41,42) were involved and were functioning independently in the restitution of RNA synthesis *in vivo*. This interpretation is further supported by the demonstration of RNA synthesis in early telophase nuclei by electron microscopic autoradiography and by the demonstration of the synthesis of non-mitochondrial 4S and 5S RNA in metaphase-arrested Chinese hamster ovary cells (43).

D. RNA Synthesis by Preparations of Metaphase Chromosomes

The persistence of RNA polymerase activities in mitotic cells was further investigated in a series of *in vitro* studies of the RNA synthetic capacity associated with isolated metaphase chromosomes. Chromosomes isolated from HeLa cells under aseptic conditions (44), in the absence of zinc ions, were used as sources of both substrate and endogenous RNA polymerase activity (or activities) in an *in vitro* assay that measured radioactive UTP incorporation. Assay conditions were selected to meet the requirements of both Mn^{+2} and Mg^{+2} dependent enzyme activities (45,46). Dialysis of preparations of chromosomes for 16 hours under conditions known to preserve RNA synthetic capacities (42,47) increased the level of $[^3H]$-UTP incorporated into an ethanol and acid precipitable product. This suggests that the protein-nucleic acid complex is modified during the course of dialysis. The data presented in Fig. 5 indicate that the incorporation of radioactive UTP directed by dialyzed preparations of metaphase chromosomes represents the synthesis of an RNA product. The addition of high levels of actinomycin D or excess unlabeled UTP to reaction mixtures prevented further time-dependent increase in tritiated UTP incorporation. Alpha-amanitin, known to inhibit only DNA-dependent RNA polymerase II (48), reduced the level of recoverable radioactivity by 20% while pancreatic ribonuclease reduced this level by 60%. The tritiated product of the reaction mixtures is also labile in alkali. These results utilizing an *in vitro* assay further support the implications of the *in vivo* data: at least two DNA-dependent RNA synthetic activities persist in metaphase cells and these may be responsible for the initiation of RNA synthesis in late mitosis. Subunit modification of RNA polymerases has been shown in prokaryotes to alter template specificity (49). The isolation and characterization of the RNA polymerases persisting in metaphase cells should be pursued in order to compare the subunit composition of the G_2, M, and G_1 forms of the enzymes.

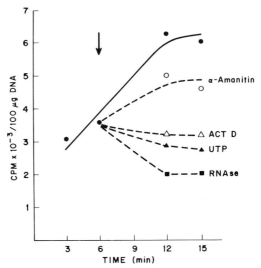

Fig. 5. Kinetics of UTP incorporation in the presence of inhibitors. Dialyzed chromosomes were incubated at 37°C in reaction mixtures adjusted to 5mM Mg Cl_2, 4 mM $MnCl_2$, 1 mM dithiothreitol, 1 µg/ml pyruvate kinase, 5mM phosphoenol pyruvate, 1 mM spermidine, 0.5 mM each of ATP, CTP, and GTP, 20 mM $(NH_4) SO_4$ and 75 mM Tris-HCl pH 7.9. The residual glycerol concentration from dialyzed preparations was approximately 10%. At 6 min (arrow) incubation, portions of the reaction mixture were transferred to tubes, pre-warmed to 37°C, each of which contained a quantity of additive such that the desired final concentration was attained: untreated control (●–●); 10 µg/ml α-amanitin (O————O); 4 µg/ml actinomycin D (△—△); 500 fold excess unlabelled UTP (▲—▲), 75 µg/ml pancreatic RNase (■—■). At 3, 6, 12, and 15 min of incubation, the ethanol and acid insoluble radioactivity in 0.5 ml aliquots of each incubation condition was determined. The data have been normalized for a concentration of 100 g of DNA per assay.

Comparison by polyacrylamide gel electrophoresis of the radioactive products of 3 min and 15 min reaction mixtures revealed that the 15 min product was larger (Fig. 5). When $[^{32}P]$-γ-ATP is used in the assay as the labeled nucleotide, radioactive phosphate was found in an acid-insoluble, RNase sensitive product. These results argue for the elongation and initiation of RNA molecules in the *in vitro* system. It is of interest that the non-mitochondrial RNAs made in metaphase-arrested cells also appear to be small in size (43).

IV. Viral RNA Synthesis in Synchronized Transformed Cells

As yet, in mammalian cells, there has been only limited success in demonstrating cell cycle related fluctuations in the synthesis of specific messenger RNAs. This has been accomplished for histone mRNA whose synthesis is confined to the DNA synthetic period of the cell cycle (50). As previously discussed, current techniques of mRNA preparation and DNA-RNA hybridization have not been able to discriminate the intermitten synthesis of mRNA transcribed from unique sequences of nuclear DNA. In order to approach a study of the regulation of mRNA synthesis in the cell cycle, one must be able to isolate and identify specific mRNA. The use of virus transformed cells facilitates the study of control of RNA synthesis and expression because the virus specific RNA sequences provide markers which can be readily identified. These markers can be used to investigate cell cycle related RNA synthetic events in the same way that specific proteins have been employed for the study of genetic "induction" and "repression" (51-53). It is conceivable that such data may also help to elucidate more about the interdependence or dependence of host functions and viral gene functions in transformed mammalian cells.

A. Viral RNA Synthesis in S, G₂ and Mitosis

Synthesis of viral RNA was measured following release from thymidine blockade through the entry of cells into mitosis during which there was a high degree of synchrony in the population. Cells were exposed to [^{3}H]-uridine (50 μCi/ml) for 1 hr intervals throughout this period. Total cellular RNA was prepared by disrupting the cells with SDS, extracting with phenol (20°C), and precipitating with NaCl and ethanol. The DNA-RNA hybridization procedure of Tsuei et al. (54) was employed using a non-limiting amount of alkali denatured adenovirus type 2 DNA (Fig. 7). When the total incorporation of uridine in acid precipitable radioactivity was compared to virus-specific radioactivity, it was clear that viral RNA synthesis coincided with total cellular RNA synthesis. It was also apparent that as cells entered mitosis, both cellular and viral RNA synthesis declined sharply.

Metaphase-arrested cells obtained from albumin density gradients were used to investigate the extent of restriction of viral RNA synthesis during mitosis. Randomly growing cells were added to the metaphase cells to form populations of defined content. The level of tritiated uridine incorporation in the mixed populations indicated that viral RNA synthesis was restricted along with cellular RNA synthesis (Fig. 8). The amount of total RNA and virus-specific RNA synthesized decreased linearly with increasing proportions of metaphase cells.

Since the G₂ period is short and is not clearly defined by thymidine synchronization, it could be argued that viral RNA synthesis occurs primarily during S, and that restriction during G₂ is masked by the decay in synchrony. A more precise examination of the G₂ phase was possible because of the nearly complete restriction of virus RNA synthesis during metaphase. A culture was synchronized for S phase and colcemid was added at the time of release from the second blockade. This culture was divided into three aliquots which were

301

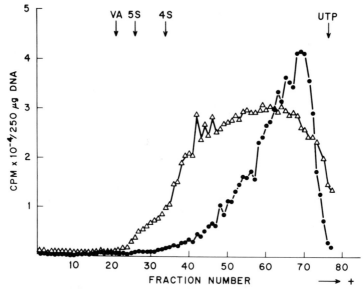

Fig. 6. Electrophoretic analysis of reaction mixture products. Products of the reaction mixtures were extracted with phenol and then analyzed by electrophoresis in 8.0 cm, 10% acrylamide gels containing 0.1% SDS at a constant current of 7 amps per gel for 5 hrs. The position of migration of adenovirus-associated RNA (VA), of cellular 5S RNA (5S), of cellular 4S RNA (4S), and of [3H]-UTP was determined in parallel gels. Each fraction represents the radioactivity of a 1 mm gel slice. Profile of the product of a reaction mixture incubated for 3 min at 37°C (●—●). Profile of the product of a reaction mixture incubated for 15 min at 37°C (△—△).

continuously exposed to [3H]-uridine for 6, 4, and 2 hrs prior to collection of metaphase cells. The advantage of this approach is that by isolating metaphase cells, the incorporation during a given period represents synthesis for a maximum of that time before mitosis. For example, cells incorporating radioactivity in the 2 hr interval immediately preceding the collection of metaphase-arrested cells represented a pure G_2 population. The data indicated that significant amounts of virus-specific RNA were made in all three populations that represented S, late S-G_2 and G_2 cells (Table II). To date only tentative data have been obtained concerning virus-specific RNA synthesis in G_1 cells. Cells synchronized with thymidine were allowed to pass through mitosis and the postmitotic incorporation of [3H]-uridine was followed as previously described.[2] Although there is much synchrony decay with this

---

[2]A variety of other techniques to obtain G_1 cells were unsuccessful. For example, selective detachment did not work because of the extreme calcium sensitivity of the Ad2-T cells. It was possible to bring the cells to stationary phase by allowing them to grow to high density by reducing the serum content of the medium (55), and by deleting isoleucine from the medium (56). But transference of these stationary phase cultures to fresh complete medium did not result in synchronous growth.

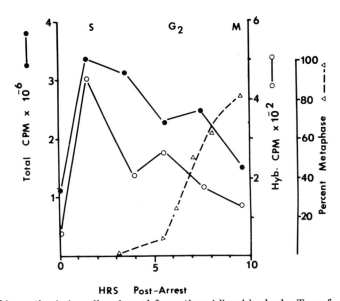

Fig. 7. RNA synthesis in cells released from thymidine blockade. Transformed rat embryo cells synchronized by the technique of double thymidine blockade were exposed to 50 µCi per ml of [^{3}H]-uridine for 1 hr intervals following release. Total cellular RNA was prepared from $2\text{-}4 \times 10^7$ cells (58) and was allowed to hybridize in 7.5 M urea to 8 µg of alkali denatured adenovirus type 2 DNA immobilized on nitrocellulose filters for 96 hr at 37°C. The total acid precipitable radioactive material (●—●) and radioactive material specifically bound to DNA filters (O————O) were determined. The data were normalized as described in Fig. 4. 4 hrs after release, colcemid was added and the accumulation of mitotic cells was monitored. The per cent metaphase refers to both prophase—and metaphase—arrested cells.

procedure, the results have indicated that viral and cellular RNA synthesis were also coincident during G_1. These experiments support the conclusion that viral RNA is synthesized along with cellular RNA throughout the cell cycle, and that viral RNA synthesis is restricted during mitosis coordinate with nuclear RNA synthesis.

B. Approaches for Future Investigations

The characteristics of this system make possible further study of the cell cycle regulation of viral RNA synthesis and function, especially since the viral RNA produced at a given time can be isolated from DNA-RNA hybrids. It is conceivable that further analyses by competitive hybridization procedures would reveal whether the same RNA sequences are synthesized at all times in

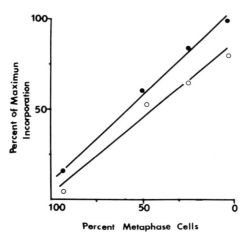

Fig. 8. Restriction of viral RNA synthesis during mitosis. Metaphase-arrested rat embryo transformed cells were collected with albumin gradients (see Cells and Methods of Synchronization). Randomly growing transformed cells and metaphase-arrested cells were mixed to form populations containing 96%, 50%, 25%, or 3% mitotic cells. [^{3}H]-uridine incorporation into total acid precipitable radioactivity and specifically hybridizable radioactivity was determined as in Fig. 7 for each of the mixed populations: total radioactivity (●——●); specifically hybridizable radioactivity (O————O).

the cell cycle. Similarly a comparison of the RNA synthesized during the same phase of the cell cycle in different clones might provide information about the site of integration of the viral genome, especially if sufficient quantities of virus-specific RNA can be obtained for limited structural analysis.

In randomly growing rat embryo transformed cells, the virus transcripts are in the large molecular weight heterogeneous RNA attached to cellular RNA sequences (54). However, such heterogeneous RNA is usually degraded in the nucleus and at most only 10% appears in the cytoplasm. Thus, even though there appears to be continuous nuclear synthesis of viral RNA it should be possible to study the cell cycle regulation of viral RNA processing. For example, does continuous synthesis mean that all viral RNA transcripts are exported at all times to the cytoplasm and become associated with polyribosomes? Competitive hybridization between nuclear and cytoplasmic viral RNA could suggest whether there is selective exit of particular viral RNA transcripts at unique times in the cell cycle. It is conceivable that there is continuous nuclear synthesis of viral RNA but exit from the nucleus only at restricted times in the cell cycle; e.g. following the addition of polyadenylic acid sequences. Another point of possible regulation in RNA metabolism during the cell cycle is the loss of significant amounts of messenger-RNA during polyribosome disaggregation at mitosis (57). It would be of interest to

Time before Metaphase	Cell Cycle Stage	Total CPM x 10^7	Hyb. CPM
2	G_2	1.4	327
4	$S - G_2$	2.6	377
6	S	5.3	550

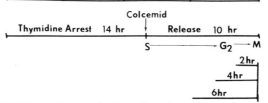

Table 2. Viral RNA syntheses during G_2. A culture of rato embryo cells synchronized with thymidine was released after the second blockade in the presence of colcemid and was divided into three equal portions. The subcultures were continuously exposed to [^{3}H]-uridine for the 2, 4, or 6 hour period preceding the time of collection of metaphase cells which occurred at 10 hrs after resuspension in thymidine-free medium (see Cells and Methods of Synchronization). The virus specific radioactivity for each subculture was determined as described in Fig. 7. For convenience the experimental details are diagramed.

determine whether viral RNA is in the fraction of m-RNA conserved or lost at this time.

V. Conclusion

These studies support the idea that synchronized cells can provide a useful approach for probing regulatory events in RNA metabolism in cultured mammalian cells. With HeLa S_3 cells tightly synchronized in the late stages of mitosis, it was possible to correlate nuclear reformation with the restitution of RNA synthesis. This restitution appears to be a membrane-associated event and to involve performed RNA polymerase activities. In other eukaryotes in which mitosis is not accompanied by nuclear envelope dissolution, it may be possible to further explore the apparent membrane-associated RNA synthesis in telophase nuclei. Further experiments with synchronized HeLa cells should explore the subunit composition of RNA synthetic enzymes during the transitions from G_2 to mitosis and from mitosis to G_1. Although subunit regulation of RNA synthetic enzymes has not been demonstrated in mammalian cells, this synchronized system would offer a logical time in which to probe for this type of regulation.

While it is too early to be certain, adeno 2 transformed rat embryo cells appear to offer a system with which the metabolism of mRNA(s) can be monitored. At least some virus-specific RNA is synthesized in all phases of the cell cycle except metaphase. The relative composition of the RNA transcripts synthesized in each phase, and the post-transcriptional processing occurring in each phase remain to be elucidated. Finally, the virus transformed cells, when synchronized, promise to provide useful information concerning the maintenance of the transformed state and the expression of the host cell genome.

Acknowledgement

These investigations were supported by United States Public Health Service Research Grant CA-12229. T. Simmons is a trainee supported by United States Public Health Training Grant AI-291. L. D. Hodge is a recipient of United States Public Health Service Research Career Development Award 5K04-GM-42-385 from the National Institutes of General Medical Sciences.

References

1. J. E. Darnell, Jr., *Bacteriol. Rev. 32*, 262 (1968).
2. R. A. Weinberg, *Ann. Rev. Biochem. 42*, 329 (1973).
3. A. Howard and S. R. Pelc, *Heredity* (suppl.) *6*, 261 (1953).
4. D. M. Prescott, *In* Synchrony in Cell Division. E. Zeuthen, editor. Wiley, New York. p. 71 (1964).
5. R. Baserga, *Cancer Res. 25*, 581 (1965).
6. R. Baserga, *Cell Tissue Kinet. 1*, 167 (1968).
7. M. D. Scharff and E. Robbins, *Nature 208*, 464 (1964).
8. L. J. Bello, *Biochim. Biophys. Acta 157*, 8 (1968).
9. S. E. Pfeiffer and L. J. Tolmach, *J. Cell. Physiol. 71*, 77 (1968).
10. S. E. Pfeiffer and L. J. Tolmach, *J. Cell. Physiol. 71*, 95 (1968).
11. J. H. Taylor, *Ann. N.Y. Acad. Sci. 90*, 409 (1960).
12. D. M. Prescott and M. A. Bender, *Exp. Cell Res. 26*, 260 (1962).
13. L. E. Feinendegen and V. P. Bond, *Exp. Cell Res. 30*, 393 (1963).
14. C. G. Konrad, *J. Cell Biol. 19*, 267 (1963).
15. A. O. A. Miller, *Arch. Biochem. Biophys. 122*, 270 (1967).
16. L. J. Bello, *Biochim. Biophys. Acta. 179*, 204 (1969).
17. G. N. Pagoulatous and J. E. Darnell, *J. Cell Biol. 44*, 476 (1970).
18. E. Robbins and P. I. Marcus, *Science 144*, 1152 (1964).
19. E. Robbins and M. D. Scharff, *In* Cell Synchrony. I. L. Cameron and G. M. Padilla, editors. Academic Press, Inc., New York. p. 353 (1966).
20. T. Terasima and L. J. Tolmach, *Exp. Cell Res. 30*, 344 (1963).
21. N. Xeros, *Nature 194*, 682 (1962).
22. D. Bootsma, L. Budke, and O. Vos, *Exp. Cell Res. 33*, 301 (1964).
23. J. J. Puck, P. I. Marcus, and S. J. Cieciura, *J. Exp. Med. 103*, 273 (1956).
24. A. E. Freeman, P. H. Black, E. A. Vanderpool, P. H. Henry, J. B. Austin, and R. J. Huebner, *Proc. Natl. Acad. Sci. 58*, 1205 (1967).

25. R. A. Erlandson and E. deHarven, *J. Cell Sci. 8*, 353 (1971).
26. J. M. Mitchison. The Biology of the Cell Cycle, Cambridge University Press, Cambridge, England, p. 115 (1971).
27. J. Farber, G. Stein, and R. Baserga, *Bioch. Biophys. Res. Comm. 47*, 790 (1972).
28. G. Stein and J. Farber, *Proc. Natl. Acad. Sci. 69*, 2918 (1972).
29. M. M. Salpeter, L. Bachmann, and E. E. Salpeter. *J. Cell Biol. 41*, 1 (1969).
30. T. Simmons, P. Heywood, and L. Hodge, *J. Cell Biol. 59*, 150 (1973).
31. L. Goldstein, *Exp. Cell Res. 61*, 218 (1970).
32. S. Fakan and W. Bernhard, *Exp. Cell Res. 67*, 129 (1971).
33. C. F. Robinow and J. Marak, *J. Cell Biol. 29*, 129 (1966).
34. D. F. Kubai and H. Ris, *J. Cell Biol. 40*, 508 (1969).
35. G. F. Leedale, *In* The Biology of Euglena. D. E. Buetow, editor. Academic Press, Inc., New York. p. 185 (1968).
36. H. C. Aldrich, *Amer. J. Bot. 56*, 290 (1969).
37. E. G. Jordan and M. B. E. Godward, *J. Cell Sci. 4*, 3 (1969).
38. R. P. Perry, *Exp. Cell Res. 29*, 400 (1963).
39. S. Penman, *In* Fundamental Techniques in Virology. K. Habel and N. P. Salzman, editors. Academic Press, Inc., New York. p. 35 (1969).
40. R. P. Perry and D. E. Kelley, *J. Cell Physiol. 76*, 127 (1970).
41. S. P. Blatti, C. J. Ingles, T. J. Lindell, P. W. Marris, R. F. Weaver, F. Weinberg, and W. J. Rutter, *Cold Spring Harbor Symp. Quant. Biol. 35*, 649 (1970).
42. S. T. Jacob, E. M. Sajdel, W. Muecke, and H. N. Munro, *Cold Spring Harbor Symp. Quant. Biol. 35*, 681 (1970).
43. E. A. Zylber and S. Penman, *Science 172*, 947 (1971).
44. J. J. Maio and C. L. Schildkraut, *J. Mol. Biol. 24*, 29 (1967).
45. R. G. Roeder and W. J. Rutter, *Nature 224*, 234 (1969).
46. S. T. Jacob, E. M. Sajdel, and H. N. Munro, *Biochem. Biophys. Res. Comm. 38*, 765 (1970).
47. R. F. Weaver, S. P. Blatti, and W. J. Rutter, *Proc. Nat. Acad. Sci. 68*, 2994 (1971).
48. T. J. Lindell, F. Weinberg, P. W. Morris, R. G. Roeder and W. J. Rutter, *Science 170*, 447 (1970).
49. A. Travers, *Nature (NB) 229*, 69 (1971).
50. T. W. Borun, M. D. Scharff, and E. Robbins, *Proc. Nat. Acad. Sci. 58*, 1977 (1967).
51. D. W. Martin, Jr., G. M. Tomkins, and M. A. Bresler, *Proc. Natl. Acad. Sci. 63*, 842 (1969).
52. D. W. Martin, Jr., G. M. Tomkins, and D. Granner, *Proc. Natl. Acad. Sci. 62*, 248 (1969).
53. R. A. Lerner and L. D. Hodge, *J. Cell Physiol. 77*, 265 (1971).
54. D. Tsuei, K. Fujinaga, and M. Green, *Proc. Natl. Acad. Sci. 69*, 427 (1972).

55. W. A. Strohl, *Virol. 39*, 653 (1963).
56. R. A. Tobey and K. D. Ley, *J. Cell Biol. 46*, 151 (1970).
57. L. D. Hodge, E. Robbins, and M. D. Scharff, *J. Cell Biol. 40*, 497 (1969).
58. R. A. Lerner and L. D. Hodge, *Proc. Nat. Acad. Sci. 64*, 554 (1969).

CHANGES IN NUCLEAR MACROMOLECULES
DURING NORMAL AND NEOPLASTIC GROWTH

Jen-Fu Chiu, K. Wakabayashi, Catherine Craddock
H. P. Morris and L. S. Hnilica*

Department of Biochemistry, The University of Texas System Cancer Center
M. D. Anderson Hospital and Tumor Institute, Houston, Texas 77025
and
*Department of Biochemistry, Howard University College of Medicine
Washington, D. C. 20001

Cells may be characterized by the efforts to reproduce themselves and to manufacture numerous specialized molecules including macromolecules in a manner determined by their differentiated state. Both of these principal functions can be traced to highly specific manipulations of the cellular memory bank, i.e., its DNA. Cumulative efforts of many investigators led to the development of methods for the isolation of chromatin in a form resembling its native state. Purified chromatin from most cells contains DNA, histones, nonhistone proteins, some RNA, and small quantities of other molecules. Since isolated chromatin migrates in zone electrophoresis as a homogeneous material, it can be concluded that its components form stable complexes. Hence, it appears that all chromatin components interact with the DNA in a meaningful fashion and that these interactions serve a primary purpose of controlling its ability to function as a gentic template (1).

Stimulated by findings that histones alone cannot restrict DNA in chromatin specifically, several investigators began an intensive search for macromolecules which, in association with histones, would direct these proteins to specific segments to DNA, thereby restricting its transcriptional capacity in a highly specific manner. Using RNA-DNA hybridization, evidence was presented that the DNA transcription in chromatin is tissue specific and that the specificity of this restriction is relatively unaffected by the isolation of chromatin (1-4). Further studies on the biochemical identity of macromolecules conferring tissue specificity to the DNA restriction by histones were considerably facilitated by the discovery that chromatin components can be dissociated with concentrated salt solutions and brought back together without a substantial loss of its tissue specific transcription, providing the dissociation and reassociation takes place in the presence of 5.0 M urea (5-8).

Dissociation experiments demonstrated that the removal of histones from chromatin increases both the rate of transcription as well as the number of RNA species transcribed (2,9). Reconstitution of pure DNA with histones produced transcriptionally inactive nucleohistones. However, if the reconstitution was performed with the dissociated mixture of histones,

309

nonhistone proteins and DNA, then the resulting artifical chromatin was capable of templating for RNA species similar to those templated by the native sample (5). It was concluded that although the histones are necessary to restrict the DNA quantitatively, the qualitative specificity of the restriction is determined by the macromolecules present in the nonhistone protein fraction.

The essential role of the nonhistone chromatin proteins in tissue specific DNA restriction was further supported by studies on "hybrid chromatins" composed of DNA and a major part of the chromatin nonhistone proteins from one tissue combined with the histones from another (8,10). The substitution of histone fraction from the chromatin of one tissue for that of another, as well as the substitution from one species (cow) to another (rat), demonstrated that the transcriptional specificity of reconstituted chromatin is determined by the DNA-associated nonhistone protein fraction. The histones thus served only as general, quantitative repressors. A similar conclusion was reached by Gilmour and Paul (11) who used chromatography of QAE Sephadex A-50 to separate the chromatin proteins from histones. Again, the specificity of the RNA synthesis, directed by reconstituted chromatins, depended on the tissue origins of the nonhistone protein fraction with the histones contributing no specificity.

According to Chytil and Spelsberg (12), dehistonized chromatin contains nonhistone protein species capable of producing tissue-specific antibodies. Wakabayashi and Hnilica (13) confirmed the tissue specificity of these NP-DNA pellets obtained by dissociation and ultracentrifugation of chromatin dissociated in 2.0 M NaCl, 5.0 M urea and 50 mM sodium phosphate buffer at pH 6.0. The macromolecular components comprising these pellets were antigenic in rabbits, and using the complement fixation method of Wasserman and Levine (14), Wakabayashi and Hnilica (13) demonstrated significant immunochemical differences between the NP-DNA pellets isolated from several normal tissues and transplantable tumors. However, individual neoplasms, when assayed with antibodies against Novikoff hepatoma NP-DNA material, did not exhibit any tissue specificity. Simply, they behaved as only one tissue type, i.e., the tumor.

Since the material used for immunization of rabbits (NP-DNA pellets) contained considerable amounts of DNA in addition to the nonhistone proteins, a possibility was tested that the tissue specific antibodies may be oriented against the complexes of some nonhistone proteins with DNA. Indeed, when the NP-DNA pellets were dissociated in 2.5 M NaCl, 5.0 M urea and 50 mM Tris-HCl buffer, pH 8.0 and the DNA separated by extensive ultracentrifugation, neither the DNA containing pellet nor the nonhistone protein containing supernatant were immunoreactive in the complement fixation assay with antibodies produced against the intact NP-DNA complex (13). The possibility that the DNA as electronegative polymer may influence the conformation of the nonhistone protein antigen, making it immunochemically active, was eliminated by assaying complexes of nonhistone proteins reconstituted with various polyanionic molecules, including DNA from another species (heterologous DNA). Additionally, when isolated nonhistone proteins without DNA were used as antigens to immunize rabbits, the resulting antibodies were immunochemically positive when assayed in the presence of their antigens but not in the presence of chromatin. Furthermore, antibodies produced against the isolated nonhistone proteins from one tissue type cross-reacted with similar proteins of another tissue, showing essentially no tissue specificity (15). It was concluded that chromatin of higher vertebrates contains species of nonhistone proteins which can form highly tissue specific complexes with the DNA of the same species (homologous) but not with the DNA of another species (heterologous).

Using this information, Wakabayashi et al (15,16) showed that the mixture of nonhistone proteins present in the pH 6.0 pellet obtained by dissociation and ultracentrifugation of rat liver chromatin contains protein species which bind to homologous (rat spleen) DNA but not to heterologous (calf thymus) DNA. Using affinity chromatography, Wakabayashi et al (15,16) isolated this protein fraction and showed that it consists of three major protein bands in polyacrylamide gel electrophoresis resolved in the presence of sodium dodecyl sulfate. The amino acid composition of these low molecular weight proteins (12-16,000 daltons) eliminated the possibility that these proteins are histones or their degradation products. When these nonhistone proteins isolated by DNA affinity chromatography were reconstituted to homologous DNA, the reconstitution product exhibited immunochemical specificity very similar to that of the reconstituted NP-DNA pH 6.0 pellet from the same tissue. It appears therefore that cells contain special proteins interacting very selectively with the DNA in chromatin, producing complexes which reflect in their specificity the differentiated state of cellular types comprising the particular tissue.

As was mentioned here, neoplastic transformation results in the formation of new NP-DNA complexes, specific for the cancerous growth (13). Individual transplantable tumors did not differ immunochemically in their NP-DNA complexes, regardless whether the tumors were of rat, mouse, or human origin. In this chapter, we show that the change in immunochemical specificity of the NP-DNA complexes can be detected relatively early during experimental azo-dye hepatocarcinogenesis in rats and that this change is different from changes produced by cellular response following partial hepatectomy.

The antigenic materials were prepared from chromatin isolated from liver nuclei or partially hepatectomized (17) rats or from livers of rats exposed to a diet containing 10% corn oil (Mazola) and 0.06% N,N-dimethyl-o-(m-tolylazo) aniline (3'-MDAB). Controls were either normal, sham-operated, Sprague-Dawley rats or Fisher rats fed standard Wayne Laboratory Meal (Allied Mills, Inc.) containing 10% corn oil. The isolation of liver nuclei and chromatin was described previously (18,19).

The NP-DNA complexes used for immunization were prepared by a modification of our earlier procedure (13). This modification yields purer complexes of DNA with the specific binding proteins. This is illustrated in

TABLE 1

FRACTIONATION SCHEME OF CHROMATIN PROTEINS

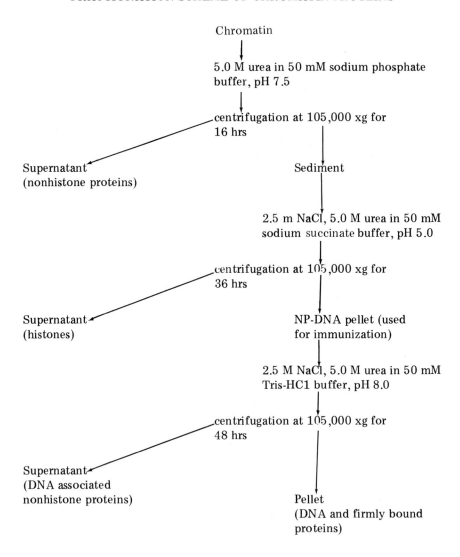

Chromatin

5.0 M urea in 50 mM sodium phosphate buffer, pH 7.5

centrifugation at 105,000 xg for 16 hrs

Supernatant
(nonhistone proteins)

Sediment

2.5 m NaCl, 5.0 M urea in 50 mM sodium succinate buffer, pH 5.0

centrifugation at 105,000 xg for 36 hrs

Supernatant
(histones)

NP-DNA pellet (used for immunization)

2.5 M NaCl, 5.0 M urea in 50 mM Tris-HC1 buffer, pH 8.0

centrifugation at 105,000 xg for 48 hrs

Supernatant
(DNA associated
nonhistone proteins)

Pellet
(DNA and firmly bound proteins)

Table I. Briefly, the nonhistone proteins are extracted by suspending isolated chromatin in 5.0 M urea, 50 mM sodium phosphate buffer, pH 7.5. The sediment resulting from a prolonged ultracentrifugation contains DNA, histones and the DNA-binding nonhistone proteins. The histones are removed by dissociation and ultracentrifugation in 2.5 M NaCl, 5.0 M urea and 50 mM succinate buffer, pH 5.0. The final pellet, containing the NP-DNA complexes, was used for immunization of rabbits (12,13).

The immunospecificity of the nonhistone protein-DNA complexes was determined by the micro-complement fixation method of Wasserman and Levine (14). The antisera were purified by chromatography on DEAE-cellulose columns and the gamma-globulin was concentrated by ammonium sulfate precipitation (20). For immunoassay, chromatin samples were prepared from the livers of rats maintained on the 3'-MDAB containing diet. Groups of animals were sacrificed at 7, 15, 111 and 175 days after the beginning of the experiment. In hepatectomized series, samples of liver chromatin were prepared 6, 12, 18, 24 and 48 hours following the hepatectomy. Finally, the immunospecificity of Morris hepatomas 7800, 7787, 3924A and 7777 was assayed to study the effect of their growth rates on their immunospecificity.

When the complement fixation of NP-DNA complexes prepared from normal or post-natal (three weeks) rat liver, Novikoff hepatoma, or livers of rats fed the 3'-MDAB containing diet for 175 days was compared using antiserum against the NP-DNA complex from livers of rats on 3'-MDAB diets, the normal and post-natal liver NP-DNA complexes exhibited no affinity for the 3'-MDB liver NP-DNA antiserum. On the other hand, the Novikoff hepatoma NP-DNA complex was highly immunoreactive (Fig. 1).

The notion that neoplasia changes the immunospecificity of the complexes of DNA with certain chromatin nonhistone proteins is further emphasized by the results shown in Figure 2. Here, the NP-DNA complexes prepared from the chromatins of either Novikoff hepatoma or livers of rats maintained on 3'-MDAB diet were assayed in the presence of antibodies against Novikoff hepatoma NP-DNA complex. While the initial seven days of 3'-MDAB feeding did not affect the immunospecificity of chromatin NP-DNA proteins, the complement fixation began to increase in two weeks reaching full reactivity of the Novikoff hepatoma antigen in 111 and 175 days.

In an attempt to correlate the immunoreactivity of various tumors to their growth rates, several Morris hepatomas were compared. The tumors were transplanted in the Department of Biochemistry at Howard University, Washington, D. C. by Dr. H. P. Morris and shipped to Houston, Texas. The tumors were harvested when reaching approximately 2-3 cm in diameter and their chromatin or NP-DNA complexes were compared using the complement fixation assay performed in the presence of Novikoff hepatoma NP-DNA antiserum. As can be seen in Figure 3, the affinity of various hepatoma NP-DNA preparations for the Novikoff hepatoma NP-DNA antiserum approximately parallels the degree of their differentiation as well as their individual growth rates. The fast growing and poorly differentiated 7777 and

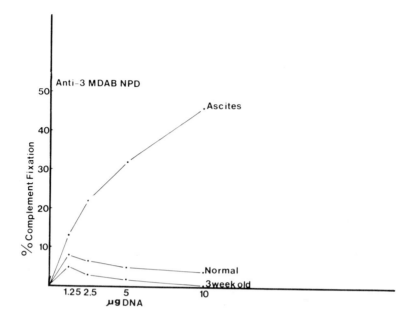

Fig. 1. Complement fixation of chromatin isolated from Novikoff hepatoma (ascites), adult rat liver (normal) and post-natal rat liver (3 week old) in the presence of antibodies against NP-DNA from livers of rats subjected to a diet containing hepatocarcinogen 3'-MDAB (Anti-3'-MDAB NPD). All experimental points were corrected for anticomplementarity. The amount of antigen is expressed as µg of DNA.

3942A hepatomas were more reactive than the slow growing and better differentiated 7800 and 7787 tumors. This correlation is only approximate, however, since the slowly proliferating tumors fixed the complement better than could be anticipated from their slower growth rates.

Since it is well documented that compensatory hypetrophy of liver produces both qualitative and quantitative changes in RNA transcription (21-23), liver chromatin preparations from hepatectomized rats were compared using antisera directed against normal liver or Novikoff hepatoma NP-DNA complexes.

As illustrated in Figure 4, 18 hours of regeneration changed the immunochemical character of NP-DNA complexes as compared with that of normal liver. However, the regenerating rat liver chromatin NP-DNA did not fix the complement more efficiently than the NP-DNA or chromatin samples prepared from Novikoff hepatoma or livers of rats maintained on 3'-MDAB diet for 111 days.

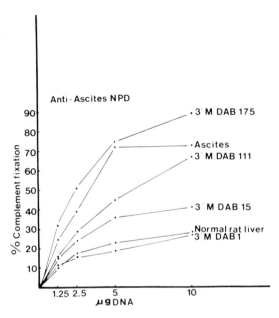

Fig. 2. Complement fixation of chromatin isolated from normal rat liver and livers of Fischer rats fed a hepatocarcinogen (3'-MDAB) containing diet for 1, 15, 111 and 175 days in the presence of antibodies against NP-DNA from Novikoff hepatoma (anti-ascites NPD). All experimental points were corrected for anticomplementarity.

Because the chromatin sample isolated 18 hrs after the hepatectomy was immunochemically different from normal or sham-operated rat liver NP-DNA, it was of interest to compare the immunospecificity of several post-operative time point preparations with that of the Novikoff hepatoma (Fig. 5). Quite unexpectedly, the NP-DNA sample isolated from livers 24 hrs after hepatectomy interacted strongly with antibodies against Novikoff hepatoma. The 6 and 12 hour samples were essentially nonreactive (similar to normal liver) while the 48 hr sample exhibited only limited immunoreactivity. It appears as if at least part of the immunochemically detectable changes characteristic for neoplasms were attributable to their rapid growth and proliferation, similar to that of the 24 hour regenerating rat liver NP-DNA sample.

It can be only speculated, at the present time, what biological roles the tissue specific DNA-binding proteins may play in living cells. It is noteworthy

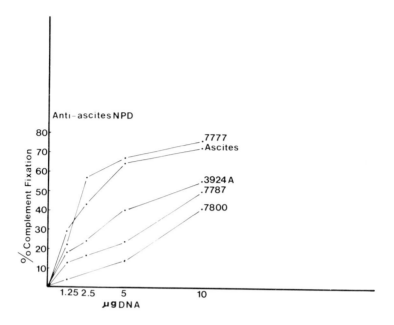

Fig. 3. Complement fixation of chromatin isolated from Novikoff hepatoma (ascites) and Morris hepatomas 7800, 7787, 3924A and 7777 in the presence of antibodies against Novikoff hepatoma NP-DNA (anti-ascites NPD). All experimental points were corrected for anticomplementarity.

that the NP-DNA pellet obtained by dissociation and ultracentrifugation of chromatin at pH 6.0 exhibits many interesting capabilities. In addition to the presence of the immunochemically tissue specific non-histone proteins interacting with DNA, it also contains traget tissue-specific receptors of steroid hormones (24), proteins regulating the transcription of tissue-specific RNA species (8,10), nonhistone protein phosphokinase enzymes (25), and possibly, the receptor sites for the interaction of certain carcinogens with chromatin (25). It can be anticipated that a detailed investigation and analysis of macromolecular species comprising this pellet may enlighten at least some of the mechanisms operating in the controls of gene activation and cell proliferation.

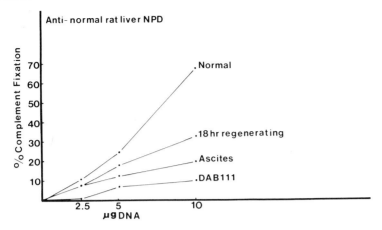

Fig. 4. Complement fixation of chromatin isolated from normal rat livers (normal), Novikoff hepatoma (ascites), livers of rats fed 3'-MDAB containing diet for 111 days (DAB 111) and livers 18 hrs after partial hepatectomy (18 hr regenerating) in the presence of antibodies against NP-DNA complexes from normal rat liver (anti-normal rat liver NPD). All experimental points were corrected for anticomplementarity.

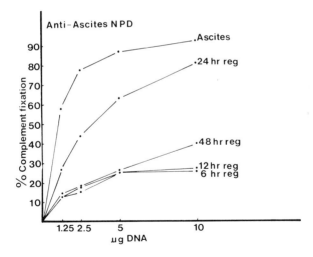

Fig. 5. Complement fixation of chromatin isolated from Novikoff hepatoma (ascites) and from livers of hepatectomized rats (6, 12, 24 and 48 hours after hepatectomy) in the presence of antibodies against Novikoff hepatoma NP-DNA (anti-ascites NPD). All experimental points were corrected for anticomplementarity.

Acknowledgements

Supported by the U. S. P. H. S. contract NIH-NCI-E-72-3269 and grants from the U. S. Public Health Service (CA 07746) and The Robert A. Welch Foundation (G-138).

References

1. L. S. Hnilica, *The Structure and Biological Function of Histones*, CRC Press, Cleveland, Ohio, 1972.
2. J. Paul and R. S. Gilmour, *J. Molec. Biol. 34*, 305 (1968).
3. K. D. Smith, R. B. Church and B. J. McCarthy, *Biochemistry 8*, 4271 (1969).
4. J. Bonner, M. E. Dahmus, D. Fambrough, R. C. C. Huang, K. Marushige and D. Y. H. Tuan, *Science 159*, 47 (1968).
5. R. S. Gilmour and J. Paul, *J. Molec. Biol. 40*, 137 (1969).
6. I. Bekhor, G. M. Kung and J. Bonner, *J. Molec. Biol. 39*, 351 (1969).
7. R. C. C. Huang and P. C. Huang, *J. Molec. Biol. 39*, 365 (1969).
8. T. C. Spelsberg and L. S. Hnilica, *Biochem. J. 120*, 435 (1970).
9. T. C. Spelsberg and L. S. Hnilica, *Biochim. Biophys. Acta 228*, 212 (1971).
10. T. C. Spelsberg, L. S. Hnilica and A. T. Ansevin, *Biochim. Biophys. Acta 228*, 550 (1971).
11. R. S. Gilmour and J. Paul, *FEBS Letters 9*, 242 (1970).
12. F. Chytil and T. C. Spelsberg, *Nature New Biol. 233*, 215 (1971).
13. K. Wakabayashi and L. S. Hnilica, *Nature New Biol. 242*, 153 (1973).
14. E. Wasserman and L. Levine, *J. Immunol. 87*, 290 (1961).
15. K. Wakabayashi, S. Wang and L. S. Hnilica, *Biochemistry*, in press.
16. K. Wakabayashi, S. Wang and L. S. Hnilica, Fogarty Internatl. Center Symposium on Poly ADP-Ribose, NIH, Bethesda, 1973, in press.
17. G. M. Higgins and R. M. Anderson, *Arch. Pathol. 12*, 186 (1931).
18. T. C. Spelsberg and L. S. Hnilica, *Biochim. Biophys. Acta 228*, 202 (1971).
19. J. A. Wilhelm, A. T. Ansevin, A. W. Johnson and L. S. Hnilica, *Biochim. Biophys. Acta 272*, 220 (1972).
20. H. J. Rapp, Immunochemical Methods (J. F. Ackboyd, ed.) p. 1, F. A. Davis Co., Philadelphia, 1964.
21. R. B. Church and B. J. McCarthy, *J. Molec. Biol. 23*, 459 (1967).
22. A. O. Pogo, V. C. Littau, V. G. Allfrey and A. E. Mirsky, *Proc. Natl. Acad. Sci. U. S. 57*, 743 (1967).
23. M. M. Thaler and C. A. Villee, *Proc. Natl. Acad. Sci. U. S. 58*, 2055 (1967).
24. T. C. Spelsberg, A. W. Steggles, F. Chytil and B. W. O'Malley, *J. Biol. Chem. 247*, 1368 (1972).
25. J. F. Chiu, Y. H. Tsai, S. Wang and L. S. Hnilica, in preparation.

POSSIBLE ROLES OF MICROTUBULES
AND ACTIN-LIKE FILAMENTS DURING CELL-DIVISION

Arthur Forer

Biology Department, York University
Downsview, Ontario M3J 1P3 Canada

In this article I discuss chromosome movements during cell-division (mitosis and meiosis). *Firstly* I consider what produces the force for chromosome movement. *Secondly* I consider how these forces are regulated such that movements of different chromosomes are coordinated.

I. Force Production

A. *Forces Come from Chromosomal Spindle Fibres*

The paradigm agreed upon by most workers is that the force for chromosome movement comes from the spindle fibres attached to the kinetochores of the chromosomes [i.e., the "chromosomal spindle fibres," as defined by Schrader (1)]. One reason for this belief is the *negative* argument that all other theories can be ruled out (2-4); for example, electrostatic or electromagnetic attractions (or repulsions) could not cause chromosome movements in anaphase because the needed electrostatic or electromagnetic energy could not exist in a living cell (3,4). As *positive* arguments, data from micromanipulation studies show that chromosomes are mechanically linked to the spindle by forces stronger than those needed to grossly stretch the chromosomes (5); data from microbeam irradiation studies show that irradiation of a chromosomal spindle fibre can stop the associated chromosome from moving, while the same irradiation doses applied to the interzonal region (i.e., between the separating chromosomes) or to the region outside the spindle do not stop chromosome motion (6).

Detailed discussions of the experimental bases for this paradigm can be found elsewhere (1,2,6-10): the data deal primarily with *anaphase* chromosome movements, but it is assumed that the prometaphase movements up and down the spindle are caused by spindle fibres as well (see, e.g., 8,11).

B. *Spindle Microtubules*

The term "microtubule" denotes objects which appear as dark, hollow cylinders, of about 25 nm outer diameter and 12 nm inner diameter, as seen in sectioned material after cells are fixed with glutaraldehyde, post-fixed with osmium tetroxide, dehydrated, embedded, sectioned, and stained (12-16).

Microtubules are universally associated with chromosomes during cell-division, attached to the chromosomal kinetochores (for reviews see 8,11). Because from light microscopic evidence one expects the force for chromosome movement to come from the chromosomal spindle fibre, and because from electron microscopic evidence microtubules are universally seen attached to kinetochores, many current hypotheses assume that microtubules produce the force for chromosome movement (e.g. 8,9,11, 17-20).

There are no data requiring one to consider *only* microtubules, however. Microtubules *per se* comprise only about 10% of the mass of isolated mitotic apparatus (21,22); *isolated* mitotic apparatus, in turn, contain only 5-10% of the mass of *in vivo* mitotic apparatus (7,23,24). Thus, microtubules *per se* comprise a very small fraction of the mass of the spindle, likewise the volume (7,25). The remainder of the spindle material is unknown, but *may* contain non-microtubule components, as suggested by the following data: Chemical analyses of isolated mitotic apparatus show several components (reviews in 7,26). Ultraviolet microbeam irradiation experiments suggest spindle components (6,8,9). Stored isolated mitotic apparatus can have 50% of the initial birefringence yet no visible microtubules (27), which suggests more than one spindle component. Cytochemical staining indicates several spindle components (7). Thus we conclude that microtubules *per se* comprise only a small fraction of the mass of a spindle, and that there is some evidence for additional components. The exact composition of the spindle is not known, however, because isolated mitotic apparatus contain only 10% of the mass of *in vivo* mitotic apparatus, and because some of this 10% is cytoplasmic contamination (24).

What do the spindle microtubules do? Despite the uncertainties regarding the exact composition of a spindle fibre, the fact that microtubules are universally seen attached to kinetochores strongly suggests that microtubules have *some* role in chromosome movement: what might this role be? Unfortunately, the evidence in this regard is unsatisfactory. The difficulty is that most of the experimental perturbations utilized so far have been treatments of entire cells. For example, entire cells have been treated with altered temperatures or with D_2O (18,19), either of which could affect many cellular processes; or cells have been treated with drugs such as vinblastine or colchicine (18,28). The difficulty in interpretation can perhaps best be illustrated by considering the colchicine data, as follows.

Colchicine binds strongly to *one* protein in the cell [called tubulin (29)], and the action of colchicine is thought to be specific on microtubules since microtubules are formed from this protein (30-35). When applied to dividing cells colchicine causes microtubule breakdown, spindle destruction, and concomitant block of chromosome movement (18,19). Thus, it is argued, since the drug specifically affects microtubules and since destruction of microtubules causes motion to stop, the motion must be caused by the microtubules. This argument need not be true, however. Firstly, colchicine binds to cytoplasmic and nuclear membranes as well as to tubulin (36-38), and colchicine alters nucleoside transport (39) as well as other cell processes (e.g., 40). Secondly, even if the colchicine effect were due solely to alterations of microtubules, the data are consistent with many roles of microtubules: for example, spindle microtubules could be strictly supportive (like girders of a building), or could be strictly guidelines (like tram tracks). The data might then be looked at as follows. Consider some men that move from the 3rd floor to the 6th floor of a building, while other men move from the 3rd floor to the basement (much as chromosomes move from the equator to the two poles). Destruction of the girders which causes the building to collapse would also cause each man's downwards or upwards movement to cease. But this would not mean that the men were pulled upwards by the girders, rather than by elevator, escalator, or the like. Likewise, destruction of the spindle prevents chromosome movement, but this does not necessarily mean that chromosomes normally move by virtue of the component destroyed.

Thus I conclude that while one does not know what microtubules do, nor whether there are other important spindle components, microtubules are universally associated with kinetochores during cell-division and must therefore have *some* role. On the other hand, I think it most likely that *actin-like filaments produce the force for chromosome movement,* and that microtubules have a different role. I shall now present the argument that the force is produced by actin-like filaments.

C. Actin-like Filaments

I shall first briefly describe actin, and the properties used for its identification. Then I will describe the apparent ubiquity of actin-like filaments in non-muscle cells, at sites where one expects force production; from the ubiquity I shall argue that actin-like filaments are likely to be involved in *all* motility, including chromosome movement.

Skeletal muscle contracts by virtue of 6-7 nm (diameter) actin filaments sliding past 14 nm myosin filaments (41), the energy coming from ATP; the regulatory proteins troponin and tropomyosin give a Ca^{++} requirement to the ATPase activity (41,42). Brief trypsin digestion of myosin forms 2 fragments, called light meromyosin (LMM) and heavy meromyosin (HMM), the latter having the ATPase activity and, unlike myosin or LMM, being soluble in low salts (43). "Arrowheads" are formed when HMM is added to actin filaments (44,45), the arrows pointing in one direction along any given filament. The formation of "arrowheads" enables one to identify actin-like filaments and to locate them intracellularly in non-muscle cells (46): cells are glycerinated (to make them permeable to HMM), HMM is added, and arrowheads are seen if the filaments in question are actin-like, nothing other than actin forming such arrowheads (45,46,47), with a further condition being that the arrowheads are removed by the addition of ATP (44,46).

Actin-like filaments are not only involved in muscle movement but are involved as well in cytoplasmic movement in non-muscle cells, as follows.

Motile cytoplasm isolated from amoebae contains filaments (48,49,50,51), of 2 size classes, both of which are necessary for movement (50). The 6-7 nm filaments are actin-like, as identified by arrowhead formation (52). Similar filaments are seen in *Physarum* cytoplasm (53,54), and the 6-7 nm filaments are similarly actin-like (55). Further, actin and myosin can be isolated *chemically* from both amoebae (56-59) and *Physarum* (60-63), and even cytoplasmic tropomyosin is found (64). All of this makes it difficult to avoid the conclusion that actin-myosin interactions cause protoplasmic movement in these amoeboid cells. An equally strong case can be made for actin involvement in amoeboid movements in higher cells as well: actin-like and myosin-like filaments are seen when *mammalian* blood platelets become amoeboid (65), and *actin*-like, *myosin*-like, and *tropomyosin*-like components have been isolated chemically from blood platelets (66-69). Actin-like and myosin-like proteins also have been isolated from human fibroblasts (70,71,72), mammalian leukocytes (73,74), and sea urchin eggs (75-77). Therefore actin-myosin interactions seem responsible for protoplasmic (amoeboid) movements in various cell types. This conclusion is strengthened by the identification of actin-like material (either by arrowhead formation or by chemical criteria) in cytoplasms of other cells as well, namely: in brain cells, neurons, and lens epithelia (78); in extending nerve processes (79,80); in macrophages (81); in rat megakaryocyte cytoplasm (82); in *Naegleria* (83); in various chick embryonic cells (46); in the brush border of chicken intestine epithelial cells (84); in synaptosomes (85-87); in *Chaos* cytoplasm (88); in rat kidney tubule and interstitial cells (89); in various motile mouse embryo and chick embryo cells (90); in mitochondrial (91-93); and in chloroplasts (94). Thus the conclusion seems inescapable that actin-like proteins are ubiquitous cytoplasmic components, and that actin-myosin interactions cause protoplasmic movements in *all* eukaryotes, from amoebae to man.

Actin-like filaments are probably involved with *other* motile systems as well, for such filaments have been identified (by *in situ* arrowhead formation) in several other places where, from previous work, one expects force production (or contraction). These places include the contractile ring causing cell cleavage of animal cells (95-97); the animal cell cortex, in which the actin-like filaments are present as a 1-2 micron thick layer of filaments around the outside of the cell (46,96-99); acrosomes of sperm (100,101); and cilia, flagella, and sperm tails (47,102-104), though some other reports find myosin-like material in cilia but no actin (105,106) and yet others find no actin-like material (107,108).

Condensation of chromatin is a further example of intracellular contractility probably mediated by actin-myosin interactions (109), as follows. Actin has been isolated from *Physarum* nuclei (110,111), actin and myosin have been isolated from calf thymus nuclei (112,113), and myosin was identified in sections of *Physarum* nuclei (114). Actin, myosin, and tropomyosin have been purified from *Physarum* nuclei, and have been identified (on SDS-polyacrylamide gels) in nuclear extracts from HeLa cells and human fibroblasts (109). That these contractile proteins have a role in condensation of chromatin is indicated by the changes in intranuclear concentrations of these components in the life of the cell (109,115,116): cells which become quiescent due to contact inhibition or starvation get condensed chromatin and at the same time the nuclei both lose tropomyosin and gain actin; when quiescent cells become active the reverse occurs (i.e. nuclei lose actin and gain tropomyosin). (The myosin concentrations are roughly the same in all conditions.) Thus, actin-myosin interactions seem to be implicated in chromatin condensation.

I argue from the data summarized above that actin-myosin interactions are universal agents of eukaryotic intracellular motility, and thus that actin-myosin interactions are probably involved in producing force for chromosome movement. There is no evidence against this hypothesis that I know of, especially since, as described above, spindle microtubules comprise only a small portion of the spindle's mass and volume, and since the composition of the non-microtubule portion is not known. The only positive evidences so far for the idea that filaments might be involved in chromosome movement include observations that filaments are sometimes seen in glutaraldehyde-fixed spindles (117,118), and that from arrowhead formation *in situ* actin-like filaments have been identified in meiotic spindles of a crane fly (102,119) and in mitotic spindles of locust spermatogonia (120). Also, fluorescent HMM binds to isolated spindles (121). Despite the so far meagre positive evidence for the hypothesis, I nonetheless expect actin-like filaments to produce force for chromosome movement.

Summarizing the argument to this point: the force for chromosome movement comes from the chromosomal spindle fibres, with which microtubules are universally associated. Based on the apparent universal use of actin-like filaments in eukaryotic motile systems, and a few reports of actin-like material in spindles, I expect actin-myosin interactions to produce the force for chromosome movement. That is to say, microtubules are universally present, but actin-myosin interactions produce the force: what then do microtubules do? Before considering this, I want to consider evidence of interactions between microtubules and actin-like filaments.

D. Are There Interactions Between Actin-like Filaments and Microtubules?

Some data (119) seem interpretable only in terms of an interaction between actin-like filaments and microtubules, as follows. *After glycerination*, crane fly spermatocytes contain neither microtubules nor actin-like filaments; *after glycerination followed by HMM treatment*, on the other hand, some spermatocytes contain microtubules and all contain actin-like filaments (see Table I). [Whether or not microtubules are present after HMM depends on the stage of division and type of microtubule: *no* microtubules are seen in prometaphase, roughly *all* microtubules are seen in metaphase, and only chromosomal microtubules are seen in anaphase (see 110 for qualifications and for all details).] These results force one to conclude that HMM somehow influences spindle microtubules: *before* addition of HMM microtubules are *not* seen, but *after* addition of HMM microtubules *are* seen, provided that ATP is

TABLE I

Results of treatment of crane fly spermatocytes [summarizing data from (119)].

Treatment	Microtubules	Actin-like filaments
glycerination followed by fixation	Absent	Absent
glycerination followed by HMM followed by fixation	Some present, some not *	Present
glycerination followed by HMM followed by ATP followed by fixation	Absent	Absent

* Described in the text.

not added (Table I). One must therefore say *either* that HMM interacts *directly* with microtubules, in an ATP sensitive reaction, *or* that HMM interacts in the usual way with actin, but *indirectly* affects microtubules because the actin interacts with microtubules. To help decide which possibility is more likely, let us look at some of the data on whether microtubules from *other* systems interact with myosin (or with HMM).

Some data suggest that microtubules do *not* interact with myosin (or HMM), as follows. Microtubules in glycerinated chick embryo mitotic cells are not altered in appearance due to addition of HMM, and therefore presumably do not bind HMM (46). *Protofilaments* of blood platelet microtubules (65) or of crane fly spermatid microtubules (47) are not altered in appearance after addition of HMM, and therefore do not appear to bind HMM. Isolated *doublet microtubules* from sea urchin sperm tails do not bind rabbit muscle HMM under conditions at which they bind dynein (122), as measured by sedimentation of the doublets out of an HMM-containing medium, and they do not activate the Mg^{++} - ATPase of rabbit muscle myosin at *low ionic strength* (123). Nor is the viscosity of a rabbit muscle myosin solution at *high ionic strength* altered by the addition of A-microtubules [prepared by thermal treatment of isolated sea urchin sperm tail doublet microtubules (123)]. All these data argue against a direct interaction between HMM and microtubules.

But the case is not clear cut, however. For example, if purified muscle actin is subjected to some of the treatments used to isolate sperm tail doublet microtubules, then the *actin* itself does *not* interact with myosin (123). Also, the distinctive appearance of the actin-HMM arrowhead complex depends on the summation of appearances of individual HMM molecules bound to actin on adjacent monomer subunits at different angles with respect to the actin filament length (45); thus it is not altogether clear that HMM binding to

324

protofilaments of microtubules would be detected morphologically, even if it occurred, and thus it is not clear whether the normal appearance of microtubules after addition of HMM unequivocally negates the possibility of HMM binding to the microtubule protofilaments. Indeed, other data suggest that microtubules *can* interact with myosin, as follows. A mixture of isolated sea urchin sperm tail doublet microtubules plus rabbit muscle myosin, at low ionic strength, undergoes ATP-induced "superprecipitation" (29,123), which pure myosin does not do. Also, tubulin purified from porcine brain both stimulates the Mg^{++} - ATPase activity of cat muscle myosin and causes an increase in viscosity when mixed with myosin (in high ionic strength solution), and the viscosity increase is sensitive to ATP (124).

Thus we have no clear-cut unequivocal answer regarding possible interactions of HMM and microtubules. On the other hand; there are many distinct clear-cut *differences* between *actin* and *tubulin* (14); to list a few, these include differences in molecular weight on SDS-acrylamide gels (79), a different bound nucleotide (125), differences in peptide patterns (79,123,126), and differences in antigenicity (127). Thus, despite the conflicting data on whether HMM can interact directly with microtubules, in view of the clear-cut differences between actin and tubulin I think that the most cautious interpretation is that HMM interacts directly only with actin, and that the HMM-actin interaction affects crane fly spermatocyte spindle microtubules only because the microtubules also interact with actin. Those anaphase spindle microtubules which interact with actin, then, would seem to be those associated with the chromosomes (119).

Other evidences also point towards interactions between actin and microtubules, as follows. Firstly, while it is clear that microtubules *can* form from *purified tubulin*, and essentially from *only* tubulin (33,34,35), it is not clear that all microtubules exist this way *in vivo*. Indeed, microtubules *isolated as microtubules* from cow, rabbit, or guinea pig brain are associated consistently with a component which on SDS-acrylamide gels co-migrates with actin (128): though no actin-like filaments are seen in the microtubule preparations themselves (128), 0.6M KC1 dissolves pelleted microtubules and then actin-like filaments are subsequently seen (129). Thus, these data suggest a possible interaction between actin and tubulin. Secondly, other data can be construed to suggest a similar interaction: microtubules *isolated as microtubules* from spores of a microsporidian are composed of 2 components which migrate on SDS-acrylamide gels with molecular weights of 46,000 and 53,000 daltons (130), which is the same migration behaviour as actin and tubulin, respectively. Here, too, a direct actin-tubulin interaction is possible.

In summary, I have argued that actin-like filaments produce force for chromosome movement, and have given evidence that actin-like material interacts with microtubules. This raises the question of what spindle microtubules do; but before considering the possible role(s) of the microtubules, and before considering how the actin-like filaments might produce the force, I want to briefly take up two points raised by the data summarized in Table I.

Firstly, regardless of whether the HMM affects spindle microtubules directly or rather via an intermediate, the data suggest that the microtubules are different in different stages of meiosis, as well as in different parts of the same spindle (119). That is to say, since after identical HMM treatment *no* microtubules are seen in *prometaphse* cells and *normal numbers* of microtubules are seen in *metaphase* cells (119), either the microtubules themselves have altered or else something associated with the microtubules has altered in that time period. Likewise for the differences between metaphase and anaphase, and between different portions of the anaphase spindle (119).

Secondly, actin-like filaments are not seen in crane fly spermatocyte spindles unless HMM is added (Table I): why are they not seen otherwise? This dilemma is not unique to crane fly spermatocyte spindles, for the same occurs in other systems (46,50,96,98). There are several possible explanations. *One* is that the filaments are always present, but that without HMM they are not preserved by the fixation methods. (This is analogous to studies before 1963 when osmium tetroxide was the sole fixative, and where microtubules were always *present* in normal cells but were not preserved.) This explanation requires the actin-HMM complex to be more stable than actin itself, and this is indeed true for the corresponding muscle proteins (131,132). *Another* possibility is that the filaments are present, and are preserved, but are just not recognized—for example because they are tightly packed and do not bind stain. Relevant to this possibility is the observation that actin filaments are seen when glycerinated spermatids are broken open and negatively stained, but are not recognized in sectioned spermatids (47). A *third* possibility is that the spindle actin exists as monomers, and that the HMM causes the monomers to polymerize: this *can* occur for both muscle actin and depolymerized spermatid actin (see 47). Thus, there are several conceivable, reasonable answers to the puzzle.

E. Roles of Actin-like Filaments and of Spindle Microtubules

If actin-like filaments *do* produce the force for chromosome movement, it seems reasonable to expect that the actin-like filaments slide past and work against other filaments: actin filaments in skeletal muscle slide past and work against myosin filaments (41), and one expects the same to apply elsewhere. To get chromosome movement, then, using the general criterion of filaments working against each other, one requires one kind of filament attached to the chromosome working against other filaments which are anchored somehow. Several possible arrangements can be imagined. *One possibility* is that one kind of filament is attached *directly* to the chromosome, and the other kind anchored either on (a) "continuous" microtubules (i.e., those not connected to the chromosomal kinetochores), or on (b) other spindle material. [With regard to the latter possibility, it is known that isolated spindles do not necessarily fall apart when microtubules are extracted (22,23), so the filaments could conceivably be anchored to the non-microtubule "matrix" material which holds the spindles together after extraction of microtubules.] *Another possibility* is that filaments are not attached directly to the chromosome, but rather are attached to microtubules which in turn are attached to the chromosome. The anchored countervailing filaments then could, as above, be anchored either to "continuous" microtubules or to matrix. A *third possibility* is that actin filaments work directly against microtubules attached to the chromosome; indeed, one or more of the microtubules attached to any given chromosome could conceivably be hollow-looking myosin filaments (see, for example, 133,134).

It is relevant to point out that the second or third possibilities are formally identical to other sliding microtubule models (8,17), with the substitutions of actin filaments for one kind of microtubule and either myosin filaments or microtubules for the other kind of microtubule.

Since I argued above that actin interacts with chromosomal microtubules at anaphase, those models are preferred which accommodate a microtubule-actin interaction. Even with this restriction, however, there are too many possibilities and too little hard data to justify detailed discussion of any models. Nonetheless, it is relevant to point out that the suggested microtubule-actin interaction does not necessarily imply that the same kind of microtubule actin interaction occurs *in vivo*: the observed association might occur *after* glycerination. For example, cytoplasmic microtubules in not-treated crane fly spermatids are generally well spaced throughout the cells, and are generally surrounded by a non-staining "clear area" (135); after glycerination, however, the same microtubules are "clumped" together and almost touch, and the clear areas are gone (47). Analogously, actin-like material need not necessarily be associated with spindle microtubules *in vivo* but could "clump" to and interact with them after glycerination removes impeding material (such as the "clear areas").

Assuming, then, that actin-like filaments produce the forces for chromosome movements, what do the microtubules do? In some of the models discussed above, the microtubules act as supports for the actin-like filaments and myosin-like filaments. This is one possible role. But I want to suggest another role; in doing so I wish to tie together 3 sets of data, which I now summarize.

(1.) The rates of chromosome movements have been carefully analyzed under conditions in which the chromosomal load pulled by the spindle fibre would be expected to vary by factors of 2 to 10 (3,8,136,137): the rates of movement were constant despite the large variations in load. From these data one concludes that *either* the chromosomal load is insignificant compared to some other rate-limiting step (3,137), *or* that the force is continuously adjusted to match different loads ["force compensation" (136,137)].

(2). From measured (or estimated) physical parameters one calculates the force needed for chromosome movement as about 10^{-9} - 10^{-8} dynes (3,4,7,8). This is a factor of 3000 *less* than that produced by one myosin and one actin filament of muscle acting in concert (7,138).

(3). The velocities of chromosome movements are much lower than those of other intracellular movements, or of cilia and flagella movements (139): chromosome velocity is around 0.017 μ/sec. [i.e., 1 μ/min. (10)], whereas actin and myosin filaments in muscle move 400 times faster [i.e., around 7 μ/sec.] and other intracellular movements have around the same velocity as muscle filaments (139).

To tie together these points I suggest that the depolymerization of microtubules is the rate-limiting step in chromosome movement; that is, actin-like filaments working in concert either with myosin-like filaments or with microtubules produce more than enough force to move chromosomes, and to move them much more quickly than the observed rate of movement, but the rate of movement is limited by microtubule breakdown: microtubules are pushed polewards (either directly or via the attached chromosomes), but need to be depolymerized at the ends before chromosomes can move. [In cells in which the spindles elongate without appreciable shortening of the chromosomal spindle fibres, the equivalent rate-limiting step might be polymerization of interpolar microtubules.] The role of the microtubules, then, is to damp the otherwise rapid movement of chromosomes, and thereby to provide mechanisms for regulating speeds of movement (137) and mechanisms for coordination of movement (discussed in the next section)* Put in another way, the polymerization-depolymerization reactions of spindle microtubules have been considered by others as the motive force for chromosome movement (18,19,20,140), but, in my suggestion, these reactions are not the motive force but rather are a means of regulating speeds and directions of movement and a means of coordinating movements (such as the start of anaphase); in my suggestion the force comes from the actin component but the speed is damped by the microtubules, and some of the coordination between chromosomes might arise through a time and space monitoring of the equilibrium between polymerized and not-polymerized components (19,140).

In proposing a "coordination" or "regulation" role for spindle microtubules I suggested that the microtubules were depolymerized at the poles during anaphase. There are no data which allow an unequivocal answer to the question of where spindle microtubules depolymerize, but since a region of reduced birefringence on an individual chromosomal spindle fibre moves poleward without change in dimensions (141), this suggests that removal of material from the spindle fibre occurs solely at the pole. Perhaps a more relevant reason for considering the fibre to be depolymerized at the pole is the expectation that some kind of geometrical order is required in order for forces to be produced by the interaction of 2 kinds of sliding filaments; one would expect that interstitial depolymerization of microtubules would interfere with maintenance of the needed geometry if indeed microtubules were one of the "filaments" involved (or if microtubules were associated with the filaments). Thus depolymerization at the ends seems more likely.

Before discussing evidences of precise coordination between chromosomes, I will briefly discuss two points relevant to considerations of spindle

---

*Others (8,17) have also considered the possibility that microtubule depolymerization may control the rates of chromosome movement.

microtubules.

Firstly, the equilibrium reaction of spindle microtubules is generally considered to be one in which the monomer directly polymerizes into a microtubule, with no intermediate steps or other possible forms of the polymer (18,19,142); this need not be so, however. There is evidence in several other systems that some tubulin exists in a form which is not microtubules but which is nonetheless centrifuged down into a pellet, and which is therefore not free monomer (e.g., 36,143,144); other experiments indicate that brain tubulin can form aggregates (with increased turbidity) which are not microtubules (145). Besides these data on "amorphous" aggregates, there is also evidence for tubulin being packed into forms other than the standard, cylindrical microtubule: individual microtubules can be converted by vinblastine into "bedspring helices" (146,147), and presumably into crystals (148), and both helices and crystals can be found within some cells even *without vinblastine treatment* (149,150,118). Other data suggest the possibility that tubulin can pack into a "macrotubule" form [review in (151)]. In sum, these data suggest that tubulin can exist in several states, and not just simply as free monomer and as microtubule.

Secondly, it is relevant to point out that all spindle microtubules need not be alike despite similarity in appearance. Not only can various microtubules which look alike respond differently to different treatments (135), but the same microtubules can be different at different points along their length (135), and other (non-tubulin) proteins can be indistinguishable in sections from microtubules (e.g., 152,153). Thus there may be several kinds of spindle microtubules.

Summing up this section, I have argued that microtubules comprise a small portion of the spindle, and that the remaining material is largely unknown. From the data showing actin-like filaments associated with various motile systems, in many cell types, I argued that actin filaments are involved with *all* intracellular force production, including that for moving chromosomes. I discussed several sliding filament geometries which might cause chromosome movement, and suggested that slow microtubule depolymerization damps the rate of chromosome movement, providing a means of regulating and perhaps coordinating chromosome movements.

II. Coordination Between Chromosomes

The basic paradigm underlying the discussion in the first section is that the forces for chromosome movements come from the chromosomal spindle fibres. A secondary paradigm is that the forces on individual chromosomes can be applied independently of those on other chromosomes (e.g. see 7,8,9,154). As an example of the data supporting this second idea, irradiation of one chromosomal spindle fibre prevents the associated chromosome from moving to the pole, yet *other* chromosomes nonetheless move normally (6): each chromosome therefore had an independent "motor."

On the other hand, as I shall argue below, the movements of various

329

chromosomes are often coordinated, and are *not* independent. This conjunction of independence with non-independence is comparable to motorists who are independent of each other in the sense that they each have separate cars, ignitions, etc. but who are at the time not-independent of each other in their movements because of traffic regulations, stop lights, the need to drive on roads, and even the need to obtain fuel from the same limited supply centre. I will now illustrate some of the coordination between chromosomes, and afterwards discuss possible mechanisms.

A. *Examples of Coordination Between Chromosomes*

As examples of not-independent chromosome movements, one could cite the following. Anaphase generally does not start until all chromosomes are at the equator (10), even though this may require all the other chromosomes to wait for a deviant "centrophilic" chromosome to leave the spindle pole (155). In some cells non-paired sex chromosomes regularly move to opposite poles during meiotic anaphase ["distance segregation" (156)], despite the absence of pairing (9). In meiosis in *Gryllotalpa* spermatocytes the larger partner of a heteromorphic bivalent always moves to the same pole as the univalent sex chromosome (157).

In all these examples the movements of different chromosomes seem somehow coordinated. Rather than continuing to list isolated examples from different cells, I will describe various examples of coordination detectable within one cell type, namely crane fly spermatocytes.

Crane fly spermatocytes contain three autosomal bivalents plus two unpaired sex-chromosomes (158,6,7,20,119,140,154). Anaphase does not begin until all five chromosomes are at the equator and oriented such that each half-bivalent has one chromosomal fibre to one of the two spindle poles and each univalent has *two* chromosomal spindle fibres, one to each pole. At anaphase the half-bivalents move to the two poles while the sex-chromosomes remain at the equator: only after the autosomes reach the poles do the sex-chromosomes move to opposite poles. This brief description illustrates coordination in *timing* of autosomal anaphase and in *timing* of sex-chromosomal anaphase, as well as coordination in *direction* of motion such that each sex-chromosome moves to a different pole.

In crane fly spermatocytes one can demonstrate further that the previous autosomal motion influences sex-chromosome movements, and that the total number of sex-chromosomes influences sex-chromosome movements as well (20,154). For example, if instead of three autosomes going to each pole the cell has four going to one pole and two to the other [this situation arising after micromanipulation (154) or after unequal disjunction from experimentally induced quadrivalents (20)], then one sex-chromosome moves poleward normally but the other sex-chromosomes does not move. Similarly, in cells with an uneven number of sex chromosomes (20) one sex-chromosome does not move. One sex-chromosome does not move either if the 2 sex-chromosomes in a cell are supplemented by one autosomal half-bivalent oriented to have one spindle fibre to each pole (20,154).

330

Not only can experimental perturbations *block* movement of one sex-chromosome, but the *direction* of sex-chromosome motion can be reversed even after it starts, by experimentally changing the position of the *other* sex-chromosome (154). This demonstrates that there must be continual "reading" of each other's position as the sex-chromosomes move to opposite poles in anaphase.

Autosomes not only influence sex-chromosomes, but they can influence each other as well. Microbeam irradiation of one spindle fibre *prior to anaphase* blocks the anaphase movement of *all* half-bivalents: even if all half-bivalents separate from each other immediately after the irradiation, *none* of them moves poleward without a delay, even though only one of the six spindle fibres in question was irradiated (141,6); when the half-bivalents *do* move, all begin move poleward at the same time (6). As another example, when one spindle fibre is irradiated in *anaphase* the associated half-bivalent may temporarily stop moving, and when it does the partner half-bivalent moving to the opposite pole also stops moving, even though other half-bivalents move normally (6); both the half-bivalent in question and the partner resume movement at the same time (6).

As all these examples illustrate, even within one and the same cell one can recognize various kinds of coordinated, interdependent chromosome movements.

B. Mechanisms of Coordination

So little is known about the mechanisms of coordination between chromosomes, that one is tempted to just shrug one's shoulders and await more data. It might be worthwhile, however, to mention two mechanisms which have been suggested to account for various aspects of coordination. These are that (a) coordination occurs by means of the amounts of polymerized *vs.* depolymerized microtubule material, considering microtubules as the motive force for movement (20,140), and that (b) coordination occurs by means of interactions between chromosomal spindle fibres (157,9,154). In the latter hypothesis one need not consider solely interactions between spindle fibres *per se*; for example, one could imagine that different spindle fibres, all of which contain actin, all compete for matrix-anchored myosin filaments, and the fibre-fibre interaction then would not be a direct one but rather would occur via an "intermediate component" (154). Micromanipulation experiments on crane fly spermatocytes favour the idea that fibre-fibre interactions (perhaps via an "intermediate component") are responsible for at least some of the coordination (154), but the case is not yet closed and some part of the *coordination* may still be due to a monitoring of the equilibrium between polymerized and depolymerized microtubules.

In *summary*, I have suggested that actin-like filaments produce the force for chromosome movement, but that microtubule breakdown is the rate-limiting step. This "damping" by the microtubules may have a role in regulating forces on chromosomes and coordinating movements of different chromosomes.

Chromosome coordination was discussed briefly, mainly to illustrate the problem and to point out our ignorance of the mechanisms involved.

Acknowledgement

I acknowledge the support of grants from the National Research Council of Canada.

References

1. F. Schrader, "Mitosis." Columbia Univ. Press, N.Y. (1953).
2. I. Cornman, *Amer. Naturalist 78*, 410 (1944).
3. E. W. Taylor, in Proc. Fourth Internat. Congress Rheology, E. H. Lee, edit., Interscience, N.Y., p. 175 (1965).
4. A. D. Gruzdev. *Tsitologiya 14*, 141 (1972).
5. R. B. Nicklas and C. A. Staehly, *Chromosoma 21*, 1 (1967).
6. A. Forer, *Chromosoma 19*, 44 (1966).
7. A. Forer, in "Handbook of Molecular Cytology," A. Lima-de-Faria, edit., North-Holland, London, p. 553 (1969).
8. R. B. Nicklas, *Adv. Cell Biol. 2*, 225 (1971).
9. P. Luykx, *Intern. Rev. Cytol. Suppl. 2*, 1 (1970).
10. D. Mazia, in "The Cell," Vol. 3, J. Brachet and A. E. Mirsky, edit., Acad. Press, N.Y. (1960).
11. A. S. Bajer and J. Mole-Bajer, *Internat. Rev. Cytol. Suppl. 3*, 1 (1972).
12. M. C. Ledbetter and K. R. Porter, *J. Cell Biol. 19*, 239 (1963).
13. O. Behnke, *Intern. Rev. Path. 9*, 1 (1970).
14. J. B. Olmsted and G. G. Borisy, *Annual Rev. Biochem. 42*, 507 (1973).
15. L. G. Tilney, in "Origin and Continuity of Cell Organelles," J. Reinert and H. Ursprung, edit., Springer-Verlag, Berlin, p. 222 (1971).
16. D. Slautterback, *J. Cell. Biol. 18*, 367 (1963).
17. J. R. McIntosh, P. K. Hepler, and D. G. Van Wie, *Nature 224*, 659 (1969).
18. S. Inoue and H. Sato, *J. Gen Physiol. 50 (No. 6, Pt. 2)*, 259 (1967).
19. S. Inoue, in "Biology and the Physical Sciences," S. Devons, edit., Columbia Univ. Press, N.Y., p. 139 (1969).
20. R. Dietz, *Naturwiss. 56*, 237 (1969).
21. W. D. Cohen and L. I. Rebhun, *J. Cell Sci. 6*, 159 (1970).
22. T. Bibring and J. Baxandall, *J. Cell Biol. 48*, 324 (1971).
23. A. Forer and R. D. Goldman, *Nature 222*, 689 (1969).
24. A. Forer and R. D. Goldman, *J. Cell Sci. 10*, 387 (1972).
25. H. Fuge, *Protoplasma* in press (1973).
26. J. F. Hartmann and A. M. Zimmerman, in "The Cell Nucleus," H. Busch, edit., Academic Press, N.Y., (1974).
27. R. D. Goldman and L. E. Rebhun, *J. Cell Sci. 4*, 179 (1969).
28. P. George, L. J. Journey, and M. N. Goldstein, *J. Natl. Cancer Instit. 35*, 355 (1965).

29. H. Mohri, *Nature 217*, 1053 (1968).
30. R. C. Weisenberg, G. G. Borisy, and E. W. Taylor, *Biochem.* 7, 4466 (1968).
31. G. G. Borisy and E. W. Taylor, *J. Cell Biol. 34*, 525 (1967).
32. G. G. Borisy and E. W. Taylor, *J. Cell Biol. 34*, 535 (1967).
33. R. C. Weisenberg, *Science 177*, 1104 (1972).
34. G. G. Borisy and J. B. Olmsted, *Science 177*, 1196 (1972).
35. M. L. Shelanski, F. Gaskin, and C. R. Cantor, *Proc. Nat. Acad. Sci. 70*, 765 (1973).
36. H. Feit and S. H. Barondes, *J. Neurochem. 17*, 1355 (1970).
37. J. Stadler and W. W. Franke, *Nature New Biol. 237*, 237 (1972).
38. W. W. Franke, J. Stadler, and S. Krien, *Beitr. Path 146*, 289 (1972).
39. S. B. Mizel and L. Wilson, *Biochem. 11*, 2573 (1972).
40. P. H. Fitzgerald and L. A. Brehaut, *Exp. Cell Res. 59*, 27 (1970).
41. H. E. Huxley, *Science 164*, 1356 (1969).
42. A. Weber and J. M. Murray, *Physiol. Rev. 53*, 612 (1973).
43. A. G. Szent-Györgyi, *Arch. Biochem. Biophys. 42*, 305 (1953).
44. H. E. Huxley, *J. Mol. Biol. 7*, 281 (1963).
45. P. B. Moore, H. E. Huxley, and D. J. De Rosier, *J. Mol. Biol. 50*, 279 (1970) and H. E. Huxley, *Nature New Biol. 243*, 445 (1973).
46. H. Ishikawa, R. Bischoff, and H. Holtzer, *J. Cell Biol. 43*, 312 (1969).
47. A. Forer and O. Behnke, *J. Cell Sci. 11*, 491 (1972).
48. C. M. Thompson and L. Wolpert, *Exp. Cell Res. 32*, 156 (1963).
49. J. Morgan, *Exp. Cell Res. 65*, 7 (1971).
50. T. D. Pollard and S. Ito, *J. Cell Biol. 46*, 267 (1970).
51. C. R. Gicquaud and P. Couillard, *Cytobiologie 5*, 139 (1972).
52. T. D. Pollard and E. D. Korn, *J. Cell Biol. 48*, 216 (1971).
53. H. Komnick, W. Stockem, and K.E. Wohlfarth-Bottermann, Z. *Zellforsch. 109*, 420 (1970).
54. D. Kessler, *J. Mechanochem. Cell motility 1*, 125 (1972).
55. A. Allera, R. Beck, and K.-E. Wohlfarth-Bottermann, *Cytobiologie 4*, 437 (1971).
56. T. D. Pollard, E. Shelton, R. R. Weihing, and E. D. Korn, *J. Mol. Biol. 50*, 91 (1970).
57. R. R. Weihing and E. D. Korn, *Biochem. 10*, 590 (1971).
58. T. D. Pollard and E. D. Korn, *J. Biol. Chem. 248*, 4682 (1973).
59. D. E. Woolley, *J. Cell Physiol. 76*, 185 (1970): *Arch. Biochem. Biophys. 150*, 519 (1972).
60. S. Hatano and F. Oosawa, *Biochim. Biophys. Acta. 127*, 488 (1966).
61. M. R. Adelman and E. W. Taylor, *Biochem. 8*, 4976 (1969).
62. M. R. Adelman and E. W. Taylor, *Biochem. 8*, 4964 (1969).
63. V. T. Nachmias, H. E. Huxley, and D. Kessler, *J. Mol. Biol.50*, 83 (1970).
64. H. Tanaka and S. Hatano, *Biochim. Biophys. Acta 257*, 445 (1972).
65. O. Behnke, B. I. Kristensen, and L. E. Nielsen, *J. Ultrastruct. Res. 37*,

351 (1971).

66. M. Bettex-Galland and E. F. Lüscher, *Adv. Prot. Chem. 20*, 1 (1965).

67. E. Probst and F. Lüscher, *Biochim. Biophys. Acta 278*, 577 (1972).

68. R. S. Adelstein, T. D. Pollard, and W. M. Kuehl, *Proc. Nat. Acad. Sci. 68*, 2703 (1971).

69. I. Cohen and C. Cohen, *J. Mol. Biol. 68*, 383 (1972).

70. Y. Yang and J. F. Perdue, *J. Biol. Chem. 247*, 4503 (1972);J. F. Perdue, *J. Cell Biol. 58*, 265 (1973).

71. R. S. Adelstein and M. A. Conti, *Cold Spring Harbor Symp. Quant. Biol. 37*, 599 (1973).

72. R. S. Adelstein, M. A. Conti, G. S. Johnson, I. Pastan, and T. D. Pollard, *Proc. Nat. Acad. Sci. 69*, 3693 (1972).

73. N. Senda, N. Shibata, N. Tatsumi, K. Kondo, and K. Hamada, *Biochim. Biophys. Acta 181*, 191 (1969).

74. N. Shibata, N. Tatsumi, K. Tanaka, Y. Okamura, and N. Senda, *Biochim. Biophys. Acta 256*, 565 (1972).

75. T. Miki-Noumara and F. Oosawa, *Exp. Cell Res. 56*, 224 (1969).

76. T. Miki-Noumara and H. Kondo, *Exp. Cell Res. 61*, 31 (1970).

77. I. Mabuchi, *J. Cell Biol. 59*, 542 (1973).

78. D. Bray, *Cold Spring Harbor Symp. Quant. Biol. 37*, 567 (1973).

79. D. E. Fine and D. Bray, *Nature New Biol. 234*, 115 (1971).

80. C. M. Chang and R. D. Goldman, *J. Cell Biol. 57*, 867 (1973).

81. A. C. Allison, P. Davies, and S. de Petris, *Nature New Biol. 232*, 153 (1971).

82. O. Behnke and J. Emmersen, *Scand. J. Haemat. 9*, 130 (1972).

83. A. J. Lastovica and A. D. Dingle, *Exp. Cell Res. 66*, 337 (1971).

84. L. G. Tilney and M. Mooseker, *Proc. Nat. Acad. Sci. 68*, 2611 (1971).

85. S. Puzkin, W. J. Nicklas, and S. Berl, *J. Neurochem. 19*, 1319 (1972).

86. S. Puzkin and S. Berl, *Biochim. Biophys. Acta 256*, 695 (1972).

87. S. Berl, S. Puzkin, and W. J. Nicklas, *Science 179*, 441 (1973).

88. L. T. Comly, *J. Cell Biol. 58*, 230 (1973).

89. J. Rostgaard, B. I. Kristensen, and L. E. Nielsen, *Z. Zellforsch. 132*, 497 (1972).

90. B. S. Spooner, J. F. Ash, J. T. Wrenn, R. B. Frater, and N. K. Wessels, *Tissue and Cell 5*, 37 (1973).

91. T. Ohnishi and T. Ohnishi, *J. Biochem. 52*, 230 (1962).

92. T. Ohnishi, H. Kawamura, K. Takeo, and S. Watanabe, *J. Biochem. 56*, 273 (1964).

93. J. C. Arcos, R. E. Stacey, J. B. Mathison, and M. F. Arcus, *Exp. Cell Res. 48*, 448 (1967).

94. T. Ohnishi, *J. Biochem. 55*, 494 (1964).

95. M. M. Perry, H. A. John, and N.S.T. Thomas, *Exp. Cell Res. 65*, 249 (1971).

96. A. Forer and O. Behnke, *Chromosoma 39*, 175 (1972).

97. T. E. Schroeder, *Proc. Nat. Acad. Sci. 70*, 1688 (1973).

98. H. Holtzer, J. W. Sanger, H. Ishikawa, and K. Strahs, *Cold Spring Harbor Symp. Quant. Biol. 37*, 549 (1973).
99. R. D. Goldman and D. M. Knipe, *Cold Spring Harbor Symp. Quant. Biol. 37*, 523 (1973).
100. H. Jessen, O. Behnke, K. G. Wingstrand, and J. Rostgaard, *Exp. Cell Res. 80*, 47 (1973).
101. L. Tilney, S. Hatano, H. Ishikawa, and M. S. Mooseker, *J. Cell Biol. 59*, 109 (1973).
102. O. Behnke, A. Forer, and J. Emmersen, *Nature 234*, 408 (1971).
103. L. G. Young and L. Nelson, *Exp. Cell Res. 51*, 34 (1968).
104. L. G. Young and L. Nelson, *J. Cell Physiol. 74*, 315 (1969).
105. S. A. Burnasheva, *Biokhimiya* (English translation) *23*, 523 (1958).
106. S. A. Burnasheva and N. V. Raskidnaya, *Dokl. Akad. Nauk USSR* (English translation) *179*, 73 (1968).
107. P. R. Burton and W. L. Kirkland, *Nature New Biol. 239*, 244 (1972).
108. R. W. Rubin and L. P. Everhart, Jr., *J. Mol. Biol. 75*, 437 (1973).
109. W. M. Le Stourgeon, A. Forer, Y.-Z. Yang, J. S. Bertram, and H. P. Rusch, submitted to *Nature* (Nov., 1973).
110. B. M. Jockusch, D. F. Brown, and H. P. Rusch, *J. Bact. 108*, 705 (1971).
111. B. M. Jockusch, D. F. Brown, and H. P. Rusch, *Biochem. Biophys. Res. Comm. 38*, 279 (1970).
112. T. Ohnishi, H. Kawamura, and T. Yamamoto, *J. Biochem. 54*, 298 (1963).
113. T. Ohnishi, H. Kawamura, and Y. Tanaka, *J. Biochem. 56*, 6 (1964).
114. B. M. Jockusch, U. Ryser, and O. Behnke, *Exp. Cell Res. 76*, 464 (1973).
115. W. M. Le Stourgeon and H. P. Rusch, *Arch. Biochem. Biophys. 155*, 144 (1973).
116. W. M. Le Stourgeon, W. Wray, and H. P. Rusch, *Exp. Cell Res. 79*, 487 (1973).
117. A. Bajer and J. Molè-Bajer, *Chromosoma 27*, 448 (1969).
118. W. Müller, *Chromosoma 38*, 139 (1972).
119. A. Forer and O. Behnke, *Chromosoma 39*, 145 (1972).
120. N. Gawadi, *Nature 234*, 410 (1971).
121. J. F. Aronson, *J. Cell Biol. 26*, 293 (1965).
122. M. Hayashi and S. Higashi-Fujime, *Biochem. 11*, 2977 (1972).
123. H. Mohri and M. Shimomura, *J. Biochem. 74*, 209 (1973).
124. S. Puzkin and S. Berl, *Nature 225*, 558 (1970).
125. R. E. Stephens, F. L. Renaud, and I. R. Gibbons, *Science 156*, 1606 (1967).
126. R. E. Stephens, *Science 168*, 845 (1970).
127. C. Fulton, R. E. Kane, and R. E. Stephens, *J. Cell Biol. 50*, 762 (1971).
128. P. Filner and O. Behnke, submitted to *Exp. Cell Res.*
129. A. Forer, unpublished observations.
130. D. M. Dwyer and E. Weidner, *Z. Zellforsch. 140*, 177 (1973).
131. M. Kikuchi, H. Noda, and K. Maruyama, *J. Biochem. 65*, 945 (1969).

132. K. Tawada and F. Oosawa, *Biochim. Biophys. Acta 180*, 199 (1969).

133. G. Hoyle and P. A. McNeill, *J. Exp. Zool. 167*, 487 (1968).

134. C. Franzini-Armstrong, *J. Cell Sci. 6*, 169 (1967).

135. O. Behnke and A. Forer, *J. Cell Sci. 2*, 169 (1967).

136. R. B. Nicklas, *Chromosoma 14*, 276 (1963).

137. R. B. Nicklas, *J. Cell Biol. 25 (No. 1, Pt. 2)*, 119 (1965).

138. J. Lowy, B. M. Millman, and J. Hanson, *Proc. Roy Soc. London B 160*, 525 (1964).

139. L. Wolpert, *Symp. Soc. Gen. Microbiol. 15*, 270 (1965).

140. R. Dietz, *Chromosoma 38*, 11 (1972).

141. A. Forer, *J. Cell Biol. 25 (No. 1, Pt. 2)*, 95 (1965).

142. R. E. Stephens, *J. Cell Biol. 57*, 133 (1973).

143. R. C. Weisenberg, *J. Cell Biol. 54*, 266 (1972).

144. R. A. Raff and J. F. Kaumeyer, *Develop. Biol. 32*, 309 (1973).

145. G. G. Borisy, J. B. Olmsted, and R. A. Klugman, *Proc. Nat. Acad. Sci. 69*, 2890 (1972).

146. R. Marantz and M. L. Shelanski, *J. Cell Biol. 44*, 234 (1970).

147. O. Behnke and A. Forer, *Exp. Cell Res. 73*, 506 (1972).

148. J. Bryan, *J. Mol. Biol. 66*, 157 (1972); B. Jockusch and J. Blessing, *Exp. Cell Res. 69*, 465 (1971).

149. H. Fuge, *Exp. Cell Res. 60*, 309 (1970).

150. M. P. Osborne, *J. Cell Sci. 11*, 295 (1972); N. Gawadi, *J. Cell Sci. 11*, 887 (1972).

151. G. E. Tyson and R. E. Bulger, *Z. Zellforsch. 141*, 443 (1973).

152. S. Dales, I. T. Schulze, and S. Ratner, *Biochim. Biophys. Acta 229*, 771 (1971).

153. J. Cronshaw and K. Essau, *J. Cell Biol. 34*, 801 (1967).

154. A. Forer and C. Koch, *Chromosoma 40*, 417 (1973).

155. R. B. Uertz, W. Bloom, and R. E. Zirkle, *Science 120*, 197 (1954).

156. S. Hughes-Schrader, *Chromosoma 27*, 109 (1969).

157. R. Camenzind and R. B. Nicklas, *Chromosoma 24*, 324 (1968).

158. H. Bauer, *Z. Zellforsch. 14*, 138 (1931).

FLUCTUATIONS IN CHROMATIN FUNCTION
AND COMPOSITION DURING CELL PROLIFERATION

M. E. McClure

Department of Cell Biology
Baylor College of Medicine
Houston, Texas

I. Introduction

The transition from a resting cell to a mitotically cycling cell necessarily entails a substantial realignment from the precisely regulated circumstances of metabolism concerned with the specialized, differentiated function of the cell to a metabolic pattern in which a major expenditure of energy is directed toward the continued chemical synthesis of those nucleic acid and protein molecules specifically required for cell growth and reproduction. Studies on the life cycle of homogeneous cell populations grown *in vitro* uniquely demonstrate that the chemical composition of a growing cell changes continuously throughout interphase in an orderly, discretely definable manner. It is clear at present that the exquisitely planned order of the biochemical sequences observed during the cell cycle reflects, in large part, a genetically directed logistical expression of anticipated macromolecular changes associated with cell proliferation. What is not clear at present is the specific role played by genome utilization (transcription) in the mechanism of cell proliferation. It is currently unknown whether the progressive biochemical events required for cell cycle traversal are programmed by a few transcriptional bursts of genomic expression or by a more continuous process of incremental transcriptional activity.

In the quest for knowledge of the manner in which a normal cell genetically controls its own growth and reproduction, the synchronously growing *in vitro* cell population represents a major tool. During the creation of an entire new genome within the cell, the developmentally compressed history of the cell type is recreated as the pattern of genomic restriction for that cell type. The materials and mechanisms for both the regions of permanent genetic restriction and potential genetic expression are synthesized anew and assembled by definable chemical events. Thus, in one system one can experimentally search for the novel molecular components involved in creating the chromosome regions associated with developmentally dictated genetic activity or inactivity. Furthermore, significant insight into the mechanism of selective gene activation can be gained by studies conducted on the definable chemical entities present in the genetic apparatus during G_1 or G_2 of the cycle when marked variations in chromosome stability occur which are related to variations in gene action.

337

Unfortunately, present knowledge is most inadequate in this important area of research. It is the purpose of the present paper to discuss the relationship between the compositional modification of genomic chromatin and the functional status of that chromatin during the cell cycle.

II. Characterization of the Chinese Hamster Fibroblast Cell Cycle

Although much has been learned from the general metabolic patterns observed during cell proliferation, it has only recently been possible to demonstrate a requisite temporal order of chemical synthetic events required for the successful entrance into and continued traversal of the mitotic cycle. It is clear that these events are a consequence of controlled chromatin function since they rely on unhampered gene activity at the biochemical level and reoccur with precise temporal order in subsequent progeny cells. In this section, we will briefly discuss the major patterns of macromolecule synthesis with respect to the importance of their proper scheduling to the successful traversal of the cell cycle by Chinese hamster fibroblast cells.

A. Variations in the DNA Content

DNA synthesis has been the most exhaustively studied of the biochemical events during the cell cycle. A number of methods (1,2) have evolved for determining the temporal profile and duration of DNA synthesis during the cell cycle. The most sophisticated technique currently available (mitotic selection) is one in which those elements of the cell population undergoing mitosis are selectively detached and isolated as an essentially pure preparation of metaphase cells which, upon culturing, initiate a synchronous wave of division and traversal of the resultant cell cycle (2). The application of this technique (3) to growing Don-C Chinese hamster cells produced synchronous populations with temporal patterns for both the rate of synthesis and the accumulation of newly synthesized DNA which were essentially the same as described by other workers (4-6) for mitotic detachment synchronization of the Don-C Chinese hamster cells. The pause in net *accumulation* of ^{3}H-Thymidine (^{3}H-TdR) labelled DNA generally occurred around 6 hours into the cell cycle (Fig. 1). Using the data obtained on the accumulation of ^{3}H-TdR into newly synthesized DNA over the cell cycle, the average rate of change of accumulated isotope per time interval (CPM_2-CPM_1 divided by t_2-t_1) was calculated. The histogramed results (Fig. 1) clearly illustrated the bimodal nature of the rate of accumulation of newly synthesized DNA in the synchronously growing cultures. We do not yet clearly understand the nature of the observed pause in net DNA accumulation associated with the depressed rate of DNA synthesis. Irrespective of the rate interpretation, the data obtained on DNA accumulation parameters demarcated the duration and temporal position of the cell cycle stages (horizontal bars) following entrance into the cycle at zero time. Thus, zero to one hour after reversal represented reinitiated mitosis, one to two and one-half hours was G_1, two and one-half to ten hours was the S period, and ten to twelve hours was the G_2 mitosis period.

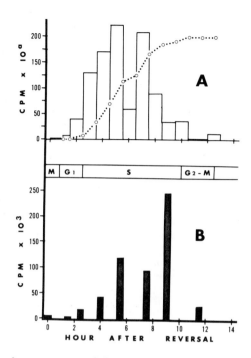

Fig. 1. Temporal parameters of DNA synthesis during synchronous Chinese hamster cell proliferation. In 1A, the open circles represent the net accumulation of ³H-TdR during *continuous* labeling with the isotopic precursor. The histogram represents the calculated average rate of change. Details are given in the text. The ordinate value for "a" is 3 for the histogram and 4 for the curve data. In 1B, the results of ³H-TdR pulse labeling of independent cultures established on different days at defined periods (after initiation of cell cycle traversal) are shown for numerically equivalent cultures.

B. *Primary Initiation of Chromatin-dependent RNA Synthesis*

Various authors have demonstrated the resumption of chromatin-directed RNA synthesis in late telophase of mitosis (3, 7-9). In the case of the Chinese hamster cell (3), metaphase populations reversed in medium containing ³H-Uridine (³H-Ur, 5 µCi/ml, 17.3 Ci/m) and prepared for radioautography at intervals of 5 minutes thereafter demonstrated an incorporation of ³H-Ur into acid-precipitable, RNAse-sensitive material which commenced only after the advent of cytokinesis (at telophase-late telophase). It has been demonstrated by others that nuclear reformation commenced at this stage of mitosis (9). The chromosomes in such cells were frequently observed to form closed loops by telomere adherence and showed regions with varying degrees of despiralization. The appearance of chromatin-directed RNA synthesis activity was evidenced in radioautographs by individual grains sparsely aligned along the chromosomal axis and, more interestingly, minor and major foci of ^{3}H-Ur incorporation over spatially restricted chromosomal regions (Fig. 2, arrows). Based on 500 cell counts, the number of major sites increased from 1.7 ± 0.4 (S. D. in all cases) for late telophase cells to 3.2 ± 1.4 in late telophase-early G_1 and 4.6 ± 0.8 for early G_1 cells. Randomly growing cultures of the same cell line showed prophase site numbers of 3.4 ± 0.3 and an interphase cell site number of 3.5 ± 0.2. The foci of ^{3}H-Ur incorporation in the latter two cases correlated spatially with nucleolar locations. These results suggested that the major sites of RNA synthesis observed gradually increased in number from late telophase to early G_1 and were associated with nucleolar function. While it was tempting to further suggest that G_1 was characterized by the existence of an additional major site of RNA synthesis, this suggestion remained unwarranted due to the inability to exclude the possibility of one or more of the active sites becoming spatially coincident following early G_1, i.e., reduced detection rather than occurrence.

C. Required Intervals of RNA Synthesis

In agreement with the work reported by Buck et al. (10), the author was unable to interrupt the traversal of mitosis in Don-C cells with inhibitors of RNA or protein synthesis. Thus, if synchronized metaphase cell populations (0.26×10^6 cells/cm^2) were allowed to resume cell division in the presence of either Cycloheximide (9.1 μg/ml) or Actinomycin-D (5.0 μg/ml), the time dependent disappearance of metaphase cells and the appearance of nucleated G_1 cells in the population (Fig. 3) were indistinguishable from the control in both cases. Since controls proved the agents active and pre-exposure (30-90 minutes) permitted adequate time for action, it was clear that neither continued RNA synthesis nor protein synthesis were required for traversal of mitosis past metaphase. Interestingly, radioautographs failed to demonstrate the appearance of the active sites of RNA synthesis described earlier (Fig. 2) when the metaphase cell population had been released in Cycloheximide or Actinomycin-D. It seemed clear from these results that no requirement for continued RNA or protein synthesis attended mitotic traversal to G_1 entry.

Interphase progression has also been surveyed using inhibitors of RNA and protein synthesis to study the obligatory requirements for RNA and/or protein synthetic activity prior to a defined cell cycle event as an end-point (11,12,13). The extent of the requirement for continuing RNA and protein synthesis during G_1 traversal in synchronized Don-C Chinese hamster fibroblasts was found to be remarkable (Table I). These studies were accomplished by reversing synchronized cells into medium containing 1 μCi/ml of either ^{14}C-Uridine, ^{3}H-Uridine (Ur) or ^{3}H-Thymidine (TdR) and adding either Actinomycin-D (5 μg/ml) or Cycloheximide (10 μg/ml) at various time points during the cell cycle. The series of cultures established in ^{14}C- or ^{3}H-Ur

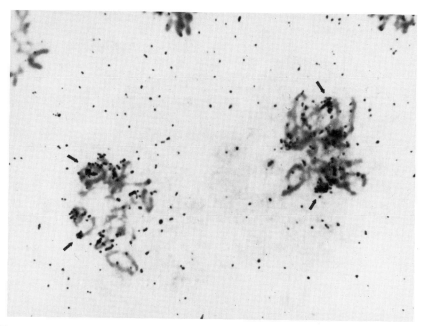

Fig. 2. Focalized initiation of RNA synthesis, in late telophase-early G_1 Chinese hamster cells. Details are given in the text. Focalized incorporation of ^{3}H-Uridine into acid precipitable, RNAse sensitive material appears at the arrows.

medium served to monitor the inhibition of RNA synthesis, while the ^{3}H-TdR labeled cultures monitored the extent of DNA synthesis which occurred. The experiment reported in Table IA was also controlled by a study of the cell kinetics of reinitiated cell division which demonstrated that the 0 and 20 minute additions of Actinomycin-D (AMD) occurred prior to the advent of *de novo* chromatin-directed RNA synthesis at late telophase. It was observed from the data of Table I that the addition of AMD during cell division (0 to 70 minutes) or at points less than 90 minutes into the cycle (early G_1) effectively prevented the traversal of G_1 and entry into the S period. The addition of AMD at points closer to the initiation of the S period (150-180 minutes) permitted an increased amount of DNA synthesis, yet still evidenced severe inhibition of net DNA synthesis during the cell cycle, e.g., the addition of

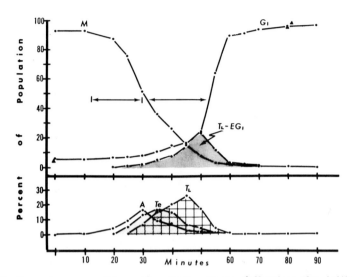

Fig. 3. Percent composition of mitotic stages following the initiation of synchronous cell cycle traversal (500 cell counts). The stages scored are metaphase (m), anaphase (A), telophase-early (Te), telophase-late (TL), late telophase-early G_1 (TL-G_1) and G_1. The triangles represent the percent nucleated cells during the initiation of synchronous growth in the presence of actinomycin-D blocked RNA synthesis. Details are given in the text.

AMD at 180 minutes restricted net DNA synthesis to 18% of the control level. It was subsequently found (Table IIA) that AMD (2 μg/ml) treated replicate cultures pulse labeled with ^{3}H-TdR (2 μCi/ml, 6.7 μCi/mM) at 6 hours post-reversal demonstrated, as expected, the nearly total inhibition of net DNA synthesis. Radioautography, however, showed the presence of labeled nuclei in numbers consistent with the 3.2% (random) interphase cells contaminating the synchronous population (96.8%) only in samples exposed to AMD prior to 80 minutes (early G_1) into the cell cycle. The increased number of labeled nuclei observed thereafter was not accounted for by contaminant interphase cells. Although the addition of AMD at 120 minutes (late G_2) inhibited 96% of the total net DNA synthesized by 6 hours into the cycle, 76% of the cell nuclei demonstrated ^{3}H-TdR incorporation and, hence, DNA synthesis characteristic of entry into the S period.

The results discussed above demonstrated the obligatory requirement of continued *de novo* RNA synthesis during G_1 with an apparent "relaxing" of this requirement during mid to late G_1 which might correlate with either a gradual achievement of a state of "competence" just prior to entry or, more likely, the existence of inadequate time for effective inhibitor action. It

TABLE I

A. EFFECTIVENESS OF AMD INHIBITION OF DNA SYNTHESIS
DURING SYNCHRONOUS GROWTH

	Post-reversal Time of AMD (5 μg/ml) Addition				Control (No AMD)
	0	20	40	60	720
Total c.p.m. (ave.)† ^{14}C-Ur	6225	5904	6693	8044	320,720
% Inhibition	98	98	97	98	0
Total c.p.m. (ave.)† ^{3}H-TdR	3094	2990	3294	3889	126,694
% Inhibition	98	98	97	97	0

†c.p.m. ^{14}C-Ur incorporated into RNA or c.p.m. ^{3}H-TdR incorporated into DNA during one 720 minute labeling period (one generation time). The ^{14}C-Ur incorporation was monitored to verify effective activity of AMD.

B. EFFECTIVENESS OF AMD INHIBITION OF DNA SYNTHESIS
DURING SYNCHRONOUS GROWTH

	30	60	90	120	180	720
Total c.p.m. (ave.)† ^{3}H-Ur	1397	1772	2310	3589	7768	91,009
%Inhibition	98	98	97	96	91	0
Total c.p.m. (ave.)† ^{3}H-TdR	3748	5047	7242	9736	17,943	102,191
%Inhibition	96	95	93	90	82	0

†c.p.m. ^{3}H-Ur or ^{3}H-TdR incorporated during one generation time (720 minutes) into RNA and DNA respectively. The ^{14}C-Ur incorporation was monitored to verify effective activity of AMD.

appeared that the failure of chromatin function during early G_1 effectively stalled traversal of the cell cycle, while the inhibition of chromatin function during late G_1 permitted the cells to make an abortive attempt to enter the S period by initiating DNA synthesis that could not be sustained.

D. *Variations in Chromatin Protein Contents*

In order to study variations in the composition and functional ability of the genomic chromatin, colcemid accumulated metaphase cells were selectively

TABLE II

A. INCOMPLETE AND INHIBITION OF DNA SYNTHESIS DURING SYNCHRONOUS GROWTH

	Post-reversal Time of AMD addition					Control (No AMD)
	0	40	60	80	120	360
Ave. c.p.m. ^{3}H-TdR	66	141	65	333	401	8870
%Inhibition	99.3	98.2	99.3	96.2	95.5	0
Labeled Nuclei	9	24	------	50	437	--------
Unlabeled	793	881	------	381	575	--------
% Labeled Nuclei	1.1	2.7	------	13.1	76	--------

B. CHM INHIBITION OF DNA SYNTHESIS DURING SYNCHRONOUS GROWTH

	Post-reversal Time of CHM addition				Control (No CHM)
	0	40	80	120	240
Ave. c.p.m. ^{3}H-TdR	1,266	945	759	1,944	18,946
%Inhibition	93.3	94.6	96.4	87.4	0

detached from rapidly growing cultures of a male Chinese hamster fibroblast cell line (Don-C) and allowed to synchronously resume traversal of the cell cycle as described earlier. Chromatin was prepared from cell populations harvested at various defined points of the cell cycle according to the method of Marushige and Bonner (14). The compositional stability of the chromatin was studied by pulse-labeling synchronously growing cell cultures for one hour with a mixture of ^{14}C-amino acids (1 μCi/ml) at defined points of the cell cycle. The various chromosomal proteins were extracted from isolated chromatin in dilute acid or base and fractionated as described elsewhere (3). The specific activities of the fractions were obtained by densitometry or Lowry analysis and liquid scintillation spectrometry. The average level of newly synthesized protein accumulation into chromatin was calculated from the fractional values and compared to the expected accumulation of newly synthesized DNA into chromatin (Fig. 4A). Assuming a continuous rate of DNA synthesis and perfect cell synchrony in the system the temporal profile of DNA accumulation into chromatin would appear as shown by the dashed lines in the figure. Theoretically, DNA synthesis should have commenced abruptly at the onset of the S period and ceased sharply at the end of the S period. The first units of DNA accumulated would be added onto the initial content of DNA and, hence, would establish the maximum specific activity permitted by the pulse-labeling interval used. As time progressed, the DNA content would gradually increase to double the initial content by the end of the S period. Thus, the pulse-labeled amount of DNA would represent a progressively decreasing proportion of the total DNA and the specific activity of the DNA should have declined to about one-half of the initial value by the end of the S period.

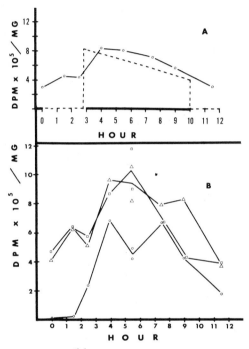

Fig. 4. Incorporation of ^{14}C-amino acids into chromosomal proteins. A. Incorporation into total chromosomal protein (open circles) and the theoretically expected incorporation (dashed line) profile. B. Incorporation into chromosomal protein fractions. Histone (open circles), CNS non-histone (triangles), and HNS non-histone proteins (squares) are shown. Details appear in the text.

Newly replicated HeLa cell DNA, like that in *Escherichia coli*, has been reported to exist in the form of small sub-units (15) which were synthesized at a rate of 1-2 microns per minute (15-18). The newly synthesized sub-units were subsequently joined tandemly into a larger continuous piece (presumedly by ligase action). It has been suggested that such sub-units complexed with concurrently synthesized chromosomal proteins to form small nucleoprotein units which became rigidly integrated into the chromatin structure (19). If the protein content of chromatin was dependently linked to DNA synthesis, it was not unreasonable to expect a temporal profile of chromosomal protein accumulation which followed that observed for DNA. The abrupt rise in the accumulation of newly synthesized protein which occurred coordinately with the onset of DNA synthesis and declined to one-half the initial increment of increase by the end of the S period clearly indicated that part of the chromosomal protein accumulation was dependent on DNA synthesis. In contrast, the accumulation of chromosomal protein during the G_1 and G_2 to mitosis indicated that another part of the chromosomal protein accumulation occurred independently of DNA synthesis.

A comparison of the accumulation profiles for the various fractions of chromosomal proteins proved of further interest. Significant histone accumulation commenced abruptly with the onset of DNA synthesis (Fig. 4B) at 2 to 3 hours into the cell cycle and declined concertedly with the cessation of DNA synthesis. During the course of the S period, the F_1 and F_3 histone fractions incorporated labeled amino acids about twice as readily as the other fractions. Despite the apparent differences in the synthetic rates, the histone fractions were found to accumulate in chromatin in nearly constant proportions (3). It thus appeared that the histone specific activities reflected limited rates of accumulation as well as synthetic rates *per se.* The "coupled" nature of the biosynthesis of histone and DNA has been studied in a number of biological systems and has been recently reviewed elsewhere (20,21). Evidence from a number of sources has demonstrated that histones accumulated in chromatin without subsequent release (turnover). This necessarily occurred in conjunction with the assembly of a new genome copy during the S period in a manner designed to maintain the tissue-specific restriction of the chromatin's genetic potential. The pattern of accumulation observed for histone was consistent with the DNA rate model described, since the abrupt appearance of histone accumulation coincided with the onset of DNA synthesis and reached a level about one-half that initially observed by the end of the S period (10 hours). The "pause" in the rate of accumulation of histone observed during the S period of Chinese hamster fibroblasts was followed by a briefly elevated rate of accumulation and coincided with variations in the rate of DNA synthesis described for this cell line (c.f. Fig. 1). Since the chromatin incorporation profiles for the non-histone (CNS and HNS) proteins also demonstrated elevated accumulation levels during the S period, the elevated incorporation level observed for chromosomal protein accumulation (Fig. 4A) was not due solely to the rate of histone accumulation during the S period.

The metabolic profile of the non-histone proteins was not as easily interpreted as the events of histone accumulation. Unlike the histones, newly synthesized non-histone proteins accumulated in chromatin throughout interphase and mitosis. Thus, the accumulation of these proteins was at least partly independent of the events of DNA synthesis. The pronounced effect of DNA synthesis on the level of non-histone protein accumulation into chromatin during the S period demonstrated, however, that the rate of synthesis and/or accumulation of the proteins was partly dependent on DNA synthesis. The level of isotopic precursor incorporation into the cold 0.1 N NaOH soluble (CNS) and hot 1 N NaOH soluble (HNS) proteins observed during G_1 (Fig. 4) sharply increased with the onset of DNA synthesis and rapidly attained a value nearly double the initial level. The CNS protein value gradually declined to a level about one-half the incremental increase from the G_1 to the initial S period value, while the elevated HNS protein value observed during the first half of the S period fell rapidly to a value similar to that observed for mitosis. The HNS incorporation level remained low from the end of the S period through G_2 to mitosis. At the end of the cell cycle both the CNS and HNS protein accumulation levels declined to a value similar to the initial metaphase value observed at zero time.

Since the incorporation of isotopic precursors of non-histone proteins into the structural complex of chromatin appeared to be partly dependent on the rate of accumulation of the newly synthesized proteins into chromatin, a dependent link to DNA synthesis would limit the rate of accumulation of such proteins into chromatin. In contrast, those proteins which appeared in chromatin independently of DNA synthesis would accumulate at a rate limited by the event of chromatin function with which they were associated. It must also be noted that in addition to limits imposed by the mechanism integrating chromosomal proteins into chromatin, the rate of synthesis for the particular proteins would also influence the specific activity determined for a protein fraction. Thus, the specific activities observed for non-histone chromosomal proteins reflected the rate of synthesis, the rate of accumulation, and dilution by unlabeled protein present in the chromatin. These results suggested that the synthesis and/or accumulation of non-histone proteins during the Chinese hamster fibroblast cell cycle was partly dependent on and partly independent of DNA synthesis. Since we were not able to discriminate between variations in non-histone protein synthetic rates and the rate limitation of integration into chromatin structure, it was difficult to extend the interpretation of non-histone protein accumulation much further than discussed above. The two fractions studied, however, clearly evidenced different mechanisms regulating their presence in chromatin.

E. Variations in In Vitro Chromatin-dependent RNA Synthesis

In order to determine the template activity level for chromatin during the cell cycle, Chinese hamster cell chromatin was prepared from various stages of the cell cycle by the method of Marushige and Bonner (14) with the addition of a second passage through the crude sucrose gradient described. The isolated chromatin was assayed for its ability to support in vitro DNA-dependent RNA synthesis using the reaction system described by Spelsberg and Hnilica (22). The RNA transcripts generated in the in vitro reaction mixture by Micrococcus lysodeikticus (= luteus) RNA polymerase from chromatin DNA templates have been shown to evidence tissue-specific similarities to RNA transcript populations generated in vivo (22-25). The in vitro enzyme reaction mixture

347

contained treated bentonite (1% $^w/v$) to retard RNAse activity and was designed to provide an ionic strength (0.12 M NaCl, 0.0025 M $MnCl_2$) and spermidine concentration (0.002 M) which was optimal for both the isotonic structure of chromatin and the propagated rate of RNA synthesis. The ionic strength was not high enough to selectively stimulate endogenous polymerase (26) and, indeed, no significant endogenous activity was detected. The rate of reaction under these conditions was linear with time significantly in excess of the incubation end-point chosen and was directly dependent on both the concentration of bacterial polymerase enzyme and template DNA concentration.

For quantitative estimations, the rate of reaction was related to known amounts of template as DNA or Chinese hamster cell chromatin per unit incubation time (10 min) and unit reaction volume (0.25 ml). Exogenously supplied *Micrococcus lysodeikticus* RNA polymerase was present in excess (32-64 mg per reaction). The rate response generally approximated a hyperbolic profile under the conditions employed. Using the Lineweaver-Burke method of linear derivation, a plot of the DNA (template) concentration with respect to the corresponding rate response observed for the standardized assay system permitted maximum velocity estimates (V max) to be obtained for each type of chromatin assayed relative to appropriate controls. The estimates achieved for simultaneous reaction series of free DNA, control known chromatin, and unknown chromatin allowed a comparison (Fig. 5) of the chromatin template *efficiency* relative to that of purified DNA (100%). The averaged estimates obtained were 17% (G_2-m). Chromatin from a non-synchronous cell population evidenced a template efficiency level of 6.3 $\pm$ 0.7% (S.D.) Calculation of the average percent efficiency expected from the various cell cycle preparations weighted for the proportional contributions of each stage to the generation time of the cell (and hence the population) predicted a value of 6.5% for the non-synchronous cell population. The agreement between the observed and the expected results provided confidence in the results achieved. It must be noted at this point that template activity per unit amount of chromatin DNA did not reflect the template activity *per cell* during the cycle due to the increasing accumulation of DNA during the S period of the cycle in this system. The net template activity value *per cell* for the cycle would be G_1 (7.6%), early-S (3.7%), mid-S (5.0%), late-S (6.6%), and G_2-mitosis (11.4%) based on DNA contents which were linearly increased 200% by the end of the S period.

In the study accomplished by McClure and Hnilica (27) on metaphase cells, the extractable histone mass yield was about 50% of that expected. Additionally, metaphase chromatin showed protein/DNA mass ratios of 0.95 to 1.18, while nucleated cell chromatin showed values of 1.35 (G_1), 1.38 (G_1-S early), 1.57 (mid-S), 1.66 (late-S), and 1.81 (G_2-mitosis). These values represented perchloric acid precipitable protein and, hence, were 10-15% lower than total protein estimates achieved directly. According to Wray and Stubblefield (28), the protein/DNA ratio of isolated Don-C metaphase

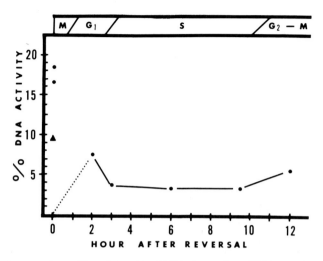

Fig. 5. Temporal profile of *in vitro* DNA-dependent RNA synthesis templated by cell cycle chromatin. The undersheared metaphse preparation (triangle) and standard preparation produced samples (closed circles) are shown. Details appear in the text. The dashed line represents the theoretically expected response predicted for metaphase cell chromatin.

chromosomes was about 2.1 to 2.2, i.e., about double that observed by McClure and Hnilica. The preparation of metaphase chromatin at a lowered shearing rate was found to produce a protein/DNA ratio of 1.53 and a template efficiency level of 9.5% relative to free DNA. In view of the decreased histone yield per cell, lowered protein/DNA ratio, elevated template efficiency, and influence of the shearing technique on the protein content and template efficiency of metaphase chromatin, the standardized isolation procedure was not considered to produce a metaphase chromatin product comparable to that obtained from nucleated cells. This did not appear to be the case with regard to interphase chromatin, since expected histone yields were observed. Additionally, the interphase chromatin with lower protein/DNA ratios obtained earlier in the cell cycle failed to show marked elevations in template efficiency. It was assumed, therefore, that the standardized isolation procedure produced comparable chromatin preparations from nucleated Chinese hamster cells.

Estimation of the "free" DNA content in chromatin by comparative enzyme rate kinetics has proven to be of a highly controversial nature due to considerations of the ionic strength effects on chromatin structure, the selective stimulation of multiple polymerase activity, the stimulation of endogenous polymerase, and the selective propagation of RNA polymer formation (initiation, elongation, termination). In addition, homologous eukaryotic RNA polymerase and bacterial RNA polymerase have not yet been proven to transcribe eukaryotic chromatin DNA in an equivalent manner. However, eukaryote chromatin was able to be transcribed by bacterial polymerases *in vitro* in a manner which produced transcripts with similarity to *in vivo* synthesized RNA (29,30). Although the extent of the RNA sequence homology in hybridization studies was judged to be significant (31-33), a recent study by Maryanka and Gould (34) demonstrated that homologous rat liver RNA polymerase (form B) generated high molecular weight (18-45 S) RNA *in vitro* only slightly smaller than that observed for native Hn-RNA transcribed *in vivo* (by endogenous enzyme) in contrast to the smaller size (10 S) generated by bacterial polymerase. Additionally, Butterworth *et al.* (26) demonstrated that rat liver chromatin was templated more efficiently by homologous rat liver RNA polymerase (form B) than by *Micrococcus lysodeikticus* RNA polymerase. It appeared that rat liver from B and *M. lysodeikticus* RNA polymerase bound to and transcribed from different sites on the chromatin DNA. Most recently, Cedar and Felsenfeld (35) reported that bacterial polymerase was able to directly measure the number of binding sites in eukaryote (calf thymus) chromatin. These workers found that chromatin bound 5-10% the number of polymerase molecules bound by free DNA, all bound enzymes were capable of chain elongation, the rate of chain elongation on chromatin only approached one-third that observed for free DNA, and, finally, chromatin solubility was not a factor in template restriction *per se*. While the methods employed permitted a determination of template concentration based on the number of chains and the rate of chain elongation, it was noted that the rate changes in chromatin template activity may be a function of the number of template sites (capacity) and the propagation rate (efficiency) at such sites. Despite these reservations, it was clear that chromatin from various cell cycle periods failed to perform uniformly as a template for *in vitro* DNA-dependent RNA synthesis under conditions of constant reaction parameters. Thus, variations in template site availability (capacity) or utilization (efficiency) occurred during the cell cycle with regard to chromatin sites templated under the reaction conditions used.

III. The Temporal Modification of Chromatin Function and Composition During the Cell Cycle

A. *Temporal Patterns of Chromatin Function* In Vivo

Based on general estimates of complementarity to DNA sites achieved by DNA-RNA molecular hybridization studies on ribosomal RNA and several other RNA species (36-40), one can calculate that about 0.39 percent of the *total* genomic DNA in the mammalian cell represented DNA sites for 45S r-RNA, t-RNA, and small (5S) RNA species. This corresponded to 2.6-7.9 percent of the *active* portion of the genomic DNA assuming the usually estimated 85-95% restriction of the total genomic DNA currently accepted (31-33, 41). Thus, the bulk of the DNA sites avilable as a template source were non-ribosomal DNA sites presumed to be (largely) Hn-RNA sites. Despite the more limited availability of r-DNA sites, most of the RNA contained in the cell (42,43) and synthesized *in vivo* by randomly dividing (47) and synchronized (5,44) mammalian cells appeared to be ribosomal RNA species. It was not unreasonable to assume, therefore, that the bulk of the labeled RNA present in the cell after the usual conditions of isotope administration (10-30 min) was ribosomal RNA.

The successful traversal of the cell cycle has been shown to be critically dependent on the continuous synthesis and translation of RNA transcripts from the genome of the cell (11-13, Tables I and II) with major periods of importance during G_1 and G_2. Except for several small RNA species synthesized in concert with DNA synthesis, RNA species such as the 4-7S RNA (45), Hn-RNA (46), and r-RNA species (47) were continuously synthesized through interphase of the cell cycle. Additionally, the processing of 45S r-RNA appeared to occur at similar rates during the same interval (43, 46). In general, the net mass increase in RNA during the cell cycle was partly linked to that of DNA synthesis (48-51) and the inhibition of DNA synthesis differentially interfered with the accumulation of RNA in the cell (5,50,52). In *Escherichia coli* (53) evidence has clearly indicated that the control of the three classes of RNA synthesized must be independently mediated during variable growth responses. Similarly, the demonstration of multiple RNA polymerase species in a variety of eukaryotic species (26, 54-58), the differential quantitative content of the various RNA polymerases in dividing mammalian cells (56), the discrepancy between DNA site numbers available and the cell content of the various RNA species, and the differential dependency on continuous DNA synthesis, argued that the balance between synthesis, turnover, and export of the nuclear RNA species was attributable to independently mediated mechanisms of synthesis for the various RNA species in eukaryotic cells. The existence of these multiple parameters and the observation that the bulk of newly synthesized RNA in the cell were r-RNA species severely complicated interpretations of the temporal patterns of RNA synthesis during the cell cycle with respect to information on genomic function during the cell cycle. Since multiple mechanisms independently regulated the synthesis of various RNA species, it was made evident that the "whole-cell" RNA synthesis observed could be affected by means other than the simple modulation of the amount of DNA template available. Hence, the level of cellular RNA synthesis might be a complicated function of both the template capacity *and* efficiency for DNA-dependent RNA synthesis. Temporal patterns of RNA synthesis have been reported for a number of cell types and circumstances (4, 5, 44, 48, 50, 59-63). A brilliant discussion of those papers prior to 1968 appeared in the work of Pfeiffer (44, 50) and more recent consideration appeared in the works of Mitchison (64, 65) and Klevecz (4, 5, 66).

In mammalian cells, an abrupt doubling in the rate of synthesis during early S of the cell cycle has been reported for whole cell RNA (4, 5, 44, 50), total cell protein (4,5), cytoplasmic "soluble" proteins (67), cytoplasmic microsomal proteins (67) and certain specifically identified enzyme proteins (64, 65, 68). These patterns suggested the occurrence of a gene-dosage effect during early S of the cell cycle that was similar to the circumstances described for bacteria and yeast cells (64,65,69) in which the capacity for the synthesis of particular enzymes appeared to double as the genes in question were duplicated during DNA synthesis. Since the bulk of the RNA polymerase in the proliferating mammalian cell has been shown to be nucleolar (AI) polymerase (56) and the bulk of the newly synthesized RNA accumulated was ribosomal RNA (5,44,47), the gene-dosage effect observed for whole cell RNA synthesis suggested that the nuclear DNA sites for ribosomal RNA doubled during early S of the cell cycle. Moreover, in agreement with earlier reports of retarded RNA synthesis at genomic sites of DNA replication (7,8) the report of Kasten and Strasser (61), which demonstrated a complete block of nucleolar RNA synthesis during early S of the cell cycle, also suggested that nucleolar DNA synthesis occurred during early S of the cell cycle. This has been confirmed for Chinese hamster cells using the techniques of molecular hybridization. Amaldi *et al.* (70) reported the replication of ribosomal DNA (r-DNA) between 23 and 46% of the way through the S period. Stambrook (71) found that half of the r-DNA replicated during the first 23% of S and most of the remainder prior to 46% of the way through the S period. Thus, the evidence described supported the conclusion that r-RNA was synthesized at rates determined by the number of r-DNA copies per cell and was amplified by: (1) quantitatively elevated levels of nucleolar-specific (AI) polymerase; (2) prolonged r-RNA stability and rapid Hn-RNA turnover; and (3) the complete utilization of the available r-DNA continuously during cell proliferation. Unfortunately, the contribution of Hn-RNA synthesis to the whole cell pattern of RNA synthesis was unable to be adequately considered due to insufficient data on this specie. However, the gene-dosage effect observed for certain mammalian cell enzymes suggested that the translation capacities of the enzymes might be dependently linked to the amount of m-RNA copies available and, assuming m-RNA species were processed from precursor Hn-RNA species, the amount of Hn-RNA synthesized. Alternately, the amount of r-DNA site transcription of r-RNA might link the supply of ribosomes to the translational capacity of the proteins noted. Conceivably, both might occur coordinately.

B. *Patterns of Chromatin Function* In Vitro

The degree of genomic restriction in isolated chromatin from a wide range of species has been found to be in the range of 80 to 95% based on comparative template efficiencies estimated by enzyme-dependent rate kinetics (33,41,72) and DNA-RNA molecular hybridization abilities (31-33). The limited RNA transcription accomplished by isolated chromatin *in vitro* produced RNA populations markedly similar to those synthesized *in vivo* prior to chromatin isolation (31-33). Thus, the mechanism of genomic restriction remained largely intact in isolated chromatin.

Variations in *in vitro* chromatin-dependent RNA synthesis have been demonstrated during both developmental (73,74) and physiological (74-82) changes in cell status. In particular, studies on semisynchronous cell proliferation induced by the removal of inhibitory conditions or mitogen action showed increased levels (120-300 percent) of *in vitro* chromatin template activity (74-77, 79, 80, 82), an enhanced content (200 percent) of nuclear RNA polymerase (76), and accelerated rates of nuclear protein synthesis and accumulation in nuclei (81,83,84, see also 20) between the time of stimulation and the advent of major DNA synthesis. While phytohaemaggluginim stimulated lymphocytes evidenced the appearance and persistence of rapidly-labeled 4S RNA one hour post-stimulation, newly synthesized r-RNA appeared at four hours (78) when a 300 percent increase in *in vitro* chromatin template activity was also observed (77). Interestingly, nuclei isolated from developing *Rana pipiens* embryos and incubated in an *in vitro* reaction mixture (85) showed unchanged levels at high ionic strength, and at low ionic strength, a 300 percent stimulation after the tail bud stage (when *in vivo* r-RNA was elevated normally). In both contact released WI-38 cell cultures (79) and isoproternol (IPR) stimulated mouse parotid glands (82), the temporal pattern of chromatin-dependent *in vitro* RNA synthesis paralleled the "whole cell" RNA synthesis pattern. As discussed by Novi and Baserga (82), the observation of continued r-RNA synthesis in IPR stimulated mouse parotid glands and the parallel increases in whole cell RNA synthesis, ctyoplasmic-RNA synthesis and the *in vitro* template activity of chromatin suggested that the control of r-RNA synthesis was involved in regulating the proliferative ability of the cells studied. An abrupt increase in RNA synthesis also preceded DNA synthesis in rabbit kidney cortex cells in primary culture (86) and was found to correlate with an increased rate of ribosomal RNA synthesis and an unchanged rate of Hn-RNA synthesis. According to the above authors, the accelerated RNA synthesis observed was thought to be primarily associted with accelerated ribosome production.

The variation of *in vitro* chromatin template activity during the cell cycle of highly synchronized cultures of eukaryote cells has been little studied at present. The incubation of isolated HeLa cell metaphase chromosomes in an *in vitro* enzyme reaction mixture which probably failed to maintain the metaphase structure demonstrated template efficiency levels which were 13-50% of the purified DNA level (42). Johnson and Holland (87) reported that the deoxyribonucleoprotein (DNP) fibers of chromatin extracted from mitotic cells were one-third to one-half as efficient as interphase chromatin with regard to template ability. More recently, Farber and coworkers (88, 89) reported that mitotic chromatin obtained from HeLa cells demonstrated one-quarter the template efficiency observed for S period chromatin. In the study by McClure and Hnilica (Sec. IIE), the interphase preparations considered comparable showed an elevated level of template efficiency (per cell) during G_1, which was depressed at the onset of DNA synthesis and gradually increased (continuously) during the S period to about the initial G_1

value. The template efficiency level was further elevated during G_2. Under the moderate ionic strength reaction conditions employed, the template efficiency pattern was not preferentially determined by r-RNA synthesis, failed to parallel the known whole cell RNA synthesis pattern, and failed to show a gene-dosage effect during early S of the cell cycle. Indeed, the template sites transcribed *in vitro* were not abruptly doubled at any point during the S period.

C. Fluctuations in Chromatin Protein Composition during the Cell Cycle

The biosynthetic parameters of chromosomal protein synthesis have been determined in synchronized non-diploid cells by Stein and coworkers (88,90,91) who reported that the residual chromatin proteins continued isotopic precursor incorporation at 96% of the S period value during G_2 and mitosis, respectively. A similar observation of continued incorporation into chromatin proteins and nuclear proteins throughout the cell cycle has been reported by McClure and Hnilica (27,92, Fig. 4) and Rotti Rotti *et al* (67). Stein and Borun (93) found an increased rate of synthesis (and accumulation) of the non-histone nuclear proteins preceding the onset of HeLa cell DNA synthesis. According to these workers, nuclear non-histone protein synthesis was insensitive to inhibitors of DNA synthesis (93), showed different fractional profiles of isotopic leucine incorporation during the various cell cycle periods (94), and evidenced different durations of retention in the nucleus (94).

While the data reported by Stein and coworkers (93,94), Rotti Rotti *et al* (67) and McClure and Hnilica (27, Fig. 4) were largely in agreement, neither the nuclear protein rate data of Rotti Rotti *et al* (67) nor the chromatin protein rate data of McClure and Hnilica (Fig. 4) demonstrated the degree of elevated incorporation during G_1 calculated by Stein (93) from the cumulative incorporation data observed in the HeLa cell system. Nonetheless, it appeared that either the rate of synthesis or rate of accumulation of the non-histone nuclear proteins changed immediately preceding the onset of DNA synthesis and was not entirely coupled to DNA synthesis.

While the data of Stein and Borun (93) showed distributional variations in the sodium dodecyl sulfate (SDS) polyacrylamide gel electropherograms with respect to both the amount of protein present and isotope incorporated, the extent of the heterogeneity was hard to discern. The recent work of Bhorjee and Pederson (95), however, demonstrated these circumstances clearly. The chromatin non-histone proteins isolated from synchronized HeLa S_3 cells at different periods of the cell cycle and electrophoresed in SDS-polyacrylamide gels showed a highly reproducible pattern of 22 bands with molecular weights of 15,000 to 180,000. Eighty five percent of these proteins exceeded a molecular weight of 40,000. Certain protein concentrations varied by nearly fifty percent and one group at about 75,000 daltons was greatly reduced just before the onset of DNA synthesis. A comparison between the protein patterns from chromatin and those reported by Stein and coworkers (93,94) from nuclei was difficult to effect.

Although the functional role and subnuclear localization of the metabolically active nonhistone nuclear proteins has yet to be clarified, the evidence described here (sections II-IV) and reivewed previously (20) provided a formidable array of evidence from diverse biological systems which supported the concept that non-histone proteins were involved in chromatin function and variably underwent turnover at such sites. The continued flux of chromatin proteins during the cell cycle argued that chromatin functional activity continued throughout the cell cycle.

IV. Concluding Remarks

Early molecular models of mechanisms of regulating RNA transcription in the cell were overly dominated by the simplicity of the sequential flow of genetic information (DNA→RNA→Protein). The initial view was more or less a "one on one" model in which an RNA polymerase molecule produced an RNA transcript from a genomic DNA site which shuttled to the cytoplasm and was translated into a protein product. It was frequently assumed that certain "signal factors" (presumedly proteins) dictated the availability of a DNA site for the relatively non-specific activity of RNA polymerase. Recent knowledge in this area emphatically demonstrated that the RNA synthesis regulatory circumstances were not as simple as first envisioned. Various authors have discussed a number of models consistent with the contemporary evidence available to them (44, 96-98). Briefly stated, studies on the interaction of bacterial RNA polymerase and certain factors with respect to DNA site selection (binding), as well as the initiation, elongation, and termination of the synthesis of RNA transcripts, have been described. Although yet unproved, such events might occur in mammalian cells as well. The recognition of the multiple polymerase system which preferentially transcribed nuclear or nucleolar DNA sites (c.f. discussion in ref. 98) suggested that mechanistic differences existed in the regulation of RNA transcription at these sites. In addition to possible regulatory effects at the level of RNA polymerase interaction with available DNA sites, molecular hybridization investigations demonstrated that genes might be tandemly transcribed as multiple rather than individual units (97). This suggestion was recently underscored by the suggestion that Hn-RNA might be transcribed from large portions of the genome with most of the RNA transcripts destroyed intranuclearly. In such a case, the excision of m-RNA from a small amount of the Hn-RNA might occur to provide m-RNA transcripts to be exported from the nucleus to the cytoplasm for translational purposes. The possibility of a post-transcriptional regulation of the release of m-RNA to the cytoplasm suggested the possibility that niehter changing quantities of genomic DNA availability (template capacity) nor qualitative changes in the Hn-RNA population transcribed would be necessarily required for the differential appearance of gene products (proteins) in the cell. Thus, in light of current knowledge of the multiple mechanisms operant in functional chromatin activity, the evidence from early works which studied the accumulation and rates of synthesis of RNA during the cell cycle permitted only a few conclusions to be offered regarding chromatin function during the cell cycle. The absence of chromatin function during mitosis showed that this activity was confined to interphase of the cell cycle. Within interphase, however, functional genomic transcription was not confined to any one period of the cell cycle. Thus, the transcriptional use of genomic DNA sites was presumed to occur continuously during the cell cycle. The variations in the rate of RNA synthesis observed, however, indicated that the chromatin template capacity (amount of available DNA template use) and/or the chromatin efficiency (frequency of available DNA template use) or both varied during the cell cycle. It was not entirely possible to discriminate between these possibilities at present.

Acknowledgement

The author wishes to acknowledge that this work was performed in the laboratory of Dr. L. S. Hnilica (The M.D. Anderson Hospital and Tumor Institute, Houston, Texas) and was supported in part by American Cancer Society grant E-388, USPHS grant CA-07746, and Robert A. Welch Foundation grant G-138 to Dr. Hnilica and an institutional grant (FR-05511-07-IN-88) to Dr. McClure.

References

1. D. J. Lehmiller, In: *Cellular and Molecular Renewal in the Mammalian Body*, I. L. Cameron and J. D. Thrasher (eds.), Academic Press, N.Y., 1971, p. 2.
2. E. Stubblefield, *Methods in Cell Physiol.*, *3*, 25 (1968).
3. M. E. McClure and L. S. Hnilica, In: *Oncology, 1970*, R. L. Clark, R. W. Cumley, J. E. McCay and M. M. Copeland (eds.), Yearbook Medical Publishers, Inc., Chicago, Ill., 1971, Vol. 1, p. 494.
4. E. Stubblefield, R. Klevecz and L. Deaven, *J. Cell Physiol. 69*, 345 (1967).
5. R. R. Klevecz and E. Stubblefield, *J. Exptl. Zool. 165*, 259 (1967).
6. R. R. Klevecz, *Science 166*, 1536 (1969).
7. J. Taylor, Ann. N.Y. Acad. Sci. *90*, 409 (1960).
8. J. Showacre, W. Cooper, and D. Prescott, *J. Cell Biol. 33*, 273 (1967).
9. B. R. Brinkley, E. Stubblefield, and T. C. Hsu, *J. Ultrastructure Res.,19*, 1 (1967).
10. C. A. Buck, G. A. Granger, and J. J. Holland, *Current Mod. Biol.*, *1*, 9 (1967).
11. R. Tobey, E. Anderson, and D. Peterson, *Proc. Natl. Acad. Sci. —U.S.*, *56*, , 1520 (1966).
12. J. Kim, A. Gelbard, and A. Perez, *Exptl. Cell Res.*, *53*, 478 (1968).
13. Y. Fujiwara, *J. Cellular Physiol.*, *70*, 291 (1967).
14. K. Marushige and J. Bonner, *J. Mol. Biol.*, *15*, 160 (1966).
15. R. Painter and A. Schaefer, *Nature*, *221*, 1215 (1969).

16. J. Huberman and A. Riggs, *J. Mol. Biol.*, *32*, 327 (1968).
17. E. Schandl and J. Taylor, *Biochem. Biophys. Res. Comm.*, *34*, 291 (1969).
18. J. H. Taylor, *J. Mol. Biol.*, *31*, 579 (1968).
19. M. Thaller and C. Villee, *Biochem. Biophys. Res. Comm.*, *29*, 273 (1967).
20. M. E. McClure and L. S. Hnilica, *Sub-cell. Biochem.*, *1*, 311 (1972).
21. L. S. Hnilica, In: *The Structure and Biological Functions of Histones*, Chemical Rubber Co. Press, Cleveland, Ohio, 1972.
22. T. C. Spelsberg and L. S. Hnilica, *Biochem. Biophys. Acta.*, *228*, 212 (1971).
23. R. S. Gilmour and J. Paul, *J. Mol. Biol.*, *40*, 137 (1969).
24. C. H. Tan and M. Miyagi, *J. Mol. Biol.*, *50*, 641 (1970).
25. V. K. Singh and S. C. Sung, *J. Neurochem.*, *19*, 2885 (1972).
26. P. Butterworth, R. Cox, and C. J. Chesterton, *European J. Biochem.*, *23*, 229 (1971).
27. M. E. McClure and L. S. Hnilica, *Biochim. Biophys. Acta*, (Submitted, 1974).
28. W. Wray and E. Stubblefield, *Exptl. Cell Res.*, *59*, 469 (1970).
29. R. Reeder and D. Brown, In: *Lepetit Colloquia on Biology and Medicine*, *1*, 249 (1970).
30. V. Daniel, S. Sarid, J. Beckmann, and U. Littawer, *Proc. Natl. Acad. Sci. U.S.*, *66*, 1260 (1970).
31. J. Paul and R. S. Gilmour, *Nature*, *210*, 992 (1966).
32. J. Paul and R. S. Gilmour, *J. Mol. Biol.*, *34*, 305 (1968).
33. J. Bonner, R. C. Huang and R. V. Gilden, *Proc. Natl. Acad. Sci. — U.S.*, *50*, 893 (1969).
34. D. Maryanka and H. Gould, *Proc. Natl. Acad. Sci. U.S. 70*, 1161 (1973).
35. H. Cedar and G. Felsenfeld, *J. Mol. Biol.*, *77*, 237 (1973).
36. R. Pedersen, *J. Exptl. Zool.*, *177*, 65 (1971).
37. A. Siegel, D. Lightfoot, O. Ward and S. Keener, *Science 179*, 682 (1973).
38. C. W. Vermeulen and K. C. Atwood, *Biochem. Biophys. Res. Comm. 19*, 221 (1965).
39. E. H. McConkey and J. W. Hopkins, *Proc. Natl. Acad. Sci. U.S.*, *54*, 1197 (1964).
40. L. Hatlen and G. Attardi, *J. Mol. Biol.*, *56*, 535 (1971).
41. J. Bonner, M. Dahmus, D. Fambrough, R. C. Huang, K. Marushige, and D. Tuan, *Science*, *195*, 47 (1968).
42. H. Ju Lin, J. D. Karkas, and E. Chargaff, *Proc. Natl. Acad. Sci. — U.S.*, *56*, 954 (1966).
43. J. D. Thrasher, In: *Cellular and Molecular Renewal in the Mammalian Body*, I. L. Cameron and J. D. Thrasher (eds.), Academic Press, N.Y., N.Y., 1971, p. 107.
44. S. E. Pfeiffer, *J. Cell Physiol.*, *71*, 94 (1968).
45. A. E. Clason and R. H. Burdon, *Nature*, *223*, 1063 (1969).
46. G. Pagoulatos and J. E. Darnell, *J. Cell Biol.*, *44*, 476 (1970).
47. M. D. Scharf and E. Robbins, *Nature*, *208*, 464 (1965).

48. J. Seed, *Nature, 198,* 147 (1963).
49. D. Killander and A. Zetterberg, *Exptl. Cell Res. 38,* 272 (1965).
50. S. E. Pfeiffer and L. J. Tolmach, *J. Cell Physiol. 71,* 77 (1968).
51. J. McLeish, *Chromosoma, 26,* 312 (1969).
52. N. P. Salzman and E. D. Sebring, *Biochim. Biophys. Acta, 61,* 406 (1962).
53. T. E. Norris and A. L. Koch, *J. Mol. Biol., 64,* 633 (1972).
54. D. Doenecke, Ch. Pfeiffer and C. Sekeris, *FEBS Letters, 21,* 237 (1972).
55. R. G. Roeder and W. J. Rutter, *Nature, 224,* 234 (1969).
56. C. J. Chesterton, S. M. Humphrey, and P. H. Butterworth, *Biochem. J. 126,* 675 (1972).
57. D. Lentfer and A. G. Lezius, *European J. Biochem. 30,* 278 (1972).
58. S. C. Froehner and J. Bonner, *Biochemistry, 12,* 3064 (1973).
59. R. R. Rueckert and G. C. Mueller, *Cancer Res., 20,* 1584 (1960).
60. J. M. Reiter and J. W. Littlefield, *Biochem. Biophys. Acta, 80,* 562 (1964).
61. F. H. Kasten and F. F. Strasser, *Nature, 211,* 135 (1966).
62. E. V. Gaffney and R. M. Nardone, *Exptl. Cell Res., 53,* 410 (1968).
63. M. D. Enger and R. A. Tobey, *J. Cell Biol., 42,* 308 (1969).
64. J. M. Mitchison, *Science, 165,* 657 (1969).
65. J. M. Mitchison, In: *The Biology of the Cell Cycle,* Cambridge University Press, N.Y., N.Y., 1971, pp. 159-175.
66. R. R. Klevecz, *Science, 166,* 1536 (1969).
67. J. L. Rotti Rotti, S. Okada, and H. Eberle, *Exptl. Cell Res., 76,* 200 (1973).
68. R. R. Klevecz, *J. Cell Biol., 43,* 207 (1969).
69. P. Hanawalt and R. Wax, *Science, 145,* 3636 (1964).
70. F. Amaldi, D. Giacomoni and R. Zito-Bignami, *Eur. J. Biochem., 11,* 419 (1963).
71. P. J. Stambrook, *J. Cell Biol., 59,* 332a (1973).
72. J. Bonner and R. C. Huang, *J. Mol. Biol., 6,* 169 (1963).
73. K. Marushige and G. Dixon, *Develop. Biol., 19,* 397 (1969).
74. M. Thaler and C. Villee, *Proc. Natl. Acad. Sci.-U.S., 58,* 2055 (1967).
75. R. B. Church and B. J. McCarthy, *J. Mol. Biol., 23,* 457 (1967).
76. S. Bannai and H. Terayama, *J. Biochem., 289* (1969).
77. R. Hirschhorn, W. Troll, G. Brittinger, and G. Weissman, *Nature, 222,* 1247 (1969).
78. J. P. P. V. Manjardino and A. J. MacGillivray, *Exptl. Cell Res., 60,* 1 (1970).
79. J. Farber, G. Rovera, and R. Baserga, *Biochem. J., 122,* 189 (1971).
80. G. Rovera, J. Farber, and R. Baserga, *Proc. Natl. Acad. Sci. — U.S., 68,* 1725 (1971).
81. G. Stein and R. Baserga, *J. Biol. Chem., 245,* 6097 (1970).
82. A. Novi and R. Baserga, *J. Cell Biol., 55,* 554 (1972).
83. R. Levy, S. Levy, S. Rosenberg, and R. Simpson, *Biochemistry, 12,* 224 (1973).
84. G. Rovera and R. Baserga, *J. Cell. Physiol., 77,* 201 (1970).

85. D. M. Kohl, J. Norman and S. Brooks, *Cell Differentiation*, *2*, 21 (1973).
86. M. J. Lee, M. H. Vaughan and R. Abrams, *J. Biol. Chem.*, *245*, 4525 (1970).
87. T. C. Johnson and J. H. Holland, *J. Cell Biol.*, *27*, 565 (1965).
88. G. Stein and J. Farber, *Proc. Natl. Acad. Sci. — U.S. 69*, 2918 (1972).
89. J. Farber, G. Stein, and R. Baserga, *Biochem. Biophys. Res. Comm.*, *47*, 790 (1972).
90. G. Stein and R. Baserga, *Biochem. Biophys. Res. Comm.*, *41*, 715 (1970).
91. G. S. Stein and D. E. Mathews, *Science*, *181*, 71 (1973).
92. M. E. McClure and L. S. Hnilica, *J. Cell Biol.*, *47*, 132a (1970).
93. G. S. Stein and T. W. Borun, *J. Cell Biol.*, *52*, 202 (1972).
94. T. W. Borun and G. S. Stein, *J. Cell Biol.*, *52*, 308 (1972).
95. J. Bhorjee and T. Pederson, *Proc. Natl. Acad. Sci. — U.S.*, *69*, 3345 (1972).
96. J. Darnell, W. Jelinek, G. Molloy, *Science*, *181*, 1215 (1973).
97. R. J. Britten and E. H. Davidson, *Science*, *165*, 349 (1969).
98. A. J. McGillivray, J. Paul, and G. Threlfall, *Adv. in Cancer Res.*, *15*, 93 (1972).

SUBJECT INDEX

A

A-state, 46, 47, 52, 56
A-B transition, 51
Acetylcholine, 92
Actin
  microtubules, 325
  properties of, 321
Actin-like filaments, 321
  and cell division, 319
  and interactions between microtubules,
    323,
  role of, 322, 326
Actin-like proteins, distribution of,
    322
Actin-myosin interactions, in intracel-
    lular motility, 323
Actinomycin D, 63, 64, 69, 94
  differential inhibition of RNA
    synthesis, 298
  effects on epidermis, 70
  effects on eukaryotic cell cycle, 68
  effect on G_2 period, 67
  enhancement of mitosis, 65
  in chick magnum, 71
  in sea urchin embryo, 70
  on intestinal crypts, 69
  superinduction, 71, 72
Adenyl cyclase, isoproterenol, 91
Age distribution, 31, 35, 37, 38
  ideal, 41
  ideal and real, 32
Age, effect on mating ability, 190
Allelic interaction, 191
Alpha-amanitin, action of, 299
Amoeba
  amputation of, 139
  cytoplasmic initiator in, 140
Antibody binding, and cell cycle, 122
Artichoke
  DNA synthesis in, 261
  enzyme activities during the cell
    cycle, 252
Aspartate transcarbamylase, 129
  expression of, 136
  pattern of accumulation, 227

regulation of, 227
state of repression of the gene, 228
ATPase
  Ca-activated, 270
  Na-K, 270

B

Bacterial chromosome, replication of,
    143
β-adrenergic mechanism, evidence for,
    89
β_1-adrenergic receptor, and adenyl
    cyclase, 88
β-galactosidase, 217
BHK cells, 51
  stimulation with serum, 53, 54
Block point
  and age distribution, 241
  concept of, 239
  effects on cell division, 240
  equation for, 241
  for cb mutants, 240, 242
  in cell cycle, 135, 136
  relationship to execution point, 239
  relationship to transition point, 239
Bone marrow, 87
  action of drugs on, 88
Bone marrow cells, drug-induced hypo-
    plasia or aplasia, 98
Bone marrow cultures, isoproterenol
    treatment of, 90
B-phase, 51
Budding yeast, *see also Saccharomyces
    cerevisiae*, 132
  cell cycle mutants in, 132
  dependence of cycle in, 133
  initiation of DNA synthesis, 132
  temperature sensitivity (TS), 132
  transition point, 132

C

Carbamycholine, 92
Carbohydrate content of membrane, in-
    crease during G_1, 120

361

369

Ecology of
Salt Marshes
and Sand Dunes

Ecology of
Salt Marshes
and Sand Dunes

D. S. RANWELL

Head of the Coastal Ecology
Research Station (Nature Conservancy)
Norwich

LONDON
CHAPMAN AND HALL

First published 1972
by Chapman and Hall Ltd
11 New Fetter Lane, London EC4P 4EE
Printed in Great Britain by
Cox & Wyman Ltd., Fakenham

SBN 412 10500 4

Contents

List of Plates

Preface

Some attempt has been made in this book to bring together recent knowledge of the ecology of salt marshes and sand dunes and to relate it to current work on associated subjects.

What could be a greater contrast than the flatness and wetness of a marsh and the hilliness and dryness of a dune? Yet in both there are interesting parallels in the ways in which plants and animals achieve mastery over these initially inhospitable environments as well as in the obvious contrasts.

The extreme differences in the two habitats have influenced approaches to the study of each in the past. This has led to emphasis on study of salinity in the salt marsh and lack of appreciation of the significance of drought near the upper limits of the marsh. Emphasis on the study of drought effects on the dunes has resulted in neglect of moisture effects in the damp slacks between them.

Individual habitats, or parts of them, have been studied in isolation in the past. Now, with increasing knowledge and better facilities, it is possible and essential to study whole systems and the relationships between salt marshes and sand dunes and their associated environments.

With the increasing pace of human activities on and near the coast it is vital for the coastal ecologist to gain a more balanced understanding of these habitats and to develop effective predictive models of the processes at work in them.

I hope that this book will help to encourage new studies where they are most needed, not only in existing marshes and dunes, but also in the design of the new coastal environments derived from the exciting coastal engineering schemes now on the drawing boards. We should no longer submit to a defensive protection of coastal wildlife resources, but insist that due regard for a rightful place for these be built into the new environments to come and ensure that adequate stepping stones are preserved to allow them to take this place alongside their human neighbours.

The book is planned in four parts: the first (Chapters 1 to 3) and last (Chapters 12 and 13), synthetic in character; the two central parts (Chapters 4 to 11) analytical. The first part concerns general relationships of both habitats, the second and the third parts contain separate treatments of the ecology of salt marshes and sand dunes, while the fourth part deals with human influences and management.

Coastal Ecology Research Station,   D.S.R.
(Nature Conservancy),
Norwich

February 1972

Acknowledgements

I am very much aware of the dependence of this book on the guidance and help I have received from teachers and colleagues in the past, but I should make it clear that the responsibility for opinions expressed is solely my own.

In particular I am grateful to Professor W. T. Williams and Professor T. A. Bennet-Clark F.R.S., who taught me respect for the discipline of plant physiology; Dr G. Metcalfe and Dr J. M. Lambert who encouraged my initial ecological interest; Professor J. A. Steers who encouraged my interest in coastal physiology; and especially Professor P. W. Richards who supervised my first studies in coastal ecology with a wisdom which I still appreciate. I would like to thank also colleagues who helped with my work and influenced my thinking: Dr M. V. Brian, Dr D. A. Ratcliffe, Mrs B. Brummitt, Mr R. E. Stebbings, Mr J. C. E. Hubbard and Dr E. C. F. Bird. I am especially grateful to Dr R. L. Jefferies for much helpful advice on Chapter 3 and to the many others too numerous to mention who have helped in many ways. Special thanks are due to Miss E. J. Reeve and Mrs S. van Piere for help in typing the manuscript, and to Mr P. G. Ainsworth, Miss S. S. Anderson and Mr B. H. Grimes for assistance with a number of the illustrations and photographs.

I should also like to thank the following for permission to use various quotations: Dr A. J. Brereton; the American Society of Limnology and Oceanography; the British Entomological and Natural History Society; the Institution of Civil Engineers; the New York Botanical Garden; Academic Press Inc. (London) Ltd.; G. Bell and Sons Ltd.; Blackwell Scientific Publications Ltd.; and the University of Chicago Press. Grateful acknowledgement is made to those who have granted permission to reproduce figures and tables; the source in each case is given at the end of the caption or table title.

Finally I would like to record my gratitude to the editor, Mr D. C. Ingram, for much helpful advice and to my wife for her constant encouragement throughout.

D.S.R.

PART ONE General Relationships

1   Climatic Restraints

Distributions and Climate

World patterns

There is a certain similarity in the appearance of the vegetation of salt marshes or of sand dunes in whatever part of the world they are found. Each is subjected to two overriding physical restraints which control the type of growth they can support. For the salt marsh these are silt and saline water; for the dune, sand and wind.

Given these physical restrictions, the kinds of plants and animals which can survive in any particular part of the world is then largely governed by prevailing climate. Whether they actually occur there or not depends on chance factors of migration or introduction. Opportunities for widespread dispersal are in fact much better for coastal plants than for other kinds because the transporting powers of the sea, migratory birds and ships all help to promote this.

It follows that we might expect to find among coastal plants some of the most widely distributed species in the world; species whose distribution most nearly reflects the absolute climatic limits they can withstand. So, we find for example that *Phragmites communis*, accredited to be perhaps the most widely distributed species in the world (Ridley 1930), is an element of the coastal flora on moderately saline coastal soils. The submerged coastal aquatics *Ruppia maritima* and *Zostera marina*, which occur respectively at the upper and lower limits of estuaries, and the brackish marsh species *Scirpus maritimus*, are among the most widely distributed species in the northern hemisphere. The recent discovery of *Zostera marina* beneath 1 m of winter ice at nearly 65°N in the Bering Sea, Alaska, is a remarkable further extension of the known range of this species (McRoy, 1969).

B

But we must bear in mind the limitations of the species concept when considering widely distributed species spanning major climatic zones. Such widely distributed plants as *Phragmites*, though relatively uniform in appearance and behaviour throughout its range, will undoubtedly be broken down eventually into a complex of closely related forms each adapted to special conditions. For example coastal *Phragmites* of the Red Sea area is known to tolerate a much higher soil salinity than temperate forms of *Phragmites* (Kassas – in litt.).

The absence of large accumulations of blown sand on tropical coasts was noted by Hitchcock (1904). He concluded that in these latitudes the long growing season is especially favourable to vegetation allowing it to colonize closer to high water mark than in more temperate zones. It therefore covers up sites from which sand dunes would normally receive their sand supply. Jennings (1964) has recently reached a similar conclusion from a study of literature on tropical coasts, but adds that the humid climate may also limit the extent to which sand can be blown. Special local conditions may account for exceptions such as the large dune accumulations in south Java. On corraline strands, the angularity of fragments may lock them into a more wind resistant surface (Oosting 1954). It seems likely that, as Van Steenis (1958) claims, genuine dunes tend to be restricted to regions with a seasonal climate in the more temperate parts of the world.

While it is true that salt marshes throughout the world are generally similar in appearance, even a layman would place himself in the tropics if set down in a mangrove marsh. These coastal marsh trees form a very distinctive group of some half a dozen genera whose main distribution falls neatly within the tropic zone. Their absolute limits are at 28°N and 25°S in the New World and 28°N and 38°S in the Old World (MacNae 1968, Good 1966, Clark and Hannon 1967, and Chapman and Ronaldson 1958). Frost damage occurs both at their north limit in Florida (Webber 1895 and Davis 1940) and at their south limit (Chapman 1944) in New Zealand. Their restriction to these limits is thought to be controlled largely by frost incidence.

The mangrove (*Rhizophora mangle*) forms dense growths on mainland and island shores as far north as Ormond (29° 22′N), Florida. It was killed during the 1894–95 winter (minimum temperature Ormond 29 December 1894, −7.7°C) as far south as Lake Worth (26° 40′N), except in cases where plants grew on the south side of large stretches of water. On the west shore of the Florida peninsula at Myers (26° 40′N) mangroves *were* killed on the south shore of the Calosabatchee river estuary (Webber 1895). The points of interest here are the width, over 200 miles (320 km), of the damage belt at the mangrove northern limit, and the pockets of survival related to local topography lying within it. This illustrates the difficulty of applying precision to the concept of a climatic limit although it may be clearly significant for the plant or animal concerned.

The latitudinal temperature gradient in the sea is very gradual because here fluctuations in climate are literally damped down. Work on the distribution of marine Angiosperms in relation to temperature has suggested that the distribution of these plants is likely to be controlled by temperature and the following main world zones have been distinguished on this basis. (Table 1)

Table 1. Marine temperature zones (after Setchell 1920).

Zone	Mean maximum temperature range °C
Upper Boreal	0–10
Lower Boreal	10–15
Temperate	15–20
Subtropical	20–25
Tropical	25–30

Crisp (see Johnson and Smith 1965) observes, 'Our success in relating intertidal population changes to climatic fluctuations may result from the fact that the main factor, temperature, is easily measured. The lack of comparable evidence from land plants may perhaps be due to the intervention of interaction of other factors (such as rainfall, sunlight, day length) which tend to obscure correlations. Are we clear about what limits the distribution of land plants?'

The answer to this is certainly no at the present time, but it does suggest that we may have a better chance of finding limits which are climatically controlled in intertidal salt marsh plants than in dune plants beyond tidal influence.

Assumptions about climatic limits are best tested by transplants to find out potential range. This is a useful preliminary to more critical work in the growth cabinet and field. Success and failure results for world-wide transplants of the salt marsh grass *Spartina anglica* probably give a true indication of the climatic limits of this species (Fig.1). The northern limit of *S. anglica* in the northern hemisphere (57°N), like mangrove, seems to be controlled by frost frequency. Frost killed 99 per cent of transplants in plantations in northern Holland; those in southern Holland were virtually unaffected. The northern limit of the species in the southern hemisphere

(35°S) may be controlled by day length as the plant does not flower under short day conditions (Hubbard 1969). However, Chapman (1964 *in litt.*) has suggested that winter temperatures may be too high at 35°S for normal development and notes that little seed is set at this latitude in New Zealand.

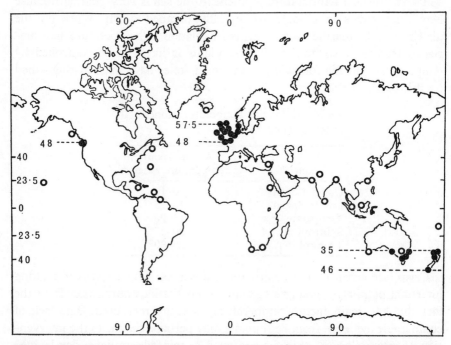

Fig. 1. Distribution of *Spartina townsendii* (*sensu lato*). Solid circles show sites where plants were known to be established in 1965. Open circles show sites where plantings are known to have failed. There are records of many other introductions, but the fate of these is unknown (from Ranwell 1967).

Ammophila arenaria, the principal dune-building grass native to European coasts, has been planted successfully in N. America, South Africa, New Zealand. It thrives from the boreal to subtropical zones (Cooper 1958). No similar study of its world distribution and climatic requirements seems to have been made in spite of its acknowledged success as a foredune builder and aid to coastal protection.

Regional patterns

We are so ignorant of what limits individual species of plants and animals to particular parts of the world that great care must be taken not to make any assumptions about the extent to which climate controls the individual components of regional floras and faunas. Indeed the very constitution of the coastal dune and salt marsh floras (let alone faunas) of many parts of the world are still very incompletely known. This does not only apply to the remoter shores of South America or Asia, but also in surprising measure to shorelines like the Mediterranean close to major centres of civilization.

Chapman (1960) distinguishes nine major regional salt marsh floras adapted to the particular climate conditions where they occur:

Arctic	Chinese, Japanese and Pacific Siberian
North European	South American
Mediterranean	Tropical
Eastern North American	Australia and New Zealand
Western North American	

Within each group are certain characteristic species of relatively wide range which impart some degree of inherent homogeneity to the group as a whole (e.g. *Puccinellia phryganodes* in Arctic salt marshes).

No similar attempt has been made to classify the dune floras of the world. These are much more variable in species composition than salt marshes partly because their distribution tends to be much more discontinuous and less freely linked by any major agent of dispersal like the sea.

Matthews (1937) showed how the British flora could be clumped into recognizable geographical groups. He found that some 600 species out of 1500 native or well naturalized species could be classified in distinctive geographical groups. The rest were generally distributed. Since his analysis was based on the known absolute distribution of the species concerned it seemed reasonable to assume that climatic factors were predominant in determining the flora of these groupings.

In fact an analysis of species reaching coastal limits in the British Isles (based on data from Perring and Walters 1962) shows a background of continuous replacement with evidence of disjunction at certain points clearly related as much to geographical barriers as to climatic zones (Fig. 2). Further analysis of the species reaching limits at points of concentration shows that less than half of them fall into Matthew's groups, the remainder being presumably capable of much wider distribution round the coast and to that extent relatively independent of climatic restraints.

Mörzer-Bruijns and Westhoff (1951) have shown how the Netherlands can be divided into climatic areas using indices based on extremes of climate over the course of a decade. These climatic areas can be correlated with biogeographical regions as determined by the distribution of insects

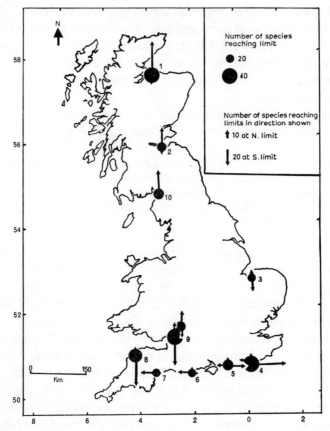

Fig. 2. Examples of some important meeting points for geographical elements of the Coastal flora in Great Britain.

1. Moray Firth	6. Poole Harbour
2. Forth of Firth	7. Exe
3. Wash	8. Taw-Torridge
4. Beachy Head	9. Severn
5. Arun	10. Solway Firth

Results are derived from records of all species recorded in coastal 10 Km squares in the Atlas of the British Flora (1962) and show numbers of species reaching north and south limits on east and west coasts, and north, east and west limits on the south coast, for only those squares in which 20 or more species reach such limits. There are two exceptions: (1) Beachy Head to Sussex Ouse, two contiguous squares each with more than 20 species at limits which have been combined to make the map record clearer. (2) Exe estuary with only 16 species at limits, included because other evidence, not considered here, suggests it is an important meeting point of eastern and western elements of the flora (from Ranwell 1968).

and plants. Beeftink (1965) has produced a comprehensive classification of European halophytic higher plant communities related to geographic regions.

Studies of this type help to narrow down the field to specific locations and particular species where intensive study of the biology of the plants or animals concerned in relation to key climatic factors is likely to prove rewarding (Plate 2). They do not in themselves tell us exactly how climatic factors control survival.

Oceanicity

Troll (1963) points out that: 'the greatest modification of the latitudinal zonation of thermal seasons in temperate latitudes is due to the distribution of oceans and continents'. Reduction of annual fluctuation of temperature is a measure of oceanicity. It is achieved by the greater heat storage capacity of water compared with land and the possibilities of convective heat exchange between warm water surface and colder land surface.

Thermophilous species such as *Samolus valerandi* are capable of surviving in inland localities of England and Wales but are restricted to the coast in Scotland (Fig. 3) and in Scandinavia. The more oceanic climate of the north-west coast of the North American continent allows *Spartina anglica* to survive at least 2° of latitude further north than on the north-east coast where the incidence of frost in the more continental climate prevents its survival.

Faegri (1958 and 1960) gives useful discussions of the climatic demands of oceanic plants. He concludes that many oceanic or maritime plants are sensitive to winter frosts, and that summer warmth for the successful ripening of seeds may be an additional limiting factor. The northernmost limit of oceanic species frequently follows winter isotherms closely up to a certain point then crosses them to follow particular summer isotherms. This suggests that a point has been reached where summer temperature has become more critical than that of winter. Rising lower altitude limits towards the south (such as *Dryas octopetala* shows in Britain) suggest that humidity may be critical for certain species. Ratcliffe's (1968) studies of Atlantic bryophytes also stress the importance of humidity and demonstrate interaction with varying soil tolerances. The concept of oceanicity is not a simple one and involves the interaction of at least three important climatic factors (temperature, wind and humidity) together with other environmental restraints.

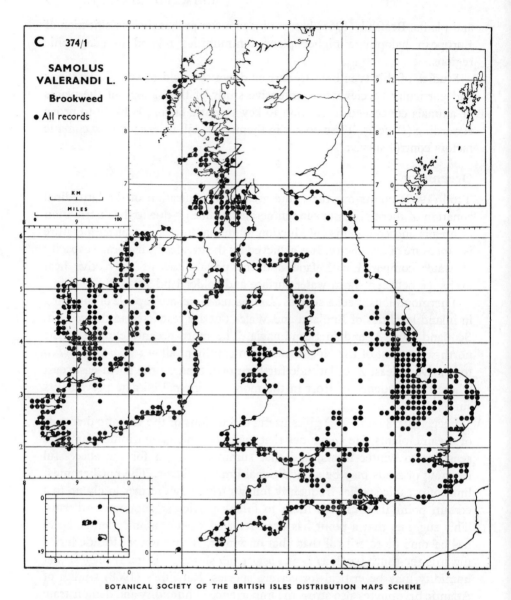

Fig. 3. Distribution of *Samolus valerandi* in the British Isles showing trend to strictly coastal distribution in the northern part of its range (from *Atlas of the British Flora*, F. H. Perring and S. M. Walter, T. Nelson & Sons Ltd., 1962).

Specific Climatic Factors

Local variations

It will be useful at this point to consider the effects of sunlight, temperature, rain and wind on dune and marsh, because they differ profoundly in each habitat. In particular, climatic factors operate much more uniformly over the flat surface of a marsh than over the broken terrain of a dune system. Climatic variability tends to be unidirectional across a marsh surface from seawards to landwards and is a contributory factor to the marked zonation of vegetation apparent in this direction (Fig. 4). Much greater diversity of microclimate is detectable in the different parts of a dune system according to local shelter effects and proximity to the water table (Fig. 5). This is reflected in much more complex mosaic patterns of the vegetation.

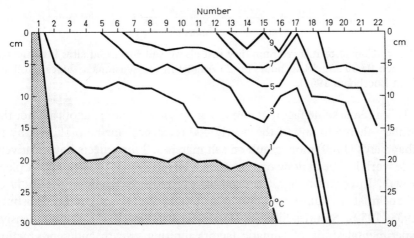

Fig. 4. Soil temperature 0–30 cm deep in different habitats at Liminka, Finland. 18. V. 1968 at 12–14 o'clock. The shaded area is below 0°C. No. 1. *Carex nigra* var. *juncea*, 2. *Phragmites communis*, 3. *arable land*, 4. wood of *Vaccinium – Myrtillus* type (VMT), shaded by spruce, 5. *Juncus gerardii – Odontites litoralis*, 6. *Carex aquatilis*, 7–9. *Carex mackenziei*, 10. *Salix phylicifolia* shrubs, 11. *Juncus gerardii – Primula finmarchica*, 12. *Eleocharis palustris*, 13. *Alnus incana* wood, 14 – 15. *Phragmites communis*, 16. *Eleocharis palustris*, 17. VMT wood (open location), 18. *Triglochin maritimum* (a depression in the Carices distigmaticae zone), 19–20. *Deschampsia caespitosa*, 21. *Betula pubescens* wood, 22. field drainage ditch. All readings correspond to littoral zones, except numbers 3, 4, 17, 19 and 22, which were taken on the epilittoral 5 km from the water boundary (from Siira 1970). N.B. Communities not in zonal sequence.

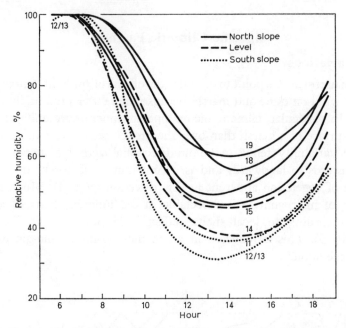

Fig. 5. Course of relative humidity just above soil surface at sites investigated between 05.30 and 18.40 hours on 23 August in the Wassenaar dunes, Holland (from Boerboom 1964).

It is a remarkable fact that climate or specific climatic factors are not the primary theme in any of the nearly 400 references in the bibliography to Chapman's (1960) monograph on salt marshes. This reflects how relatively little work has been done on climatic factors in relation to salt marsh plant growth. In contrast, significant advances have been made in the study of microclimate in dune vegetation by a number of authors, notably Salisbury (1933, 1952), Stoutjesdijke (1961) and Boerboom (1964). Comprehensive experimental study of climatic factors limiting growth and reproduction still remains to be carried out for the majority of the commoner salt marsh and dune species which make up the bulk of the floras of these two habitats.

Stoutjesdijke (1961) discusses thermo-electric measuring techniques suitable for use on dunes and in dune vegetation. He obtained simultaneous measurements of temperature and absolute humidity of the air, wind velocity, radiation intensity and soil temperature in dune grass and scrub vegetation.

Miniaturization at the sensor end and direct linkage to computer analysis at the other would be the best way to reveal the complex micro-meteorological patterns prevailing in salt marsh and sand dune communities.

Light

Cottam and Munro (1954) record that *Zostera marina var. latifolia* survives at the 'almost incredible' depths of 100 ft (30m) on the slopes of La Jolla submarine canyon in the clear waters off southern California. Beneath the relatively turbid waters over mudflats around European shores *Zostera marina* is usually confined to water depths of less than 20 ft (6m). Transplants of *Zostera marina var. angustifolia* made in the upper reaches of Poole Harbour a few years ago died gradually *in situ* at levels normally within the tidal range of this species, apparently for lack of adequate light. Johnson and York (1915) have shown that the anatomical leaf structure of *Zostera* and *Ruppia* is similar to that of shade tolerant plants. Day (1951) notes that in the turbid waters of the Berg estuary, South Africa, *Zostera* (*Z. capensis*) is limited to inter-tidal banks, but in the clear waters of the Knysa estuary it extends at least 3 ft (1m) below low tide mark. He summarizes techniques of measuring turbidity and gives a useful brief discussion of this factor.

Much work has been done to show how salt marsh plant zonation may be correlated with duration of tidal submergence (Chapman 1938, Adams 1963, Hinde 1954), but practically none on the survival of individual species subjected to submergence experimentally. *Spartina anglica* has been shown to survive continuous submergence in clear sea water in greenhouse conditions without apparent harm for $4\frac{1}{2}$ months. This is far beyond the maximum possible duration of submergence it receives in its normal intertidal habitat. Growth chamber studies have demonstrated that this species tillers but does not flower in short days (Hubbard 1969) and flowering may be inhibited by limitations on the supply of light during submergence at its lower limit. Flowering is generally poor at this limit, but this may be due to mechanical damage as well as reduced light.

The effects of reduced light on salt marsh plants is very noticeable beneath the shade of overhanging oaks in estuaries like the Fal, Cornwall or Beaulieu estuary, Hampshire. Here the growth of *Glaux maritima, Aster tripolium* or *Triglochin maritima* becomes very straggly and attenuated. It would be interesting to measure light values, morphological responses and the sequence of survival as shade increases in such situations.

Not all salt marsh plants require high light intensities, however, and *Althaea officinalis* appears to thrive best in the half shade of oaks near high water mark on wooded shores of Hampshire and Sussex. Lack of light may limit growth at both ends of the salt marsh. At the seaward end turbidity may be operative, and at the landward end the growth of taller plants shades out shorter plants where the marshes are ungrazed.

Thus in a mixed boundary where *Phragmites* was invading *Spartina anglica* marsh in Poole Harbour, it was found that *Spartina* was first drawn up in the taller growth, then ceased to flower, then lost its ability to tiller, became aetiolated, and finally died out some 25 m landward of the seaward limit of *Phragmites*. Merely cutting the stems of *Phragmites* near the extinction point of *Spartina* in no more than 1 metre square was sufficient to enable *Spartina* to recommence both tillering and flowering again.

It might be assumed that the light requirements of pioneer plants on the open sand of dunes would be high. In fact this is not necessarily so. It is intriguing to find for example that *Carex arenaria*, which we associate with the very high light conditions of open sand, is one of the last species to survive beneath the quite dense shade of mature pines on the dunes at Holkham in Norfolk. Once again no study appears to have been made of the shade-tolerance of dune plants and this would be of particular interest in view of the current trend towards scrub on dunes since 1954 when rabbit populations in Britain were greatly reduced by the virus disease known as myxomatosis.

Temperature

Little information is available about the field temperature controls on the germination, establishment, growth and flowering of salt marsh plants. There must certainly be far lower temperature ranges and smaller rates of fluctuation in lower marshes subject to regular inundation by the sea than in higher marshes free from its influence for days or even weeks at a time. Duff and Teal (1965) provide evidence supporting this in *Spartina alterniflora* marsh. *Cochlearia anglica* grows on both lower and higher shore in the Orne estuary, France. Binet (1965*a*) shows that on the lower shore seed germination is delayed and growth slowed down in winter in this species compared to germination and growth on the higher shore, but he does not relate these differences to temperature differences. Species like *Halimione portulacoides* and *Agropyron pungens* are often generally abundant in English and Welsh salt marshes, but absent from Scotland north of a line from the Solway to the Forth and in the more northerly parts of Europe. The abrupt disappearance of common species such as these at certain latitudes strongly suggests that their limits are controlled by some climatic factor, but just how and at what stage in the life cycle this happens we do not know. Iverson (1954) records the rapid spread of *Halimione portulacoides* at Skallingen, Denmark between 1931 and 1954 and this may be partly in response to ameliorated climatic conditions. The species was clearly damaged by frost in Holland in 1963. *Limonium vulgare* on the other hand is not particularly frost sensitive (Boorman 1968) and it seems more likely that day length requirements for flowering may be a controlling factor near its northern limit in Scandinavia.

Binet (1965*b*) has shown that seeds of *Cochlearia anglica* germinate best at relatively low temperatures between 5° – 15°C while those of *Plantago maritima* will scarcely germinate at all at these temperatures, but will do so at 25°C, though even better in diurnal alternations of 5°C and 25°C.

It is extremely difficult to get a clear indication of the way in which temperature may limit survival because of its complex interactions with photoperiod and salinity which is itself affected by rainfall. Tsopa (1939), Chapman (1942) and more recently, Binet (1964 *a* and *b*, 1965 *a, b, c, d, e,* and 1966) have made important contributions in this field in relation to germination requirements of salt marsh species, while Seneca (1969) has studied the germination responses of dune grasses. Chapman (1960, p. 315) points out that the germination of seeds of most salt marsh species occurs at times of reduced salinity, that is when low temperatures combine with high rainfall.

Stubbings and Houghton (1964) have shown that the cooling effect in winter and heating effect in summer which takes place in the harbour shallows of Chichester Harbour, Sussex, is of the order of 2°C below or above the open surface water temperature of the Harbour.

In contrast to the ameliorating effect of more or less permanently moist ground on temperature fluctuation in salt marshes, extremely wide and rapid temperature fluctuations are characteristic of the dune habitat. For example at Newborough Warren in Anglesey the temperature of the sand surface rose 14°C between 4.30 and 10.0 a.m. on a clear day in August. Salisbury (1952) notes that temperatures of over 60°C occur at the surface of bare sand on a hot summer day and diurnal fluctuations of 30°C were common. Such fluctuations are much reduced in damp dune slacks compared with dunes. In the generally cooler climate of damp slacks leafing of *Salix repens* may be delayed up to a fortnight after that on nearby dunes. Boerboom (1964) noted the persistence of late frosts in dune slacks in Dutch dunes in June 1957. *Juncus acutus* suffered over 80 per cent mortality in open drier slacks, but less than 50 per cent mortality locally in closed wetter slacks to landward in the 1962–1963 cold winter at Braunton Burrows, Devon (Hewett 1971).

Soil temperature fluctuation is markedly reduced beneath dune scrub, compared with bare sand (Fig. 6). Stoutjesdijk (1961) notes that the heat storage capacity of the soil beneath open sand was able to compensate for the net radiation loss from the surface during the night and there was no dew formation. By contrast, beneath *Hippophaë* scrub the much lower heat storage of the soil was only able to compensate for about one third of the net radiation loss so that the remainder was compensated by heat taken up from the air either as sensible heat or as heat of condensation resulting in strong dew formation on the *Hippophaë* surfaces. Further studies related to dew formation are considered in Chapter 9.

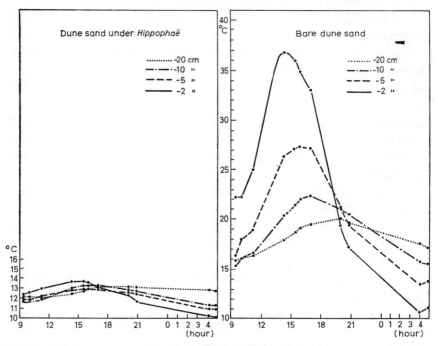

Fig. 6. Soil temperatures on Oostvoorne dunes, Holland between 9 a.m. on 16 June and 5 a.m. on 17 June (from Stoutjesdijke 1961).

The sand dune flora is notable for its high proportion of annual plants. These, unlike those of arable land, tend to pass through the hot summer period as seed and grow during cooler months, some germinating in autumn to grow as winter annuals. Salisbury (1934) has shown that the air close to leaves of annual dune species in April may be around 10°C higher in temperature than that of the general ambient air temperature. This could allow rapid photosynthesis in annuals at a time when low temperature was preventing photosynthesis in other species. Following the earlier work of Salisbury there is renewed interest in the temperature and light controls on the germination and growth of these species at the present time (Bakker *et al*, 1966). The variable topography of dunes produces striking climatic contrasts according to aspect. For example *Tortula ruralis* carpets may differ in temperature as much as 9°C on north and south aspects of Dutch dunes (Boerboom 1946). Such contrasts provide a juxtaposition of almost regionally distinct climatic zones in the form of mosaic units a few metres apart.

Rainfall

There is a very characteristic difference in the appearance of salt marsh vegetation as one goes south from North Europe to the sub-tropics. This is shown by the increasing openness of the vegetation dependent on the degree to which evapo-transpiration exceeds precipitation for substantial periods of time. In general the vegetation of north European marshes forms a more or less continuous sward (except in the pioneer zone), interrupted only by discontinuities in the surface. In the Camargue, France, Bigot (1958) notes that cover in the main salt marsh areas is often less than 80 per cent and some times as low as 50 per cent and here evapo-transpiration exceeds precipitation during much of the summer (Plate 1). On the borders of the Red Sea vegetation cover rarely exceeds 70 per cent and is mostly below 50 per cent, though this is partly due to the incidence of grazing in addition to the desert-type climate (Kassas 1957). In Gambia (in the tropical zone) where the rainy season is confined to June to October, the remaining months are in general hot and very dry and Giglioli and Thornton (1965) note that this results in extensive areas of barren mud within the mangrove swamps. There is marked interaction between seasonal and season to season rainfall and salinity. Indeed it is the incidence and amount of rainfall and not tidal influence which dominates salinity concentrations in upper salt marsh levels (Ranwell *et al* 1964).

Plants of the higher salt marsh are generally tolerant of the highest salinities likely to obtain in the climate of north European coasts. Growth room studies have shown that *Spartina anglica* for example can survive salinities up to twice that of sea water, a value much in excess of any recorded in 162 field samples measured. *Glaux maritima* and *Limonium vulgare* are also known to be tolerant of salinities in excess of sea water. Yet, in periods of summer drought these plants may die back extensively as they did on the northern Irish coast for example in the dry summer of 1968. Little study seems to have been made of the drought tolerance of salt marsh plants and this would be well worth investigation for it might help to explain the successive replacement of species in progressively higher zones of marsh. Boorman (1967) has demonstrated by transplant experiments that drought is a limiting factor of growth for *Limonium vulgare* on sandy soil at the upper levels of a marsh.

As mentioned earlier, species characteristic of the montane element in the British flora such as *Dryas octopetala, Trollius europaeus* and *Saxifraga aizoides* descend to the coast and are able to survive on north and west Scottish dunes. Further south they are confined to montane regions. These plants are characteristic of high rainfall in their mountain habitats and it is probably the more humid climate on the northern dunes which allows them to survive there.

Mosses normally found in dune slacks further south, grow on dunes at Luskentyre in the Outer Hebrides and Gimingham *et al* (1948) suggest this may be correlated with the relatively uniformly rainy climate of the area. Similar moisture-loving mosses occur in *Ammophila* tussocks in north-west Sutherland. In both of these sites the average annual potential water deficit is less than 0·5 ins (13 mm), among the lowest for any coastal site in Great Britain, one of the wettest parts of Europe (Green 1964).

Wind

There is much historical evidence (from the burial of human settlements), of pronounced dune activity on European coasts in Mediaeval times during periods of exceptionally stormy weather, e.g. from Penhale, Cornwall (Steers 1964); South Wales (Higgins, 1933); Newborough, Anglesey (Ranwell, 1959); Forvie, Aberdeenshire (Barkley, 1955), and many other sites on the European coast from the Baltic dunes to those of The Landes in France. Attempts have been made to link exceptional periods of storminess with recurrent astronomical events affecting tidal maxima (Petterssen, 1914; Brooks, 1949) and there does seem to be some evidence in the above of pronounced dune activity at times of major (1700 year cycle) and minor (90 year cycle) tidal maxima. According to Petterssen's hypothesis the last major maximum was in 1433 A.D. and we should be reaching the next minor maximum in the 1970 to 1980 period, when it might be expected that the storminess noted at the end of the nineteenth century may be repeated. Lamb (1970 *a* and *b*) has recently drawn together evidence of climatic periodicity which suggests that cycles of 5, 10, 20, 90, 200 years may also be operative.

The direct effect of wind on plant growth in salt marshes has not been investigated, but it must curtail both height and extent of growth in more exposed sites. There is a noticeable reduction in height of *Phragmites* growth at the windward edge of clones invading *Spartina* marsh in Poole Harbour, Dorset, but this may be partly due to competition for water or nutrients. Lines (1957) has shown how wind exposure may be compared from site to site using tatter flags.

Wind exerts a profound influence on the growth of dune plants, on the redistribution of organic and inorganic nutrients, on the amount of salt received, on the distribution of propagules, and above all shapes the very ground on which they grow. These different influences are discussed in Chapter 8 and subsequently.

Response to Climatic Changes

While it is true as Major (in Shaw 1967) points out that: '. . . ecologists ascribe to climate major importance in differentiating the kinds of vegetation on earth' it is also paradoxically true as he says, '. . . it is difficult even to rank the relative importance of a climatic difference in relation to other site factors.'

The problem becomes less acute perhaps if individual species rather than vegetation types are considered. Especially those species which by their powers of regeneration, superior height or dominating influence in one form or another gain ascendancy and exert such influence as to control to a large extent the type of habitat and even survival of accompanying plants and animals. In fact one searches the literature in vain for critical studies of the performance of such species near the limits of their range. Experimental transplants beyond such limits have scarcely been attempted let alone monitored to see how prevailing climate affects their growth.

Many botanists are reluctant to deliberately move plants outside their existing range. They are keenly aware of the risks of liberating potentially invasive species in a new environment. However the case against introductions is frequently based on the argument that they may interfere with studies on geographical distributions. In fact such studies often reveal only correlations of doubtful significance between range and climatic factors. Properly controlled introductions provide information of real predictive value and are a powerful tool for those who are asked to make a practical contribution to advice on vegetation management.

Lethal and sub-lethal damage.

Wholesale damage to *Spartina anglica* and *Halimione portulacoides* was noted in Holland near the northern limit of range of these species after the cold winter of 1963. But in fact despite this setback *Halimione* has apparently extended its range northwards in Denmark.

The 1963 winter froze *Spartina anglica* marsh soil to a depth of 10 cm

c

(the main rooting level) at the height of the frost, and surface growth was much diminished in the following summer at Poole Harbour, Dorset. There was full recovery in lower *Spartina* marsh in 1964, but near the upper limit of *Spartina* growth species such as *Agropyron pungens* and *Aster tripolium* invaded and persisted, occupying much of the ground formerly covered by *Spartina*. This differential effect of severe frost on plants in optimum, compared with sub-optimum growth conditions, underlines the need to examine climatic effects on the same species in different habitats.

Suaeda fruticosa near the northern limit of its range in Europe at Blakeney Point, Norfolk lost all its leaves after the 1962–1963 cold winter, but some buds on higher twigs survived to put out new growth which however mostly came from the base of plants (White 1967). This illustrates how extremely local even severe climatic effects can be.

Zoologists use the level at which 50 per cent of the population is killed by any particular factor operating over unit time as an indication of lethal dose (L.D. 50). Crisp (1965) for example gives L.D. 50's for low temperature operating for 18 hours for a range of shore organisms many of which were severely affected by 1962–1963 frosts. It would be of great interest to have similar information on low temperature lethal limits for salt marsh and dune plants of predominantly southern distribution in the British Isles. *Festuca arundinacea*, though by no means confined to these habitats, does occur in dune grassland and in upper estuarine brackish marsh grassland. It also shows a marked coastal trend in northern Scotland. Recent work by Robson and Jewiss (1968) demonstrates an inverse relationship between winter growth and winter hardiness in forms of this species. In culture the L.D. 50 for Mediterranean varieties occurs at temperatures of $-13°C$ and for more northerly varieties at $-16°C$. Their work illustrates the further complexity of interaction between climate and genetic adaptation.

Population response to climatic changes

The nature of the limit in relation to a plant population must be considered as the combination of internal and external environmental factors which regularly result in L.D. 50, rather than in reference to any one factor in particular.

As we saw in the case of the Florida mangrove, local pockets of the species may persist in sheltered refuges (perhaps for a century or more), so a frequency value needs to be added to the L.D. 50 to make it meaningful for a climatically controlled limit.

Study of microclimates at refuge limits could characterize climatic tolerances where these are believed to be limiting to growth, reproduction or survival. They would be representative of the effective adaptive limits but not necessarily of the potential adaptive limits of the species as a whole. For example, Sakai (1970) has recently drawn attention to the latent genetic potential of tropical willows to withstand freezing to very low temperatures.

There is little evidence of climatically induced population changes in the British sand dune flora, with a few notable exceptions. Prior to 1930 *Otanthus maritimus* occurred as far north as Anglesey and Suffolk, but now survives only in Cornwall in Great Britain. *Eryngium maritimum*, formerly in the Shetlands, now survives only south of a line from the Hebrides to south Yorkshire, while *Glaucium flavum* appears to have retreated south from its Scottish habitats. The latter is a continental southern species which may require hot dry summers and therefore not be favoured by the increasing trend towards higher rainfall evident up to at least the 1950's, (Lamb in Johnson and Smith, 1965). *Parnassia palustris* on the other hand, a plant of the continental northern flora, has disappeared from several coastal habitats including dune slacks in southern England, possibly in response to the generally warmer winters which have characterized our climate in the first half of the twentieth century.

It is now becoming evident (Lamb 1969) that climatic trends may be quite short-lived and that records over periods as short as a decade may give a better indication of current trends, than longer runs of records. We are only beginning to see how rapidly there may be an adaptive adjustment in species populations where other factors related to establishment and selection operate in favour or against a species. Adaptation to climatic change is clearly a product of inherent variability and where this is low in a rare species, ground lost by the species may not be regained for very long periods of time, even though climatic trends once more favour its growth in the site.

2 Physiography and Hydrology

As the sum of knowledge grows in different disciplines it becomes possible to break down the isolation between them, to see relevant connections and work towards a synthesis which in turn opens up new lines of study and unsuspected possibilities for the practical use of knowledge gained. Understandably in the past both ecological and physiographic studies have tended to concentrate on the immediate influences at work in particular sites, and this has led to a distinctly piecemeal approach to advice on problems of coastal protection or coast land reclamation. For example the benefits in terms of raised shore levels obtained by groyning at one site may be gained at the expense of lowered shorelines in another. Again, a low water training bank to hold a deep water channel for shipping on one side of an estuary may result in silting on the other side of the estuary and loss of beach recreational facilities.

Now there are welcome signs that we are moving towards a synthesis of coastal physiographic knowledge and in this chapter some attempt is made to link up the complex of forces which controls the disposition, type and development of salt marshes and sand dunes on the British coast as a whole. Sources of sediment supply come from erosion of the land surface, the edge of the coast or the sea bed mainly during periods of high wind or high rainfall; deposition occurs mainly in calm weather conditions.

Physiographic Influences

Rock-type influences

Primary geological effects on the distribution of marshes and dunes are particularly evident in a country like Great Britain because its small size and unusually diverse series of rock types bring out striking differences between one part of the coast and another.

There is an obvious natural division for example between highland and lowland Britain. To the north and west of a line from the Tees to the Exe hard rocks predominate in relatively high-lying country. To south and east of this line soft rocks predominate in relatively low-lying country. This is reflected in the predominance of coarse sediment deposits in the north and west which form the building material of dunes. Finer sediments have accumulated to form the extensive marshlands of the south and east and also in the three isolated soft rock outlets on the west coast: the Bristol Channel, the Dee-Mersey area, and the Solway.

Apart from any contributions of sedimentary material brought down by rivers which we shall consider shortly, the erodibility and lime status of coastal rocks partly governs the particle size and particle type of adjoining sedimentary deposits. It is not easy to sort out the various contributions from country rock, glacial material and soil erosion which go to make up marsh and dune sediments and this is a field that would repay much further study. Clues to the origins of sediments can be obtained from heavy mineral analysis, X-ray diffraction and differential thermal analysis of clay minerals (Guilcher and Berthois, 1957), and more recently from electron micrography studies of particle surface textures (e.g. Biederman, 1962; Porter, 1962 and Krinsley and Funnell, 1965). There are one or two sites where direct connections between adjoining rock strata and dune type can be made with some confidence. The Spurn Head dunes, Yorkshire, are clearly influenced in their high lime status by the extensive exposures of chalk on the Yorkshire coast. The lime deficient dunes of Studland, Dorset (an area not subject to glaciation), must owe their relatively recent origin to erosion of the adjoining lime-deficient Bagshot sand deposits of Poole Bay.

Indented cliff coasts of lime-deficient rocks, though not themselves a source of lime, provide innumerable sites for molluscs which derive lime for their shells from the adequate supplies in sea water. Much of the beach material for building dunes on these hard coasts is therefore in the form of broken shell material often comprising more than 50 per cent of the sand, as at Penhale, Cornwall, the dunes of the Outer Hebrides, or the shell sand beaches of the west coast of Scotland.

The legacy of glacial deposits

Superimposed on the rock formations of northern Europe are varying quantities of glacial material of mixed and derived origin. Its deposition occurred rapidly in the form of out-wash fans from the main centres of glaciation as the glaciers themselves began to melt and retreat northwards some 20,000 years ago. These glacial deposits, re-worked by the sea at the time of the last major land and sea level adjustment some 7,000 years ago, probably form the structural basis of many of our present-day marshes and dunes. They also form a bank of material, easily eroded at the coast and in the immediately offshore zones, from which supplies are drawn to help feed accreting coastal systems today. We do not know the size of this bank, nor, what is more important, its capacity to supply withdrawals, but both are clearly limited.

Long term changes in the relationship of land and sea level modify the distribution of salt marshes and sand dunes by exposing or submerging the foreshore deposits from which they are built according to varying amounts of wind or wave action. The pattern of the present configuration of the coast and its adjoining inter-tidal flats was established about 6,000 years ago when sea level is believed to have attained its present level. Since that time there have been only minor oscillations (Fairbridge 1961). Land subsidence or elevation operates more locally and it may increase or reduce the relative sea level rise in any particular part of the coast.

It is not certain whether the sand in Caernarvon Bay is of glacial or pre-glacial origin, but there is evidence (Ranwell 1955) that there has been a steady deepening of the bay during the past 250 years. During that time very considerable quantities of sand have been blown from the shore onto land beyond reach of tides. This clearly must have come from the bay.

Studies in the Danish Waddensea have shown that there is a net input of sedimentary material to the system and the bulk of this is believed to be derived from off-shore deposits of glacial or pre-glacial material. The movement of submarine material can be studied indirectly by means of sea bed drifters (Perkins *et al* 1963), or more directly by the use of radio-active tracers (Perkins and Williams (1965)), but much of this work has been confined to littoral deposits and we are largely ignorant about the extent to which storms can stir up and transport deeper sea bed deposits, and incidentally, potential pollutants dumped off-shore.

Topographic relationships

Coastal sedimentary material derives ultimately from aerial or marine erosion of the land surfaces and forms a discontinuous belt of various thickness and width on either side of the shoreline.

The disposition of material is related to the vertical angles and horizontal distributions of the 'rocks' across and along the shoreline which in turn is governed by the nature of these rocks. These topographical relationships control rates of river and sea-water flow in which loose sediments are carried when it is fast, or deposited when it is slow.

The type of material available for transport or deposition varies according to the texture of the rocks eroded: harder rocks producing coarser-particled material and softer rocks finer-particled material.

It follows that of the two habitats with which we are concerned, sand deposits are characteristic of the inlets and embayments of the hard rock cliffed coast of Northern Scotland; silt deposits of the graded soft coastline of south and east England. Elsewhere the conjunction of both types of rock provides both sand and silt for deposition side by side according to local shelter and the sorting action imposed by differential rates of water flow.

Only at the extreme heads of the most sheltered inlets on the rugged north and west coasts of Scotland (e.g. Kyle of Tongue, Sutherland), can one find vestigial salt marshes on silt. This contrasts with the widespread occurrence of salt marsh in the estuaries of lowland coasts further south and even extending onto the open coasts of Essex shores (e.g. at Dengie and Foulness). These are the two extremes. More commonly estuaries are found with silt in the sheltered upper reaches and sand accumulations in the form of various types of spits and cusps on one or both sides of the estuary mouth.

Before considering the different types of dune and marsh systems we need to consider external hydrological factors operating on them from the larger systems of which they form a part: on the landward side, the catchment area, and to seawards, its effective counterpart, the tidal basin.

Hydrological Influences

Influences from the land catchment

In the past, as we have seen, the bulk of European coastal deposits are believed to have derived from glacial deposits distributed by glacial melt-water erosion of the land surface. At the present time available evidence and deductions from river volume and flow rates, suggests that relatively small amounts of material are derived from the land surface (excluding for the moment the contribution from coastal erosion), and these mainly at times of bank-full river conditions during heavy rain. Nevertheless catchment type, size, vegetative cover and management, together make discernible contributions, particularly to upper estuarine marshes. The catchment influences deserve much fuller study than they have yet received, and a few examples will help to make this clear.

Catchment types with predominantly basic rocks and soils derived from them in the south of England provide nutrients which favour the growth of *Artemisia maritima, Carex divisa, Inula crithmoides* and many other species absent from salt marsh soils deficient in lime. At Langstone Harbour, Hampshire, chalk bed-rock outcrops in the inter-tidal zone and freshwater springs from it help to keep the chalk surface exposed and free from mud locally. These lime-rich waters, quite brackish at low tide, favour growths of *Zostera angustifolia* and some of the biggest marsh populations of *Inula crithmoides* found in Britain. Permeable sandy marshes and those with only a thin layer of silt overlying permeable deposits or bedrock may be strongly influenced in this way by sub-surface fresh water drained from the catchment.

A relatively high-lying catchment commanding high rainfall such as that of the River Tay in Scotland, which produces a greater discharge than any other in Britain, markedly reduces the estuarine salinity. This allows species such as *Scirpus maritimus* (normally a pioneer of upper estuarine muds), to extend its range as a pioneer to much nearer the estuarine mouth than normal.

A large catchment like that of the Mississippi produces great quantities of silt and characteristic deltaic marshes. Some tendency towards deltaic marsh development is evident in the Thames with its reclaimed marsh islands (Canvey, Grain), lying at the mouth.

Vegetative cover in the form of blanket bog and hard rocks resistant to erosion provide little sediment in the rivers of north and west Scotland. Much more sediment is transported from catchments with a high proportion of agricultural land, notably catchments from the Humber to the Thames. Little quantitative information is available in this country on catchment erosion, but Gottschalk and James (1955) estimate that soil loss from arable land is about 500 times that from grassland. Recent estuarine barrage feasibility studies (Anon 1966 *a, b,* and *c*) for freshwater reservoirs provide rough comparative estimates of the river-borne sediment supplies to the Solway and Morecambe Bay. Those to the Solway are sufficiently high to influence decisions in favour of selecting Morecambe Bay as a potential barrage site rather than the Solway.

Currently the enormous increase in the use of fertilizers, herbicides and pesticides in agricultural catchments is also changing the chemical quality of water and silt coming from the land. Westlake (1968) calculates that 500 metric tons per year of nitrate nitrogen at an average concentration of 2 mg/1, and about 20 metric tons per year of phosphate phosphorus at an average concentration of 0·09 mg/1, are currently discharged to the sea by the River Frome, Dorset, a small well-graded river. Relationships with catchment activities on the one hand and contributions of silt and nutrients from these sources to the marshes on the other, have yet to be made. However it is recorded that the amount of nitrogen applied as fertilizer to agricultural land in England and Wales has doubled in the 7 years from 1962 to 1969 (Anon 1970 c).

Tidal basin influences

One way of considering the marine contribution to the hydrology and physiography of marsh and dune systems is to think in terms of the tidal basin. Like the catchment for the land, this is a convenient, though much less discrete unit of the sea. Superimposed on both, of course, are the effects of aerial weather systems and one can foresee that with better understanding it may eventually be possible to relate contributions from land, sea and air in some unit form to the modelling of coastal systems.

As an example of this type of approach, Phillips' (1964) studies on the Yorkshire coast is of outstanding interest. She studied beach features known as 'ords' which develop on the Yorkshire coasts between Flamborough Head and Hornsea. Where protection of Flamborough Head is no longer effective, strong northerly winds directly on shore generate big waves which may initiate upper beach erosion to form an 'ord'. 'Ords' are about 45 to 55 m long and may lower the top beach by as much as 3 m. They travel southwards as units at an average rate of about 1·6 km per year and enable storm waves to erode embryo dunes or coast defence works, whichever is present. Their movement as a unit is only initiated when winds of over 15 knots (7·7 m/s) blow from the northerly quarter for at least several hours. Such winds are usually associated with a deep depression centred over southern Scandinavia. These conditions produce not only powerful waves developed in the long fetch available to the north west, but also a storm surge which causes a temporary oscillatory rise in sea level in the North Sea basin. Under most other wind conditions the beach tends to become built up in form. Once developed, 'ords' and the higher sections of beach between them, move as a unit since, as Phillips points out, the rate of longshore movement on this only slightly curving stretch of coast would tend to be uniform. The importance of this work lies in the link-up and interaction of major environmental systems, the distinctive threshold effect that produces such beach features, and the very short time periods that may be involved in beach re-structuring. Concerning the latter, Groves (in Steers 1960) records that even under relatively calm conditions en found that the level of a shore feeding a dune system may alter uch as 0·3 m over the greater part of the profile within as little as rs. Clearly beach profiles should be sampled in relation to weather ations rather than at regular time intervals.

Tides are produced in the oceans by the gravitational pull of planetary bodies, primarily the sun, the moon and the earth. Because these are moving in relation to each other the gravitational forces change. Constraints to movement formed by land boundaries result in water basins within which oscillations are set up and rotated by Coriolis forces about nodal or amphidromic points, points or areas of no significant tidal rise or fall (Defant, 1964). Around these points tidal range increases concentrically. Fig. 7 shows the distribution of tidal basins around British coasts. The minimum tidal ranges of the Solent area are, as one might expect, close to an amphidromic point in that region. Maximum tidal ranges on the west coast in the Bristol Channel region lie towards the outer edge of the much larger Atlantic Ocean tidal basin.

The vertical range in level of a salt marsh is primarily related to tidal range and secondarily to turbidity of the water. If the turbidity of the water is high, this reduces the potential vertical range of growth of salt marsh plants. In the maximum tide range of 12 m in the Bristol Channel, salt marsh growth ranges vertically over 4 m; in the small tidal range of 1·8 m in Poole Harbour salt marsh growth is telescoped into a vertical range of about 1 m. Marshes within a large tidal range tend to be more steeply sloping and consequently have more clearly zoned vegetation and sharper drainage systems normal to the shore. Marshes within a small tidal range have less clearly zoned vegetation and sluggish drainage on the ebb which produces a more complex network of winding and much-branched creeks. The Poole Harbour to Chichester Harbour salt marshes in a minimum tide zone have these distinctive features of complex drainage patterns and mosaics of indistinctly zoned vegetation. It is true also that long established high level marshes tend to flatten near the limit at which tidal submergence (and therefore silting), becomes insignificant. In such regions zonation again becomes indistinct and vegetation mosaics are common.

Water in the tidal basin moves in response to tidal streams and currents which may have their origin within or outside the basin and, to winds which generate waves. While the former may transport material in the vicinity of coasts, it is waves under the influence of wind which predominate in moving material along the shore or add material to the shore from supplies immediately off-shore or by erosion of the coastline (Steers, 1964).

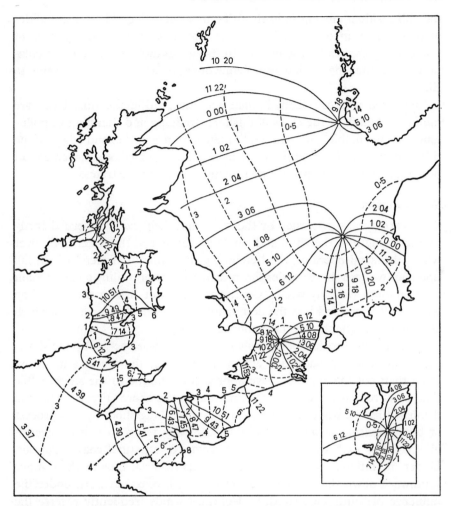

Fig. 7. Co-tidal and range lines for the North Sea tides. (Based on BA chart no 5058 with the sanction of the Controller, H.M. Stationery Office and of the Hydrographer of the Navy, from *Beaches and Coasts*, C. King, Edward Arnold Ltd. 1972.)

Robinson (1966) concluded from marine surveys, observations of current movements and drogue runs, that residual ebb/flood currents were important in determining off-shore bottom topography on the East Anglian coast. In addition to waves, tidal streams also influenced coastal topography. The tidal basin of the southern North Sea gives a decrease of tidal range along the Norfolk coast southwards leading to a tendency for the flood residual to be dominant. South of Yarmouth the situation is reversed and the ebb residual current is likely to be more effective. These inequalities of tidal range affect the tidal streams which have velocities capable of transporting the sea-bed sediment. In this area sand is in constant circulation throughout the shallow water zone down to at least 18 m according to Robinson.

Compared with the land, human activities within the tidal basin are minimal, though in the cases of offshore dredging and dumping of pollutants, not negligible. By contrast, human activities at the land edges of the tidal basin may have profound effects on the longshore movement of beach material as a glance at any groyne system demonstrates.

Wave action

Waves are usually generated by the wind, and their size is governed partly by the strength of the wind and partly by the size of the water body over which the wind operates, the 'fetch', e.g. the Atlantic 'rollers' which fall upon our western shores under the influence of the prevailing and dominant west or south-westerly winds. On the east coast of Britain the prevailing wind is off-shore and it is the dominant winds from the north-east which occur most frequently on these shores.

When a wave breaks upon the shore it has a forward action up the shore, the 'swash', and a backward action downwards, the 'backwash'. More often than not the angle of wave approach is oblique to the shore, while that of its retreat under gravity is normal to the shore (Fig. 8). Material disturbed by the wave is consequently carried along shore by this process of beach drifting. Material carried in this way tends to pile up against objects in its path whether this be a groyne or a natural coastal deposit. Under the influence of north easterly winds material thus tends to move southward down the east coast or on the south coast eastward under the influence of south-westerly drift. Contrary winds frequently reverse the directional flow.

Certain types of waves throw up more material than they comb down off the beach (constructive waves), others remove more material than they supply (destructive waves). The conditions under which these different types of waves occur are not yet fully understood, further discussion of the subject is given in King (1972) and Steers (1964). Certainly storm waves can be of either type, and undoubtedly the shore profile on their approach path (whether well or poorly stocked with material) must influence this. Efficient sampling methods for estimating mean particle size of beach samples, are discussed by Krumbein and Slack (1956). Where the whole sample population is used, careful choice of size grades enables the results

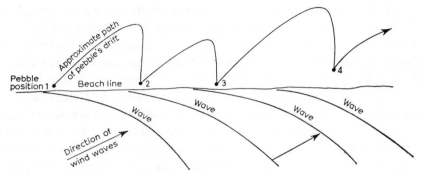

Fig. 8. Beach-drifting (from Steers 1964).

to be plotted on a logarithmic basis and departures from the smooth curve give clues to the presence and possible origins of unsorted material newly deposited on the shore.

Kestner (1961) points out that two groups of processes, sorting and mixing are continually going on in estuaries. Suspended load travel sorts particle sizes so that fine material is transported up or downstream in tidal flow and can only settle in fairly still water. Mixing processes include side erosion, scour and bed movement. From his studies on the Lune estuary, Lancashire and elsewhere, Kestner distinguishes three types of loose boundary transport.

Silt (particles below 0·1 mm diameter) forms loose boundary surfaces which are flat and unrippled. The material has arrived in suspension, is highly cohesive and resistant to surface scour; it is most likely to be eroded by some form of side erosion.

Fine sand between 0·1 and 0·2 mm can be very highly mobile and capable of moving with equal ease either in suspension or along the bed. Changing current velocities rather than high velocity itself may be responsible for starting this material into movement. Since more of this material is carried in the flood than on the ebb tide it may be responsible for rapid upstream shoaling especially where fresh water flow is reduced.

Medium and coarser sands above 0·2 mm may not be in suspension and sand banks of this material tend to move along the bed as units.

Kestner (1963) has demonstrated from suspended load sampling in the Wash that the important material for mudflat and therefore salt marsh development is a coarse silt or fine sand around 0·64–0·128 mm particle size. He notes that quite a small addition of finer particles makes it very cohesive. Clearly these threshold effects are highly relevant to both animal and plant establishment.

Experiments with radio-active tracers have demonstrated the rate at which silt or sand can be moved in estuaries. For example silt has been shown to move 10 miles (16 km) downstream in a matter of hours in the Thames estuary (Inglis and Allen 1957). Sand may move upstream up to 500 ft (152 m) per tide in the Mersey estuary (Price and Kendrick 1963). The speed of such changes and the distances involved underline the difficulties of investigating these movements and the need to extend studies (and plans for dealing with coastal erosion problems) far beyond the immediate vicinity of the coastal area being investigated. Just as whole catchment studies are necessary to understanding changes in an estuary, so studies based on whole tidal basins are likely to be needed to understand forces moulding estuary mouths and adjoining open coasts.

Habitat Series

The salt marsh series

Both salt marsh and dune physiographic series are best classified in relation to their maritime or terrestrial affinities.

The most maritime of all salt marshes are those which develop on relatively open coast conditions. Those in the lee of small islands offshore from low coasts backed by low hills are bathed in almost full strength sea water since the island fresh water catchment is negligible and drainage from the land catchment minimal. Some of the best examples of these marshes in Europe are found in association with the shingle and dune islands and spits off the north Norfolk coast, e.g. at Scolt Head Island. Marshes attached to the open coastline are developed on relatively coarse-particled sediments, as one might expect in the conditions of relatively strong water flow. They occur where broad expanses of inter-tidal sediments are found e.g. off the North Norfolk coast or on the Danish, German and Dutch Wadden coasts.

Next in the series are estuarine mouth marshes. These usually form in the lee of coastal spits, tend to be more coarse-particled than those further up estuary, and subject to stronger saline influence though less so than the open coast marshes. Estuarine mouth marshes such as those on the Dovey or Burry estuaries in Wales occupy about one fifth of the marsh area in any estuary.

Marshes of the more maritime type are characteristically rich in algae including free-living diminutive forms of *Fucus* species derived from normal forms attached to rocks or stones in the neighbourhood of the marshes.

Embayed marshes themselves form series according to the depth of the embayment and degree of fresh water flushing they receive. Shallowly embayed marshes are often extensive in area and developed on relatively freely-draining sandy silt of relatively high salinity as at Morecambe Bay in Lancashire for example. Marshes at the head of lochs tend to be small in area and subject to much fresh water flushing for silt is in short supply, but land drainage water abundant in the rocky terrain where they occur. Narrow-mouthed deep embayments are more sheltered and allow deposition of finer particles, producing salt marshes with intricate drainage systems on poorly drained clay and silt soils as at Hamford Water, Essex, Poole Harbour, Dorset or Chichester Harbour, Sussex. The salinity of embayed marshes depends on the amount of fresh water inflow. At Hamford Water and Chichester Harbour very little surface-drained fresh water reaches the marshes. At Poole Harbour two sizable rivers enter the upper reaches and reduce marsh salinity.

This brings us to the final group of marshes associated with the mid and upper reaches of estuaries which are progressively more terrestrial in affinity and increasingly influenced by fresh, rather than salt water tidal flooding.

Mid estuary marshes are extensive in area, occupying some three fifths of the total marsh area of an estuary. They are usually developed on silt and much subject to alternate periods of advance or retreat on any mid-estuarine shore under the influence of swings of the low water channel where this has not been trained in a particular direction. Studies on the Wyre-Lune estuary provide evidence that these fluctuations are self maintaining (Kestner and Inglis 1956). The important thing to bear in mind here is the relative impermanence of at least the lower levels of mid-estuarine marshes.

Upper estuarine marshes are the most sheltered of all and develop on clay-silt in regions of reduced salinity.

Any particular marsh may be intermediate in character between the three main types distinguished: spit-associated marshes; embayed marshes; estuarine marshes. So long as the various gradients: exposure; soil particle size; salinity; are kept in mind there should be no difficulty in placing the marsh in the ecosystem of which it forms a part. It is essential to consider the system as a whole when interpreting results obtained from the study of any part of it.

Transitional habitats

Before considering the dune series it is logical at this point to refer briefly to habitats transitonal between salt marshes and dunes, and between either of these two and any other coastal habitats.

The important point to bear in mind is that special conditions obtain in the transition zone dependent on interactions between the two or more coastal habitats represented.

Take the case of the horizontal transition between high level sand flats and high level mudflats. The primary colonist of such regions on European coasts is *Puccinellia maritima* which characteristically forms dome-shaped hummocks under the joint influence of contributions of wind-borne sand and water-borne sand and silt (Plate 3). These transitions usually occur either in sand-floored bays with silt deposits in the more sheltered inlets to landward (e.g. Morecambe Bay, Lancashire or Baie de la Frenaye, Côtes du Nord, France) or near the mouths of estuaries where the mud deposits of upper and mid estuary give way to sand deposits at the estuary mouth (e.g. Burry estuary, Glamorgan or the Dovey estuary, Cardiganshire). In all these cases the resultant marshes are rich in pans whose development is clearly traceable to a primary origin dependent on the formation of *Puccinellia maritima* hummocks, and the associated hollows between them.

The development of vegetation on these transitional tidal flats increases surface roughness, reduces rates of flow of tidal water and encourages deposition of increasing amounts of fine-particled material. Consequently the original sandy flat becomes overlaid with silt and gradually converted to a more uniform level salt marsh. In the early stages when only a thin layer of silt overlies sand the habitat is transitional not only in the horizontal, but also in the vertical plane and certain species (e.g. *Armeria maritima*), favoured by relatively well-aerated marsh soil conditions, are characteristic of this phase. In open bays with limited supplies of silt, marshes of the transitional type may persist through to maturity. In estuaries, gradually increasing silt accretion insulates the marsh from sub-surface sand effects (e.g. improved drainage, aeration, and percolation of fresh water from land drainage) and a typical silt or silt and clay marsh flora develops.

Similar horizontal and vertical transitions occur between salt marsh or sand dune and shingle to produce special relationships of drainage and aeration and characteristic floras and faunas. The tendency to fresh water flooding in dune slacks is usually presumptive evidence of an impermeable silt or clay sub-surface below the sand. However, in the case of island dunes on shingle there may be a tidal influence on the water table due to the high permeability of the underlying shingle.

Study of these transitional habitats and the complexities resulting from interaction of different coastal habitat systems is still in its infancy but it is central to an understanding of much of the pattern of diversity which distinguishes minor variants in the salt marsh or sand dune series.

In the case of the sand-silt flat transition one useful line of laboratory investigation would be to study seedling displacement from sand and silt mixtures in relation to rates of water and wind flow. In the field it might be possible to estimate strain forces at work using anchored fibres of different breaking-strain strengths attached to seedlings.

The dune series

Six main physiographic types of dune system are distinguished on the basis of position in relation to the shoreline. Three of these types: offshore islands, spits, and nesses, project seawards from the shoreline and are generally of a prograding nature. They are best collectively described as Frontshore systems and are more characteristic of sheltered shores including those where the prevailing and dominant winds are at least partially in opposition.

The other three types of dune systems distinguished here are bay dunes, hindshore dunes (usually with well developed slacks) and hindshore sand plains, or 'machairs' as they are known in western Scotland where the latter characteristically occur. These last three systems, progressively more terrestrial in position, may be collectively called Hindshore systems though their point of origin is of course on the shoreline itself. Bay dunes may have equally developed frontshore and hindshore components and form an intermediate link between the more maritime frontshore systems and the more terrestrial hindshore systems. The latter are particularly associated with more exposed shorelines where prevailing and dominant winds reinforce each other.

For a dune to form at all there must be some obstacle, natural or artificial around which wind blown sand can accumulate. In the absence of significant amounts of tidal litter or other wind-stilling barriers, but where abundant supplies of sand at the backshore level occur, the backshore flats may become colonized *in situ* and the whole system develops as a rapidly prograding vegetated sand plain with minimal dune formation. Where sections of lower-lying backshore levels become enclosed by higher-lying backshore levels to seawards, primary slacks develop in parallel with the shore line and they often carry elements of the sandy salt marsh facies due to occasional incursions of sea water or a brackish water table. The low-relief vegetated sand plain, Morrich More, Ross is a good example of this rather unusual rapidly prograding Frontshore system.

D

The effect of a solitary obstacle placed on backshore flats just high enough to accommodate the growth of strandline plants is well illustrated at Holkham, Norfolk. Here an abandoned military vehicle left on the backshore is said to have accumulated an isolated dune some 5 or 6 m high in about 20 years.

More usually tidal litter initiates sand accumulation at the top of backshore. This becomes colonized first by annual species and then by perennial grasses capable of growing up through sand accretion to form dunes of varying height according to the level at which sand loss exceeds sand supply.

Dunes up to 100 m high occur in the Coto Doñana, Spain. On British coasts they rarely exceed 30 m though sand may be blown to higher levels over rock outcrops in the hinterland e.g. at Penhale in Cornwall. Dunes with low relief may either occur in more sheltered areas with rapid progradation and stabilization, or in sites so exposed that strong winds never allow growth of high dunes at all e.g. Shetland Island dunes.

Sand or shingle spits built by beach drifting form a foundation on which dunes can accumulate, fan-wise by apposition of new material to seawards, or terminally as the spit lengthens alongshore. Storm conditions may throw material at the exposed tip landwards intermittently so that the linear growth alongshore is interrupted by a series of spur-like recurves. Systems such as these at Scolt Head Island or Blakeney, Norfolk tend to achieve stabilization in the position of formation of main dune ridges. This results in a series of dune zones of increasing age from the distal to the proximal end of a spit, or in the case of nesses, a series of apposition ridges of increasing age from seaward to landward.

This tendency to stabilize *in situ* is dependent on reduced wind strengths found on sheltered coasts and reduced frequency of higher wind strengths in any one particular direction. In other words coasts where the stronger winds tend to cancel out each others effects.

Bay dunes frequently occur as a narrow strip of dunes at the head of the bay with some penetration back onto the land surface. Relatively small bay systems set in cliffed coast are much influenced by fixed shelter factors of the surrounding topography. This greatly modifies the prevailing and dominant wind pattern of the general area so that dune systems only a few miles apart may be controlled by dominant winds from totally different directions. Ritchie and Mather (1969) have calculated the Exposure Index (pattern of wind incidence and direction) for a number of Scottish dune sites (Figs 9 and 10). Melvich Bay and Strathy Bay, both facing north on the north shore of Sutherland and, only 5 km apart, have dominant winds from the south and north west respectively. Sand supply

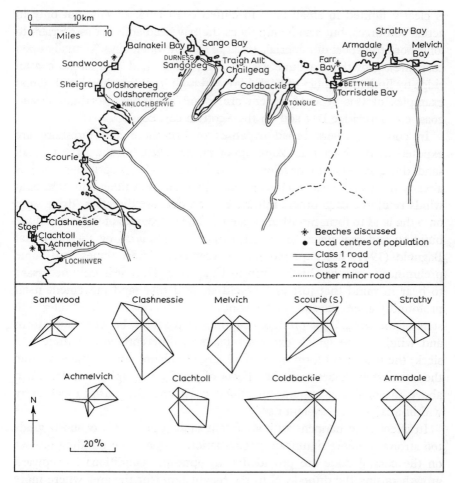

Fig. 9. Location of beaches in Sutherland, Scotland where Ritchie and Mather calculated exposure index as given in Fig. 10. (After Ritchie and Mather 1971)

Fig. 10. Exposure index for a selection of sites in Sutherland, Scotland as given in Fig. 9 (from Ritchie and Mather 1969)

is clearly limited in small bays in cliffed coast such as this, not only in actual resources, but also in supply as the bay headlands act as barriers to longshore transport of material. In large shallow open bays (virtually open coastline), growth of a dune system, may be limited to a single coastal ridge by limited supplies of backshore and high level foreshore sand. Good examples of this type of system are found along the Northumberland coast e.g. Druridge Bay and on the Scottish east coast.

In contrast, where broad expanses of backshore and foreshore are exposed in the inter-tidal zone opportunities exist for the largest of all sand dune systems to develop. In addition to the prograding systems already discussed, such shores orientated so that prevailing and dominant winds reinforce each others influence in an onshore direction, drive sand onto the land to form hindshore systems. The mechanics of vegetated dune building and movement are much in need of study along the lines that Bagnold (1941) has explored for un-vegetated desert dunes, but some preliminary work on this is given in Chapter 8. Here it is only necessary to bear in mind that just as Bagnold showed that sand movement itself profoundly alters the state of the wind, so the dune system itself alters the state of the wind. This produces continuous feedback between dune form and wind regime which expresses itself in differentiation of dune and slack, the basic land form units of all large hindshore dune systems. Note the contrast here between the effects of *fixed* topography as referred to above in small bay systems in cliffed coast and the *plastic* topography within a large dune system itself.

In effect what happens with abundant sand supplies but onshore winds too strong to allow significant progradation is that dune building grasses on the coastal dune are provided with optimum conditions for upward growth raising the dune level to the height limit for the area where more sand is lost than gained at the crest from wind erosion, and at the seaward toe, from wave erosion. Thereafter, the mature coastal dune, unable to repair the continuously eroding vertical faces on its seaward side, is eroded back to make way for new embryo dunes to develop on tidal litter cast to the limit of storm tides. Such a dune once set in motion continues inland until it is sufficiently flattened and sheltered to stabilize while a new coast dune builds vertically again. Between the two, erosion persists to the damp sand level where it stops and a slack surface is available for colonization. The ridge structure may be broken up into a series of parabola-shaped mobile dunes. Landsberg (1956) has developed formulae which give good correlation between orientation of the long axis of these parabolas and calculated wind resultant.

Perhaps the most significant distinction between these active hindshore systems and prograding systems stabilized *in situ* is the periodic re-exposure of bare sand at intervals in the mobile part of the system and the tendency to cyclic alternation of dune and slack at any particular point within it. Braunton Burrows, Devon, Newborough Warren, Anglesey and Ainsdale, Lancashire are among the best examples of hindshore systems in Britain. Full stabilization is only achieved when the landwardmost ridges are deflated to an almost flat plain at the dry slack level.

The last system in the dune series is really an extreme example of the hindshore system and has close affinity with its stable stage. Under extreme conditions of exposure on the north west coast of Scotland where the full force of Atlantic gales reach low lying islands, the height limit of dune building is so low that much of the sand from the shore is swept inland to form low lying sandy plains. The universal influence of open range sheep grazing or of cultivation of the lime-rich sandy pastures further limit the possibilities of dune development. Towards the landward side of these sandy plains or 'machairs' lime-rich sand thinly overlies lime-deficient moorland peat providing contrasting conditions for great variety of plant species.

Many salt marshes and sand dune systems are of course intermediate between the types distinguished on the maritime to terrestrial gradient. Familiarity with the series as a whole soon enables intermediate and apparent anomalies to be placed in the general framework described. We begin to see these plastic deposits of the coast moulded into recognizable forms by the four great agents, wind, water, land and living things, operating simultaneously upon them.

3 Mineral Nutrient Relations

The essential nutritional problem for plants that grow on salt marsh or calcareous dune soils is one of adaptation to growth on soils, with elevated levels of certain ionic species. Such adaptation is intimately concerned with osmotic effects. These partly control ionic concentrations either side of cell membranes and also the capacity of plants to function in the presence of elevated ionic levels, particularly of sodium. Although these adaptations can be studied at a variety of organizational levels; within the cell, within the whole plant, or within the community, they must be related to the capacity of a plant to survive, grow and reproduce in the habitat where selection operates at all levels in the life cycle; on individuals, populations and at the species level.

While elevated levels of sodium ions are the main problem for salt marsh plant growth, there is increasing evidence that high levels of calcium ions in addition exert profound effects on the tolerance of certain species to sodium and on the actual species composition of the salt marsh flora.

The sodium ion problem is only critical for a small but important element of the dune flora near the coast where significantly most of the species confined to the dune habitat occur. Elsewhere in the dune system the main nutritional problems for dune plants are either, growth in elevated levels of calcium ions or, the more widespread problem of general mineral deficiency.

The presence of high ionic levels makes it essential to discuss a number of soil physical parameters related to soil evolution, development, and structure since they have big effects on availability of ions for plants. It must be remembered also that ionic exchange is not confined to the soil and root boundary, but also to the leaf and water, or leaf and air boundaries. For example ionic exchange in *Zostera* occurs primarily through the leaf cuticle and salt may be exuded through leaf glands as in *Spartina* in addition to exchange through cuticle and stomata. Consideration of osmotic effects follows logically discussion of soil physical parameters. This in turn sets the scene for discussion of chemical parameters associated with nutrient uptake, salt tolerance mechanisms and ionic balance. This chapter concludes with a brief account of nutrient levels and availability in the two habitats. Nutrient supply and nutrient cycling at the community level are mentioned in passing but considered in more detail in appropriate places in the sections dealing with salt marshes and sand dunes separately.

Physical Constraints on the Chemical System

Effects of particle size composition

Kelley (1951) notes that the amount of growth of plants is much more seriously affected by exchangeable sodium in the soil than by equal amount in true solution. The yield of beans (*Phaseolus vulgaris*) for example has been shown to be much reduced by quite low levels of exchangeable sodium in the soil, but is maintained in the presence of soluble sodium at much higher concentrations (Bernstein and Pearson 1956). Now Lopez-Gonzales and Jenny (1959) have demonstrated with the aid of cation-exchange resin discs in contact and apart, that contact exchange of ions occurs much more rapidly than that by mass transfer of ions through the solution when discs were separated. As Heimann (1958) points out this suggests an explanation for the different effects of sodium in soil and solution on plant growth. In soil it seems likely that contact exchange mechanisms may operate additionally to mass flow under the influence of transpiration or diffusion mechanisms. If so, then ionic readjustments of equilibrium disturbed at the root surface by selective uptake may well be more rapidly compensated by the common pool of ions where solution mechanisms are at work than where less readily compensated contact exchange mechanisms are in operation. The ecologist might advance knowledge in this field by comparing the salinity tolerance of clonal material of different species in mud, sand and culture solution. It is clear that we should beware of concluding that tolerances in culture are immediately relevant to field conditions. Equally, in reporting field tolerances it is evidently important to specify the clay and organic matter content of the soil. The lengthy procedures associated with mechanical analysis have been a barrier to this, but water holding capacity of the soil can be quite rapidly measured and Glopper (1964) has shown this is closely correlated with the clay and organic matter content.

It is relevant at this point to recall the early experiments of Joseph and Oakley (1929) on the structural improvements to soil with impeded drainage produced by charging their exchange complexes with calcium or potassium ions. With either, this results in at least a halving of the water retaining capacity and a capillary rise of water increased by at least a factor of 40 compared with sodium-charged exchange complexes. These effects may not be important in lower salt marshes subject to regular inundation by sea-water, but they are likely to be important in improving both the nutrient environment and the amelioration of salinity at higher marsh levels.

Soil macrostructure

Salt marsh and sand dune soils form an extreme contrast so far as texture is concerned. Improvements in texture (and nutrient availability) in the salt marsh soil depend on factors *increasing* pore space; in contrast in sand dune soil factors *decreasing* pore space are instrumental in improving texture and nutrient availability.

Reduced incidence of flooding in higher marsh levels does not automatically improve texture as dried out sodium clays are mechanically difficult for roots to penetrate. Cracking of the surface helps, but only in a very localized way. Far more important are the biological effects of roots and the larger soil animals in opening up cavities. Green and Askew (1965) have investigated the activities of roots, ants, and earthworms in reclaimed marsh soils at Romney Marsh, Sussex, using infiltration techniques. They attribute the high fertility of these soils to improvements in drainage caused in this way.

Olson (1958 *a*) notes the improvements of exchange capacity with buildup of the clay fraction in an inland dune system and incidentally notes the value of dune soils for studies on dust fall and weathering.

Base exchange characteristics of salt marsh and sand dune soils

In young salt marsh muds there may be very little organic matter (less than 5 per cent) and the bulk of adsorbed ions will be associated with the clay mineral lattices which would be fully saturated with ions, presumably mainly sodium ions. The cation exchange capacity of different clay minerals varies considerably. For kaolinite it is approximately 3 to 15 mE; for illite and chlorite 10 to 40 mE; for montmorillonite 80 to 150 mE. (Grim 1953). Equivalence (E) is defined here as 1 mole of electronic charge (i.e. 96 487 coulombs) on the clay lattice complex. There have been very few analyses of the clay mineral composition of salt marsh muds, but Guilcher and

Berthois (1957) investigating the possible origin of salt marsh muds in Brittany showed that illite was the dominant clay mineral in that area. Unpublished work by Stebbings and Ball indicated that in muds at Poole Harbour, Dorset, and Bridgwater Bay, Somerset, illite and chlorite were dominant while in the upper reaches of the Fal estuary, Cornwall, the fine clay fraction was practically pure kaolinite. Now it is interesting that on the Fal estuary, *Scirpus maritimus* exhibits much poorer growth and flowering than in Bridgwater Bay, and Poole Harbour. At Bridgwater Bay it is true that the exceptionally strong growths there are also no doubt influenced by greatly increased nutrient supply and abundant calcium both of which may be limiting factors in the other two environments. However, the point to be made is that salt marsh muds do differ by as much as a factor of 10 in cation exchange capacity and this may well be partly responsible for differences in productivity.

Nutrient-deficient muds like those of the Fal estuary, could profitably be used as an experimental medium for studies on the effects of adding nutrients in relation to salt marsh plant growth. It would also be interesting to know the total cation exchange capacity of these muds in relation to the amount of organic matter as this might shed light on that part of the exchange capacity dependent on organic matter.

Sand dune soils with free carbonate are invariably base-saturated, but the total cation exchange capacity is itself low due to the very low clay and organic matter contents of the soil. In dune soils it is of the order of 10–15 mE. per 100 g and in dune slacks from 15–30 mE. per 100 g (Ranwell 1959). In dune sands devoid of free carbonate, either initially or through leaching, the exchange complexes are frequently partly unsaturated and this may occur even where the soil reaction is neutral (Table 2). We can see these figures in perspective against Renger's (1965) observations derived from over 1,500 soil samples estimating the mE. per 100 g of: organic matter (168 to 249); clay (38 to 51); silt (2 to 22) and sand (0·7 to 6·5).

Ecologists can make a useful contribution by studying field sites where a particular environmental factor is likely to exert an extreme influence. So far as nutrient deficiencies are concerned, the kaolin muds of the Fal estuary, Cornwall and the base deficient dunes of Winterton, Norfolk or Studland, Dorset are therefore recommended for study.

Osmotic Effects

Plant cells must maintain internal osmotic potential lower than the external osmotic potential or they lose water. The osmotic effect only becomes

Table 2. Base status of stable dune and slack soils at Newborough Warren, Anglesey beneath Salix repens associes. Equivalence (E) is defined as 1 mole of electronic charge (i.e. 96,487 coulombs) on the soil/organic complex (from Ranwell 1959).

Sample site	Sample depth (cm)	Exch. bases in mE/100 g oven-dry soil	Exch. H_2 in mE/100 g oven-dry soil	Total cation exch. cap. in mE/100 g oven-dry soil	% Base saturation	CaO as % of total exch. bases	CO_3 as % of weight of oven-dry soil	pH value Water	KCl	Water table (free) depth (cm)
Fixed dune	8	12·02	0	12·02	100	79·07	0·03	7·8	7·3	>200
Salicetum	30	19·09	0	19·09	100	6C·82	0·23	8·8	8·2	
Fixed dry slack turf	8	2·76	0·61	3·37	81·9	67·11	0	5·3	5·0	104
	30	7·10	0·08	7·18	98·9	89·04	0	7·1	7·2	
Fixed dry slack Salicetum	8	2·85	4·38	6·23	45·7	61·58	0	4·4	4·1	84
	30	17·78	0·76	18·54	95·9	7·71	0	4·9	4·7	
Fixed wet slack Salicetum	8	18·84	0	18·84	100	63·16	0·22	7·2	7·1	64
	30	27·36	0	27·36	100	39·66	0·77	7·3	8·1	
Fixed wet-dry slack Salicetum	8 (i)	17·09	2·46	19·55	87·4	86·22	0	6·1	6·3	79
	(ii)	16·12		18·58	86·8	89·76				
(+ Calluna)	30 (i)	27·54	0	27·54	100	48·51	0·61	7·5	8·0	
	(ii)	32·75		32·75	100	59·91				

serious for non-halophytes at high external concentrations of salt in the order of 0·05 M (e.g. one tenth sea water strength) or more, corresponding to osmotic potentials around two bars (Epstein 1969). It is of interest that concentrations of salt of this order occur in the water flooding the seaward limit of tidal woodland in Europe marking a major change in the flora from few halophytes to many non-halophytes (see Stebbings 1971). Obligate halophytes subject to salinities at full strength sea water need to develop osmotic pressures greater than 20 bars to survive. Arnold's (1955) results (Table 3) show in fact that they do. The two exceptions (*Glaux maritima* and *Scirpus maritimus*) are facultative halophytes and persist for centuries in effectively non-saline habitats e.g. in dune slacks or old reclaimed marshland.

Table 3. Osmotic pressures of cell-sap of various halophytes and proportion due to chloride ion. One atm. is equal to 101325 Nm^{-2} (from Arnold 1955).

Species	No. of determinations	O.P. sap (atm.)	Proportion O.P. sap due to Cl$^-$ (atm.)	Cl$^-$ as per cent O.P. sap
Atriplex patula var. hastata	6	31·6	13·3	42
Suaeda fruticosa	15	35·2	15·3	43
Glaux maritima	2	14·6	7·4	51
Distichlis spicata	10	29·3	15·1	52
Juncus gerardi	3	27·8	15·5	56
Iva frutescens var. oraria	4	23·9	13·8	58
Suaeda nigra	2	41·1	24·7	60
Spartina glabra	4	27·0	16·9	62
Triglochin maritima	10	24·6	16·1	66
Scirpus maritimus	2	14·7	10·4	71
Salicornia rubra	11	44·3	31·5	71
Limonium carolinianum	4	29·2	21·4	73
Spartina patens	4	20·9	15·7	75
Salicornia ambigua	6	42·5	34·1	80
Salicornia stricta (herbacea)	10	39·7	35·9	91
Salicornia mucronata	3	34·0	31·5	93
	Av.	32·6	21·1	65 ± 11·2

There are four main effects of high external osmotic potential according to Slatyer (1967).

(1) It depresses growth and therefore yield of either the whole plant or parts of it and this effect occurs even with halophytes.

(2) It may depress transpiration initially, but not to the extent that growth is depressed. The extent to which this occurs depends in part on stomatal behaviour.

(3) It may have similar effects to low soil water content and reduce water availability.

(4) It induces excess ion accumulation in tissues which may combine with reduced uptake of essential mineral elements.

In the past workers have tended to give undue weight to the effect of osmotic pressure of the soil solution and pay little attention to the effect of soil matric pressure. The soil water potential (soil water stress) is now recognized to be a combination of these forces. However, the osmotic relations within the plant are modified by uptake of soil solutes and entry of excess salts. Therefore the effects of osmotic pressure of the soil solution and those of the soil matric pressure do not have equivalent effects on the internal osmotic relations of the plant tissue.

Experimental studies suggest that some reduction of metabolic function may be attributable to direct osmotic effects on internal water deficits, but this does not appear to be as pronounced as that caused by a similar reduction in substratum water potential caused by a reduction in soil water content. However these are short term responses to imposed salinity. The main long term effects of salinity are associated with ion accumulation (and therefore mineral uptake disturbance or toxic ion effects) in the plant rather than with reduced water availability in the substratum, as earlier workers tended to assume.

It follows that tolerance to elevated ionic levels therefore requires a high degree of selectivity in ion uptake. Excess accumulation of electrolytes in plant cells, particularly of sodium or chloride ions, is likely to result in progressive changes in protein hydration and conformation and enhancement or depression of enzyme activity resulting in gradual dislocation of metabolism.

Less tolerant species with less efficient ion discriminatory mechanisms may expend more respiratory energy, resulting in reduction of net assimilation rate and growth suppression, than more tolerant species with more efficient discriminatory mechanisms.

There is faster recovery from soil water stress than from salinity stress not only because of the more serious metabolic disturbances induced by the latter, but also as a result of the time required for excess ion accumulation within the plant to be diluted by new growth.

Mannitol has a large organic molecule which enters the cells of higher plants extremely slowly. A solution of mannitol can therefore be used to simulate osmotic effects similar to those produced by a solution of electrolytes, but without an immediate direct chemical effect on cell physiology.

However, mannitol is slowly metabolized and this must be kept in mind in the interpretation of experimental results. Parham (1970) has used solutions of mannitol to separate osmotic and ionic effects of saline solutions on the germination of halophyte seeds and this seems a very promising application of the technique. His results indicated that germination of *Plantago coronopus*, *P. lanceolata* and *P. major* seeds was inhibited by high ionic concentrations rather than high osmotic pressures. Germination of the halophytes *Plantago maritima* and *Triglochin maritima* and the non-halophyte *Triglochin palustris* occurred in high ionic concentrations, but was apparently inhibited by osmotic effects at salinities equivalent to a 70 per cent solution of sea water. However, it was shown that continued growth of the halophyte seedlings was better in solutions of sodium chloride than mannitol, while the reverse was found with the non-halophyte *Triglochin palustris*.

Chemical Problems of Nutrition in Elevated Ionic Environments

Nutrients essential for the growth of plants

At least 16 elements are required for the growth of all higher plants. They include: hydrogen, oxygen, carbon, potassium, calcium, magnesium, nitrogen, phosphorus, sulphur, iron, manganese, zinc, copper, chlorine, boron and molybdenum.

Cobalt is required for symbiotic nitrogen fixation in root nodules of legumes (Bollard and Butler 1966), and also for nitrogen fixation in blue-green algae. Diatoms have a specific requirement for silicon and there is evidence that the presence of silicon increases resistance to blast fungus disease in rice (Okuda and Takahashi 1965).

This is not the place to consider specific roles of mineral nutrients and Evans and Sorger (1966) have produced an excellent review of this subject. However, it is of some interest to see how concentrations of physiologically important elements are distributed in marine organisms compared with their distribution in sea water and Kalle's (1958) results for this are given in Table 4.

Other elements may be accumulated by sand dune or salt marsh plants without having any obvious nutritional role. For example *Lycium* species which occur on dunes and in other coastal habitats though not confined to the coast, accumulate especially high levels of lithium even when lithium is not abundant in the soil and the levels remain high throughout the growth period (Bollard and Butler 1966).

Table 4. Distribution of physiologically important elements in sea water organisms (from Kalle 1958).

Plastic Elements	Content of			Catalytic elements	Content of		
	100 g organisms (dry) N	1 ml sea water 35% S, A	Ratio A/N		100 g organisms (dry) N	1 ml sea water 35% S, A	Ratio A/N
Hydrogen	7 g			Copper	5 mg	10 mg	2
Sodium	3 g	10·75 kg	3,600	Zinc	20 mg	5 mg	4
Potassium	1 g	390 g	390	Boron	2 mg	5 g	2,500
Magnesium	0·4 g	1·3 kg	3,300	Vanadium	3 mg	0·3 mg	0·1
Calcium	0·5 g	416 g	830	Arsenic	0·1 mg	15 mg	150
Carbon	30 g	28 g	1	Manganese	2 mg	5 mg	2·5
Silicon a	0·5 g	500 mg	1	Fluorine	1 mg	1·4 g	1,400
Silicon b	10 g	500 mg	0·05	Bromine	2·5 mg	66 g	26,000
Nitrogen	5 g	300 mg	0·06	Iron a	1 g	50 mg	0·05
Phosphorus	0·6 g	30 mg	0·05	Iron b	40 mg	50 mg	1·3
Oxygen as O_2 and CO_2	47 g	90 g	2	Cobalt	0·05 mg	0·1 mg	2
				Aluminium	1 mg	120 mg	120
				Titanium	100 mg		
Sulphur	1 g	900 g	900	Radium	$4 \cdot 10^{-12}$ g	10^{-10} g	25
Chlorine	4 g	19·3 kg	4,800				

Iodine has been shown to be incorporated in amino acids in *Salicornia perennis* and *Aster tripolium* (Fowden 1959) and it is well known that iodine often accumulates in marine algae, but it is not known to be essential for growth. The mangrove *Rhizophora harrisonii* accumulates aluminium and up to 10 per cent of the aluminium in the mud is in the form of a complex with organic matter beneath stands of this species.

Certain halophytes have been shown to have a specific requirement for sodium. *Atriplex vesicaria* grown in the absence of sodium in water culture showed severe growth retardation and chlorosis but recovered when sodium was added to the culture solution (Brownell 1965). *Salicornia perennis* and *Suaeda maritima* grow larger with sodium well in excess of amounts needed by non-halophytes (Pigott in Rorison 1969). These effects seem to be characteristic of mature plants for, as mentioned in Chapter 1, in general halophytes germinate more freely in lower than in high salinities.

Salt tolerance mechanisms

The study of nutrition in coastal plants is still in its infancy. Much more work has been done on the nutrition of crop plants grown with the aid of irrigation in the very extensive salinized soils of central continental areas. A vast literature on this subject exists and perhaps because of this, the problem of salinity in relation to coastal plants tends to be given undue prominence at the expense of other nutritional aspects equally vital to their growth.

Much of the work on crop plants is not directly relevant to halophytes, because the crop species have only a marginal as opposed to full salinity tolerance. Much of the crop plant work also rests on short period studies and does not shed light on the adaptive phenomena associated with halophytism. Of far greater relevance is work associated with extreme forms of salt tolerance.

There are certain bacteria (*Halobacteria*) which cannot survive in sodium chloride concentrations *less than* 2 M and show optimum growth at the extremely high concentration of 4·3 M sodium chloride. They have been the subject of intensive study and it is of particular interest that even these highly specialized organisms are believed to have metabolic pathways not basically different from those in non-halophilic bacteria (Larsen 1967). It follows that we would not expect to find the basic metabolism of salt marsh plants very different from that of plants in non-saline environments and this seems to be so. But it is remarkable the number of ways in which halophytes have become adapted to high salinity environments. Since they shed light on the central problem of nutrition, the maintenance of balanced ionic environment in the plant cell in the face of temporary or permanent imbalance with relation to the external medium, they are worth considering in more detail. Without this background, the ecologist cannot fully appreciate for example the significance of measurements of salinity tolerance in relation to soil physical and chemical structure, or narrow down his selection of field situations for experimental study to sites where nutritional factors are paramount for the survival of particular species.

There are at least four ways in which halophytes have become adapted to enable normal metabolic functioning in high sodium environments: ion selection; ion extrusion; ion accumulation and ion dilution. Many halophytes exhibit more than one of these adaptations simultaneously and the relative importance of each in any particular species may vary according to the stage of growth and environmental conditions.

As we have seen, the extreme form of adaptation in which organisms can only function in the presence of abnormally high sodium levels occurs in the *Halobacteria* and marine algae, both of which are capable of a high degree of selective ion absorption (of potassium for example) in the presence of high sodium concentrations in the external medium.

Recently Parham (1970) has shown that in this respect certain flowering plant halophytes exhibit such marine algal-type properties. Maximum uptake of potassium into the roots of *Triglochin maritima* was shown to occur only when the concentration of both sodium and potassium in the external solution was high and moreover this uptake was shown to be metabolically mediated. This species, like *Salicornia* and *Suaeda maritima*, is therefore in the true sense an obligate halophyte.

Many species such as *Avicennia* and *Spartina* have long been known to practise ion extrusion via special salt glands on the shoots and there is evidence of sodium ion 'out pumps' in submerged halophytes and algae (Bollard and Butler 1966).

Ion accumulation in parts of the plant where concentrations may be stored away from active metabolic sites is probably a feature of most halophytes. For example *Agropyron elongatum* accumulates chlorides in the roots which, as in most grasses, are shed annually together with accumulated ions. Similarly it is commonly observed that in periods of drought on high marsh levels the leaves of *Limonium* die off prematurely shedding accumulated ions as they do so.

The development of succulence is another feature common to many halophytes (e.g. *Aster*) and this has the effect of increasing ion dilution by increasing the volume to surface area ratio of the plant.

Ionic balance

Gutnecht and Dainty (1968) suggest that the evolution of a sodium 'out-pump' may have developed in the earliest living systems in response to the need to control osmosis and that subsequently, with the development of a mechanically resistant cell wall, the sodium pump remained primarily for nutritional purposes. With elegant reasoning they go on to postulate that since extrusion of sodium ions results in high intracellular concentrations of potassium ions this would lead to enzyme adaptation to a high potassium ion environment and evolution of the inward potassium ion pumps. Sodium may also stimulate the growth of many organisms under conditions where potassium is deficient (Evans and Sorger 1966).

Several recent authors have drawn attention to the importance of calcium in enabling plants to grow in saline conditions (e.g. *Avicennia* (Macnae 1966) and *Agropyron* species (Elzam and Epstein 1969)). The sodium concentrations in leaves of *Phaseolus* fell by a factor of 16 over a range of calcium sulphate from 1×10^{-4}M to 3×10^{-3}M in 5×10^{-2}M sodium chloride solutions indicating a massive breakthrough of sodium to leaves in calcium sulphate concentrations below 3×10^{-3}M (i.e. 0·01 per cent calcium). It is suggested that calcium is essential for the maintenance of the integrity of selective ion transport mechanisms (especially selective absorption of potassium in the presence of sodium) at the surface of the absorbing cells of the roots where the entry of ions into the roots is governed.

Ionic balance and hydrogen ion concentration

In salt marsh soils the ionic balance is dominated by sodium. In dune soils landward of significant sea spray influence, the concentration of cations in solution is determined, either by a dominant calcium inflow (from shell fragments) or (in calcium-deficient sands), by the concentrations of nitrate, sulphate and bicarbonate anions resulting from (in part) microbial activity (Black 1968). The salt spray effect imposes a very localised nutritional regime at the dune coast well illustrated by the results of Gorham (1958), Willis *et al* (1959) and Sloet van Oldruitenborgh (1969) in respect of sodium, magnesium, and chloride (Tables 5 and 6). Its effects on the dune flora are masked by the severe limitations on the number of species that can occur because of high soil mobility. The nutritional effect of salt spray is much more readily appreciated on low coastal cliffs on acid soil where heath is replaced by a narrow band of grassland in the spray-dominated zone (e.g. in low offshore islands like the Isles of Scilly).

The salt effect is negligible in dune soils further inland from the coast.

E

Table 5. Soluble materials in dune sands from Blakeney Point, Norfolk. Equivalence (E) is defined as 1 mole of electronic charge (i.e. 96,487 coulombs) (from Gorham 1958 a).

	1 Embryo dunes	2 Sand hills	3 Face of main ridge	4 Laboratory dunes	5	6 Long Hills	7 The Hood	8 Lifeboat House well 1957	9 Lifeboat House well 1955 (analysed 1957)
		(Ammophila)		(Ammophila-Festuca)	(Carex arenaria)	(Ammophila-Festuca)	(Ammophila-Carex arenaria)		
pH (original)	8·2	8·1	7·9	6·2	5·7	6·1	5·7	7·9	8·6
(aerated)	7·9	7·9	7·8	7·7	7·0	7·6	7·1	8·6	8·7
Specific conductivity (micromhos at 20°C)	110	53	54	45	28	51	50	551	656
Sum of cations	1·18	0·60	0·61	0·52	0·29	0·57	0·52	6·57	7·73
Sodium	0·46	0·08	0·08	0·04	0·04	0·07	0·15	3·65	4·94
Potassium	0·05	0·03	0·04	0·04	0·04	0·07	0·11	0·20	0·24
Calcium	0·49	0·42	0·43	0·21	0·08	0·23	0·08	1·53	1·30
Magnesium	0·17	0·07	0·05	0·15	0·07	0·09	0·08	1·18	1·25
Ammonium (mE/200 g dry sand)	0·006	0·004	0·008	0·08	0·06	0·11	0·10	0·009	nil
Bicarbonate	0·49	0·47	0·47	0·35	0·12	0·29	0·15	3·56	3·95
Chloride	0·42	0·05	0·06	0·01	0·02	0·04	0·08	2·16	2·91
Sulphate	0·21	0·07	0·06	0·07	0·08	0·12	0·17	0·80	0·87
Phosphate	0·004	0·004	0·006	0·05	0·05	0·11	0·11	0·004	<0·001
Nitrate	0·011	0·005	0·004	0·001	<0·001	<0·001	<0·001	0·019	0·009
Silica (p.p.m. SiO_2)	1·6	1·6	1·5	1·0	1·0	1·1	1·0	8·1	8·1
Optical density ($\log \frac{I_0}{I}$ at 320 mμ, 10 cm cells)	0·5	0·5	0·6	2·0	2·4	2·7	3·9	0·5	0·5
Na/Cl ratio	1·1	1·6	1·3	4·0	2·0	1·8	1·9	1·7	1·7
Mg/Cl ratio	0·4	1·4	0·8	15·0	3·5	2·3	1·0	0·6	0·4
Ca/HCO_2 ratio	1·0	0·9	0·9	0·6	0·7	0·8	0·5	0·4	0·3
*$CaCO_3$ (% dry wt. of soil)	0·28-0·61			nil-0·64		nil-0·05	nil-0·05		
*Ignition loss (% dry wt. of soil, corrected for carbonates)	0·2-0·5			0·3-0·9		0·6-2·7	0·6-6·3		

*Soil data from Salisbury (1922).

Table 6. Chemical analyses of the soils from different types of vegetation at Braunton Burrows, Devon (from Willis *et al* 1959)

The results are expressed per gm dry weight of soil.

		Dry dunes							Slacks and hollows						
		Ammophila fore dunes	Pure Ammophiletum (main dunes)	Mixed Ammophiletum	Dry dune pasture	Stable pasture with lichens	Pteridietum	Dune scrub	Slack near sea	Plantago-Leontodon community	Festuca-Agrostis pasture	Festuca-Carex flacca pasture	Caricetum nigrae	Salicetum repentis	Salicetum atrocinereae
pH		9·05	9·06	8·79	8·70	8·66	8·60	8·18	8·99	8·73	8·42	8·22	8·12	8·11	8·06
Organic Carbon	mg	0·52	0·19	0·94	0·74	2·44	4·47	12·60	0·41	1·39	5·82	9·23	22·93	13·55	19·47
Total Nitrogen	mg	0·18	0·11	0·20	0·23	0·41	0·67	2·15	0·15	0·26	0·81	1·36	2·38	1·38	2·82
Calcium	mg	70·4	69·5	62·3	57·3	60·6	49·2	33·9	64·4	66·1	49·1	62·0	46·8	50·4	45·2
Magnesium	μg	2,270	990	1,070	1,060	880	980	220	1,570	1,470	770	970	570	280	240
Sodium	μg	528	14	11	6	11	14	55	15	25	17	24	50	26	81
Potassium	μg	50	6	9	7	7	12	13	11	14	18	6	26	15	17
Carbonate	mg	119·6	115·4	103·9	92·8	98·2	78·4	51·0	109·7	112·1	75·9	96·6	73·5	78·3	68·4
Phosphate-P	μg	109	110	98	107	59	91	148	112	110	103	108	133	110	131
Chloride	μg	845	14	7	3	3	3	10	10	25	14	17	29	22	39

In these regions, the hydrogen ion concentrations of the soil solution exerts profound effects on both nutrient availability and on concentrations of ions which may have toxic effects in the soil solution.

At high pH values for example there may be a decrease in availability of potassium, phosphorus and iron due in part to competition for sites on soil ion exchange complexes. The solubilities of metallic cations vary markedly at different hydrogen ion concentrations (Sparling 1967). In general they tend to increase with increasing hydrogen ion concentrations (i.e. at lower pH values). This effect is particularly marked with aluminium which may be released in soluble form in toxic concentrations at pH values around 4. There is also a marked increase in the solubilities of many other metallic cations around pH 7.

High levels of soluble aluminium may interfere with uptake of calcium, phosphorus and iron at low pH values. In toxic concentrations aluminium inhibits root growth, particularly in seedlings (Bollard and Butler 1966), so it would be interesting to experiment with calcicole members of the dune flora to see how their seedlings respond to varying aluminium levels where these species become eliminated from the more leached parts of a dune system.

Nutrient Levels and Availability

We are now in a better position to appreciate the significance of the general nutrient levels in salt marsh and dune soils in relation to availability. It must be stressed however that much of the older data is of limited value because of inadequacies of sampling and methodology. In view of this, emphasis is given to more recent work.

General levels of the commoner nutrients in salt marsh soils are given by Chapman (1960); Zonneveld (1960); Goodman and Williams (1961); Ranwell (1964 a, b) and Pigott (in Rorison 1969), and in sand dune soils by Gimingham (1951); Gorham (1958 a); Olsen (1958 a); Ranwell (1959); Willis *et al* (1959); Willis and Yemm (1961); and Freijsen (1964). The nutrient supply from rainfall, a source of particular importance to dune plants, has been analysed by Gorham (1958 b), Allen *et al* (1968), and Parham (1970).

Macro-nutrients

Both sea water and shore sand are very deficient in nitrogen, but so far as pioneer salt marsh plants are concerned this may be a distinct advantage as lush, brittle growth induced by high nitrogen levels would be subject to severe damage from wave action. Supplies in marsh mud are probably augmented by fixation of atmospheric nitrogen by blue-green algae the precursors of salt marsh growth. Gorham (1958 *a*) shows that nitrate nitrogen is distinctly higher in the embryo dunes subject to spray than further landward, but organic matter in tidal litter is undoubtedly an important source of nitrogen in this region. Also, Metcalfe (*pers. comm.*) has shown that *Azotobacter* is widely distributed in dune soils and again fixation of atmospheric nitrogen by this means is probably an important nitrogen source for the younger dune communities (see Chapter 9).

It is well known that organic carbon, and with it total nitrogen, augments with increasing age and increasing density of plant cover on dunes. The rate of increase in damp slacks due to inflow of nutrients from leaching and reduced rate of organic matter breakdown in the wetter soils, is faster than on dunes (Ranwell 1959; Willis *et al* 1959). Little is known about the changes in total nitrogen from seaward to landward in salt marshes, but evidence from newly accreted surface layers at Bridgwater Bay show an increasing trend in nitrogen (Table 7) from bare mud to marsh levels near high water mark. A similar trend was also found in Poole Harbour marshes. Since both marshes are dominated by *Spartina anglica* we can see also how the much higher soil moisture levels at Poole dependent on a high clay content result in almost twice as much organic carbon and total nitrogen compared with the more freely draining coarser-particled marsh soil at Bridgwater Bay.

Table 7. Nutrient content of newly accreted silt, 0 to 0·5 cm depth, from *Spartina anglica* marsh, Bridgwater Bay, Somerset, sampled 7 Nov. 1962 (from Ranwell 1964 *b*).

Sampling site	Distance seaward (m)	% oven dry weight of silt					% volume of wet silt				
		K	Ca	P	N	Organic carbon	K	Ca	P	N	Organic carbon
Level *Spartina*	30	2·2	5·11	0·11	0·30	5·66	1·27	2·94	0·06	0·17	3·26
Level *Spartina*	150	2·1	5·17	0·10	0·29	6·39	1·26	3·11	0·02	0·17	3·85
Spartina on ridge	230	1·3	5·62	0·08	0·14	4·68	1·38	5·55	0·08	0·14	4·63
Bare mud on ridge	230	1·1	6·11	0·07	0·11	3·80	1·03	5·74	0·07	0·10	3·57

Hesse (1961) however has shown that organic matter type may override soil conditions, for example he finds greater accumulation of organic matter beneath *Rhizophora* than *Avicennia* even though the soil of the latter had a higher clay content.

Total nitrogen does not of course tell us much about availability and inorganic nitrogen added to *Rhizophora* mud was rapidly immobilized (Table 8), presumably through bacterial consumption since the carbon/nitrogen ratio of 36 was extremely high. In dune soils the C/N ratio is usually less than 10 (Willis *et al* 1959) though Olson (1958) did find a ratio of 20 in ancient inland dune forest soils. In salt marsh soils the ratio usually lies between 10 and 20 but it may be higher at least for short periods where tidal litter accumulates. In spite of the frequent references to nitrophilous species being associated with tidal litter deposits, one of the commonest (*Atriplex hastata*) has been shown to grow on highly nitrogen-deficient substrata (Weston 1964).

Table 8. The immobilization of nitrogen by fibrous mangrove-swamp soil. Results on an oven-dry basis (from Hesse 1961).

Soil	ppm NH$_4$-N		ppm NO$_3$-N	
	Initial	After 15 days	Initial	After 15 days
Mud alone	12·5	14·0	0	1·6
Mud + (NH$_4$)$_2$SO$_4$ at 2 cwt/acre 6″	32·0	10·2	0	2·2

Some extremely interesting studies have been carried out recently on nitrogen availability to *Suaeda maritima* in Conway estuary marshes, North Wales (Stewart *et al* 1972). They have shown that total nitrogen, soluble amino acids, and nitrate show a well marked gradient which decreased in *Suaeda* tissue from seaward to landward up the marsh. Using an enzyme bio-assay technique they showed that nitrate reductase levels in *Suaeda maritima* were found to be as much as 40 to 50 times higher in plants from low-lying seaward parts, as opposed to higher-lying landward parts, of the salt marsh (Table 9). This adaptive behaviour of the enzyme system is an intriguing example of physiological plasticity. We know very little about this in salt marsh plants. The results also suggest that the actual through-put of nitrate as a result of daily incursions of the tide in lower marsh levels is very much higher than the supply at higher marsh levels rarely reached by the tide, even though the total nitrogen values of the soil are highest in the latter.

Table 9. Nitrate reductase activity in samples of *Suaeda maritima* from different levels in a salt marsh on the Conway estuary, Caernarvonshire. Figures in brackets refer to number of samples taken randomly at each site (from Stewart *et al* 1972).

Site	μ moles NO_2/h/g fresh wt. Average value	Range of values	μ moles NO_2/h/mg protein Average value
1 (Seaward)	5·55 (8)	4·86–6·14	0·62
2	2·45 (6)	2·01–2·69	0·40
3	1·84 (12)	1·09–2·49	0·21
4	0·81 (5)	0·41–1·19	0·11
5 (Landward)	0·11 (8)	0·08–0·16	0·01

Hesse (1963) has studied the forms in which phosphate occurs and their distribution in mangrove swamp mud (Fig. 11). He gives a useful discussion of the complexities associated with phosphate availability in saline muds.

The responses of salt marsh species from low marsh and high marsh levels to nitrogen and phosphate fertilisers have been examined by Pigott (in Rorison 1969). *Salicornia* species and *Suaeda maritima* grown on their own marsh soils with and without fertilizer all showed response to nitrogen

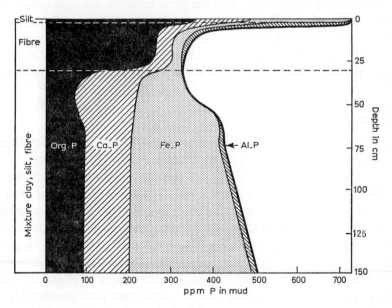

Fig. 11. Vertical distribution of phosphorus in mangrove-swamp mud (from Hesse 1963).

and phosphate fertilizer at higher marsh levels. In the case of the pioneer species *Salicornia dolichostachya*, growth on cores of bare mud was rapid with or without the addition of nitrogen. It was also shown that *Halimione portulacoides* taken from sub-optimal growth sites responded to nitrogen and phosphate. The apparent nutrient deficiency symptoms found in high marsh plants might be due in part to deficient water supply and soil mechanical restrictions on rooting since high marsh levels may dry out considerably in summer as Pigott points out.

Willis and Yemm (1961) used turf transplants and tomato culture to reveal that nitrogen, phosphate, and potassium were deficient in the soils of a calcareous dune system. Big responses to complete fertilizer applications were also found in both dune and slack vegetation resulting in a marked increase in grasses at the expense of lower-growing herbs and bryophytes (Willis 1963). The changes were in fact remarkably similar to those occurring in dune vegetation following myxomatosis, (Ranwell 1960 *b*). This illustrates the improved nutritional status that conservation of organic matter beneath taller vegetation can bring to these relatively arid soils. One notable effect of mineral deficiency is to restrain growth of fast growing species resulting in the maintenance of high species diversity.

Both Brown and Hafenrichter (1948) and Augustine *et al* (1964) found that ammonium compounds were the most effective nitrogen fertilizers in experiments concerned with promoting growth of *Ammophila* plantations on dunes. The controlled release fertilizer magnesium ammonium phosphate with potassium added, gave highly significant increases in growth over controls in *Ammophila arenaria* and *A. breviligulata*. (Augustine *et al* 1964).

Micro-nutrients

Willis (1963) found no significant response to trace element fertilization (iron, sodium, manganese, zinc, copper or molybdenum) in dunes.

Gorham and Gorham (1965) record that ash from salt marsh plants was 2 to 10 times greater than in plants from other habitats. However, iron was from $\frac{1}{3}$ to $\frac{1}{15}$, and manganese from $\frac{1}{7}$ to $\frac{1}{25}$ that of plants from other habitats. It is suggested that ion antagonism may prevent iron and manganese ions, mobilized in the reduced state, entering salt marsh plants freely in an ion-rich environment. Adams (1963) found that plants of *Spartina alterniflora* became chlorotic in iron-deficient soils where high level marsh species such as *Distichalis spicata* and *Juncus roemerianus* were unaffected. The *Spartina* responded to experimental foliar applications of ferrous sulphate or the addition of iron to a saline nutrient solution. Now *Spartina alterniflora* is restricted to low marsh where the mean soluble iron content is in excess of 4 p.p.m. Adams suggests that iron deficiency may limit the growth of *Spartina* in high level marsh. He also notes that iron becomes available through the action of anaerobic iron-reducing bacteria and observed that *Spartina alterniflora* shows the greatest tendency towards chlorosis in better-drained, rather than wetter sites at the high marsh level.

PART TWO   Salt Marshes

4 Salt Marshes: Tidal Influence

Salt marsh formation normally begins at a level subject to salt water tidal inundation twice daily. The upper level of the tidal influence, where salt marsh is replaced by fresh-water marsh, is subject to tidal inundation only a few times a year at the spring and autumn equinoxes, and then by almost fresh water. Between these two extremes organisms are zoned according to the range of conditions they can withstand; conditions dominated by the tide at lower levels, but almost independent of them at the highest levels.

It follows that efforts to explain zonation as a whole in relation to any one factor such as duration of tidal submergence or salinity is bound to fail not only in relation to different species, but even in relation to upper and lower limits of the same species where these have wide vertical ranges.

So far as survivial, growth and reproduction of organisms in the intertidal zone is concerned, tidal factors of particular importance include: intensity and frequency of *mechanical disturbance* due to tidal action; the *vertical range* over which the tide operates which controls tidal flooding depths and the vertical extent of salt marsh; the *form of tidal cycle* which controls frequency and duration of submergence and emergence, and *water quality* which controls the amount of light reaching submerged growths and the salinity to which they are subjected.

Tidal Parameters

Tidal disturbance

Forward colonization of a salt marsh only takes place in relatively sheltered inlets or behind broad expanses of tidal flats where much of the energy of waves has already been dissipated. Under these conditions waves are relatively low in height and do not exert much force in their downward plunge on breaking. On these relatively level shores the swash, or landward motion of the water, tends to be more powerful than the backwash. The aerial parts of a pioneer plant are first pulled landward strongly and then pulled seaward less strongly as a wave completes its breaking.

There is much evidence (Ragotskie 1959, Møller 1964 and Pestrong 1965) that ebb flow reaches higher velocities than flood flow and controls drainage patterns in tidal channels on salt marshes and this is a subject we shall return to in Chapter 5. However, it is of interest to note here that Bradley (in Ragotskie 1959) found that tidal velocities of flood currents 3 cm above a tidal flat were consistently 15 to 20 per cent higher than those of the ebb.

No attempts appear to have been made to measure the strength of the water-borne forces required to break salt marsh plants or uproot them and this would be a profitable line of study of fundamental importance to an understanding of conditions required for salt marsh formation.

Clearly, as well as the force, the frequency with which a plant is rocked back and forth by the water affects its chances of survival. Wave force and frequency varies with both distant and local weather conditions but the maximum incidence of wave break tends to occur on a salt marsh at the

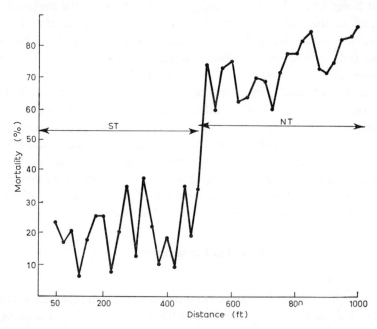

Fig. 12. Percentage mortality of *Salicornia* seedlings in the spring tide (ST) zone and the neap tide (NT) zone (From Chapman 1960, after Wiehe 1935).

level of mean high water neap tides because most tides pass over or reach this point and more waves have time to break there while the tide is on the turn.

Wiehe (1935) has shown that the density of *Salicornia* seedlings in the spring tide zone increases from 2 to 6 times that in the neap tide zone and that a two fold increase in density occurs within a 30 m distance above the top of the neap tide zone. In his work on the Dovey estuary in Wales he found *Salicornia* seedling mortalities of 60 per cent or more in the neap tide zone but less than 40 per cent in the spring tide zone (Fig. 12). As Chapman (1960) points out in relation to these observations, the sharp rise in seedling mortality occurs at the level flooded daily by the tides and mortality is a direct result of tidal water washing out a high percentage of the seedlings.

Brereton (1965) found that during the first half of the period March to June *Salicornia* seeds continue to germinate at a rate sufficient to offset losses due to up-rooting by the tide so that seedling density increases. In

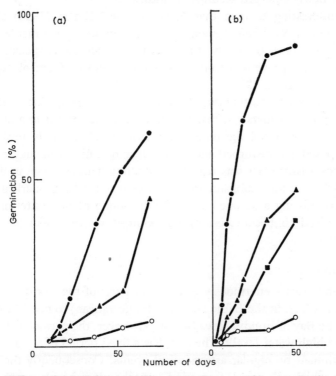

Fig. 13. Effect of (a) light and (b) temperature on the germination of seed of *Spartina anglica*. Light: ●, dark; ▲, 8 h; ○, 16 h. Temperature: ●, 25°C; ▲, 20°C; ■, 15°C; ○, 7°C. (from Hubbard 1970).

the second half of this period the rate of germination falls off but the uprooting effect of the tide does not and seedling density decreases.

Chapman (1960) suggests that pioneer species need a specific period of time of continuous emergence of the marsh surface so that seedlings are not swept away by the tides. In fact newly germinated seedlings of *Spartina anglica* at the seaward edge of the marsh at Bridgwater Bay, Somerset were invariably found with seed cases and/or stem collars from 2 to 10 cm below the mud surface. The seeds must have germinated below the surface and Hubbard (1970) has shown that both rate and amount of germination of this species is greater in the dark than in the light (Fig. 13). It follows that the period of continuous submergence is more likely to be related to the light requirements of the emerging seedling shoot than to mud surface stability given that wave action is sufficiently moderate to allow establishment to occur at all. Experimental work on the establishment of pioneer species from ripe seed buried at different depths and on the light requirements of newly emerged seedlings is indicated.

It is interesting to note that Stevenson and Emery (1958) obtained germination of *Salicornia bigelowii* in seeds completely submerged in sea water. Also tests on the seeds of five Australian salt marsh species showed that none were prevented from germination beneath 5 cm of tap water (Clarke and Hannon 1970).

The density of *Spartina anglica* seedlings in the neap tide zone at the seaward edge of a marsh in Bridgwater Bay, Somerset was at a maximum opposite the centre of the marsh and diminished regularly towards the more exposed and disturbed ends of the marsh. This again suggested that mechanical disturbance was regulating seedling density. Even in the most favourable zone for establishment 40 per cent of seedlings marked in July 1964 were missing after one month. Of the young plants wintering 50 per cent survived to become established clumps by the following August (Table 10).

Tide range

Tidal range varies enormously in different parts of the world from a few centimetres (e.g. in the Baltic and the Mediterranean) to nearly 19 metres (e.g. in the Bay of Fundy, Nova Scotia). However, the period of tidal cycle from one high water level to the next has a fixed period of about 12 hours predetermined by regular astronomical events. Consequently the vertical range in which salt marsh formation can take place is very much greater in the big tide range areas than in the smallest ones. Further in the big tide range areas there is room for a series of communities zoned according to the tidal conditions. In the smallest tide range (e.g. in the Baltic) tidal movements may be strongly modified by over-riding weather effects, nevertheless distinctive vegetational zonation related to duration of submergence is still found in such situations (Tyler 1971). Changes in atmospheric pressure alone are sufficient to depress or increase water level by as much as 30 cm in the Danish Wadden sea (Jacobsen, N.K. *pers. comm*); on west Swedish coasts where the tidal range is about 30 cm, changes in wind direction and atmospheric pressure give non-periodic water level ranges up to 170 cm (Gillner 1965).

Table 10. Survival of newly germinated *Spartina anglica* seedlings on open mud at the seaward limit of the Bridgwater Bay marsh, Somerset. Seedlings were searched for within a mudflat area 450 × 150 m (long axis parallel to the shore) and marked with a bamboo cane 1 m away.

Date	No. of marked sites (out of 50 originally marked) located	No. of marked seedlings still present	Survival %
13/7/64	50	50	100
17/8/64	48	30	62
23/9/64	47	22	46
14/6/65	45	10	22
27/8/65	45	9	20

Tide curve modifications

It is an observed fact that the tidal curve steepens up estuary though its ultimate amplitude of course diminishes eventually to zero. It was noted that the lower limits of *Spartina anglica* marsh in Poole Harbour, Dorset were higher at successive intervals up estuary (Ranwell *et al* 1964), though the differences were small and only just significant. However, Beeftink (1966) has shown on the Scheldt estuary in Holland that different species do retreat to higher levels upstream apparently proportionate to the steepening of the tidal curve.

Contrasted Habitats in Submergence and Emergence Zones

Tidal submergence

If the duration of tidal submergence per annum is plotted against vertical level within the tide range, the curve obtained is typically S-shaped (Fig. 14). In Poole Harbour, Dorset a difference in marsh level of 10 cm near high water spring tides gives difference of only about 100 hours submergence per year. In contrast, a difference in marsh level of 10 cm near high

F

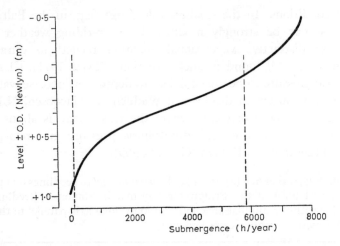

Fig. 14. Duration of submergence per annum in relation to O.D. levels. Broken lines mark upper and lower limits of *Spartina*. Derived from tide gauge records at Poole Bridge, Poole Harbour, July 1962 to July 1963 (from Ranwell *et al* 1964).

water neap tides gives a difference of about 700 hours submergence per year. In other words level is much more sensitively related to submergence at the marsh seaward limit than at its landward limit.

Now it is common knowledge that many salt marsh plants can survive much longer periods of submergence in the quiet waters of sheltered brackish lagoons than they are ever likely to experience at their seaward limits of growth in open marsh conditions. Hubbard (1969) has shown that *Spartina anglica* may be submerged throughout daylight hours during certain neap tide periods (Fig. 15) and it can also withstand submergence in the laboratory for at least 4½ months. So submergence itself is not critical for survival of this, the lowest growing salt marsh species.

Apart from any mechanical effect, the big difference between quiet saline water and tidal water flooding an open marsh is that the former tends to be clear while the latter is turbid with silt in suspension and this cuts down the length of time available for photosynthesis. It is remarkable that virtually no work seems to have been done on light as a limiting factor for establishment, growth, and survival of salt marsh plants, though shading effects are obvious wherever a salt marsh is overhung by trees.

The most turbid non-toxic tidal waters in Britain are found in Cornish estuaries clouded by sediment derived from china clay mining activities in the catchment. In the upper estuary of the River Fal this results in a

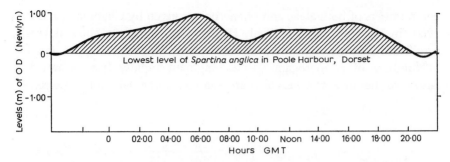

Fig. 15. Tidal curve showing the lowest level of *Spartina anglica* in Poole Harbour and the extent of its immersion (hatched area) by a neap tide on 5 November 1962. Readings obtained from the tide gauge on Poole Harbour, Dorset (from Hubbard 1969).

truncation of salt marsh zonation which precludes entirely the development of lower marsh levels. *Puccinellia maritima* growth is confined to the extreme upper limit of its vertical range where it shares a pioneer role with *Triglochin maritima* and other species of higher marsh levels.

Effects of turbidity on photosynthetic activity might be investigated using algal culture techniques (see Hopkins 1962).

Pioneer salt marsh plants capable of withstanding regular submergence must also withstand extremely low oxygen levels especially in marsh soils with a high clay content. However even in the exceptionally poorly drained soils associated with 'die-back' in *Spartina anglica* marsh, Goodman and Williams (1961) were unable to find direct evidence of damage primarily due to anaerobiosis.

Submergence and Emergence marshes

From extensive studies in British and American marshes on frequency and duration of submergence, Chapman (1960) concludes there is a fundamental distinction between lower marshes and higher marshes with the line of demarcation at about mean high water level. He notes that 'lower marshes commonly undergo more than 360 submergences per annum, their *maximum* period of continuous exposure never exceeds nine days, whilst there is more than 1·2 hours submergence *per diem* in daylight. Upper marshes on the other hand undergo less than 360 submergences annually, their *minimum* period of continuous exposure exceeds ten days, and there is less than one hour's submergence daily during daylight.'

However, as Chapman points out, this does not hold for the San Francisco Bay marshes studied by Hinde (1954). Moreover data on zonation and numbers of both algae and phanerogams given by Chapman (in Steers 1960) in his study area at Scolt Head Island, Norfolk, show no really significant disjunction at the proposed demarcation line (Figs 16 and 17).

Numbers of phanerogam species augment rapidly from seaward to landward, because only very few are adapted to regular submergence.

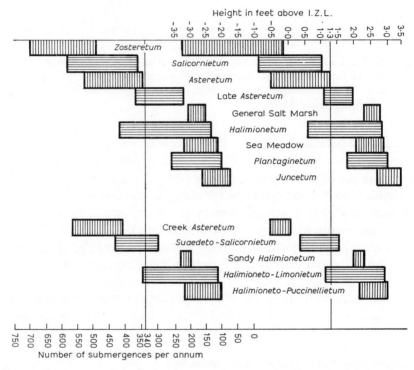

Fig. 16. Zonation of salt marsh communities at Scolt Head Island, Norfolk. Levels and number of tidal submergences per annum (0·0ft I.Z.L. = + 7ft O.D. Newlyn). The lines at 1·30 ft and 340 submergences per annum mark the approximate division between Submergence and Emergence marsh (from Chapman in Steers, 1960).

It is of particular interest that Chapman (1940) found that it was characteristic of American salt marsh species that 'the further south they spread on the continent the more tolerant they are of submergence (i.e. the lower they go in the tidal plane).' Although he adduces temperature as the factor possibly responsible for this it seems more likely that stronger light intensities and temperature act together to allow photosynthesis at deeper levels of submergence further south.

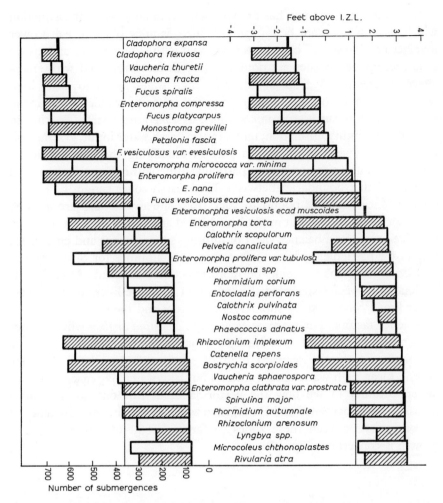

Feet above I.Z.L.

| Cladophora expansa |
| Cladophora flexuosa |
| Vaucheria thuretii |
| Cladophora fracta |
| Fucus spiralis |
| Enteromorpha compressa |
| Fucus platycarpus |
| Monostroma grevillei |
| Petalonia fascia |
| F. vesiculosus var. evesiculosis |
| Enteromorpha micrococca var. minima |
| Enteromorpha prolifera |
| E. nana |
| Fucus vesiculosus ecad caespitosus |
| Enteromorpha vesiculosis ecad muscoides |
| Enteromorpha torta |
| Calothrix scopulorum |
| Pelvetia canaliculata |
| Enteromorpha prolifera var. tubulosa |
| Monostroma spp |
| Phormidium corium |
| Entocladia perforans |
| Calothrix pulvinata |
| Nostoc commune |
| Phaeococcus adnatus |
| Rhizoclonium implexum |
| Catenella repens |
| Bostrychia scorpioides |
| Vaucheria sphaerospora |
| Enteromorpha clathrata var. prostrata |
| Spirulina major |
| Phormidium autumnale |
| Rhizoclonium arenosum |
| Lyngbya spp. |
| Microcoleus chthonoplastes |
| Rivularia atra |

Number of submergences

Fig. 17. Zonation of salt marsh algal species at Scolt Head Island, Norfolk. Levels and number of tidal submergences per annum (0·0 ft I.Z.L. = + 7 ft O. D. Newlyn). The lines at 1·30 ft and 340 submergences per annum mark the approximate division between Submergence and Emergence marsh (from Chapman in Steers 1960).

Nevertheless, Chapman has drawn attention to an important distinction which, even if it cannot be critically defined, is none the less real. Lower marshes are dominated by factors associated with submergence and, as we shall see, upper marshes are dominated by factors associated with emergence. Lower marsh from about mean high water neaps to mean high water is best described as *Submergence marsh*, and upper marsh from mean high water to mean high water springs, as *Emergence marsh*. Bearing in mind that turbidity, insolation, temperature, fresh water input, grazing and adaptive ecotypes will all play a part in modifying the exact limits of species both at a site and at any particular time within a site it should be recognized that boundaries between these marsh zones cannot be exactly defined since they are bound to vary from site to site and are anyway oscillatory in nature. What might be achieved and would be most useful in practice would be a balance sheet of limiting submergence and emergence factors for particular species so that their true affinities for Submergence and Emergence marsh could be more effectively assessed. Already it is evident that species spanning the transition zone are likely to have their lower limits controlled by submergence factors and their upper limits by emergence factors. The extreme example is *Spartina anglica* whose clones are known to persist for at least 50 years without significant seedling recruitment from the seaward limit of salt marsh growth to the zone around mean high water spring tides. The Bridgwater Bay marsh gives us an opportunity to see how the performance of this one species varies across the Submergence and Emergence zones. There is some evidence of discontinuity between these zones so far as height of *Spartina anglica* growth is concerned (Fig. 18).

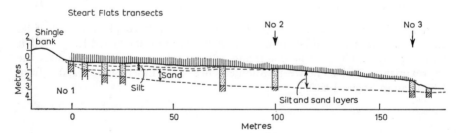

Fig. 18. Vertical height of shoots of *Spartina anglica* from landward (left) to seaward (right) at Bridgwater Bay, Somerset. Note the quite distinct change in height (indicated by closely spaced vertical lines) just to the left of arrow No. 2. Growth in Emergence marsh is much taller than in Submergence marsh (after Braybrooks 1958).

Mechanical effects on Emergence marsh

The incidence of mechanical disturbance from the tide is much reduced on Emergence marsh compared with that affecting establishment and growth on Submergence marsh. It is greatest at higher marsh levels around mean high water spring tides where wave presence often persists, prolonged by onshore winds delaying the ebb. Tidal litter accumulates at this level, crushes weak growths, and temporarily reduces, or cuts off the light supply (Plate 4). This produces the characteristic open ground of the strandline near the landward limit of the marsh favoured by temporary colonists like the annual species of *Atriplex*. Kidson and Carr (1961) recorded disturbance of shingle markers at Bridgwater Bay, Somerset and showed that significant movement was limited to a narrow zone near high water mark. Measurements are needed of mechanical disturbance in different salt marsh zones and fracture devices involving threads of different breaking strains might be employed. It is evident from the gradient of normal tide curves that flood and ebb tides pass relatively swiftly over marsh levels between mean high water neap and mean high water spring tides and mechanical disturbance should be reduced there.

Submergence, or reduced light due to submergence, is unlikely to be a limiting factor for plants of Emergence marsh except at the very lowest levels, because the duration of submergence is too short to seriously impede photosynthesis.

Field Salinity in Relation to Tidal Action

Salinity in Submergence and Emergence marshes

The one common factor that does affect all plants of Emergence salt marsh and precludes the growth of non-halophytic terrestrial species is of course the high level of salinity maintained persistently by the tides. Salinities in Submergence marsh rarely rise above that of the tidal water with which it is regularly and frequently bathed, Emergence marsh however can develop much higher salinities in the soil solution than that of the tidal water as a result of water evaporation during dry inter-tidal periods. Hannon and Bradshaw (1968) have in fact demonstrated that upper marsh populations of *Festuca rubra* are more salt tolerant than those of lower marsh.

Gillham (1957) took advantage of a drought in 1955 to record extreme high salinities in the soil solution of many species regarded as non-halophytes in Emergence marsh on the shores of islands off the west coast of Scotland. Her results (Fig. 19) show that in addition to the normal halophytes, non-

Chlorides as % by weight of soil solution

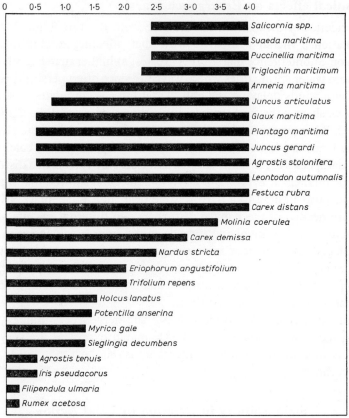

0   0·5   1·0   1·5   2·0   2·5   3·0   3·5   4·0

Salicornia spp.
Suaeda maritima
Puccinellia maritima
Triglochin maritimum
Armeria maritima
Juncus articulatus
Glaux maritima
Plantago maritima
Juncus gerardi
Agrostis stolonifera
Leontodon autumnalis
Festuca rubra
Carex distans
Molinia coerulea
Carex demissa
Nardus stricta
Eriophorum angustifolium
Trifolium repens
Holcus lanatus
Potentilla anserina
Myrica gale
Sieglingia decumbens
Agrostis tenuis
Iris pseudacorus
Filipendula ulmaria
Rumex acetosa

Range of tolerance of soil salinity

Fig. 19. Distribution of common lochside species in relation to salt-tolerance. Combined data from three west coast lochs, Mull. Samples taken during the drought of August 1955. Note the downshore penetration of moorland plants into soils having as much as 3.5 per cent salt in the soil solution (i.e. full sea strength). Moorland dominants which are apparently intolerant of salt are *Calluna, Erica* spp., *Trichophorum, Juncus acutiflorus* and *Sphagnum* spp. (from Gillham 1957).

halophytes such as *Carex demissa, Molinia coerulea* and *Nardus stricta* were surviving chlorinity levels of 2·1 per cent by weight of soil solution (equivalent to full strength sea water salinities), without apparent harm to growth. In addition facultative halophytes like *Agrostis stolonifera, Carex distans* and *Juncus articulatus* were surviving hyper-saline conditions of up to 4 per cent chlorinity. Judging from the distribution of these species at the extreme upper limits of Emergence marsh where they are more likely to come under the influence of regular fresh water flushing than these exceptionally high salinities, they would be unlikely to survive the latter for more than brief periods. We need to know much more about the time factor in relation to salinity tolerance than we do at present.

Salinity control of *Phragmites* growth

Field measurements of salinity rarely catch these extreme conditions where levels rise close to a species limit of tolerance. It follows that for much of the time, salinity is not limiting for the growth of many salt marsh plants or facultative halophytes. Only where persistently high levels occur close to a species limit of tolerance does salinity become limiting. Since these conditions occur in quite a narrow sector of the estuarine gradient, where the species may become discontinuous due to other limiting factors, the location of a site where salinity is an operative limiting factor is difficult to find or may not occur at all in particular estuarine marsh systems. If we take the case of *Phragmites communis* in Poole Harbour, the great majority of 87 samples taken in dry summer conditions at the seaward limit of *Phragmites* growth gave salinities well within the tolerance of this species. Limiting values were only found at the point where the seaward limit of growth coincided with the limit of penetration down-estuary of *Phragmites* in salt marsh. The field limit of 1·3 per cent chlorinity of the soil solution at 10 cm depth (Ranwell *et al* 1964) was very close to the experimentally determined limit of 1·25 per cent found by Taylor (1939).

There is little evidence that salinity limits the growth of true halophytes in Cool Temperate zone marshes. It is more likely to do so in Warm Temperate and Sub-Tropical zones where evaporation regularly exceeds precipitation. Purer (1942) records salinities of 17 to 23 per cent on Californian marshes and notes that plant growth is inhibited where salt efflorescence occurs.

However, there is growing evidence that the seaward penetration of facultative halophytes in the Cool Temperate zone is often limited by salinity although other limiting factors such as grazing may prevent a species growing to the limit of its salinity tolerance.

Inversions of zonation

Complexities arise in the highest part of estuaries where inversions of zonation occur (see Gillham 1957 and Beeftink 1966). A similar phenomenon has been observed in the upper parts of the Humber estuary Yorkshire. *Puccinellia maritima* zoned below *Agrostis stolonifera* in the lower estuary retreats to the tops of hummocks above *Agrostis stolonifera* in the upper parts of the estuary where it seems likely that temporarily high salinity or drought in dry weather may exclude *Agrostis*.

Clearly whatever conclusions may be drawn about the relation of species zonation to tidal submergence in the more marine bay marshes or lower estuary marshes, these do not necessarily hold in the special conditions existing in the up-estuary regions.

As marsh levels rise under the influence of tide-borne sediments many factors other than those directly dependent on tidal action control the zonation and growth of the increasing number of species which colonize Emergence marsh. In particular physical soil factors: structure, particle size, aeration and moisture become significant. It is convenient to deal with sedimentation in association with these factors in the next chapter and to discuss biological factors in the two that follow.

5 Salt Marshes: Sedimentation, Drainage and Soil Physical Development

Salt Marsh Precursors

Biological activity on and in tidal mud flats starts off the processes of soil formation from mineral sediments, and salt marsh development accelerates these processes. We have little quantitative information about the role played by tidal flat organisms but there are a number of ways in which these precursors favour salt marsh formation once the flats are formed in sufficiently sheltered conditions, and high enough in relation to tides, for salt marsh plants to grow.

Ginsberg and Lowenstam (1958) review evidence of the stabilizing and silt trapping powers of filamentous algae and marine phanerogams in shallow lagoonal waters. Ginsberg *et al* (1954) also found that a culture of the algae *Phormidium* could re-establish a surface mat through as much as 4 mm of sediment in 24 hours. The burrows of invertebrate animals improve soil aeration, for example those of worms (e.g. *Nereis* sp.) are often lined with a layer of oxidized iron, reddish in colour in contrast to adjoining grey-blackish reduced mud. Colonies of molluscs like the Horse Mussel in San Francisco Bay (Gillham 1957) provide surface roughness which promotes sedimentation. On the sandy-silt flats adjoining Holy Island, Northumberland, *Enteromorpha* first colonizes on loose shells of *Cardium* in more mobile sites. Growth encourages silt deposition. The filament bases of the algae become embedded and ultimately lose their attachment to the now buried shells. Stronger growths of algae on the silt increase stability and accretion for the subsequent establishment of salt marsh species.

Kamps (1962) has shown that molluscs reconstitute clay into faecal pellets which settle quicker than unmodified clay flakes and help to promote accretion. His paper deserves much wider recognition than it has achieved for it gives much valuable quantitative information about the movement of clay in inter-tidal waters and its dependence on wind-induced waves and on the effects of inter-tidal organisms on silt movement in the Dutch Wadden sea.

Sedimentation

Techniques of measurement of salt marsh accretion

Kestner has developed an effective method for measuring changes in surface level of tidal flats using as a marker a creamed suspension of silica flour poured into cored out holes in the mudflats (Inglis and Kestner 1958). Where very high levels of accretion (several cm per year) occur, deeply embedded bamboo canes with about 15 cm sticking out of the surface enable repeated measurements to be made without destruction of the sampling site. They give reliable results when checked against natural markers like depth to previous years rosettes in perennials or successive seed case layers of annual species in the marsh soil profile (Ranwell 1964 a). In marshes with very low accretion rates it is necessary to lay down coloured marker layers (sand, coal dust, brick dust etc.) and record their depth of burial after several years (Jakobsen et al 1955, Steers 1964).

Rates of accretion

True accretion is the depth of sediment deposited in unit time minus the reduction in thickness of the accreted layer due to settlement factors. The latter become increasingly important at the higher levels of Emergence marsh where accretion diminishes, organic matter increases, and drying out occurs more frequently. Stearns and MacCreary (1957) for example found that settlement factors effectively compensated for an annual accretion of 0·64 cm on a high level marsh carrying *Scirpus olneyi*, *Spartina patens* and *Eleocharis rostellata* so that over a 20 year period there was no significant rise in marsh level. However, even over this sort of period, changes in land and sea level can be important and net increase in marsh level has to be considered in relation to them.

Chapman and Ronaldson (1958) found less than 0·2 mm per year accretion on a tropical mangrove marsh where roots and rhizomes are too widely spaced to retain much mud washed between them. In the pioneer zone of temperate marshes *Salicornia* accretion may be as much as 3 cm per year (Oliver 1929) and *Puccinellia maritima* can accrete sandy silt at a rate of 10 cm per year (Jakobsen et al 1955). Pioneer *Spartina anglica* can

Table 11. Values of correlation coefficients and their statistical significance for relationships between accretion, topography and vegetation at Bridgwater Bay, Somerset (from Ranwell 1964 *a*).

* P = 0·05, r = 0·2839
** P = 0·01, r = 0·3676
*** P = 0·001, r = 0·4594
With 46 degrees of freedom

	Dependent variate				Independent variates						
	Mean annual accretion 1960–61	Mean annual accretion 1961–62	Mean annual accretion 1960–62	Mean height of site	Distance seaward	Creek depth	Shoot density	Mean height of vegetation	Air dry weight of vegetation	Dry weight/Density (mean shoot weight)	Dry weight per shoot/Mean height
Mean annual vertical accretion (cm) 1960–61		+0·5317***	+0·8799***	+0·4133**	−0·4101**	−0·3316*	N.S.	+0·5098***	+0·5030***	+0·3825**	N.S.
1961–62			+0·8668***	+0·4074**	−0·3955**	N.S.	−0·3279*	+0·4372**	+0·3025*	+0·3408*	N.S.
1960–62				+0·4751***	−0·4692***	−0·3115*	−0·3277**	+0·5418**	+0·4626***	+0·4181**	N.S.
Mean height of site above O.D. (m)					−0·9724***	−0·7571***	−0·4120**	+0·6898***	+0·5062***	+0·7081***	N.S.
Distance seaward from landward limit of marsh (m)						+0·6742***	+0·4511**	−0·6474***	−0·4866***	−0·7006***	N.S.
Creek depth (highest minus lowest O.D. level at site) (m)							N.S.	−0·4585**	−0·3286*	−0·4118**	N.S.
Shoot density of vegetation (per m²)								−0·4366**	N.S.	−0·5143***	N.S.
Mean height of vegetation (cm)									+0·7635***	+0·8644***	N.S.
Total air dry weight of vegetation (g/m²)										+0·7350***	N.S.
Dry weight/Density (mean shoot weight)											+0·2992*
Dry weight per shoot/Mean height											

accrete 5 to 10 cm per year and at higher levels has been recorded as regularly accreting 15 cm of silt per year under exceptionally favourable conditions (Ranwell 1964 a). In general however accretion on temperate European and North-East American marshes in the Emergence zone is between 0·2 and 1 cm per year, and is unlikely to be a limiting factor for survival of salt marsh plants.

Pattern of accretion

In a study of accretion on an exceptionally rapidly accreting marsh at Bridgwater Bay, Somerset it was possible to show how annual accretion varied in different parts of the marsh (Ranwell 1964 a). Maximum accretion occurred in the centre section of the marsh which is not grazed; accretion diminished towards the ends of the marsh which were more exposed and grazed. The seaward to landward pattern (Fig. 20) showed maximum accretion about 50 m seawards of the landward limit of the marsh and minimum accretion at the seaward edge. Richards (1934) found that maximum accretion rates on the Dovey marshes occurred nearer the seaward limit. Now the Bridgwater Bay marsh was young and actively developing at all levels while the Dovey marsh was mature and more or less static at its seaward limit. Flattening out of marsh level occurs just landward of the zone of maximum accretion (evident in Fig. 20). This zone must migrate seawards as the marsh reaches maturity. Analysis of the Bridgwater Bay results suggested that rise in accretion rate was positively correlated with marsh height and the height or weight of the vegetation, and negatively correlated with distance seaward and the density of the vegetation (Table 11). The individual effects of the variables could not be separated and this emphasises the intimate nature of the phytogeomorphological processes at work. The positive correlation of accretion with marsh height only holds for young actively developing marshes. Evidently one significant feature of marsh system development is a change in surface angle from a slope to seaward to a more level surface and consequently a less readily drained one.

Mechanism of accretion

Because of the high rate of accretion at Bridgwater Bay it also proved possible to identify distinct seasonal changes in the pattern of mud supply to the marsh (Fig. 21). It seems that mud builds up at the seaward edge of the marsh during spring and summer, accretion is at a maximum over the whole marsh in the autumn, while in winter there is either no change or a slight trend towards erosion.

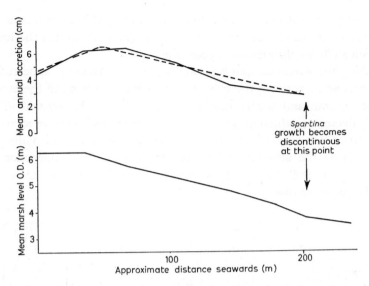

Fig. 20. Annual accretion and marsh-level curves derived as the means of results from five transects combined, Bridgwater Bay, Somerset (from Ranwell 1964 *a*).

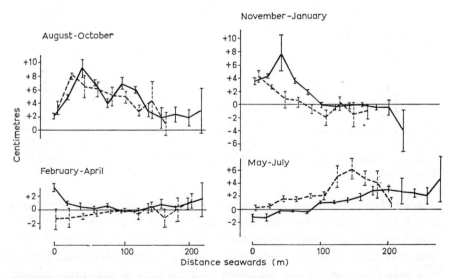

Fig. 21. Seasonal pattern of accretion and erosion on Transects 3 (solid line) and 4 (broken line), 1960–61. Transect 4 results have been slightly offset to the right for clarity. Vertical lines show plus or minus twice the standard error of the mean. *Spartina* growth becomes discontinuous at about 200 m seaward. Bridgwater Bay, Somerset (from Ranwell, 1964 *a*).

It is of particular interest here to note Kamps (1962) results on a study of variations in the clay content of the surface layers of mud in the Eastern Wadden shallows throughout a year (Fig. 22):

'Taking the whole sampling period into consideration it appeared that there was an increase in the clay content from the spring to the autumn. In the autumn and winter months it seemed that the amount of mud rapidly decreases, which means that during the spring and summer months large quantities of mud are deposited along the coast which disappear again during the autumn and winter months. In this connection it is interesting to recall that the former owners of the Banks called the autumn months the warp months.'

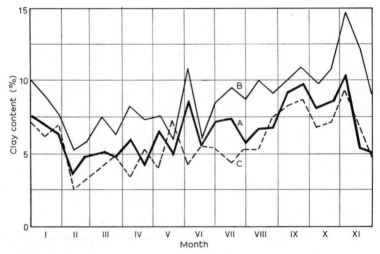

Fig. 22. The variations in the clay content of the upper layer (0·5 cm) of the Westpolder Shallow, Eastern Wadden, Netherlands in the course of a year. A = mean of 15 measuring points; B and C two individual measuring points (from Kamps, 1962).

Riley (1963) examined the seasonal distribution of organic aggregates in sea-water in the inshore waters of Long Island Sound and concluded: 'The level of non-living particulate matter was moderately high in mid-winter when the quantity of phytoplankton was small. It increased further to a peak that more or less coincided with the peak of the spring diatom flowering and thence declined to the lowest point of the year at about the time of the termination of the flowering. The quantity was relatively small during most of the summer and autumn.'

One further relevant factor is the flocculation effect of salt water on silt which Price and Kendrick (1963) believe may explain seasonal changes in suspended silt in the Thames and other estuarine areas:

'In the summer of 1959 the water became unusually free of silt and visibility increased from 6 in to 2 ft (15 to 60 cm). During the late autumn and early winter the deposits again became muddy. This may be explained as follows: In the dry summer of 1959 fresh water flow was minimal, hence salinity upstream was higher than usual. Charges on the silt particles neutralized by the electrolyte salt water caused flocculation and deposition of silt higher upstream than usual. When the fresh water flow returned to normal it is possible that the charges on the deposited silt could be restored mobilizing it again.'

The pattern of seasonal events in relation to marsh accretion seems to be as follows: In spring and summer large quantities of mud are deposited adjoining seaward edges of sheltered salt marshes as a combined result of flocculation due to high salinity and coagulation of particles through increased biological activity. In autumn the reverse effects occur as salinity declines with increased rainfall and biological activity declines with lowered temperatures. Under the influence of autumn storms mobilized silt arrives over the marsh and settles in maximum amount because the fully grown vegetation is at maximum trapping capacity. In winter fresh supplies of silt are mobilized under conditions of minimal salinity and biological activity, to settle once more against marsh edges in the following spring.

It is worth noting that Jakobsen (1961) using a simple siphon sampler found that under dense salt marsh vegetation not more than half the material suspended in tidal water was deposited.

If the top-most layer of newly accreted autumn silt is analysed for its chemical constituents in samples taken successively from landward to seaward, trends are apparent, at least on a dry weight basis (see Table 7). This suggests that the mud supply to higher marsh zones is derived from those successively lower. In other words the source of silt for accretion seems to be immediately adjacent and to seaward of the site concerned.

If the above interpretation is correct we have here some at least of the elements for a mathematical model of the accretion system. It will not be easy to get measurements of some of the parameters in this complex physico-biological system, but one begins to see the need for continuous monitoring of salinity and temperature and for critical studies on silt coagulation through biological activity in order to understand it.

G

Meanwhile it may be of interest to note that a regression equation has been calculated for the Bridgwater Bay marsh as follows:

Accretion = 0·643 (mean height of site above O.D.) + 0·0462 (mean height of vegetation) + 0·00135 (air dry weight of vegetation) — 1·143.

Stratigraphy and the Age of Marshes

Deep level stratigraphy of marshes is a study more in the province of the physiographer than the ecologist and has little relevance to current ecological processes at the surface which is the concern of this book. But it does have relevance to past land and sea level changes and calculations of the age of marshes.

Rising coastlines

On rising coastlines (e.g. such as parts of the Swedish and Scottish coasts are believed to be) there is little opportunity for salt marsh development to occur and the vertical extent of marsh deposits may be less than the tidal range would permit on a stable coastline. Where the coast is stable, sheltered inlets silt up with marsh deposits to a depth equivalent to that part of the tidal range where salt marsh growth is possible i.e. from about mean high water neaps to extreme high water spring tides.

Falling coastlines

On a subsiding coastline (e.g. south-east England) or one on which the sea level is rising, marsh formation can continue so long as its rate is not exceeded by the isostatic adjustment for as long as this occurs. In this way depths of marsh sediment far in excess of that permissible within the tidal range can accumulate.

The history of the marsh changes can be read from fragments preserved in the marsh soils. The reconstruction of the ontogeny of the Barnstaple marsh, New England by Redfield (1965) is an elegant example, and that of Jacobsen (1960 and 1964) in Denmark another.

Techniques for measuring the age of marshes

Three ways of measuring marsh age have been used: direct records, calculation from existing vertical zonation of plants and accretion rates

Table 12. Minimum rates of development of salt marshes to maturity.

Site	Type of marsh	Method of age estimation	Minimum age to maturity in years	Author
Malltraeth, Anglesey	Part enclosed	Direct observation	10	Ranwell (*pers comm.*)
Bridgwater Bay, Somerset	Open	Accretion rates	40	Ranwell (*pers comm.*)
St. Cyrus, Scotland	Part enclosed	Documentary records	70	Gimingham (1953)
Newport Bay, New England	Part enclosed	D. records	90	Stevenson and Emery (1958)
Fal, Cornwall	Part enclosed	Map records	100	Ranwell (*pers comm.*)
Baltrum Island	Open	Map records	100	Tuxen (1956)
Skallingen, Denmark	Open	Accretion rates	100	Nielsen (1935)
North Bull Island, Eire	Part enclosed	Documentary records	125	Chapman (1960)
Scolt Head Island, Norfolk	Open	Accretion rates	200	Chapman (1960)
Dovey, Wales	Open	Accretion rates	330	Chapman (1960)

corrected for subsidence and settlement, and radio-carbon dating. We need to distinguish between minimum age for reaching maturity and marshes whose life is extended indefinitely by the rejuvenating effect of subsidence at a rate less than that of the accretion rate.

Some minimum ages for marshes reaching maturity (which might be defined as the point where settlement tends to balance accretion) are given in Table 12. In general, part-enclosed marshes develop more rapidly than open coast marshes and the minimum order of time to reach maturity on the majority of marshes is about 100 years. What these estimates do not take into account is the level of the tidal flats when salt marsh formation originally took place. If for example tidal flats are significantly above the submergence limit for the pioneer species to begin with, and salt marsh formation is prevented by too high mobility, they will have a head start on salt marsh formation starting on lower level but less mobile flats if mobility is suddenly reduced by an artificial or natural barrier partly enclosing them. The extremely rapid development of the marsh on the Malltraeth estuary, Anglesey is a case in point. Here high level sand flats suddenly became more stable due to the artificially aided growth of a sand barrier and only 10 years were needed to build from pioneer *Puccinellia maritima* to a mature stage with *Juncus maritimus* (Packham and Liddle 1970). In the case of the Bridgwater Bay marsh, exceptionally silt-laden tidal water combined with the vigorous growth of *Spartina anglica* to speed up marsh development. At Scolt Head Island, Norfolk silt supplies are limited and marsh growth occurs at a much slower rate than normal (Chapman 1960).

Radio-carbon dating of deep level salt marsh deposits has been used by Godwin *et al* (1958) and Newman and Rusnak (1966) to estimate the eustatic rise in sea level and by Redfield and Rubin (1962) in connection with his studies of the Barnstaple marsh.

Superficial stratigraphic studies are of more immediate concern to the ecologist. They enable comparisons to be made between the vertical zonation of the living plant at the surface and the depth of its organic remains in the soil profile (Fig. 23). Such studies also provide direct evidence of successional relationships that have occurred in the past and that are likely to be occurring at the present time (Fig. 24). Any species present in quantity is likely to leave distinctive remains in the profile. An end can often be put to speculation about successional relationships with the aid of a trowel and a sieve. Subsurface discontinuities in sediments which may profoundly affect soil moisture supplies for growth are also revealed by stratigraphic work.

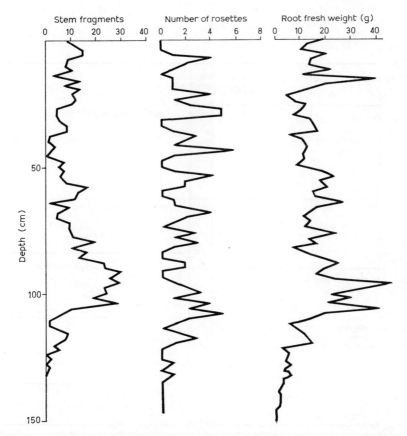

Fig. 23. Distribution of organic remains (*Spartina*) in the marsh profile near
the site of the first plantings of *Spartina* in 1929, Bridgwater Bay. Samples were
taken every 2 cm down the profile in summer 1962. Note that stem fragments,
rosettes and root remains lie progressively deeper at the bottom of the profile as
death in the natural position would be expected to leave them. Also, high values
obtained near the bottom of the profile and the sudden tailing off of all three
items measured near the same general level suggests that the true base level of
Spartina has been reached and that the results are not significantly affected by
loss of material due to decay in the lower half of the profile (from Ranwell 1964 *a*).

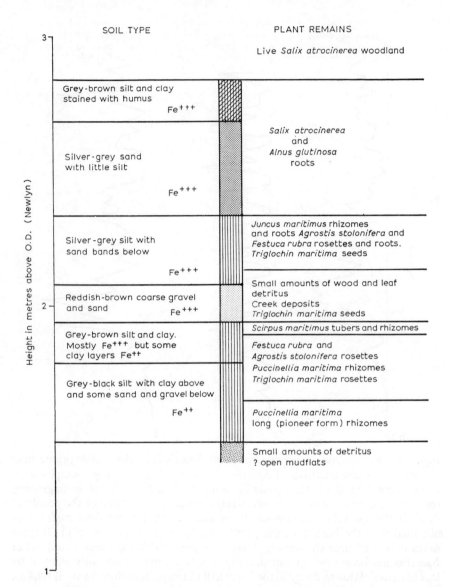

Fig. 24. Soil profile 150 m to landward of the present seaward limit of tidal woodland with evidence from plant remains of direct succession from pioneer salt marsh to tidal woodland. Fal estuary, Cornwall.

Marsh Formation, Drainage and Re-cycling

Marsh relief

Marsh morphology is dominated initially by the surface relief and hydrology of the tidal flats on which it takes place. Localized flood and ebb tide effects in relation to growth form of pioneer species and sedimentation has been studied by Jakobsen (1964) and Møller (1963 and 1964). In general it seems that flood tide effects may dominate under storm conditions while ebb tide effects are dominant in calmer conditions. Flood tide features in higher order creeks with two-way flow are also confined normally to the seaward edge of the marsh, for example the flood bars

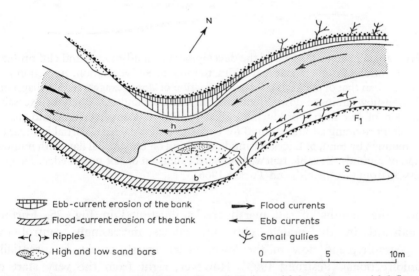

⊞⊞⊞ Ebb-current erosion of the bank		➡ Flood currents	
⫽⫽⫽ Flood-current erosion of the bank		⬅ Ebb currents	
⤙( )⤚ Ripples		⅄ Small gullies	
⬬ High and low sand bars		0 5 10m	

Fig. 25. The successive flood-bars F_1 and F_2 in a tidal creek on a tidal flat at Hojer. The salt marsh plants appear on the tidal flats, and the tidal creek is changing into a salt marsh creek. The flood-bar F_1 was formed in the original wadden gully. The flood-bar has made a barring for the flood-current in the flood-channel which is now a swamp (S). The flood-current now runs round the flood-bar forming ripples and eroding the outside of the bar. In the lower part of the original flood-channel the right side is eroded by the flood-current which forms an isolated flood-bar F_2 to the left. Because of the closing of the original flood-channel and the forming of a new bar the ebb-current is forced to erode the right side of the wadden gully forming a bigger ebb-channel meander. 1. Ebb-current erosion of the bank. 2. Flood-current erosion of the bank. 3. Ripples. 4. High and low sand bars. 5. Flood-currents. 6. Ebb-currents. 7. Small gullies. (from Jakobsen 1964).

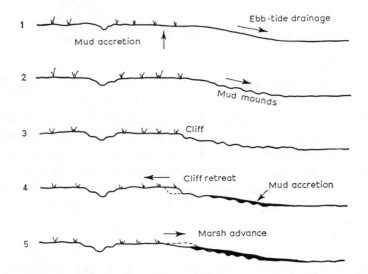

Fig. 26. Theoretical stages in the development of mud-mound and cliff profiles. 1. Accretion of mud on salt marsh due to influence of vegetation. Accretion rate higher than that of adjacent tidal flats. Creation of slope (exaggerated in diagram). 2. Ebb-flow drainage down slope creating mud-mounds. 3. Phase of excessive erosion of salt marsh producing cliffed profile. 4. Retreat of cliff and marsh erosion providing additional mud for deposition on tidal flats. Partial inundation of mounds by mud. 5. Excessive mud deposition on upper tidal flats with inundation of mounds. Possible renewed seaward extension of salt marsh edge. Dengie, Essex (from Greensmith and Tucker 1966).

near the mouths of salt marsh creeks illustrated in Fig. 25. Further landward in the marsh, ebb tide effects increasingly control the topography, and flow in the lower order marsh creeks is essentially unidirectional (Pestrong 1965). However, right from the very start of marsh formation the two opposing processes of accretion and erosion are at work side by side, and one after another, in fashioning the form of the marsh.

Usually one or other of these processes is clearly in the ascendancy. The change from a predominantly accreting situation to one of erosion and vice versa is usually quite rapid. However, Greensmith and Tucker (1966) have made a fascinating study of an Essex marsh where the processes of erosion and accretion are much more neatly balanced than usual (though erosion is believed to be dominant at the present time) and their theoretical conclusions on the cycle of events is summarized in Fig. 26.

Several recent authors (Ragotskie (1959), Redfield (in Ragotskie 1959) and Beeftink (1966)) have described the general changes in relief which accompany marsh genesis as follows:

(1) A change from the convex contours of the open mud flat surface to a concave marsh surface due to better growth at edges and hence levee formation through accretion.
(2) A general levelling of the relief as the marsh approaches maturity and its higher parts receive less and less accretion while the lower parts with more catch up to their level.
(3) A tendency to dissection of large marsh units into small ones, dependent on creek head ramification induced by ebb run-off.

Drainage

Pestrong (1965) has shown that hydraulic velocities and discharge are highest just after the onset of the ebb, and notes that vegetation dominates the hydraulic geometry of the marsh. Using multivariate analysis he concludes that moisture content and particle size exert most control over erodibility.

As studies on loose boundary hydraulics like those of Kestner (1961) become integrated with those on tidal marsh hydraulic geometry (e.g. Myrick and Leopold 1963, Ragotskie 1959 and Pestrong 1965) the approach to mathematical model analogues of the tidal marsh system comes closer. The contributions of physiographer and ecologist still tend to be largely descriptive though much more comprehensive in scope than in the past as shown for example in a study of marsh structure, mud origin, and re-cycling of deposits in Breton estuaries (Guilcher and Berthois 1951). There seems no reason why the various vegetation types and their trapping power for silt and the distinct physiographic units with which they are associated: flats, levees pans, flood bars, slumped blocks and cliff edges should not be analysed (e.g. with the aid of air photographs) separately, quantified, and then re-associated with hydraulic parameters like drainage density to create useful models for predictive purposes.

Re-cycling of sediments

Apart from the small scale re-cycling that accompanies marsh genesis it is a common feature of middle estuary regions that periodic swings of the river channel cause large scale re-cycling of marsh deposits over relatively short periods of time. Studies on a 1,500 acre salt marsh at Caerlaverock, Dumfries led to the conclusion that virtually the whole of it had developed, 'in something less than 140 years, large parts of it in much less' (Marshall

1963). Because of this relatively short period of time, Marshall concludes there would not have been long enough to allow isostatic change to be primarily responsible for the terracing characteristic of these marshes at one time believed to be an indication of a rising coastline.

Soil Composition, Aeration and Water-logging

Soil physical variation and growth

The majority of common salt marsh plants can grow on an extremely wide range of salt marsh soils of highly varied physical structure and composition. For example it is possible to find *Aster tripolium, Plantago maritima,* or *Triglochin maritima* equally on the clay-rich muds of south English coast harbours or on almost purely sandy gravel marshes in western Scottish coast sites. Leaving aside for the moment the possibilities of genetically controlled adaptation, it would be wrong to conclude from this that soil physical factors were of minor importance in relation to the salt marsh flora. While each species may have considerable latitude in survival on a wide range of soil types, performance and abundance are very markedly affected by physical factors and their dependent variates such as soil aeration. For this reason, floristic analyses which depend on presence and absence records alone may often give a misleading impression of uniformity in any comparison of marsh vegetation types.

Soil aeration

Adriani (1945) found air contents of 2–4 per cent in *Salicornia – Spartina* marsh and 25–45 per cent in *Salicornia fruticosa – Halimione* marsh. Chapman (1960) found that by poking a hole through the mud surface of submerged salt marsh, gas with significant oxygen trapped below the mud surface was released. His results (Table 13) also show an increase in oxygen content of marsh soils going from pioneer to longer-established and higher-lying plant communities.

Chapman noted that at root depth soil biological remains were abundant and that the aerated layer was especially associated with them. Green and Askew (1965) have examined soil macropores in reclaimed marshland at Romney marsh in Kent using latex infiltrant techniques. They found fine pores up to 1 mm in diameter were especially associated with roots, *Enchytraeid* worms and *Gammarus* sp. Small cavities and tunnels 2–3 mm or more in diameter were associated with ant activities (*Laesius flavus*). The pore system suggested long periods of activities rather than intensity of current use and the pores and cavities were evidently highly persistent.

Table 13. Composition of salt marsh soil gas (from Chapman 1960).

Region	Marsh	% CO_2*	% O_2	% residual
	Aster marsh a^1	2·99	1·61	95·4
	Aster marsh a^2	2·53	0·82	96·63
Norfolk	Aster marsh a^3	3·26	0·71	96·03
(England)	Aster marsh b	4·22	1·42	94·36
	Limonium marsh 1	1·46	10·5	88·04
	Limonium marsh 2	0·93	17·5	81·57
Romney	Spartina glabra 1	1·79	3·42	94·79
	Spartina glabra 2	3·23	8·28	88·49
(New England)	Spartina patens	0·58	17·3	82·12
	Distichlis	1·17	17·3	81·53
New Zealand	Juncus, Leptocarpus	2·95	17·34	max.
marshes	Juncus, Leptocarpus	0·49	10·05	min.
	Juncus, Leptocarpus	1·75	12·7	av. (12 values)

* The method of analysis (absorption by KOH) also measures any hydrogen sulphide that may be present. This, however, is only likely to be a small percentage.

The mud of pioneer zones of salt marsh contains few large cavities due primarily to activities of *Arenicola* and *Nereid* worms, molluscs (especially *Cardium*), and finer more superficial cavities associated with *Corophium* and *Hydrobia*. A marked increase in finer cavity structure develops as a result of root growth. The persistence of biological effects on soil structure is immediately apparent wherever salt marsh soils show erosion faces.

Air is replenished in these cavities by biological activity (e.g. *Arenicola* air holes) and at higher levels by fissuring of clay soils in dry weather in summer. Air is trapped in these cavities by surface sealing with deposited clay (especially in autumn), and mucilaginous algal growths at the surface in winter and spring.

Clarke and Hannon (1967) have demonstrated with ring infiltrometers just how low the infiltration rate of even highly sandy marsh soils can be (Table 14). They suggest that sodium in the soil may disperse the small amounts of silt and clay so impeding infiltration. However, in a later paper Clarke and Hannon (1969) report that the aerated soil layer is not universal and tends to occur mainly in finer grained, less permeable sediments. Possibly this is because macropores are more abundant and persistent than in sandy soils. Tidal flooding was found to cause water table rise to surface in sandier soils. Penetration of tidal water may be either through solution of crystalline salt, reduction of soil surface salinity, and clay flocculation and/or by lateral seepage due to saturated flow.

Table 14. Infiltration rates in sandy mangrove and salt marsh soils (from Clarke and Hannon 1967).

Species	Infiltration rate 100 cm/hour		% Coarse + fine sand at surface
	Dry period	Wet period	
Avicennia marina	0–8	2–4	79
Arthrocnemum australasicum	25–67	2–42	78
Juncus maritimus var. australis	46–149	14–36	69
Casuarina glauca	240	150	70

Stevenson and Emery (1958) note the large bulk storage of air in shoots of *Spartina* (see also Baker 1970 *a*) and suggest this may account for its success in colonizing frequently submerged zones of fine-particled mud. The same may be true for *Aster tripolium*.

Using Poel's (1960) polarographic apparatus to measure oxygen diffusion rate in sandy salt marsh soils, Brereton (1965) showed that in spite of the increasing water retaining capacity of soils with increasing silt contents associated with rise in altitude, conditions with respect to waterlogging improved. Oxygen diffusion rate rose from 6·1 μA in pioneer *Salicornia* to 8·6 to 13·2 μA in *Puccinellia* zones.

According to Brereton (1965), 'initially population structure is a reflection of the dominant influence of a high water-table which produces a highly plastic marsh surface accompanied by water-logged conditions. Later as the water-table falls population structure is a reflection of point to point variations in soil composition.'

He concludes 'an examination of environmental features show that soil water relations as controlled by drainage (through altitude), and soil physical characters, are primarily responsible for producing differences in species performance between stands and within stands respectively'.

Effects of water-logging

Very little experimental work has been done on the effects of water-logging on salt marsh plants, but Brereton (1971) concludes from his work on *Salicornia* that the main factor affecting *Salicornia* during the succession appears to be aeration. This controlled both germination and growth rates, but there was interaction with salinity. While *Salicornia* shows a preference for soils having high redox levels and shows tolerance of high salt levels, *Puccinellia maritima* shows the opposite. *Puccinellia* performance is improved in water-logged soils of relatively low salt status. *Salicornia* shows the opposite. Field and laboratory culture data confirmed these relationships.

More recently Clarke and Hannon (1970) found that none of the Australian species (*Arthrocnemum australasicum, Suaeda australis, Triglochin striata, Juncus maritimus* var *australiensis* and *Casuarina glauca*) were prevented from germination by submergence in 4 mm of water, but coverage by 5 cm of water retarded and reduced germination. These species and *Sporobolus virginicus* were grown experimentally at three water levels from water-logged to free-drained but damp. All except *Suaeda* and *Casuarina* grew satisfactorily under water-logged conditions, but there is evidence that *Arthrocnemum* and *Aegiceras* seedlings are more intolerant of water-logging than mature plants.

As mentioned in Chapter 4, Goodman and Williams (1961) studying *Spartina* 'die-back' soils found no direct evidence of plant damage due primarily to anaerobiosis. Mineral studies moreover showed no evidence of nutrient unbalance. Since ion accumulation requires oxygen this is further evidence that anaerobiosis is not seriously limiting. They conclude that death is brought about by a toxic reduced ion, but could not definitely establish sulphide as the responsible ion. It might be interesting to compare the tolerance of salt marsh species to different sulphide concentrations side by side with treatments providing a similar level of anaerobiosis, but without sulphide.

While at the present time we seem to be faced again and again with the close interaction of significant factors in the study of causative phenomena in ecology (as illustrated all too apparently in this account of aeration in relation to saltmarsh plant growth) we are at least beginning to understand how the parts of the system are connected even if we do not know exactly how it works.

6 Salt Marshes: Species Strategies

Autecological Limits

Each species on a salt marsh has evolved its own particular strategy for dispersal, establishment and growth; each has its own dimensional limits of age, height and potential clonal size.

Strategies are controlled by the range of environments in which a species can survive and the kinds of change the environments have undergone in the past and are undergoing now acting on the somatic and genetic material of which the species is composed. Perhaps we should remind ourselves right from the start that 'for sexual organisms, it is the local interbreeding population and not the species that is clearly the evolutionary unit of importance' (Ehrlich and Raven 1969).

The salt marsh habitat as a whole imposes certain limits on the kinds of plant that survive in it: all for example must be tolerant of high salinity and some of the ways in which this has been achieved have already been discussed. In addition each sub-habitat of the salt marsh imposes specific additional limits so that certain species survive in them preferentially.

So in this chapter we will see how life in different zones on the salt marsh and in the types of habitats within them has been solved by different species. Autecological studies help us to understand how the various species dovetail into the complex tapestry of populations of which salt marsh plant communities are composed.

Grime (1965) has drawn attention to the necessity of studying and experimenting with events in the field rather than trying to draw nebulous conclusions from data correlations. He also emphasizes the value of looking for susceptibility limits and trying to discover '. . . of what adaptation is the susceptibility a consequence.' This approach is being increasingly adopted with success by workers concerned with the practical application of autecological knowledge and examples are selected for discussion here with this in mind.

Algal Strategies

A small number of red, filamentous green, and dwarf brown algae have become adapted to the salt marsh habitat. Chapman (in Steers 1960) has made a special study of the distribution of some of these in time and space on an English marsh at Scolt Head Island, Norfolk (Fig. 27 and 28).

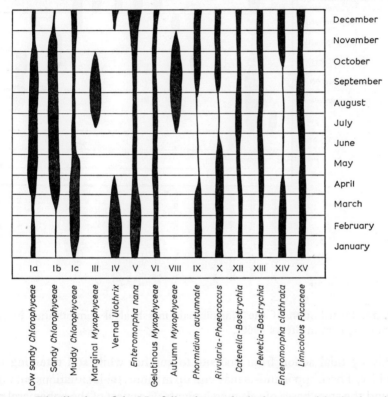

Fig. 27. Distribution of the Norfolk salt marsh algal communities in time (from Chapman in Steers, 1960).

Little is yet known about the autecology of these species and indeed their taxonomic relations are still in some cases little understood. But Chapman's studies reveal clearly some of the distinctive strategies, which enable algae to colonize salt marshes.

Mobile strategies

Certain *Enteromorpha* species (*E. prolifera*) which occur on sandy ground in the pioneer marsh zone have adopted a mobile strategy. They are moved

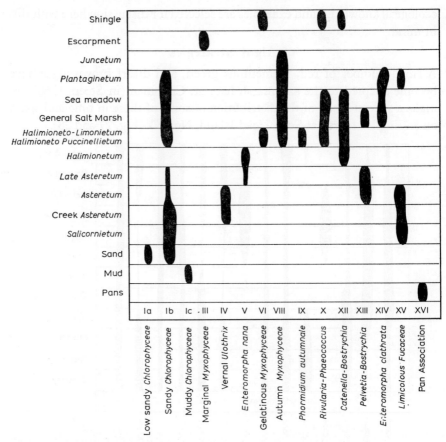

Fig. 28. Distribution of the Norfolk salt marsh algal communities in space (from Chapman in Steers 1960).

bodily by tidal action from one site to another while still retaining their viability. These opportunists may end up in more stable situations entangled around the stem bases of salt marsh plants or as part of the ephemeral algal mixtures stranded in temporarily flooded salt marsh pans. *Pelvetia canaliculata* also produces free-floating forms which accumulate in pans e.g. at Tyninghame marsh, East Lothian, Scotland.

Fixed strategies

Then there are species like *Vaucheria thurettii* and other filamentous blue-green algae whose gelatinous secretions are sufficiently binding on mud surfaces to enable growth to occur on the steep banks of creeks where no other plants can grow, what one might call an adhesive strategy.

Many algae (e.g. *Ulothrix speciosa* in spring, and *Oscillatoria sancta* in autumn) have developed optimal growth patterns outside the season favourable to the growth of most salt marsh plants. This enables them to take advantage of improved light conditions beneath taller salt marsh plants at the ends of the normal growth season when it is still warm enough for algal metabolism. This is effectively an off-peak seasonal growth strategy.

Yet others adopt a shelter-hold strategy (e.g. *Bostrychia scorpioides*, *Catenella repens*, dwarf *Fucus* species) embedded in mud accreted in the shelter formed by salt marsh plant growth. A few show an epiphyte strategy (e.g. *Catanella repens* or *Enteromorpha nana*), growing on the stems of other plants. This confers advantages of improved light compared with those growing in more shaded conditions at ground level.

Flowering Plant Strategies

Salt marshes are often thought of as an open habitat. If this were true one would expect a high proportion of annuals in the flora, but in fact this is not so. Certainly the pioneer zone is open and is kept open by high incidence of wave break since it is located around mean high water neap tides. Similarly at mean high water spring tides the high incidence of wave break and local smothering of vegetation by tidal litter maintains an open habitat. A few species of annuals (e.g. *Salicornia* sp. in the pioneer zone; *Atriplex* sp. on the strandline) are abundant in these two zones but the zones where annual species are dominant are relatively narrow in width. Between them the bulk of the salt marsh vegetation, whether grazed or ungrazed covers most of the ground when viewed from above and shows little more well-lit bare ground than in an inland grazed pasture for example.

It is not surprising to find therefore that the majority of salt marsh species are perennial and in fact relatively few species of annuals have become adapted to the true salt marsh habitat. Many casual species not specific to the salt marsh flora occur on the strandline and this might account for the relatively high values for Therophytes quoted in Chapman's analyses (Chapman 1960, Table 30). However, he does conclude that 'on the basis of the percentage of total species it is clear that the salt marsh is fundamentally a Hemicryptophyte area'.

Annual Strategies

The majority of salt marsh annuals are members of one family, the Chenopodiaceae. Many of the 500 species comprising this family show a high

H

degree of tolerance to high salinities, but most of them are adapted to relatively stable inland salines: only a few to the very specialized saline *and* tidally-disturbed salt marsh conditions. The genera *Atriplex, Suaeda,* and *Salicornia* have all produced annual species adapted to survival on salt marshes.

Salt tolerance

Hunt (1965) has shown that the improvement in salt tolerance of selected seedlings of *Agropyron intermedium* averaged 100 per cent and exceeded 500 per cent in several of the 20 clonal lines examined, indicating it was a highly heritable character. Gutnecht and Dainty (1968) have shown how the appropriate ion systems to utilize high sodium environments could have evolved. In the case of *Salicornia,* selection appears to have run its course to produce species which not only tolerate high salinity, but ones in which the capacity for persistence in fresh water conditions has been either 'bred-out' or is so low that re-invasion of open fresh-water muds is no longer a possibility. It would be interesting to measure any residual capacity for fresh water growth adaptation to see just how low the improvement capacity might be as compared with Hunt's figures for the improvement of salt tolerance as given above.

Resistance to mechanical damage

Boyce (1954) and Oosting (1954) point out the relationship between salt, succulence, and resistance to mechanical damage. It seems there may be a syndrome in which a succulent leaf shape (characteristic of salt marsh Chenopodiaceae and other salt marsh plants) is better resistant to mechanical damage (e.g. by wind and tide on coastal flats) and reduced mechanical damage in turn reduces injury from salt (or in particular the Chloride ion), which stimulates the development of succulence.

The annual strategy in the pioneer zone on a salt marsh demands high resistance to mechanical damage. *Salicornia* shows reduction to a phylloclade form and this presents minimum leaf appendanges for tearing by wave action, but an adequate photosynthetic surface in the high-light open habitats where it occurs.

Tolerance of water-logged conditions

The lowest zones of salt marshes are characteristically water-logged and as Brereton (1965) has shown have low oxygen diffusion rates. But he also points out that *Salicornia* shows a preference for soils having a high redox potential, compared for example with *Puccinellia maritima*. In fact high level tidal flats where *Salicornia* can grow are frequently colonized by algal blooms in spring and summer and it is not uncommon to see a silvery texture on the surface of water-logged mud formed by millions of tiny oxygen bubbles produced by the diatom *Pleurosigma* in early summer. Newly germinating *Salicornia* in April and May may take advantage of this oxygen source and the annual growth strategy in fact takes advantage of the potentially better aerated conditions in summer and passes the unfavourable winter season in seed.

Dispersal

Annual salt marsh species commonly produce seeds of from 1 to 3 mm diameter and not uncommonly the fruit itself (e.g. in *Atriplex*) or even the fruiting head forms the dispersal propagule. Thus Dalby (1963) has shown that the fruiting head with from 4–10 seeds may be dispersed as a whole in *Salicornia pusilla* and can float for up to 3 months in sea water before germination. Stevenson and Emery (1958) found that 10 per cent of seeds of *Salicornia bigelowii* (a Californian species) floated for at least 19 days although Praeger (1913) found that Irish *Salicornia* seeds sank within a minute. Ball and Brown (1970) noted that in *Salicornia europaea* and *S. dolichostachya* some ripe seeds fell out of plants, but in many cases seeds were retained on plants and germinated *in situ*. It seems that dispersal strategy is variable in *Salicornia*. Retention of at least a proportion of propagules occurs in the pioneer zone, some are strewn throughout the marsh by the tide, while a high proportion are carried to the strandline at the upper limits of the marsh. It is of interest here to note that Ball and Brown (1970) found that *Salicornia dolichostachya*, characteristic of the pioneer zone, has larger seeds and a more rapid rate of elongation of the radicle on germination than *S. europaea*, characteristic of more closed marsh habitats (Fig. 29). Moreover *S. dolichostachya* was not capable of maturing seeds in the more shaded sites where *S. europaea* could still achieve maturity.

Phenotypic and genotypic variability

There is a high degree of plasticity within species of *Salicornia* and Ball and Brown (1970) could not find any single character of 14 examined (other than chromosome number) to distinguish the *Salicornia europaea* and *S. dolichostachya*. *Salicornia ramosissima* can survive as isolated depauperate plants beneath tall *Spartina anglica* growth. Extremely vigorous plants of this species developed when the *Spartina* growth was killed in

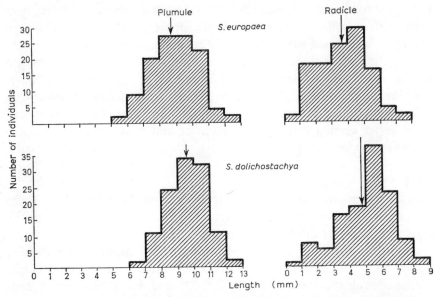

Fig. 29. Frequency distribution of the length of the plumule and radicle of *Salicornia* seedlings 4 days after germination (After Ball and Brown 1970).

herbicide experiments at Bridgwater Bay, Somerset (Ranwell and Downing 1960).

Dalby (1962) has brought together evidence showing that both diploid and tetraploid species occur among British *Salicornia* and that the tetraploids (e.g. *S. dolichostachya*) tend to be heavier seeded than the diploids. There is strong evidence (Fig. 30) that wind-pollination occurs, but plants have been shown to set a high proportion of seed by self-fertilization. This pattern of variation could result in the propagation of micro-species and also segregates from occasional crosses between lines to produce new lines.

Short-Lived Perennial Strategies

Flowering behaviour in *Aster* in relation to clines

Aster tripolium is a short-lived perennial widespread on most European and Mediterranean salt marshes. It is also found in inland salines in Europe and Western Asia. Gray (1971) has distinguished geographical and topographical clines in flowering behaviour. In the southern part of its range flowering occurs in the first year; in the northern part of its range flowering may be delayed up to 4 years or more. The topographical cline expressed by flowering behaviour is associated with a tendency for peren-

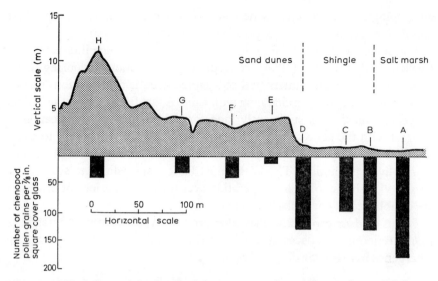

Fig. 30. Wind dispersion of pollen along transect from salt marsh to sand dunes, Blakeney, Norfolk. The lines marking the sampling points, A to H are not drawn to scale vertically (After Dalby 1962).

nial (and therefore earlier-maturing) plants, to occur in Submergence marsh, while annual (and therefore later-maturing) plants, occur more frequently in Emergence marsh.

Perhaps the most striking feature of *Aster* is its remarkable morphological plasticity ranging from tiny depauperate non-flowering plants one or two cm high in grazed salt marsh swards to robust freely flowering plants up to 180 cm high in optimum growing conditions. The latter seem to be associated with lime-rich marshes on soils rich in silt where salinity is generally not above 2 per cent, as in upper estuarine marshes on the Humber estuary, Yorkshire. The early studies of Montfort and Brandrup (1927) and Schratz (1934) indicated that the salinity of the parental habitat influenced seed size and seedling growth rates.

Close environmental 'tracking'

Chapman (1960) found that *Aster* seeds germinated most freely in fresh water, at about half the fresh-water rate in cultures at 2 per cent salinity, and not at all in seawater. However, Chater (in Clapham *et al* 1942) reports plants in inland salines surviving 4·6 per cent salinity and there is evidently genetic adaptation to high salinity in some races of this highly variable species. Indeed Gray (1971) has demonstrated marked physiological and morphological distinctions which are very localized in different *Aster* populations according to their position on a particular marsh.

The short-lived perennial exemplified by *Aster tripolium* illustrates well Ehrlich and Raven's (1969) conception of selection favouring close 'tracking' of the environment and co-adapted genetic combinations related to environmental factors changing with time. Apart from survival selection, its generation time is sufficiently long for environmental modification to interact with the gene complex, but sufficiently short to enable it to adapt more closely to specific environmental conditions at a time and place than longer-lived perennials. The latter must be less specialized to survive the greater vicissitudes of change operating over the longer periods of time they are established at particular locations.

Forms of *Aster* especially adapted to growth on submergence marsh have more numerous air spaces in the rhizome than those adapted to growth on emergence marsh according to Iverson (1936).

Response to grazing

The succulent leaves of *Aster* are attractive to both herbivorous wildfowl like the Brent goose (Ranwell and Downing 1959), and sheep (Ranwell 1961), but plants are capable of surviving in the semi-hemicryptophyte state (normally adopted during the winter), all the year round in grazed pastures. Vegetative propagation by detachable axillary buds still enables plants to reproduce although non-flowering, but continued persistent grazing probably accounts for the absence of this species on some Scottish salting pastures e.g. in some Argyllshire marshes.

Dispersal

Although seeds of *Aster tripolium* are supplied with pappus hairs, they tend to stick together in the capitulum and only some are partially dispersed over relatively short distances by wind; the remainder fall to the ground where they are dispersed by water.

Bearing in mind the limited extent of the salt marsh habitat and the wastage which would occur to unsuitable habitats either side of the shore-line if wind dispersal were not restricted, this compromise seems a very effective adaptation for providing *Aster* with the means of rapidly coloniz-ing open ground within the confines of the salt marsh. Both this and the compromise between the annual and perennial strategy equip *Aster* to behave like a particularly well-adapted weed species (Gray 1971) and in emergence marsh it is frequently a temporary colonist of gaps created by drought (Fig. 31) or by tidal litter. As with most weeds it seems intolerant of shade

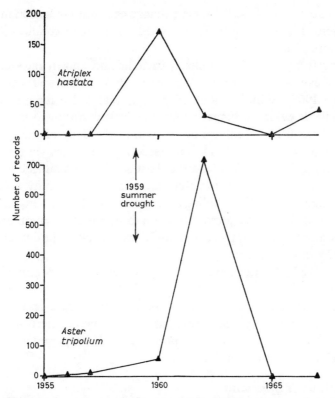

Fig. 31. Changes in salt marsh flora associated with summer drought. Note the peaking of the annual (*Atriplex hastata*) and the short-lived perennial species one year, and two years respectively, after the drought in 1959. Results derived from point quadrat records over a 12 year period in a grazing exclosure at the *Festuca rubra* marsh level, Bridgwater Bay, Somerset.

and is usually sparse in dense or tall communities such as *Spartina* or *Phragmites* marsh. Although an abundant colonist of reclaimed agricultural land temporarily salinized at the time of a sea flood in S.E. England (Hughes 1953), it was unable to persist as a weed element once the salt had leached and the land was again restored to agriculture.

Herbaceous Perennial Strategies

Consequence of longevity

Unlike the more ephemeral strategies discussed above, the herbaceous perennial strategy must cater for persistence at a location on the salt marsh for several decades. During this time the initial conditions operating on the seedling at the time of establishment change as a result of accretion and soil maturation. These produce distinctly different living conditions for the aged mature plant.

Judging from easily visible clonal patterns and known rates of growth individual clones of *Spartina anglica* commonly survive for 50 years or more (Hubbard 1965). While undoubtedly many plants live for much shorter time, available evidence suggests that very little recruitment of perennials by newly established seedlings occurs within the body of the salt marsh. For example, Boorman (1967) found seedlings of *Limonium vulgare* only very occasionally in the field and these usually in areas outside its optimal range, although mature plants were abundant in his study area. Again although *Spartina anglica* seedlings occur scattered beneath mature swards in early spring they do not survive beneath the shade of tall summer growths at Bridgwater Bay in Somerset, and rarely occur at all in the denser growths of Poole Harbour, Dorset.

Grasses are by far the commonest type of herbaceous perennial in most salt marshes. The relatively few species found such as *Spartina anglica, Puccinellia maritima, Festuca rubra, Agrostis stolonifera* and *Phragmites communis* are extremely widespread, at least in the northern hemisphere. In spite of their importance they have received comparatively little autecological study until recently.

Adaptations of *Spartina*

Very few species have succeeded in spanning survival from the absolute seaward limit of salt marsh growth on open high level mudflats to the landward limit of salt marsh growth at high water equinoctial tides, and from fully saline to brackish water conditions. The recently evolved *Spartina anglica* has achieved this. Longer established species of *Spartina* on American marshes (Mobberly 1956) have become specialized for growth in specific zones within this range but only *Spartina anglica* has the capacity to span the whole range. It has achieved this through a type of polyploidy (see Marchant 1967) which confers vigorous growth, large size, and high fertility, enabling it to dominate other species and reproduce extremely rapidly. At the same time it has developed unusually high phenotypic plasticity. This enables it to elongate stems as much as 10 or 15 cm a year to grow up through accreting mud in pioneer marsh and at a later stage in its life to adjust to accretion between one tenth and one twentieth of this rate in fully mature marsh. As Bradshaw (1965) has pointed out, 'where changes in environment occur over very short distances, adaptation by the formation of genetically different populations may be precluded. In these conditions very spectacular types of plasticity may be evolved'. The salt marsh habitat and *Spartina anglica* exemplify this.

Adaptations to survival at the seaward limit of submergence marsh include: a relatively large seed with considerable food reserves; rapid rate of shoot and root growth; development of deep, stout, anchor roots, and shoots well supplied with air spaces.

Taylor and Burrows (1968) have shown that some establishment does occur by fragments, but this seems to be of minor importance compared with seedling establishment except perhaps in long distance establishment via the tide-line to new sites. One of the putative parent species, *Spartina maritima*, sets little or no fertile seed (Marchant 1968). Even in favoured sites (e.g. Foulness, Essex), only widely scattered isolated clones develop from the occasional established fragment. *Spartina anglica* by contrast, rapidly fills in gaps with seedlings once pioneer clones have established. This enables it to develop a continuous sward over large areas in about 20 years.

Experimental studies have shown that *Spartina anglica* plants are tolerant of salinities up to 9 per cent in culture. Even in Emergence marsh where higher salinities than sea water develop salinity is unlikely to limit growth. Summer drought probably limits seedling establishment at higher levels, but it is often the density of its own tidal litter which kills out plants near the strandline and finally defeats its growth (Ranwell 1964 *b*). Grazing reduces flowering, but increases tillering to produce short dense swards (Ranwell 1961).

Adaptation of *Puccinellia*

Puccinellia maritima also spans a very wide range of the salt marsh habitat, but it is distinguished from *Spartina anglica* by much smaller size and much greater reliance on vegetative, as opposed to sexual reproduction. In common with *Spartina anglica* it is tolerant (but to a lesser extent), of water-logging, high salinity and high accretion rates of up to 5 cm per year. It too exhibits considerable phenotypic plasticity and pioneer forms are charac-terized by stolons up to 50 cm long, while under intense sheep grazing a tight mat-like growth little more than 1 cm high develops.

A vegetative propagation strategy is favoured in this species by tendencies to apomictic seed development (Hubbard 1968) and by sheep grazing which produces quantities of discarded fragments (much in evidence on the strandline of salting pastures), which root readily when heeled into damp marsh surfaces by sheep treading. In an experimental study of sheep grazing at Bridgwater Bay, Somerset, tread-planting was particularly noticeable where a period of grazing was followed by a big tide flooding the plots so irrigating the newly 'planted' *Puccinellia* and aiding its establishment.

A distinctive growth of narrow-leaved, widely spaced, upright shoots of *Puccinellia maritima* is characteristic of ungrazed marsh at Dengie, Essex, a marsh which is subject to surface scour erosion. This and other distinctive forms are in cultivation at the Coastal Ecology Research Station, Norwich to discover their phenotypic and genotypic relationships. *Puccinellia maritima* is not entirely apomictic and some sexual reproduction occurs, so one would expect greater variation in genotype on ungrazed marshes where seed propagation is likely to occur and greater uniformity on grazed marshes where vegetative reproduction would result in propagation of the few initially established types.

Adaptations in *Agrostis* and *Phragmites*

Spartina anglica and *Puccinellia maritima* have adapted to pioneer growth on water-logged saline silt liable to accretion, in ungrazed and grazed conditions respectively. Their counterparts in the brackish zone, *Phragmites communis* in ungrazed, and *Agrostis stolonifera* in grazed salt marsh, play a similar role and show similar adaptations related to tolerance of water-logging, rapid powers of horizontal spread, and capacity for vertical adjustment. They might be expected to differ principally from the first two species in their tolerance of saline conditions, and this proves to be so. Field measurements (Ranwell 1964) and culture experiments (Taylor 1939 and Gray 1971) indicate that both *Agrostis stolonifera* and *Phragmites communis*, in temperate zones at least, are restricted to estuarine marshes where chlorinity of the soil solution does not rise much above 1 per cent. This does not necessarily mean that chlorinity is the limiting factor at the seaward limit for these species and Hannon and Bradshaw (1968) give evidence suggesting that it is not in the case of *Agrostis stolonifera*.

All four species are likely to be limited to landward by competition for soil moisture with more drought tolerant species or by shade in the case of the shorter species. Experimental studies are needed to determine these tolerance limits and in the case of drought, with and without association with other species.

Woody Perennial Strategies

Adaptations in *Limonium*

We can take *Limonium* species as representative of the woody perennial strategy as recent studies by Boorman (1967, 1968) have added considerably to knowledge of the autecology of *Limonium vulgare* and *L. humile*. They both possess a stout woody rootstock and deep tap root and are tolerant of salinities in excess of sea water. In fact pre-treatment with sea-water followed by fresh water may actually enhance germination in *Limonium humile* compared with that in fresh water alone, (Fig. 32). Boorman suggests that sea water pre-conditions the embryo for germination and has an osmotic shock effect which weakens the seed coat and stimulates subsequent germination in fresh water. As in the case of *Aster*, seed size and early environmental history play a part (in addition to genetic control), in variations in germination response to different treatments. Germination is reduced by low oxygen levels and the seaward limit of *Limonium vulgare* is related to tidal flooding frequency, while the landward limit of both species seems to be controlled by competition for light or soil moisture with other species.

Both species are very susceptible to trampling damage by grazing animals, and sheep bite young buds off *L. vulgare* in spring. This rapidly leads to its disappearance. Both species are insect-pollinated and clearly likely to

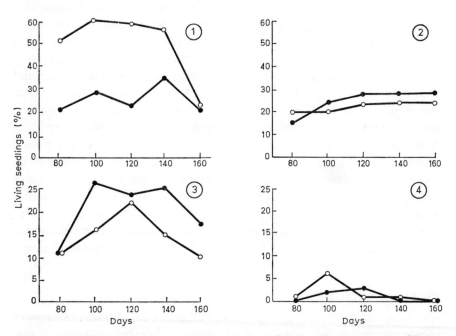

Fig. 32. Seedling establishment of *L. vulgare* and *L. humile*. Percentage of living seedlings after 160 days in an unheated greenhouse. 1. Fresh water and sand. 2. Fresh water and mud. Sea water/fresh water and sand. 4. Sea water/fresh water and mud. *L. vulgare* ○, *L. humile* ● (from Boorman 1968).

be left unpollinated if populations are so reduced that bees and other insects are no longer attracted to them. It may be a combination of intense sheep grazing and depressed insect activity in the cooler climate of Scotland, rather than a simple direct climatic factor alone, which has prevented both species from penetrating into more northerly Scottish marshes.

Advantages of high density growth in Emergence marsh

Boorman (1967) found that the vertical range of *L. humile* was restricted to one third of its potential range when in competition with *L. vulgare* and to one half its potential range when other species were present in high density. Rhizomes are little more than 5 cm long and it is evident that in contrast to Submergence marsh, where long rhizomatous species or ephemeral species occur, Emergence marsh favours short rhizome plants and dense or tufted growths. Most Emergence marsh species have dense woody rootstocks (e.g. *Armeria maritima, Halimione portulacoides, Inula crithmoides, Limonium* sp. and *Plantago maritima*). Others while not woody, tend to have compact short rhizomatous growth (e.g. *Festuca rubra, Juncus gerardii, Juncus maritimus* and *Triglochin maritima*). The latter are less sensitive to trampling than the brittle woody perennials and persist better in grazed marsh.

There is obviously selective advantage in relatively impenetrable dense growth in Emergence marsh. This is presumably related to the increasing competition that develops as the marsh vegetation becomes diversified and closes up. It is clearly shown in air photographs of *Spartina anglica* where the original rounded clone structure becomes compressed into polygonal clonal boundaries in more mature marsh. Marchant (1967) notes that *Spartina anglica* derivatives thought to be of polyhaploid origin have a much higher tiller density, about double that of the parent. It is noteworthy that these high density *Spartina* forms are particularly found in the upper Emergence marsh levels in Poole Harbour, Dorset and elsewhere. There is little vertical root layering in salt marsh soils because of the limits on respiration imposed by water-logging not far below the surface. Competition for moisture and nutrients at the 5 to 10 cm level is likely to be keen at the upper limits of Emergence marsh especially on more coarse particled substrata.

Problems of distinguishing limiting factors

Autecological studies of the type described above help to sort out master factors controlling the strategies appropriate to different salt marsh zones. Within the major zones, limiting factors may sometimes be too closely confounded for statistical approaches like the partial component analysis employed by Dalby 1970, to distinguish them. The solution of these problems must lie in laboratory study and field experimental study at critical seasons of growth and at both actual and potential limits of growth.

Limited knowledge of animal autecology on salt marshes

It should not be assumed that animals are of negligible importance in salt marshes because they receive so little mention here. This is due mainly to lack of knowledge.

The dominant influence of animals on salt marshes is vertebrate grazing but very little is known about animal behavioural patterns influencing this. Invertebrate influences tend to be localized to particular sub-habitats of limited area such as creeks, pans, tidal litter and the rather limited epifauna associated with certain plant species. With some exceptions such as Dahl's (1959) outstanding autecological studies on Scandinavian Ephydridae (Diptera), much of the work on salt marsh invertebrates remains at the taxonomic and distribution stage of study. There is great scope for autecological study of invertebrate adaptations to life on salt marshes and speciation arising from this.

Some attempt to balance the largely plant ecological emphasis of this chapter, by reference to animal population studies, is made in the next chapter concerning synecology.

7 Salt Marshes:
Structure and Function of Communities

The primary constraint on the types and numbers of organisms found on a salt marsh is the availability of the organisms themselves, that is the composition of evolved biological material at any particular place at any particular time.

The next constraint is the energy level imposed particularly by the climatic factors temperature and light. As we have seen in Chapter 1 these are primarily responsible for determining the regional type of marsh vegetation found.

Subordinate to these higher orders of restraint are water and particularly water quality (especially salinity) and oxygen availability. In other words this amounts to position on landward to seaward gradients and degree of submergence or emergence in relation to that position.

The ultimate level in the hierarchy of environmental constraints is nutritional and dependent not only on physical and chemical factors but also on the biological material which is itself energetic and can in turn generate the evolution of new biologic material.

This chapter concerns the arrangement of biological material, its movement, and the energy flowing through it in the salt marsh ecosystem. It is not concerned with the floristic or faunistic composition of different salt marsh communities except so far as they illustrate relation between structure and function. The presence or absence of most species in a habitat is irrelevant to the great majority of other species (including man) in the habitat. The dimensions, population size, and behaviour of a few species, and the degree of diversity of the remaining species (rather than their particular type), seem to be the more significant biological elements of the ecosystem. It is worth keeping in mind in the approach to the complexities of community study that, as Odum (1961) points out, 'basic work which is functional in approach is almost immediately practical . . . description alone, no matter how detailed, does not bring understanding.'

Plant Populations

Haline zones and plant distributions

Beeftink (1962) has produced a conspectus of phanerogamic salt plant communities in the Netherlands and considers that their zonation coincides very well with the classification of saline waters known as the Venice system (Final Resolution of the Symposium on the Classification of Brackish Waters, 1959) i.e.:

Zone	% Chlorinity (mean values at limits)
Euhaline	1·65 – 2·2
Polyhaline	1·0 – 1·65
Mesohaline	0·3 – 1·0
α - mesohaline	0·55 – 1·0
β - mesohaline	0·3 – 0·55
Oligohaline	0·03 – 0·3
Fresh water	0·03 – or less

To the extent that some of these limits coincide approximately with boundaries of some of the more abundant species controlling the character of certain communities these divisions seem of practical value in relation to British habitats also.

In particular there is a small group of highly salt tolerant species (e.g. *Halimione portulacoides, Limonium vulgare*) which do not normally penetrate beyond the up-estuary limit of the polyhaline zone. There are others which penetrate seaward just so far as the down estuary limit of the Polyhaline zone (e.g. *Agrostis stolonifera* and *Phragmites communis*). The limit of tidal woodland growth (e.g. *Alnus glutinosa, Salix cinerea* spp. *atrocinerea*) lies close to the down estuary limit of the oligohaline zone. Too many other considerations affect distribution to press these relationships too hard, but a competent ecologist should be able to deduce the halinity zone from the spectrum of species or vice versa quite reliably.

Sub-habitats of the salt marsh ecosystem

There are 10 quite distinctive sub-habitats (Fig. 33) found in most salt marshes and each provides distinctive growing conditions for the communities that occupy them. Teal (1962) has attempted to sum the proportions of sub-habitats on a complete marsh and find out how they equate with the total populations of the more abundant species occupying them. High quality air photographs which are now available should enable this type of analysis to be carried out quite readily, and it seems a useful basis for rapid survey of wildlife resources over large areas comparable to that developed by Poore and Robertson (1964).

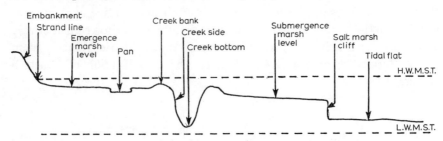

Fig. 33. Sub-habitats of the salt marsh ecosystem.

While it is true that a sizeable area of a salt marsh consists of level marsh, the other component sub-habitats may together make up an even greater area on much dissected marshlands. Yet the distinctive communities of these sub-habitats have been subject to very little individual study. Much of what follows therefore concerns level marsh communities. However, as Odum (1961) points out, because of the importance of tidal action in nutrient cycling and production, the entire estuarine system must eventually be considered as one ecosystem or productive unit. This means that the sub-systems based on plankton, benthos, and marsh-detritus food economies and the sub-habitats with which they are associated, must be analysed separately and ultimately related, before the system as a whole can be understood.

Transitional habitats

Salt marshes normally develop as relatively narrow belts adjoining sheltered coasts. Consequently they have extensive boundaries adjoining other systems where transitional habitats between the two are developed. In smaller marsh systems, transitional habitats may be more extensive in area than the pure salt marsh system itself.

Transitions to wet land, dry land, shingle and cliff tend to be relatively sharp with fairly steep salinity gradients near the top of the marsh in which many of the less common and more locally distributed species find a home. Transitions to sand dune may be much less sharp owing to the mobility of sand in wind and water.

In coastal sectors which have been reworked by wave action, horizontal and vertical mixtures of contrasting particle size; clay, silt, sand or shingle, impose patterns of transitional habitat type on the more typical salt marsh habitat.

Formation of the plant communities

The threshold level for salt marsh plant growth on a tidal flat often develops quite suddenly, and new colonization over a comparatively large surface area may occur very rapidly. West (1956) also finds that 'colonization of a mud flat (by mangrove) is not gradual but sudden and this results in stepwise bands of even age growth adjoining the coast'.

This threshold level varies in height in relation to the tidal regime and in accordance with stability, turbidity, and the light climate as we have seen. If this threshold happens to be low in level there is likely to be rapid colonization by a very pure community of the lowest growing pioneer species (e.g. *Salicornia* or *Spartina*); if it is higher then the amount of emergence may be sufficient for most of the main marsh species to colonize almost simultaneously and a mixed marsh community develops. In the intermediate case it depends very much on the proportions of annuals and perennials as pioneers whether diversification takes place early or late in marsh development, for the entry of new species depends very much on available space.

The subsequent vegetational history of the marsh is a product of interacting physical and biological boundaries. These are developed in horizontal and vertical planes and become diversified in time. Consequently it is of vital importance, to measure the rate at which the various interfaces and boundaries are changing to understand the plant and animal community transformations which both accompany, and at the same time, modify them.

Types of population change

What is becoming evident is that whether a pure dominant marsh community (like *Spartina anglica* marsh) or a mixed marsh community (e.g. *Triglochin, Plantago, Limonium, Puccinellia* marsh) develops initially, it may be relatively stable in composition for decades. Changes occur either in the vicinity of significant vertical changes in level (e.g. cliffs, creeks, pan edges or accretion to some new critical threshold level) or, where vertical changes in height of the vegetation are imposed by grazing, cutting or the invasion of taller species which can over-shade shorter ones.

Minor population changes are associated with physiographic development

I

of creeks, pans and cliffing. More far-reaching ones are associated with management interferences with the surface growths. The primary population changes which are perhaps of more fundamental interest will depend on the pioneer species growing and accreting silt to a level at which their reproductive performance is sub-optimal, and a new threshold for change in the flora has been reached.

There are surprisingly few studies of population changes on salt marsh though there are many assumptions in the literature implying that zonation can be equated with succession. Chapman (1959) has provided a valuable series of maps showing plant community changes over a period of 25 years at Scolt Head Island, Norfolk. But one has only to look at the slumped clods of main marsh level communities doomed to die in the bottom of a creek to realize that the probabilities of any particular square metre of salt marsh turf taking part in uninterrupted text-book succession may be very low indeed. Just what these probabilities are from site to site remains to be worked out.

Successional processes

Studies of population changes in mature *Spartina* marshes have helped to illustrate some of the general points made above and perhaps at the same time throw a little light on the successional processes.

In the Bridgwater Bay marsh it was found that at least three processes were involved in replacement of *Spartina* marsh by other species in mesohaline zones near the landward limit of *Spartina*, (1) suppression of *Spartina* growth by accumulation of its litter, (2) accretion towards a less saline and drier zone which favoured reproduction and growth of such invading species as *Agropyron pungens, Scirpus maritimus* and *Phragmites communis* and depressed reproduction and growth of the *Spartina* and, (3) shading out of *Spartina* beneath vegetatively expanding taller growths (Ranwell 1964 b). It was found that *Spartina* retained dominance for about 20 years, but in the subsequent 12 years about 50 per cent of the *Spartina* had been replaced by the invading species along the 2 mile (1·6 km) landward edge of the marsh. It was noted also that the build-up of levels suitable for growth of the invading species occurred more rapidly than they could be utilized. This was demonstrated by experimental transplants of the invading species well beyond their natural limits within *Spartina* marsh where they survived in competition with *Spartina*.

Yet another successional process, involving frost damage to *Spartina* growth near its upper limit, became apparent after the 1962–1963 cold winter in the Keysworth marsh in Poole Harbour. Patches of

Spartina killed back by frost together with suppression of growth locally by patches of litter, opened up the marsh surface for colonization by invaders similar to those at Bridgwater Bay (Hubbard and Stebbings 1968).

The younger marsh at Bridgwater Bay is still relatively lower in relation to tidal flooding than that at Keysworth and invasion is still largely linear in association with litter along the landward boundary. By contrast at Keysworth, litter is more widely scattered over an older and relatively higher marsh and *Spartina* is being invaded in irregular patches over the whole surface because it has all reached the new threshold level for such a transformation. It looks as though succession occurs in sudden jumps after long periods of relative stability and that it is the loss of vigour of the original species in sub-optimal growth conditions that results in its withdrawal to make room for the next phase.

Invasion boundaries

Fig. 34 shows a typical profile of one of many linear invasion boundaries subsequently mapped over a number of years by plane table mapping. The profile shows a distinctive pattern in which the *invading* species (*Phragmites*) grades quite gradually from high to low in height or biomass from landward to seaward i.e. from near optimal to sub-optimal growth conditions. By contrast the *invaded* species (*Spartina*) shows a characteristic

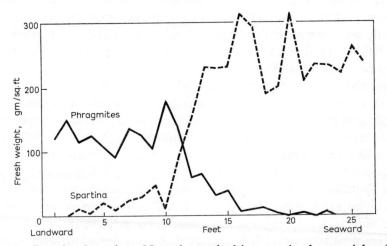

Fig. 34. Invasion boundary. Note the gradual increase in shoot weight of the invading species (*Phragmites*) and the sudden decrease in shoot weight of the invaded species (*Spartina*) from seaward to landward across the boundary. Poole Harbour, Dorset.

sudden decline and eventual extinction, within the taller growths of the invading species. This pattern was similar in boundaries between *Spartina* and other invaders such as *Scirpus maritimus* or *Agropyron pungens*.

It was found that if shoots were removed reciprocally at points across a mixed boundary of *Spartina* and *Phragmites*, *Spartina* maintains, but does not increase, its density in the absence of *Phragmites*, but can be extinguished by *Phragmites* growths of density 150 shoots per metre or more. By contrast, the removal of *Spartina* had no significant effect on *Phragmites* at the mid-point of the boundary.

Crude as this experiment is, it illustrates the susceptibility of *Spartina* near its landward survival limit. It is of interest that reciprocal shoot removal experiments in main marsh communities recently carried out at Scolt Head Island (Woodall *pers comm.*) show very little adjustment between species which implies much greater stability in these populations, at present in the middle rather than near the limit of their vertical range of growth.

Influence of grazing and fire

Grazing and exclosure experiments (Ranwell 1961 and 1968) show that very rapid readjustments in population balances can be induced by such wholesale alterations of the habitat. Recent studies in oligohaline marshes on the Fal estuary have demonstrated that it is possible for ungrazed salt marsh to succeed to tidal woodland without major isostatic change, but conditions for tree growth in the badly drained ground are poor and trees are frequently undermined by flooding and tend to be short-lived. It is relevant that Reid (1913) found trees to be mostly quite young in age in submerged forest beds around the British coast.

Peat accumulation (e.g. up to 2 ft (60 cm) of slowly decaying litter in 8 years), occurs in *Spartina patens* marsh in south east Louisiana. This usually gets burnt in dry weather and fire prevents tree establishment on these marshes so they remain grass-dominated (Lynch *et al* 1947). It is interesting that under the highly nutrient-rich conditions at Bridgwater Bay, *Spartina anglica* marshes are now developing a distinct peat formed from *Spartina* litter at the top of the shore. This recently took fire and was difficult to extinguish (Morley (*in litt.*)). Fire has also occurred in the *Phragmites* beds invading *Spartina* marsh in Poole Harbour.

More usually however it is grazing that maintains some form of grass marsh whether it be by sheep and cattle in Europe; cattle in North America; camels in the Red Sea marshes (Kassas 1957), or marsupials and rabbits in Australia or New Zealand.

Truncated development

Polyhaline marshes tend to form abrupt saline to non-saline boundaries with adjoining coasts since by their nature they are remote from ameliorating influences of fresh-water. They are more prone to disturbance from wave action and tend to be coarse-particled as they are also remote from river-borne silt sources. There is some evidence (Guilcher and Berthois (1957)) that their more readily erodable sandy-silt may tend to re-cycle and plant communities on them may not undergo significant directional succession. Human activities of course frequently truncate the development of marshland before it can reach its final stages.

Animal Populations

Numbers and biomass

Apart from vertebrate grazers there seems to have arisen a general belief that animals are of minor significance in the salt marsh habitat. Chapman (1960) for instance devotes less than two pages to them in a 350-page book on salt marshes and salt deserts. In fact it is only in the last decade or so since Chapman wrote that the abundance and significance of animals in salt marshes has come to be realized.

Paviour-Smith (1956) obtained the astounding figure of 7,631,460 animals per m^2 from a closely rabbit-grazed salt marsh turf in New Zealand with a soil (admittedly highly organic in character) only 20 cm deep over almost pure sand. Her figures for biomass show that animals represent only about 2 per cent of the value for plant biomass:

	mg/m^2 dry wt.
Total zoomass (max.)	32,436
Phytomass (bacteria)	10
(higher plants)	1,680,000
	1,712,446

Of particular interest also are her figures for organic matter. They immediately suggest the importance of detritus and hence of detritus feeders in the salt marsh economy:

	g/m^2 organic matter dry wt
Dead	17,374·4
Live plants	760·9
Live animals	25·6
	18,160·9

Origins of animal groups

Teal (1962) examined the terrestrial and aquatic macro-invertebrate fauna of a salt marsh in Georgia, U.S.A. (Fig. 35) and found it was distributed as follows:

(a) Terrestrial species	(1) General marsh levels
	(2) Upper limits of marsh
(b) Aquatic species	(1) Seaward edge of marsh
	(2) Creek sides
	(3) General marsh levels
(c) Marsh species (aquatic derived origin)	(1) Planktonic in larval stage
	(2) Marsh-living throughout the life cycle

Terrestrial groups formed nearly 50 per cent of the marsh fauna, but aquatic groups were found to be more important in the energetics of the system. He also found there was only slight adaptation to marsh conditions

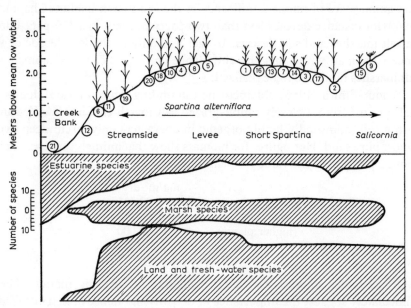

Fig. 35. Representative section of a Georgia salt marsh with horizontal scale distorted non-uniformly. Sample sites indicated by circled numbers. Site 2 represents the beginning of a drainage channel, not an isolated low spot. Symbols for grass are drawn to correct height for average maximum growth at those sites. The number of species of animals of 3 groups listed in Table 1* are plotted against sample sites. Names of marsh types used herein are also indicated (from Teal, 1962, by permission of the Duke University Press).

* Not given here.

Table 15. Ordinal composition of sets of ten samples taken in five types of salt marsh vegetation in the summer of 1960. Newport river estuary, North Carolina, U.S.A. From Davis and Gray, 1966, by permission of Duke University Press).

Stations	Mean percentage composition							
	Homoptera	Diptera	Hemiptera	Orthoptera	Coleoptera	Hymenoptera	Other orders	Average number of insects per sample
Spartina alterniflora								
Bogue Banks*	75·15	9·27	12·31	2·05	0·46	0·61	0·08	1,316
Piver's Island	53·47	40·28	0·70	0·70	2·78	2·08		288
Beaufort Channel	55·09	30·94	5·28	3·02	3·02	2·26	0·38	265
Bell Creek*	87·74	10·20	0·16	1·22	0·26	0·37	0·05	1,893
Harlowe Creek**	95·73	3·07	0·68	0·13	0·21	0·16	0·02	11,095
Lennoxville Point	31·65	52·23	3·16	7·91	2·22	2·53	0·32	316
All stations	90·11	6·72	1·76	0·66	0·38	0·37	0·04	2,529
Spartina-Salicornia-Limonium								
Bogue Banks	78·10	13·41	3·89	2·67	0·98	0·73	0·24	411
Juncus roemerianus								
Bogue Banks	85·61	5·30	1·52	5·30		2·27		132
Core Creek	30·00	15·00	20·00	20·00		10·00	2·27	20
Nelson Bay	73·17	7·32	1·22	15·85	1·22	1·22		82
Bell Creek	16·66	50·00		22·22	11·11	11·11		18
All stations	72·22	9·92	2·78	11·11	0·79	2·78	0·40	63
Distichlis spicata								
Lennoxville Point	60·83	13·64	21·83	2·35	0·51	0·78	0·06	1,782
Core Creek	72·29	14·13	9·45	0·39	3·04	0·62	0·08	1,281
Nelson Bay	38·41	40·71	14·60	1·50	3·54	1·15	0·09	1,130
Harlowe Creek	54·32	10·12	30·19	2·36	1·85	1·18	0·08	1,186
All stations	57·32	18·66	19·18	1·17	2·04	0·91	0·07	1,345
Spartina patens								
Bogue Banks	41·06	25·17	17·22	2·65	9·27	3·31	1·32	115
Piver's Island	19·90	61·22	4·59	2·04	3·07	7·65	1·53	196
Beaufort Channel	28·65	40·00	10·81	3·24	5·41	10·81	1·08	185
Harlowe Creek	43·90	31·22	6·83	4·39	1·95	11·22	0·49	205
Lennoxville Point	22·04	54·69	6·94	2·45	3·67	9·39	0·80	245
All stations	30·41	43·88	8·78	2·96	4·39	8·78	1·02	196

★ *Prokelisia marginata* (Homoptera) estimated in two samples.
★★ *P. marginata* estimated in four samples.

under flood, most climbing to escape (e.g. spiders) or trapping air (e.g. ants). Certainly one of the most dramatic and enlightening experiences the author has had was in watching the mass escape of terrestrial animals by flying, crawling up stems, swimming, or walking over the water surface supported by surface tension, one quiet evening on the equinoctial flood of the Fal marshes in Cornwall.

The basically terrestrial nature of the fauna of these marshes was confirmed by Stebbing's (1971) findings that, 'in general the faunal species recorded were representative of any marshland ecosystem in southern Britain and were not indicative of saline or brackish conditions'.

Distribution of insects on marshes

Davis and Gray (1966) found some 250 species of insects on North Carolina salt marshes and concluded they were abundant in both variety and quantity. They studied the distribution of insect groups above ground with respect to plant zones of increasing elevation (Table 15). Most of the insect species spend the winter in the egg state either in dead *Spartina* stems or in the ground. They found that shelter and food are factors that affect the size of insect groups more than tidal influences. For example the grass *Distichalis spicata* was rich in insects as it provided much cover and food, while the slender tough stems of the rush *Juncus roemarianus* attracted *Orthoptera* which are characteristic of open stands of coarse vegetation and can utilize tough plant tissues for food better than most insects. *Homoptera* decreased with increasing elevation, but ants, common in *Spartina patens*, were excluded by the tide at lower levels.

Dahl (1959) has made a special study of the distribution of species and numbers within species of Diptera Brachycera in salt marsh and dune habitats on the coasts of Norway and Sweden. Six sample surfaces of each sub-habitat were recorded and proportions of the more abundant species compared to determine in which sub-habitat particular species were dominant. Studies on the biology of the different species reveal their preferences and adaptations. Work on the species – habitat relationships of wide ranging groups of closely related species of this type are of particular value because they show how the capacity for dominance may be altered by climate or other environmental factors.

It will be of absorbing interest to watch the build up of the marsh fauna on the relatively new *Spartina anglica* marshes in Europe. Preliminary studies on one of the oldest of these new marshes in Poole Harbour, Dorset show that the principal species at present include a herbivorous bug (*Euscelis obovata*), an omnivorous grasshopper (*Conocephalus dorsalis*) and a carnivore (*Dolichonabis lineatus*), a rather nice illustration of the balanced way in which this new animal community seems to be developing. *Euscelis* feeds on *Spartina*, *Conocephalus* on *Euscelis* and *Spartina*, and *Dolichonabis* on *Euscelis* (Payne 1972).

Trophic Levels and Relationships

Paviour-Smith (1956) first outlined some basic trophic relationships on a salt marsh (Fig. 36).

Marples (1966) has recently utilized radio-isotopes to clarify the food

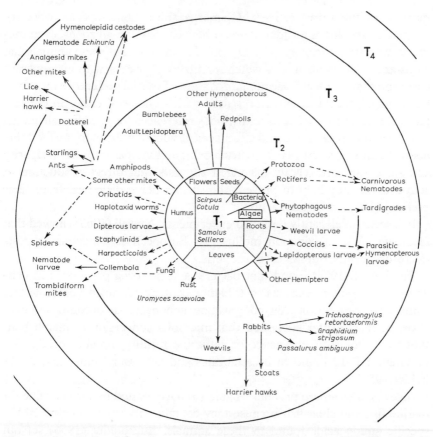

Fig. 36. Foodweb of the salt meadow community showing trophic (T$_1$, T$_2$ etc.) levels. The producer organisms are enclosed in boxes, and succeeding trophic levels are enclosed in succeeding concentric circles. Solid lines represent known relationships and broken lines assumed relationships. Hooper's Inlet, Otago Peninsula, New Zealand. (After Paviour-Smith 1956.)

chains of Arthropods. He labelled *Spartina* and detritus-rich sediment surfaces with the Phosphorus isotope P^{32}. He sampled standing crops and used
sweep netting to capture insects to record their distribution and changes in
their radioactivity with time (Fig. 37).

Results showed that four species of insects were dominant grazing
organisms (one *Orthoptera*, two *Hemiptera* and one *Homoptera*). Species in
two families of Diptera (*Dolichopodidae* and *Ephydridae*) and *Littorina*
snails were dominantly detritus feeders. Spiders were the important carni-
vores and obtained their energy from both the detritus and the grazing
food chains.

Luxton (1964) has examined the zonation of *Acarina* on the grazed salt
marshes of the Burry estuary, Glamorganshire. Some species were restricted to specific intertidal zones, others occurred in all zones, even into
the seawardmost *Spartina anglica* zone. He showed that these animals can
withstand immersion in sea water for up to 12 weeks without apparent
harm and that neither larvae nor adults showed special preference for salt or
fresh water conditions when presented with a choice of either in culture.
However these mites did have distinct preferences for specific salt marsh
fungi as food, so salinity could affect acarine zonation through the food. He
noticed that many salt marsh Acarina exhibited viviparity or ovoviviparity
and since eggs are readily dislodged suggests that direct production of
active larvae may help to prevent the species being dislodged from their
preferred habitats by tidal action.

In a series of elegantly designed experiments, Newell (1965) showed that
Hydrobia ulvae (a common mollusc on European salt marshes) digests
micro-organisms, but not organic debris.

Taschdjian (1954) found that bacteria and protozoa participate in conversion of *Spartina* extracts to a higher content of mixed vegetable and
animal protein. Studies on composition and bacterial decomposition of
Spartina marsh litter suggested that microbial conversion in marsh benthos was a key to maintenance of estuarine fertility, notably production
of Vitamin B 12 found in invertebrates and fish (Burkholder and Burkholder 1956).

Spartina alterniflora proteins (Table 16) have limited biological value for
marine fish and shell-fish because they contain only small amounts of the
specific amino acids found in these animals. Burkholder (1956) set out
cages of *Spartina* marsh litter in creeks and found that about 50 per cent of
the dry matter disappeared after a period of about 6 months. About 11 per
cent (dry wt. basis) of the annual crop of marsh grass may be rapidly converted to bacteria, but microbial utilization of crude fibre takes place more

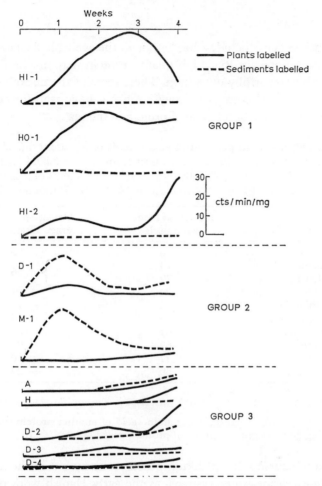

Fig. 37. Uptake curves of 10 animal populations based on mean weekly activity density for 4 weeks following the labelling of plants. (*Spartina alterniflora*) and sediments with P^{32}.

Group 1 includes three species (in descending order, *Trigonotylus* sp., *Prokelisia marginata* and *Ischnodemus badius*) which became highly labelled within 1 or 2 weeks after the grass was labelled. These species were judged to be primarily grazing herbivores.

Group 2 includes the group of highly related Dolichopodid flies and the *Littorina* snails which became highly labelled when the sediments were labelled, but less so when the grass was labelled.

Group 3 includes those Anthropods (Arachnida, Hymenoptera: Ichneumonidae, Braconidae, Chalcidae and Scelionidae, *Oscinella insularis*, *Chaetopsis apicalis* and *C. aenea* and *Hoplodicta*) which did not become highly labelled in either quadrant or became labelled only after 3 to 4 weeks. These groups were judged not to be actively feeding on either the growing grass or the detritus, or to be predators, in the case of delayed uptake. Sapelo Island marshes, Georgia (from Marples, 1966, by permission of the Duke University Press).

slowly. Burkholder concludes that 'although the available data are too few, still the indications are that high quality protein may not be formed by either marsh grass or phytoplankton. There remains the unexplored possibility that microbial conversion may act like a huge transformer to step up the potential value of the pool of protein in the sea'.

Table 16. Deviation* in per cent of amino acids of mature terminal leaf stalks of *Spartina alterniflora* (collected in August from Sapelo Island area, Georgia, U.S.A.), above (+) or below (—) the amino acids in proteins of fish and average *Graminae*. Data from Block, 1945 and Lugg 1949 (after Burkholder 1965).

Amino acids	Deviation from fish muscle	Deviation from average Gramineae
Arginine	— 78	— 90
Leucine	— 73	} — 56
Isoleucine	— 79	
Lysine	— 24	+ 7
Methionine	— 90	— 76
Phenylalanine	— 75	— 55
Tryptophane	— 41	— 60
Histidine	— 63	— 93
Valine	— 77	— 60
Threonine	— 50	— 24

* Per cent deviation $= \dfrac{x - y}{y}$. 100, where x is the amino acid of one protein (*Spartina*) and y is the corresponding amino acid of another protein the values of x and y being given in gm. per 100 g of protein, based upon N = 16%.

The sea of course has a plankton-based economy but in the marsh and near-shore estuarine water systems, as the early Danish work on *Zostera* (Petersen 1915, 1918) suggested, detrital food chains are likely to be of greater importance. These unlike the direct plant/herbivore/carnivore food chain are much more complex. Both of these food chains are operative in most ecosystems, but often in widely different proportions (Odum 1963). It is largely thanks to Odum and his co-workers that we are at last beginning to understand how a salt marsh nutritional system really works, largely through the concept of energy flow.

Energy Flow

Odum and Smalley (1959) showed how numbers tend to overemphasize, and biomass to underemphasize, the importance of small organisms in the community (see Paviour-Smith's figures above), while the reverse tends to be true of large organisms. Numbers and biomass can be integrated by a consideration of energy flow. These relationships were compared in two species (*a*) a herbivore and (*b*) an omnivore in *Spartina* marsh.

These workers obtained data by seasonal sampling, respirometry and calorimetry, of *Spartina alterniflora*, the herbivorous grasshopper *Orchelimum fidicinium*, and the omnivorous mollusc *Littorina irrorata*. They found that energy flow fluctuated only two-fold while numbers and biomass fluctuated five- or six-fold (Fig. 38). They noted that there was synchronization of the energy flow peak with medium numbers of median sizes (i.e. stages of active growth) of the secondary producers rather than with maximum numbers or maximum biomass. Food availability was related to periods of high energy flow and operated over a longer period for the mollusc than the grasshopper (Fig. 39).

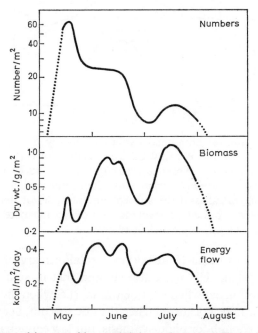

Fig. 38. Numbers, biomass (dry weight), and energy flow per square metre (1 kcal. = 4186 J) in a population of salt marsh grasshoppers (*Orchelimum fidicinium*) living in low-level *Spartina alterniflora* marsh, Sapelo Island, Georgia, U.S.A. (from Odum and Smalley 1959).

Role of benthic algae

Now benthic algae have a high production all the year round in Georgia, U.S.A. at least and these and detritus were considered to be the principal food of the mollusc *Littorina*. Pomeroy (1959) found that net production of

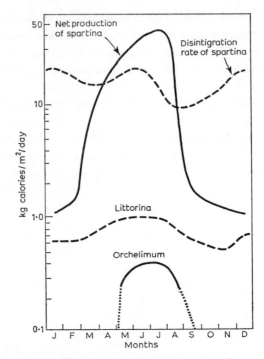

Fig. 39. Comparison of the annual pattern of energy flow (1 kcal = 4186 J) in *Littorina* and *Orchelimum* populations in relation to certain potential food sources. Net production of *Spartina alterniflora* in low level marsh is the sole source of food energy for *Orchelimum* while the disintegration of dead *Spartina* through the entire marsh and the subsequent transport of detritus (including associated microflora) to the high level marsh provides one potential food source for *Littorina* (from Odum and Smalley 1959).

'mud-algae' makes a major contribution to total primary production of the salt marsh ecosystem. His work also showed that in summer, maximum photosynthesis occurred when the tide was in, (i.e. when the marsh was not desiccated); in winter it occurred when the tide was out, i.e. when temperatures were high enough for photosynthesis, but low enough not to desiccate the algae.

It is of interest to note here in relation to the above that the herbivorous Brent Goose was found to synchronize its feeding at Scolt Head Island, Norfolk on tidal flat algae in winter (where temperatures even at these latitudes favour active growth of algae) and on the salt marshes in spring (Ranwell and Downing 1959). We begin to see why the tidal flats are so productive for waders and wildfowl during winter months in the northern hemisphere.

On a larger scale of course, productivity is also related to regional climate. One study on the mollusc faunas of *Spartina-Salicornia* marshes illustrates this. The local, seasonal and latitudinal variations in their faunas were analysed quantitatively in 11 marshes ranging over 20° of latitude on the North American Pacific Coast by Macdonald (1969). He found that the standing crop of the living animals increases considerably from north to south, suggesting that available resources increase at lower latitudes.

Grazing and detrital pathways

Odum (1962) and Teal (1962) found that major energy flow between autotrophic and heterotrophic levels on a marsh is by way of the detritus food chain rather than the grazing food chain. In Georgia estuaries dominated by *Spartina alterniflora*, organic detritus (more than 90 per cent of *Spartina* origin) is the chief link between primary and secondary produc-

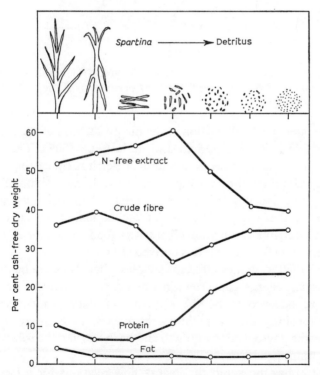

Fig. 40. The nutritive composition of successive stages of decomposition of *Spartina alterniflora* marsh grass, showing increase in protein and decrease in carbohydrate with increasing age and decreasing size of detritus particles. Sapelo Island, Georgia, U.S.A. (from Odum and Cruz 1967).

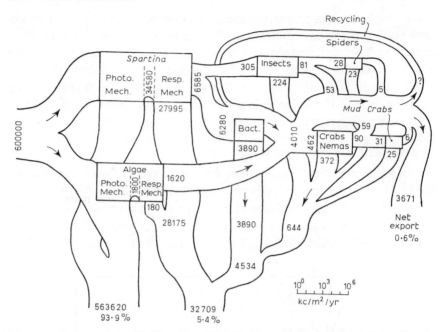

Fig. 41. Energy flow (1 K cal = 4186 J) diagram for a salt marsh. Sapelo Island, Georgia, U.S.A. (from Teal 1962, by permission of the Duke University Press).

tivity because only a small portion of the net production of the marsh grass is grazed while it is alive (Odum and de la Cruz 1967). These authors studied the seasonal changes in amounts of detritus by netting at a creek mouth, they examined the distribution of size and composition of particles, their origin, the decomposition rates of litter, its nutritive value and the metabolism of detritus particles under incubation. They conclude that bacteria-rich detritus is nutritionally a better food source for animals than the fresh *Spartina* from which it is derived (Fig. 40).

Pomeroy (1959) concludes that in water less than 2 m deep benthos is the more important energy converter; in water greater than 2 m deep phytoplankton are believed to be the principal primary producers. In fact, increasing turbidity due to pollution in inshore waters may reduce the possibilities for phytoplankton growth and be increasing the role of benthos in inshore water productivity.

Teal (1962) has measured the energy flow relations for a Georgia salt marsh (Fig. 41) and these are summarized in Table 17 which shows that 45 per cent of the net production of the marshes is exported to the estuarine waters where it is believed to largely support harvestable shrimps and crabs in waters too turbid for significant phytoplankton production. Clearly increasing turbidity may not merely modify the lower limit of salt marsh growth but also affect profoundly the balance in energy paths at critical junctions between ecosystems.

Table 17. Summary of salt marsh energetics derived from studies at Sapelo Island, Georgia, U.S.A. One kc = 4186 J (from Teal 1962 by permission of the Duke University Press).

Input as light	600,000 kcal/m²/year
Loss in photosynthesis	563,620 or 93·9%
Gross production	36,380 or 6·1% of light
Producer Respiration	28,175 or 77% of gross production
Net Production	8205 kcal/m²/year
Bacterial respiration	3,890 or 47% of net production
1° consumer respiration	596 or 7% of net production
2° consumer respiration	48 or 0·6% of net production
Total energy dissipation by consumers	4,534 or 55% of net production
Export	3,671 or 45% of net production

K

PART THREE　Sand Dunes

8 Sand Dunes: Formation and Differentiation of the Habitat

One thing that salt marsh and sand dune plants have in common, in spite of the striking differences between these two habitats, is the problem of establishing in a soil which is initially unstable. Foreshore sand washed daily by the tides on an open shore is in fact so unstable that no flowering plants or even macro-algae have yet succeeded in colonizing it. Even when sand is exposed for days at a time, it soon dries out in the wind and may still be blown about too frequently for plants to establish. Moreover, unlike silt on which salt marsh plants grow, sand is very deficient in plant nutrients and moisture (at least in the surface layers). Without nutrient income from the tide, sand would be unlikely to support much plant growth for long even if it were stable.

All three of these deficiencies; instability, lack of nutrients, and lack of soil moisture, are ameliorated by tidal litter, deposited at the top of the fore-shore in strandlines, and it is here that the process of dune formation can start.

The Strandline

Sand feeding to the strandline

Krumbein and Slack (1956) recognize four zones on sandy shores as shown in Table 18. To the extent that each zone feeds sand to the zone above, adequate sand supply in all of them is essential to the continued long term growth of dunes. But width, height and orientation of the backshore zone is of more immediate concern, as it is from this zone that the bulk of the sand is derived by wind action for dune building (Plate 9).

Coastal physiographers have carried out intensive studies of the sweep zone of open shores (see King 1972), but so far there has been little attempt to link these changes with rates of supply to dune systems, though studies on this are now in progress in Northern Ireland.

Sandy shores bordering estuaries or sounds are usually narrow, and being well protected from strong wave action tend to accumulate considerable quantities of tidal litter. This enables strandline plants to establish but there is usually insufficient sand supply to lead to significant dune formation.

In contrast sandy shores on the open coast are often wide and what tidal litter there is forms temporary surface accumulations above mean high water spring tide level. Litter sticking up from the sand reduces windflow near the sand surface and wind blown sand is deposited until the litter is buried. No further sand is likely to accumulate once a smooth surface is restored again, and the deposit is likely to be re-worked by the next tide to reach it.

Table 18. Sand shore zones (Krumbein & Slack 1956)

Shore Zone	Limits	Tidal relations	Agents of sand movement
1. Nearshore bottom	Mean low water to minus 9 m	Nearly always submerged	Currents and breaking waves
2. Foreshore	Mean low water to high tide line	Alternately submerged and exposed	Currents, breaking waves, occasional wind action
3. Backshore	High tide line to dunes	Nearly always exposed but occasionally submerged during storms or exceptionally high water	Breaking waves, wind action
4. Dunes	Above highest tide limit	Always exposed	Wind action

Colonization of the strandline

Hulme (1957) has shown (Fig. 42) how after the high tides of the spring equinox, falling high water levels leave a zone up to 11·5 m wide seaward of maximum high spring tide level at Longniddry, East Lothian. Annual strandline plants can colonize in this zone. The tidal litter contains or traps varying quantities of viable seeds. Tidal litter reduces daily temperature fluctuations in summer at the sand surface from 25°C in open sand to 7°C

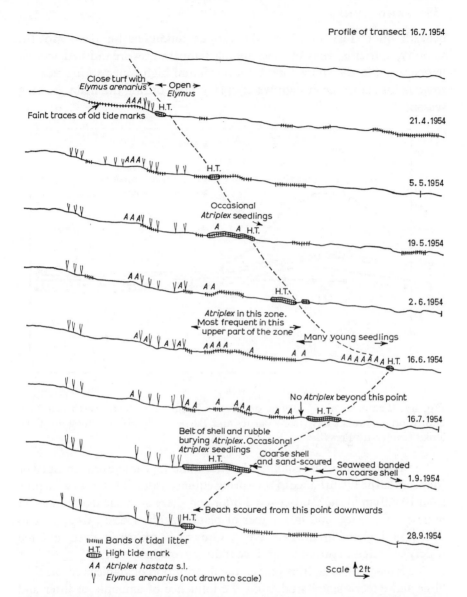

Profile of transect 16.7.1954

Close turf with
Elymus arenarius ← Open → *Elymus*
AAAyyy H.T.
Faint traces of old tide marks

21.4.1954

H.T.

5.5.1954

Occasional
Atriplex seedlings
A A H.T.

19.5.1954

H.T.

2.6.1954

Atriplex in this zone.
Most frequent in this
upper part of the zone →
← Many young seedlings
AAAA A A A AAAAAAA H.T. 16.6.1954

No *Atriplex* beyond this point
H.T.

16.7.1954

Belt of shell and rubble
burying *Atriplex*. Occasional
Atriplex seedlings
H.T.
Coarse shell
and sand-scoured → ← Seaweed banded
on coarse shell

1.9.1954

← Beach scoured from this point downwards
H.T.

28.9.1954

ıɪɪɪɪ Bands of tidal litter
H.T. High tide mark
AA *Atriplex hastata* s.l.
ɣ *Elymus arenarius* (not drawn to scale)

Scale ↑2ft→

Fig. 42. A permanent transect across the foreshore at Longniddry, East Lothian, charted at intervals throughout the summer. Note the downward extension of colonization by *Atriplex*, as the zone free from tidal scour (to left of dotted line) increases during the early part of the summer. (After Hulme, 1957).

beneath litter (Fig. 43). Thus the various conditions for plant growth; stability, nutrients, moisture and suitable temperatures are satisfied, at least temporarily, in the shore zone from maximum high water spring tides to some point above mean high water spring tides during the main growing season.

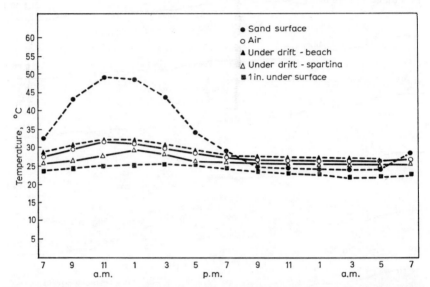

Fig. 43. Temperature variations on Piver's Island, Beaufort, North Carolina, U.S.A. on a summer day (from Barnes and Barnes, 1954, by permission of the Duke University Press).

The annual plants that colonize the strandline appear late (e.g. in April or May), after the disturbance of the spring equinox tides. They root at depths from 10–40 cm in the sand-covered litter and there are wide fluctuations in number of species, and in number of plants within species, from year to year (Gimingham in Burnett 1964). Growth is too short-lived and too widely scattered to promote significant dune growth.

The biology of strandline plants, the distribution of viable seeds in tidal litter and experimental studies on the influence of amounts of litter and seeds on the thresholds for embryo dune formation, all deserve further study.

The Embryo Dune

Perennial grass pioneers and accretion

Tidal litter and strandline plants accumulate sand to a level just above H.W.M.O.S.T. This is sufficient for perennial grasses like *Agropyron junceiforme* and *Elymus arenarius* to establish, and their growth laterally and vertically is sufficiently persistent to raise a dune a metre or two high. Bond (1952) found evidence of continued vegetative growth in *Elymus arenarius* in all months of the year except January at Aberdeen, but Nicholson (1952) found that *Agropyron junceiforme* was dormant during winter at the same latitude. Nevertheless even in winter the dead shoots of *Agropyron* are persistent and help to retain sand trapped in summer.

The inherent capacity of these plants and others like them elsewhere in the world (e.g. *Uniola paniculata*, North Carolina, U.S.A.) to bind sand lies in the ability to perennate and to develop extensive horizontal and vertical rhizome systems. Bond (1952) recorded vertical rhizomes 150 cm long in *E. arenarius* and noted that viable buds survived at depths of 60 cm. Viable buds of *Agropyron junceiforme* occur at depths of 60 cm but this species is not tolerant of accretion rates of more than 60 cm per year (Nicholson 1952). This is probably about the limit for both species, but it needs to be determined experimentally.

Agropyron junceiforme propagates readily by seed and within 10 days the seedling root has elongated 7 cm to more or less permanently humid sand. Rhizome fragments are also important in reproduction and according to Nicholson (1952) propagation from broken rhizomes is common. *Elymus arenarius* propagates rather less freely from seed in the field than *Agropyron* according to Bond (1952).

Seedling regeneration only becomes significant in *Elymus* when plants are abundant. Germination tends to be delayed by the presence in seeds and glumes of a water soluble germination inhibitor until spring, when seedlings have a better chance of survival. Seedlings can not withstand more than about 7·6 cm of sand burial (Clarke 1965).

Growth of *Agropyron* embryo dune

Gimingham (in Burnett 1964) gives a good account of the way in which *Agropyron junceiforme* forms a fore dune from a newly established seedling:

'The single primary root quickly extends to a depth of about 15 cm where a level of moisture content rather higher than that of surface layers is often maintained. The first lateral roots, however, extend horizontally closely below the sand surface. After a rosette of tillers has been established, short rhizomes are formed extending obliquely for distances of between 5 and 30 cm from the original plant giving rise to new groups of tillers (Fig. 44 A to C). This type of growth may continue for two seasons, but in time long horizontal rhizomes are produced greatly increasing the vegetative spread of the plant (Fig. 44 D). Their tips normally turn upwards in autumn, first breaking the surface, ready to produce a new group of shoots in spring. Development may continue indefinitely in this way if sand accumulation is only slight, for elongation of the shoots can bring them to the surface through layers not exceeding about 23 cm in depth. Where however, burial is more rapid, shoots are killed and rhizomes instead of extending laterally assume a vertical direction until the new surface is reached, when again tillering takes place. This sympodial development may keep pace with repeated sand deposition, often up to heights of 1·8 m and considerably more.'

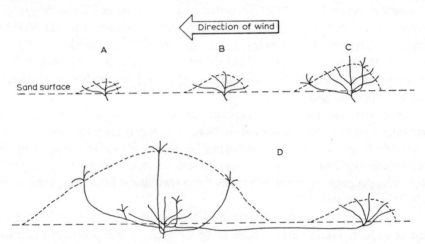

Fig. 44. Stages in colony development and formation of 'embryonic dunes' by *Agropyron junceiforme*. The upper broken line represents the surface of the dune. (from Nicholson, 1952).

Plant growth habit and dune formation

Although the occasional bud at depths of 60 cm in *Agropyron junceiforme* and *Elymus arenarius* may produce vertical growth to reach the surface, continuous accretion of this order per year would result in such sparse surface growth that plants would be unlikely to survive. Effectively they are limited to annual accretion zones of about 30 cm a year, and are incapable of building the really high dunes produced by *Ammophila* growth, and certain other species.

Cowles (1899) recognized two contrasting growth habits among dune-forming plants which are of fundamental importance in understanding their capacities and limitations. *Ammophila* species and *Agropyron junceiforme* both produce horizontal rhizomes of potentially unlimited growth. Others like *Elymus arenarius, Salix repens* and *Populus* species seem to lack the capacity for rapid horizontal spread, but do readjust to sand burial with oblique or vertical rhizome growth. In the case of the two shrub species the vertical rhizome growth may be virtually unlimited at acceptable accretion rates. Their tight growth form tends to produce rather steep-sided hummock dunes in contrast to the much broader dune units formed by the more widely spreading growth of *Ammophila*.

Limitations of the pioneer grasses

It is not clear whether or not *Agropyron junceiforme* and *Elymus arenarius* are simply overwhelmed by high accretion rates and/or shading (see Nicholson 1952) which *Ammophila* induces, or whether their moisture or nutrient requirements cannot be satisfied in high dunes. Gimingham (in Burnett 1964) concludes that since *Agropyron junceiforme* seldom has active roots below 60 cm and the water table may be over 1·2 m below the surface the roots are independent of it. However, they may be more dependent on organic matter and spray-borne nutrients than *Ammophila*. It may be significant also that *Elymus arenarius* does survive in high *Ammophila* dune at Durness, Sutherland where rainfall and spray are high compared with more southerly dunes. Clearly there is scope for experimental transplant studies and studies of moisture and nutrient requirements to help solve this problem.

We are now in a position to appreciate the special advantages which *Ammophila* species possess in dune building, namely potentially unlimited horizontal *and* vertical rhizome growth. No other species combine these two vital attributes and throughout the world it is either *Ammophila arenaria* or *Ammophila breviligulata* which have created the really high dune landscapes.

Establishment and Growth Patterns of *Ammophila*

Pioneer studies by Gemmell, Greig-Smith and Gimingham (1953) showed how *Ammophila arenaria* initiates dune building. They found that establishment by seedlings on the higher parts of the backshore at Luskentyre in the Outer Hebrides was sporadic. At Ainsdale, Lancashire establishment from rhizome fragments from the eroded coast dune occurred more commonly than seedling establishment.

Shoot tufts from rhizome fragments or seedlings were initially un-branched and created smooth dunes parallel with the upper shore like those formed by *Agropyron*. These leafy shoots were capable of growth through moderate accretion by leaf elongation. If leaves were buried, axillary buds developed to form vertical shoots with long internodes which produced new leafy shoots at the surface. Adventitious roots formed on the vertical rhizomes just below the surface and deeper horizontal rhizome connections gradually died. As the leafy shoots occasionally produced more than one vertical rhizome branching increased and dome-shaped tussocks developed.

Laing (1954) confirmed and amplified this picture with detailed studies of the American species *Ammophila breviligulata* on the inland dunes around Lake Michigan. He found this species regenerating mainly from eroded rhizome fragments on the beach. Seedling regeneration was confined to damp hollows or protected sites on the lee slopes of eroding dunes.

The evidence suggests that seedling establishment on shore or dunes depends on periods of heavy rainfall which gives both moisture and temporary stability and that where coastal dunes are eroding, rhizome fragments from toppled clumps are the principal means of re-establishment of a new embryo dune. Where erosion of a coast dune cuts back to forested dunes beneath which *Ammophila* has been shaded out, such regeneration may be prevented as at Holkham, Norfolk.

Laing (1954) found that shoot elongation occurs in spring, especially from buried vertical stems which at the onset of dormancy in the previous autumn had a fully formed blade but an incompletely elongated sheath and internode. Newly formed internodes of the current spring also develop and elongate. Buds of the continuous development type form only on vigorous shoots from depositing surfaces and only on those internodes which mature from late April to early June. Dormant buds form elsewhere and may be dormant for months or years. Branching of the vertical shoots creates a cluster of shoots around the parent shoot; loose open clumps in accreting surfaces, compact tufts on stable areas.

Dune stratigraphy

Both Laing (1954) and Olson (1958 *c*) have shown how past depositional patterns can be determined through measurement of internodal lengths which occur in response to burial (Fig. 45). Olson (1958 *c*) found that foredunes accrete only about 30 cm a year probably because shoot burial removes their power to hold more sand until new growth appears in the following year. He notes how the appearance of a new foredune effectively cuts off sand supply to the one to landward where dead growths persist, clumping becomes more compact and flowering diminishes markedly within 3 or 4 years of stabilization. These very swift reactions of *Ammophila* to changes in sand accretion indicate how greatly its vigour depends upon them. Both *Ammophila breviligulata* and *A. arenaria* have been shown (Laing 1954 and Ranwell 1958) to just tolerate an absolute limit of sand burial of 1 m per year, but density diminishes rapidly if these conditions persist. These high rates of accretion are especially characteristic of the higher lee slopes of dunes.

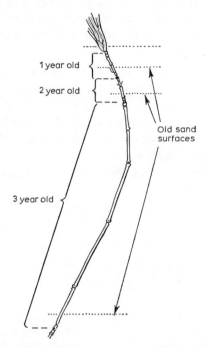

Fig. 45. Three annual growth cycles of a buried marram grass stem. Wide spacing of internodes occur in years following rapid deposition of sand in winter. Stabilization of the dune shown by sudden decrease in annual growth elongation (from Olsen 1958 *c*).

In order to understand the mechanics of dune building and development of the dune landscape it is necessary at this point to consider some fundamental principles relating to the effect of surface roughness in the form of vegetation on wind flow and sand movement.

Windflow, Sand Movement and Dune Vegetation

Bagnold (1941) has shown that sand moves by saltation, a process whereby the first grains moved by a sufficiently strong wind fall under gravity to the loose sand surface and bounce back into the wind at the same time setting other grains in motion by their impact. He showed that the impact gradient threshold varies approximately as the square root of the grain diameter. Above the impact velocity, about 10 miles per hour (4·5 m/s), rate of sand flow varies as the cube of the wind velocity. It follows that really substantial sand movement is only accomplished by high wind velocities.

Much remains to be learned about the frequency of movement of shore sand in relation to the incidence of rainfall and the special conditions operating on drying out sandflats with or without variously spaced pioneer growths of algae or higher plants.

Air flow over a smooth level surface decreases in velocity in a regular manner near to the surface. Over a curved smooth surface, the rate of velocity decrease increases near the surface especially at the point of maximum curvative. Surface roughness interferes with the smooth laminar flow of air and creates turbulence. As a result the profile of mean velocity $(\overline{V})$ as a function of height (Z) is not linear, as it is in laminar flow, but logarithmic. The logarithmic relationship implies that velocity decreases to zero at some height greater than zero (Z_0).

Bagnold (1941) showed that Z_0 can be related to height and spacing of

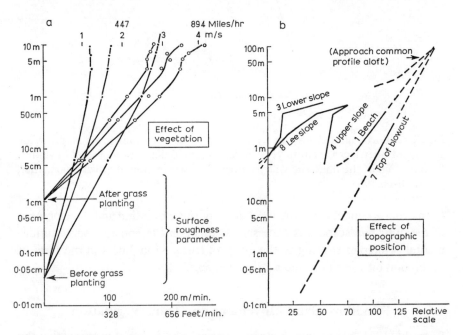

Fig. 46. Dune wind profiles. *a*, Before and after grass planting on an Indiana dune; *b*, comparable plotting of key profiles for a bare Michigan dune, after Landsberg and Riley 1943 (from Olsen 1958 *b*).

surface irregularities and he called it 'the surface roughness parameter'. For a given roughness (Z_0), the slope of velocity with respect to log Z increases as velocities aloft increase. This slope is proportional to the 'drag velocity' (V_*) which equals the square root of shear stress over air density. The proportionality constant is 5·75 when log Z is to base 10. Hence the relationship:

$$\overline{V} = 5\cdot75 \, V_* \, (\log Z - \log Z_0)$$
$$= 5\cdot75 \, V_* \log \frac{Z}{Z_0}$$

Using this relationship on anemometer measurements of wind velocity at series of heights above dune surfaces, Olson (1958 b) showed that the upper level of calm air near the surface (i.e. Z_0) was raised approximately thirty-fold from the bare sand value to the new value after the sand was planted with *Ammophila breviligulata* (Fig. 46).

A fairly low surface roughness (Z_0) around 0·03 cm seems to be a general characteristic of bare dune surfaces (Bagnold 1941, Landsberg and Riley 1943 and Olson 1958 b).

Now the threshold velocity for sand movement, 4·5 m/s must be exceeded at the 1 cm Z_0 level for sand to move from a vegetated surface. In fact Olson's results show that most dune-building vegetation reduces wind velocity at the sand surface well below this value, indeed to zero in an *Ammophila* plantation. As Olson points out we can see why most sand is trapped within a few metres of a vegetated edge. The effect is maintained by the fact that unlike other obstacles, dune-building vegetation regenerates surface roughness by growth.

Wind profiles over a dune

Ammophila keeps pace with sand accretion in the building phase as we have seen, but changes in dune shape themselves modify the overall air flow patterns. It is known from aerodynamics that laminar flow over an aerofoil crowds the streamlines and accelerates flow near the point of curvature. This produces the well-known half Venturi effect on which the lift factor of an aerofoil depends. Now an *Ammophila* dunelet or a mature coast dune, especially in regions where the prevailing and dominant wind is on shore, approximates an aerofoil shape and measurements over the profile show that wind profiles (Fig. 47) behave somewhat similarly to those over the aerofoil. The anatomy of a dune system with dunes and intervening slacks (Plate 12) is explicable in terms of these relationships.

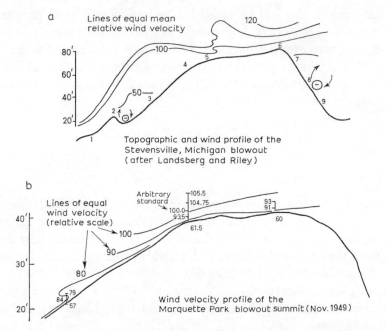

Fig. 47. Effects of topography on wind profiles over dunes. *a*, Michigan profile shows negative velocities behind small foredune ridge and large blowout dune, and crowding of high velocities very near dune surface at the upper slope (position 5); *b*, Indiana profile of present study (see table 1) shows similar crowding of high velocity near surface on windward slope. Velocities here are given relative to 3·34+ meter anemometer on the main tower, labelled 'arbitrary standard' (from Olson 1958 *b*).

Critical zones

Critical zones to bear in mind are (1) Protection in front of a big dune due to a 'stalling' effect (suggested by the spreading of the streamlines in Fig. 47); (2) Maximum wind velocity near the surface leading to maximum erosion near the crest of the windward face of the dune which according to exposure, wind and rain climate controls the height to which dunes can grow in any particular region; (3) the vortex behind the lee slope which creates calm for deposition in relation to winds flowing over the dune (and where non-prevailing winds have greater influence); (4) the extended shelter to leeward of the dune where a dune slack may be developed and (because the wind has deposited most of its sand load already on the lee slope) accretion is minimal and (5) the point beyond the shelter of the dune to landward (which is a function of the dune height) where higher wind velocities approach the sand surface carving down to non-erodable damp

1 *Arthrocnemum* marsh (sansouire) in the Camargue (Rhone delta). France. Note the open nature of the vegetation in this fully mature marsh in a climate where evapo-transpiration exceeds precipitation.

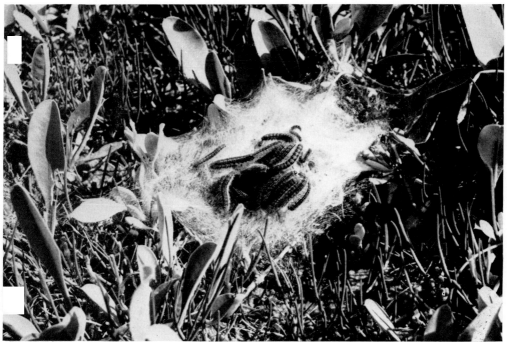

2 A 'nest' of caterpillars of the Ground Lackey moth (*Malacosoma castrensis*), one of the larger terrestrial invertebrates specifically adapted to living on salt marsh vegetation at the higher Emergence marsh level. This species has a very restricted distribution in Britain and is confined to ungrazed *Halimione* and *Limonium* marsh in south-east England. Havergate Island, Suffolk.

3 Pioneer growth form of *Puccinellia maritima* with stolons radiating onto open silt at Morecambe Bay, Lancashire. Sheep-grazing prevents flowering over most of these marshes and it is only in the more remote pioneer zones to seaward that sexual reproduction becomes important, and plants are more inherently variable. Note also the hummock form of growth which right at the start of marsh formation imposes local topographic point to point variation, later expressed as soil, moisture, and plant species variation in the fully colonized marsh.

4 Tidal litter (foreground on right) from *Spartina* marsh accumulates at the upper limit of the marsh and open up the dense *Spartina anglica* sward (foreground on left) allowing invaders like *Typha latifolia* (centre) to colonize. Bridgwater Bay, Somerset.

5 Oil pollution experiments on *Puccinellia maritima* marsh in the Burry estuary, Glamorganshire, South Wales. The plot shows the effect of chronic pollution with persistent applications of oil. Occasional pollution with oil has much less dramatic effects on the growth of salt marsh plants.

6 Sheep grazing experiments in progress at the upper limits of *Spartina anglica* marsh, Bridgwater Bay, Somerset. This is a critical transition zone where the *Spartina* marsh has reached maturity. Grazing favoured invasion by *Puccinellia maritima*, a short grass palatable to both sheep and wildfowl. The ungrazed marsh became invaded by tall growths of *Scirpus maritimus* and *Phragmites communis*.

7 Cattle walkway and flooded borrow pits in Gulf Coast marshland, Louisiana, U.S.A. These walkways are built so that man and animals can get in and out of marshes when they are flooded at high tide. The borrow pits from which silt has been dug to build the walkway provide habitat for wildfowl. (*Photograph by R. E. Williams.*)

8 Turf cutting experiment in *Festuca rubra* salting pasture at Bridgwater Bay, Somerset. This marsh has been undergrazed and sheep and wildfowl tend to avoid the tussocky *Festuca*. Turf has been cut to see if the succession can be put back to the earlier *Puccinellia* stage, as this grass is more palatable. Small amounts of *Puccinellia* are still present in the *Festuca* marsh, and in addition to the complete turf-cutting treatment, strips of marsh from which regeneration can take place are left in another treatment. The uncut control plots are also clearly visible. (*Photograph by P. G. Ainsworth.*)

9 Backshore zone of a sandy shore with abundant tidal litter. Unlike the foreshore (top left) which remains damp, the backshore sand readily dries out and it is from this zone that most shore sand is blown by the wind to feed dune vegetation (right). The tidal litter supplies nutrients for the growth of pioneer species *Agropyron junceiforme* and *Elymus arenarius*. Holy Island, Northumberland.

10 Coast dunes exposed to the maximum onslaught of prevailing and dominant winds at Morfa Dyffryn, Merionethshire, Wales. Once the dunes have grown to their maximum height they erode back from the shore. In the gaps so formed, strandline vegetation and embryo dunes start the cycle of dune building again. This shore is much trampled by tourists, but prickly Saltwort (*Salsola kali*) is avoided and scattered patches are developing in spite of the trampling.

11 *Salix repens* hummocks ('hedgehogs') at Newborough Warren, Anglesey. These very characteristic bio-topographic units are a common sub-habitat type in dunes in many parts of the world, although formed by different plant species. They are produced by alternating periods of accretion and erosion.

12 Dune slack colonized by *Salix repens* formed in the wake of a landward-migrating parabola dune (background) at Newborough Warren, Anglesey. Low-lying hollows like that in the foreground flood with fresh water in the winter.

13 Vigorous and free-flowering *Ammophila arenaria* among mobile dunes at Newborough Warren, Anglesey. This grass and the related *Ammophila breviligulata* are responsible for the build-up of dunes in many parts of the world.

14 De-pauperate growths of non-flowering *Ammophila arenaria* on stabilized dunes at Ross Links, Northumberland. The upright habit of growth persists but sand accretion has stopped, shoots topple and are very susceptible to damage from trampling at this stage.

15 A plantation of *Ammophila arenaria* protected by fishing netting pegged about 30 cm above the ground surface at Gullane dunes, East Lothian, Scotland. Sand movement at the surface is instantly stilled in the plantation. Nets prevent people trampling on sites planted to heal erosion scars and are visually unobtrusive.

16 A graded and hydraulically seeded dune at Camber, Sussex, showing good growth of introduced turf grasses. Hydraulic seeding is carried out by machine spray. Chopped straw is sprayed dry onto the bare sand to still the sand surface and provide a mulch. Seed and fertilizer are then sprayed on in a water mix. In this way a compensated environment for germination and growth is provided. Chestnut pale fencing protects the plantations and is an ideal permeable barrier for catching blown sand and helping to heal erosion gaps at the coast.

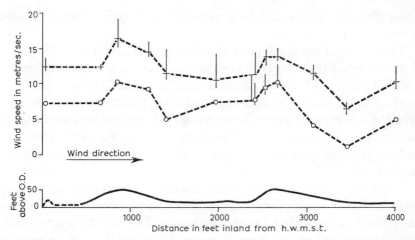

Fig. 48. Wind speeds recorded at 5 cm. (0—0) and 1m. (+—+) above the ground surface on T1 - T4 during a south-west gale of 40 knots (20 m/sec.) at Newborough Warren. The verticals on the curves show the range of speeds recorded at individual sites. The point where the dotted lines cut the verticals is the estimated average speed at the site (from Ranwell 1958).

sand just above the water table. The wind profile in Fig. 48 illustrates these relationships well.

Formation of the Dune and Slack System

Displacement of the coast dune

As we saw in Chapter 2, sand supply, orientation in relation to prevailing wind, and local topography, all greatly influence the type and extent of any particular dune system. But whatever type is considered, growth of the coast dune is initially in a linear or curvi-linear manner parallel with the strandline.

Here it may either stabilize at a low level, erode and re-cycle as in a small bay, or continue to accumulate in open coast sites where sand supply is abundant. One of two things can then happen to the coastal dune, either (1) a new coast dune forms to seaward e.g. where there are broad and high backshore levels and moderate onshore winds or (2) the coast dune grows to its maximum height and then erodes moving landwards e.g. where there are narrow backshore levels and prevailing and dominant winds are on shore.

The first or prograding type of system may simply add ridge on ridge which become stable *in situ* or, if there is more than one backshore

L

zone, alternate ridges and hollows which may again become stabilized *in situ.*

The second or eroding type of system (which of course may also develop from the first if the vegetation is destroyed in any way) may undergo centuries of instability before the sand is fixed permanently by vegetation.

Much of the earlier work on dune ecology in this country was concerned with the first type of system. Special emphasis is given here to the second or more plastic system because it has been the subject of more recent study and has much to teach us about problems of combating instability in dune landscapes.

As the coast dune builds up to the wind limit it takes on the characteristic form in relation to the prevailing wind which as we have seen approximates the streamline form of half an aerofoil section. In effect the vegetated dune has a relatively short and steeply sloping windward slope and a relatively long and gently sloping leeward slope. Now the non-vegetated barchan dune common in deserts has exactly the opposite. Taken together these facts imply that the natural tendency of a sand mound as big as a dune is to develop a shallow windward slope, but vegetation, and the growth of *Ammophila* in particular, modifies this tendency in the direction of a dune with a steep windward slope. Bearing in mind that the windward face of a vegetated coastal dune is rarely completely closed by vegetation it is clear that the vegetated coastal dune form is inherently unstable. This of course has been tacitly recognized in classical techniques of dune restoration where the aim is to create a more stable gently sloping windward face (see Steers 1964 p. 513 Fig. 108).

Seaward growth of the coast dune is restricted by the height of storm tides which can undercut the dune to form a near vertical seaward face. Once this condition develops in regions where there are strong onshore winds the coast dune windward face continues to erode sand which accumulates on the grass-covered leeward slopes. Thus the dune effectively moves back from the shoreline while still continuing to build to its maximum height. Ultimately the crest of the dune may reach a critical height in exposed areas where *Ammophila* can no longer hold the sand and the entire seaward face and crest becomes bare eroding sand. Whole coastal dune ridges may move landward in this way (Plate 10) and on the western coast of Britain where prevailing Atlantic winds are on shore this is a natural phenomenon which occurs quite independently of human disturbance.

The highest dunes are usually found some way inland in sites where prevailing winds are directly onshore (Oosting 1954; Willis *et al* 1959). On the Atlantic coast of France or Spain the inland dunes may reach 70 or

80 m in height. On coasts where the prevailing wind is offshore maximum heights are likely to occur at the coast dune (e.g. on the east coast of the British Isles at Strathbeg, Aberdeen).

Rates of dune building and travel

In seeking to understand complex phenomena it is especially valuable to study situations in which particular effects are maximized. For example the seasonal mechanics of salt marsh accretion (see Chapter 5), only became readily apparent from the studies at Bridgwater Bay where accretion was at a maximum. In just the same way, the mechanics of mobile dune system development can be more readily understood by studying a maximal erosion situation where the prevailing and dominant wind has uninterrupted flow to a coastline exactly normal to its direction. Such a situation occurs at Newborough Warren, Anglesey. Landsberg (1956) found perfect correlation at this site between a calculated wind resultant and the orientation of parabola or U-shaped dunes (Plate 12) of which many parts of it are composed. Subsequently the rate of dune building and dune travel in a region where entire dune ridges were moving landward successively were measured by means of repeatedly levelled transects at this site (Ranwell 1958).

The theoretical point of maximum erosion was confirmed and in this case occurred about 18 m to windward of the crest of 15 m high dunes. Zones of maximum accretion varied from 0 to 18 m behind the crest in low stable dune sections to as much as 164 to 183 m to leeward of the crest in high unstable sections.

It was calculated that the coastal dune must take at least 50 years to build

Table 19. Some recorded rates of dune movement (from Ranwell 1958).

Place	Rate m/annum	Authority
Inland		
Indiana	1–2	Cowles (1911)
Lake Michigan	2–4	Gates (1950)
Coastal		
Kurische Nehrung	5·5–6·1	Care & Oliver (1918)
Gascony	9·1 (mean)	
Wales, Morfa Harlech	3·7 (max.)	Steers (1939)
Morfa Dyffryn	6·1 (max.)	
South Lancashire		
Great Crosby	1·1	Salisbury (1952)
Freshfield	5·5–7·3	
Norfolk coast	1·5	

to maximum height and its mean rate of travel inland near the coast was estimated at 6·7 m per year. It would therefore take at least another 20 years or so for the coast dune to travel landwards sufficiently for a new embryo dune system to develop. So the cycle could take some 70 or 80 years to complete. Failure to understand such time factors may result in costly, unnecessary and undesirable attempts to stabilize a system which ultimately achieves its own stabilization.

Some recorded rates of dune movement are given in Table 19.

The water table and slack differentiation

The wind ceases to erode a bare dune surface when some underlying non-erodable surface such as shingle, rock, clay or wet sand is reached. Most usually, dune slacks have either a freely-drained shingle base or a damp sand base.

As Willis *et al* (1959) have pointed out, a big dune system perched on low lying ground acts itself as an isolated catchment. They showed that water percolated through the dunes at Braunton Burrows, Devon and accumulated over impermeable sub-surface deposits to form a dome-shaped water table (Fig. 49).

They point out that steeper water gradients at the margins of the system

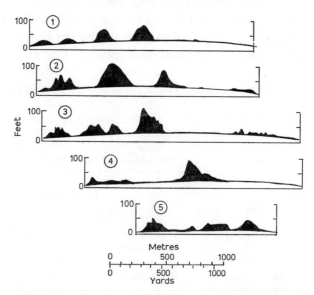

Fig. 49. Profiles across the dune system, Braunton Burrows, Devon. Heights obtained by survey are given in feet above O.D. (Newlyn). The sand above the water table of June 1952 is shown in black (from Willis *et al* 1959).

result from the fact that greater volumes of water (dependent on the greater catchment involved) percolate through the margins of the system compared with the centre of the system. They note the close correspondence of slack ground level with water table level and similar results were obtained at Newborough Warren, Anglesey (Ranwell 1959).

Parabolic dune units and cyclic alternation between dune and slack

It is rare for whole ridges to erode uniformly in the way described earlier for Newborough. Even on this system, ideally orientated for maximum uniform erosion, parts of the coast dune reach maximum height more rapidly than adjoining parts locally and this produces irregular erosion of ridges in the form of parabolic or U-shaped dune units (Fig. 50).

From the direct measurements of dune movement and studies of the growth and age patterns of *Salix repens* either side of mobile dunes, it was possible to demonstrate that there must have been cyclic alternation of dune

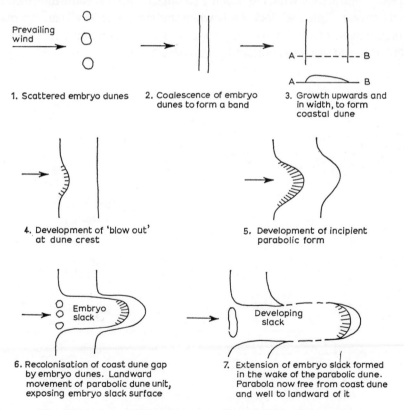

Fig. 50. The formation and development of a parabolic dune unit.

and slack at many points among the more mobile parts of the system. The period of this cycle was estimated at about 80 years and clearly corresponds to the period of the dune building cycle at the coastline (Ranwell 1960 *a*).

There is similar *a priori* evidence in N. American dunes as Oosting's (1954) graphic account testifies: 'When a dune moves over a forest the trees are buried as they stand and are preserved in the dune. Again and again "ghost" or "grave-yard" forests are reported where such long-dead trees have again been exposed as a dune moves on.' However there seems to have been no critical study of the dune and slack relationships in these American sites.

It is not implied of course that any particular point will alternate regularly, because once the dune ridge structure has become broken up complex local shelter effects play a part, and development becomes irregular.

Ultimately dunes are worn down and are sufficiently remote from coastal exposure and strong winds to become stabilized, often at some intermediate level between dune and slack. Only when this stage is reached can one really compare plant successional changes with those found on dunes that stabilize more or less *in situ* on a sheltered prograding coast.

9 Sand Dunes: Sand, Water Relations and Processes of Soil Formation

If initially the problem for growth of plants on salt marshes is too much water, for dune plant growth it is the reverse. Dune sand is not only remote from the permanent water table but also very low compared with silt, for example, in water-holding capacity. Plants and their dead remains increase the water-holding capacity of sand. Increasing organic matter increases the capacity of the soil to retain nutrients. This chapter concerns the properties of sand and especially its water relations. It is also concerned with the way sand is transformed into soil capable of supporting the many hundreds of plant and animal species found on dunes. The basic ideas on these subjects are still largely derived from the pioneer work of Salisbury (1952) and his results are quoted extensively. However, it is evident that on some of the more controversial points more data and more critical work with modern techniques is badly needed in both the study of water and nutrient relations on dune systems.

Properties of Sand

Particle size in dune sands

Salisbury (1952) concluded that the average particle size of dune sand has a general relationship to the average strength of the prevailing onshore winds, but this conclusion was derived from very limited data.

Subsequent studies (Cooper 1958 and Ritchie and Mather 1969) show that there is as much variation in either average or median particle size from site to site within a particular climatic region, or even within a dune system (Fig. 51), as there is from one major climatic region to another. The original source and sedimentation history of the shore sand of any particular dune system is probably of far greater significance in relation to particle size, than the wind strengths in the area. Ritchie and Mather (1969) define sand sizes in general terms as follows:

Median particle diameter in microns

Coarse sand	> 600 to < 1100
Medium sand	< 600 to > 200
Fine sand	< 200

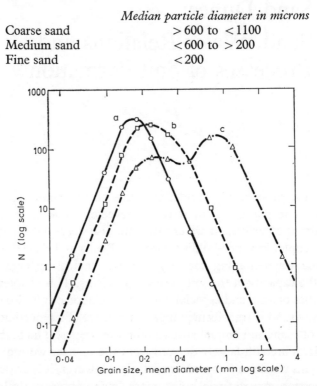

Fig. 51. Grading diagrams for sand samples from three different situations at Braunton Burrows. For each of the three samples the graphs show the proportion of sand having a particular grain size. This proportion is expressed by the parameter N, which is the percentage of the total sample which falls in a given fraction, divided by the difference between the logarithms of the maximum and minimum particle sizes in that fraction (*see* Bagnold 1941, p. 113 et seq.). The samples illustrated are as follows: (*a*) Fine sand from among *Ammophila* shoots on the lee slope of the main line of dunes. (*b*) Medium sand from the shore above high water mark. (*c*) Sand from the erosion face of a blow-out and containing two components, a coarse fraction left on the surface by wind sorting and a medium sand from below (from Willis *et al* 1959).

In their study of 16 dune sites in N.W. Scotland they found the median ranged from 192 to 496 microns and other studies confirm that most dune systems are composed of medium grade sand. At Newborough Warren, Anglesey particles of less than 20 microns (clay/ silt fraction) only contributed two per cent to the total. Particulate air pollution may well be augmenting this finer fraction on some of our dune systems now and this could well have far reaching effects on their water and nutrient holding capacity and consequently their flora.

Pore space

Measurements on dune soils at Braunton Burrows, Devon and Harlech, Merionethshire indicated a pore space around 40 per cent of the soil volume (Salisbury 1952). In the closest packing array the pore space between spherical particles would be 26 per cent of the volume. The difference is a measure of sand particle angularity and the extent to which particles are held apart by interstitial live and dead organic material. It is this organic matter which plays such a vital part in improving the water-holding capacity of sand.

Water Relations in the Dune

Field moisture capacity and availability

It is essential to keep in mind the changing weight to volume ratios which occur with soil maturation on a dune system. Old dune soils weigh only about half as much per unit volume as young dune soils.

Field moisture capacity may vary from 7 per cent (by volume) in young dune sand to 33 per cent in old dune sand (Salisbury 1952). In dry slack sand the field moisture capacity was between 25 and 30 per cent and in surface humus horizons of wet slacks as high as 50 per cent (by volume) at Newborough Warren, Angelsey (Fig. 52). Salisbury (1952) records minimum water contents of 1 per cent or less in dune sand and notes that water is no longer available to plants when it falls to values of 0·5 per cent, when wilting ensues. However, this does not necessarily mean the death of plants, for wilting cuts down the surface for evapo-transpiration and at least some dune plants can survive daily wilting (Oosting 1954).

Water content and plant requirements

The water content (by weight) to loss on ignition (corrected for carbon dioxide loss from carbonate) ratio is rather constant; usually from 1 to 2·5 (full range 0·15 to 3·48) Salisbury (1952). In general the water content of old dune soils is about twice that of young dune soils when the soils are at field capacity. This is clearly another indication of the changing weight to volume ratio referred to above and evidence that the water content is closely dependent on the organic matter content of dune soil.

The water content of the soil exploited by a plant of *Trifolium arvense*

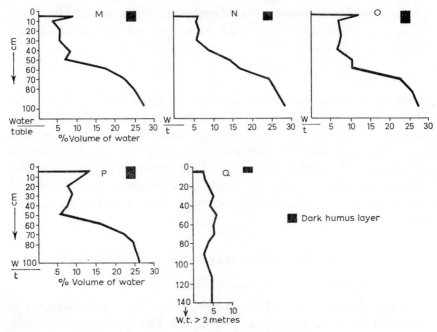

Fig. 52. Soil moisture profiles, Newborough Warren.
M & N beneath fixed dry slack *Agrostis tenuis* – *Festuca rubra* turf associes.
O & P beneath fixed dry slack *Salix repens* associes.
Q beneath semi-fixed dune *Salix repens* associes (Ranwell 1959).

would be used up in less than 4 days by evapo-transpiration in dry weather but the species can survive apparently unharmed on dunes up to 6 weeks without rain (Salisbury 1952). In general the water content at any one time in young dune soils is only enough for plants growing in it to survive for 2 to 5 days.

To resolve this apparent anomaly we must consider the distribution of soil moisture within the dune and the ways in which this moisture can be augmented other than by rainfall.

Soil moisture distribution within the dune

Several studies in open dune communities have shown that soil moisture augments to depths of about 60 cm below the dune surface and then tends to fall off to a more constant level (Salisbury 1952, Ranwell 1959 and Willis *et al* 1959). This is usually at about 1 m below the surface (Fig. 52). Live roots of plants characteristic of the open dune such as those of *Euphorbia portlandica* or *Ammophila* penetrate to depths of about 1 m, but not significantly below this.

In closed dune communities light rainfall is absorbed and held near the surface by organic matter to give a reversal of the soil moisture gradient characteristic of open dune soils. The water content of sand at depths between 60 to 90 cm in dry seasons is lower beneath dry dune pasture than beneath *Ammophila* on a high dune. In August 1955 at Braunton Burrows for example the values were 1 per cent and 4·9 per cent respectively (Willis *et al* 1959). However, it is still true that the establishment of seedlings in the surface layers of an open dune is very much dependent on the incidence of rainfall because the top few centimetres heat up and dry out daily in hot dry weather. In artificially stabilized mobile dunes where there has not been time for organic matter to build up it seems likely that serious moisture deficiencies might occur limiting the establishment of other species, and this would be worth investigating.

Sources of water for dune plants

Olsson-Seffer (1909) showed that the capillary rise of water from a free water surface even in very fine sand 30 to 50 microns particle size was not more than about 40 cm. The water table in a dune only 3 or 4 m high can therefore make no direct contribution to the moisture requirements of plants rooting to depths of only 1 m. In really high dunes the water table lies many metres below the surface and has no significance for plant growth at the dune surface.

The primary source of water for dune plants comes from rainfall, and in particular that proportion of it held as pendular water dependent on the moisture-holding capacity of the sand. But as Salisbury's studies have shown there must be some other source of water to carry plants through long periods of dry weather. Olsson-Seffer (1909) was the first to suggest a possible source: 'It must be remembered that the diurnal and nocturnal temperature variations are considerable on an open sand formation, on which the radiation factor is one of considerable moment. Such fluctuations in soil temperature . . . are sufficient to cause periodical condensation of water vapour in the soil.'

Salisbury (1952) demonstrated that the average water increment from dew was 0·9 ml per 100 ml soil per night in cloudless conditions, and transpiration measurements showed that this was sufficient to maintain plants exploiting that soil volume in rainless periods. Salisbury believed that warm moisture-laden air from above the adjacent sea after sunny days was drawn into the pore spaces of the sand 'as a concomitant of the upward convection currents maintained especially on the southern face and crest of the dune after dusk', where it was deposited on cold grains as internal dew.

Willis *et al* (1959) point out that at night the temperature gradient in the soil (Fig. 53) is favourable for an *upward* movement of water vapour from the warm and wetter layers of the sand below and this is so even when the upper layers do not fall below the dew point of the external air. The actual moisture contents found by Salisbury (Table 20) seem to preclude the possibility of drainage of dew from surface condensation. But as Willis *et al* (1959) point out, the greatest increase of moisture content after dew formation recorded by Salisbury (Table 20) occurred at a depth of 36 in (0·9 m) where the temperature is very unlikely to fall below the dew point. The problem clearly requires further study, perhaps with the aid of tracers, and one would like to know more about the distribution of moisture in the various components of the soil including that imbibed by micro-organisms which possibly undergo diurnal/nocturnal migration in the soil.

Table 20. Water content of sand after and before dew formation. Blakeney Point, Norfolk (from Salisbury 1952).

	Sample depth (cm)	Water contents by weight		
		27/7/38 Night samples %	27/7/38 Day samples %	Gain night-day
Single samples from the side of a pit	7	0·59	0·18	+ 0·41
	30	0·99	0·35	+ 0·64
	90	3·80	1·94	+ 1·86
Composite samples	7	0·71	0·53	+ 0·18
	30	1·31	1·18	+ 0·13

Water Relations in the Slack

Very little work has been published on the ecology of dune slacks or low-lying flat areas where growth is influenced by the proximity of the water table although they may occupy up to half the area of some dune systems.

Shape of the water table and drainage

As we have seen, the overall shape of the water table in a large isolated dune system is dome-shaped (Fig. 49). This means that peripheral slack communities are particularly likely to have nutrient enriched ground water derived from lateral seepage outwards from the centre of the dune system. It also explains why permanent dune lakes often occur at the landward side of the dune system where hinterland and dune system drainage meets.

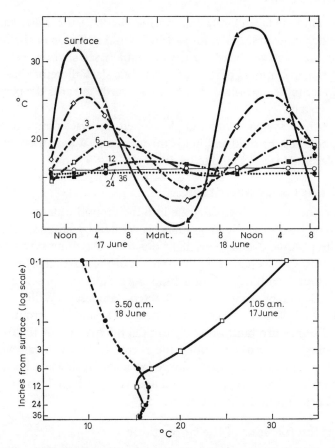

Fig. 53. Diurnal fluctuations in temperature down a sand profile. An undisturbed profile about 40 in (1 m) deep was exposed by digging a pit in the top of a bare dune. Resistance thermometers were inserted without delay into the profile at 36, 24, 12, 6, 3 and 1 in from the surface, and the excavated sand was restored as far as possible in its original position. After 2 days, periodic measurements of temperature were made, and are shown in the upper diagram. The temperature at the surface was recorded by means of a mercury-in-glass thermometer. The lower diagram shows the maximum extent of the fluctuation against a logarithmic scale of depth. Times are given in G.M.T. Braunton Burrows, Devon (from Willis *et al* 1959).

The ground surface closely follows the shape of the water table from slack to slack indicating the limit to which wind deflation of the dunes can occur. At the edge of a slack, the water table rises slightly, but does not follow the steep contours of the dune. Any one slack therefore has a saucer-shaped water table unit tilted slightly downwards towards the periphery of the dune system.

In most dune systems with well developed wet slacks impermeable clay or rock underlies the sand and holds the water up. Where they are built on permeable shingle slacks tend to be dry.

Tidal effects on the water table

Sea water does not readily penetrate into the ground water zone through a coastal dune (unless a reverse gradient is created by ground water extraction). The positive drainage gradient out from the dune system normally prevents this. Consequently, no fluctuation of the water table with respect to tide is found within the larger hindshore dune systems like Braunton Burrows, Devon or Newborough Warren, Anglesey (Willis et al 1959, Ranwell, 1959). Where a large part of the dune system is surrounded by sea water and dunes are built on shingle, tidal fluctuations of the fresh water table floating on salt water among the permeable shingle can occur (Hill and Hanley 1914). Brown (1925) showed that the fresh water body beneath small pervious islands capped with dunes is lens-shaped, and floats upon the convex salt ground water surface.

Seasonal fluctuations of the water table

In a wet year there may be widespread flooding in dune slacks at Newborough Warren, Anglesey from November to April. The water table falls from April to August and recovers the high winter levels during autumn rains. As Fig. 54 shows there is close correlation in this seasonal pattern with the distribution of rainfall. The rapid fall of the water table in April and May coincides with the leafing of deciduous plants like *Salix repens* (dominant in the slacks) and the autumn rise with the leaf fall in mid-October.

A ten year study of water level and rainfall fluctuations at Braunton Burrows, Devon enabled Willis et al (1959) to estimate the extent and duration of recent flooding at any point on the site from a knowledge of its land height and water level. They found that the slopes of lines for the regression of water level on rainfall were nearly proportional to the mean heights of the water table at each sampling point. They also calculated an index of flooding which could be related to different plant communities.

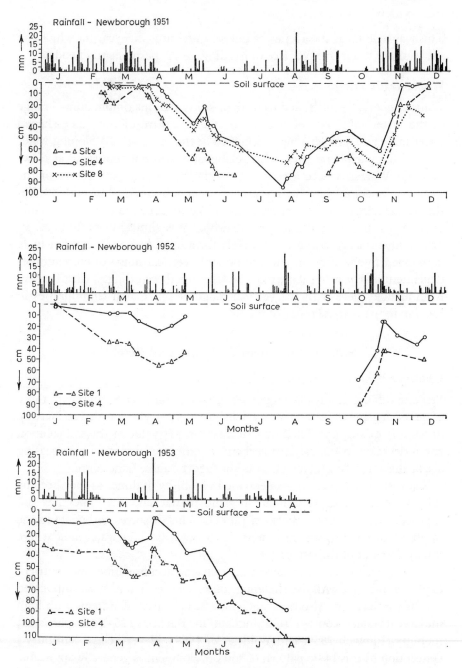

Fig. 54. Seasonal fluctuation of the free water table at selected sites in wet slacks with *Salix repens* associes, and daily rainfall 1951–1953 at Newborough Warren, Anglesey (from Ranwell 1959).

This was defined as the average numbers of months in a year for which the site is under water, from 0, free of flooding, to 12, permanently flooded.

Water table range

Willis *et al* (1959) concluded from their study that maximum annual range of the water table occurred near the flatter centre of the dome-shaped water table and smaller ranges occurred at the more steeply sloping water table of the peripheral sites. This relationship was not confirmed at Newborough Warren where a rock ridge running through the centre of the system may complicate the pattern of water movements. In both studies the annual range of the water table was of the order of 1 m and this probably depends on the total annual rainfall which was similar in both areas. In drier climates as at Winterton, Norfolk the annual range of the water table is reduced to about 0·5 m and the possible vertical range of communities dependent on the water table must be telescoped accordingly. Clearly there is an analogy here with the vertical range of salt marsh communities in relation to the vertical range of the tide.

Soil Development in Dunes and Slacks

Changes in soil properties with time

Striking differences in soil properties of young dunes compared with those of increasing age were detected by Salisbury (1925). He recorded an increase in organic matter paralleled by a decrease in pH value of the soil solution and a decrease in the calcium carbonate content of dunes with increasing age at Blakeney Point, Norfolk and Southport dunes, Lancashire.

These trends depend upon decrease in mobility of dunes as they become fully vegetated and accumulate organic matter and the time they are exposed to the leaching effects of rainfall. This dissolves carbonate in the soluble bicarbonate form and carries it downwards to the water table where it may flow out of the system.

The opposing agencies favouring nutrient accumulation and nutrient depletion interact with the changing soil properties in a rather complicated way as we shall see. Also the trends may be obscured in systems with high initial carbonate content (Gimingham in Burnett 1964), or persistent mobility (Ranwell 1959, and Willis *et al* 1959). Nevertheless, Salisbury's conception of this basic pattern of soil development is generally applicable to dune systems wherever they occur.

Differences in soil moisture in young and old dune and slack soils have already been discussed. Right from the start this effectively controls the oxidation of organic matter so that young slack soils start with twice the amount of organic matter that comparable aged dune soils contain (Ranwell 1959). In addition, gravity favours the accumulation of nutrients in the slacks at the expense of the bank of nutrients in dunes. As Olson (1958 *a*) succinctly puts it, 'the rich system gets richer and the poor system poorer', at least until wind deflation brings the two systems together so closely at the dry slack level that the systems interact as in a chalk heath situation where shallow-rooted plants live in an acid soil side by side with deeper rooted plants tapping base rich soil below (Fig. 55 Profile EF).

Particle size changes

Perhaps because the initial silt and clay contents of dune sand are so low as to be almost unmeasurable, little attention has been given to changes in the finest fractions of dune and slack soil. However Olson's (1958 *a*) work on the ancient inland dune systems of Indiana, U.S.A., has shown that these finer particle fractions do accumulate with time as a result of weathering *in situ* and accumulation of airborne dust. The exchange capacity of the sand will be increased by clay accumulation but is of course decreased by hydrogen ion replacement as leaching proceeds with age.

Rates of change in carbonate content

The evidence from a number of dune systems with an initial carbonate content of not more than 5 per cent (by weight) shows that most free carbonate is lost from the first decimetre of surface dune soil within 300 to 400 years (Salisbury 1952, Olson 1958 *a*, Ranwell 1959). Under the prevailing weather conditions, Olson (1958 *a*) found that it takes about 1,000 years for carbonate to leach out of the first 2 metres of dune soil in the Indiana dunes. The rate of carbonate loss was found to be proportional to the amount of cabonate left at any point in time. Salisbury (1952) points out that calcium may be replenished at the surface in three ways: by wind blow of shell fragments from mobile dunes, by re-cycling via plant roots and leaves, and by the burrowing activity of rabbits. Ultimately however it seems from Olson's studies that leaching wins in the end, even where forest cover is developed as on the Indiana dunes. Similarly Ovington (1950) found in studies of the afforested Culbin dunes that nutrients are lost from the soil at a greater rate than they are being made available. He points out that if an allowance were made for the nutrients in the trees there would be an overall increase of nutrients in the afforested areas, but these of course will be removed at felling.

M

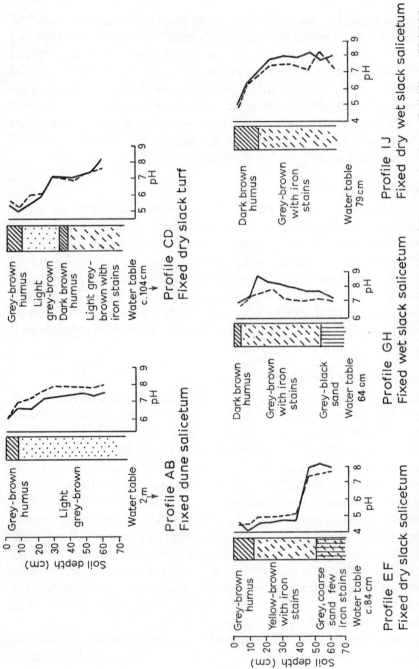

Fig. 55. Soil profiles and pH values found in the most mature soils of Newborough Warren, Anglesey. Solid lines pH (KCl), broken lines pH (Water) (from Ranwell 1959).

Rates of organic matter and nitrogen changes

Salisbury (1925) found that organic matter augments slowly at first, but appreciably faster after about 200 years in the dunes especially in the higher rainfall climate at Southport, Lancashire.

In contrast Wilson (1960) found very rapid increase in organic matter in the very lime-deficient dunes at Studland, Dorset. This he attributes to early invasion by *Calluna* which is largely responsible for the rapid litter accumulation, and possibly also accelerated leaching of what little carbonate there is present initially, by means of humic acids as well as carbonic acid. Under the rapidly developing acid conditions litter breakdown is inhibited and organic matter accumulation promoted (Fig. 56). Optimum organic matter accumulation was found at the limit of winter flooding at Newborough Warren. Olson (1958 *a*) found that organic carbon increases about three times faster at the surface compared with 10 cm depth on the Indiana dunes. Even in soils believed to be about 500 years old not more than 2 per cent organic matter accumulated at depths of 20 to 25 cm at Newborough Warren (Ranwell 1959).

Carbon to nitrogen ratios were found to vary from 10:1 in young dune

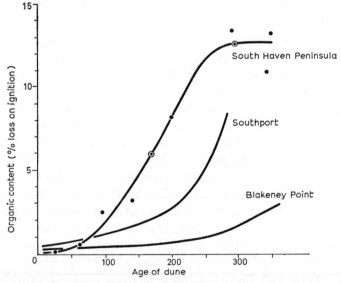

Fig. 56. Changes in the organic content of dune surface soil (0 to 5 cm depth) with age at South Haven Peninsula, Dorset. The ringed points represent overall means for the second and third dune ridges from the coast. Salisbury's (1925) curves for the Southport, Lancashire and Blakeney Point, Norfolk dune systems have been inserted for comparison (from Wilson 1960).

grassland soils, to 20:1 in old forest dune soils in the Indiana dunes (Olson 1958 *a*). A balance between gain and loss of nitrogen seems to be reached at about 1000 years following stabilization of the dune surface.

Cation exchange capacities

The cation exchange capacity of the slack soils at Newborough rose from 2–3 m E per cent in young dry slack soils to 15 to 20 m E per cent in older wet slack soils (Ranwell 1959). Olson (1958 *a*) found low values in very ancient dunes at Indiana: 5·3 m E per cent in 8,000 year old dunes and 6·5 to 8 m E per cent in 10,000-year-old dunes. In contrast Ovington (1950) found quite high values in the humus layers at Tentsmuir, Fife up to 25–30 m E per cent and these were probably due to the presence of *Calluna* litter, the dominant species in the area examined.

Biological influences on soil formation

The initial organic matter supplied to the shore in tidal litter forms a relatively rich nutrient medium for the growth of strandline plants. It is processed by temporary faunas including abundant amphipods and mites (McGrorty *pers comm.*). Brown (1958) found little evidence of fungi in tide-washed sand, but Gunkel (1968) has recently demonstrated that oil-decomposing bacteria are present in shore sand and pollutant products decomposed by bacteria may well be enriching not only the shore, but also the sand being blown onto coastal dunes at the present time.

Prior to the advent of serious shore pollution, the switch from the relatively rich tidal litter-based nutrient economy of the shore to the nutrient deficient high level *Ammophila* dune sand must clearly have presented serious nutritional problems for the inhabitants of the open dune community. Webley *et al* (1952) developed a technique whereby the microfloras of the open sand, the rhizosphere, and the root surface of *Agropyron* and *Ammophila* could be compared. They showed that the rhizosphere of these sand dune plants contains abundant bacteria which are present only in very low numbers in intervening sand areas. Hassouna and Wareing (1964) have demonstrated by experimental cultures that nitrogen is limiting for the growth of plants of the open dune and that the growth of root-surface sterilized *Ammophila* seedlings exhibited much poorer growth than those inoculated with bacteria capable of fixing nitrogen (providing carbon sources, possibly derived from exudates, are adequate). The regular association of fungi e.g. *Inocybe* species with *Ammophila* suggests that mycorrhizal symbiosis may also be important in the supply of nitrogen.

Webley *et al* (1952) also demonstrated that the bacterial flora of the early stages of dune fixation increased to the dune pasture stage, but decreased markedly where dune heath was developed. At this stage the fungal flora, which like the bacterial flora augmented with increasing fixation, became more abundant than the bacterial flora (Table 21). However, it should be borne in mind that the values are given on a weight basis.

Table 21. Microbiological analyses of dune soils on a transect through the Newburgh dunes, Aberdeenshire, 8 September 1948 (from Webley *et al* 1952).

Sample (depth 5–15 cm)	Moisture content (% of dry wt. of sample)	pH	Bacteria per g oven-dried material	Fungi per g oven-dried material
(1) Open sand	3·5	6·80	18,000	270
(2) Yellow dunes, *Ammophila arenaria*	3·7	6·68	1,630,000	1,700
(3) Early fixed dunes *Ammophila arenaria* with grasses, etc.	1·5	5·14	1,700,000*	69,470
(4) Dune pasture	7·1	4·80	2,230,000	109,780
(5) Dune heath	16·8	4·27	127,000	148,190

* Just over 50 % were actinomycetes.

Gimingham (in Burnett 1964) notes that legumes increase in the later stages of stabilization and shrubs like *Hippophaë rhamnoides* also develop root nodules with bacteria which have been shown to be capable of nitrogen fixation (Bond *et al* 1956).

We are still very ignorant of the nutrient pathways in sand dune habitats, but it does seem evident that there is a distinct switch from the bacterial-based economy of the younger or more base-rich dune soils to a fungal-based economy on the older or more base-deficient soils dominated by *Calluna* or conifer plantations where litter accumulates. No information seems to be available on the biological influences on soil formation in dune slacks, but those that flood late into spring have an additional source of production from algal blooms contributing to their enrichment. Shields (1957) has demonstrated high levels of amino nitrogen in algal lichen surface crusts compared with sand at 15 cm depth in New Mexico desert soils. Similar crusts formed by congeneric species occur in young lime-rich dune slacks and probably engender a continually renewable supply of soil nitrogen. In fact it has been demonstrated with the use of labelled nitrogen that blue-green algae in dune slacks fix atmospheric nitrogen and that such labelled nitrogen is taken up by mosses and higher plants from the soil and assimilated (Stewart 1967).

10 Sand Dunes: Structure and Function of Dune Communities

Dunes are immensely rich in species of plants and animals in contrast to salt marshes. Unlike a salt marsh with exacting constraints imposed by salt-dominated relatively level ground, dune habitats are highly variable. They range from mobile to fixed dunes of varying aspects and from ground-water dependent, to ground-water independent levels. This complex mosaic of habitats is further diversified at the surface by the plant and animal components which both inhabit and help to create the environment in which they live.

Ecologists were rather mesmerized by this wealth of diversity and, until recently devoted much of their time to lengthy descriptive studies and rather little to behavioural study and experiment. As Elton (1966) puts it, 'One wishes that some of the very elaborate European investigations of dune faunas had given less time to statistical abstractions and cryptic classical terminologies, and more to finding out what the animals do and where they do it, in relation to the rather obvious patterns of habitat structure and climate.' This is equally true of the study of dune floras and dictates limits to the accounts of the dune and slack communities given in this and the next chapter. The emphasis is therefore primarily on structure and rather less on function, but it is hoped that the accounts will encourage future study on function of the component parts of the dune system, such as the ecology of the *Ammophila* tussock, the *Salix repens* hummock, or *Hippophaë* scrub.

The Dune Flora and Fauna

Species diversity

Taking British dune systems as a whole, i.e. dune and slack complexes, Salisbury (1952) notes there are about 400 species of vascular plants. It is clear that this represents only the truly native flora, for a recent tally of native and introduced vascular plants on 43 of the more important dune systems in Great Britain shows that over 900 species occur on them. Many of these additional species are associated with the forestry plantations developed on dunes over the past 100 years. More than half the species of vascular plants growing on British dunes at the present time were probably introduced there directly or indirectly by man or by birds (especially gulls, see below). It has always been accepted of course that the dune flora contains a large element derived from the weed flora of agricultural land. This element of heterogeneity from site to site in the dune flora presents classification problems for classical phytosociologists. Floristic assemblages are often as unstable as those on a rubbish dump. Bearing out Darwin's conclusion that the indigenous plants of a district are not necessarily those best suited to it, Westhoff (1952) points out that no less than 31 species in the *Hippophaë – Ligustrum* community on Dutch dunes are exotics.

There are in addition several hundred species of lichens, bryophytes, fungi and algae on our dune systems and no doubt an equally rich microflora. It would be a formidable task to develop floristic classification of the dune floras of the world let alone the fact that species have not even been listed for such major European dune systems as the Coto Doñana in Spain for example.

No figures are available for the animal species found on a whole sand dune system. Heerdt and Morzer Bruyns (1960) found 368 species of arthropods in the open dune communities on the island of Terschelling off the Dutch coast. Hincks *et al* (1951–1954) recorded over 2,000 species of invertebrates (between one fifth and one quarter of the British fauna) on the sand dune and salt marsh complex at Spurn Head, Yorkshire in an area of less than 1 square mile (less than 200 ha). Duffey (1968) found no less than 188 species of spiders alone in systematic sampling on the lime-rich dunes of Whiteford Burrows, Glamorgan and the more lime-deficient dunes at Tentsmuir, Fife.

He points out that the relatively low total of only 54 species of spiders found on Terschelling by Heerdt and Morzer Bruyns (1960) is partly due to the fact that the collections were limited to a particular time of the year (August) and partly to the limited area within each vegetation type sampled. The sampling intensity is particularly important, because even with spring and summer sampling at Whiteford Burrows, Cotton (1967) found only 25 per cent of Duffey's total of spider species for that site.

Coleoptera, Hymenoptera and Diptera seem to be the best represented insect groups in the dune habitat.

Communities

Biological spectra

Raunkiaer's system of life form groups as modified by Braun-Blanquet (1932) is based on the position of the perennating organs during the unfavourable season, i.e. during cold winter or hot dry summer, and it is a useful way of analysing complex floras.

Böcher (1952) gives results for a series of sand ridges on shingle at Isefjord, Denmark (Table 22). The flora of dunes and slacks were analysed separately at Newborough (Table 23).

Both analyses clearly emphasize the dominance of the hemicryptophyte habit i.e. plants with perennial shoots and buds close to the earth's surface, and this is reinforced by the fact that the abundant cryptogams of dune systems also are to be classified as hemicryptophytes. Also notable is the importance of therophytes (annual species) in the open habitats of the early stages of succession in Böcher's figures and the high preponderance of this group in both dunes and slacks at Newborough, a west coast dune system with a high degree of mobility and hence extensive areas of open ground.

Nobuhara (1967) has studied the way in which proportions of life form types change zonally and seasonally in the Japanese strand flora.

Analytical studies

Much of the European literature on sand dune ecology is dominated by subjective classification of vegetation types which appear relatively homogeneous to the observer. Often they are accompanied by somewhat arbitrarily chosen measurements of habitat factors, but the information contained in them is not readily applicable to functional aspects of the vegetation groups or usable in relation to modern management requirements. More recently their value as a basis for vegetation mapping, which as Westhoff (1952) points out is of value to the dune landscape architect in enabling him to see what will grow where and maintain itself in competition with existing species, has been demonstrated by such studies as those of Martin (1959) in America and Boerboom (1960) and Maarel and Westhoff (1964) in Holland. At the same time attempts are being made to draw together classifications of dune associes and selected environmental factors for dune systems in several European countries (Wiemann and Domke 1967).

More objective methods of purely floristic analysis such as the association analysis developed by Williams and Lambert (1960) have been applied to dune floras in Holland (Maarel 1966), in Australia (Welbourn and Lange

Table 22. Biological spectra of dune ridges at Isefjord, Denmark.
Figures refer to numbers of species in each group (from Böcher 1952)

	Phanerophyte	Chamaephyte	Hemicryptophyte	Geophyte	Therophyte
Strandline	0	0	33	18	48
Coast ridge	3	0	56	19	22
2nd ridge	8	0	68	11	13
3rd ridge	12	2	65	12	9
4th ridge	10	9	70	5	5
5th ridge	17	5	63	5	9
Landward dune	11	7	64	7	11
Heath	13	13	70	4	0

Table 23. Biological spectra of dunes and slacks at Newborough Warren, Anglesey.
Figures refer to numbers of species in each group (from Ranwell 1959).

	Phanerophytes	Chamaephytes	Hemicryptophytes	Geophytes	Helophytes	Therophytes	Total species
Dunes	4	11	40	5	0	40	122
Slacks	0	9	57	9	9	16	109

1969) and in England (Moore 1971). In addition Maarel (1966) has sought to combine classification and ordination techniques and to correlate directly vegetation and environmental types (analysed separately) in a unified system.

Study of the Island beach dune spit in New Jersey led Martin (1959) to conclude that the dune system consisted of a zoned mosaic dominated by topographically determined environmental features. Topography and vegetation were interrelated and interacting. Most community types could occur on more than one topographic facet and topographic facets could support more than one community type.

The hummock is a good example of a biotopographic unit which maintains integrity as a habitat unit in dune systems in many parts of the world. Thus Osborn and Robertson (1939) writing of New South Wales dunes, Australia note 'that mat plants (*Mesembryanthemum aequilaterale, Scaveola suaveolens* and *Stackhausia spathulata*) may persist, building the sand into rounded hummocks a metre or more in height, long after wind erosion has removed the low dune on which they had been growing'. Ramaley (1918) describing San Francisco dunes in California, U.S.A. notes that, 'Distributed throughout the dune area are many small mounds 2 to 4 m high capped with willow (*Salix lasiolepis*).' Yano (1962) in Japan speaks of clonal hummocks in unfixed dune, and Duffey (1968) notes that at Whiteford Burrows, South Wales *Salix repens* hummocks 2 to 3 m high form small steep-sided dunes which have earned the local name 'hedgehogs' (Plate 11).

The point to be made here is that clonally based pattern is of such universal significance as a module of plant and animal community structure that it must become more widely utilized in the study of dune ecology if functional aspects are to be successfully unravelled.

Duffey (1968) points out that 'phytosociological concepts are of limited value to the zoologist studying animal associations on dunes and habitat classification is proposed based mainly on structural characteristics of the vegetation cover' (Fig. 57). He finds that the more distinctive spider faunas

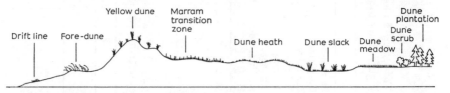

Fig. 57. Diagrammatic representation of habitats on a dune system (from Duffey 1968).

are associated with the more specialized open dune and dune heath habitats compared with other dune habitat types and notes close similarity in dune heath and inland heath spider faunas.

Ardö (1957) found that the dipterous fauna of sand dunes was 'overwhelmingly rich'. He distinguished a eurytope group of species and a stenotope group. In the latter a close relationship was demonstrated between thermal preference and choice of zone. Drought resistance and light preference were also important.

The main factors affecting distribution of invertebrates include: mechanical damage by flying sand particles; physical injury by desiccation and salt content of onshore wind; shelter; and food according to Heerdt and Morzer Bruyns (1960). They note that few species can survive seaward of the main coast dune ridge and that this is therefore a principal boundary for invertebrates of the open dune areas they studied.

Primary biotopographic units of the dunes

The primary biotopographic units of dunes are those associated with orientation, particularly in regard to incidence of direct insolation and incidence of dominant winds. Boerboom (1964) as we saw in Chapter 1, has recorded the microclimatic differences due to aspect of dune slopes. In many continental studies plant community analyses are related to aspect in a general way. However, we are a long way from understanding precisely how *Empetrum nigrum* becomes dominant on a shade slope and how its establishment seems to be prevented completely on a sun slope as on Terschelling, for example. Primary topographic units include:

(1) Dune plateaux
(2) Solar slopes
(3) Shade slopes
(4) Windward slopes
(5) Leeward slopes
(6) Intermediate slopes
(7) Dune hollows

The ecology of any one of these major units would be worth studying on its own in relation to the zonal gradients of a dune system.

Secondary biotopographic units of dunes

Superimposed on the plateaux or orientated slopes of dunes are secondary units ranging from physiographically dominated units like blow-outs, through biotopographic units like tussocks, and hummocks, gulleries, warrens, burrow mouths, ant hills and pathways to patterned plant shape and height arrangements which themselves create diverse topography on otherwise level ground.

Once again one would like to see studies orientated more specifically on these individual secondary biotopographic units of the dune system and already the trend towards this type of study is starting. For example, Ardö (1957) studied the microclimate of *Elymus* dunes in Scandinavia in relation to Diptera, Heerdt and Morzer Bruyns (1960) the microclimate of *Ammophila* tussocks in Holland in relation to the fauna, while Elton (1966) gives us a fascinating insight into the oasis-like character of this habitat from observations on the Daymer dunes, Cornwall.

The clonal hummock habitat is of particular interest because the plants that create it frequently can only originate in a slack where there is adequate moisture for germination. Subsequently they keep pace with accretion from the advancing dune and ultimately come to lie at the dune summit (e.g. *Salix repens* at Newborough Warren, *Populus alba* in the Coto Doñana, Spain). This fundamental form has received very little specialized study and offers particularly attractive opportunities for comparative study in discrete units in a great variety of orientations, vertical heights, and horizontal zones from seaward to landward.

Animal-controlled biotopographic units are just beginning to receive detailed study. Work is in progress for example on the South Walney Island dunes, Lancashire where the largest Lesser Black-backed Gull colony in Europe has transformed the dune vegetation over considerable areas from open *Ammophila* dune to a distinctive lush carpeting growth of annual weed species such as *Stellaria media*. Trampling by gulls has been observed to destroy robust *Ammophila* tussock growth at Newborough Warren, Anglesey but to so enrich the dune sand with guano locally that *Dactylorchis* species (normally confined to slacks), grow in lush turf among the high level dunes. Gillham (1964) records that the commonest means of arrival of viable weed seeds in off-shore island sites with gulleries was in the pellets which gulls eject from their crops. Single plants of weed species such as *Brassica sinapis*, *Plantago lanceolata* and *Stellaria media* were found on perching hummocks where regurgitation pellets also occurred at Newborough Warren.

The activities of grazing animals (especially rabbits) on dunes have not yet received the systematic study they deserve. We are aware in a general way of the profound effects that rabbit grazing has had on the dune flora, especially following the striking changes that have occurred since myxomatosis. But we do not have any critical experimental data and nor has there been any serious attempt to relate specific population densities of rabbits to plant community types. Rabbit population density and grazing intensity varied enormously from site to site on any particular dune system prior to myxomatosis. Intensively warrened areas were particularly associated with the better vegetated more landward parts of the dune system which are frequently worn down to a level intermediate between the dune and dry slack level. On larger dune systems with extensive areas of highly mobile dunes to seaward, rabbit density and grazing intensity was much reduced in these seaward parts. On smaller more stable systems such as that at Blakeney, Norfolk, rabbit grazing evidently did occur extensively at the coast and White (1961) records much improved growth and flowering of *Agropyron junceiforme* in embryo dunes following loss of rabbits from myxomatosis. The general changes he recorded on transect studies at Blakeney are similar to those found in dry slack transect studies at Newborough Warren, Anglesey described in the next chapter.

Both grazing and wind exposure have profound effects on the structure of vegetation of dunes and it is useful at this point to consider the vertical layering of vegetation.

The Vegetation Layers

Boerboom (1960) points out that deviations in the correlated occurrence of the respective vegetation layers may lead to anomalies in classification and suggests that more attention should be given to their separate analysis. It is interesting here to note that no correlation was found between the distribution of species of filamentous algae on a salt marsh and the species of higher plants in a recent (unpublished) study of a Norfolk coast marsh. Stewart and Pugh (1963) found a similar lack of correlation between blue-green algae and the dominant flora at Gibraltar Point, Lincolnshire. Similarly one would expect that shade tolerant bryophytes with wide edaphic tolerance might well prove to be indifferent to what species of plant provides the shade so long as it has the right structure. Experimental work with artificial structures could soon clear up this point. Such studies are incidentally of direct relevance to the colonization of erosion scars after their treatment with dead brushwood.

The ground carpet layer

One of the most noticeable features of dune vegetation in the later stages of stabilization is the abundance of bryophytes and lichens. There has been very little study of the growth of dune lichens, but interesting work has been carried out on bryophytes. Richards (1929), and Ducker (in Steers 1960), demonstrated their zonation at Blakeney Point, and Scolt Head Island, Norfolk respectively, and Birse and Gimingham (1955) have studied growth form in relation to increasing stability. Pioneer species like *Tortula ruraliformis* on open sand are of the acrocarpous type with an upright growth habit capable of growing through small amounts of accreting sand. In more stable areas pleurocarpous mosses with a spreading habit of growth like *Hypnum cupressiforme* are found.

In moss transplant experiments (Birse *et al* 1957) there was no emergence of 8 species transplanted into *Ammophila* zones subject to 4 to 5 cm sand accretion in 10 months. In more stable *Ammophila*, 3 species emerged through 3 cm accretion and only where accretion was less than 1 cm in 10 months did all 8 species survive. In pot burial experiments several species emerged through 4 cm burial and none from deeper burial. The ability to produce rhizoids in overlying sand and the rapidity of upward growth were important in the ability to survive burial.

Robertson (1955) demonstrated the importance of shade by transplant experiments and found, with the exception of two species which were indifferent, the order in which transplants to a sunny site died out was in the same order as the level of shading in which they normally grew. Northfield (1968) found that pioneer bryophytes at Studland, Dorset were always in well protected sites associated with other plants and only began to move out into open sand with the invasion of *Calluna* which occurs at an early stage on this lime-deficient dune system. Lichens tend to become more abundant in dry rabbit-disturbed ground than bryophytes. With the general growth of grasses since myxomatosis, bryophytes at first increased in the formerly lichen-rich areas, but are now decreasing as shade increases and litter isolates them from the sand surface.

Alvin (1960) has recorded the lichen zonation on Studland dunes (Fig. 58) and compares his results with Böcher's (1952) study of lichen zonation on the island dunes at Laesø, Denmark. Differences were believed to be primarily due to climatic factors as both are lime-deficient dune systems. Brown and Brown (1969) found that lichens at Blakeney, Norfolk were well represented in arid areas less favourable to higher plant growth and Robertson (1955) found that on dry eminences at Ross Links, Northumberland, lichen mats break into polygons especially where turf death attributed to drought due to rabbit burrows close to the surface occurs. Such areas were especially prone to erosion.

Dune annuals are a very characteristic element of the dune flora occurring on the ground layer among lichens and especially in the sheltered mouths of old rabbit burrows. Salisbury (1952) has shown that these are

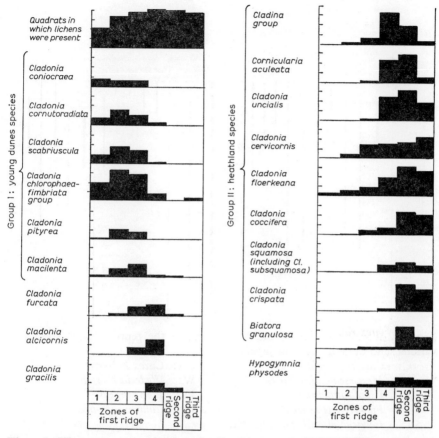

Fig. 58. Histograms representing the distribution of the main lichen species in six zones of the dunes as determined by their occurrence in 100 random quadrats in each zone. One division of the vertical scale represents 25 occurrences. The top left histogram shows the number of quadrats in which any lichens occurred. South Haven Peninsula, Dorset (from Alvin, 1960).

shallow-rooted plants which mostly germinate in autumn and are able to grow in the warmer days of winter at a season when moisture is adequate. Lichen rich sites are 'hot-spots' and winter temperatures are adequate for growth. In summer the winter annuals pass the unfavourable season in seed.

The grass/herb layer

It is of particular interest that in addition to many of the hummock species, such dominant plants of the dunes as *Carex arenaria* can also only establish from seed in damp hollows. Unlike the hummock formers which

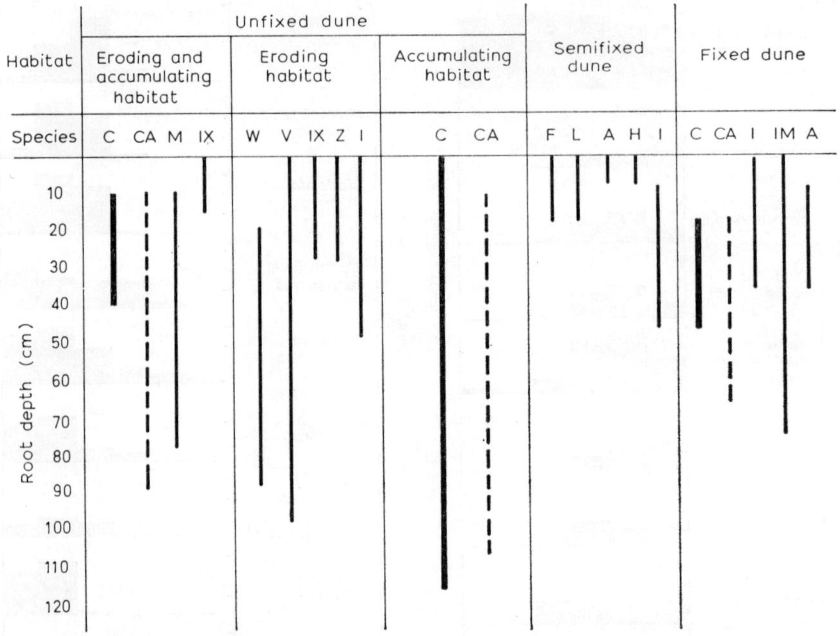

Fig. 59. Working ranges of the root system of the representative species in various sand dune habitats in Japan.

C: *Calystegia soldanella*
M: *Messerschmidia sibirica*
V: *Vitex rotundifolia*
F: *Fimbristylis sericea*
I: *Ischaemum anthephroides* var. *eriostachyum*
IM: *Imperata cylindrica* var. *koenigii*
A: *Artemisia capillaris*

CA: *Carex kobomugi*
W: *Wederia prostrata*
IX: *Ixeris repens*
L: *Linaria japonica*

Z: *Zoysia macrostachya*
H: *Heteropappus arenarius*

(from Yano 1962)

have rather limited horizontal, as opposed to vertical powers of growth, this species spreads widely onto dunes by means of rapidly extending horizontal growth. In relatively dry systems like that at Holkham or Blakeney, Norfolk, *Carex arenaria* may well be of uniform clonal origin over very large areas. It would be interesting to apply Harberd's (1961) technique of clonal analysis to see if this can be demonstrated. Certainly White's (1961) observation that *Carex arenaria* is holding its own, but not spreading to new areas at Blakeney Point since myxomatosis implies problems of establishment.

Those biennial or perennial species which do regenerate at the dune level to become part of the herb layer like *Euphorbia paralias* or *Cyno-*

glossum officinalis tend to be large seeded and capable of very rapid root extension down to moist layers at 15 cm depth within a week of germination. Thus their occupancy as seedlings of the tricky arid ground layer is cut down to a minimum. Yano (1962) has studied root formation of Japanese dune plants in relation to depth, root area, morphology, and propagation across the dune zones. He found that plants with well developed rhizomes rooting deeply over a wide area, formed large clones on mobile dunes. On fixed dunes plants had poorly-developed rhizome systems, were generally shallow-rooted over more limited areas, and formed small clones (Fig. 59).

The striking change in the growth form of *Ammophila* species from the vigorous close tussock habit of mobile dunes (Plate 13) to the sparse depauperate isolated shoots of fixed dunes (Plate 14) has been variously interpreted as due to drought, mineral deficiency, toxicity or senescence (Marshall 1965). Olson (1958 *c*) was the first to suggest that normal internodal elongation in *Ammophila breviligulata* on a stable surface raises the meristem into surface dry sand which prevents normal development of adventitious roots. However Laing (1954) produced experimental evidence that mineral deficiency and toxicity were not associated with decline in vigour of *Ammophila breviligulata* and found that sand burial alone was sufficient to restore vigour. Marshall (1965) reached similar conclusions (see also Hope-Simpson and Jefferies 1966) with regard to decline of both *Ammophila arenaria* and *Corynephorus canescens* (Fig. 60). In a nice demonstration using split root-culture technique he found that new adventitious roots in *Corynephorus canescens* were more efficient in water and nutrient uptake than nine-month-old roots. The interest of these studies lies in the emphasis they give to inherent morphological, growth, and senescence characteristics of dune dominants in relation to successional changes of plant pattern. This tendency for *Ammophila* to literally grow out of the ground in stable dune surfaces is well illustrated in the toppling of *Ammophila* shoot tufts in response to trampling (Plate 14).

With the development of really fine, detailed studies of the types just mentioned one is forcibly reminded of Watt's (1947) vision in his classical paper on 'Pattern and process in the plant community'. We begin to see the possibilities of, as he puts it, 'fusing the shattered fragments into the original unity.' But we have a long way to go in understanding the complex interacting patterns of the dune grass/herb layer. As Olson (1958 *a*) points out, 'successions in the dunes are going off in different directions and have different destinations according to the many possible combinations of independent variables which determine the original site and subsequent

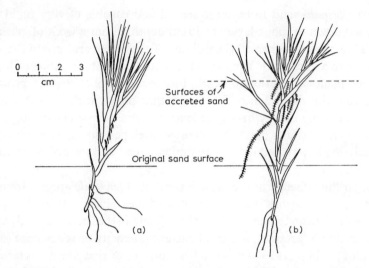

Fig. 60. Tillers from two plants of *Corynephorus canescens* growing approximately 20 cm apart. In plant (a) the site of adventitious root production was above the sand surface and the roots failed to develop. Plant (b) was partially covered with sand by rabbits. The site of adventitious root production was below the sand surface and the new roots, which have not been drawn to their full extent, had developed. Note the difference in diameter of the new roots in comparison with the older roots and also the presence of root hairs on the former (from Marshall, 1965).

conditions for development.' He emphasizes the value of multivariate techniques of analysis such as Maarel (1966) has recently applied to the Voorne dune environments in Holland, in helping us to understand these complex relationships.

The shrub and woodland layers

Partly for reasons of space and partly for lack of detailed study it is not possible here to deal adequately with the woody components of the dune flora. In Europe at the present time we are witnessing a resurgence of scrub on dune systems as a result of myxomatosis. In many areas these developing communities have not yet achieved a stable equilibrium with the environment, and especially with the human part of it.

An attempt to analyse some of the problems associated with the development of a particular type of dune scrub, *Hippophaë rhamnoides*, which is transforming many areas of dune grassland in Europe has recently been made by a group of Nature Conservancy biologists (Ranwell (ed.) 1972).

Some discussion of the dwarf shrub community formed by *Salix repens* in dune slacks is given in the next chapter, and of afforestation in the last chapter. Dwarf shrub communities anyway tend to be associated with dune slacks and those landward parts of the dune system where erosive and accumulative processes draw together the dune and slack levels to a plateau level somewhat intermediate between the two. Native woodland is a rather rare phenomenon on dunes and the accounts that do exist are primarily of a general descriptive character (e.g. Osborn and Robertson 1939; Westhoff 1947 and 1952; Lambinon 1956; Olson 1958; Boerboom (1960)). If the study of the ecology of dune scrub is in its infancy, that of the afforested dunes has scarcely yet begun and it offers challenging opportunities for work relating to the changing communities associated with fire and felling-recovery sequences of considerable practical interest.

Perhaps one might end this chapter with a plea that instead of directing attention to the generalized zonal trends that are more or less universal and fairly well known in dunes throughout the world but which tend to obscure what goes on in time at a particular spot in a particular dune system, we should turn our attention to the more detailed study of the unit 'fragments' of the ecosystem, the better to eventually understand the whole.

11 Sand Dunes: Structure and function of slack communities

Surprisingly little work has been carried out on the ecology of dune slacks outside the British Isles and Holland. This may be because these damp or wet hollows among the dunes are much more restricted in distribution than dunes and are sparsely represented in the more arid parts of the Mediterranean climatic zone for example. Slacks are however a very characteristic feature of large dune systems underlain by impervious deposits in the more humid Temperate zones, and it would be valuable to have fuller accounts than we have at present of slack systems in American and Australasian sites to say nothing of the still largely undescribed magnificent slack communities of the Coto Doñana in Spain.

Apart from their intrinsic interest as floristically and faunistically rich sites, their study is vital to an understanding of dune communities because many species of the dunes regenerate either more freely or solely at damp slack levels. The study of plant growth in the vicinity of the water table is also of considerable significance to both agriculturalists and water engineers concerned with cropping water-logged land, or water extraction.

In many ways a dune slack with its level surface; often influenced by salt in the early stages of formation; generally influenced by moderate accretion; and subject to the opposing influences of submergence or drought seasonally or at different stages of its development over longer periods of time, has much in common with a salt marsh. Certainly in transitional zones between the two, the sandy salt marsh and the saline dune slack merge so completely that they become indistinguishable.

It is not possible with the present state of knowledge to give a balanced account of dune slack ecology for we know little about the biology of most dune slack plants and have scarcely begun to study their animal communities. It is hoped that the somewhat fragmentary account given here will nevertheless stimulate further study of this neglected habitat.

Species Diversity

The biological spectrum given in Table 23 shows that in a site where slacks form nearly half the area of the dune system, they are almost as rich in vascular plants as the dunes. When we take into account the cryptogamic flora, and especially bryophytes which are particularly abundant, there is little doubt that in some sites the dune slack flora is the richer of the two. This forms a striking contrast with the relative poverty of the salt marsh flora and indicates optimum conditions in the dune slack for plant establishment, though not necessarily as we shall see, for plant growth.

We have no data on the relative species diversities of dunes and slacks so far as animals are concerned, but from general observations, the cooler climate of slacks, and the unpredictable tendencies towards flooding or surface desiccation from season to season it seems likely that fewer species of animals will have become adapted to them than to the dunes. Arthropods and molluscs (on lime-rich systems) are the most abundantly represented groups of the larger invertebrates, and annelids form a small but important section of the soil fauna in older slacks as at Newborough Warren. On machair, annelids are especially found in association with dung pats (Boyd 1957). The soil macrofauna of fully vegetated slacks at Newborough is sufficient to attract moles in quite large numbers in winter (e.g. 80 fresh mole hills noted in an area of less than $350\,m^2$ in December 1953 at Newborough.

Prior to myxomatosis, rabbits exerted a controlling influence on the vegetation of dry slacks and a more indirect influence on that of wet slacks (see section on grazing below).

Analytical Studies of Slack Vegetation Gradients

Much of what has already been said about the analysis of the dune flora in Chapter 10 is applicable also to the flora of slacks, but there is one important series of papers (Crawford and Wishart 1966, 1967 and 1968), devoted to the analysis of the dune slack flora of Tentsmuir, Fife. This provides an objective analysis which is enlightening. A grid of quadrats were examined for floristic cover and abundance, analysed by association analysis, and the groupings compared for similarity (Crawford and Wishart 1966). Results gave a spatially orientated series which was meaningful in relation to the

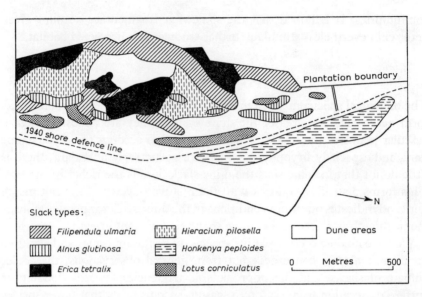

Fig. 61. Distribution of slack types determined by association analysis. Data from Tentsmuir, Fife (from Crawford and Wishart 1966).

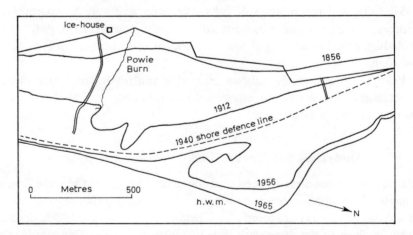

Fig. 62. Coastal changes at Tentsmuir from 1856 to 1965. All the coastlines except the most recent are taken from the unpublished report of Grove, A.T. (1953) Tentsmuir, Fife; soil blowing and coastal changes. Nature Conservancy, Edinburgh (from Crawford and Wishart 1966).

temporal development of this prograding dune system (Fig. 61 and 62). It was found that seasonally high salinity was correlated with the floristic groups associated with seawardmost slacks, while water level and soil activity were correlated with distinctive groupings in landwardmost slacks.

In a further treatment of this data (Crawford and Wishart 1967), a multivariate technique based on coincidence of occurrence and species interaction distinguished fewer groups than the original analysis and showed that many were ecotonal in character (Fig. 63) as might be expected in rapidly changing vegetation of this type. In their final paper (Crawford and Wishart 1968), an agglomerative method is used following the original divisive process to check for misclassification and secondly a means of representing the variance both within and between the terminal groups of an ordination procedure. It is then shown how the potential of any quadrat for membership of any classified type can be used to give computer maps of varying group potential, rather than discrete vegetation boundaries. Significantly the authors point out that, 'While it is always possible to draw a boundary marking the distribution of one particular species . . . no boundary can be drawn with any precision for any vegetation type that is defined on the basis of the probable occurrence of a number of species.' Objective studies of this type provide valuable confirmation of the

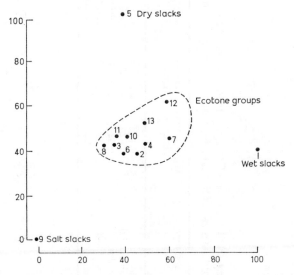

Fig. 63. Ordination of slacks obtained by group analysis; x axis based on types 9 and 1, y axis on types 9 and 5. Data from Tentsmuir, Fife (from Crawford and Wishart 1967).

Table 24. Soil analyses of samples from five different stands in the transition from sand dune to salt marsh. Each value represents the mean of five samples as expressed as g/100g oven dry weight. Mairiut District, Egypt (after Rezk 1970).

Soil type	Hygroscopic moisture	Loss on ignition	Org. C	Loss on acid treatment	Exchangeable Ca	Total soluble salts	Soluble chlorides	Soluble sulphates
Dune sand	0·42	2·7	0·033	99·0	20·29	0·11	0·004	0·001
Partly stabilized dune	0·62	2·8	0·080	98·1	18·31	0·10	0·004	0·001
Stabilized dune	0·62	5·0	0·156	95·3	15·71	0·14	0·004	0·007
Transition to salt marsh	1·12	5·7	0·138	86·1	16·29	0·23	0·015	0·009
Salt marsh	1·82	15·1	0·306	76·5	23·14	1·31	0·424	0·351

significance of trends hitherto distinguished primarily by more subjective approaches.

Westhoff *et al* (1961), Freijsen (1967) and Rezk, M. R. (1970) have studied dune slack to salt marsh transitions including relationships between plant associes and physico-chemical soil properties (Table 24).

On the dune to slack gradient most of the environmental changes are dependent solely on the change in water relations, and the gradient contains a mixture of species from each habitat, but none confined to the gradient zone. In contrast, at the salt marsh to sand dune slack gradient environmental changes are much more profound and involve changes in soil type, in salinity, *and* in water relations. Many species are more or less confined to the intermediate gradient zone (e.g. *Blysmus rufus, Centaurium littorale, Frankenia laevis, Limonium bellidifolium* and, though also associated with damp shingle, *Suaeda fruticosa*). This 'sandwich zone' as it might be called from the characteristic alternating soil types deserves special study in its own right as a separate habitat complex.

Maarel and Leertouwer (1967) used ordination and classification of contiguous quadrats on a dune gradient at Schiermonnikoog, Holland based on an index distinguishing floristic difference between adjacent quadrats. This showed continuous variation of vegetation which was correlated with variation in pH value and vertical height of the ground (Fig. 64). The remarkable pH gradient they found from *Drosera* areas on soil with pH 4 to *Schoenus* areas on soil of pH 7 over a very short distance is clearly similar to that in the dune heath situation studied at Newborough (Ranwell 1959).

While it is true that these studies emphasize the continuum nature of dune slack transitions, Leeuwen (1965) points out that the presence of a fluctuating water table near the surface marks out a zone on the gradient between dune and slack which is characterized by instability (in regard to the water factor) in time. This gives a specialized habitat which appears to suit orchids for example which are characteristic of (though not necessarily confined to) gradient situations like this and path sides. Freijsen's (1967) work (see below) shows that *Centaurium littorale* is best developed where the fluctuating influence of the water table is maximal.

For practical purposes it is necessary to draw boundaries and preferably ones which can be readily distinguished on the ground or on aerial photographs. Crawford and Wishart (1968) note the distinctive appearance of certain physiognomic dominants like *Erica tetralix, Glyceria maxima, Carex nigra* or *Juncus effusus* on photographs of the Tentsmuir dune slacks. Individual species of this type have special value for mapping when it comes to monitoring the rapid vegetation changes that occur in coastal systems.

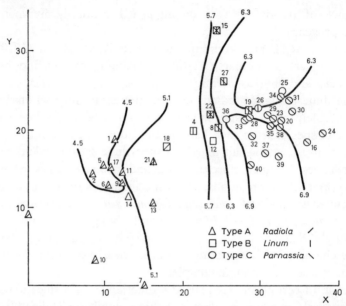

Fig. 64. Ordination of 40 quadrats in a 10 × 4 m transect laid down in the transition between *Schoenus nigricans* associes on base-rich ground at low elevations, and *Drosera rotundifolia* associes on base-deficient ground at slightly higher elevations, with isonomes of pH superimposed. Species number was taken as a measure of species diversity and was found to be related to range of pH and height. Results suggest a continuum-like variation in vegetation types mainly governed by pH variation itself governed by variation in height (from Maarel and Leertouwer 1967).

Similarly we need practical and workable divisions of the apparent continuum from the aquatic habitat to the dune in the vertical plane, and the next section outlines a system suitable for this purpose.

Water Table Limits as a Basis for Defining Slack Plant Community Limits

Studies at Newborough Warren and subsequent observations suggest that four distinctive levels of plant growth in relation to the water table can be distinguished (Ranwell 1959 and 1960 *a*). They are levels which appear to be significant in a number of ways which are biologically meaningful, but it is highly desirable that their delineation should be based on more objective criteria. In particular, one would like to see comparative study of the rooting depths of all species and study of the nutrient economy within and between the different zones. At the present time the literature on sand dune vegetation seems especially confusing in relation to the levels variously referred to as 'dune pasture', 'dune grassland', 'fixed dune', and 'grey dune'. Judging from the species composition and descriptions, many of these landscape types have more in common with what is defined below as the dry slack level than they do with the dune level. Perhaps the following classification may serve as a useful basis for clearing up such problems:

(1) *Semiaquatic habitat*

The water table is never more than 0·5 m below the soil surface and floods the surface from autumn to spring or later; plant roots are almost permanently waterlogged. Amphibious hydrophytes like *Littorella uniflora*, *Polygonum amphibium* or *Ranunculus aquatilis* are represented in the flora.

(2) *Wet slack habitat*

The water table never falls below 1 m below the soil surface and plants have their roots within reach of adequate moisture supplies at all seasons. Bryophytes are particularly abundant, the bulk of the flora is mesophytic and relatively few grass species occur.

(3) *Dry slack habitat*

The water table lies between 1 and 2 m below the surface at all seasons. Shallower rooted species are beyond the influence of the water table, but deeper rooted species benefit from its influence in summer drought. Phreatophytes and grasses are especially abundant and under rabbit-grazed conditions lichens may be locally abundant.

(4) *Dune habitat*

The water table never rises above 2 m below the surface and most of the plant growth is independent of it and wholly dependent on pendular water. Xerophytes and therophytes are common and vegetation tends to remain open.

In connection with the paucity of grasses in wet slacks, Jones and Etherington (1971) found from pot experiments that the dune grasses (*Agrostis stolonifera* and *Festuca rubra*) showed reduced tiller production and stunted roots in water-logged conditions. Growth of sedges (*Carex flacca and Carex nigra*) was less affected by water-logging.

It is true that Boerboom (1963) found that the presence of moisture indicator plants was found to be ordered by the quality (humus content) of the topmost soil layer rather than by the depth of the groundwater level and concluded that the field moisture capacity of the soil surface layers during drought was a better standard for the presence of moisture indicators than total pore space or the moisture content at various tensions. It is also true that as Vose *et al* (1957) showed there may be a higher moisture retentiveness in the surface layers in more humid areas such as the Tiree machairs off the western seaboard of Scotland. Also species like *Empetrum nigrum* normally dependent on groundwater survive at higher levels on north slopes (Westhoff 1947). Nevertheless, the broad distinctions outlined above seem to hold in European dunes at least, and it is very much easier to determine the approximate position of the water table within 2 m of the surface at any season with an auger than it is to measure field moisture capacities in drought periods.

Goethart *et al* (1924) have worked out the vertical ranges in relation to the water table of 91 species of flowering plants in several Dutch dune systems, but this work has been overlooked by most later workers.

Minor slack habitats

Within the main slack levels which are generally flat, secondary biotopographic units may be developed locally. They include: wet flushes; manmade pools and turf or peat cuts; low hummocks associated with clonal patches of such species as *Agropyron junceiforme* and *Salix repens*; molehills; ant-hills; and rabbit-disturbed ground. Each of these surface irregularities may show zonal sequences with characteristic communities varying according to salinity, moisture and physico-chemical soil factors. They are minor habitats found again and again in different dune systems wherever dune slacks occur, and they are worthy of detailed comparative study. To take one example, Freijsen (1967) has made a special study of the low *Agropyron junceiforme* hummocks of the Boschplat on the island of Terschelling. These form characteristic circular or horseshoe-shaped isolates on the floor of coastal slacks. He investigated the performance and regeneration of *Centaurium littorale*, a plant characteristic of the wet-dry gradient, on these hummocks. His results (see below) are of considerable interest in helping us to understand the problems of plant establishment and growth in this sub-habitat of dune slacks. These *Agropyron* hummocks are especially characteristic of rapidly prograding phases of dune systems and are also found at Tentsmuir, Fife; Morrich More, Ross; Towyn Burrows, Carmarthen, and Newborough Warren, Anglesey.

Establishment and Growth in the Slack Habitat

Migration into the slack

Salisbury (1952) concluded that the flora of wet slacks is chiefly composed of marsh plants commonly found outside the dune system. Very few species are confined to the dune slack habitat. Apart from being a relatively rare habitat, slacks are isolated among the drier dunes and only in direct communication with other wet habitats such as a salt marsh during relatively short periods when they are formed at the coast or with other fresh-water bodies via temporary streams during flooding. Once the slack is isolated by the growth of dunes, migration into it must come largely via wind-borne seeds, by birds, or by human activities.

The open damp slack surface exposed in the wake of eroding dunes becomes warm and moist in late spring or early autumn and ideal for seedling establishment, but only for very short and irregular periods. Where the slack surface is level, Martin (1959) failed to find any obvious environmental gradients of soil moisture, salinity, or soil physico-chemical factors in Island Beach, New Jersey. Yet there were striking differences from point to point in the vegetation (e.g. locally pure stands of *Dryopteris thelypteris* up to 40 m in diameter). Blanchard (1952) found similar point to point variation and large pure stands of a variety of species in a relatively young slack at Ainsdale, Lancashire. Martin concluded that chance factors of migration and establishment must play a big part in the colonization of slacks to account for the variety found. At Newborough Warren a few large clones of *Juncus maritimus* occur in certain landward slacks. During a 20 year period of observation only one new seedling of this species became established in these slacks. The seeds of this species are small enough to be carried in mud on the feet of birds, but unlikely to be distributed far by wind. Now there is a 200 acre (80 ha) *Juncus maritimus* marsh within a distance of 3 flight miles (4·48 km) of these slacks. This example suggests that bird-induced migration of plant species into slacks is likely to be of very rare occurrence indeed, a conclusion also reached by Westhoff (1947) in relation to colonization of the West Friesian Islands off the Dutch coast. Westhoff concluded that cattle were an important agent of dissemination within dune systems in addition to the main agents, water and the wind.

Establishment problems

The presence of such relict salt marsh species as *Glaux maritima* which establish in seaward slacks and persist in landward ones long since cut off from the sea, often gives the impression that there is a persistent saline influence in landward slacks operating through the water table. This is not necessarily so and many authors (Lambinon 1956, Ranwell 1959, Martin 1959, Duvigneaud 1947) have confirmed that the saline groundwater influence is limited to coastal slacks affected by tidal influence either by occasional tidal flooding or via percolation through a permeable shingle base. Pioneer species like *Centaurium littorale, Agrostis stolonifera* and *Juncus articulatus* are able to tolerate temporary periods of high salinity and even *Salix repens* can germinate in 25 per cent sea water (Ranwell 1960 *a*).

Submergence in wet slacks limits the period of establishment of most species to late spring or summer when the flood waters recede. Freijsen (1947) found that *Centaurium littorale* occurs on a level just reached by the capillary fringe (c. 55 cm above the water table). Germination in this species occurred when the mean afternoon temperature exceeded the critical temperature for germination (13°C) and was at a maximum when the temperature reached 24°C. The effect of the sun's altitude is important in autumn when slack surfaces become warmer than north slopes. Shallow water acts as a heat reservoir during the night and warms slacks to slope temperatures in spring. Experimental sowings of *Centaurium littorale* showed that germination was inhibited where the surface soil was saturated (Freijsen).

The lower limit of many species in slacks appears to be controlled by submergence though precisely how is not clear. Blanchard (1952) made detailed observations of the duration of fresh-water submergence in dune slacks at Ainsdale, Lancashire.

Birse (1958) carried out experiments on the tolerance of mosses to flooding and showed that species characteristic of dry slacks like *Ceratodon purpureus* and *Climacium dendroides* could tolerate up to 4 months flooding and survive. Westhoff (1947) found that in dune slacks with open water the strip along the shores is almost always water-logged, while elsewhere the soil is drier even though the water table is at the same depth. As Freijsen (1967) points out this is because the capillary rise is higher in wet than in dry soil where large pore spaces limit the rise of capillary threads. Westhoff observed that species requiring water-logged soil (e.g. *Littorella uniflora* can only thrive in the neighbourhood of open water and hence are particularly sensitive to drainage.

The density of vegetation in fully-vegetated slacks is persistently very high and this makes it difficult for other plants to establish. The only sites where *Calluna vulgaris* seedlings were found at Newborough were in turf cut areas where the vegetation had been temporarily opened up. *Calluna* does not enter the succession until a late phase at Newborough where leaching provides a sufficiently acid soil. There is of course no such problem for this species in lime-deficient dune slacks such as Good (1935) described at Studland, Dorset as *Calluna* can enter the slacks in the open phase.

Freijsen (1967) demonstrated how populations of *Centaurium vulgare* (syn. *Centaurium littorale*) oscillated in their vertical distribution from year to year according to the depth of the soil water table in spring (Fig. 65). Drought was clearly limiting both germination and establishment at the upper vertical limit of the populations. At the wetter, lower limit *Centaurium* showed delayed development up to three years though it is normally a biennial plant. As in *Aster* we have here another good example of the 'close-tracking' of the environment (see Chapter 6) shown by a short-lived perennial.

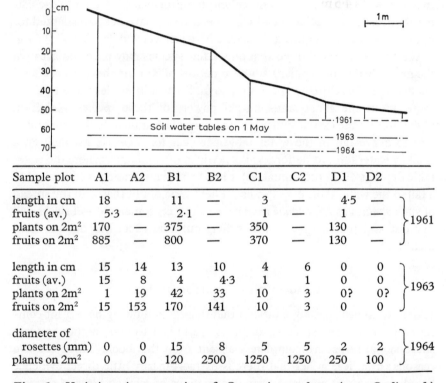

Sample plot	A1	A2	B1	B2	C1	C2	D1	D2	
length in cm	18	—	11	—	3	—	4·5	—	
fruits (av.)	5·3	—	2·1	—	1	—	1	—	1961
plants on 2m²	170	—	375	—	350	—	130	—	
fruits on 2m²	885	—	800	—	370	—	130	—	
length in cm	15	14	13	10	4	6	0	0	
fruits (av.)	15	8	4	4·3	1	1	0	0	1963
plants on 2m²	1	19	42	33	10	3	0?	0?	
fruits on 2m²	15	153	170	141	10	3	0	0	
diameter of rosettes (mm)	0	0	15	8	5	5	2	2	1964
plants on 2m²	0	0	120	2500	1250	1250	250	100	

Fig. 65. Variations in properties of *Centaurium vulgare* (syn. *C. littorale*) populations across a transect in a dune slack on Terschelling, West Friesian Islands, Netherlands (from Freijsen 1967).

It is of interest that these gradient environments favour the development of hybrids between *Agrostis stolonifera* (which occurs in wet slacks and semi-stable dune) and *Agrostis tenuis* (confined to drier sites) on the older dry slacks at Newborough (Bradshaw 1958). Anderson and Stebbins (1964) have pointed out the significance of habitat gradients for survival of hybrids.

Species of dune slack plants show very extreme morphological modifications in reaction to excess or deficiency of water. Non-flowering and very flaccid-leaved forms of *Myosotis scorpioides* occur in slacks submerged until summer. Reduced terrestrial forms of aquatic species like *Polygonum amphibium* occur in areas normally flooded but exposed in drought. As we have seen, submergence causes dwarfing in *Centaurium littorale* and *Salix repens* while as Salisbury (1952) demonstrates, drought has a similar effect on *Samolus valerandi* and *Plantago coronopus*. It is not surprising that with the two opposed extremes of too much and too little water constantly oscillating that very few species like *Centaurium littorale* have become adapted for growth specifically in the gradient zone between wet and dry slacks.

Westhoff (1946) observed that only a few species may have contact with the groundwater or capillary zones to depths of 3 m. At the dry slack level (as defined above) many species do so and it would be desirable to know more about the evapo-transpiration powers of these species. Robinson (1952) quotes the extraordinarily high figure of an annual discharge equivalent to a 2 m fall in the water table per year for *Tamarix* growing on a shallow water table at Safford Valley, Arizona. By contrast where the water table depth is 4·5 m the annual water loss from transpiration is equivalent to a fall in the water table of only 5 cm. There is no doubt that with the spread of such species as *Hippophaë rhamnoides* since myxomatosis, wet slacks will dry out and marsh species will suffer, but this effect has not yet been measured.

Community Transformations

It may seem a surprising omission that diagrams showing how plant communities of salt marsh or sand dunes are linked to one another to form successional series are completely absent from this book. There are two reasons: first, it is a fact that there is extraordinarily little direct evidence based on frequent and long term observations of marked plots in support of the assumptions made in such diagrams; second, two-dimensional diagrams are inadequate for the expression of the complexity of communities that can arise at any one location according to its history and subsequent treatment.

Successional diagrams not only oversimplify the many directions in which a particular community can develop, but they tend to falsify the reality of the situation in the minds of student ecologists. Successions do not end at the point where man's influence becomes dominant, they are simply modified by it. Many of the earlier accounts of dune succession stop at the first fence-line and ignore the fact that beyond it in the sandy pastures, the golf links or the conifer plantations, a high proportion of the dune flora and fauna continues to exist and develop under the imposed conditions of management. Some account of these various forms of more intensive management are discussed in the last chapter.

Physiographic changes (Chapter 8) and soil changes (Chapter 9) in space and time have already been discussed. These exert primary control on the type of communities present at any particular time and in any particular place on a dune system. As we have seen, the seaward edge of a dune system tends to undergo alternating change from the strandline community through embryo to open dune community and back again. Within the mobile dunes another type of alternation or cycle may occur from the open wet slack community to a more species rich, but still open dune community which erodes right down to damp bare sand to start the next phase in the cycle. And finally only when the dunes are worn down at the landward side of the dune system to near the dry slack level is there persistent stability. Here fully closed vegetation maintained as turf with dwarf shrubs under grazing, or developing towards some type of woodland, can develop.

Thus as in an estuarine series of marshes, we must think in terms of several quite distinct successional series operating persistently at different points in the system and not in terms of units of the spatial sequence from seaward to landward succeeding one another in time.

Coastal slacks

These are usually very transient features liable to sea water flooding or to obliteration by the growth of embryo dunes. They usually contain very open communities of scattered strandline species and a few salt marsh plants tolerant of well-drained soils. Where the dune system is prograding in alternating bands of low dunes and intervening coastal slacks they can develop more stable communities in which individual clones of halophytic and wet dune slack perennials close up to form a mosaic. The development of these communities has received little study and it would be interesting to establish long term observations on such sites as Morrich More, Ross or Tentsmuir, Fife to see how persistent the halophytic elements are with time and to find out just how perennials characteristic of later stages establish.

Blanchard (1952) made a detailed map of semi-aquatic plant communities associated with a coastal slack in Ainsdale dunes, Lancashire. She

o

found these formed a mosaic, the elements of which were evidently laid down at the time of establishment and subsequently formed a pattern of very persistent form. It is interesting that although this slack has been cut off from the sea for many years, now some 20 years after Blanchard's study the sea is threatening to break through again. If this is allowed to happen the existing vegetation of tall fresh water marsh species may well be destroyed leaving a bare surface for a new pattern, stemming from chance factors affecting establishment, to be set up.

Wet slacks

Where these are created by the erosion of a mobile dune as at Newborough, Anglesey, each new crescent of newly exposed wet sand in the wake of the eroding dune carries a slightly different seed complement from that of the previous year. Consequently banded communities commonly occur across the slacks often at slightly different levels according to the depth of the water table and the intensity of wind erosion at the time of their formation.

Among mobile dunes these slacks may persist with relatively open vegetation for 50 years or more. It was interesting to find at Newborough Warren that even with the reduction in rabbit grazing following myxomatosis that the existing vegetation of these wet slacks remain short although the flowering of many species (especially terrestrial orchids) was much improved. Evidently the adverse effects of summer drought for some species and winter flooding for others and not grazing are the main factors restricting growth.

However Westhoff (1946) records that since about 1910 the Dutch government has been active in controlling rabbits, and scrub has come up in the dune slacks. This led Westhoff to conclude that biotic rather than climatic factors have limited the spread of dune woodland in the past. Tansley (1949) suggested that the lack of tree seed parents in the neighbourhood of the coast was mainly responsible for absence of native forest on coastal dunes.

Now of course we are witnessing the post-myxomatosis transformation and *Alnus* and *Salix* are developing extensively in wet slacks and especially where they are close to afforested areas where rabbits were controlled prior to myxomatosis. Nevertheless in some of the bigger systems still remote from tree seed parents, tree seedlings have not yet appeared in wet slacks.

Dry slacks

Open vegetation communities at the dry slack level, like coastal slacks, are usually very transient habitats. They occur either at the base of the lee slope of advancing mobile dunes or occasionally over wider areas where shifts in the dune contours have led to re-erosion of a low dune area. Hummocks of *Salix repens*, *Ligustrum vulgare* or other shrub species are especially characteristic of such sites and they usually carry a rather sparse associated flora in which annual species are common, at least in the more stable areas.

The dune system comes to rest at the closed dry slack to dune level, and it is here that biological, as opposed to other environmental influences, become paramount in controlling community changes.

Prior to myxomatosis it was shown that there was strong evidence from sequences in the mosaic of vegetation that turf and dwarf shrub communities alternated with one another in time at Newborough Warren, Anglesey (Ranwell 1960 *a*). In this particular example a defoliating beetle (*Lochmea capreæ*) and the drought effect from rabbit-burrowing beneath *Salix repens* were believed to be responsible for its death. The occasional chance establishment of *Salix* seedlings in rabbit-disturbed turf started the cycle of shrub growth and ultimate death and decay again.

In this more stable zone it is legitimate to equate seral relations in time and space in local areas. From a study of serally related transects on dry slack turf at Newborough before and three years after myxomatosis, it was shown that while no seral trends were reversed, significant changes did occur in the frequency of species which were serally static. In particular there was a marked increase in the growth and flowering of most grasses and sedges. Turf 1 cm high grew to 15 cm in 3 years, low growing herbs and lichens were much reduced, but mosses remained abundant (Ranwell 1960 *b*). Now 15 years later the grassland has formed a 'rough' 40 to 50 cm high, mosses are much reduced, lichens are absent and scattered shrubs are overtopping the grassland.

Even more dramatic effects are evident on the dunes where rabbits no longer graze in any numbers. The vastly increased seed output has filled up the gaps in many formerly mobile dune areas and effectively locked the moving dunes into place.

The future clearly lies with the newly developing shrub communities, especially those associated with *Hippophaë rhamnoides* in European systems, unless active management takes a hand.

PART FOUR   Human Influences

12 Management of Salt Marsh Wildlife Resources

Salt marshes are the product of land erosion and therefore an expanding resource. It is no accident that the greater part of the world's population derives its food from the great deltas, largely in the form of fish and rice. No other habitat has sufficient natural fertility to support it. According to Grist (1959) possibly over 600 million people in Asia derive 50 per cent or more of their food calories from rice.

For centuries human beings have settled at the head of deltas and estuaries and expanded cropping and port facilities in pace with the seaward thrust of siltation. Yet it is extraordinary how even today in the most highly civilized countries this elementary geographical process seems to catch unawares the local authorities or other coastal landowners who suddenly find that their creek or tidal flat frontages are no longer open but clothed with salt marsh vegetation. Similarly, the significance of the dredger off-shore or the coastal engineering works on the other side of the estuary is rarely grasped by those whose coastal facilities will suffer in 10, 20 or 50 years time, until it is too late to do anything about it.

Rather more subtle changes associated with isostatic adjustment may be equally significant over periods of 50 or 100 years in those areas where the rate of coastal sinking is nearly balanced by the rate of salt marsh accretion. As we have seen this appears to be happening on parts of the south and south-east coasts of England at the present time.

Human influence on the salt marsh environment is increasing. The ecologist and the physiographer have an important responsibility to inform themselves of the directions and time scales of change relating to salt marsh formation, development and destruction. They also have the responsibility to pass on their knowledge in intelligible form to those who need to act on it.

An excellent account of human influences on general estuarine processes and the animals inhabiting estuaries is given by Cronin (in Lauff 1967); the following is chiefly concerned with human influence on salt marsh vegetation, and its management.

External Human Influences

Catchment activities

Land cultivation and mining activities have big effects on water and silt inflow into estuarine basins and hence on the life cycle of salt marshes.

It is claimed that the hydraulic mining in operation from 1850 until it was banned in 1884 added a metre of silt to the Suisan and San Pablo parts of San Francisco Bay (Gilliam 1957).

If silt is added in quantity to an estuary it increases the turbidity and raises the lower vertical limit to which salt marsh plants can grow. At the same time the rate of accretion of the levels which can support marsh growth will increase. These opposing effects result in a tendency to extensive cliffing at the seaward edge of the marsh which is very persistent even after the silt input is subsequently reduced. No one has measured these effects yet recent changes in land cultivation have been on a vast scale and must have had a profound influence on salt marsh development. Equally if cultivation demands improved drainage, then a greater volume of fresh water enters the estuary and this on meeting tidal waters will cause an increase in the height to which the tide will rise and consequently a reduction in the depth at which salt marsh can establish.

The need for data relating to these input factors is now recognized in relation to current studies on estuarine barrage schemes (Anon 1966 *a* to *c*, 1967, and 1970 *a*).

Pollution

In addition to the relatively innocuous effects of increased water and sediment input there may come along with it a frighteningly complex array of chemical substances derived from agricultural operations, industry, and sewage. This pollution approaches the salt marshes from landward and seaward. There is also a vertical component from air pollution.

In industrialized estuaries the sediment of which marshes are built may contain a high proportion of man-made detritus. Cinders, siliceous and metallic fly ash, slag, and coal were found in the sand-size fractions from the top 35 cm of a core in bottom sediment from the centre section of the Hudson estuary, New York State (McCrone 1966).

With the possible exception of oil, we are again very ignorant of both the nutritive or toxic effects of these substances on the life of salt marshes. No one has measured for example changes in the deathline for salt marsh growth in heavily polluted estuaries due to the combined effect of all these influences close to centres of civilization. It has been shown that overall productivity of macro-algal communities on the Adriatic coast remains unimpaired right up to the deathline where macro-algal growth suddenly fails (Golubic 1970). Significantly as this point is approached, the species diversity is reduced from many to only two algal species (*Ulva lactuca* and *Hypnea musciformis*) and one larger animal, the sea hare *Aplysia fasciata*. Beyond this point persistently anaerobic organic-rich mud forms a foul-smelling 'bacterial soup' virtually devoid of higher forms of life.

One of the most noticeable changes in southern English salt marshes over the past 20 years is the extensive growths of green algae (*Enteromorpha* and *Ulva* species) which have developed around the seaward edges of salt marshes. They occur particularly in the more sheltered bays where sewage or industrial effluents in built up areas, or fertilizer outwash in arable farming areas, are likely to accumulate. Now these algae are capable of utilizing nitrogen in the ammonium form and these various effluents must contribute substantial quantities of organic or ammonium nitrogen, which would normally be converted to readily assimilated nitrite and nitrate. But it is significant that under anaerobic conditions, the conversion of organic nitrogen stops with the step of ammonium formation (Black 1968). It seems likely that accumulation of ammonium nitrogen may preferentially benefit algal, rather than salt marsh plant growth. The growths are so extensive that in sheltered bays which could act as nutrient traps in the Blackwater estuary, Essex, or Poole Harbour, Dorset, for example, they appear to smother salt marsh growth and replace it locally. Studies are in progress to test the truth of this hypothesis and to determine the ultimate fate of the algal growths in chronic pollution conditions.

Chronic pollution from oil refinery effluent has much the same effect and the line between apparently normal live *Spartina anglica* marsh and dead marsh at Fawley in Southampton Water for example was found to be very sharp indeed when the site was visited in 1962. Boorman (*pers. comm.*) notes that *Limonium* species disappear from salt marsh in heavily polluted estuaries. It is important that studies should be made of the more subtle effects of pollution: sub-lethal damage, nutritional disturbance and, in the case of invertebrates, behavioural disturbance due to chemo-sensitivity, but there is much to be said for concentrating first on gross effects of total pollution as described above.

Heavy but isolated oil pollution may be tolerated without serious harm by salt marsh plants like *Spartina anglica* (Ranwell and Hewett 1964), indeed marsh growth forms a valuable trapping surface for oil in estuaries and strains it from the tidal water where it is so harmful to birds. However, most vegetation including salt marsh is rapidly killed by emulsifiers used to disperse oil (Ranwell and Stebbings 1967), so it is pointless to use them on an oiled salt marsh.

A recent bibliography by Nelson-Smith (1968) gives a valuable key to the literature on oil pollution and outstanding contributions have been made on the effects of oil (Plate 5) and emulsifiers on salt marsh plants and salt marshes by Baker (1970 *a* to *i*). This work is incidentally a model example of the experimental approach and its presentation. It has demonstrated the relatively high resistance of *Puccinellia* marsh turf to repeated oil sprayings, clearcut differences in the tolerance of different salt marsh species to oil pollution, and a (possibly indirect) nutritive effect of oil on salt marsh vegetation.

Baker (1970 *c*) also tested the effect of emulsifiers used to disperse oil pollution on *Puccinellia maritima/Festuca rubra* turves and found that emulsifiers in current use killed plants in concentrations above 10 per cent (Fig. 66). In a field trial where emulsifier (B.P. 1002) was used to clean oil, no decrease in damage to *Puccinellia* or *Spartina* marsh was noted and it was concluded that oiled salt marshes are best left to recover naturally (Baker 1970 *i*).

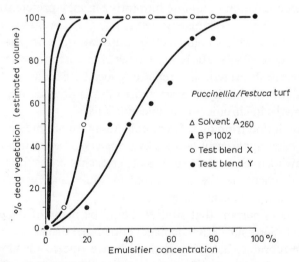

Fig. 66. Effects of emulsifiers and a solvent on *Puccinellia maritima/Festuca rubra* turf (from Baker 1970 *c*).

The effects of air pollution on salt marshes are probably minimal because of regular tidal flooding. However, there is evidence (McCrone 1966) that algae and silt accumulate radio-isotopes and Preston (1968) has shown that radionuclide concentrations decrease exponentially with depth from the surface in silt cores taken from the Ravenglass estuary, Cumberland. By contrast there was little significant change in radionuclide concentrations, in cores taken in nearby beach sands.

Introduced species

One of the most powerful human influences on salt marshes in Europe has been the deliberate introduction of *Spartina anglica* to particular sites and its subsequent uncontrolled spread to other sites from them (Ranwell 1967 a). This plant has also been established in Australia, Tasmania and New Zealand and recently planted in Puget Sound, Washington State, U.S.A. where it seems to be flourishing (Table 25). It has partially replaced *Zostera* and algal communities on high level mudflats and temporarily reduces variety where it becomes dominant in salt marshes. The rationale of many of the introductions is questionable and many attempts have been made to eradicate it locally. In most cases they have failed because of the high reproductive potential of this species. Currently work is in progress in Northern Ireland and elsewhere to determine the population level at which effective control can be achieved at reasonable cost.

Table 25. World resources of *Spartina anglica* marsh (from Ranwell 1967).

| Country | Date of first record* | Area | |
		Acres	Hectares
Great Britain	1870	30,000	12,000
Ireland	1925	500–1,000	200–400
Denmark	1931	1,230	500
Germany	1927	1,000–2,000	400–800
Netherlands	1924	9,800–14,300	4,000–5,800
France	1906	10,000–20,000	4,000–8,000
Australia	1930	25–50	10–20
Tasmania	1927	50–100	20–40
New Zealand	1913	50–100	20–40
United States	1960	< 1	< 1
Total		52,400–68,500	21,000–27,700

* Dates refer to the first recorded appearance, or first known introduction, to a country. All British material before 1892, when the fertile form was first recorded, is of the sterile form. Area estimates are of ground covered 50 % or more by *Spartina* and must be considered as very approximate.

The introduction of *Tamarix gallica* from the Mediterranean to the salt marshes of the southern United States has had more serious consequences. Martin (1953) records that its uncontrolled spread from wind and water dispersed seed has interfered with drainage, promoted flooding, reduced the value of grazing and waterfowl habitat, and resulted in extensive losses of irrigation water through evapo-transpiration. The latter has the effect of increasing ground water salinity in brackish areas. This limits the extent of rice cultivation and other salt-sensitive crops in the delta areas. The growth of deep-rooted phreatophytes like *Tamarix* is primarily a problem associated with marshlands in the warmer and more arid parts of the world.

Much of the literature on this subject relates to inland growths of *Tamarix*. Fletcher and Elmendorf (1955) for example quote annual water losses of up to 5 acre – feet due to *Tamarix* in the Pecos river delta, New Mexico. They give a useful review of the significance of phreatophytes in water control and the effect of attempts to limit their growth.

Internal Human Influences

The mildest forms of human influence on salt marshes result from sporadic direct cropping of the plants and animals which live on them (e.g. the gathering of 'samphire' (*Salicornia* sp.) or wildfowling). Indirect cropping with domestic grazing animals has a stronger influence as it changes the physical environment and the species composition. Draining and spraying activities produce even more profound changes in hydrological, chemical and biological parts of the environment. Both the creation and extermination of salt marshes may be caused by coast protection activities, reclamation, or other coastline modifying activities such as estuarine barrage or airport construction.

It is useful to consider these activities as given above in the sequence of increasingly strong human influence and we have to bear in mind there is often a hidden legacy from past activities, such as salt panning or derelict reclamation schemes which have left their mark on the marsh. Directly or indirectly, the character of most salt marshes throughout the world today has been largely determined by human activities past or present just as in other habitats.

Direct cropping

Spartina marshes on the north-east coast of North America were formerly mown for hay and Burkholder (1956) records an amusing energy chain from Georgia, U.S.A. where 'in former years marsh grass was harvested as the sole feed for mules that were used to haul fuel for the wood-burning steam locomotives on the Old Brunswick and Florida Railroad.' *Spartina patens* was formerly cut for hay in Delaware and *Spartina pectinata* (said to be dominant over some 28,000 acres (11,3000 ha) of marsh around the Bay of Fundy, Nova Scotia) is under investigation at the present time for use as hay and pasture (Nicholson and Longille 1965). Hubbard and Ranwell (1966) demonstrated that it was possible to cut *Spartina anglica* marsh in dry weather using a light tractor at Bridgwater Bay, Somerset and to make palatable and digestible silage for sheep. No information has been found on the effect of regular mowing of salt marshes on their botanical composition. It seems unlikely that mowing was practised in the native short grass or herb-rich marshes on European coasts, though reed (*Phragmites*) cutting is still an important activity in the larger deltas like that of the Danube.

Most of the *Spartina anglica* planting stocks used in different parts of the world were derived from a small bay in Poole Harbour, Dorset (Ranwell 1967) and seed stocks of *Festuca rubra* are currently harvested from Lancashire marshlands. Both this species and *Agrostis stolonifera* are being propagated for use on embankments on the German North Sea coast (Wohlenberg 1965). The selection and breeding of coastal grasses for use in specialized habitats of this type has hardly begun, and there is great scope for further work in this field and for their use as pasture plants on inland salinized soils.

Turf cutting is practised on the sheep-grazed Lancashire and Solway marshes. Usually 2 in (5 cm) strips are left between cuts to improve regeneration and within 5 years the same areas may be cut again. Experiments are in progress in the Bridgwater Bay National Nature Reserve, Somerset to determine the botanical changes of this cycle and to see if it is possible to reverse succession from coarse and relatively unpalatable *Festuca rubra* marsh to *Puccinellia* marsh, more palatable for wildfowl (Plate 8).

Wintering flocks of wildfowl on tidal marshes have been cropped for centuries. While in general this has developed in an uncontrolled manner particularly in Europe, the controlled cropping of wildfowl and muskrat on Delaware marshes has been combined with the application of sophisticated habitat management techniques. These include the excavation of flight pools, the planting of wildfowl foods, and spraying or burning to control tall marsh growths (Lynch 1941; Schmidt 1950 and Steenis *et al* 1954).

There is another less tangible 'crop' of increasing importance from salt marshes and that is the education and recreation derived by students, naturalists, yachtsmen and anyone who seeks to explore the life of the marshes or derives pleasure from the subtle contours and colours of their remote landscapes.

Indirect cropping

By far the most widespread use of salt marshes has involved indirect cropping by open range grazing with domestic animals. On the Gulf and Atlantic coasts of North America cattle graze the coarse *Spartina* marsh growths and older breeds of sheep do so in the British Isles. Williams (1955, 1959) has shown how access to the marshes can be improved by provision of cattle walkways (Plate 7). These are ridges of spoil bull-dozed from the marsh and spaced half a mile (0·8 km) apart where they provide refuge for cattle at high tide. The borrow pits from which the soil is dug, flood, and attract wildfowl. *Spartina* and *Distichalis* marshes will support a cow for every 2 to 4 acre (0·8 to 1·6 ha) during the 6 month grazing season. Burning is practised widely by stockmen to stimulate succulent new growth, but during drought, fire can reach plant crowns and severely damage the marsh vegetation (Williams 1955).

Salt marshes of the European seaboard are used for cattle, but more generally for sheep-grazing. Very extensive salting pastures are found in Northern France (e.g. the Baie de St. Michel) and on a smaller scale in most of the estuaries and bays of the west coast of England and Wales from the Solway to the Bristol Channel. European saltings are composed of three principal grasses, *Agrostis stolonifera*, *Festuca rubra*, and *Puccinellia maritima*. They support 2 to 3 sheep to the acre (0·4 ha) for most of the year when the marshes are free of tides. It has been shown that high level *Spartina anglica* marsh can be converted to *Puccinellia maritima* marsh by sheep grazing (Ranwell 1961). Experimental studies (Plate 6) indicate that on high level *Spartina* marsh growing on firm silt this can be achieved in about 5 to 10 years. With the development of intensive agriculture and loss of inland pastures for folding sheep at high tide, or in mid winter, sheep-grazing has declined on many coast marshes in southern England. This has led to the spread of unpalatable tufted growths of *Festuca rubra* or *Agropyron pungens* which accrete silt more rapidly than close-grazed salting, quickly replacing succulent *Puccinellia* marsh. Chippindale and Merricks (1965) have shown how gang-mowing can help to maintain reclaimed salting pasture at times when sheep are in short supply. Wohlenberg (1965) records that turf cutting on salting pasture on the West German coast may enable *Agropyron pungens* to establish. Once established this coarse, unpalatable species can rapidly invade high level salting pasture.

There is a very critical stage near the upper limit of *Puccinellia* growth where coarser grasses can invade but at this level sheep-trampling helps to offset the very small but significant accretion brought by the few tides that reach these high level salting pastures. This compaction, aided by the normal settlement due to drying, can hold the marsh at a level suitable for *Puccinellia* growth for decades longer than it would otherwise be able to survive, providing it continues to be hard-grazed by sheep (Ranwell 1968 *a*).

Drainage

Large scale residential development near tidal marshes in the warmer parts of the world invariably promotes activities associated with mosquito control; in particular drainage and spraying.

Bourn and Cottam (1950) record that ditching for mosquito control began in New Jersey in 1912; greatly expanded in 1933 when relief labour (organized as a result of the economic depression) became available, and by 1938 had encompassed 90 per cent (562,500 acre (227,700 ha)) of the original tidewater lands lying between Maine and Virginia. Inevitably there was a clash with sporting and conservation interests, but by the time this became vociferous enough to achieve action most of the marshes had been criss-crossed by a network of drainage channels.

Taylor (1937) testified to the effectiveness of mosquito control on newly ditched marshes in comparison with unditched marshes and concluded from a superficial study of the vegetation that the only significant changes were in the development of secondary vegetation on ditch banks, notably the spread of *Iva ovaria*. However Bourn and Cottam (1950) carried out a much more detailed study over a period of 12 years on ditched Delaware marshes and found that shrubby growths of *Baccharis halimifolia* and *Iva Frutescens* had largely replaced the marshes natural grass associations and resulted in serious reductions in populations of marshland inverte-brates important as food for wildfowl and waders. There is no mention of reduction in grazing or hay cutting on these marshes with the advent of ditching, but inevitably this would result from reduced access due to the ditches and must have contributed to the spread of taller vegetation. In addition to the serious loss of wildlife habitat, Bourn and Cottam (1950) note that mosquito control has not been effective in many areas due to lack of maintenance on the ditches.

A great deal of literature exists relating to the use of brackish water for crop irrigation in coastal areas (see Gallatin *et al* 1962 and numerous publications of the U.S. Salinity Laboratory, Riverside, California), but very little study has been given to the effects of changing water quality on the wildlife of tidal marshlands. However, valuable studies have been carried out on this subject in the Camargue marshes of the Rhône delta, France by biologists at the Tour du Valat Biological Station. For example, Aguesse and Marazanof (1965) have studied changes in salt marsh and brackish water populations of invertebrates over a period of some 30 years in relation to climate, the effects of irrigation for rice cultivation, and habitats modified by salt production. Of particular interest is their conclusion that all the changes observed are reversible. But one should not assume that this would be true for larger vertebrates. As we have seen the spread of phreatophytes like *Tamarix* species may contribute to increasing the salinity of ground water, while the development of desalination plants will increase the salinity of superficial waters flooding estuarine marshes. Presumably this will result in a partial reversal of the normal salinity gradient in estuaries and encourage the spread of more salt tolerant species further up the estuary.

Spraying

In the past 30 years insecticide or herbicide spraying has been used increasingly on coastal marshes for management purposes in relation to wildlife cropping. Spraying has been used also for mosquito control, for oil pollution decontamination purposes, and in mangrove swamp for clearance for military purposes.

In general the use of herbicides for wildlife management is a responsible activity carried out or supervised by well-informed people who are primarily interested in protecting wildlife rather than destroying it. Certain herbicides like Dalapon (sodium salt, 2, 2 – dichloropropionic acid) are not known to be significantly harmful to life on tidal marshes other than the grass species like *Spartina* or *Phragmites* they are used to control. However, it is not possible to use sprays effectively in marshland washed daily by the tides and in the control of *Spartina anglica* in such sites the use of pelleted substituted urea compounds such as Fenuron (3-phenyl-i, i-dimethylurea) has been found effective (Ranwell and Downing 1960). These of course are soil sterilents and non-specific. It would be desirable to know more about their side effects on invertebrates and the risks of promoting erosion before they are used on a wide scale.

The aerial spraying of tidal marshland for mosquito control has had serious consequences and Springer and Webster (1951) have demonstrated the more immediate effects of aerial spraying on experimental plots in the New Jersey marshes. Plots were 50 acre (20 ha) or 100 acre (40 ha) and were treated with dosages ranging from 0·2 (0·09 kg) to 1·6 lb (0·7 kg) DDT per acre (0·4 ha) and results measured against untreated controls. Birds were not obviously affected, but heavy losses of fish were recorded in dosages above 0·8 lb per acre (0·36 kg/ha) and crabs were almost completely killed, these effects being greater in ponds than in creeks or ditches. Effects on smaller invertebrates were variable; shrimps and amphipods were seriously affected, insects, spiders and worms less so, and mites and molluscs not apparently harmed.

Now of course, we are aware of the more subtle dangers that accrue through the build up of chlorinated hydrocarbon residues from substances like DDT in food chains. Haderlie (1970) records the death of hundreds of fish-eating marine birds and some sea lions, believed to have accumulated lethal doses of DDT off the Monterey Bay coast, California. He is currently studying the accumulation of this substance and its derivatives in the estuary of the Salinas River draining the Salinas valley. Here, during the past 10 years it is estimated that 125,000 lb (56,700 kg) of DDT has been sprayed on agricultural land each year.

Tschirley (1969) estimates that the regeneration of mangrove forest to its original condition following defoliant spraying in Vietnam for military purposes with 2, 4–D and 2,4,5–T (normal butyl esters of 2,4–dichlorophenoxy – acetic acid and 2,4,5 – trichlorophenoxyacetic acid) or with triisopropanolamine salts of 2,4–D and picloram (4 – amino – 3,5,6 – trichloropicolinic acid) will require about 20 years. Fish yields have increased during a period of intensive defoliation, but this could be a temporary phenomenon due to release of nutrients.

Reclamation and coastal transformation

Reduction in the tidal area of Poole Harbour, Dorset through natural siltation and reclamation is estimated to have increased within the past 150 years to a rate 12 times that of the previous 6,000 years (May 1969). This gives some idea of the accelerated pace at which salt marshes are being diminished. In San Francisco Bay less than a quarter of the original marshland survives (Harvey 1966). Much depends on how the reclamation is achieved, and there is convincing evidence that embankment of marshland around the Wash in England has in the past stimulated the formation of new salt marsh to replace that reclaimed (Inglis and Kestner 1958 a and Kestner 1962). Dalby (1957) estimates that some 80,000 acre (32,400 ha) have been embanked around the Wash since the seventeenth century and estimated that embankment could continue at a rate of some 15 000 acre (6 100 ha) per century. This reclaimed land has produced some of the most highly fertile agricultural soils in the world, but only at a controlled rate of reclamation which does not exceed the rate of new marsh formation. In fact,

P

Inglis and Kestner (1958 b) give evidence which suggests that supplies of silt which had taken thousands of years to accumulate in the Wash may already be so depleted as a result of reclamation that little of the progressive silting expected seaward of a recent embankment has occurred.

The pace of salt marsh formation has been increased on the Dutch, German and Danish wadden coasts by means of ditched and groyned sedimentation fields and an excellent account of the techniques involved is given by Kamps (1962).

Unfortunately, in many industrialized estuaries, land prices are so high and the need for new land so urgent that it becomes economic and expedient to infill marshes with rubble and rubbish directly to make up the level at a rate faster than there is time for new marsh to form. Obviously this brings a serious risk of pollution especially if the tipping is not done behind bunds which effectively keep the sea from re-working the rubbish.

Reclaimed marshland used as pasture and intersected by drainage ditches may retain elements of the salt marsh flora for at least 100 years (Petch 1945). It provides grazing and roost for wildfowl at high tide and the dykes and ditches extend the habitat of many rare species normally localized at the salt marsh upper limit and in brackish flushes. This habitat has never received proper ecological study although it probably carries almost as high species diversity as the sand dune and slack gradient. The present trend towards arable farming is rapidly destroying reclaimed pastureland at a time when its wildlife potential is only beginning to be recognized and valued.

The needs of coastal protection, improved navigation and wholesale transformation for fresh water reservoirs behind estuarine barrages, or coastally sited airports, all result in re-structuring of coastal sediments and the marshes derived from them. It is not always appreciated that foreshore amenity may be lost in a few decades as a result of siltation and marsh formation at sites apparently remote from newly constructed works. For example the training of the low water channel to the south shore of the Dee Estuary, Flint is the indirect cause of the loss of coastal waterfront at Parkgate, Cheshire on the north shore.

Kestner (in Thorn 1966) has reviewed the effects of dredging, barrages and training walls, on the tidal and siltation regime in estuaries. He concludes that the most successful schemes have been those in which the estuary as a whole has been modified. Half measures have usually not been successful and have produced undesirable side effects.

Gilson (in Lowe-McConnell 1966) has discussed some of the biological implications of proposed barrages in Morecambe Bay, Lancashire and the

Solway Firth to the north of it. More specifically Gray (in Perring 1970) who has completed a two year study of the Morecambe Bay salt marshes, draws attention in an interim report to the hazards of ecological prediction and to the probably ill-founded assumption that the present ecological behaviour of a given species is a reliable guide to its reaction to new situations. Most likely it is not, and the explosive spread of *Typha* in possibly new genetic combinations on the pseudo-delta of the Niger estuary (Trochain *pers. comm.*) may well be a pointer to the sort of biological reactions we should expect.

Integrated Management

It should be apparent from this account that the human impact on the salt marsh environment has not in general been based on informed understanding or consideration for the wildlife resources it contains. It follows that we are not fully aware of the value of these resources. Somewhat frantic efforts are being made at the present time to bring to the attention of authorities a fuller understanding of what is being lost and what might be gained by combined planning for the use of existing resources and the deliberate design of new salt marsh resources. For example, the creation of the Rømo dam joining the mainland to the island of Romø in Denmark, has been foreseen to encourage the formation of new salt marshes in its sheltered angles. Their formation is actively aided by ploughing drainage ditches in high level mudflats to seaward so that salt marsh growth is improved on the intervening ridges.

The activities of the San Francisco Bay Conservation and Development Commission (Harvey 1966 *a* and *b*) are spreading wider understanding of the value of existing wildlife resources to the people that live around the shores of the Bay. Steenis *et al* (1954) have done the same for the Delaware marshes and Goodwin *et al* (1961) for Connecticut's coastal marshes where significant advances in legislation have provided valuable protection to these habitats.

Fresh water reservoir proposals in the inter-tidal zone of estuaries are under joint investigation by engineers, hydrologists, fisheries, biologists, limnologists and all who are directly concerned with the protection and production of wildlife. All these activities are leading towards integrated management proposals which should enable the living things on salt marshes space to exist and should no longer allow the marsh to be treated as a convenient potential rubbish dump.

But they cannot only exist. They must be made to produce in common with other land for our crowded societies. Work in the larger nature reserves must evolve new management techniques, the full value of marshes in coastal protection must be assessed and the value of a fully utilized marsh set against any reclamation proposals for other purposes.

One example of the seasonal cycle of use that might be more fully developed is as follows. In the spring when migratory wildfowl have left, marshes may be rested for a few weeks to allow vegetation to recover and resident marshland birds to breed. Turf cutting could commence on suitable sites and stock return to graze. In summer, marshland areas could be increasingly used for recreation and education at a time when least harm will be done to wildlife resources. Mowing can be carried out to preserve the quality of salting pasture and in preparation for autumn turf cutting. In winter the migratory wildfowl will take up residence and could be cropped on a regulated permit system as at Caerlaverock National Nature Reserve, Dumfries, or fully protected as in the case of diminishing species such as the Brent Goose as at Scolt Head Island, Norfolk.

Only when we have tried to dove-tail these various forms of management can we hope to set a proper value on the salt marsh.

13 Management of Sand Dune Wildlife Resources

Sand dunes, unlike salt marshes, are effectively a diminishing resource around the coasts of lowland Europe and North America. Not only is their regeneration limited by what is believed to be a diminishing bank of off-shore sand supplies, but their rate of destruction under development of various kinds is almost certainly exceeded by their rate of formation. Expansion in area of a dune system is a much slower process than that associated with salt marsh formation. No figures are available for the proportion of sandy prograding coasts where the rate of formation is maximal as opposed to systems where the coastline is static or eroding, but it seems likely that if dune coastline lengths were scored for these properties, prograding sandy coasts would be in the minority.

While a certain amount of re-cycling of material goes on, this is primarily of a very local nature and most of the sand of a dune system being above the inter-tidal zone is out of circulation anyway. So, while there may be considerable internal mobility, dune systems as a whole are much more static in position on the coastline than salt marshes. They also tend to be more isolated one from another than salt marshes and this accounts for the distinctiveness of each individual dune flora. This is well illustrated by the colour variants of *Viola tricolor* sub-species *curtisii* on European dunes. Only yellow-flowered forms may occur on one system, on another, both yellow and blue-flowered forms are found, presumably evidence of isolation in terms of gene flow.

As we have seen, the low fertility of dune soils coupled with much open ground for casual colonization produces an immensely rich flora. This, combined with the distinctive landscape and shorelines ideal for recreation, attracts people in ever increasing numbers.

Sand dunes were among the earliest of sites settled by primitive man.

They have often been used with little understanding and disastrous results when the dunes, re-mobilized by over-cropping, have overwhelmed adjoining land settlements. More enlightened management policies followed and the value of dunes in coast protection was recognized. Some of the larger systems were afforestated in the eighteenth and nineteenth century. As land became scarce dune systems were levelled for industry, housing, and airport needs. Now we are beginning to realize the special virtues of the diminishing dune landscape for recreation and the need for protecting these resources from further despoliation.

External Influences

Water extraction

The effect of water extraction on the dune flora has received little study except in the Netherlands. Here, Westhoff (1964) records that the dune area has 'to a large extent been dried up by the extraction of drinking water'. The Wassenaar dunes near the Hague have been exploited as a catchment area since 1874 and from about 1885 onward this has caused a serious fall in the level of the ground water table (Boerboom 1960). Many moisture-loving plants disappeared and the plant communities dependent on a high water table level were almost destroyed except in a few small man-made hollows formerly used as wells. Even a small permanent fall in the water table of about 10 cm can be fatal especially to the plants and animals of sub-aquatic and wet dune slacks (Voo 1964). In their place, common species such as *Molinia caerulea* or *Calamagrostis canescens* have spread over these habitats in the Netherlands. Uncontrolled water extraction from sites close behind the coastal dune can also lead to contamination of fresh-water supplies with brackish water.

Fortunately the dangers have been recognized in time and artificial fresh-water infiltration has been started in the Wassenaar dunes since 1955. Boerboom (1960) has been recording the floristic changes as the water table began to rise again. These changes are not necessarily a straight reversal to the original damp and wet slack communities, partly because of loss of parent material and partly because rabbit-grazing has diminished so altering the floristic balance. Studies on changes due to falling water tables are urgently needed in British dunes and especially those where there is little immediate prospect of new slack formation at the coast (as at Ainsdale, Lancashire for example). Any new slacks would of course result from wind excavation down to the new water table level.

Pollution

In a low-lying country like the Netherlands, there is a serious problem in maintaining oligotrophic communities like those found in lime-deficient dune slacks for they are enriched by nutrients from fertilizer residues washed out of agricultural land. Westhoff (1964) points out that a high proportion of the rare flowering plant species found in European dune slacks are characteristic of leached soils developed in the Atlantic climate zone. These are the first to diminish as soil enrichment progresses. This enrichment effect is proceeding only slowly in the Netherlands according to Voo (1964). Nevertheless from samples of about 900 oligotrophic waters throughout the Netherlands, it was found that significant changes in communities attributed to enrichment occurred in 42 per cent of them. Much depends on the way drainage from the agricultural catchment impinges on the dune water table. High-lying arable land directly to landward of the dune system is likely to have the most serious effects. In mesotrophic dune systems, incipient oligotrophic dune heath develops at the landward edge of the system. Such areas, lying closest to cultivated land, are particularly susceptible. For this reason it is essential to control drainage or at least to have a buffer zone to landward where high fertility cultivation is discouraged if oligotrophic systems are to be preserved as nature reserves.

Oil pollution does not have such serious effects on dune systems as on salt marshes, but where it does reach coastal dune slacks such species as *Euphorbia paralias* may be damaged by combinations of oil and emulsifiers (Ranwell 1968 *b*). It has been observed that up to 10 per cent of wind blown sand grains may be contaminated with oil and emulsifiers (Elliston – *pers. comm.*) after a serious pollution incident. We have no measure of the background oil contamination levels on sandy shores, but this will be maximal at the backshore, the source of dune sand, where conditions may often be too dry for effective bacterial breakdown of oil residues.

An increasing quantity of litter of all types is brought to the shores by tides and into the dune system by tourists. Teagle (1966) has analysed the weekly quantities of litter collected at Studland dunes, Dorset over a two-year period and finds the summer values about ten times greater than the winter quantities with peak values of thirty times the winter values on public holidays. The bulk of the litter is paper, but food remains attract gulls, and empty milk bottles trap small mammals in alarming quantities e.g. 48 mammals in 15 bottles in 1 year.

The possible consequences of air pollution on the mineral deficient soils of dune systems has been referred to earlier, but remains open for study.

Voo (1964) notes that shelter belts have been planted along the borders of nature reserves in the Netherlands to reduce the effects of airborne pollution and Bernatsky (1969) has demonstrated the importance of design of protective plantations in reducing air pollution.

Introduced species

Because of the need to control sand dune movement species like *Ammophila arenaria* have been deliberately introduced from Europe to the United States, South Africa, Australia and New Zealand. *Hippophaë rhamnoides* has now been planted on more dune systems within the British Isles than there are in its native range. The presence of this species (frequently planted in gardens), within a radius of about 5 miles (8 km) of a dune system brings a persistent risk of invasion via the agency of birds. A vigilant management policy is needed to check sporadic appearances and subsequent spread if the dune flora is not to be shaded out by its growth. This is a serious problem in some sand dune native reserves (e.g. Ainsdale Lancashire and Gibraltar Point, Lincolnshire). The rare sterile hybrid grass *Ammocalamagrostis baltica* was widely planted to new stations on Norfolk and Suffolk coasts during a dune re-planting programme following damage to the coast by floods in 1953 (Ellis 1960).

Afforestation is by far the most powerful agent for introduction of new species on to sand dunes and Holder (1953) records that following afforestation at Ainsdale, Lancashire the flora became far richer than it was originally. Similarly at Newborough, Anglesey the introduction of trees and shrubs for stabilization purposes, the use of roadside verge cuttings and forestry 'brash' to still the sand, and the introduction of weeds with hop manure in nursery beds, increased the flowering plant species total of the system by at least one third in about 10 years.

Teagle (1966) found little evidence that increasing numbers of tourists had added to the introduced flora at Studland, Dorset but since 1953 the New Zealand alien, *Acaena anserinifolia* has become well established in car park areas and in heavily trampled pathways. This species has hooked burrs on the fruiting head and is readily transported on the fur of animals or on clothes. The burrs may so clog the feathers of fledgling ground-nesting birds that they are unable to move effectively and die of starvation. This has been observed at Holy Island dunes, Northumberland where *Acaena* is abundant in *Ammophila* dunes.

Garden rubbish dumping adds to the dune flora and bulb cultivation on the Isles of Scilly has produced a remarkable assortment of aliens on many of the small dune systems there.

Internal Influences

Direct cropping of sand

The sand itself is mined locally for mineral extraction or building purposes. Mineral-bearing sands are widespread on the shores and dunes throughout the New South Wales coast and also on the south and central Queensland coasts of Australia (Sless 1956). Sand-winning for building purposes occurs sporadically around the British coast e.g. at Ainsdale, Lancashire and Druridge Bay, Northumberland. In many areas it has been discontinued (e.g. at Rock dunes, Cornwall) because of the risks to coast protection and loss of amenity beaches.

Other direct cropping

With the exception of the cranberry bogs associated with some of the North America dune systems, there is little available evidence that dune floras have so far yielded plants of any significant economic value. However, it is interesting that *Elymus arenarius* has been successfully hybridized with wheat (Pissarev and Vinogradova 1944), and with barley (Tsitsin 1946). Tsitsin considers that the hybrids thus obtained are 'of very great importance indeed' and should lead to big increases in crop yield.

In the past, marram grass (*Ammophila arenaria*) was regularly cut for thatching as at Newborough Warren, Anglesey (Ranwell 1959), but with increased availability of straw and development of plastics this is now discontinued in most areas.

Sand dune-building plants (especially *Ammophila* species) are cropped for stabilization purposes, but only on a small scale as a few strong tussocks will produce a great many planting units (Plate 15).

Rabbit cropping

Sand dunes were used extensively in medieval times as rabbit warrens, at least in Britain. They have also been used for centuries as open range grazing for stock. Warrens were effectively managed at first, but wild populations established and spread without control. Tansley (1949) records that rabbits were little known in Scotland until as late as the nineteenth century when their numbers rapidly increased. The structure of sand dune communities in Europe prior to myxomatosis was effectively the product of intensive rabbit-grazing.

Stock cropping

There is little information about the effect of stock grazing on dunes and this has never received experimental study. Frame (1971) records that a cow's hoof exerts a pressure of 40 to 60 lb per in^2 and that an acre of pasture would be trodden some three or four times in a year at normal stocking. By contrast sheep hooves exert a pressure of about 25 to 35 lb per in^2 and tread an acre of pasture six to ten times in a year. It becomes immediately clear from this why sheep have been found to be particularly damaging to dune pasture. However the low-lying lime-rich dune pastures (machairs) of Scotland have a relatively high moisture-holding capacity and have supported sheep for centuries without serious erosion. Elsewhere the uncontrolled mobility of many European dune systems which developed in stormy periods was undoubtedly triggered off by over-grazing by rabbits and stock in the past. All forms of large mammal grazing have now declined in many dune areas though it is still possible to see the typical grazed sward flora in pockets where rabbits have survived. At Whiteford Burrows, Glamorgan, ponies graze the dunes (in addition to sheep) and with little apparent harm to the dune turf which they crop almost as closely as rabbits.

Golf links

The use of sand dunes as golf links involves heavy local fertilizing, extensive mowing, some drainage, and local shrub clearance. Small areas are intensively managed as greens, tees or bunkers, but for the most part a modified, fairly varied dune flora and fauna survives unharmed. Wallace (1953) recorded some 350 species of flowering plants on Dawlish Warren, Devon a small dune system of about 100 acre ($c.40$ ha) partly used as a golf links. Beeftink (1966) found no less than 220 species on only one hectare of the Heveringen dunes formerly grazed by horses and goats and afterwards used as a golf links. Experimental studies on the effects of mowing dune vegetation are now in progress in the Newborough Warren (Anglesey), and Holkham and Winterton (Norfolk) National Nature Reserves.

No one has yet brought together the very considerable practical experience obtained by golf links management on sand dunes. Ecologists need to relate this knowledge to the modified, but locally species – rich plant and animal communities produced. It may well help in the design of field experimental studies required for effective management of sand dune nature reserves. It is probably true to say that this relatively benign use has done more to preserve the dune flora and fauna near built up areas than any other factor. In doing so it has helped to keep open the lines of migration between one dune system and another for recruitment of flora and fauna.

Afforestation

The primary reason for planting trees on sand dunes has always been to protect the surface of shifting sands which in the past have overwhelmed coastal settlements in many parts of the world on more exposed dune coasts. Plantings have also been made for amenity purposes, as at Holkham, Norfolk. Afforestation is not the best means of protecting dunes from coastal erosion as the trees shade out *Ammophila* and other plants capable of recruiting new coastal dunes at the strandline. Timber production is only significant on the very largest dune plantations, and then only behind the shelter of protection forest consisting of wind-deformed trees, themselves useless for timber production.

At Les Landes in France 250,000 acre (101,075 ha) of dune were afforested during the nineteenth century mainly with *Pinus maritima* (Macdonald 1954). English (1969) has described the technique developed on this coast by the French engineer Bremontier. A shallow sloping littoral dune is created with fences and *Ammophila* planting, and behind this lies the protection forest itself protecting the production forest. In 1949 forest fires destroyed 200,000 acre (80,940 ha) of this woodland and 82 people died. However, the speed of the fire was so rapid that seeds survived intact in cones and pine regenerated in the burnt areas (English 1969).

In Denmark 75,000 acre (30,352 ha) of coastal dune have been afforested and are managed by the State primarily as amenity woodlands. *Pinus mugo* is used as both *Pinus nigra* and *Pinus maritima* (widely used on dunes elsewhere) were found to be attacked and destroyed by the fungus *Crumenula pinea* after 15 to 20 years growth. Careful attention is paid to thatching felled areas with cut heather or to planting with *Ammophila* before new plantations are started (Thaarup 1954). Trees on dunes are very deep-rooted and not readily subject to wind throw.

Some 10,000 acre (4,047 ha) have been planted chiefly with *Pinus nigra* var. *calabrica* and some *P. maritima* and *P. sylvestris* in Great Britain (Macdonald 1954).

The immediate effect of afforestation is to increase the diversity of flowering plant species largely through introduction as mentioned earlier. As the trees mature they shade out the ground flora almost completely although certain species like *Goodyera repens* at Culbin, Moray (not present in the unplanted dunes), are widespread in the plantations.

Ovington (1950 and 1951) has studied changes in the soil environment due to afforestation on dunes. He found that the water table was lowered by 17 cm in 20 year old conifer plantations compared with unplanted areas at Tentsmuir, Fife. At both Culbin and Tentsmuir, the nutrient content decreased with afforestation and the soil acidity increased (Fig. 67) while the organic matter at the surface and the manganese content increased in

plantations over a 20 year period. Nutrients are bound up in the tree crop
and when this is removed the impoverished soil is highly vulnerable to
erosion.

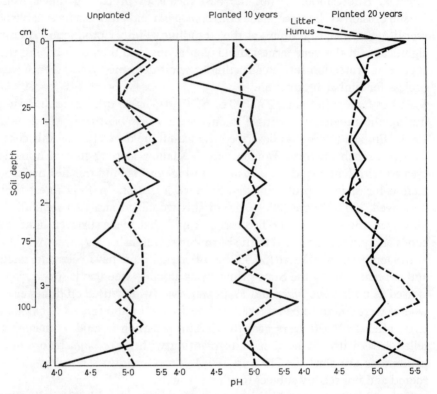

Fig. 67. The effect of Conifer planting upon soil pH at various depths in dune
soil from Culbin Sands, Morayshire. Two profiles for each area are shown
(from Ovington 1950).

Wright (1955) extended Ovington's studies to a wider variety of tree
species and age classes at Culbin and recorded soil moisture and temperature in plantation soils using gypsum soil moisture blocks and thermistor
techniques. The growth of trees dried out upper sand layers considerably
although the surface organic layers of older plantations had a high moisture-
holding capacity.

Little is known about the ecology of these dune forests or the young
plantations. Where seedling pines invade dune nature reserves as at Tentsmuir, Fife they have to be cut out regularly, to protect the native fauna and
flora.

Coast protection and amenity use

Ever since the stormy periods of the fourteenth and fifteenth centuries *Ammophila* has been planted to stabilize sand surfaces. Brown and Hafenrichter (1948) in an important series of papers describe experiments on the influence of date of planting, density, and different combinations of fertilizers on the growth of *Ammophila breviligulata*, *A. arenaria* and *Elymus mollis*. Charlton (in Anon 1970 *b*) has also carried out fertilizer trials on the growth of *Ammophila arenaria* and *Elymus arenarius* in Scotland. Thornton and Davis (1964) have selected and propagated genotypes of *Ammophila breviligulata* and studied germination of this species. Organic material of various kinds (e.g. forestry trimmings, roadside verge cuttings) is regularly used to protect bare sand from erosion. Haas and Steers (1964) describe a latex spray technique for stilling sand surfaces and Zak (1967) experimented with the use of hydraulically sprayed mulches and seed mixtures for stabilization purposes (Plate 16).

There is now a very extensive world literature on dune stabilization techniques and as this is being reviewed elsewhere it is not appropriate to deal with it here. Perhaps of more direct interest to the ecologist is the effect of the treatments on the plants and animals, the effect that the recent upsurge in tourist use of dunes is having upon them, and the techniques being used to study this.

Hewett (1970) has recorded the re-establishment of the dune flora within a 100 acre (40 ha) *Ammophila* plantation at Braunton Burrows, Devon and found that 53 species of flowering plants, two mosses and one lichen had colonized the bare sand in 15 years or less. These were all plants characteristic of the existing dune system and included several of the less common species. *Festuca rubra* was beginning to close up the gaps in the plantations and leguminous species were becoming increasingly common 10 to 15 years after stabilization.

No studies have yet been made on grasslands established on dunes by hydraulic seeding using cultivated grass seed. Where this has been done successfully, at Camber, Sussex for example, the dunes have been artificially graded before seeding to produce uniform slopes. Some habitat diversity has therefore already been lost and it may be many years before native species are able to reinvade. In fact this system of some 250 acre (100 ha), part of which is used as a Golf Links, may receive up to 17,000 people a day in summer and it seems likely there will be a constant need for repair of trampled turf by seeding and fertilizing. Shrub planting is now in progress and the system is clearly moving towards a very artificial habitat specifically designed for recreational use. Nevertheless it may still harbour many native species alongside the introduced ones.

Where human population pressures are rather lower as on parts of the East Lothian sand dune coast, dune restoration with native species, *Elymus arenarius* and *Ammophila arenaria*, and the judicious clearance of pathways through invasive growths of *Hippophaë rhamnoides*, help to retain much of the original character of the dune systems while still allowing large numbers of people access to the shores. (Tindall 1967 and Anon 1970 *b*.)

Low level aerial photography from a captive balloon coupled with quadrat ground survey has been used to record the distribution of people and plants at Whitesands and Yellowcraig, East Lothian (Duffield in 1970 *b*). Here a very detailed picture is being built up from which future changes can be measured.

Quinn (in Anon 1970 *b*) used conventional air photography to study recreational use over a 230 acre (85 ha) dune system at Brittas Bay, Wicklow, Ireland and recorded a maximum of 250 people per 50 m², falling to 120 to 80 people per 50 m² at 100 to 150 m distance from paths.

Goldsmith *et al* (1970) have studied the effect of trampling on dune vegetation in the Isles of Scilly. Schofield (in Duffey 1967) used electronic counters for monitoring the movement of people on dunes at Gibraltar Point, Lincolnshire, while Bayfield (1971) has shown how short soft metal wires can be set in paths and used to measure trampling by the proportion of wires that get bent flat.

These studies are just beginning, but it is clear that the ecologist is at last beginning to treat man as a highly significant animal in the dune landscape worthy of objective study.

Integrated Management

Because dune systems are now recognized as a valuable and limited resource not only for wildlife but also as recreational land, each country with dune resources clearly needs to develop a national plan for their protection and use. Such a plan would record the distribution and size of dune resources and designate those areas in which there is a priority for coast protection, recreational use or for protection of wildlife.

Any dune sites which are managed primarily for coast protection or recreation will still contain significant wildlife resources. Their protection is clearly relevant to the maintenance of the system for both these uses. Nevertheless the initiative to protect specific populations of rare or local species on such systems must lie with local naturalists or voluntary bodies. They can do much to help avoid unnecessary destruction by providing owners and planners with maps showing location of sites for which protection is desirable or by actually transplanting species to nearby safe areas from those which must be disturbed. The recording of such transplanting activities is clearly desirable.

Management for coast protection

Where coast protection is the primary aim any sand-winning activities should be gradually phased out, growth of trees should not be allowed to shade out dune-forming grasses near the coast and a regular maintenance commitment accepted at the coast. Air photography should be commissioned on a regular 3 to 5 year basis to record the success of management activities and to help understand the structural development of the system in relation to adjoining coastal changes. Instant stabilization can be employed in serious trouble spots with modern techniques referred to above. Elsewhere the principle of developing an aerodynamically stable shallow-sloping seaward face to the coast dune should be followed with conventional fencing and planting techniques using selected strains of appropriate grasses with use of fertilizers to aid establishment.

Management for recreation

Where recreational use is the primary object it is essential to provide convenient access to the shore where most people want to be, via specified pathways which effectively protect the sand from erosion. Where visitors are few natural vegetated pathways can be protected by rotational use. With increasing numbers of visitors fertilizing and regular repair of pathways by seeding becomes essential. Where large numbers of people need access to the shore artificially surfaced paths (plastic netting, wood, shells or gravel) must be provided.

The extent to which people penetrate back into the dunes from the shore is under active study at the present time. Many factors are involved here including the orientation of the dune coast, the freedom of the shore from pollution, the weather, and the type of vegetation on or behind the coast dune. Much could be done by judicious management of the grass/shrub balance to accommodate more people within dunes in relative privacy from one another, but there is a limit and at the present time we do not have the facts and figures which will tell us the optimum design for maximum acceptable densities.

While coast protection is primarily concerned with dune maintenance at the coast itself, a more comprehensive dune maintenance programme over the whole dune system is required where it is under intensive recreational use. Air photo monitoring may be required more frequently to detect changes in pathway patterns pointing to the need for closer access control or urgent restoration activities. Car park capacities must be related to holding capacity of the dune system for people and uncontrolled parking on dune turf is bound to lead to expensive control measures or ultimate abandonment of the site. The siting of car parks and caravan sites to the landward of the dunes should be designed to avoid releasing large numbers of people where the shore is narrow and sand supply for dune building minimal.

Management for wildlife protection

The objectives of management for coast protection and recreation are simply defined. They are to keep the sea out and to enable as many people as possible to enjoy the dune amenity without destroying it.

The objectives of management for wildlife protection are more complex and less easily defined and achieved. The British series of dune National Nature Reserves has been chosen to include representative physiographic and soil and climate-determined biological types throughout the country. Broadly speaking the objectives of management in these systems is to maintain the plant and animal communities for which they were originally selected, to utilize them for educational and research purposes and, where there is scope for this, to increase the diversity of habitats within them by controlled disturbance.

The maintenance of dune communities produced by a long history of intensive rabbit grazing, now much reduced since myxomatosis, presents special problems. It may be possible on one or two larger systems to enclose a captive rabbit population within rabbit-proof fencing. This would in fact be a return to the way in which these communities presumably started in specially created warrens. However, because of costs and the general undesirability of building up rabbit populations this could not be a universally acceptable policy. Cutting out of invasive trees and shrubs at least in selected areas is more generally acceptable and practicable where invasion is still in an early stage. This policy is already in practice in a number of reserves e.g. at Whiteford Burrows, Glamorgan where *Hippophaë* is removed and at Tentsmuir, Fife where unwanted pines are cut out.

Mowing could only be used on a limited scale and while it may help to maintain populations of certain low-growing species threatened by under-grazing. it will produce different communities to those characteristic of rabbit grazing. Carefully controlled sheep and pony grazing may be more generally applicable, but imply control of dogs which is not easily achieved near centres of population.

It seems inevitable that we must accept major changes towards scrub and woodland communities in some reserves where invasion is already well-advanced as at Ainsdale, Lancashire or Gibraltar Point, Lincolnshire. These new communities may eventually become as diverse and interesting as the species-rich dune grasslands they replace. However, it seems likely that they will themselves tend to be controlled by fire, as already happens at Studland, Dorset. It follows therefore, that specially designed fire-breaks will have to be created and maintained if fire control is not to become too destructive and lead to massive erosion.

As we have seen the problem of falling water tables can and has been tackled in Dutch dune reserves by active management of the water table. This again is costly and probably only applicable to few selected areas. The deliberate creation of pools by excavation can help to recreate late stage hydroseres where these have been lost by drying up or where as at Newborough Warren, Anglesey it was done to diversify the system. But this does bring the risk of attracting Herring Gulls which may in turn attack the young of other ground-nesting species such as terns. Wherever possible the natural formation of new slacks at the coast should be allowed to proceed unhindered to replenish those which in the normal course of development become drier as they mature.

Unlike the salt marsh habitat there is little opportunity for zoning different activities in time on dune systems, but there is a great deal to be gained from spacial zonation. The principle of resting sections from intensive educational or recreational use with temporary fencing can improve both coast protection and restoration of a trampled strand-line flora. Afforestation designed for recreational use can develop side by side with undisturbed wildlife sectors running from the coast to the landward limits in the bigger dune systems. Educational use can be separated from remoter research sites to cut down disturbance to the latter to a minimum.

It is already apparent in the British sand dune reserve series that systems showing a high recreational use are more appropriate for intensive educational use (e.g. Studland, Dorset) those readily accessible to research stations with lighter recreational use are more appropriate for research use (e.g. Holkham, Norfolk), while small remote systems with climatically distinctive communities (e.g. Invernaver, Sutherland) are best left undisturbed as much as possible.

We begin to see a pattern emerging which should be applicable not just to nature reserves, but to dune systems generally whatever their use: appropriate use of different systems; appropriate use of parts within the systems, joint care for the needs of people and of wildlife.

Q

References

ADAMS, D. A. (1963), 'Factors influencing vascular plant zonation in North Carolina salt marshes', *Ecology*, **44**, 445–456.

ADRIANI, E. D. (1945), 'Sur la Phytosociologie, la Synécologie et le bilan d'eau de Halophytes de la région Néerlandaise Méridionale, ainsi que de la Mediterranee Française.' *S.I.G.M.A.*, *Groningen*, **88**, 1–217.

AGUESSE, P. and MARAZANOF, F. (1965), 'Les modifications des milieux aquatiques de Camargues au cours des 30 dernières annés', *Ann. de Limiol.*, **1**, 163–190.

ALLEN, S. E., CARLISLE, A., WHITE, E. J. and EVANS, C. C. (1968), 'The plant nutrient content of rainwater'. *J. Ecol.*, **56**, 497–504.

ALVIN, K. L. (1960), 'Observations on the lichen ecology of South Haven Peninsula, Studland Heath, Dorset', *J. Ecol.*, **48**, 331–339.

ANDERSON, E. and STEBBINS, G. L. (1954), 'Hybridization as an evolutionary stimulus', *Evolution*, **8**, 378–388.

ANON (1966 *a*), *Solway Barrage*, Water Resources Board Report, London: H.M.S.O.

ANON (1966 *b*), *Morecambe Bay Barrage*, Water Resources Board Report, London: H.M.S.O.

ANON (1966 *c*), *Morecambe Bay and Solway Barrages*, Water Resources Board Report, London: H.M.S.O.

ANON (1967), *Dee crossing study. Phase 1*. Ministry of Housing and Local Government Report, London: H.M.S.O.

ANON (1970 *a*), *The Wash: estuary storage*, Water Resources Board Report, London: H.M.S.O.

ANON (1970 *b*) *Dune conservation 1970*, North Berwick Study Group Rep, North Berwick: East Lothian County Council.

ANON (1970 *c*), *Modern Farming and the Soil*, London: H.M.S.O.

ARDÖ, P. (1957), 'Studies in the marine shore dune ecosystem with special reference to the dipterous fauna', *Opusc. ent. Suppl.*, **14**, 1–255.

ARNOLD, A. (1955), Die Bedeutung der Chlorionen für die Pflanze. Bot. Stud. 2. Jena.

AUGUSTINE, M. T., THORNTON, R. B., SANBORN, J. M. and LEISER, A. T. (1964),

'Response of American Beachgrass to fertilizer', *J. Soil and Water Consvn.*, **19**, 112–116.

BAGNOLD, R. A. (1941), *The Physics of Blown Sand and Desert Dunes*, London; Methuen.

BAKER, J. M. (1970 *a*), 'Oil and salt marsh soil', Institute of Petroleum Symposium on the ecological effects of oil pollution on littoral communities. London, Morning Session 1–10.

BAKER, J. M. (1970 *b*), 'Growth stimulation following oil pollution', *Ibid*, 11–16.

(1970 *c*), 'Comparative toxicities of oils, oil fractions and emulsifiers', *Ibid*, 17–26.

(1970 *d*), 'The effects of oils on plant physiology', *Ibid*, 27–37.

(1970 *e*), 'The effects of a single oil spillage', Institute of Petroleum Symposium on the ecological effects of oil pollution on littoral communities. London, Afternoon Session, 1–5.

(1970 *f*), 'Successive spillages', *Ibid*, 7–18.

(1970 *g*), 'Refinery effluent', *Ibid*, 19–29.

(1970 *h*), 'Seasonal effects', *Ibid*, 31–38.

(1970 *i*), 'Effects of cleaning', *Ibid*, 39–44.

BAKKER, D., TER BORG, S. J. and OTZEN, D. (1966), 'Ecological research at the Plantecology Laboratory, State University, Groningen', *Wentia*, **15**, 1–24.

BALL, P. W. and BROWN, K. G. (1970), 'A biosystematic and ecological study of *Salicornia* in the Dee estuary', *Watsonia*, **8**, 27–40.

BARKLEY, S. Y. (1955), 'The morphology and vegetation of the sands of Forvie with reference to certain related areas', Ph.D. Thesis, Aberdeen.

BARNES, B. M. and BARNES, R. D. (1954), 'The ecology of the spiders of maritime drift lines', *Ecology*, **35**, 25–35.

BAYFIELD, N. G. (1971), 'A simple method for detecting variations in walker pressure laterally across paths', *J. appl. Ecol.*, **8**, 533–535.

BEEFTINK, W. G. (1962), 'Conspectus of the phanerogamic salt plant communities in the Netherlands', *Biol. Jaarb. Antwerp*, 325–362.

BEEFTINK, W. G. (1965), De zoutvegetatie van ZW – Nederland beschouwd in Europees Verbaud. Wageningen.

BEEFTINK, W. G. (1966), 'Vegetation and habitat of the salt marshes and beach plains in the south-western part of the Netherlands', *Wentia*, **15**, 83–108.

BERNATZKY, A. (1969), Die Bedeutung von Schutzpflanzungen gegen Luftverunreinigungen. Air Pollution. Proc. 1st. Europ. Congr. on Influence of Air Pollution on Plants and Animals, Wageningen 1968, 383–395.

BERNSTEIN, L. and PEARSON, G. A. (1956), 'Influence of exchangeable sodium on the yield and chemical composition of plants. 1. Green beans, garden beans, clover and alfalfa', *Soil Sci.*, **82**, 247–258.

BIEDERMAN, E. W. Jr. (1962), 'Distinction of shoreline environments in New Jersey', *J. Sediment. Petrol.*, **32**, 181–200.

BIGOT, M. L. (1958), Les grands caractères écologiques des milieux terrestes de Camargue, 3° Congr. Soc. Sav., 533–539.

BINET, P. (1964 *a*), 'Action de la température et de la salinité sur la germination des graines de *Plantago maritima* L.', *Bull. Soc. bot. Fr.*, **111**, 407–411.

BINET, P. (1964 *b*), 'La germination des semences des halophytes', *Bull. Soc. Fr. Physiol. Vég.*, **10**, 253–263.

BINET, P. (1965 *a*), Etudes d'écologie expérimentale et physiologique sur *Cochlearia anglica* L. I Etudes dans l'estuaire de l'Orne.' *Oecol Planta.*, **1**, 7–38.

BINET, P. (1965 *b*), 'Action de la température et de la salinité sur la germination des graines de *Cochlearia anglica* L.', *Revue gen. Bot.*, **72**, 221–236.

BINET, P. (1965 *c*), 'Action de divers rhythmes thermiques journaliers sur la germination des semences de *Triglochin maritima* L.', *Bull. Soc. Linn. Normandie Series* 10, **6**, 99–102.

BINET, P. (1965 *d*), 'Action de la température et de la salinité sur la germination des graines de *Glaux maritima* L'., *Bull. Soc. bot. Fr.*, **112**, 346–350.

BINET, P. (1965 *e*), 'Aptitude a germer en milieu salé de trois espèces de *Glyceria*: *G. borreri* Bab., *G. distans*. Wahlb. et *G. maritima* Wahlb.', *Bull Soc. bot. Fr.*, **113**, 361–367.

BIRSE, E. L. and GIMINGHAM, C. H. (1955), 'Changes in the structure of bryophytic communities with the progress of succession on sand dunes', *Trans. Br. Bryol. Soc.*, **2**, 523–531.

BIRSE, E. L., LANDSBERG, S. Y. and GIMINGHAM, C. H. (1957), 'The effects of burial by sand on dune mosses', *Trans. Br. Bryol. Soc.*, **3**, 285–301.

BIRSE, E. M. (1958), 'Ecological studies on growth-form in Bryophytes. III. The relationship between the growth-form of mosses and ground-water supply', *J. Ecol.*, **46**, 9–27.

BLACK, C. A. (1968), *Soil-plant Relationships*. 2nd edn. Chichester: J. Wiley.

BLANCHARD, B. (1952), An ecological survey of the vegetation of the sand dune system of the South West Lancashire coast, with special reference to an associated marsh flora. Ph.D. Thesis, Liverpool.

BLOCK, R. J. (1945), 'Amino acid composition of food proteins', Adv. Protein Chemistry. **2**, 119–134.

BÖCHER, T. W. (1952), 'Vegetationsudvikling iforhold til marin akkumulation', *Bot. Tidsskr.*, **49**, 1–32.

BOERBOOM, J. H. A. (1960), 'De plantengemeenschappen van de Wassenaarse duinen', *Meded. LandbHoogesch. Wageningen*, **60**, 1–135.

BOERBOOM, J. H. A. (1963), 'Het verband tussen bodem en vegetatie in de Wassenaarse duinen', *Boor en Spade*, **13**, 120–155.

BOERBOOM, J. H. A. (1964), 'Microclimatological observations in the Wassenaar dunes', *Meded. LandbHoogesch. Wageningen*, **64**, 1–28.

BOLLARD, E. G. and BUTLER, G. W. (1966), 'Mineral nutrition of plants', *A. Rev. Pl. Physiol.*, **17**, 77–112.

BOND, G., MacCONNELL, J. T. and McCULLUM, A. H. (1956), 'The nitrogen nutrition of *Hippophaë rhamnoides*, L.', *Ann. Bot. (N.S.)*, **20**, 501–512.

BOND, T. E. T. (1952), '*Elymus arenarius*. Biological Flora of the British Isles', *J. Ecol.*, **40**, 217–227.

BOORMAN, L. A. (1967, '*Limonium vulgare* Mill. and *L. humile* Mill, Biological flora of the British Isles', *J. Ecol.*, **55**, 221–232.

BOORMAN, L. A. (1968), 'Some aspects of the reproductive biology of *Limonium vulgare* Mill. and *Limonium humile* Mill.', *Ann. Bot.*, **32**, 803–824.

BOURN, W. S. and Cottam, C. (1950), Some biological effects of ditching tide-water marshes. U.S. Fish and Wildlife Service, Rep., **19**, 1–30.

BOYCE, S. G. (1954), 'The salt spray community', *Ecol. Mongr.*, **24**, 29–67.

BOYD, J. M. (1957), 'The Lumbricidae of a dune – machair soil gradient in Tiree, Argyll', Ann. Mag. nat. Hist., Ser. 12, **10**, 274–282.

BRADSHAW, A. D. (1958), 'Natural hybridization of *Agrostis tenuis* Sibth. and *A. stolonifera* L.', *New Phyt*, **57**, 66–84.

BRADSHAW, A. D. (1965), 'Evolutionary significance of phenotypic plasticity in plants', *Adv. Genet.*, **13**, 115–155.

BRAUN-BLANQUET, J. (1932), *Plant Sociology*, London: McGraw-Hill.

BRAYBROOKS, E. M. (1958), The general ecology of *Spartina townsendii* (*sic. S. anglica*) with special reference to sward build-up and degradation. M.Sc. Thesis, Southampton.

BRERETON, A. J. (1965). Pattern in salt marsh vegetation. Ph.D. Thesis, Univ. of Wales.

BRERETON, A. J. (1971), 'The structure of the species populations in the initial stages of salt-marsh succession', *J. Ecol.*, **59**, 321–338.

BROOKS, C. E. P. (1949), *Climate Through the Ages*. London: Ernest Benn.

BROWN, D. H. and BROWN, R. M. (1969), 'Lichen communities at Blakeney Point, Norfolk', *Trans. Norfolk Norwich Nat. Soc.*, **21**, 235–250.

BROWN, J. C. (1958), 'Soil fungi of some British sand dunes in relation to soil type and succession', *J. Ecol.*, **46**, 641–664.

BROWN, J. S. (1925), A study of coastal ground water. U.S. Geol. Survey Water Supply Paper 537, 16–17.

BROWN, R. L. and HAFENRICHTER, A. L. (1948), 'Factors influencing the production and use of beach-grass and dune-grass clones for erosion control. I. Effect of date of planting. II. Influence of density of planting. III. Influence of kinds and amounts of fertilizer on production', *J. Am. Soc. Agron.*, **40**, 512–521; 603–609; 677–684.

BROWNELL, P. F. (1965), 'Sodium as an essential micronutrient element for a higher plant (*Atriplex vesicaria*)', *Pl. Physiol.*, **40**, 460–468.

BURKHOLDER, P. R. (1956), Studies on the nutritive value of *Spartina* grass growing in the marsh areas of coastal Georgia', *Bull. Torrey bot. Club*, **83**, 327–334.

BURKHOLDER, P. R. and BURKHOLDER, L. M. (1956), 'Vitamin B_{12} in suspended solids and marsh muds collected along the coast of Georgia', *Limnol. Oceanogr.*, **1**, 202–208.

BURKHOLDER, P. R. and BORNSIDE, G. H. (1957), 'Decomposition of marsh grass by aerobic marine bacteria', *Bull. Torrey bot. Club*, **84**, 366–383.

BURNETT, J. H. (ed.) (1964), *The Vegetation of Scotland*. Edinburgh: Oliver and Boyd.

CAREY, A. E. and OLIVER, F. W. (1918), *Tidal Lands*. London: Blackie.

CHAPMAN, V. J. (1938), 'Studies in salt marsh ecology. Sections I–III', *J. Ecol.*, **26**, 144–179.

CHAPMAN, V. J. (1940), Succession on the New England salt marshes. *Ecology*, **21**, 279–282.

CHAPMAN, V. J. (1942), 'The new perspective in the Halophytes, *Q. Rev. Biol.*, **17**, 291–311.

CHAPMAN, V. J. (1944), 'Cambridge University expedition to Jamaica', *J. Linn. Soc.* (Bot), **52**, 407–533.

CHAPMAN, V. J. (1959), 'Studies in salt marsh ecology. IX. Changes in salt marsh vegetation at Scolt Head Island, Norfolk', *J. Ecol.*, **47**, 619–639.

CHAPMAN, V. J. (1960), *Salt Marshes and Salt Deserts of the World*. London: Leonard Hill.

CHAPMAN, V. J. and RONALDSON, J. W. (1958). The mangrove and salt marsh flats of the Auckland Isthmus. N.Z. Dept. Sci. and Indust. Res., Bull. 125, 1–79.

CHIPPINDALE, H. G. and MERRICKS, R. W. (1965), 'Gang-mowing and pasture management', *J. Br. Grassld Soc.*, **11**, 1–9.

CLAPHAM, A. R., PEARSALL, W. H. and RICHARDS, P. W. (1942), '*Aster tripolium.* Biological flora of the British Isles'. *J. Ecol.*, **30**, 385–395.

CLARKE, L. D. and HANNON, N. J. (1967), 'The mangrove swamp and salt marsh communities of the Sydney district. I Vegetation, soils and climate', *J. Ecol.*, **55**, 753–771.

CLARKE, L. D. and HANNON, N. J. (1969), 'The mangrove swamp and salt marsh communities of the Sydney district. II The Holocoenotic complex with particular reference to physiography', *J. Ecol.*, **57**, 213–234.

CLARKE, L. D. and HANNON, N. J. (1970), 'The mangrove swamp and salt marsh communities of the Sydney district. III. Plant growth in relation to salinity and waterlogging', *J. Ecol.*, **58**, 351–369.

CLARKE, S. M. (1965), Some aspects of the autecology of *Elymus arenarius* L. Ph.D. Thesis, Hull.

COTTON, M. J. (1967), 'Aspects of the ecology of sand dune arthropods', *Entomologist*, **100**, 157–165.

COOPER, W. S. (1958), 'Coastal sand dunes of Oregon and Washington', Geol. Soc. America Memoir, **72**. Baltimore.

COTTAM, C. and MUNRO, D. A. (1954), 'Eelgrass status and environmental relations', *J. Wildl. Mgmt*, **18**, 449–460.

COULL, J. (1968), 'Crofting townships and common grazings', *Agr. Hist. Rev.*, 16.

COWLES, H. C. (1899), 'The ecological relations of the vegetation on the sand dunes of Lake Michigan', *Bot. Gaz.*, **27**, 95–117; 167–202; 281–308; 361–391.

COWLES, H. C. (1911), 'A fifteen year study of advancing sand dunes', Rep. Br. Ass. 1911, 565.

CRAWFORD, R. M. M. and WISHART, D. (1966), 'A multivariate analysis of the development of dune slack vegetation in relation to coastal accretion at Tentsmuir, Fife', *J. Ecol.*, **54**, 729–743.

CRAWFORD, R. M. M. and WISHART, D. (1967), 'A rapid multivariate method for the detection and classification of groups of ecologically related species', *J. Ecol.*, **55**, 505–524.

CRAWFORD, R. M. M. and WISHART, D. (1968), 'A rapid classification and ordination method and its application to vegetation mapping', *J. Ecol.*, **56**, 385–404.

DAHL, R. G. (1959), 'Studies on Scandinavian *Ephydridae (Diptera, Brachycera)*', Opusc. ent. suppl., **15**, 1–224.

DALBY, D. H. (1962), 'Chromosome number, morphology and breeding behaviour in the British *Salicorniae*', *Watsonia*, **5**, 150–162.

DALBY, D. H. (1963), 'Seed dispersal in *Salicornia pusilla*', *Nature*, **199**, 197–198.

DALBY, D. H. (1970), 'The salt marshes of Milford Haven, Pembrokeshire', *Field Studies*, **3**, 297–330.

DALBY, R. (1957), 'Problems of land reclamation. 5. Salt marsh in the Wash', *Agric. Rev.*, **2**, 31–37.

DAVIS, J. H. (1940), The ecology and geologic role of mangroves in Florida. Carnegie Inst. Publ. 517, 303.

DAVIS, L. V. and GRAY, I. E. (1966), Zonal and seasonal distribution of insects in North Carolina salt marshes. Ecol. Monogr., **36**, 275–295.

DAY, J. H. (1951), 'The ecology of South African estuaries Pt. I. A review of estuarine conditions in general', *Trans. Roy. Soc. S. Afr.*, **33**, 53–91.

DEFANT, A. (1964), *Ebb and Flow*. New York: University of Michigan Press.

DUFF, S. and TEAL, J. M. (1965), Temperature change and gas exchange in Nova Scotia and Georgia salt marsh muds. Woods Hole Oceanographic Inst. Contrib. No. 1501, 67–73.

DUFFEY, E. (1967), The biotic effects of public pressure on the environment. Nature Conservancy, Monks Wood Experimental Station Symposium, **3**, 1–178.

DUFFEY, E. (1968), 'An ecological analysis of the spider fauna of sand dunes', *J. Anim. Ecol.*, **37**, 641–674.

DUVIGNEAUD, P. (1947), 'Remarques sur la végétation des pannes dans les dunes littorales entre La Panne et Dunkerque', *Bull. Soc. roy. Bot. Belg.*, **79**, 123–140.

EHRLICH, P. R. and RAVEN, P. H. (1969), 'Differentiation of populations', *Science*, **165**, 1128–1232.

ELLIS, E. A. (1960), 'The purple (hybrid) Marram, *Ammocalamagrostis baltica* (Fluegge) P. Fourn. in East Anglia', *Trans. Norfolk Norwich Nat. Soc.*, **19**, 49–51.

ELTON, C. S. (1966), '*The pattern of Animal Communities*'. London: Methuen.

ELZAM, O. E. and EPSTEIN, E. (1969), 'Salt relations of two grass species differing in salt tolerance. I. Growth and salt content at different salt concentrations', *Agrochimica*, **13**, 187–195.

ENGLISH, N. (1969), 'Les Landes', Nature Conservancy unpubld. typescript, 1–20.

EVANS, H. J. and SORGER, G. J. (1966), 'Role of mineral elements with emphasis on the univalent cations', *A. Rev. Pl. Physiol.*, **17**, 47–76.

FAEGRI, K. (1958), 'On the climatic demands of Oceanic plants', *Bot. notiser*, 3, 325–332.

FAEGRI, K. (1960), *The Distribution of Coast Plants*. Oslo: Oslo University Press.

FAIRBRIDGE, R. W. (1961), 'Eustatic changes in sea level', *Phys. Chem. Earth*, **4**, 99–185.

FLETCHER, H. C. and ELMENDORF, H. B. (1955), 'Phreatophytes – a serious problem in the West. U.S.', *Yearb. Agric.*, **1955**, 423–429.

FRAME, J. (1971), 'Fundamentals of grassland management. 10. The grazing animal', *Scottish Agric.*, **50**, 1–17.

FREIJSEN, A. H. J. (1967), *A Field Study of the Ecology of Centaurium vulgare* Rafn., Tilburg: H. Gianotten.

FOWDEN, L. (1959), *Physiologia Pl.*, **12**, 657–664.

GALLATIN, M. H., LUNIN, J. and BATCHELDER, A. R. (1962), 'Brackish water sources for irrigation along the eastern seaboard of the United States', *U.S. Dept. Agric. Prod. Resour Rep.*, **61**, 1–28.

GARRET, P. (1971), The sedimentary record of life on a modern tropical tidal flat, Andros Island, Bahamas. Ph. D. thesis John Hopkins Univ., Baltimore.

GATES, F. C. (1950), 'The disappearing Sleeping Bear Dune', *Ecology*, **31**, 386–392.

GEMMELL, A. R., GREIG-SMITH, P. and GIMINGHAM, C. H. (1953), 'A note on the behaviour of *Ammophila arenaria* (L.) Link in relation to sand-dune formation', *Trans. bot. Soc. Edinb.*, **36**, 132–136.

GIGLIOLI, M. E. C. and THORNTON, I. (1965), 'The mangrove swamps of Keneba, Lower Gambia river basin. I. Descriptive notes on the climate, the mangrove swamps and the physical conditions of their soils', *J. appl. Ecol.*, **2**, 81–103.

GILLHAM, M. E. (1957), 'Coastal vegetation of Mull and Iona in relation to salinity and soil reaction', *J. Ecol.*, **45**, 757–778.

GILLHAM, M. E. (1964), 'The vegetation of local coastal gull colonies', *Trans. Cardiff Nat. Soc.*, **91**, 23–33.

GILLIAM, H. T. (1957), *San Francisco Bay*. New York: Doubleday.

GILLNER, V. (1965), 'Salt marsh vegetation in Southern Sweden', *Acta Phytogeogr. Suecica*, **50**, 97–104.

GIMINGHAM, C. H. GEMMELL, A. R. and GREIG-SMITH, P. (1948), 'The vegetation of a sand dune system in the Outer Hebrides', *Trans. Proc. Bot. Soc. Edinb.*, **35**, 82–96.

GIMINGHAM, C. H. (1951), 'Contributions to the maritime ecology of St. Cyrus, Kincardineshire. Part II. The sand dunes', *Trans. Proc. Bot. Soc. Edinb.*, **35**, 387–414.

GIMINGHAM, C. H. (1953), 'Contributions to the maritime ecology of St. Cyrus, Kincardineshire. III. The salt marsh', *Trans. Proc. Bot. Soc. Edinb.*, **36**, 137–164.

GINSBURG, R. N., ISHAM, L. B., BEIN, S. J. and KUPERBERG, J. (1954), Laminated algal sediments of south Florida and their recognition in the fossil record: unpublished Rep. No. 54–21, Marine Laboratory, University of Miami, Coral Gables, Florida.

GINSBURG, R. N. and LOWENSTAM, H. A. (1958), 'The influence of marine bottom communities on the depositional environment of sediments', *J. Geol.*, **66**, 310–318.

GLOPPER, R. J. de (1964), 'About the water content and shrinkage of some Dutch lacustrine and marine sediments', *Neth. J. agric. Sci.*, **12**, 221–226.

GODWIN, H. SUGGATE, R. P. and WILLIS, E. H. (1958), 'Radiocarbon dating of the eustatic rise in ocean level', *Nature*, **181**, 1518–1519.

GOETHART, J. W. C., TESCH, P., HESSELINK, E. and DIJT, M. D. (1924), 'Cultuuren waterleidingbelangen wittreksel uit het rapport inzake het verband tusschen wateronttrekking en plantengroei', *Meded. Rijksboschb Proefstn. s'Gravenhage*, 1/3, 5, 5–28.

GOLDSMITH, F. B., MUNTON, R. J. C. and WARREN, A. (1970), 'The impact of reacreation on the ecology and amenity of semi-natural areas: methods of investigation used in the Isles of Scilly', *Biol. J. Linn. Soc.*, **2**, 287–306.

GOLUBIC, S. (1970), 'Effect of organic pollution on benthic communities', Marine Pollut. Bull. 1 (N.S.), 56–57.

GOOD, R. (1935), 'Contributions towards a survey of the plants and animals of South Haven Peninsula, Studland Heath, Dorset. II General ecology of the flowering plants and ferns', *J. Ecol.*, **23**, 361–405.

GOOD, R. (1964), *The Geography of the Flowering Plants*. (3rd edn.) London: Longman.

GOODMAN, P. J. and WILLIAMS, W. T. (1961), 'Investigations into 'die-back' in *Spartina townsendii* agg. III. Physiological correlates of 'die-back', *J. Ecol.*, **49**, 391–398.

GOODWIN. R. H. (ed.) (1961), 'Connecticut's coastal marshes. A vanishing resource', *The Connecticut Arboretum Bull.*, **12**, 1–36.

GORHAM, A. V. and GORHAM, E. (1965), 'Iron, manganese, ash and nitrogen in some plants from salt marsh and shingle habitats', *Ann. Bot.*, **19**, 571–577.

GORHAM, E. (1958 *a*), 'Soluble salts in dune sands from Blakeney Point in Norfolk', *J. Ecol.*, **46**, 373–379.

GORHAM, E. (1958 *b*), 'The influence and importance of daily weather conditions in the supply of chloride, sulphate and other ions to fresh waters from atmospheric precipitation', Phil. Trans. Roy. Soc. Series B, **241**, 147–178.

GOTTSCHALK, L. C. and JONES, V. H. (1955), 'Valleys and hills, erosion and sedimentation', *Yearb. U.S. Dep. Agric.*, 135–143.

GRAY, A. J. (1971), 'Variation in *Aster tripolium* L., with particular reference to some British populations'. Ph.D. Thesis, University of Keele.

GREEN, F. H. W. (1964), 'A map of annual average potential water deficit in the British Isles', *J. appl. Ecol.*, **1**, 151–158.

GREEN, R. D. and ASKEW, G. P. (1965), 'Observations on the biological development of macropores in soils of Romney Marsh', *J. Soil. Sci.*, **16**, 342–349.

GREENSMITH, J. T. and TUCKER, E. V. (1966), 'Morphology and evolution of inshore shell ridges and mud-mounds on modern intertidal flats, near Bradwell, Essex', *Proc. Geol. Ass.*, **77**, 329–346.

GRIM, R. E. (1953), *Clay Mineralogy*. New York: McGraw-Hill.

GRIME, J. P. (1965), 'Comparative experiments as a key to the ecology of flowering plants', *Ecology*, **46**, 513–515.

GRIST, D. H. (1959), *Rice*, London.

GUILCHER, A. and BERTHOIS, L. (1957), 'Cinq années d'observations sédimentologiques dans quartre estuaires-témoins de l'ouest de la Bretagne', *Rev. de Géomorph. Dynamique*, **5–6**, 67–86.

GUNKEL, W. (1968), 'Bacteriological investigations of oil-polluted sediments from the Cornish coast following the Torrey Canyon disaster. The biological effects of oil pollution on littoral communities', *Field Studies 2, supp*l., 151–158.

GUTNECHT, J. and DAINTY, J. (1968), 'Ionic relations of marine algae', *Oceanogr. Mar. Biol. Ann. Rev.*, **6**, 163–200.

HAAS, J. A. and STEERS, J. A. (1964), 'An aid to stabilization of sand dunes: experiments at Scolt Head Island', *Geogr. J.*, **130**, 265–267.

HADERLIE, E. C. (1970), 'Influence of pesticide run-off in Monterey Bay', *Mar. Pollut. Bull.* **1** (*N.S.*), 42–43.

HANNON, N. and BRADSHAW, A. D. (1968), 'Evolution of salt tolerance in two co-existing species of grass', *Nature*, **220**, 1342–1343.

HARBERD, D. J. (1961), 'Observations on population structure and longevity of *Festuca rubra*', *New Phytol.*, **60**, 184–206.

HARVEY, H. T. (1966 *a*), 'Marshes and mudflats of San Francisco Bay', San Francisco Bay Conserv. and Dev. Comm., San Francisco.

HARVEY, H. T. (1966 *b*), 'Some ecological aspects of San Francisco Bay', San Francisco Bay Conserv. and Dev. Comm., San Francisco.

HASSOUNA, M. G. and Wareing, P. F. (1964), 'Possible role of rhizosphere bacteria in the nitrogen nutrition of *Ammophila arenaria*', *Nature*, **202**, 467–469.

HEERDT, P. F. VAN and MÖRZER BRUYNS, M. F. (1960), 'A biocoenological investigation in the yellow dune region of Terschelling', *Tijdschr. Ent.*, **103**, 225–275.

HEIMANN, H. (1958), 'Irrigation with saline water and the ionic environment', 'Potassium-symposium'. 1958. Berne, 173–220.

HESSE, P. R. (1961), 'Some differences between the soils of *Rhizophora* and *Avicennia* mangrove swamps in Sierra Leone', *Pl. Soil*, **14**, 335–346.

HESSE, P. R. (1963), 'Phosphorus relationships in a mangrove swamp mud with particular reference to aluminium toxicity', *Pl. Soil*, **19**, 205–218.

HEWETT, D. G. (1970), 'The colonization of sand dunes after stabilization with Marram grass ' (*Ammophila arenaria*), *J. Ecol.*, **58**, 653–668.

HEWETT, D. G. (1971), 'The effects of the cold winter of 1962/63 on *Juncus acutus* at Braunton Burrows, Devon'. Devon Assoc. Adv. Sci. Lit. Art Rep. Trans., 1970, **102**, 193–201.

HIGGINS, L. S. (1933), 'An investigation into the problem of the sand dune areas on the South Wales coast', *Arch. Camb.*, June 1933.

HILL, T. G. and HANLEY, J. A. (1914), 'The structure and water content of shingle beaches', *J. Ecol.*, **2**, 21–38.

HINCKS, W. D., MICHAELIS, H. N., SHAW, S., BRAHAM, A. C., MURGATROYD, J. H. and BUTLER, P. M. (1951–4), 'The entomology of Spurn Peninsula', *Naturalist*, 1951: 75–86, 139–46, 183–90; 1952: 131–8, 169–76; 1953: 125–40, 157–72; 1954: 74–8, 95–109.

HINDE, H. P. (1954), 'Vertical distribution of salt marsh phanerogams in relation to tide levels', *Ecol. Mon.*, **24**, 209–225.

HITCHCOCK, A. S. (1904), 'Methods used for controlling and reclaiming sand dunes', U.S. Department of Agriculture, Bureau of Plant Industry Bull. No. 57.

HOLDER, F. W. (1953), 'Changing flora of the South Lancashire dunes', *N.West. Nat.*, **1** (N.S.), 451–452.

HOPE-SIMPSON, J. F. and JEFFERIES, R. L. (1966), 'Observations relating to vigour and debility in Marram grass (*Ammophila arenaria* (L.) Link)', *J. Ecol.*, **54**, 271–274.

HOPKINS, B. (1962), 'The measurement of available light by the use of *Chlorella*', *New Phyt.*, **61**, 221–223.

HUBBARD, C. E. (1968), *Grasses*. (2nd edn.), Harmondsworth: Penguin.

HUBBARD, J. C. E. (1965), '*Spartina* marshes in Southern England. VI. Pattern of invasion in Poole Harbour', *J. Ecol.*, **53**, 799–813.

HUBBARD, J. C. E. (1969), 'Light in relation to tidal immersion and the growth of *Spartina townsendii* (s.l.)', *J. Ecol.*, **57**, 795–804.

HUBBARD, J. C. E. (1970), 'Effects of cutting and seed production in *Spartina anglica*', *J. Ecol.*, **58**, 329–334.

HUBBARD, J. C. E. and RANWELL, D. S. (1966), 'Cropping *Spartina* salt marsh for silage', *J. Br. Grassld Soc.*, **21**, 214–217.

HUBBARD, J. C. E. and STEBBINGS, R. E. (1968), '*Spartina* marshes in Southern England. VII Stratigraphy of the Keysworth marsh, Poole Harbour', *J. Ecol.*, **56**, 707–722.

HUGHES, G. P. (1953), 'The effect on agriculture of the East Coast floods', Unpublished report by the National Agricultural Advisory Service, Ministry of Agriculture Fisheries and Food, 1–262.

HULME, B. A. (1957), 'Studies on some British species of *Atriplex* L.', Ph.D. Thesis, University of Edinburgh.

HUNT, O. J. (1965), 'Salt tolerance in intermediate wheatgrass *Agropyron intermedium*', *Crop Sci.*, **5**, 407–409.

INGLIS, C. C. and ALLEN, F. H. (1957), 'The regimen of the Thames Estuary as affected by currents, salinities and river flow', *Proc. Instn Civ. Engrs*, **7**, 827–878.

INGLIS, C. C. and KESTNER, F. J. T. (1958 a), 'Changes in the Wash as affected by training walls and reclamation works', *Proc. Instn Civ. Engrs.*, **11**, 435–466.

INGIS, C. C. and KESTNER, F. J. T. (1958 b), 'The long-term effects of training walls, reclamation, and dredging on estuaries', *Proc. Instn Civ. Engrs*, **9**, 193–216.

IVERSON, J. (1936), *Biologische Pflanzentypen als Hilfsmittel in der Vegetationsforschung*. Copenhagen: Medd. Fra. Skalling-Labor. Bd., 4.

IVERSON, J. (1954), 'The zonation of the salt marsh vegetation of Skallingen in 1931–4 and in 1952', *Meddr Skalling – Lab.*, **14**, 113–118.

JACOBSEN, N. K. (1960), 'Types of sedimentation in a drowned delta region', *Geogr. Tidsskr.*, **59**, 58–69.

JACOBSEN, N. K. (1964), 'Troek af Tøndermarskens naturgeografi med saerligt henblik på morfogenesen', *Folia Georgr. Dan.*, **7**, 1–350.

JAKOBSEN, B., JENSEN, K. M. and NIELSEN, N. (1955), 'Forlag til landvindingsarbejder langs den sømderjyske vadehavskyst', *Geogr. Tidsskr.*, **55**, 62–87.

JAKOBSEN, B. (1961), 'Vadehavets sedimentomsoetning belyst ved kvantitative målinger', *Geogr. Tidsskr.*, **60**, 87–103.

JAKOBSEN, B. (1964), 'Vadehavets morfologi', *Folia Geogr. Dan.*, **11**, 1–176.

238   REFERENCES

JENNINGS, J. N. (1964), 'The question of coastal dunes in tropical, humid climates', *Z. Geomorph.*, **8**, 150–154.

JOHNSON, C. G. and SMITH, L. P. (Eds.) (1965). *The Biological Significance of Climatic Changes in Britain*. London: Institute of Biology and Academic Press.

JOHNSON, D. S. and YORK, H. H. (1915), The relation of plants to tide levels, Carnegie Institute Washington Publication, 206.

JONES, R. and ETHERINGTON, J. R. (1971), Comparative studies of plant growth and distribution in relation to water-logging IV. The growth of dune and slack plants, *J. Ecol.*, **59**, 793–801.

JOSEPH, A. F. and OAKLEY, H. B. (1929), 'The properties of heavy alkaline soils containing different exchangeable bases', *J. agric. Sci.*, **19**, 121–131.

KALLE, K. (1958), Sea water as a source of mineral substances for plants, Nature Conservation (London) Translation No. 11.

KAMPS, L. F. (1962), 'Mud distribution and land reclamation in the eastern Wadden shallows', *RijkswatSt. Commun.*, No. 4, 1–73.

KASSAS, M. (1957), 'On the ecology of the Red Sea coastal land', *J. Ecol.*, **45**, 187–203.

KELLEY, W. P. (1951), *Alkali Soils: their Formation, Properties and Reclamation*. Monograph No. 111. New York: Rheinhold.

KESTNER, F. J. T. and INGLIS, C. C. (1956), 'A study of erosion and accretion during cyclic changes in an estuary and their effect on reclamation of marginal land', *J. Agric. Engng. Res.*, **1**, 63–67.

KESTNER, F. J. T. (1961), 'Short term changes in the distribution of fine sediments in estuaries', *Proc. Instn. Civ. Engrs.*, **19**, 185–208.

KESTNER, F. J. T. (1962), 'The old coastline of the Wash', *Geogr. J.*, **128**, 457–478.

KESTNER, F. J. T. (1963), The supply and circulation of silt in the Wash. 10th Congr. International Association Hydraulic Research, London, 231–238.

KIDSON, C. and CARR, A. P. (1961), 'Beach drift experiments at Bridgwater Bay, Somerset', *Proc. Bristol Nat. Soc.*, **30**, 163–180.

KING, C. A. M. (1972), *Beaches and Coasts* (2nd edn.), London: Edward Arnold.

KRINSLEY, D. H. and FUNNELL, B. M. (1965), 'Environmental history of quartz sand grains from the Lower and Middle Pleistocene of Norfolk, England', *Q J. Geol. Soc. Lond.*, **121**, 435–461.

KRUMBEIN, W. C. and SLACK, H. A. (1956), 'The relative efficiency of beach sampling methods', *Tech. Memo. Beach Eros. Bd U.S.*, **90**, 1–34.

LAING, C. (1954), The ecological life history of the marram grass community on Lake Michigan dunes. Ph.D. dissertation, University of Chicago.

LAMB, H. H. (1969), 'The new look of climatology', *Nature*, **223**, 1209–1215.

LAMB, H. H. (1970), The variability of climate. Met. Office, Bracknell, unpublished typescript, 1–22.

LAMBINON, J. (1956), 'Aperçu sur les groupements végétaux du district maritime Belge entre La Panne et Coxyde', *Bull. Soc. Roy. Bot. Belg.*, **88**, 107–127.

LANDSBERG, S. Y. (1956), 'The orientation of dunes in relation to wind', *Geogr. J.*, **122**, 176–189.

LANDSBERG, H. and RILEY, N. A. (1943), Wind influences on the transportation of sand over a Michigan sand dune. Proceedings 2nd. Hydraulics Conference Bulletin 27, Univ. Iowa Studies in Engineering.

LARSEN, H. (1967), 'Biochemical aspects of extreme Halophilism', *Adv. Microb. Physiol.*, **1**, 97–132.

LAUFF, G. H. (ed.) (1967), Conference on estuaries. Jekyll Island (Ga.), 1964. Washington.

LEEUWEN, C. G. van (1965), 'Het verband tussen naturrlijke en anthropogene landschapsvormen, bezien vanuit de betrekkingen in grenzmilieu's', *Gorteria*, **2**, 93–105.

LINES, R. (1957), 'Estimation of exposure by flags', Report on Forestry Research, H.M.S.O. London 1957, 47–48.

LOPEZ-GONZALEZ, J. de and JENNY, H. (1959), 'Diffusion of strontium in ion-exchange membranes', *J. Colloid Sci.*, **14**, 533–542.

LOWE-MCCONNELL, R. H. (ed.) (1966), *Man-made Lakes*. London: Institute of Biology and Academic Press.

LUGG, J. W. H. (1949), 'Plant Proteins', *Adv. Protein Chem.*, **5**, 230–295.

LUXTON, M. (1964), 'Some aspects of the biology of salt marsh Acarina', *Acaralogia*. C.R. 1er Congrès Int. d'Acaralogie, Fort Collins, Colorado, U.S.A. 1963, 172–182.

LYNCH, J. J. (1941), 'The place of burning in management of the Gulf Coast refuges', *J. Wildl. Mgmt.*, **5**, 454–458.

LYNCH, J. J., O'NEIL, T. and LANG, D. W. (1947), 'Management significance of damage by geese and muskrats to Gulf Coast marshes', *J. Wildl. Mgmt.*, **2**, 50–76.

MAAREL, E. VAN DER (1966), 'Dutch studies on coastal sand dune vegetation, especially in the delta region', *Wentia*, **15**, 47–82.

MAAREL, E. VAN DER and LEERTOUWER, J. (1967), 'Variation in vegetation and species diversity along a local environmental gradient', *Acta. Bot. Neerl.*, **16**, 211–221.

MAAREL, E. VAN DER and WESTHOFF, V. (1964), 'The vegetation of the dunes near Oostvoorne', *Wentia*, **12**, 1–61.

MACDONALD, J. (1954), 'Tree planting on coastal sand dunes in Great Britain', *Adv. Sci.*, **11**, 33–37.

MACDONALD, K. B. (1969), 'Quantitative studies of salt marsh mollusc faunas from the North American Pacific coast', *Ecol. Mongr.*, **39**, 33–60.

MACNAE, W. (1966), 'Mangroves in eastern and southern Australia', *Aust. J. Bot.*, **14**, 67–104.

MACNAE, W. (1968), 'A general account of the fauna and flora of mangrove swamps and forests in the Indo-West-Pacific Region', *Adv. Mar. Biol.*, **6**, 73–270.

MARCHANT, C. J. (1967), 'Evolution in *Spartina* (Graminae) I. The history and morphology of the genus in Britain', *J. Linn. Soc. (Bot.)*, **60**, 1–24.

MARCHANT, C. J. (1968), 'Evolution in *Spartina* (Graminae) II. Chromosomes, basic relationships and the problem of *S. x townsendii* agg.,' *J. Linn. Soc. (Bot.)*, **60**, 381–409.

240 REFERENCES

MARPLES, T. G. (1966), 'A radionuclide tracer study of Arthropod food chains in a *Spartina* salt marsh ecosystem', *Ecology*, **47**, 270–277.

MARSHALL, J. K. (1965), '*Corynephorus canescens* (L.) P. Beauv. as a model for the *Ammophila* problem', *J. Ecol.*, **53**, 447–463.

MARTIN, A. C. (1953), 'Improving duck marshes by weed control. U.S. Fish and Wildlife Service Circular', **19**, 1–49.

MARTIN, W. E. (1959), 'The vegetation of Island Beach State Park, New Jersey', *Ecol. Mongr.*, **29**, 1–46.

MATTHEWS, J. R. (1937), 'Geographical relationships of the British Flora', *J. Ecol.*, **25**, 1–90.

MAY, V. J. (1969), 'Reclamation and shore line change in Poole Harbour, Dorset', *Proc. Dorset Nat. Hist. Archaeol. Soc.*, **90**, 141–154.

McCRONE, A. (1966), 'The Hudson river estuary. Hydrology, sediments and pollution', *Geogr. Rev.*, **56**, 175–189.

McROY, C. P. (1969), 'Eelgrass under Arctic winter ice', *Nature*, **224**, 818–819.

MOBBERLEY, D. G. (1956), 'Taxonomy and distribution of the genus *Spartina*', *Iowa St. J. Sci.*, **30**, 471–574.

MØLLER, J. T. (1963), 'Accumulation and abrasion in a tidal area', *Geogr. Tidsskr.*, **62**, 56–79.

MØLLER, J. T. (1964), *Fladkystems og Flodens Morfologiske Elementer*, Copenhagen: K. G. Wingstrand.

MONTFORT, C. and BRANDRUP, W. (1927), 'Physiologische und Pflanzengeographische Seesalzwirkungen II. Okologische Studien über Keimung und erste Entwicklung bei Halophyten', *Jb. Wiss. Bot.*, **66**, 902–946.

MOORE, P. D. (1971), 'Computer analysis of sand dune vegetation in Norfolk, England, and its implications for convservation', *Vegetatio*, **23**, 323–338.

MÖRZER BRUYNS, M. F. and WESTHOFF, V. (1951), The Netherlands as an environment of insect life. 9th Int. Congr. Entom., Amsterdam.

MYRICK, R. M. and LEOPOLD, L. B. (1963), 'Hydraulic geometry of a small tidal estuary', U.S. Geol. Survey Professional Paper, 422–B, 1–18.

NELSON-SMITH, A. (1968), *A Classified Bibliography of Oil Pollution*, Swansea: University College (typescript 1–51).

NEWELL, R. C. (1965), 'The role of detritus in the nutrition of two marine deposit feeders, the Prosobranch *Hydrobia ulvae* and the bivalve *Macoma balthica*, *Proc. Zool. Soc. Lond.*, **144**, 25–45.

NEWMAN, W. S. and RUSNAK, G. A. (1965), 'Holocene submergence of the Eastern shore of Virginia', *Science*, **148**, 1464–1466.

NICHOLSON, I. A. (1952), A study of *Agropyron junceum* (Beauv.) in relation to the stabilization of coastal sand and the development of sand dunes. M.Sc. Thesis, University of Durham.

NIELSEN, N. (1935), 'Eine methode zur exakten sedimentations-messung studien über die marschbildung auf der halbinsel Skalling. Danske Videnskabernes Selskab', *Biol. Meddr.*, **12**, 1–96.

NOBUHARA, H. (1967), 'Analysis of coastal vegetation on sandy shore by biological types in Japan', *Jap. J. Bot.*, **19**, 325–351.

NORTHFIELD, J. (1960), 'The bryophyte flora of Studland Heath', *Durham Colleges Nat. Hist. Soc. J.*, **7**, 38–45.

ODUM, E. P. (1961), 'The role of tidal marshes in estuarine production', N.Y. State Conservationist Information Leaflet.

ODUM, E. P. (1962), 'Relationship between structure and function in the ecosystem', *Jap. J. Ecol.*, **12**, 108–118.

ODUM, E. P. (1963), 'Primary and secondary energy flow in relation to ecosystem structure', Proceedings of XVI International Congr. Zool., **4**, 336–338.

ODUM, E. P. and CRUZ, A. A. de la (1967), 'Particulate organic detritus in a Georgia salt marsh – estuarine ecosystem'. Estuaries. Publication No. 83 American Association for the Advancement of Science, Washington, 383–388.

ODUM, E. P. and SMALLEY, A. E. (1959), 'Comparison of population energy flow of a herbivorous and deposit-feeding invertebrate in a salt marsh ecosystem', *Proc. Natn. Acad. Sci., U.S.A.*, **45**, 617–622.

OKUDA, A. and Takahashi, E. (1965), The role of silicon, in *Mineral Nutrition of the Rice Plant*, Ch. 10, 123–146 (Proc. Intern. Conf. Rice Res. Inst., Los Bañjos, Philippines, 1964. John Hopkins Press, Baltimore).

OLIVER, F. W. (1929), 'Blakeney Point Reports', *Trans. Norfolk Norwich Nat. Soc.*, **12**, 630–653.

OLSON, J. S. (1958 *a*), 'Rates of succession and soil changes on Southern Lake Michigan sand dunes', *Bot. Gaz.*, **119**, 125–170.

OLSON, J. S. (1958 *b*), 'Lake Michigan dune development. 1. Wind-velocity profiles', *J. Geol.*, **66**, 254–263.

OLSON, J. S. (1958 *c*), 'Lake Michigan dune development. 2. Plants as agents and tools in geomorphology', *J. Geol.*, **66**, 345–351.

OLSSON-SEFFER, P. (1909), 'Hydrodynamic factors influencing plant life on sandy sea shores', *New Phytol.*, **8**, 37–49.

OOSTING, H. J. (1954), Ecological processes and vegetation of the maritime strand in the United States. *Bot. Rev.*, **20**, 226–262.

OSBORN, T. G. B. and ROBERTSON, R. N. (1939), 'A reconnaissance survey of the vegetation of Myall Lakes', *Proc. Linn. Soc. N.S.W.*, **64**, 279–296.

OVINGTON, J. D. (1950), 'The afforestation of the Culbin sands', *J. Ecol.*, **38**, 303–319.

OVINGTON, J. D. (1951), 'The afforestation of Tentsmuir Sands', *J. Ecol.*, **39**, 363–375.

PACKHAM, J. R. and LIDDLE, M. J. (1970), 'The Cefni salt marsh, Anglesey and its recent development', *Fld Stud.*, **3**, 331–356.

PARHAM, M. R. (1970), A Comparative study of the mineral nutrition of selected halophytes and glycophytes. Ph.D. Thesis University of East Anglia.

PAVIOUR-SMITH, K. (1956), 'The biotic community of a salt meadow in New Zealand', *Trans. Roy. Soc. N.Z.*, **83**, 525–554.

PAYNE, K. T. (1972 – in press), 'A survey of the *Spartina* feeding insects in Poole Harbour, Dorset', *Entomologist's mon. Mag.*

PERKINS, E. J., WILLIAMS, B. R. H. and BAILEY, M. (1963), 'Some preliminary notes on the bottom currents of the Solway Firth and North East Irish sea', *Trans. Dumfries and Galloway Nat. Hist. and Antiq. Soc. Ser.* 3, **41**, 45–51.

PERKINS, E. J. and WILLIAMS, B. R. H. (1965), 'Some results of an investigation of the biology of the Solway Firth in relation to radioactivity', *Trans. J. Dumfries. Galloway nat. Hist. Antiq. Soc., Ser.* **3, 42**, 1–5.

PERRING, F. (1970), 'The flora of a changing Britain', *Bot. Soc. British Isles Rep.* No. 11. Hampton, Middlesex.

PERRING, F. H. and WALTERS, S. M. (1962), 'Atlas of the British Flora', London: Nelson.

PESTRONG, R. (1965), 'The development of drainage patterns on tidal marshes', Stanford University Publication Geological Science, 10, 1–87.

PETCH, C. P. (1945), 'Reclaimed lands of West Norfolk', *Trans. Norfolk Norwich Nat. Soc.*, **16**, 106–109.

PETERSEN, C. G. J. (1915), 'A preliminary result of the investigations on the valuation of the sea', *Rep. Dan. biol. Stn.*, **23**, 29–32.

PETERSEN, C. G. J. (1918), 'The sea bottom and its production of fish food', *Rep. Dan. biol. Stn.*, **25**, 1–62.

PETTERSSON, O. (1914), 'Climatic variations in historic and prehistoric time', *Svenska hydrogr.-biol. Kommn. Skr.*, **5**.

PHILLIPS, A. W. (1964), 'Some observations on coast erosion studies at South Holderness and Spurn Head', *Dock Harb. Auth.*, **45**, 64–66.

PISSAREV, V. E. and VINOGRADOVA, N. M. (1944), 'Hybrids between wheat and *Elymus*', C.r. *Dokl. Proc. Acad. Sci. U.S.S.R.*, **45**, 129–132.

POEL, L. W. (1960), 'The estimation of oxygen diffusion rates in soils', *J. Ecol.*, **48**, 169–177.

POMEROY, L. E. (1959), 'Algal productivity in salt marshes of Georgia', *Limnol. Oceanogr.*, **4**, 386–395.

POORE, M. E. D. and ROBERTSON, V. C. (1964), An approach to the rapid description and mapping of biological habitats. International Biological Programme Publication. London.

PORTER, J. J. (1962), 'Electron microscopy of sand surface textures', *J. Sedim. Petrol.*, **32**, 124–135.

PRAEGER, R. L. (1913), 'On the buoyancy of the seeds of some Brittanic plants', *Proc. Roy. Dub. Soc.*, **14**, 13–62.

PRESTON, A. (1968), Radioactive waste disposal, radioecology and radiobiology, Annual Report Fisheries Laboratory, Ministry of Agriculture Fisheries and Food, London. 108.

PRICE, W. A. and KENDRICK, M. P. (1963), 'Field model investigation into the reasons for silting in the Mersey estuary', *Proc. Inst. Civ. Engrs.*, **24**, 273–518.

PURER, E. A. (1942), 'Plant ecology of the coastal salt marshes of San Diego County, California', *Ecol. Mon.*, **12**, 81–111.

RAGOTSKIE, R. A. (1959), Proc. Salt Marsh Conf. Marine Inst. University of Georgia Publication, Athens, Georgia.

RAMALEY, F. (1918), 'Notes on dune vegetation at San Francisco, California', *Pl. Wld*, **21**, 191–201.

RANWELL, D. S. (1955), Slack vegetation, dune system development and cyclical change at Newborough Warren, Anglesey. Ph.D. Thesis, University of London.

RANWELL, D. S. (1958), 'Movement of vegetated sand dunes at Newborough Warren, Anglesey', *J. Ecol.*, **46**, 83–100.

RANWELL, D. S. (1959), 'Newborough Warren, Anglesey. I. The dune system and dune slack habitat', *J. Ecol.*, **47**, 571–601.

RANWELL, D. S. (1960 *a*), 'Newborough Warren, Angelsey. II. Plant associes and succession cycles of the sand dune and dune slack vegetation', *J. Ecol.*, **48**, 117–141.

RANWELL, D. S. (1960 *b*), 'Newborough Warren, Anglesey. III. Changes in the vegetation on parts of the dune system after the loss of rabbits by myxomatosis', *J. Ecol.*, **48**, 385–395.

RANWELL, D. S. (1961), '*Spartina* salt marshes in Southern England. I. The effects of sheep grazing at the upper limits of *Spartina* marsh in Bridgwater Bay', *J. Ecol.*, **49**, 325–340.

RANWELL, D. S. (1964 *a*), '*Spartina* salt marshes in Southern England. II. Rate and seasonal pattern of sediment accretion', *J. Ecol.*, **52**, 79–94.

RANWELL, D. S. (1964 *b*), '*Spartina* salt marshes in Southern England. III. Rates of establishment, succession and nutrient supply at Bridgwater Bay, Somerset', *J. Ecol.*, **52**, 95–105.

RANWELL, D. S. (1967), 'World resources of *Spartina townsendii* (*sensu lato*) and economic use of *Spartina* marshland', *J. Appl. Ecol.*, **4**, 239–256.

RANWELL, D. S. (1968 *a*), Coastal marshes in perspective. Regional studies Group Bull. Strathclyde No. 9, 1–26.

RANWELL, D. S. (1968 *b*), 'Extent of damage to coastal habitats due to the Torrey Canyon incident', Fld Stud., **2** (Suppl,), 39–47.

RANWELL, D. S. (ed.) (1972), The Management of Sea Buckthorn (*Hippophaë rhamnoides* L.) on selected sites in Great Britain. Nature Conservancy Report. London.

RANWELL, D. S., BIRD, E. C. F., HUBBARD, J. C. E. and STEBBINGS, R. E. (1964), '*Spartina* salt marshes in Southern England. V. Tidal submergence and chlorinity in Poole Harbour', *J. Ecol.*, **52**, 627–641.

RANWELL, D. S. and DOWNING, B. M. (1959), 'Brent goose (*Branta bernicla* L.) winter feeding pattern and *Zostera* resources at Scolt Head Island, Norfolk', *Anim. Behav.*, **7**, 42–56.

RANWELL, D. S. and DOWNING, B. M. (1960), 'The use of Dalapon and Substituted Urea herbicides for control of seed-bearing *Spartina* (Cord-grass) in inter-tidal zones of estuarine marsh', *Weeds*, **8**, 78–88.

RANWELL, D. S. and HEWETT, D. (1964), 'Oil pollution in Poole Harbour and its effect on birds', *Bird Notes*, **31**, 192–197.

RANWELL, D. S. and STEBBINGS, R. E. (1967), Report on the effects of Torrey Canyon oil pollution and decontamination methods in Cornwall and Brittany, March to April 1967. Nature Conservancy (London) Unpublished rep., 1–12.

RATCLIFFE, D. A. (1968), 'An ecological account of Atlantic Bryophytes in the British Isles', *New Phytol.*, **67**, 365–439.

REDFIELD, A. C. (1965), 'Ontogeny of a salt marsh estuary', *Science*, **147**, 50–55.

REDFIELD, A. C. and RUBIN, M. (1962), 'The age of salt marsh peat and its relation to recent changes in sea level at Barnstable, Massachusetts', *Proc. natn. Acad. Sci. U.S.A.*, **48**, 1728–1735.

REID, C. (1913), *Submerged Forests*. Cambridge: University Press.

RENGER, M. (1965), 'Berechnung der Austanchkapazität der organischen und anorganischen Anteile der Böden', *Z. Pfl-ernähr. Düng. Bodenk.*, **110**, 10–26.

REYNOLDSON, T. B. (1955), 'Observations on the earthworms of North Wales', *N. West Nat.*, Sept./Dec., 291–304.

REZK, M. R. (1970), 'Vegetation change from a sand dune community to a salt marsh as related to soil characters in Mariut District, Egypt', *Oikos*, **21**, 341–343.

RICHARDS, F. J. (1934), 'The salt marshes of the Dovey estuary. IV. The rates of vertical accretion, horizontal extension and scarp erosion', *Ann. Bot.*, **48**, 225–259.

RICHARDS, P. W. (1929), 'Notes on the ecology of the bryophytes and lichens at Blakeney Point, Norfolk', *J. Ecol.*, **17**, 127–140.

RIDLEY, H. N. (1930), *The Dispersal of Plants throughout the World*, Ashford: L. Reeve.

RILEY, G. A. (1963), 'Organic aggregates in sea water and the dynamics of their formation and utilization', *Limnol. Oceanogr.*, **8**, 373–381.

RITCHIE, W. and MATHER, A. (1969), *The Beaches of Sutherland*, Aberdeen: Dept. Geography, Univ. of Aberdeen.

RITCHIE, W. and MATHER, A. S. (1971), 'Conservation and use: Case-study of Beaches of Sutherland, Scotland', *Biol. Consvn.*, **3**, 199–207.

ROBERTSON, D. A. (1955), The ecology of the sand dune vegetation of Ross Links, Northumberland with special reference to secondary succession in the blow-outs. Ph.D. Thesis, Durham University.

ROBINSON, A. H. W. (1966), 'Residual currents in relation to shoreline evolution of the East Anglian coast', *Mar. Geol.*, **4**, 57–84.

ROBINSON, T. W. (1952), 'Phreatophytes and their relation to water in Western United States. Symposium on Phreatophytes', *Trans. Am. geophys. Un.*, **33**, 57–61.

ROBSON, M. J. and JEWISS, O. R. (1968), 'A comparison of British and North African varieties of tall Fescue (*Festuca arundinacea*) II. and III.', *J. appl. Ecol.*, **5**, 179–190 and 191–204.

RORISON, I. H. (ed.) (1969), *Ecological Aspects of the Mineral Nutrition of Plants*, Oxford: Blackwell.

SAKAI, A. (1970), 'Freezing resistance in willows from different climates', *Ecology*, **51**, 485–491.

SALISBURY, E. J. (1925), 'Note on the edaphic succession in some dune soils with special reference to the time factor', *J. Ecol.*, **13**, 322.

SALISBURY, E. J. (1934), 'On the day temperatures of sand dunes in relation to the vegetation at Blakeney Point, Norfolk', *Trans. Norfolk Norwich Nat. Soc.*, **13**, 333–355.

SALISBURY, E. J. (1952), *Downs and Dunes*, London: Bell.

SCHMIDT, F. V. (1950), 'An evaluation of practical tidal marsh management on state and private marshes. Proceedings of the North-East Fish and Wildlife Conference.

SCHRATZ, E. (1934), 'Beiträge zur Biologie der Halophyten I. Keimungsphysiologie', *Jb. wiss. Bot.*, **80**, 112–142.

SENECA, E. D. (1969), 'Germination response to temperature and salinity of four dune grasses from the Outer Banks of North Carolina', *Ecology*, **50**, 44–53.

SETCHELL, W. A. (1920), 'Geographical distribution of the marine spermatophytes', *Bull. Torrey bot. Club.*, **47**.

SHAW, R. H. (Ed.) (1967), *Ground Level Climatology*. Baltimore: American Assocn. for Advancement of Science.

SHIELDS, L. M. (1957), 'Algal and lichen floras in relation to nitrogen content of certain volcanic and arid range soils', *Ecology*, **38**, 661–663.

SIIRA, J. (1970), 'Studies in the ecology of the sea shore meadows of the Bothnian Bay with special reference to the Limnika area', *Aquila Ser. Bot.*, **9**, 1–109.

SLATYER, R. O. (1967), *Plant-water Relationships,* London: Academic Press.

SLESS, J. B. (1956), 'Control of sand drift in beach mining', *J. Soil. Conserv. Serv. N.S. W.*, **12**, 164–176.

SLOET V. OLDRUITENBORGH, C. J. M. (1969), 'On the contribution of air-borne salt to the gradient character of the Voorne dune area', *Acta bot. neerl.*, **18**, 315–324.

SPARLING, J. H. (1967), 'The occurrence of *Schoenus nigricans* L. in blanket bogs. II. Experiments on the growth of *S. nigricans* under controlled conditions', *J. Ecol.*, **55**, 15–31.

SPRINGER, P. F. and WEBSTER, J. R. (1951), 'Biological effects of DDT application on tidal salt marshes', *Mosquito News*, **2**, 67–74.

STEARNS, L. A. and MACCREARY, D. (1957), 'The case of the vanishing brick dust', *Mosquito News*, **17**, 303–304.

STEBBINGS, R. E. (1971), 'Some ecological observations on the fauna in a tidal marsh to woodland transition', *Proc. Trans. Brit. Entom. Soc.*, **4**, 83–88.

STEENIS, C. G. G. J. VAN (1958), Discrimination of tropical shore vegetation. Proceedings of the symposium on humid tropics vegetation, 215–217, New Delhi, UNESCO.

STEENIS, J. H., WILDER, N. G., COFER, H. P. and BECK, R. A. (1954), 'The marshes of Delaware, their improvement and preservation. Delaware Board of Game and Fish Commissioners', *Pittman-Robertson Bull.*, **2**, 1–42.

STEERS, J. A. (1939), 'Sand and shingle formations in Cardigan Bay', *Geogrl. J.*, **94**, 209–227.

STEERS, J. A. (Ed.) (1960), *Scolt Head Island*, Cambridge: Heffer.

STEERS, J. A. (1964), *The Coastline of England and Wales.* (2nd edn.), Cambridge: University Press.

STEVENSON, R. E. and EMERY, K. O. (1958), Marshlands at Newport Bay, California. Allan Hancock Foundation Publication No. 20. Los Angeles.

STEWART, G. R., LEE, J. A. and GREBAMJO, T. O. (1972), *Nitrogen Metabolism of Halophytes.* I. Nitrate reductase activity in *Suaeda* maritima,' *New Phyt.* **71**, 263–167.

STEWART, W. D. P. (1967), 'Transfer of biologically fixed nitrogen in a sand dune slack region', *Nature*, **214**, 603–604.

STUBBINGS, H. G. and HOUGHTON, D. R. (1964), 'The ecology of Chichester Harbour', *Int. Rev. ges. Hydrobiol., Syst. beih.*, **49**, 233–279.

STOUTJESDIJKE, Ph. (1961), 'Micrometeorological measurements in vegetations of various structures', *Proc. K. ned. Akad. Wet., Amsterdam. section C,* **64,** 1–207.

TANSLEY, A. G. (1949), *The British Islands and their Vegetation,* Cambridge: University Press.

TASCHDJIAN, E. (1954), 'A note on *Spartina* protein', *Econ. Bot.,* **8,** 164–165.

TAYLOR, M. C. and BURROWS, E. M. (1968), 'Studies on the biology of *Spartina* in the Dee estuary, Cheshire', *J. Ecol.,* **56,** 795–809.

TAYLOR, N. (1937), 'A preliminary report on the relation of mosquito control ditching to Long Island salt marsh vegetation', *Proc. New Jers. Mosq. Exterm. Ass.,* **24,** 211–217.

TAYLOR, N. (1939), Salt tolerance of Long Island salt marsh plants. New York State Museum Circular No. 23, 1–42.

TEAGLE, W. G. (1966), Public pressure on South Haven Peninsula and its effect on Studland Heath National Nature Reserve, Nature Conservancy, Unpublished typescript, 1–89.

TEAL, J. M. (1962), 'Energy flow in the salt marsh ecosystem of Georgia', *Ecology,* **43,** 614–624.

THAARUP, P. (1954), 'The afforestation of the sand dunes of the western coast of Jutland', *Advmt. Sci. Lond.,* **11,** 38–41.

THORN, R. B. (1966), *River Engineering and Water Conservation Works,* London: Butterworth.

THORNTON, R. B. and DAVIS, A. G. (1964), Development and use of American Beachgrass for dune stabilization. Paper to American Society of Agronomy Meeting, Missouri, 1–21.

TINDALL, F. P. (1967), 'The care of a coastline', *J. Tn Plan. Inst.,* **53,** 387–392.

TROLL, C. (1963), *Seasonal Climates of the Earth. The Seasonal Course of Natural Phenomena in the Different Climatic Zones of the Earth. World maps of Climatology,* Berlin: Springer.

TSCHIRLEY, F. H. (1969), 'Defoliation in Vietnam', *Science,* **163,** 779–786.

TSITSIN, A. N. (1946), 'Perennial wheats', *Discovery,* **7,** 180.

TSOPA, E. (1939), La végétation des halophytes du nord de la Roumanie en connexion avec celle du reste du pays', *SIGMA,* **70,** 1–22.

TÜXEN, R. (1957), 'Die Pflanzengesellschaften des Aussendiechslandes von Neuwerk', *Mitt. flor.-soz. Arbgemein,* **6/7.**

TYLER, G. (1971), 'Studies in the ecology of Baltic sea-shore meadows III. Hydrology and salinity of Baltic sea shore meadows', *Oikos,* **22,** 1–20.

VOO, E. E. van der (1964), Danger to scientifically important wetlands in the Netherlands by modification of the surrounding environment. Proceedings of the MAR Conference, I.U.C.N. Publication. N.S. **3,** 274–278.

VOSE, P. B., POWELL, H. G. and SPENCE, J. B. (1957), 'The machair grazings of Tiree, Inner Hebrides', *Trans. Proc. bot. Soc. Edinb.,* **37,** 89–110.

WALLACE, T. J. (1953), 'The plant ecology of Dawlish Warren Pt. 1, *Rep. Trans. Devon. Ass. Advmt. Sci. Lit. Art,* **85,** 86–94.

WATT, A. S. (1947), 'Pattern and process in the plant community', *J. Ecol.*, **35**, 1–22.

WEBBER, H. J. (1895), The two freezes of 1894–95 in Florida, and what they teach. U.S. Department of Agriculture Year Book 1895, 159–174.

WEBLEY, D. M., EASTWOOD, D. J. and GIMINGHAM, C. H. (1952), 'Development of a soil microflora in relation to plant succession on sand dunes, including the 'rhizosphere' flora associated with colonizing species', *J. Ecol.*, **40**, 168–178.

WELBOURN, R. M. E. and LANGE, R. T. (1969), 'An analysis of vegetation on stranded coastal dune ranges between Robe and Naracoorte, South Australia', *Trans. Roy. Soc. S. Aust.*, **92**, 19–25.

WEST, R. C. (1956), 'Mangrove swamps of the Pacific coast of Colombia', *Ann. Ass. Am. Geogr.*, **46**, 98–121.

WESTHOFF, V. (1947), The vegetation of dunes and salt marshes on the Dutch islands of Terschelling, Vlieland and Texel. S. Gravenhage.

WESTHOFF, V. (1952), Gezelschappen met houtige gewassen in de duinen en langs de binnenduinrand. Dendrol. Jaarbk. 1952, 9–49.

WESTHOFF, V. (1962), Plant species characteristic of wetland habitats in the Netherlands. Proceedings of the MAR Conference I.U.C.N. Publication. N.S., **3**, 122–129.

WESTHOFF, V., LEEUWEN, C. G. van and ADRIANI, M. J. (1961), 'Enkele aspecten van vegetatie en bodem der duinen van Goeree, in het bizonder de contact-gordels tussen zout en zoet milieu', *Jaarb. Wet. Genot. Goeree-Overflakke*, 46–92.

WESTLAKE, D. F. (1968), The biology of aquatic weeds in relation to their management. Proceedings of the 9th British Weed Control Conference, 372–381.

WESTON, R. L. (1964), 'Nitrogen nutrition in *Atriplex hastata* L.', *Pl. Soil*, **20**, 251–259.

WHITE, D. J. B. (1961), 'Some observations on the vegetation of Blakeney Point, Norfolk, following the disappearance of the rabbits in 1954', *J. Ecol.*, **49**, 113–118.

WHITE, D. J. B. (1967), *An Annotated List of the Flowering Plants and Ferns on Blakeney Point, Norfolk*. (2nd edn) Norwich: The National Trust.

WIEHE, P. O. (1935), 'A quantitative study of the influence of the tide upon populations of *Salicornia europaea*', *J. Ecol.*, **23**, 323–333.

WIEMANN, P. and Domke, W. (1967), 'Pflanzengesellschaften der ostfriesischen Insel Spiekeroog', *Mitt. Staatsinst. Allg. Bot. Hamburg*, **12**, 191–353.

WILLIAMS, R. E. (1955), Development and improvement of coastal marsh ranges. *U.S. Year Book of Agriculture 1955*, 444–449.

WILLIAMS, R. E. (1959), Cattle walkways. U.S. Department of Agriculture Leaflet, 459, 1–8.

WILLIAMS, W. T. and LAMBERT, J. M. (1959), 'Multivariate methods in plant ecology I. Association-analysis in plant communities', *J. Ecol.*, **47**, 83–101.

WILLIS, A. J. (1963), 'Braunton Burrows: the effects on the vegetation of the addition of mineral nutrients to the dune soils', *J. Ecol.*, **51**, 353–374.

WILLIS, A. J., FOLKES, B. F., HOPE-SIMPSON, J. F. and YEMM, E. W. (1959),

Braunton Burrows: the dune system and its vegetation. I and II. *J. Ecol.*, **47**, 1–24 and 249–288.

WILLIS, A. J. and YEMM, E. W. (1961), 'Braunton Burrows: Mineral nutrient status of the dune soils', *J. Ecol.*, **49**, 377–390.

WILSON, K. (1960), 'The time factor in the development of dune soils at South Haven Peninsula, Dorset', *J. Ecol.*, **48**, 341–359.

WOHLENBERG, E. (1965), 'Deichbau und Deichpflege auf biologischer Grundlage', *Die Kuste*, **13**, 73–103.

WRIGHT, T. W. (1955), 'Profile development in the sand dunes of Culbin Forest, Morayshire I. Physical properties', *J. Soil Sci.*, **6**, 270–283.

YANO, N. (1962), 'The subterranean organ of sand dune plants in Japan', *J. Sci. Hiroshima Univ. Ser. B. Div. 2 (Botany)*, **9**, 139–184.

ZAK, J. M. (1967), 'Controlling drifting sand dunes on Cape Cod', *Massachusetts Agric. Expt. Stn. Bull.*, **563**, 1–15.

ZONNEVELD, I. S. (1960), 'The Brabantse Biesboch. A study of soil and vegetation of a freshwater tidal delta', *Bodenik Stud. No. 4. Wageningen.*

Additional References

These, for the most part, most recent references (not included in the text) are given to render the bibliography as up to date as possible.

BIRD, E. C. F. (1965), *A geomorphological study of the Gippsland Lakes*. Dept. of Geography Pubn. G/1. Australian National University, Canberra.

BIRD, E. C. F. (1971), 'Mangroves as land-builders', *Victorian Nat.*, **88**, 189–197.

BOSTON, K. G. (1971), The physiography of Anderson's Inlet, Victoria, with special reference to early stages in the establishment of *Spartina*. M. A. Thesis, University of Melbourne.

BOUGHEY, A. S. (1957), 'Ecological studies of tropical coastlines. I. The Gold Coast', *West Africa J. Ecol.*, **45**, 665–687.

CAMERON, G. N. (1972), 'Analysis of insect trophic diversity in two salt marsh communities', *Ecology*, **53**, 58–73.

CLARKE, L. D. and HANNON, N. J. (1971), 'The Mangrove swamp and salt marsh communities of the Sydney district. IV. The significance of species interaction', *J. Ecol.*, **59**, 535–553.

DENIEL, J. (1971), 'Un example d'utilisation de l'écologie et de la biometriesur un boisement de protection de l'environnement: La plantation de dunes de Cléder (Finistère)', *Penn Bed*, **8**, 147–159.

EBERSOLE, W. C. (1971), 'Predicting disturbances to the near and offshore sedimentary regime from marine mining'. *Water, Air, and Soil Pollution*, **1**, 72–88.

EVANS, G. (1965), 'Intertidal flat sediments and their environments and deposition in the Wash', *Q. Jl geol. Soc. Lond.*, **121**, 209–245.

FREIJSEN, A. H. J. (1971), 'Growth-physiology, salt tolerance and mineral nutrition of *Centaurium littorale* (Turner) Gilmour: Adaptations to its oligotrophic and brackish habitat', *Acta bot. neerl.*, **20**, 577–588.

GLUE, D. E. (1971), 'Saltmarsh reclamation stages and their associated bird-life', *Bird Study*, **18**, 187–198.

GRAY, A. J. (1972), 'The ecology of Morecambe Bay. V. The salt marshes of Morecambe Bay'. *J. appl. Ecol.*, **9**, 207–220.

GRAY, A. J. (1972), 'The ecology of Morecambe Bay. VI. Soils and vegetation of the salt marshes: A multivariate approach', *J. appl. Ecol.*, **9**, 221–234.

GRIMES, B. H. and HUBBARD, J. C. E. (1971), 'A comparison of film type and the importance of season for interpretation of coastal marshland vegetation', *Photogramm. Rec.*, **7**, 213–222.

JEFFRIES, R. L. (1972) *Aspects of salt-marsh ecology with particular reference to inorganic plant nutrition.* The Estuarine Environment, Barking, England: Applied Science Publishers Ltd., 61–85.

JENNINGS, J. N. (1965), 'Further discussion of factors affecting coastal dune formation in the tropics', *Aust. J. Sci.*, **28**, 166–167.

JONES, R. (1972), 'Comparative studies of plant growth and distribution in relation to waterlogging. V. The uptake of iron and manganese by dune and dune slack plants', *J. Ecol.*, **60**, 131–139.

JONES, R. (1972), *Ibid.* VI. 'The effect of manganese on the growth of dune and dune slack plants', *J. Ecol.*, **60**, 141–145.

JONES, R. and ETHERINGTON, J. R. (1971), *Ibid.* IV. 'The growth of dune and dune slack plants', *J. Ecol.*, **59**, 793–801.

LANGLOIS, J. (1971), 'Influence du rythme d'immersion sur la croissance et le métabolisme proteique de *Salicornia stricta* Dumort'. *Oecologia Plantarum*, **6**, 227–245.

McGUINESS, J. L., HARROLD, L. L. and EDWARDS, W. M. (1971), 'Relation of rainfall energy streamflow to sediment yield from small and large watersheds', *J. Soil Wat. Conserv.*, **26**, 233–234.

MORTON, A. (1970), A study of some factors affecting the structure of grassland vegetation, Ph.D. Thesis, University of Bangor, N. Wales.

MORTON, J. K. (1957), 'Sand dune formation on a tropical shore', *J. Ecol.*, **45**, 495–497.

NEWTON, L. E. (1965), Taxonomic studies in the British species of *Puccinellia*, M.Sc. Thesis, University of London.

OWEN, M. (1971), 'The selection of feeding site by White-fronted geese in winter', *J. appl. Ecol.*, **8**, 905–917.

OWEN, M. (1972), 'Some factors affecting food intake and selection in White-fronted geese', *J. Anim. Ecol.*, **41**, 79–92.

OWEN, M. (1971), 'On the autumn food of Barnacle geese at Caerlaverock National Nature Reserve', *Rep. Wildfowl Trust*, **22**, 114–119.

PEAKE, J. F. (1966), 'A salt marsh at Thornham in N.W. Norfolk', *Trans. Norfolk Norwich Nat. Soc.*, **19**, 36–62.

PETHICK, J. S. (1966), The ecology of the Tamar estuary, M.Sc. Thesis, University of Cambridge.

PETHICK, J. S. (1971), Salt marsh morphology, Ph.D. Thesis, University of Cambridge.

PONS, L. Z. and ZONNEVELD, I. S. (1965), *Soil ripening and soil classification.* Initial soil formation in alluvial deposits and a classification of the resulting soils. Wageningen: H. Veenman and Zonen, 1–128.

SIIRA, J. and HAAPALA, H. (1969), 'Studies in the distribution and ecology of *Puccinellia phryganodes* (Trin.) Scribn. and Merr. in Finland', *Aquilo, Serie bot.* **8**, 1–24.

SMITH, E. R. (1970), *Evaluation of a leveed Louisiana marsh.* Trans. N. Am. Wildl. Conf., No. 35, 265–275.

THALEN, D. C. P. (1971), 'Variation in some salt marsh and dune vegetations in the Netherlands with special reference to gradient situations', *Acta. bot. neerl.*, **20**, 327–342.

TYLER, G. (1971), 'Studies in the ecology of Baltic sea-shore meadows. IV. Distribution and turnover of organic matter and minerals in a shore meadow ecosystem', *Oikos*, **22**, 265–291.

VAN STRAATEN, L. M. J. (1961), 'Sedimentation in tidal flat areas'. *J. Alberta Soc. Petrol. Geol.*, **9**, 203–226.

WILLIS, A. J. and JEFFERIES, R. L. (1961), 'Investigations on the water relations of sand-dune plants under natural conditions. The water relations of plants', *Brit. Ecol. Soc. Symp. No. 3*. Oxford, 168–189.

Index

This index should be used in conjunction with the contents list at the beginning of the volume.

Index

Species	Habitat[a]	Adult Size (mm)	Site of Deposition[c]	Number of Eggs[d]
B. woodhousei	1G	130	1	25,600/
B. compactilis	1G	70	1	
B. microscaphus	1B	65	1	several thousand
B. regularis	6G	65[b]	1	23,000
B. rangeri	6G	105[b]	1,2a	
B. carens	6G	74–92[b]		10,000
B. angusticeps	6	65		650–850
B. gariepensis	6	55		100+
B. vertebralis	6	30		
Ansonia muellari	5D	31[b]		150

Phrynomeridae

Species	Habitat[a]	Adult Size (mm)	Site of Deposition[c]	Number of Eggs[d]
Phrynomerus bifasciatus bifasciatus	6G	65	1,2a	400–1,500

Microhylidae

Species	Habitat[a]	Adult Size (mm)	Site of Deposition[c]	Number of Eggs[d]
Gastrophryne carolinensis	1,2	20	2a	850
G. mazatlanensis	3	20	4	175–200
Breviceps adspersus adspersus	6	38	4a,3b	20–46

[a]Habitat: 1 = North America   A = Temporary ponds
2 = Central America B = Permanent water, xeric areas
3 = South America   C = Permanent water, mesic areas
4 = Europe   D = Permanent streams
5 = Asia   E = Caves
6 = Africa   F = Mesic   F+ = Cloud or tropical rain forest
7 = Australia   G = Grasslands, savannahs, or subhumid corridor
8 = New Zealand
[b]Size of adult female.

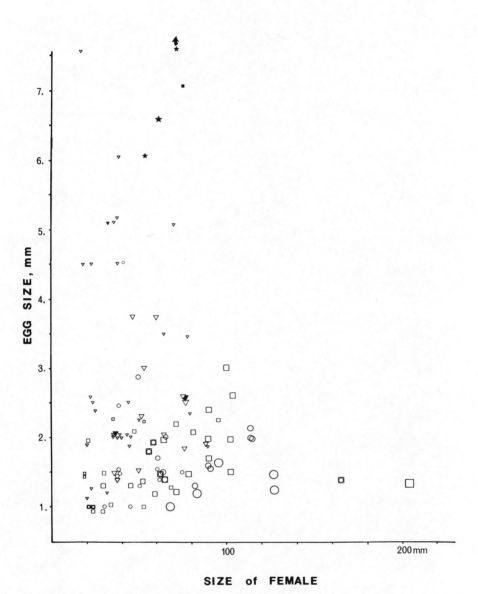

Index

Evolution of Desert Biota

Evolution of
Desert Biota

Edited by David W. Goodall

University of Texas Press Austin & London

Publication of this book was financed in part by
the Desert Biome and the Structure of Ecosystems programs of
the U.S. participation in the International
Biological Program.

Library of Congress Cataloging in Publication Data
Main entry under title:

Evolution of desert biota.

Proceedings of a symposium held during the First
International Congress of Systematic and Evolution-
ary Biology which took place in Boulder, Colo.,
during August, 1973.
Bibliography: p.
Includes index.
1. Desert biology—Congresses. 2. Evolution—
Congresses. I. Goodall, David W., 1914–
II. International Congress of Systematic and Evolu-
tionary Biology, 1st, Boulder, Colo., 1973.
QH88.E95   575'.00915'4   75-16071
ISBN 0-292-72015-7

Contents

Evolution of Desert Biota

1. Introduction   David W. Goodall

In the broad sense, "deserts" include all those areas of the earth's surface whose biological potentialities are severely limited by lack of
water. If one takes them as coextensive with the arid and semiarid
zones of Meigs's classification, they occupy almost one-quarter of the
terrestrial surface of the globe. Though the largest arid areas are to be
found in Africa and Asia, Australia has the largest proportion of its
area in this category. Smaller desert areas occur in North and South
America; Antarctica has cold deserts; and the only continent virtually
without deserts is Europe.

When life emerged in the waters of the primeval world, it could hardly have been predicted that the progeny of these first organisms
would extend their occupancy even to the deserts. Regions more different in character from the origin and natural home of life would be
hard to imagine. Protoplasm is based on water, rooted in water. Some
three-quarters of the mass of active protoplasm is water; the biochemical reactions underlying all its activities take place in water and depend on the special properties of water for the complex mechanisms
of enzymatic and ionic controls which integrate the activity of cell
and organisms into a cybernetic whole. It is, accordingly, remarkable
that organisms were able to adapt themselves to environments in
which water supplies were usually scanty, often almost nonexistent,
and always unpredictable.

The first inhabitants of the deserts were presumably opportunistic.
On the margins of larger bodies of water were areas which were alternately wetted and dried for longer or shorter periods. Organisms living there acquired the possibility of surviving the dry periods by drying
out and becoming inactive until rewetted, at which time their activity
resumed where it had left off. While in the dry state, these organisms

—initially, doubtless, Protista—were easily moved by air currents and thus could colonize other bodies of water. Among them were the very temporary pools formed by the occasional rainstorms in desert areas. Thus the deserts came to be inhabited by organisms whose ability to dry and remoisten without loss of vitality enabled them to take advantage of the short periods during which limited areas of the deserts deviate from their normally arid state.

Yet other organisms doubtless—the blue green algae among them—similarly took advantage of the much shorter periods, amounting perhaps to an hour at a time, during which the surface of the desert was moistened by dew, and photosynthesis was possible a few minutes before and after sunrise to an organism which could readily change its state of hydration.

In the main, though, colonization of the deserts had to wait until colonization of other terrestrial environments was well advanced. For most groups of organisms, the humid environments on land presented less of a challenge in the transition from aquatic life than did the deserts. By the time arthropods and annelids, mollusks and vertebrates, fungi and higher plants had adapted to the humid terrestrial environments, they were poised on the springboard where they could collect themselves for the ultimate leap into the deserts. And this leap was made successfully and repeatedly. Few of the major groups of organisms that were able to adapt to life on land did not also contrive to colonize the deserts.

Some, like the arthropods and annual plants, had an adaptational mechanism—an inactive stage of the life cycle highly resistant to desiccation—almost made to order to match opportunistically the episodic character of the desert environment. For others the transition was more difficult: for mammals, whose excretory mechanism assumes the availability of liquid water; for perennial plants, whose photosynthetic mechanism normally carries the penalty of water loss concurrent with carbon dioxide intake. But the evolutionary process surmounted these difficulties; and the deserts are now inhabited by a range of organisms which, though somewhat inferior to that of more favored environments, bears testimony to the inventiveness and success of evolution in filling niches and in creating diversity.

The most important modifications and adaptations needed for life in the deserts are concerned with the dryness of the environment there.

But an important feature of most desert environments is also their unpredictability. Precipitation has a higher coefficient of variability, on any time scale, than in other climatic types, with the consequence that desert organisms may have to face floods as well as long and highly variable periods of drought. The range of temperatures may also be extreme—both diurnal and seasonal. Under the high radiation of the subtropical deserts, the soil surface may reach a temperature which few organisms can survive; and, in the cold deserts of the great Asian land mass, extremely low winter temperatures are recorded. Sand and dust storms made possible by the poor stability of the surface soil are also among the environmental hazards to which desert organisms must become adapted.

Like other climatic zones, the deserts have not been stable in relation to the land masses of the world. Continental drift, tectonic movements, and changes in the earth's rotation and in the extent of the polar icecaps have led to secular changes in the area and distribution of arid land surfaces. But, unlike other climatic zones, the arid lands have probably always been fragmented—constituting a number of discrete areas separated from one another by zones of quite different climate. The evolutionary process has gone on largely independently in these separate areas, often starting from different initial material, with the consequence that the desert biota is highly regional. Elements in common between the different main desert areas are few, and, as between continents or subcontinents, there is a high degree of endemism. The smaller desert areas of the world are the equivalent of islands in an ocean of more humid environments.

These are among the problems to be considered in the present volume. It reports the proceedings of a symposium which was held on August 10, 1973, at Boulder, Colorado, as part of the First International Congress of Systematic and Evolutionary Biology.

2. The Origin and Floristic Affinities of the South American Temperate Desert and Semidesert Regions   Otto T. Solbrig

Introduction

In this paper I will attempt to summarize the existent evidence regarding the floristic relations of the desert and semidesert regions of temperate South America and to explain how these affinities came to exist.

More than half of the surface of South America south of the Tropic of Capricorn can be classed as semidesert or desert. In this area lie some of the richest mineral deposits of the continent. These regions consequently are important from the standpoint of human economy. From a more theoretical point, desert environments are credited with stimulating rapid evolution (Stebbins, 1952; Axelrod, 1967) and, further, present some of the most interesting and easy-to-study adaptations in plants and animals.

Although, at present, direct evidence regarding the evolution of desert vegetation in South America is still meager, enough data have accumulated to make some hypotheses. It is hoped this will stimulate more research in the field of plant micropaleontology in temperate South America. Such research in northern South America has advanced our knowledge immensely (Van der Hammen, 1966), and high rewards await the investigator who searches this area in the temperate regions of the continent.

The Problem

If a climatic map of temperate South America is compared with a phytogeographic map of the same region drawn on a physiognomic basis and with one drawn on floristic lines, it will be seen that they do not coincide. Furthermore, if the premise (not proven but generally held to be true) is accepted that the physical environment is the determinant of the structure of the ecosystem and that, as the physical environment (be it climate, physiography, or both) changes, the structure of the vegetation will also change, then an explanation for the discrepancy between climatic and phytogeographic maps has to be provided. Alternative explanations to solve the paradox are (1) the premise on which they are based is entirely or partly wrong; (2) our knowledge is incomplete; or (3) the discrepancies can be explained on the basis of the historical events of the past. It is undoubtedly true that floristic and paleobotanical knowledge of South American deserts is incomplete and that much more work is needed. However, I will proceed under the assumption that a sufficient minimum of information is available. I also feel that our present insights are sufficient to accept the premise that the ecosystem is the result of the interaction between the physical environment and the biota. I shall therefore try to find in the events of the past the answer for the discrepancy.

I shall first describe the semidesert regions of South America and their vegetation, followed by a brief discussion of Tertiary and Pleistocene events. I shall then look at the floristic connections between the regions and the distributional patterns of the dominant elements of the area under study. From this composite picture I shall try to provide a coherent working hypothesis to explain the origin and floristic affinities of the desert and semidesert regions of temperate South America.

Theory

Biogeographical hypotheses such as the ones that will be made further on in this paper are based on certain theoretical assumptions. In most cases, however, these assumptions are not made explicit; consequently, the reader who disagrees with the author is not always certain whether he disagrees with the interpretation of the evidence or with the assumptions made. This has led to many futile controversies. The fundamental assumptions that will be made here follow from the general theory of evolution by natural selection, the theory of speciation, and the theory of geological uniformitarianism.

The first assumption is that a continuous distributional range reflects an environment favorable to the plant, that is, an environment where it can compete successfully. Since the set of conditions (physical, climatical, and biological) where the plant can compete successfully (the realized niche) bounds a limited portion of ecological space, it will be further assumed that the range of a species indicates that conditions over that range do not differ greatly in comparison with the set of all possible conditions that can be given. It will be further assumed that each species is unique in its fundamental and realized niche (defined as the hyperspace bounded by all the ecological parameters to which the species is adapted or over which it is able to compete successfully). Consequently, no species will occupy exactly the same geographical range, and, as a corollary, some species will be able to grow over a wide array of conditions and others over a very limited one.

When the vegetation of a large region, such as a continent, is mapped, it is found that the distributional ranges of species are not independent but that ranges of certain species tend to group themselves even though identical ranges are not necessarily encountered. This allows the phytogeographer to classify the vegetation. It will be assumed that, when neighboring geographical areas do not differ greatly in their present physical environment or in their climate but differ in their flora, the reason for the difference is a historical one reflecting different evolutionary histories in these floras and requiring an explanation.

Disjunctions are common occurrences in the ranges of species. In a strict sense, all ranges are disjunct since a continuous cover of a species over an extensive area is seldom encountered. However, when similar major disjunctions are found in the ranges of many species whose ranges are correlated, the disjunction has biogeographical significance. Unless there is evidence to the contrary, an ancient continuous range will be assumed in such instances, one that was disrupted at a later date by some identifiable event, either geological or climatological.

It will also be assumed that the atmospheric circulation and the basic meteorological phenomena in the past were essentially similar to those encountered today, unless there is positive evidence to the contrary. Further, it will be assumed that the climatic tolerances of a living species were the same in the past as they are today. Finally, it will be assumed that the spectrum of life forms that today signify a rain forest, a subtropical forest, a semidesert, and so on, had the same meaning in the past too, implying with it that the basic processes of carbon gain and water economy have been essentially identical at least since the origin of the angiosperms.

From these assumptions a coherent theory can be developed to re-construct the past (Good, 1953; Darlington, 1957, 1965). No general assumptions about age and area will be made, however, because they are inconsistent with speciation theory (Stebbins, 1950; Mayr, 1963). In special cases when there is some evidence that a particular group is phylogenetically primitive, the assumption will be made that it is also geologically old. Such an assumption is not very strong and will be used only to support more robust evidence.

The Semidesert Regions of South America

In temperate South America we can recognize five broad phytogeo-graphical regions that can be classed as "desert" or "semidesert" regions. They are the Monte (Haumann, 1947; Morello, 1958), the Patagonian Steppe (Cabrera, 1947), the Prepuna (Cabrera, 1971), and the Puna (Cabrera, 1958) in Argentina, and the Pacific Coastal Desert in Chile and Peru (Goodspeed, 1945; Ferreyra, 1960). In addi-tion, three other regions—the Matorral or "Mediterranean" region in Chile (Mooney and Dunn, 1970) and the Chaco and the Espinal in Argentina (Fiebrig, 1933; Cabrera, 1953, 1971), although not semi-deserts, are characterized by an extensive dry season. Finally, the high mountain vegetation of the Andes shows adaptations to drought tolerance (fig. 2-1).

The Monte

The Monte (Lorentz, 1876; Haumann, 1947; Cabrera, 1953; Morello, 1958; Solbrig, 1972, 1973) is a phytogeographical province that ex-tends from lat. 24°35′ S to lat. 44°20′ S and from long. 62°54′ W on the Atlantic coast to long. 69°50′ W at the foothills of the Andes (fig. 2-1).

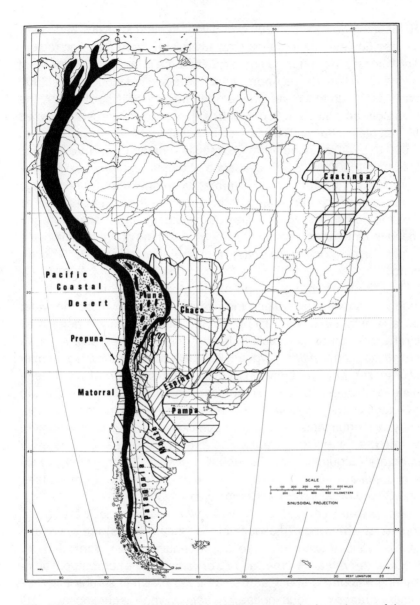

Fig. 2-1. Geographical limits of the phytogeographical provinces of the Andean Dominion (stippled) and of the Chaco Dominion (various hatchings) according to Cabrera (1971). The high cordillera vegetation is indicated in solid black. Goode Base Map, copyright by The University of Chicago, Department of Geography.

Rains average less than 200 mm a year in most localities and never exceed 600 mm; evaporation exceeds rainfall throughout the region. The rain falls in spring and summer. The area is bordered on the west by the Cordillera de los Andes, which varies in height between 5,000 and 7,000 m in this area. On the north the region is bordered by the high Bolivian plateau (3,000–5,000 m high) and on the east by a series of mountain chains (Sierras Pampeanas) that vary in height from 3,000 to 5,000 m in the north (Aconquija, Famatina, and Velazco) to less than 1,000 m (Sierra de Hauca Mahuida) in the south. Physiographically, the northern part is formed by a continuous barrier of high mountains which becomes less important farther south as well as lower in height. The Monte vegetation occupies the valleys between these mountains as a discontinuous phase in the northern region and a more or less continuous phase from approximately lat. 32° S southward.

The predominant vegetation of the Monte is a xerophytic scrubland with small forests along the rivers or in areas where the water table is quite superficial. The predominant community is dominated by the species of the creosote bush or *jarilla* (*Larrea divaricata*, *L. cuneifolia*, and *L. nitida* [Zygophyllaceae]) associated with a number of other xerophytic or aphyllous shrubs: *Condalia microphylla* (Rhamnaceae), *Monttea aphylla* (Scrophulariaceae), *Bougainvillea spinosa* (Nyctaginaceae), *Geoffroea decorticans* (Leguminosae), *Cassia aphylla* (Leguminosae), *Bulnesia schickendanzii* (Zygophyllaceae), *B. retama*, *Atamisquea emarginata* (Capparidaceae), *Zuccagnia punctata* (Leguminosae), *Gochnatia glutinosa* (Compositae), *Proustia cuneifolia* (Compositae), *Flourensia polyclada* (Compositae), and *Chuquiraga erinacea* (Compositae).

Along water courses or in areas with a superficial water table, forests of *algarrobos* (mesquite in the United States) are observed, that is, various species of *Prosopis* (Leguminosae), particularly *P. flexuosa*, *P. nigra*, *P. alba*, and *P. chilensis*. Other phreatophytic or semiphreatophytic species of small trees or small shrubs are *Cercidium praecox* (Leguminosae), *Acacia aroma* (Leguminosae), and *Salix humboldtiana* (Salicaceae).

Herbaceous elements are not common. There is a flora of summer annuals formed principally by grasses.

The Patagonian Steppe

The Patagonian Steppe (Cabrera, 1947, 1953, 1971; Soriano, 1950, 1956) is limited on its eastern and southern borders by the Atlantic Ocean and the Strait of Magellan. On the west it borders quite abruptly with the *Nothofagus* forest; the exact limits, although easy to determine, have not yet been mapped precisely (Dimitri, 1972). On the north it borders with the Monte along an irregular line that goes from Chos Malal in the state of Neuquen in the west to a point on the Atlantic coast near Rawson in the state of Chubut (Soriano, 1949). In addition, a tongue of Patagonian Steppe extends north from Chubut to Mendoza (Cabrera, 1947; Böcher, Hjerting, and Rahn, 1963). Physiognomically the region consists of a series of broad tablelands of increasing altitude as one moves from east to west, reaching to about 1,500 m at the foot of the cordillera. The soil is sandy or rocky, formed by a mixture of windblown cordilleran detritus as well as *in situ* eroded basaltic rocks, the result of ancient volcanism.

The climate is cold temperate with cold summers and relatively mild winters. Summer means vary from 21°C in the north to 12°C in the south (summer mean maxima vary from 30°C to 18°C) with winter means from 8°C in the north to 0°C in the south (winter mean minima 1.5°C to −3°C). Rainfall is very low, averaging less than 200 mm in all the Patagonian territory with the exception of the south and west borders where the effect of the cordilleran rainfall is felt. The little rainfall is fairly well distributed throughout the year with a slight increase during winter months.

The Patagonian Steppe is the result of the rain-shadow effect of the southern cordillera in elevating and drying the moist westerly winds from the Pacific. Consequently the region not only is devoid of rains but also is subjected to a steady westerly wind of fair intensity that has a tremendous drying effect. The few rains that occur are the result of occasional eruptions of the Antarctic polar air mass from the south interrupting the steady flow of the westerlies.

The dominant vegetation is a low scrubland or else a vegetation of low cushion plants. In some areas xerophytic bunch grasses are also common. Among the low (less than 1 m) xerophytic shrubs and cushion plants, the *neneo*, *Mulinum spinosum* (Umbelliferae), is the dominant form in the northwestern part, while *Chuquiraga avellanedae* (Compositae) and *Nassauvia glomerulosa* (Compositae) are dominant over extensive areas in central Patagonia. Other important shrubs are *Trevoa patagonica* (Rhamnaceae), *Adesmia campestris* (Compositae), *Colliguaja integerrima* (Euphorbiaceae), *Nardophyllum obtusifolium* (Compositae), and *Nassauvia axillaris*. Among the grasses are *Stipa humilis*, *S. neaei*, *S. speciosa*, *Poa huecu*, *P. ligularis*, *Festuca argentina*, *F. gracillima*, *Bromus macranthus*, *Hordeum comosus*, and *Agropyron fuegianum*.

The Puna

The Puna (Weberbauer, 1945; Cabrera, 1953, 1958, 1971) is situated in the northwestern part of Argentina, western and central Bolivia, and southern Peru. It is a very high plateau, the result of the uplift of an enormous block of an old peneplane, which started to lift in the Miocene but mainly rose during the Pliocene and the Pleistocene to a mean elevation of 3,400–3,800 m. The Puna is bordered on the east by the Cordillera Real and on the west by the Cordillera de los Andes that rises to 5,000–6,000 m; the plateau is peppered by a number of volcanoes that rise 1,000–1,500 m over the surface of the Puna.

The soils of the Puna are in general immature, sandy to rocky, and very poor in organic matter (Cabrera, 1958). The area has a number of closed basins, and high mountain lakes and marshes are frequent.

The climate of the Puna is cold and dry with values for minimum and maximum temperatures not too different from Patagonia but with the very significant difference that the daily temperature amplitude is very great (values of over 20°C are common) and the difference between summer and winter very slight. The precipitation is very irregular over the area of the Puna, varying from a high of 800 mm in the northeast corner of Bolivia to 100 mm/year on the southwest border in Argentina. The southern Puna is undoubtedly a semidesert region, but the northern part is more of a high alpine plateau, where the limitations to plant growth are given more by temperature than by rainfall.

The typical vegetation of the Puna is a low, xerophytic scrubland formed by shrubs one-half to one meter tall. In some areas a grassy steppe community is found, and in low areas communities of high mountain marshes are found.

Among the shrubby species we find *Fabiana densa* (Solanaceae), *Psila boliviensis* (Compositae), *Adesmia horridiuscula* (Leguminosae), *A. spinossisima, Junellia seriphioides* (Verbenaceae), *Nardophyllum armatum* (Compositae), and *Acantholippia hastatula* (Verbenaceae). Only one tree, *Polylepis tomentella* (Rosaceae), grows in the Puna, strangely enough only at altitudes of over 4,000 m. Another woody element is *Prosopis ferox*, a small tree or large shrub. Among the grasses are *Bouteloua simplex, Muhlenbergia fastigiata, Stipa leptostachya, Pennisetum chilense*, and *Festuca scirpifolia*. Cactaceae are not very frequent in general, but we find locally abundant *Opuntia atacamensis, Oreocerus trollii, Parodia schroebsia*, and *Trichocereus poco*.

Although physically the Puna ends at about lat. 30° S, Puna vegetation extends on the eastern slope of the Andes to lat. 35° S, where it merges into Patagonian Steppe vegetation.

The Prepuna

The Prepuna (Czajka and Vervoorst, 1956; Cabrera, 1971) extends along the dry mountain slopes of northwestern Argentina from the state of Jujuy to La Rioja, approximately between 2,000 and 3,400 m. It is characterized by a dry and warm climate with summer rains; it is warmer than the Puna, colder than the Monte; and it is a special formation strongly influenced by the exposure of the mountains in the region.

The vegetation is mainly formed by xerophytic shrubs and cacti. Among the shrubs, the most abundant are *Gochnatia glutinosa* (Compositae), *Cassia crassiramea* (Leguminosae), *Aphyllocladus spartioides, Caesalpinia trichocarpa* (Leguminosae), *Proustia cuneifolia* (Compositae), *Chuquiraga erinacea* (Compositae), *Zuccagnia punctata* (Leguminosae), *Adesmia inflexa* (Leguminosae), and *Psila boliviensis* (Compositae). The most conspicuous member of the Cactaceae is the cardon, *Trichocereus pasacana*; there are also present *T. poco* and species of *Opuntia, Cylindropuntia, Tephrocactus, Parodia*, and *Lobivia*. Among the grasses are *Digitaria californica, Stipa leptostachya, Monroa argentina*, and *Agrostis nana*.

The Pacific Coastal Desert

Along the Peruvian and Chilean coast from lat. 5° S to approximately lat. 30° S, we find the region denominated "La Costa" in Peru (Weber-bauer, 1945; Ferreyra, 1960) and "Northern Desert," "Coastal Desert," or "Atacama Desert" in Chile (Johnston, 1929; Reiche, 1934; Goodspeed, 1945). This very dry region is under the influence of the combined rain shadow of the high cordillera to the east and the cold Humboldt Current and the coastal upwelling along the Peruvian coast. Although physically continuous, the vegetation is not uniform, as a result of the combination of temperature and rainfall variations in such an extended territory. Temperature decreases from north to south as can be expected, going from a yearly mean to close to 25°C in north-ern Peru (Ferreyra, 1960) to a low of 15°C at its southern border. Rainfall is very irregular and very meager. Although some localities in Peru (Zorritos, Lomas de Lachay; cf. Ferreyra, 1960) have averages of 200 mm, the average yearly rainfall is below 50 mm in most places. This has created an extreme xerophytic vegetation often with special adaptations to make use of the coastal fog.

Behind the coastal area are a number of dry valleys, some in Peru but mostly in northern Chile, with the same kind of extreme dry condi-tions as the coastal area.

The flora is characterized by plants with extreme xerophytic adapta-tions, especially succulents, such as *Cereus spinibaris* and *C. co-quimbanus*, various species of *Echinocactus*, and *Euphorbia lacti-folia*. The most interesting associations occur in the so-called *lomas*, or low hills (less than 1,500 m), along the coast that intercept the coastal fog and produce very localized conditions favorable for some plant growth. Almost each of these formations from the Ecuadorian border to central Chile constitutes a unique community. Over 40 per-cent of the plants in the Peruvian coastal community are annuals (Ferreyra, 1960), although annuals apparently are less common in Chile (Johnston, 1929); of the perennials, a large number are root perennials or succulents. Only about 5 percent are shrubs or trees in the northern sites (Ferreyra, 1960), while shrubs and semishrubs constitute a higher proportion in the Chilean region. From the Chilean region should be mentioned *Oxalis gigantea* (Oxalidaceae), *Helio-tropium philippianum* (Boraginaceae), *Salvia gilliesii* (Labiatae), and

Proustia tipia (Compositae) among the shrubs; species of *Poa*, *Eragrostis*, *Elymus*, *Stipa*, and *Nasella* among the grasses; and *Alstroemeria violacea* (Amaryllidaceae), a conspicuous and relatively common root perennial. In southern Peru *Nolana inflata*, *N. spathulata* (Nolanaceae), and other species of this widespread genus; *Tropaeolum majus* (Tropaeolaceae), *Loasa urens* (Loasaceae), and *Arcythophyllum thymifolium* (Rubiaceae); in the *lomas* of central Peru the amancay, *Hymenocallis amancaes* (Amaryllidaceae), *Alstroemeria recumbens* (Amaryllidaceae), *Peperomia atocongona* (Piperaceae), *Vicia lomensis* (Leguminosae), *Carica candicans* (Caricaceae), *Lobelia decurrens* (Lobeliaceae), *Drymaria weberbaueri* (Caryophyllaceae), *Capparis prisca* (Capparidaceae), *Caesalpinia tinctoria* (Leguminosae), *Pitcairnia lopezii* (Bromeliaceae), and *Haageocereus lachayensis* and *Armatocereus* sp. (Cactaceae). Finally, in the north we find *Tillandsia recurvata*, *Fourcroya occidentalis*, *Apralanthera ferreyra*, *Solanum multinterruptum*, and so on.

Of great phytogeographic interest is the existence of a less-xerophytic element in the very northern extreme of the Pacific Coastal Desert, from Trujillo to the border with Ecuador (Ferreyra, 1960), known as *algarrobal*. Principal elements of this vegetation are two species of *Prosopis*, *P. limensis* and *P. chilensis*; others are *Cercidium praecox*, *Caesalpinia paipai*, *Acacia huarango*, *Bursera graveolens* (Burseraceae), *Celtis iguanea* (Ulmaceae), *Bougainvillea peruviana* (Nyctaginaceae), *Cordia rotundifolia* (Boraginaceae), and *Grabowskia boerhaviifolia* (Solanaceae).

Geological History

The present desert and subdesert regions of temperate South America result from the existence of belts of high atmospheric pressure around lat. 30° S, high mountain chains that impede the transport of moisture from the oceans to the continents, and cold water currents along the coast, which by cooling and drying the air that flows over them act like the high mountains.

The Pacific Coastal Desert of Chile and Peru is principally the result of the effect of the cold Humboldt Current that flows from south to north; the Patagonian Steppe is produced by the Cordillera de los Andes that traps the moisture in the prevailing westerly winds; while the Monte and the Puna result from a combination of the cordilleran rain shadow in the west and the Sierras Pampeanas in the east and the existence of the belt of high pressure.

The high-pressure belt of mid-latitudes is a result of the global flow of air (Flohn, 1969) and most likely has existed with little modification throughout the Mesozoic and Cenozoic (however, for a different view, see Schwarzenbach, 1968, and Volkheimer, 1971). The mountain chains and the cold currents, on the other hand, are relatively recent phenomena. The latter's low temperature is largely the result of Antarctic ice. But aridity results from the interaction of temperature and humidity. In effect, when ambient temperatures are high, a greater percentage of the incident rainfall is lost as evaporation and, in addition, plants will transpire more water. Consequently, in order to reconstruct the history of the desert and semidesert regions of South America, we also have to have an idea of the temperature and pluvial regimes of the past.

In this presentation I will use two types of evidence: (1) the purely geological evidence regarding continental drift, times of uplifting of mountain chains, marine transgressions, and existence of paleosoils and pedemonts; and (2) paleontological evidence regarding the ecological types and phylogenetical stock of the organisms that inhabited the area in the past. With this evidence I will try to reconstruct the most likely climate for temperate South America since the Cretaceous and deduce the kind of vegetation that must have existed.

Cretaceous

This account will start from the Cretaceous because it is the oldest period from which we have fossil records of angiosperms, which today constitute more than 90 percent of the vascular flora of the regions under consideration. At the beginning of the Cretaceous, South America and Africa were probably still connected (Dietz and Holden, 1970), since the rift that created the South Atlantic and separated the two continents apparently had its origin during the Lower Cretaceous. The position of South America at this time was slightly south (approximate-

ly lat. 5°–10° S) of its present position and with its southern extremity tilted eastward. There were no significant mountain chains at
that time.

Northern and western South America are characterized in the Cretaceous by extensive marine transgressions in Colombia, Venezuela,
Ecuador, and Peru (Harrington, 1962). In Chile, during the middle Cretaceous, orogeny and uplift of the Chilean Andes began (Kummel,
1961). This general zone of uplift, which was accompanied by active
volcanism and which extended to central Peru, marks the beginning of
the formation of the Andean cordillera, a phenomenon that will have its
maximum expression during the upper Pliocene and Pleistocene and
that is not over yet.

Although the first records of angiosperms date from the Cretaceous
(Maestrichtian), the known fossil floras from the Cretaceous of South
America are formed predominantly by Pteridophytes, Bennettitales,
and Conifers (Menéndez, 1969). Likewise, the fossil faunas are
formed by dinosaurs and other reptilian groups. Toward the end of the
Cretaceous (or beginning of Paleocene) appear the first mammals
(Patterson and Pascual, 1972).

Climatologically, the record points to a much warmer and possibly
wetter climate than today, although there is evidence of some aridity,
particularly in the Lower Cretaceous.

All in all, the Cretaceous period offers little conclusive evidence of
extensive dry conditions in South America. Nevertheless, during the
Lower Cretaceous before the formation of an extensive South Atlantic
Ocean, conditions in the central portion of the combined continent
must have been drier than today. In effect, the high rainfall in the
present Amazonian region is the result of the condensation of moisture from rising tropical air that is cooling adiabatically. This air is
brought in by the trade winds and acquires its moisture over the North
and South Atlantic. Before the breakup of Pangea, trade winds must
have been considerably drier on the western edge of the continent
after blowing over several thousand miles of hot land. It is interesting
that some characteristic genera of semidesert regions, such as *Prosopis* and *Acacia*, are represented in both eastern Africa and South
America. This disjunct distribution can be interpreted by assuming
Cretaceous origin for these genera, with a more or less continuous

Cretaceous distribution that was disrupted when the continents separated (Thorne, 1973). This is in accordance with their presumed primitive position within the Leguminosae (L. I. Nevling, 1970, personal communication). There is some geomorphological evidence also for at least local aridity in the deposits of the Lower Cretaceous of Córdoba and San Luis in Argentina, which are of a "typical desert phase" according to Gordillo and Lencinas (1972).

Cenozoic

Paleocene. The marine intrusions of northern South America still persisted at the beginning of the Paleocene but had become much less extensive (Haffer, 1970). The Venezuelan Andes and part of the Caribbean range of Venezuela began to rise above sea level (Liddle, 1946; Harrington, 1962). In eastern Colombia, Ecuador, and Peru continental deposits were laid down to the east of the rising mountains, which at this stage were still rather low. The sea retreated from southern Chile, but there was a marine transgression in central eastern Patagonia.

At the beginning of the Paleocene the South American flora acquired a character of its own, very distinct from contemporaneous European floras, although there are resemblances to the African flora (Van der Hammen, 1966). The first record of Bombacaceae is from this period (Van der Hammen, 1966).

There are remains of crocodiles from the Paleocene of Chubut in Argentina, indicating a probable mean temperature of 10°C or higher for the coldest month (Volkheimer, 1971), some fifteen to twenty degrees warmer than today. The early Tertiary mammalian fossil faunas consist of marsupials, edentates of the suborder Xenarthra, and a variety of ungulates (Patterson and Pascual, 1972). These forms appear to have lived in a forested environment, confirming the paleobotanical evidence (Menéndez, 1969, 1972; Petriella, 1972).

The climate of South America during the Paleocene was clearly warmer and more humid than today. With the South Atlantic now fairly large and with no very great mountain range in existence, probably no extensive dry-land floras could have existed.

Eocene. During the Eocene the general features of the northern Andes were little changed from the preceding Paleocene. The northern extremity of the eastern cordillera began to be uplifted. In western Colombia and Ecuador the Bolívar geosyncline was opened (Schuchert, 1935; Harrington, 1962). Thick continental beds were deposited in eastern Colombia-Peru, mainly derived from the erosion of the rising mountains to the east. In the south the slow rising of the cordillera continued. There was an extensive marine intrusion in eastern Patagonia.

The flora was predominantly subtropical (Romero, 1973). It was during the Eocene that the tropical elements ranged farthest south, which can be seen very well in the fossil flora of Río Turbio in Argentine Patagonia. Here the lowermost beds containing *Nothofagus* fossils are replaced by a rich flora of tropical elements with species of *Myrica*, *Persea*, *Psidium*, and others, which is then again replaced in still higher beds by a *Nothofagus* flora of more mesic character (Hünicken, 1966; Menéndez, 1972).

However, in the Eocene we also find the first evidence of elements belonging to a more open, drier vegetation, particularly grasses (Menéndez, 1972; Van der Hammen, 1966).

The Eocene was also a time of radiation of several mammalian phyletic lines, particularly marsupials, xenarthrans, ungulates, and notoungulates (Patterson and Pascual, 1972). Of particular interest for our purpose is the appearance of several groups of large native herbivores (Patterson and Pascual, 1972). More interesting still is "the precocity shown by certain ungulates in the acquisition of high-crowned, or hyposodont, and rootless, or hypselodont, teeth" (Patterson and Pascual, 1972). By the lower Oligocene such teeth had been acquired by no fewer than six groups of ungulates. Such animals must have thrived in the evolving pampas areas. True pampas are probably younger, but by the Eocene it seems reasonable to propose the existence of open savanna woodlands, somewhat like the llanos of Venezuela today.

The climate appears to have been fairly wet and warm until a peak was reached in middle Eocene, after which time a very gradual drying and cooling seems to have occurred.

Oligocene. The geological history of South America during the Oligocene followed the events of the earlier periods. There were further uplifts of the Caribbean and Venezuelan mountains and also the Cordillera Principal of Peru. In Patagonia the cordillera was uplifted and the coastal cordillera also began to rise. At the same time, erosion of these mountains was taking place with deposition to the east of them.

In Patagonia elements of the Eocene flora retreated northward and the temperate elements of the *Nothofagus* flora advanced. In northern South America all the evidence points to a continuation of a tropical forest landscape, although with a great deal of phyletic evolution (Van der Hammen, 1966).

The paleontological record of mammals shows the continuing radiation and gradual evolution of the stock of ancient inhabitants of South America. The Oligocene also records the appearance of caviomorph rodents and platyrrhine primates, which probably arrived from North America via a sweepstakes route (Simpson, 1950; Patterson and Pascual, 1972), although an African origin has also been proposed (Hoffstetter, 1972).

Miocene. During Miocene times a number of important geological events took place. In the north the eastern cordillera of Colombia, which had been rising slowly since the beginning of the Tertiary, suffered its first strong uplift (Harrington, 1962). The large deposition of continental deposits in eastern Colombia and Peru continued, and by the end of the period the present altiplano of Peru and Bolivia had been eroded almost to sea level (Kummel, 1961). In the southern part of the continent one sees volcanic activity in Chubut and Santa Cruz as well as continued uplifting of the cordillera. By the end of the Miocene we begin to see the rise of the eastern and central cordilleras of Bolivia and the Puna and Pampean ranges of northern Argentina (Harrington, 1962).

During the Miocene the southern *Nothofagus* forest reached an extension similar to that of today. By the end of this time the pampa, large grassy extensions in central Argentina, became quite widespread (Patterson and Pascual, 1972). We also see the appearance and radiation of Compositae, a typical element in nonforested areas today (Van der Hammen, 1966). Among the fauna no major changes took place.

The climate continued to deteriorate from its peak of wet-warm in the middle Eocene. It still was more humid than today, as the presence of thick paleosoils in Patagonia seem to indicate (Volkheimer, 1971).

Nevertheless, the southern part of the continent, other than locally, was no longer occupied by forest but most certainly by either grassland or a parkland. The reduced rainfall, together with the ever-increasing rain-shadow effect of the rising Andes, must have led to long dry seasons in the middle latitudes. Indirect evidence from the evolutionary history of some bird and frog groups appears to indicate that the Nothofagus forest was not surrounded by forest vegetation at this time (Hecht, 1963; Vuilleumier, 1967). It is also very likely that semidesert regions existed in intermountain valleys and in the lee of the rising mountains in the western part of the continent from Patagonia northward.

Pliocene. From the Pliocene we have the first unmistakable evidence for the existence of more or less extensive areas of semidesert. Geologically it was a very active period. In the north we see the elevation of the Bolívar geosyncline and the development of the Colombian Andes in their present form, leading to the connection of South and North America toward the end of the period (Haffer, 1970). In Peru we see the rising of the cordillera and the bodily uplift of the altiplano to its present level, followed by some rifting. In Chile and Argentina we see the beginning of the final rise of the Cordillera Central as well as the uplift of the Sierras Pampeanas and the precordillera. All this increased orogenic activity was accompanied by extensive erosion and the deposition of continental sediments to the east in the Amazonian and Paraná-Paraguay basins (Harrington, 1962).

The lowland flora of northern South America, particularly that of the Amazonas and Orinoco basins, was not too different from today's flora in physiognomy or probably in floristic composition. However, because of the rise of the cordillera we find in the Pliocene the first indications of the existence of a high mountain flora (Van der Hammen, 1966) as well as the first clear indication of the existence of desert vegetation (Simpson Vuilleumier, 1967; Van der Hammen, 1966).

With the disappearance of the Bolívar geosyncline in late Pliocene, South America ceased to be an island and became connected to North America. This had a very marked influence on the fauna of the continent (Simpson, 1950; Patterson and Pascual, 1972). In effect, extensive faunistic interchanges took place during the Pliocene and Pleistocene between the two continents.

By the end of the Pliocene the landscape of South America was essentially identical to its present form. The rise of the Peruvian and Bolivian areas that we know as the Puna had taken place creating the dry highlands; the uplift of the Cordillera Central of Chile and the Sierras Pampeanas of Argentina had produced the rain shadows that make the area between them the dry land it is; and, finally, the rise of the southern cordillera of Chile must have produced dry, steppelike conditions in Patagonia. Geomorphological evidence shows this to be true (Simpson Vuilleumier, 1967; Volkheimer, 1971). The coastal region of Chile and Peru was probably more humid than today, since the cold Humboldt Current probably did not exist yet in its present form (Raven, 1971, 1973). However, although the stage is set, the actors are not quite ready. In effect, the Pleistocene, although very short in duration compared to the Tertiary events just described, had profound effects on species formation and distribution by drastically affecting the climate. Furthermore, because of its recency we also have a much better geological and paleontological record and therefore knowledge of the events of the Pleistocene.

Pleistocene

The deterioration of the Cenozoic climate culminated in the Pleistocene, when temperatures in the higher latitudes were lowered sufficiently to allow the accumulation and spread of immense ice sheets in the northern continents and on the highlands of the southern continents. Four major glacial periods are usually recognized in the Northern Hemisphere (Europe and North America), with three milder interglacial periods between them, and a fourth starting about 10,000 B.P. (Holocene), in which we are presently living. It is generally agreed (Charlesworth, 1957; Wright and Frey, 1965; Frenzel, 1968) that the Pleistocene has been a time of great variations in climate, both in temperature and in humidity, associated with rather significant changes in sea level (Emiliani, 1966). In general, glacial maxima correspond to colder and wetter climates than exist today; interglacials to warmer and often drier periods. But the march of events was more complex, and the temperature and humidity changes were not necessarily correlated (Charlesworth, 1957). Neither the exact series of events nor their ultimate causes are entirely clear.

Simpson Vuilleumier (1971), Van der Hammen and González (1960), and Van der Hammen (1961, 1966) have reviewed the Pleistocene events in South America. In northern South America (Venezuela, Colombia, and Ecuador) one to three glaciations took place, corresponding to the last three events in the Northern Hemisphere (Würm, Riss, and Mindel). In Peru, Bolivia, northern Chile, and Argentina there were at least three, in some areas possibly four. In Patagonia there were three to four glaciation events (table 2-1). All these glaciations, with the possible exception of Patagonia (Auer, 1960; Czajka, 1966), were the result of mountain glaciers.

The alternation of cold, wet periods with warm-dry and warm-wet periods had drastic effects on the biota. During glacial periods snow lines were lowered with an expansion of the areas suitable for a high mountain vegetation (Van der Hammen, 1966; Simpson Vuilleumier, 1971). At the same time glaciers moving along valleys created barriers to gene flow in some cases. During interglacials the snow line moved up again, and the areas occupied by high mountain vegetation no doubt were interrupted by low-lying valleys, which were occupied by more mesic-type plants. On the other hand, particularly at the beginning of interglacials, large mountain lakes were produced, and later on, with the rise of sea level, marine intrusions appeared. These events also broke up the ranges of species and created barriers to gene flow. To these happenings have to be added the effects of varying patterns of aridity and humidity. Let us then briefly review the events and their possible effects on the semidesert areas of temperate South America.

Patagonia. Glacial phenomena are best known from Patagonia (Caldenius, 1932; Feruglio, 1949; Frenguelli, 1957; Auer, 1960; Czajka, 1966; Flint and Fidalgo, 1968). Three or four glacial events are recorded. Along the cordillera the *Nothofagus* forest retreated north. The ice in its maximum extent covered probably most of Tierra del Fuego, all the area west of the cordillera, and some 100 km east of the mountains. Furthermore, during the glacial maxima, as a result of the lowering of the sea level, the Patagonian coastline was situated almost 300 km east of its present position. The climate was definitely colder and more humid. Studies by Auer (1958, 1960) indicate, however, that the *Nothofagus* forest did not expand eastward to any con-

Table 2-1. *Summary of Glacial Events in South America*

Localities	Glaciations (no.)	Age of Glaciations Relative to Europe	Present Snow Line (m)
Venezuela Mérida, Perija	1 or 2	Würm or Riss & Würm	4,800–4,900
Colombia Santa Marta and Cordillera C.	Variable, 1 to 3	Mindel to Würm	4,200–4,500
Peru All high Andean peaks	3	Mindel to Würm	5,800(W); 5,000(E)
Bolivia All NE ranges; high peaks in SE; few peaks in SW	3 or 4	Günz or Mindel to Würm	5,900(W); ca. 5,300(E)
Argentina and Chile Peaks between lat. 30° and 42°S; all land to the west of main Andean chain; to the east only to the base of the cordillera	3 or 4	Günz or Mindel to Würm	Variable: above 5,900 m in north to 800 m in south
Paraguay, Brazil, and Argentina Paraná basin		no glaciations	
Brazil Mt. Itatiaia	1 or 2	Würm or Riss & Würm	none
Brazil Amazonas basin		no glaciations	

Source: Modified from Simpson Vuilleumier, 1971.

Glacial Snow Line (m)	Glacial Climate	Interglacial Climate
2,700–3,300		
4,500(W); 4,200(E)	wet, temp. 4° to 11° lower than present	dry, temp. 2° to 3° higher than present
4,500(W); 4,200(E)	wet, temp. 7° lower than present	
5,000–5,300(W); 4,600–5,000(E)	wet, temp. 6° lower than present	
500 m at Santiago, Chile; sea level south of lat. 42°	wet	more genial than present
	cool, dry	humid, warm
2,300		
	cool, dry	humid, warm

siderable extent. It must be remembered that, even though the climate was more humid, the prevailing winds still would have been wester-lies and they still would have discharged most of their humidity when they collided with the cordillera as is the case today. The drastically lowered snow line and the cold-dry conditions of Patagonia, on the other hand, must have had the effect of allowing the expansion of the high mountain flora that began to evolve as a result of the uplift of the cordillera in the Pliocene and earlier.

Monte. The essential semidesert nature of the Monte region was probably not affected by the events of the Pleistocene, but the extent of the area must have fluctuated considerably during this time. In ef-fect, during glacial maxima not only did some regions become covered with ice, such as the valley of Santa María in Catamarca, but they also became colder. On the other hand, during interglacials there is evidence for a moister climatic regime, as the existence of fos-sil woodlands of *Prosopis* and *Aspidosperma* indicates (Groeber, 1936; Castellanos, 1956). Also, the present patterns of distribution of many mesophytic (but not wet-tropical) species or pairs of species, with populations in southern Brazil and the eastern Andes, could probably only have been established during a wetter period (Smith, 1962; Simpson Vuilleumier, 1971). On the other hand, geomorpholog-ical evidence from the loess strata of the Paraná-Paraguay basin (Padula, 1972) shows that there were at least two periods when the basin was a cool, dry steppe. During these periods the semidesert Monte vegetation must have expanded northward and to the east of its present range.

Puna. It has already been noted that during glacial maxima the snow line was lowered and the area open for colonization by the high mountain elements was considerably extended. Nowhere did that be-come more significant than in the Puna area (Simpson Vuilleumier, 1971). During glacial periods a number of extensive glaciers were formed in the mountains surrounding it, particularly the Cordillera Real near La Paz (Ahlfeld and Branisa, 1960). Numerous and extensive glacial lakes were also formed (Steinmann, 1930; Ahlfeld and Bran-isa, 1960; Simpson Vuilleumier, 1971). However, the basic nature of the Puna vegetation was probably not affected by these events. They must, however, have produced extensive shifts in ranges and isolation of populations, events that must have increased the rate of evolution and speciation.

Pacific Coastal Desert. The Pacific Coastal Desert is the result of the double rain shadow produced by the Andes to the east and the cold Humboldt Current to the west. The Andes did not reach their present size until the end of the Pliocene or later. The cold Humboldt Current did not become the barrier it is until its waters cooled considerably as a result of being fed by melt waters of Antarctic ice. The coastal cordillera, however, was higher in the Pleistocene than it is today (Cecioni, 1970). It is not possible to state categorically when the conditions that account for the Pacific Coastal Desert developed, but it was almost certainly not before the first interglacial. Consequently, it is safe to say that the Pacific Coastal Desert is a Pleistocene phenomenon, as is the area of Mediterranean climate farther south (Axelrod, 1973; Raven, 1973).

During the Pleistocene the snow line in the cordillera was considerably depressed and may have been as low as 1,300 m in some places (Simpson Vuilleumier, 1971). Estimates of temperature depressions are in the order of 7°C (Ahlfeld and Branisa, 1960). Although the ice did not reach the coast, the lowered temperature probably resulted in a much lowered timber line and expansion of Andean elements into the Pacific Coastal Desert. There is also evidence for dry and humid cycles during interglacial periods (Simpson Vuilleumier, 1971). The cold glacial followed by the dry interglacial periods probably decimated the tropical and subtropical elements that occupied the area in the Tertiary and allowed the invasion and adaptive radiation of cold- and dry-adapted Andean elements.

Holocene

We finally must consider the events of the last twelve thousand years, which set the stage for today's flora and vegetation. Evidence from Colombia, Brazil, Guyana, and Panama (Van der Hammen, 1966; Wijmstra and Van der Hammen, 1966; Bartlett and Barghoorn, 1973) indicates that the period started with a wet-warm period that lasted for two to four thousand years, followed by a period of colder and drier weather that reached approximately to 4,000 B.P. when the forest retreated, after which present conditions gradually became established. The wet-humid periods were times of expansion of the tropical vegetation, while the dry period was one of retreat and expansion of savannalike vegetation which appears to have occupied extensive areas of what is today the Amazonian basin (Van der Hammen, 1966; Haffer, 1969; Vanzolini and Williams, 1970; Simpson Vuilleumier, 1971). Unfortunately, no such detailed observations exist for the temperate regions of South America, but it is likely that the same alternations of wet, dry, and wet took place there, too.

The Floristic Affinities

Cabrera (1971) divides the vegetation of the earth into seven major regions, two of which, the *Neotropical* and *Antarctic* regions, include the vegetation of South America. The latter region comprises in South America only the area of the *Nothofagus* forest along both sides of the Andes from approximately lat. 35° S to Antarctica and the subantarctic islands (fig. 2-1). The Neotropical region, which occupies the rest of South America, is divided further into three dominions comprising, broadly speaking, the tropical flora (Amazonian Dominion), the subtropical vegetation (Chaco Dominion), and the vegetation of the Andes (Andean-Patagonian Dominion). The Chaco Dominion is further subdivided into seven phytogeographical provinces. Two of these are semidesert regions: the Monte province and the Prepuna province. The remaining five provinces of the Chaco Dominion are the Matorral or central Chilean province, the Chaco province, the Argentine Espinal (not to be confused with the Chilean Espinal), the region of the Pampa, and the region of the Caatinga in northeastern Brazil. With exception of the Matorral and the Caatinga, the other provinces of the Chaco Dominion are contiguous and reflect a different set of temperature, rainfall, and soil conditions in each case. The other dominion of the Neotropical flora that concerns us here is the Andean-Patagonian one, with three provinces: Patagonia, the Puna, and the vegetation of the high mountains. We see then that, of the five subdesert temperate provinces, two have a flora that is subtropical in origin and three a flora that is related to the high mountain vegetation. We will now briefly discuss the floristic affinities of each of these regions.

The Monte

The vegetation, flora, and floristic affinities of the Monte are the best known of all temperate semidesert regions (Vervoorst, 1945, 1973; Czajka and Vervoorst, 1956; Morello, 1958; Sarmiento, 1972; Solbrig, 1972, 1973). There is unanimous agreement that the flora of the Monte is related to that of the Chaco province (Cabrera, 1953, 1971; Sarmiento, 1972; Vervoorst, 1973).

Sarmiento (1972) and Vervoorst (1973) have made statistical comparisons between the Chaco and the Monte. Sarmiento, using a number of indices, shows that the Monte scrub is most closely related, both floristically and ecologically, to the contiguous dry Chaco woodland. Vervoorst, using a slightly different approach, shows that certain Monte communities, particularly on mountain slopes, have a greater number of Chaco species than other more xerophytic communities, particularly the *Larrea* flats and the vegetation of the sand dunes. Altogether, better than 60 percent of the species and more than 80 percent of the genera of the Monte are also found in the Chaco.

The most important element in the Monte vegetation is the genus *Larrea* with four species. Three of these—*L. divaricata*, *L. nitida*, and *L. cuneifolia*—constitute the dominant element over most of the surface of the Monte, either singly or in association (Barbour and Díaz, 1972; Hunziker et al., 1973). The fourth species, *L. ameghinoi*, a low-creeping shrub, is found in depressions on the southern border of the Monte and over extensive areas of northern Patagonia. Of the three remaining species, *L. cuneifolia* is found in Chile in the area between the Matorral and the beginning of the Pacific Coastal Desert, known locally as Espinal (not to be confused with the Argentine Espinal). *Larrea divaricata* has the widest distribution of the species in the genus. In Argentina it is found throughout the Monte as well as in the dry parts of the Argentine Espinal and Chaco up to the 600-mm isohyet (Morello, 1971, personal communication). However, there is some question whether the present distribution of *L. divaricata* in the Chaco is natural or the result of the destruction of the natural vegetation by man since *L. divaricata* is known to be invasive. This species is also found in Chile in the central provinces and in two isolated localities in Bolivia and Peru: the valley of Chuquibamba in Peru and the region of Tarija in Bolivia (Morello, 1958; Hunziker et al., 1973). Finally, *L. divaricata* is found in the semidesert regions of North America from Mexico to California (Yang, 1970; Hunziker et al., 1973).

The second most important genus in the Monte is *Prosopis*. Of the species of *Prosopis* found there, two of the most important ones (*P. alba*, *P. nigra*) are characteristic species in the Chaco and Argentine Espinal where they are widespread and abundant. A third very characteristic species of *Prosopis*, *P. chilensis*, is found in central and northern Chile, in the Matorral where it is fairly common and in some interior localities of the Pacific Coastal Desert, as well as in northern Peru. The records of *P. chilensis* from farther north in Ecuador and Colombia, and even from Mexico, correspond to the closely related species *P. juliflora*, considered at one time conspecific with *P. chilensis* (Burkart, 1940, 1952). *Prosopis alpataco* is found in the Monte and in Patagonia. Most other species of the genus have more limited distributions.

Another conspicuous element in the Monte is *Cercidium*. The genus is distributed from the semidesert regions in the United States where it is an important element of the flora, south along the Cordillera de los Andes, with a rather large distributional gap in the tropical region from Mexico to Ecuador. *Cercidium* is found in dry valleys of the Pacific Coastal Desert and in the cordillera in Peru and Chile. In Argentina it is found, in addition to the Monte, in the western edge of the Chaco, in the Prepuna, and also in the Puna (Johnston, 1924).

Bulnesia is represented in the Monte by two species, *B. retama* and *B. schickendanzii*. The first of these species is found also in the Pacific Coastal Desert in the region of Ica and Nazca; *B. schickendanzii*, however, is a characteristic element of the Prepuna province. Other interesting distributions among characteristic Monte species are the presence of *Bougainvillea spinosa* in the department of Moquegua in Peru (where it grows with *Cercidium praecox*). The highly specialized *Monttea aphylla* is endemic to the Monte, but a very closely related species, *Monttea chilensis*, is found in northern Chile. *Geoffroea decorticans*, the *chañar*, which is an important element both in the Chaco and in the Monte, is also found in northern Chile where it is common. These are but a few of the more important examples of Monte species that range into other semidesert phytogeographical provinces, particularly the Pacific Coastal Desert.

In summary, the Monte has its primary floristic connection with the Chaco but also has species belonging to an Andean stock. In addition, a number of important Monte elements are found in isolated dry pockets in southern Bolivia (Tarija), northern Chile, and coastal Peru and are hard to classify.

The Prepuna

There are no precise studies on the flora or the floristic affinities of the Prepuna. However, a look at the common species indicates a clear affinity with the Chaco and the Monte, such as *Zuccagnia punctata* (Monte), *Bulnesia schickendanzii* (Monte), *Bougainvillea spinosa* (Monte), *Trichocereus tertscheckii* (Monte), and *Cercidium praecox* (Monte and Chaco). Other elements are clearly Puna elements: *Psila boliviensis*, *Junellia juniperina*, and *Stipa leptostachya*. Although the Prepuna province has a physiognomy and floral mixture of its own, it undoubtedly has a certain ecotone nature, and its limits and its individuality are most probably Holocene events.

The Puna and Patagonia

Although the floristic affinities of the Puna and Patagonia have not been studied in as much detail as those of the Monte, they do not present any special problem. The flora of both regions is clearly part of the Andean flora. This important South American floristic element is relatively new (since it cannot be older than the Andes). This is further shown by the paucity of endemic families (only two small families, Nolanaceae [also found in the Galápagos Islands] and Malesherbiaceae, are endemic to the Andean Dominion) and by the large number of taxa belonging to such families as Compositae, Gramineae, Verbenaceae, Solanaceae, and Cruciferae, considered usually to be relatively specialized and geologically recent. The Leguminosae, represented in the Chaco Dominion mostly by Mimosoideae (among them some primitive genera), are chiefly represented in the Andean Dominion by more advanced and specialized genera of the Papilionoideae.

The Patagonian Steppe is characterized by a very large number of endemic genera, but particularly of endemic species (over 50%, cf. Cabrera, 1947). Of the species whose range extends beyond Patagonia, the great majority grow in the cordillera, a few extend into the *Nothofagus* forest, and a very small number are shared with the Monte. This is surprising in view of some similarities in soil and water stress between the two regions and also in view of the lack of any obvious physical barrier between the two phytogeographical provinces.

The Pacific Coastal Desert

The flora of the Pacific Coastal Desert is the least known. The relative lack of communications in this region, the almost uninhabited nature of large parts of the territory, and the harshness of the climate and the physical habitat have made exploration very difficult. Furthermore, a large number of species in this region are ephemerals, growing and blooming only in rainy years. Our knowledge is based largely on the works of Weberbauer and Ferreyra in Peru and those of Philippi, Johnston, and Reiche in Chile.

One of the characteristics of the region is the large number of endemic taxa. The only two endemic families of the Andean Dominion, the Malesherbiaceae and the Nolanaceae, are found here; many of the genera and most of the species are also endemic.

The majority of the species and genera are clearly related to the Andean flora. The common families are Compositae, Umbelliferae, Cruciferae, Caryophyllaceae, Gramineae, and Boraginaceae, all families that are considered advanced and geologically recent. In this it is similar to Patagonia. However, the region does not share many taxa with Patagonia, indicating an independent history from Andean ancestral stock, as is to be expected from its geographical position.

On the other hand, contrary to Patagonia, the Pacific Coastal Desert has elements that are clearly from the Chaco Dominion. Among them are *Geoffroea decorticans, Prosopis chilensis, Acacia caven, Zuccagnia punctata*, and pairs of vicarious species in *Monttea* (*M. aphylla, M. chilensis*), *Bulnesia* (*B. retama, B. chilensis*), *Goch-*

natia, and *Proustia*. In addition there are isolated populations of *Bulnesia retama*, *Bougainvillea spinosa*, and *Larrea divaricata* in Peru. Because the Monte and the Pacific Coastal Desert are separated today by the great expanse of the Cordillera de los Andes that reaches to over 5,000 m and by a minimum distance of 200 km, these isolated populations of Monte and Chaco plants are very significant.

Discussion and Conclusions

In the preceding pages a brief description of the desert and semi-desert regions of temperate South America was presented, as well as a short history of the known major geological and biological events of the Tertiary and Quaternary and the present-day floristic affinities of the regions under consideration. An attempt will now be made to relate these facts into a coherent theory from which some verifiable predictions can be made.

The paleobotanical evidence shows that the Neotropical flora and the Antarctic flora were distinct entities already in Cretaceous times (Menéndez, 1972) and that they have maintained that distinctness throughout the Tertiary and Quaternary in spite of changes in their ranges (mainly an expansion of the Antarctic flora). The record further indicates that the Antarctic flora in South America was always a geographical and floristic unit, being restricted in its range to the cold, humid slopes of the southern Andes. The origin of this flora is a separate problem (Pantin, 1960; Darlington, 1965) and will not be considered here. Some specialized elements of this flora expanded their range at the time of the lifting of the Andes (*Drimys*, *Lagenophora*, etc.), but the contribution of the Antarctic flora to the desert and semidesert regions is negligible. The discussion will be concerned, therefore, exclusively with the Neotropical flora from here on.

The data suggest that at the Cretaceous-Tertiary boundary (between Maestrichtian and Paleocene) the Neotropical angiosperm flora covered all of South America with the exception of the very southern tip. The evidence for this assertion is that the known fossil floras of that time coming from southern Patagonia (Menéndez, 1972) indicate the existence then of a tropical, rain-forest-type flora in a region that today supports xerophytic, cold-adapted scrub and cushion-plant vegetation. The reasoning is that if at that time it was hot and humid enough in the southernmost part of the continent for a rain forest, undoubtedly such conditions would be more prevalent farther north. Such reasoning, although largely correct, does not take into account all the factors.

If we accept that the global flow of air and the pattern of insolation of the earth were essentially the same throughout the time under consideration (see "Theory"), it is reasonable to assume that a gradient of increasing temperature from the poles to the equator was in existence. But it is not necessarily true that a similar gradient of humidity existed. In effect, on a perfect globe (one where the specific heat of water and land is not a factor) the equatorial region and the middle high latitudes (around 40°–60°) would be zones of high rainfall while the middle latitudes (25°–30°) and the polar regions would be regions of low rainfall. This is the consequence of the global movements of air (rising at the poles and middle high latitudes and consequently cooling adiabatically and discharging their humidity, falling in the middle latitudes and the poles and consequently heating and absorbing humidity). But the earth is not a perfect globe, and, consequently, the effects of distribution of land masses and oceans have to be taken into account. When air flows over water, it picks up humidity; when it flows over land, it tends to discharge humidity; when it encounters mountains, it rises, cools, and discharges humidity; behind a mountain it falls and heats and absorbs humidity (which is the reason why Patagonia is a semidesert today).

As far as can be ascertained, at the beginning of the Tertiary there were no large mountain chains in South America. Therefore the expected air flow probably was closer to the ideal, that is, humid in the tropics and in the middle low latitudes, relatively dry in mid-latitudes. I would like to propose, therefore, that at the beginning of the Tertiary South America was not covered by a blanket of rain forest, but that at middle latitudes, particularly in the western part of the continent, there existed a tropical (since the temperature was high) flora adapted to a seasonally dry climate. This was not a semidesert flora but most likely a deciduous or semideciduous forest with some xerophytic adaptations. I would further hypothesize that this flora persisted with extensive modification into our time and is what we today call the flora of the Chaco Dominion. I will call this flora "the Tertiary-Chaco paleo-

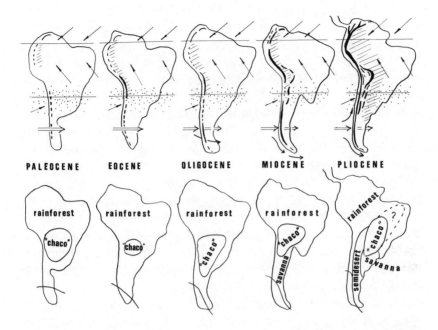

PALEOCENE   EOCENE   OLIGOCENE   MIOCENE   PLIOCENE

Fig. 2-2. Reconstruction of the outline of South America, mountain chains, and probable vegetation during the Tertiary. Solid line in Patagonia indicates the extent of the Nothofagus forest. Arrows indicate main global wind patterns.

flora" (fig. 2-2). I would like to hypothesize further that *Prosopis*, *Acacia*, and other Mimosoid legumes were elements of that flora as well as taxa in the Anacardiaceae and Zygophyllaceae or their ancestral stock. My justification for this claim is the prominence of these elements in the Chaco Dominion. Another indication of their primitiveness is their present distributional ranges, particularly the fact that many are found in Africa, which was supposedly considerably closer to South America at the beginning of the Tertiary than it is now. Furthermore, fossil remains of *Prosopis*, *Schinopsis*, *Schinus*, *Zygophyllum* (=*Guaiacum* [?], cf. Porter, 1974), and *Aspidosperma* have been reported (Berry, 1930; MacGinitie, 1953, 1969; Kruse, 1954; Axelrod, 1970) from North American floras (Colorado, Utah, and Wyoming) of Eocene or Eocene-Oligocene age (Florissant and Green River). This would indicate a wide distribution for these elements already at the beginning of the Tertiary. Verification of this hypothesis can be obtained from the study of the geology of Maestrichtian and Paleocene deposits of central Argentina and Chile as well as from the study of microfossils (and even megafossils) of this area.

Axelrod (1970) has proposed that some of the genera of the Monte and Chaco that have no close relatives in their respective families, such as *Donatia, Puya, Grabowskia, Monttea, Bredemeyera, Bulnesia,* and *Zuccagnia,* are modified relicts from the original upland angiosperm stock, which he feels is of pre-Cretaceous origin. The question of a pre-Cretaceous origin for the angiosperms is a speculative point due, to large measure, to the lack of corroborating fossil evidence (Axelrod, 1970). Assuming Axelrod's thesis were correct, not only would the Tertiary-Chaco paleoflora be the ancient nucleus of the subtropical elements adapted to a dry season, but presumably it would also represent the oldest angiosperm stock on the continent. However, some of the genera cited (*Bulnesia, Monttea, Zuccagnia*) are not primitive but are highly specialized.

The Tertiary in South America is characterized by the gradual lifting of the Andean chain. The process became much accelerated after the Eocene. The Tertiary is also characterized by the gradual cooling and drying of the climate after the Eocene, apparently a world-wide event (Wolfe and Barghoorn, 1960; Axelrod and Bailey, 1969; Wolfe, 1971). As a result, during the Paleocene and Eocene the Tertiary-Chaco paleoflora must have been fairly restricted in its distribution. However, after the Eocene it started to expand and differentiate. During the Eocene the first development of the steppe elements of the pampa region occurred. (The present pampa flora is part of the Chaco Dominion.) Evidence is found in the evolution of mammals adapted to eating grass and living in open habitats (Patterson and Pascual, 1972) and in the first fossil evidence of grasses (Van der Hammen, 1966). Since Tertiary deposits exist in the pampa region that have yielded animal microfossils (Padula, 1972), a study of these cores for plant microfossils may produce uncontroversial direct evidence for the evolution of the pampas. Less certain is the evolution of a dry Chaco or Monte vegetation from the Tertiary-Chaco paleoflora during the latter part of the Tertiary. The present-day distribution of *Larrea, Bulnesia, Monttea, Geoffroea, Cercidium,* and other typical dry Chaco or Monte elements makes me think that by the Pliocene a semidesert-type vegetation, not just one adapted to a dry season, was in existence in western middle South America. In effect, all these elements, so characteristic of extensive areas of semidesert in Argentina east of the Andes, are represented by small or, in some cases (*Geoffroea*, *Prosopis*), fairly abundant populations west of the Andes in a region (the Pacific Coastal Desert) dominated today by Andean elements that originated at a later date. Furthermore, the Mediterranean region of Chile is formed in part by Chaco elements, and we know that a Mediterranean climate probably did not evolve until the Pleistocene (Raven, 1971, 1973; Axelrod, 1973). It is, consequently, probable that toward the end of the Tertiary a more xerophytic flora was evolving from the Tertiary-Chaco paleoflora, perhaps as a result of xeric local conditions in the lee of the rising mountains. It should be pointed out, too, that the coastal cordillera in Chile rose first and was bigger in the Tertiary than it is today (Cecioni, 1970).

The Pliocene is characterized by a great increase in orogenic activity that created the Cordillera de los Andes in its present form (Kummel, 1961; Harrington, 1962; Haffer, 1970). In a sea of essentially tropical vegetation an alpine environment was created, ready to be colonized by plants that could withstand not only the cold but also the great daily (in low latitudes) or seasonal climate variation (in high latitudes). The rise of the cordillera also interfered with the free flow of winds and produced changes in local climate, creating the Patagonian and Monte semideserts as we know them today.

Most of the elements that populate the high cordillera were drawn from the Neotropical flora, although some Antarctic elements invaded the open Andean regions (*Lagenophora*) as well as some North American elements, such as *Alnus*, *Sambucus*, *Viburnum*, *Erigeron*, *Aster*, and members of the Ericaceae and Cruciferae (Van der Hammen, 1966). Particularly, elements in the Compositae, Caryophyllaceae, Umbelliferae, Cruciferae, and Gramineae radiated and became dominant in the newly opened habitats. The Puna is an integral part of the Andes and consequently must have become populated by the Andean elements as it became uplifted in the Pliocene, displacing the original tropical Amazonian elements present there in Miocene times (Berry, 1917) which were ill-adapted to the new climatic conditions. A similar but less-evident pattern must have taken place in Patagonia. The tropical flora present in Patagonia in the Paleocene and Eocene and, on mammalian evidence, up to the Miocene (Patterson and Pascual, 1972) had migrated north in response to the cooling climate and had been apparently replaced by an open steppe presumably of Chaco origin. This flora and that of the evolving dry Chaco-Monte vegetation should have been able to adapt to the increasing xerophytic environment of Patagonia in the Pliocene and perhaps did. Only better fossil evidence can tell. But it is the events of the Pleistocene that account for the present dry flora of Patagonia. The same can be said for the Pacific Coastal Desert. The lifting of the cordillera created a disjunct area of Chaco vegetation on the Pacific coast. Before the development of the cold Humboldt Current, presumably a Pleistocene event (in its present condition), the environment must have been more mesic, especially in its southern regions, and the Chaco-Monte flora should have been able to adapt to those conditions. Again it is the Pleistocene events that explain the present flora.

The Pleistocene is marked by a series (up to four) of very drastic fluctuations in temperature as well as some (presumably also up to four) extreme periods of aridity. During the cold periods permanent snow lines dropped, mountain glaciers formed, and the high mountain vegetation expanded. During the last glacial event, the Würm, which seems to have been the most drastic in South America, ice fields extended in Patagonia from the Pacific Coast to some 100 km or more east of the high mountain line, and the Patagonian climate was that of an arctic steppe with extremely cold winters. During those periods any existing subtropical elements disappeared and were replaced by cold-adapted Andean taxa. As conditions gradually improved, phyletic evolution of that cold-adapted flora of Andean origin produced today's Patagonian flora. The same is true in the Pacific Coastal Desert. Only here some elements of the old Chaco flora managed to survive the Pleistocene and account for the range disjunctions of such species as *Larrea divaricata*, *Bougainvillea spinosa*, or *Bulnesia retama*. Some elements of the old Chaco flora of the Pacific coast moved north and are found today in the *algarrobal* formation of northern Peru and Ecuador and in dry inter-Andean valleys of Colombia and Venezuela. Others moved south and are part of the Espinal and Matorral formations of Chile.

The Pleistocene also affected the Monte region. The northern inter-

Andean valleys and *bolsones* became colder, and the mountain slope vegetation was replaced by a vegetation of high-mountain type, so that the Monte vegetation was compressed into a smaller area farther south and east, principally in Mendoza, La Pampa, San Luis, and La Rioja. Cold periods were followed by warmer and wetter periods, characterized by more mesic subtropical elements which expanded their range from the east slope of the Andes to Brazil (Smith, 1962; Simpson Vuilleumier, 1971). After these mesic periods came very extensive dry periods, marked by expansion of the Monte flora and the breakup of the ranges of the subtropical elements, many of which have now disjunct distributions in Brazil and the eastern Andes.

Acknowledgments

This paper is the result of my long-standing interest in the flora and vegetation of the Monte in Argentina and of temperate semidesert and desert regions in general. Too many people to name individually have aided my interest, stimulated my curiosity, and satisfied my knowledge for facts. I would like, however, to acknowledge my particular indebtedness to Professor Angel L. Cabrera at the University of La Plata in Argentina, who first initiated me into floristic studies and who, over the years, has continuously stimulated me through personal conversations and letters, and through his writings. Other people whose help I would like to acknowledge are Drs. Humberto Fabris, Juan Hunziker, Harold Mooney, Jorge Morello, Arturo Ragonese, Beryl Simpson, and Federico Vervoorst. With all of them and many others I have discussed the ideas in this paper, and no doubt these ideas became modified and were changed to the point where it is hard for me to state now exactly what was originally my own. I further would like to thank the Milton Fund of Harvard University, the University of Michigan, and the National Science Foundation, which made possible yearly trips to South America over the last ten years. I particularly would like to acknowledge two NSF grants for studies in the structure of ecosystems that have supported my active research in the Monte and Sonoran desert ecosystem for the last three years. Sergio Archangelsky, Angel L. Cabrera, Philip Cantino, Carlos Menéndez,

Bryan Patterson, Duncan Porter, Beryl Simpson, and Rolla Tryon read the manuscript and made valuable suggestions for which I am grateful.

Summary

The existent evidence regarding the floristic relations of the semi-desert regions of South America and how they came to exist is reviewed.

The regions under consideration are the phytogeographical provinces of Patagonia and the Monte in Argentina; the Puna in Argentina, Bolivia, and Peru; the Espinal in Chile; and the Pacific Coastal Desert in Chile and Peru. It is shown that the flora of Patagonia, the Puna, and the Pacific Coastal Desert are basically of Andean affinities, while the flora of the Monte has affinities with the flora of the subtropical Chaco. However, Chaco elements are also found in Chile and Peru. From these considerations and those of a geological and geoclimatological nature, it is postulated that there might have existed an early (Late Cretaceous or early Tertiary) flora adapted to living in more arid —although not desert—environments in and around lat. 30° S.

The present flora of the desert and semidesert temperate regions of South America is largely a reflection of Pleistocene events. The flora of the Andean Dominion that originated the flora that today populates Patagonia and the Pacific Coastal Desert, however, evolved largely in the Pliocene, while the Chaco flora that gave origin to the Monte and Prepuna flora had its beginning probably as far back as the Cretaceous.

References

Ahlfeld, F., and Branisa, L. 1960. *Geología de Bolivia*. La Paz: Instituto Boliviano de Petroleo.
Auer, V. 1958. The Pleistocene of Fuego Patagonia. II. The history of the flora and vegetation. *Suomal. Tiedeakat. Toim. Ser. A 3.* 50:1–239.

————. 1960. The Quaternary history of Fuego-Patagonia. *Proc. R. Soc. Ser. B.* 152:507–516.

Axelrod, D. I. 1967. Drought, diastrophism and quantum evolution. *Evolution* 21:201–209.

————. 1970. Mesozoic paleogeography and early angiosperm history. *Bot. Rev.* 36:277–319.

————. 1973. History of the Mediterranean ecosystem in California. In *The convergence in structure of ecosystems in Mediterranean climates*, ed. H. Mooney and F. di Castri, pp. 225–284. Berlin: Springer.

Axelrod, D. I., and Bailey, H. P. 1969. Paleotemperature analysis of Tertiary floras. *Paleogeography, Paleoclimatol. Paleoecol.* 6:163–195.

Barbour, M. G., and Díaz, D. V. 1972. *Larrea* plant communities on bajada and moisture gradients in the United States and Argentina. *U.S./Intern. biol. Progr.: Origin and Structure of Ecosystems Tech. Rep.* 72–6:1–27.

Bartlett, A. S., and Barghoorn, E. S. 1973. Phytogeographic history of the Isthmus of Panama during the past 12,000 years. In *Vegetation and vegetational history of northern Latin America*, ed. A. Graham, pp. 203–300. Amsterdam: Elsevier.

Berry, E. W. 1917. Fossil plants from Bolivia and their bearing upon the age of the uplift of the eastern Andes. *Proc. U.S. natn. Mus.* 54:103–164.

————. 1930. Revision of the lower Eocene Wilcox flora of the southeastern United States. *Prof. Pap. U.S. geol. Surv.* 156:1–196.

Böcher, T.; Hjerting, J. P.; and Rahn, K. 1963. Botanical studies in the Atuel Valley area, Mendoza Province, Argentina. *Dansk bot. Ark.* 22:7–115.

Burkart, A. 1940. Materiales para una monografía del género *Prosopis*. *Darwiniana* 4:57–128.

————. 1952. *Las leguminosas argentinas silvestres y cultivadas*. Buenos Aires: Acme Agency.

Cabrera, A. L. 1947. La Estepa Patagónica. In *Geografía de la República Argentina*, ed. GAEA, 8:249–273. Buenos Aires: GAEA.

————. 1953. Esquema fitogeográfico de la República Argentina. *Revta Mus. La Plata (nueva Serie), Bot.* 8:87–168.

————. 1958. La vegetación de la Puna Argentina. *Revta Invest. agríc., B. Aires* 11:317–412.

————. 1971. Fitogeografía de la República Argentina. *Boln Soc. argent. Bot.* 14:1–42.

Caldenius, C. C. 1932. Las glaciaciones cuaternarias en la Patagonia y Tierra del Fuego. *Geogr. Annlr* 14:1–164.

Castellanos, A. 1956. Caracteres del pleistoceno en la Argentina. *Proc.IV Conf. int. Ass. quatern. Res.* 2:942–948.

Cecioni, G. 1970. *Esquema de paleogeografía chilena.* Santiago: Editorial Universitaria.

Charlesworth, J. K. 1957. *The Quaternary era.* 2 vols. London: Arnold.

Czajka, W. 1966. Tehuelche pebbles and extra-Andean glaciation in east Patagonia. *Quaternaria* 8:245–252.

Czajka, W., and Vervoorst, F. 1956. Die naturräumliche Gliederung Nordwest-Argentiniens. *Petermanns geogr. Mitt.* 100:89–102, 196–208.

Darlington, P. J. 1957. *Zoogeography: The geographical distribution of animals.* New York: Wiley.

————. 1965. *Biogeography of the southern end of the world.* Cambridge, Mass.: Harvard Univ. Press.

Dietz, R. S., and Holden, J. C. 1970. Reconstruction of Pangea: Breakup and dispersion of continents, Permian to present. *J. geophys. Res.* 75:4939–4956.

Dimitri, M. J. 1972. Consideraciones sobre la determinación de la superficie y los limites naturales de la región andino-patagónica. In *La región de los Bosques Andino-Patagonicos*, ed. M. J. Dimitri, 10:59–80. Buenos Aires: Col. Cient. del INTA.

Emiliani, C. 1966. Isotopic paleotemperatures. *Science, N.Y.* 154: 851.

Ferreyra, R. 1960. Algunos aspectos fitogeográficos del Perú. *Publnes Inst. Geogr. Univ. San Marcos (Lima)* 1(3):41–88.

Feruglio, E. 1949. *Descripción geológica de la Patagonia.* 2 vols. Buenos Aires: Dir. Gen. de Y.P. F.

Fiebrig, C. 1933. Ensayo fitogeográfico sobre el Chaco Boreal. *Revta Jard. bot. Mus. Hist. nat. Parag.* 3:1–87.

Flint, R. F., and Fidalgo, F. 1968. Glacial geology of the east flank of the Argentine Andes between latitude 39°-10′ S and latitude 41°-20′ S. *Bull. geol. Soc. Am.* 75:335–352.

Flohn, H. 1969. *Climate and weather*. New York: McGraw-Hill Book Co.

Frenguelli, J. 1957. El hielo austral extraandino. In *Geografía de la República Argentina*, ed. GAEA, 2:168–196. Buenos Aires: GAEA.

Frenzel, B. 1968. The Pleistocene vegetation of northern Eurasia. *Science, N.Y.* 161:637.

Good, R. 1953. *The geography of the flowering plants*. London: Longmans, Green & Co.

Goodspeed, T. 1945. The vegetation and plant resources of Chile. In *Plants and plant science in Latin America*, ed. F. Verdoorn, pp. 147–149. Waltham, Mass.: Chronica Botanica.

Gordillo, C. E., and Lencinas, A. N. 1972. Sierras pampeanas de Córdoba y San Luis. In *Geología regional Argentina*, ed. A. F. Leanza, pp. 1–39. Córdoba: Acad. Nac. de Ciencias.

Groeber, P. 1936. Oscilaciones del clima en la Argentina desde el Plioceno. *Revta Cent. Estud. Doct. Cienc. nat., B. Aires* 1(2):71–84.

Haffer, J. 1969. Speciation in Amazonian forest birds. *Science, N.Y.* 165:131–137.

———. 1970. Geologic-climatic history and zoogeographic significance of the Uraba region in northwestern Colombia. *Caldasia* 10: 603–636.

Harrington, H. J. 1962. Paleogeographic development of South America. *Bull. Am. Ass. Petrol. Geol.* 46:1773–1814.

Haumann, L. 1947. Provincia del Monte. In *Geografía de la República Argentina*, ed. GAEA, 8:208–248. Buenos Aires: GAEA.

Hecht, M. K. 1963. A reevaluation of the early history of the frogs. Pt. II. *Syst. Zool.* 12:20–35.

Hoffstetter, R. 1972. Relationships, origins and history of the Ceboid monkeys and caviomorph rodents: A modern reinterpretation. *Evol. Biol.* 6:323–347.

Hünicken, M. 1966. Flora terciaria de los estratos del río Turbio, Santa Cruz. *Revta Fac. Cienc. exact. fís. nat. Univ. Córdoba, Ser. Cienc. nat.* 27:139–227.

Hunziker, J. H.; Palacios, R. A.; de Valesi, A. G.; and Poggio, L. 1973. Species disjunctions in *Larrea*: Evidence from morphology, cytogenetics, phenolic compounds, and seed albumins. *Ann. Mo. bot. Gdn* 59:224–233.

Johnston, I. 1924. Taxonomic records concerning American spermatophytes. 1. Parkinsonia and Cercidium. *Contr. Gray Herb. Harv.* 70:61–68.

———. 1929. Papers on the flora of northern Chile. *Contr. Gray Herb. Harv.* 85:1–171.

Kruse, H. O. 1954. Some Eocene dicotyledoneous woods from Eden Valley, Wyoming. *Ohio J. Sci.* 54:243–267.

Kummel, B. 1961. *History of the earth*. San Francisco: W. H. Freeman & Co.

Liddle, R. A. 1946. *The geology of Venezuela and Trinidad*. 2d ed. Ithaca: Pal. Res. Inst.

Lorentz, P. 1876. Cuadro de la vegetación de la República Argentina. In *La República Argentina*, ed. R. Napp, pp. 77–136. Buenos Aires: Currier de la Plata.

MacGinitie, H. D. 1953. Fossil plants of the Florissant beds, Colorado. *Publs Carnegie Instn* 599:1–180.

———. 1969. The Eocene Green River flora of northwestern Colorado and northeastern Utah. *Univ. Calif. Publs geol. Sci.* 83:1–140.

Mayr, E. 1963. *Animal species and evolution*. Cambridge, Mass: Harvard Univ. Press.

Menéndez, C. A. 1969. Die fossilen floren Südamerikas. In *Biogeography and ecology in South America*, ed. E. J. Fittkau, J. Illies, H. Klinge, G. H. Schwabe, and H. Sioli, 2:519–561. The Hague: Dr. W. Junk.

———. 1972. Paleofloras de la Patagonia. In *La región de los Bosques Andino-Patagonicos*, ed. M. J. Dimitri, 10:129–184. Col. Cient. Buenos Aires: del INTA.

Mooney, H., and Dunn, E. L. 1970. Convergent evolution of Mediterranean-climate evergreen sclerophyll shrubs. *Evolution* 24:292–303.

Morello, J. 1958. La provincia fitogeográfica del Monte. *Op. lilloana* 2:1–155.

Padula, E. L. 1972. Subsuelo de la mesopotamia y regiones adyacentes. In *Geología regional Argentina*, ed. A. F. Leanza, pp. 213–236. Córdoba: Acad. Nac. de Ciencias.

Pantin, C. F. A. 1960. A discussion on the biology of the southern cold temperate zone. *Proc. R. Soc. Ser. B.* 152:431–682.

Patterson, B., and Pascual, R. 1972. The fossil mammal fauna of South America. In *Evolution, mammals, and southern continents*, ed. A. Keast, F. C. Erk, and B. Glass, pp. 247–309. Albany: State Univ. of N.Y.

Petriella, B. 1972. Estudio de maderas petrificadas del Terciario inferior del área de Chubut Central. *Revta Mus. La Plata (Nueva Serie), Pal.* 6:159–254.

Porter, D. M. 1974. Disjunct distributions in the New World Zygophyllaceae. *Taxon* 23:339–346.

Raven, P. H. 1971. The relationships between "Mediterranean" floras. In *Plant life of South-West Asia*, ed. P. H. Davis, P. C. Harper, and I. C. Hedge, pp. 119–134. Edinburgh: Bot. Soc.

———. 1973. The evolution of Mediterranean floras. In *The convergence in structure of ecosystems in Mediterranean climates*, ed. H. Mooney and F. di Castri, pp. 213–224. Berlin: Springer.

Reiche, K. 1934. *Geografía botánica de Chile*. 2 vols. Santiago: Imprenta Universitaria.

Romero, E. 1973. Ph.D. dissertation, Museo La Plata Argentina.

Sarmiento, G. 1972. Ecological and floristic convergences between seasonal plant formations of tropical and subtropical South America. *J. Ecol.* 60:367–410.

Schuchert, C. 1935. *Historical geology of the Antillean-Caribbean region*. New York: Wiley.

Schwarzenbach, M. 1968. Das Klima des rheinischen Tertiärs. *Z. dt. geol. Ges.* 118:33–68.

Simpson, G. G. 1950. History of the fauna of Latin America. *Am. Scient.* 1950:361–389.

Simpson Vuilleumier, B. 1967. The systematics of Perezia, section Perezia (Compositae). Ph.D. thesis, Harvard University.

———. 1971. Pleistocene changes in the fauna and flora of South America. *Science, N.Y.* 173:771–780.

Smith, L. B. 1962. Origins of the flora of southern Brazil. *Contr. U.S. natn. Herb.* 35:215–250.

Solbrig, O. T. 1972. New approaches to the study of disjunctions with special emphasis on the American amphitropical desert disjunctions. In *Taxonomy, phytogeography and evolution*, ed. D. D. Valentine, pp. 85–100. London and New York: Academic Press.

———. 1973. The floristic disjunctions between the "Monte" in Argentina and the "Sonoran Desert" in Mexico and the United States. *Ann. Mo. bot. Gdn* 59:218–223.

48   OTTO T. SOLBRIG

Soriano, A. 1949. El limite entre las provincias botánicas Patagónica y Central en el territorio del Chubut. *Revta argent. Agron.* 17:30–66.

———. 1950. La vegetación del Chubut. *Revta argent. Agron.* 17:30–66.

———. 1956. Los distritos floristicos de la Provincia Patagónica. *Revta Invest. agríc., B. Aires* 10:323–347.

Stebbins, G. L. 1950. *Variation and evolution in plants*. New York: Columbia Univ. Press.

———. 1952. Aridity as a stimulus to evolution. *Am. Nat.* 86:33–44.

Steinmann, G. 1930. *Geología del Perú*. Heidelberg: Winters.

Thorne, R. F. 1973. Floristic relationships between tropical Africa and tropical America. In *Tropical forest ecosystems in Africa and South America: A comparative review*, ed. B. J. Meggers, E. S. Ayensu, and D. Duckworth, pp. 27–40. Washington, D.C.: Smithsonian Instn. Press.

Van der Hammen, T. 1961. The Quaternary climatic changes of northern South America. *Ann. N.Y. Acad. Sci.* 95:676–683.

———. 1966. Historia de la vegetación y el medio ambiente del norte sudamericano. In *1° Congr. Sud. de Botánica, Memorias de Symposio*, pp. 119–134. Mexico City: Sociedad Botánica de Mexico.

Van der Hammen, T., and González, E. 1960. Upper Pleistocene and Holocene climate and vegetation of the "Sabana de Bogotá." *Leid. geol. Meded.* 25:262–315.

Vanzolini, P. E., and Williams, E. E. 1970. South American anoles: The geographic differentiation and evolution of the *Anolis chrysolepis* species group (Sauria, Iguanidae). *Archos Zool. Est. S Paulo* 19:1–298.

Vervoorst, F. 1945. *El Bosque de algarrobos de Pipanaco (Catamarca)*. Ph.D. dissertation, Universidad de Buenos Aires.

———. 1973. Plant communities in the bolsón de Pipanaco. *U.S./ Intern. biol. Progr.: Origin and Structure of Ecosystems Prog. Rep.* 73-3:3–17.

Volkheimer, W. 1971. Aspectos paleoclimatológicos del Terciario Argentina. *Revta Mus. Cienc. nat. B. Rivadavia Paleontol.* 1:243–262.

Vuilleumier, F. 1967. Phyletic evolution in modern birds of the Patagonian forests. *Nature, Lond.* 215:247–248.

Weberbauer, A. 1945. *El mundo vegetal de los Andes Peruanos*. Lima: Est. Exp. La Molina.

Wijmstra, T. A., and Van der Hammen, T. 1966. Palynological data on the history of tropical savannas in northern South America. *Leid. geol. Meded.* 38:71–90.

Wolfe, J. A. 1971. Tertiary climatic fluctuations and methods of analysis of Tertiary floras. *Paleogeography, Paleoclimatol. Paleoecol.* 9:27–57.

Wolfe, J. A., and Barghoorn, E. S. 1960. Generic change in Tertiary floras in relation to age. *Am. J. Sci.* 258A:388–399.

Wright, H. E., and Frey, D. G. 1965. *The Quaternary of the United States*. Princeton: Princeton Univ. Press.

Yang, T. W. 1970. Major chromosome races of *Larrea divaricata* in North America. *J. Ariz. Acad. Sci.* 6:41–45.

3. The Evolution of Australian Desert Plants   John S. Beard

Introduction

As an opening to this subject it may be well to outline briefly the where-
abouts of the Australian desert, its climate and vegetation. The desert
consists, of course, of the famous "dead heart" of Australia, covering
the interior of the continent; and it has been defined on a map together
with its component natural regions by Pianka (1969a). An important
characteristic of this area is that, while certainly arid and classifiable
as desert by most, if not all, of the better-known bioclimatic classifications and indices, it is not as rainless as some of the world's deserts
and is correspondingly better vegetated. The most arid portion of the
Australian interior, the Simpson Desert, receives an average rainfall
of 100 mm, while most of the rest of the desert receives around 200
mm. The desert is usually taken to begin, in the south, at the 10-inch,
or 250-mm, isohyet. In the north, in the tropics under higher temperatures, desert vegetation reaches the 20-inch, or 500-mm, isohyet.

Plant Formations in Australian Deserts

As a result of the rainfall in the Australian desert, it always possesses
a plant cover of some kind, and we have no bare and mobile sand
dunes and few sheets of barren rock. There are two principal plant
formations: a low woodland of *Acacia* trees colloquially known as mul-
ga, which covers roughly the southern half of the desert south of the
tropic, and the "hummock grassland" (Beadle and Costin, 1952) colloquially known as spinifex, which covers the northern half within the
tropics. Broadly the two formations are climatically separated, al-

though the preference of each of them for certain soils tends to obscure this relationship; thus, the hummock grassland appears on sand even in the southern half. The *Acacia* woodland is to be compared with those of other continents, but few Australian species of *Acacia* have thorns and few have bipinnate leaves. The hummock grassland, on the other hand, is, I think, a unique product of evolution in Australia. It is comparable with the grass steppe vegetation of other continents, but the life form of the grasses is different. Two genera are represented, *Triodia* and *Plectrachne*. Each plant branches repeatedly into a great number of culms which intertwine to form a hummock and bear rigid, terete, pungent leaves presenting a serried phalanx to the exterior. When flowering takes place in the second half of summer, given adequate rains, upright rigid inflorescences are produced above the crown of the hummock, rising 0.5 to 1 m above it. The flowers quickly set seed, which is shed within two months, although this is then the beginning of the dry season. The size of the hummock varies considerably according to the site from 30 cm in height and diameter on the poorest, stoniest sites up to about 1 m in height and 2 m in diameter on some deep sands. Old hummocks, if unburnt, tend to die out in the center or on one side, leading to ring or crescentic growth. At this stage the original root has died and the outer culms have rooted themselves adventitiously in the soil. Individual hummocks do not touch, and there is much bare ground between them.

The hummock grassland normally contains a number of scattered shrubs or scattered trees in less-arid areas where ground water is available. All of these must be resistant to fire, by which the grassland is regularly swept. After burning, the grasses regenerate from the root or from seed.

The *Acacia* woodlands, in which *A. aneura* is frequently the sole species in the upper stratum, contain a sparse lower layer of shrubs most frequently of the genera *Eremophila* and *Cassia*, 1–2 m tall, and an even sparser ground layer mainly of ephemerals and only locally of grasses.

These Australian desert formations are given distinctive character by the physiognomy of their commonest plants, that is:

Trees. Evergreen, sclerophyll. Leaves pendent in *Eucalyptus*; linear, erect, and glaucescent in *Acacia aneura*; vestigial in *Casuarina decaisneana*. Bark white in most species of *Eucalyptus*.

Shrubs. The larger shrubs are sclerophyll, typically phyllodal species
of *Acacia*; the smaller shrubs, ericoid (*Thryptomene*).

Subshrubs. Many soft perennial subshrubs typically with densely
pubescent or silver-tomentose stems and leaves, e.g., *Crotalaria
cunninghamii*, and numerous Verbenaceae (*Dicrastyles*, *New-
castelia*, *Pityrodia* spp.). Also, suffrutices with underground rootstocks and ephemeral or more or less perennial shoots, often also
densely pubescent or silver-tomentose, e.g., *Brachysema cham-
bersii*, many *Ptilotus* spp., *Leschenaultia helmsii*, and *L. striata*.
Some are viscid—*Goodenia azurea* and *G. stapfiana*.

Ephemerals. Many species of Compositae, *Ptilotus*, and *Goodenia*
appear as brilliant-flowering annuals in season. Colors are predominantly yellow and mauve, with some white and pink. Red is absent.

Grasses. Grasses of the "short bunch-grass" type in the sense of
Bews (1929) occur only on alluvial flats close to creeks or on plains
of limited extent developed on or close to basic rocks. In these
cases there is a fine soil with a relatively high water-holding capacity
and probably also high-nutrient status. On sand, laterite, and rock in
the desert, grasses belong almost entirely to the genera *Triodia*
and *Plectrachne*, which adopt the hummock-grass form as previously described. This growth form appears to be peculiar to Australia and to be the only unique form evolved in the Australian
desert.

It will therefore be seen that the Australian desert possesses special
vegetative characters of its own which can be supposed to be of some
adaptive significance, particularly *glaucescence* of bark and leaves,
pubescence frequently in association with glaucescence, *suffrutes-
cence*, the presence of vernicose and viscid leaf surfaces, and the
spinifex habit in grasses. Other characters, such as tree and shrub
growth forms and sclerophylly, are not peculiar to the desert Eremaea
but are shared with other Australian vegetation.

Growth Forms

In most of the world's deserts special and peculiar growth forms have
evolved which confer advantage in the arid environment. In North and

Central America the family Cactaceae has produced the well-known range of forms based on stem succulence, closely replicated by the Euphorbiaceae in Africa. In southern African deserts leaf succulence is a dominant feature that has been developed in many families, notably the Aizoaceae and Liliaceae. Leaf-succulent rosette plants in the Bromeliaceae are a feature of both arid northwest Brazil and the cold Andean Puna. In all cases we are accustomed to look also for deciduous, thorny trees and plants with underground perennating organs, especially bulbs and corms. In Australia there is an extraordinary lack of all these forms; where some of them exist they are confined to certain areas.

Leaf- and stem-succulent plants belonging to the family Chenopodiaceae in fact characterize two other important plant formations, less widespread than the principal formations described above and confined to certain soils. These I have named "succulent steppe" (Beard, 1969) following the usage of African ecologists; they comprise, first, saltbush and bluebush steppe dominated by species of *Atriplex* and *Kochia* respectively, and, second, samphire communities with *Arthrocnemum*, *Tecticornia*, and related genera. The former are small soft shrubs whose leaves are fleshy or semisucculent, associated with annual grasses and herbs, and sometimes with a sclerophyll tree layer of *Acacia* or *Eucalyptus*. The formation is confined to the southern half of the desert region and occupies alkaline soils, most commonly on limestone or calcareous clays. In the northern half such soils normally carry hummock grassland on limestone and bunch grassland on clays. The samphire communities, however, range throughout the region on very saline soils in depressions, usually in the beds of playa lakes or peripheral to them. The samphires are subshrubs with succulent-jointed stems. These formations are the only ones with a genuinely succulent character and are essentially halophytes.

On the siliceous soils of the desert, sclerophylly is the dominant characteristic, and stem succulence is represented in only a handful of species of no prominence, such as *Sarcostemma australe* (Asclepiadaceae), a divaricate, leafless plant found occasionally in rocky places. Others are *Spartothamnella teucriiflora* (Verbenaceae) and *Calycopeplus helmsii* (Euphorbiaceae). Likewise, leaf succulence is found in a variety of groups but is often weakly developed and never a conspicuous feature. *Gyrostemon ramulosus* (Phytolaccaceae) has somewhat fleshy foliage, which the explorers noted as a favorite feed of camels. The Aizoaceae in Australia are mostly tropical herbs, and the most genuinely succulent member, *Carpobrotus*, is not Eremaean. The Portulacaceae are a substantial group with twenty-seven species in *Calandrinia*, of which about twelve are Eremaean, and eight in *Portulaca*, which belong to the Northern Province. *Calandrinia* is herbaceous and leaf succulent, and several species are not uncommon, but it will be noted that they are not essentially desert plants. A weak leaf succulence can be seen in *Kallstroemia, Tribulus*, and *Zygophyllum* of the Zygophyllaceae and in *Euphorbia* and *Phyllanthus* of the Euphorbiaceae. Few of these are plants of any ecological importance.

Evolutionary History

The evolutionary significance of these different growth forms must now be discussed. Our view of the past history of biota has been transformed by the development of the theory of plate tectonics in quite recent years, with sanction given to the previously heretical ideas of continental drift. As long ago as 1856, in his famous preface to the *Flora Tasmaniae*, J. D. Hooker suggested that the modern Australian flora was compounded of three elements—an Indo-Malaysian element derived from southeast Asia, an autochthonous element evolved within Australia itself, and an Antarctic element comprising forms common to the southern continents which in some way should be presumed to have been transmitted via Antarctica. The trouble was that, while the reality of this Antarctic element could not be doubted, no means or mechanism save that of long-range dispersal could be used to account for it—unless one were very daring and, after Wegener and du Toit, were prepared to invoke continental drift. The thinking of those years of fixed-positional geology is typified by Darlington's book *Biogeography of the Southern End of the World* (1965), in which the southern continents are seen as refuges where throughout time odd forms from the Northern Hemisphere have established themselves and survived. Our Antarctic element would then become only a random selection of forms long extinct in the other hemisphere. This view is now discredited.

Although the breakup of Gondwanaland is dated rather earlier than the origin of the angiosperms, many of the continents do seem to have remained sufficiently close or, in some cases, in actual contact in such a way that explanations of the distribution of plant forms are materially assisted. Where Australia is concerned in this discussion of desert biota we need only go back to Eocene times, some 40 to 60 million years ago, when our continent was joined to Antarctica along the southern edge of its continental shelf and lay some 15° of latitude farther south than now (Griffiths, 1971). In middle Eocene times a rift occurred in the position of the present mid-oceanic ridge separating Australia and Antarctica; the two continents broke apart and drifted in opposite directions: Antarctica to have its biota largely extinguished by a polar icecap, Australia to move toward and into the tropics, passing in the process through an arid zone in which much of it still lies. The evolution of the desert flora of Australia has therefore occurred since the Eocene *pari passu* with this movement.

In discussions of Tertiary paleoclimates it is commonly assumed that the circulation of the atmosphere has always been much the same as it is today, so that the positions of major latitudinal climatic belts have also been fairly constant, even though there may have been cyclic variations in temperature and in quantity of rainfall. At the time, therefore, when Australia was situated 15° farther south, it would have lain squarely in the roaring forties; and it seems likely that a copious and well-distributed rainfall would have been received more or less throughout the continent. This is borne out by the fossil record which predominantly suggests a cover of rain forest of a character and composition similar to that found today in the North Island of New Zealand (Raven, 1972).

Paleontological evidence suggests rather warmer temperatures prevailing at that time and in those latitudes than exist there today. When the break from Antarctica took place, the southern coastline of Australia slumped and thin deposits of Eocene and Miocene sediments were laid down upon the continental margin. Fossils indicate deposition in seas of tropical temperature, continuing as late as Miocene times (Dorman, 1966; Cockbain, 1967; Lowry, 1970). This is consistent with the evidence of tropical flora extending to lat. 50° N in North America in the Eocene (Chaney, 1947) and to Chile and Patagonia (Skottsberg, 1956).

Evidence from the soil supports the concept of both high temperature and high rainfall. In the Canning and Officer sedimentary basins in Western Australia, the parts of the country now occupied by the Great Sandy and Gibson deserts, an outcrop of rocks of Cretaceous age has been very deeply weathered and thickly encrusted with laterite. Farther south than this an outcrop of Miocene limestone in the Eucla basin exhibits relatively little weathering or development of typical karst features and is considered to have been exposed to a climate not substantially wetter than the present since its uplift from the sea at the end of Miocene times (Jennings, 1967).

The laterization would indicate subjection for a long period to a warm, wet climate, which must therefore be early Tertiary in date. The present surface features of all of these sedimentary basins are in accord with presumed climatic history based on known latitudinal movement of the continent.

From Eocene times, therefore, Australian flora had to adapt itself to progressive desiccation. It is frequently assumed that it also had to adapt to warmer temperatures in moving northward, but I believe that this is a mistake. We have fossil evidence for warmer temperatures already in the Eocene, followed by a progressive cooling of the earth through the later Tertiary; and the northward movement of Australasia largely provided, I think, a compensation for the latter process. I do not concur with Raven and Axelrod (1972), for example, that we have to assume a developed adaptation to tropical conditions in those elements in the flora of New Caledonia which are of southern origin. Australasian flora, however, had to adapt to the greater extremes of temperature which accompany aridity, even though mean temperatures may not have greatly altered.

From my own consideration of the paleolatitudes and an attempt to map the probable paleoclimates (which I cannot now go into in detail), I believe that the first appearance of aridity may have been in the northwest in the Kimberley district of Western Australia in later Eocene times, expanding steadily to the southeast. The first Mediterranean climate with its winter-wet, summer-dry regime seems likely to have become established in the Pilbara district of Western Australia in the Oligocene and to have been progressively displaced south-ward.

The Roles of Fires and Soil

In addition to the climatic adaptations required, Australian flora also had to adapt itself to changes in soil which have accompanied the desiccation and to withstand fire. In the early Tertiary rain forests fire was probably unknown or a rarity. Such forests are able to grow even on a highly leached and impoverished substratum in the absence of fire, as a cycle of accumulation and decomposition of organic matter is built up and the forest is living on the products of its own decay. It has been shown that intense weathering and laterization occurred in the early Tertiary in some areas of Western Australia, and this may be observed elsewhere in the continent.

This process would have occurred initially under the forest without provoking significant changes, but with desiccation two things happen: fire ruptures the nutrient cycle leading to a collapse of the eco-system, and the laterites are indurated to duricrust. After burning and rapid removal of mineralized nutrients by the wind and the rain, a depauperate scrub community with a low-nutrient demand replaces the rain forest. In the absence of fire a slow succession back to the rain forest will ensue, but further fires stabilize the disclimax. This process may be seen in operation today in western Tasmania. It is intensified where laterite is present since induration of laterite by desiccation is irreversible and produces an inhospitable hardpan in the soil, usually followed by deflation of the leached sandy topsoil to leave a surface duricrust which is even more inhospitable.

Arid Australia is situated in those central and western parts of the continent where there has been little or no tectonic movement during the Tertiary to regenerate systems of erosion, so that after desiccation set in there was mostly no widespread removal of ancient weathered soil material or the rejuvenation of the soils. Great expanses of inert sand or surface laterite clothe the higher ground and offer an inhospit-able substratum to plants, poor in nutrients and in water-holding capacity. Leaching has continued, and its products have been deposited in the lower ground by evaporation where soils have been zonally accumulating calcium carbonate, gypsum, and chlorides.

Biogeographical Elements

Evolutionary adaptation to these changed conditions during the later Tertiary produced the autochthonous element in the Australian flora mostly by adaptation of forms present in the previously dominant Antarctic element. The Indo-Malaysian element is a relatively recent arrival and, as may be expected, has colonized mainly the moister tropical habitats. It has not contributed very significantly to the desert biota, but there are a few species whose very names betray their origin in that direction:*Trichodesma zeylanicum*, an annual herb in the Boraginaceae; *Crinum asiaticum*, a bulbous Amaryllid, bringing a life form (the perennating bulb) which is almost unknown in Australian desert biota in spite of its apparent evolutionary advantages.

Herbert (1950) pointed out that the autochthonous element is essentially one adapted to subhumid, semiarid, and desert conditions which has been evolved within Australia from forms whose relatives are of world-wide distribution. Evolution, said Herbert, took place in three ways: from ancestors already adapted to these drier climates, by survival of hardier types when increasing aridity drove back the more mesic vegetation, and by recolonization of drier areas by the more xerophytic members of mesic communities.

Burbidge (1960) examined the question more closely and acknowledged a suggestion made to her by Professor Smith-White of the University of Sydney that many of the elements in the desert flora may have developed from species associated with coastal habitats. Burbidge considered that such an opinion was supported by the number of genera in the desert flora of Australia which elsewhere in the world are associated with coastal areas, sand dunes, and habitats of saline type. It is certainly a very reasonable assumption that, in a well-watered early Tertiary continent, source material for future desert plants should lie in the flora of the littoral already adapted to drying winds, sand or rock as a substratum, or salt-marsh conditions. Burbidge went on to say that it is not until the late Pleistocene or early

Recent that there is any real evidence in the fossil record for the existence of a desert flora. However, this does not prove it was not there, and the evidence for the northward movement of Australia into the arid zone suggests strongly that it must have begun its evolution at least as early as the Miocene. Pianka (1969b) in discussing Australian desert lizards found that the species density was too great to permit evolution proceeding only from the sub-Recent. An identical argument is bound to apply to flora also. Speciation is too great and too diversified to have originated so recently.

Morphological Evolution

In addition to the systematic evolution of the desert flora, we may usefully discuss also its morphological evolution. It has been shown that some of the life forms considered most typical of desert biota in other continents are inconspicuous or lacking in Australia, for example, deciduousness, spinescence, and underground perennating organs. Other life forms, especially succulence, are limited to particular areas. Morphologically, there is a dualism in Australian desert flora. The typical plant forms of poor, leached siliceous soils are radically different from those of the base-rich alkaline and saline soils. The former are essentially sclerophyllous in the particular manner of so many Australian plant forms from all over the continent which are not confined to the desert. There has even been the evolution of a unique form of sclerophyll grass, the spinifex or hummock-grass form. On the other hand, succulent and semisucculent leaves replace the sclerophyll on base-rich soils. It is evident that aridity alone is not responsible for sclerophylly in Australian plants as has so often been thought. This evidence seems strongly to support the views of Professor N. C. W. Beadle, expressed in numerous papers (e.g., Beadle, 1954, 1966). Beadle has argued for a relationship between sclerophylly and nutrient deficiency, especially lack of soil phosphate. It certainly seems true to say that the plant forms of nutrient-deficient soils in the Australian desert have had the directions of their evolution dictated not only by aridity but by soil conditions as well, soil conditions largely peculiar to Australia as a continent so that this section of the Australian desert flora has acquired a unique character. It has evolved, we may say, within a straitjacket of sclerophylly. This limitation, however, has not been imposed on the ion-accumulating bottom-land soils where plant forms more similar to those of deserts in other continents have evolved.

To look back to what has been said about the taxonomic evolution of the desert flora, limitations are also imposed by the nature of the genetic source material. A subtropical and warm temperate rain forest is not a very promising source area for forms which will have the necessary genetic plasticity for adaptation to great extremes of temperature and aridity, as well as to extremes of soil deficiency. Certain Australian plant families have possessed this faculty, especially the Proteaceae, and this has resulted in a proliferation of highly specialized and adapted species in a relatively limited number of genera. This phenomenon is remarked especially on the soils which have the most extreme nutrient deficiencies or imbalances under widely differing climatic conditions, notably on the Western Australian sand plains, the Hawkesbury sandstone of New South Wales, and the serpentine outcrops in the mountains of New Caledonia, in all of which different species belonging to the same or related Australian genera can be seen forming a similar maquis or sclerophyll scrub. The sclerophyll desert flora has drawn heavily upon this source material, while the nonsclerophyll flora has been influenced particularly by the ability of the family Chenopodiaceae to produce forms adaptable to the particular conditions.

Summary

The Australian desert, covering the interior of the continent, receives an average rainfall of 100 to 250 mm annually and is well vegetated. There are two principal plant formations, *Acacia* low woodland and *Triodia-Plectrachne* hummock grassland, characteristic broadly of the sectors south and north of the Tropic of Capricorn. Component species are typically sclerophyll in form, even the grasses. Non-sclerophyll vegetation of succulent and semisucculent subshrubs locally occupies alkaline soils, in depressions or on limestone and cal-

careous clays. There is otherwise a notable absence of such xerophytic life forms as stem and leaf succulents, rosette plants, deciduous thorny trees, and plants with bulbs and corms.

Australian desert flora evolved gradually from the end of Eocene
times as the continent moved northward into arid latitudes. As the
previous vegetation was mainly a subtropical rain forest, it has been
suggested that the source material for this evolution came largely
from the littoral and seashore. Species had to adapt not only to aridity
but also to soils deeply impoverished by weathering under previous
humid conditions and not rejuvenated. It is believed that the siliceous,
nutrient-deficient soils have been responsible for the predominantly
sclerophyllous pattern of evolution; succulence has only developed
on base-rich soils.

References

Beadle, N. C. W. 1954. Soil phosphate and the delimitation of plant
communities in eastern Australia. *Ecology* 25:370–374.
———. 1966. Soil phosphate and its role in moulding segments of the
Australian flora and vegetation with special reference to xero-
morphy and sclerophylly. *Ecology* 47:991–1007.
Beadle, N. C. W., and Costin, A. B. 1952. Ecological classification
and nomenclature. *Proc. Linn. Soc. N.S.W.* 77:61–82.
Beard, J. S. 1969. The natural regions of the deserts of Western Aus-
tralia. *J. Ecol.* 57:677–711.
Bews, J. W. 1929. *The world's grasses*. London: Longmans, Green &
Co.
Burbidge, N. T. 1960. The phytogeography of the Australian region.
Aust. J. Bot. 8:75–211.
Chaney, R. W. 1947. Tertiary centres and migration routes. *Ecol.
Monogr.* 17:141–148.
Cockbain, A. E. 1967. Asterocyclina from the Plantagenet beds near
Esperance, W.A. *Aust. J. Sci.* 30:68.
Darlington, P. J. 1965. *Biogeography of the southern end of the
world*. Cambridge, Mass.: Harvard Univ. Press.
Dorman, F. H. 1966. Australian Tertiary paleotemperatures. *J. Geol.*
74:49–61.

Griffiths, J. R. 1971. Reconstruction of the south-west Pacific margin of Gondwanaland. *Nature, Lond.* 234:203–207.

Herbert, D. A. 1950. Present day distribution and the geological past. *Victorian Nat.* 66:227–232.

Hooker, J. D. 1856. Introductory Essay. In *Botany of the Antarctic Expedition, vol. III flora Tasmaniae*, pp. xxvii–cxii.

Jennings, J. N. 1967. Some karst areas of Australia. In *Land form studies from Australia and New Guinea*, ed. J. N. Jennings and J. A. Mabbutt. Canberra: Aust. Nat. Univ. Press.

Lowry, D. C. 1970. Geology of the Western Australian part of the Eucla Basin. *Bull. geol. Surv. West. Aust.* 122:1–200.

Pianka, E. R. 1969a. Sympatry of desert lizards (*Ctenotus*) in Western Australia. *Ecology* 50:1012–1013.

———. 1969b. Habitat specificity, speciation and species density in Australian desert lizards. *Ecology* 50:498–502.

Raven, P. H. 1972. An introduction to continental drift. *Aust. nat. Hist.* 17:245–248.

Raven, P. H., and Axelrod, D. I. 1972. Plate tectonics and Australasian palaeobiogeography. *Science, N.Y.* 176:1379–1386.

Skottsberg, C. 1956. *The natural history of Juan Fernández and Easter Island. I(ii) Derivation of the flora and fauna of Easter Island.* Uppsala: Almqvist & Wiksell.

4. Evolution of Arid Vegetation in Tropical America

Guillermo Sarmiento

Introduction

More or less continuous arid regions cover extensive areas in the middle latitudes of both South and North America, forming a complex pattern of subtropical, temperate, and cold deserts on the western side of the two American continents. They appear somewhat intermingled with wetter ecosystems wherever more favorable habitats occur. These two arid zones are widely separated from each other, leaving a huge gap extending over almost the whole intertropical region (see fig. 4-1). South American arid zones, however, penetrate deeply into intertropical latitudes from northwestern Argentina through Chile, Bolivia, and Peru to southern Ecuador. But they occur either as high-altitude deserts, such as the Puna (high Andean plateaus over 3,000 m), or as coastal fog deserts, such as the Atacama Desert in Chile and Peru, the driest American area. This coastal region, in spite of its latitudinal position and low elevation, cannot be considered as a tropical warm desert, because its cool maritime climate is determined by almost permanent fog. In fact, in most of tropical America, either in the lowlands or in the high mountain chains, from southern Ecuador to southern Mexico, more humid climates and ecosystems prevail. In sharp contrast with the range areas of western North America and the high cordilleras and plateaus of western South America, the tropical American mountains lie in regions of wet climates from their piedmonts to the highest summits. The same is true for the lower ranges located in the interior of the Guianan and Brazilian plateaus.

Upon closer examination, however, it is apparent that, although warm, tropical rain forests and mountain forests, as well as savannas, are characteristic of most of the tropical American landscape, the arid

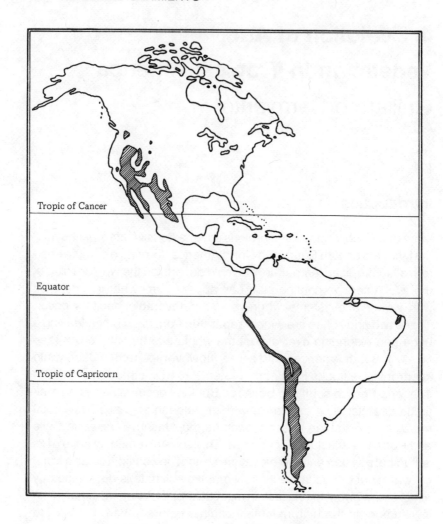

Fig. 4-1. American arid lands (after Meigs, 1953, modified).

ecosystems are far from being completely absent. If we look at a generalized map of arid-land distribution, such as that of Meigs (1953), we will notice two arid zones in tropical South America: one in northeastern Brazil and the other forming a narrow belt along the Caribbean coast of northern South America, including various small nearby islands. These two tropical areas share some common geographical features:

1. They are quite isolated from each other and from the two principal desert areas in North and South America. The actual distance between the northeast Brazilian arid Caatinga and the nearest desert in the Andean plateaus is about 2,500 km, while its distance from the Caribbean arid region is over 3,000 km. The distance from the Caribbean arid zone to the nearest South American continuous desert, in southern Ecuador, and to the closest North American continuous desert, in central Mexico, is in both cases around 1,700 km.

2. They appear completely encircled by tropical wet climates and plant formations.

3. The two areas are more or less disconnected from the spinal cord of the continent (the Andes cordillera), particularly in the case of the Brazilian region. This fact surely has had major biogeographical consequences.

Recently, interest in ecological research in American arid regions has been renewed, mainly through the wide scope and interdisciplinary research programs of the International Biological Program (Lowe et al., 1973). These studies give strong emphasis to a thorough comparison of temperate deserts in the middle latitudes of North and South America, with the purposes of disclosing the precise nature of their ecological and biogeographical relationships and also of assessing the degree of evolutionary convergence and divergence between corresponding ecosystems and between species of similar ecological behavior. Within this context, a deeper knowledge of tropical American arid ecosystems would provide additional valuable information to clarify some of the previous points, besides having a research interest per se, as a particular case of evolution of arid and semiarid ecosystems of Neotropical origin under the peculiar environmental conditions of the lowland tropics.

The aim of this paper is to present certain available data concerning tropical American arid and semiarid ecosystems, with particular reference to their flora, environment, and vegetation structure. The geographical scope will be restricted to the Caribbean dry region, of which I have direct field knowledge; there will be only occasional further reference to the Brazilian dry vegetation. The Caribbean dry region is still scarcely known outside the countries involved; a review book on arid lands, such as that of McGinnies, Goldman, and Paylore (1968), does not provide a single datum about this region.

In order to delimit more precisely the region I am talking about, a climatic and a vegetational criterion will be used. My field experience suggests that most dry ecosystems in this part of the world lie inside the 800-mm annual rainfall line, with the most arid types occurring below the 500-mm rainfall line. Figure 4-2 shows the course of these two climatic lines through the Caribbean area. Though some wetter ecosystems are included within this limit, particularly at high altitudes, few arid types appear outside this area except localized edaphic types on saline soils, beaches, coral reefs, dunes, or rock outcrops. Only in the Lesser Antilles does a coastal arid vegetation appear under higher rainfall figures, up to 1,200 mm, and this only on very permeable and dry soils near the sea (Stehlé, 1945).

This climatically dry region extends over northern Colombia and Venezuela and covers most of the small islands of the Netherlands Antilles—Aruba, Curaçao, and Bonaire—reaching a total area of about 50,000 km². The nearest isolated dry region toward the northwest is in Guatemala, 1,600 km away; in the north, a dry region is in Jamaica and Hispaniola, 800 km across the Caribbean Sea; while southward the nearest dry region is in Ecuador, 1,700 km away.

From the point of view of vegetation, only the extremes of the Seasonal Evergreen Formation Series and the Dry Evergreen Formation Series of Beard (1944, 1955) will be considered here, including the following four formations: Thorn Woodland, Thorn Scrub, Desert, and Dry Evergreen Bushland. Several papers have dealt with the vegetation of this dry area, but they analyze either only a restricted zone inside this whole region, as those of Dugand (1941, 1970), Tamayo (1941), Marcuzzi (1956), Stoffers (1956), and several others, or they are generalized accounts of plant cover for a whole country that include a short description of the arid types, like those of Cuatrecasas (1958) or Pittier (1926). The aim of this paper is to go one step further than previous investigations—first, considering the entire

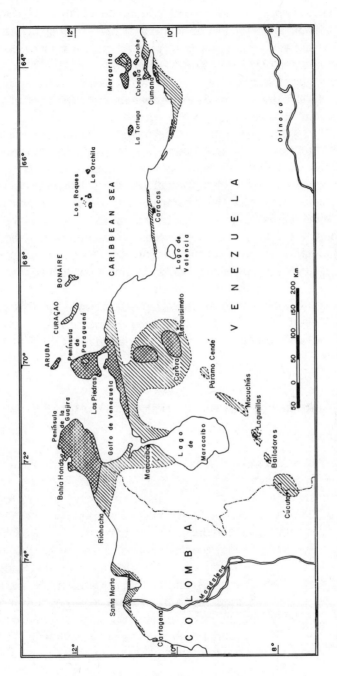

Fig. 4-2. Caribbean arid lands. Semiarid (500–800 mm rainfall) and arid zones (less than 500 mm rainfall) have been distinguished.

Caribbean dry region and, second, comparing it to the rest of American dry lands. My previous paper (1972) had a similar approach. The thorn forests and thorn scrub of tropical America were included in a floristic and ecological comparison between tropical and subtropical seasonal plant formations of South America. I will follow that approach here, but will restrict my scope to the dry extreme of the tropical American vegetation gradient.

To avoid a possible terminological misunderstanding as to certain concepts I am employing, it is necessary to point out that the words *arid* and *semiarid* refer to climatological concepts and will be applied both to climates and to plant formations and ecosystems occurring under these climates. *Dry* will refer to every type of xeromorphic vegetation, either climatically or edaphically determined, such as those of sand beaches, dunes, and rock pavements. *Desert* will be used in its wide geographical sense, that is, a region of dry climate where several types of dry plant formations occur, among them semidesert and desert formations. In this way, for instance, the Sonora and the Monte deserts have mainly a semidesert vegetation, while the Chile-Peru coastal desert shows mainly a desert plant formation. Each time I refer to a *desert vegetation* in contrast to a *desert region*, I shall clarify the point.

The Environment in the Caribbean Dry Lands

Geography

The Caribbean dry region, as its name suggests, is closely linked with the Caribbean coast of northern South America, stretching almost continuously from the Araya Peninsula in Venezuela, at long. 64° W, to a few kilometers north of Cartagena in Colombia, at long. 75° W. Along most of this coast the dry zone constitutes only a narrow fringe between the sea and the forest formation beginning on the lower slopes of the contiguous mountains: the Caribbean or Coast Range in Venezuela and the Sierra Nevada of Santa Marta in Colombia. In many places this arid fringe is no more than a few hundred meters wide. But in the two northernmost outgrowths of the South American continent, the Guajira and Paraguaná peninsulas, the dry region widens to cover these two territories almost completely (see fig. 4-2).

Besides these strictly coastal areas, dry vegetation penetrates deeper inside the hinterland around the northern part of the Maracaibo basin as well as in the neighboring region of low mountains and inner depressions known as the Lara-Falcón dry area of Venezuela. In this zone the aridity reaches more than 200 km from the coast.

Besides this almost continuous dry area in continental South America, the Caribbean dry region extends over the nearby islands along the Venezuelan coast, from Aruba through Curaçao, Bonaire, Los Roques, La Orchila, and other minor islands to Margarita, Cubagua, and Coche. The islands farthest from the continental coast lie 140 km off the Venezuelan coast. Dry vegetation entirely covers these islands, except for a few summits with an altitude over 500 m. The Lesser Antilles somehow connect this dry area with the dry regions of Hispaniola, Cuba, and Jamaica, because almost all of them show restricted zones of dry vegetation (Stehlé, 1945).

Both on the continents and in the islands dry plant formations occupy the lowlands, ranging in altitude from sea level to no more than 600–700 m, covering in this low climatic belt all sorts of land forms, rock substrata, and geomorphological units, such as coastal plains, alluvial and lacustrine plains, early and middle Quaternary terraces, rocky slopes, and broken hilly country of different ages. In the islands dry vegetation also occurs on coral reefs, banks, and on the less-extended occurrences of loose volcanic materials.

Apart from the nearly continuous coastal region and its southward extensions, I should point out that a whole series of small patches or "islands" of dry vegetation and climate occurs along the Andes from western Venezuela across Colombia and Ecuador to Peru. These small and isolated arid patches may be divided into two ecologically divergent types according to their thermal climate determined by altitude: those occurring below 1,500–1,800 m that have a warm or megathermal climate and those appearing above that altitude and belonging then to the meso- or microthermal climatic belts. The latter, such as the small dry islands in the Páramo Cendé and the upper Chama and upper Mocoties valleys of the Venezuelan Andes, even though they have low rainfall, have a less-unfavorable water budget because of their comparatively constant low temperature. Therefore, their vegetation has few features in common with the remaining dry Caribbean areas. On the other hand, the lower-altitude dry patches, like the middle Chama valley, the Tachira-Pamplonita depression, and the lower Chicamocha valley, are quite similar to the dry coastal regions in ecology, flora, and vegetation and will be considered in this study as part of the Caribbean dry lands. I shall point out further the biogeographical significance of this archipelago of Andean dry islands connecting the Caribbean dry region with the southern South American deserts.

Throughout the dry area of northern South America, dry plant formations appear bordered by one or other of three different types of vegetation units: tropical drought-deciduous forest, dry evergreen woodland, or littoral formations (mangroves, littoral woodlands, etc.). In the lower Magdalena valley, as well as in certain other partially flooded areas, marshes and other hydrophytic formations are also common, intermingled with thorn woodland or thorn scrub.

Climate

I propose to analyze the prevailing climatic features of the region enclosed within the 800-mm rainfall line, with particular reference to the main climatic factor affecting plant life, that is, the amount of rainfall and its seasonal distribution, but without disregarding other climatic elements that sharply differentiate tropical and extratropical climates, like minimal temperatures, annual cycle of insolation, and thermo- and photoperiodicity. Lahey (1958) provided a detailed discussion about the causes of the dry climates around the Caribbean Sea, and I shall refer to that paper for pertinent meteorological and climatological considerations on this topic. Porras, Andressen, and Pérez (1966) presented a detailed study of the climate of the islands of Margarita, Cubagua, and Coche, some of the driest areas of the Caribbean; some of the climatic data I will discuss have been taken from that paper.

As pointed out before, a major part of the region with annual rainfall figures below 800 mm is located in the megathermal belt, below 600–700 m, and has an annual mean temperature above 24°C. A few small patches along the Andes reach higher elevations, up to 1,500–1,800 m, and their annual mean temperatures go down to 20°C, fitting within what has been considered as the mesothermal belt. However, this temperature difference does not seem to introduce significant changes in vegetation physiognomy or ecology.

Mean annual temperatures in coastal and lowland localities range from a regional maximum of 28.7°C in Las Piedras, at sea level, to 24.2°C in Barquisimeto, a hinterland locality at 566-m elevation. Mean temperatures show very slight month-to-month variation (1° to 3.5°C), as is typical for low latitudes. The annual range of extreme temperatures in this ever-warm region is not so wide as in subtropical or temperate dry regions. The recorded absolute regional maximum does not reach 40°C, while the absolute minima are everywhere above 17°C. As we can see, then, in sharp contrast with the case in extratropical conditions, in the dry Caribbean region low temperatures never constitute an ecological limitation to plant life and natural vegetation.

I have already pointed out that, using natural vegetation as a guideline for our definition of aridity, an annual rainfall of 800 mm roughly separates semiarid and arid from humid regions in this part of the world. Excluding edaphically determined vegetation, the most open and sparse vegetation types appear where rainfall figures do not reach 500 mm. The lowest rainfall in the whole area has been recorded in the northern Guajira Peninsula (Bahía Honda: 183 mm) and in the island of La Orchila, which has the absolute minimum rainfall for the region, 150 mm. Rainfall figures below 300 mm also characterize the small islands of Coche and Cubagua and the central and driest part of Margarita. As we can see, these figures are really very low, fully comparable to many desert localities in temperate South and North America, but in our case these rainfall totals occur under constantly high temperatures and, therefore, represent a less-favorable water balance and a greater drought stress upon plant and animal life.

Concerning rainfall patterns, figure 4-3 shows the rainfall regime at eight localities, arranged in an east-to-west sequence from Cumaná at long. 64°11' W to Pueblo Viejo at long. 74° 16' W. The rainfall pattern varies somewhat among the localities appearing in the figure; some places show a unimodal distribution, with the yearly maximum slightly preceding the winter solstice (October to December), while other localities show a bimodal distribution, with a secondary maximum during the high sun period (May to June). It is clear, nevertheless, that all localities have a continuous drought throughout the year, with ten to twelve successive months when rainfall does not reach 100 mm and five to eight months with monthly rainfall figures below 50

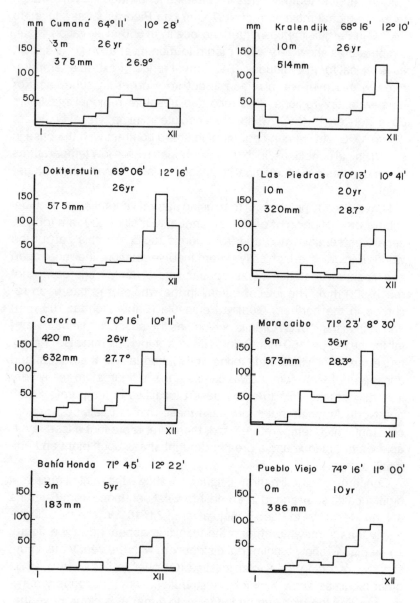

Fig. 4-3. Rainfall regimen for eight stations in the dry Caribbean region.
Each climadiagram shows longitude, latitude, altitude, years of recording,
mean annual rainfall, and mean temperature.

mm. The total number of rain days ranges over the whole area from forty to sixty.

As is typical for dry climates, rainfall variability is very high, reaching values of 40 percent and more where rainfall is less than 500 mm. This high interannual variability maintains the dryness of the climate and the drought stress upon perennial organisms.

In contrast to most other dry regions in temperate America, relative humidity in the Caribbean dry region is not as low, showing average monthly figures of 70 to 80 percent throughout the year and minimal monthly values of around 55 to 60 percent. Annual pan evaporation, however, is very high, generally exceeding 2,000 mm, with many areas having values as high as 2,500–2,800 mm. Potential evapotranspiration, calculated according to the Thornthwaite formula, reaches 1,600–1,800 mm.

According to its rainfall and temperature, this Caribbean dry region falls within the BSh and BWh climatic types of Koeppen's classification, that is, *hot steppe* and *hot desert* climates. We should remember that in this tropical region the rainfall value setting apart dry and humid climates in the Koeppen system is around 800 mm, which is precisely the limit I have taken according to natural vegetation. In turn, the 400-to-450-mm rainfall line separates BS, or semiarid, from BW, or arid, climates. Following the second system of climatic classification of Thornthwaite, this region comes within the DA' and EA' types, that is, *semiarid megathermal* and *arid megathermal* climates respectively. It is interesting that in both systems the climates corresponding to the dry Caribbean area are the same as those found in the dry subtropical regions of America, such as the Chaco and Monte regions of South America and the Sonoran and Chihuahuan deserts of North America.

We will now consider certain rhythmic environmental factors which influence both climate and ecological behavior of organisms, such as incoming solar radiation and length of day. The sharp contrast between incoming solar radiation at sea level (maximal theoretical values disregarding cloudiness) in low and middle latitudes is well known. At low latitudes daily insolation varies only slightly throughout the year, forming a bimodal curve in correspondence with the sun passing twice a year over that latitude. The total variation between the extremes of maximal and minimal solar radiation during the year is in the order of 50 percent. At middle latitudes the annual radiation curve is unimodal in shape and shows, at a latitude as low as 30° north where North American warm deserts are more widespread, a seasonal variation between extremes in the order of 300 percent.

Photoperiodicity is also inconspicuous in tropical latitudes. At 10° north or south the difference between the shortest and the longest day of the year is around one hour, while at 30° it is almost four hours. Summarizing the climatic data, we can see that the Caribbean dry region has semiarid and arid climates partly comparable to those found in middle-latitude American deserts, particularly insofar as permanent water deficiency is concerned; but these tropical climates differ from the subtropical dry climates by more uniform distribution of solar radiation, higher relative humidity, higher minimal temperatures, slight variation of monthly means, and shorter variation in the length of day throughout the year.

Physiognomy and Structure of Plant Formations

General

Beard (1955) has given the most valuable and widely used of the classifications of tropical American vegetation. Arid vegetation appears as the dry extreme of two series of plant formations: the Seasonal Evergreen Formation Series and the Dry Evergreen Formation Series. Seasonal formations were arranged in Beard's scheme along a gradient of increasing climatic seasonality (rainfall seasonality because all are isothermal climates), from an ever-wet regime without dry seasons to the most highly desert types. The successive terms of this series, beginning with the Tropical Rain Forest *sensu stricto* as the optimal plant formation in tropical America are Seasonal Evergreen Forest, Semideciduous Forest, Deciduous Forest, Thorn Woodland, Thorn or Cactus Scrub, and Desert. The first two units appear under slightly seasonal climates; Deciduous Forest together with savannas appear under tropical wet and dry climates; while the last three members of this series, Thorn Woodland, Thorn Scrub, and Desert, occur under dry climates with an extended rainless season and as such are common in the dry Caribbean area.

The Dry Evergreen Formation Series of Beard's classification, in contrast with the Seasonal Series, occurs under almost continuously dry climates but where monthly rainfall values are not as low as during the dry season of the seasonal climates. The driest formations on this series are the Thorn Scrub and the Desert formations, these two series being convergent in physiognomic and morphoecological features according to Beard and other authors, such as Loveless and Asprey (1957). The remaining less-dry type, next to the previous two, is the Dry Evergreen Bushland formation, which also occurs under the dry climates of the Caribbean area. In summary, dry vegetation in tropical America has been included in four plant formations: Dry Evergreen Bushland, Thorn Woodland, Thorn or Cactus Scrub, and Desert. Their structures according to the original definitions are represented in figure 4-4. All of them have open physiognomies, where the upper-layer canopy in the more structured and richer types does not surpass 10 m in height and a cover of 80 percent, decreasing then in height and cover as the environmental conditions become less favorable.

Plant formations occurring in the arid Caribbean region fit closely with Beard's classification and types, though it seems necessary to add a new formation: Deciduous Bushland, structurally equivalent to the Dry Evergreen Bushland, but with a predominance of deciduous woody species. Before going into some details about each dry plant formation in the Caribbean area, let me add a final remark about the evident difficulty met with when some of the vegetation is classified in one or another type, particularly in the case of some low and poor associations of the Tropical Deciduous Forest whose features overlap with those of the Dry Evergreen Bushland or the Thorn Woodland. Human interference, through wood cutting and heavy goat grazing, frequently makes subjective conclusions difficult, and in many instances only a thorough quantitative recording of vegetational features could allow an objective characterization and classification of the stand. At a preliminary survey level these doubts remain. A detailed study of dry plant formations, such as that of Loveless and Asprey (1957, 1958) in Jamaica, will emphasize the need for quantitative data on vegetation structure and species morphoecology in order to classify these difficult intermediate dry formations.

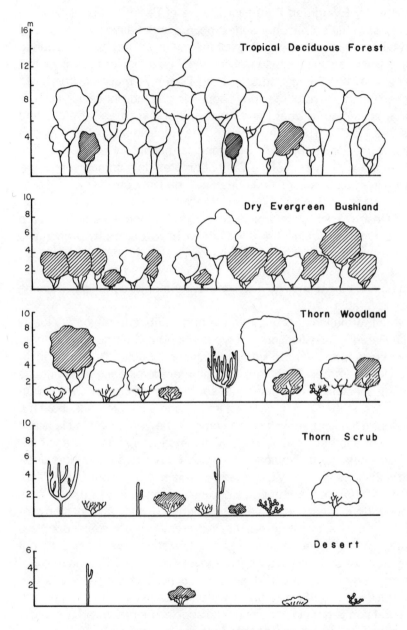

Fig. 4-4. Vegetation profiles of tropical American dry formations. Tropical Deciduous Forest has been included for comparison.

I will now very briefly consider each of the five dry plant formations as they occur nowadays in the Caribbean dry region.

Dry Evergreen Bushland

The Dry Evergreen Bushland formation has a closed canopy of low trees and shrubs at a height of about 2 to 4 m. Sparse taller trees and cacti, up to 10 m high, may emerge from this canopy. The two essential physiognomic features of this plant formation are, first, the closed nature of the plant cover, leaving no bare ground at all, and, second, the predominance of evergreen species, with a minor proportion of deciduous and succulent-aphyllous elements. The dominant woody species are evergreen low trees and shrubs, with sclerophyllous medium-sized leaves.

Floristically it is a rather rich plant formation, taking into account its dry nature, with an evident differentiation into various floristic associations. The most important families of this formation are the Euphorbiaceae, Boraginaceae, Capparidaceae, Leguminosae (Papilionoideae and Caesalpinoideae), Rhamnaceae, Polygonaceae, Rubiaceae, Myrtaceae, Flacourtiaceae, and Celastraceae. Cacti and agaves are also frequent, interspersed with a rich subshrubby and herbaceous flora.

This formation is widespread in the whole Caribbean dry region, being frequent in the islands, in the mainland coasts, and in the small Andean arid patches. Its physiognomy clearly differentiates this evergreen bushland from all other tropical dry formations; and from this physiognomic and structural viewpoint it looks more like the temperate scrubs in Mediterranean climates, such as the low chaparral of California and the garigue of southern France, than most other tropical types.

Deciduous Bushland

The Deciduous Bushland is quite similar in structure to the Dry Evergreen Bushland, but it differs mainly by the predominance of deciduous shrubs, while evergreen and aphyllous species only share a secondary role. This gives a highly seasonal appearance to the Deciduous Bushland, with two acutely contrasting aspects: one during the leafless period and the other when the dominant species are in full leaf. It also differs from other seasonal dry formations, such as the Thorn Woodland and the Thorn Scrub, because of its closed canopy of shrubs and low trees that leaves no bare ground. The most common families in this formation are the Leguminosae (mainly Mimosoideae), Verbenaceae, Euphorbiaceae, and Cactaceae. Floristically this type is not well known, but apparently it differs sharply from the Thorn Woodland and Thorn Scrub. Up to date the Deciduous Bushland has only been reported in the Lara-Falcón area (Smith, 1972).

Thorn Woodland

The distinctive physiognomic feature of the Thorn Woodland is a lack of a continuous canopy at any height, leaving large spaces of bare soil between the sparse trees and shrubs, particularly during dry periods when herbaceous annual cover is lacking. The upper layer of high shrubs and low trees and succulents is from 4 to 8 m high, with a variable cover, from less than 10 percent to a maximum of around 75 percent. A second woody layer 2 to 4 m high is generally the most important in cover, showing values ranging from 30 to 70 percent. The shrub layer of 0.5 to 2 m is also conspicuous, inversely related in importance to the two uppermost layers. The total cover of the herb and soil layers varies during the year because of the seasonal development of annual herbs, geophytes, and hemicryptophytes; the permanent biomass in these lowest layers is given by small cacti, like *Mammillaria*, *Melocactus*, and *Opuntia*.

As for the morphoecological features of its species, this formation is characterized by a high proportion of thorny elements, by many succulent shrubs, and by a total dominance of the smallest leaf sizes (lepto- and nanophyll), with a smaller proportion of aphyllous and microphyllous species together with rare mesophyllous elements; the last mentioned are generally highly scleromorphic. The relative proportion of evergreen and deciduous species is almost the same, with a good proportion of brevideciduous species.

From the floristic aspect this formation has a very characteristic flora, scarcely represented in wetter plant types. Among the most important families are the Leguminosae (particularly Mimosoideae and

Caesalpinoideae), Cactaceae, Capparidaceae, and Euphorbiaceae. Many floristic associations can be distinguished on the basis of the dominant species, but their distribution and ecology are scarcely known. The most important single species in this formation, distributed over its area, is undoubtedly *Prosopis juliflora*. When it is present, this low tree usually shares a dominant role in the community. This may probably be due, among other reasons, to its noteworthy ability for regrowth after cutting, as well as to its unpalatability to all domestic herbivores.

Thorn Scrub or Cactus Scrub

The Thorn Scrub, equivalent to the Semidesert formation of arid temperate areas, is still lower and more sparse than the Thorn Woodland, leaving a major part of bare ground, particularly during the driest period of the year. Low trees and columnar cacti from 4 to 8 m high appear widely dispersed or are completely lacking. Shrubs from 0.5 to 2 m high, though they form the closest plant layer, are also widely separated, as well as the subshrubs and herbs that form the scattered lower layer. Floristically the Thorn Scrub seems to be an impoverished Thorn Woodland, without significant additions to the flora of that formation. Cactaceae, Capparidaceae, Euphorbiaceae, and Mimosoideae continue to be the best-represented taxa. Even by its morphoecology and functionality this formation resembles the Thorn Woodland, showing a heterogeneous mixture of evergreen, deciduous, brevideciduous, and aphyllous species, with the smallest leaf sizes frequently being of sclerophyllous texture. Succulent species, particularly cacti, appear here at their optimum, frequently being the most noteworthy feature in the physiognomy of the plant formation.

This Thorn Scrub physiognomy is not so widely found in the Caribbean arid region as in the temperate deserts of North and South America. By structure and biomass it is comparable to the Semidesert formations of those arid regions, though the most extended associations of temperate American deserts, those formed by nonspiny low shrubs such as *Larrea*, are completely absent from the tropical American area. Thorn Scrub occurs in the Lara-Falcón region of Venezuela, in the northernmost part of the Guajira Peninsula, and in the driest islands like Coche and Cubagua.

Desert

Extremely desertic vegetational physiognomies are not uncommon in the Caribbean arid zone, but most of them seem determined by substratum-related factors and not primarily by climate. Thus, for example, one of the most widespread types of Desert formation occurs in the Lara-Falcón area, on sandstone hills of Tertiary age. Only four or five species of low shrubs grow there, such as species of *Cassia*, *Sida*, and *Heliotropium*, very widely interspersed with some woody *Capparis* and various Cactaceae and Mimosoideae. The total ground cover is less than 2 or 3 percent. To explain this extremely desertic vegetation in an area with enough rainfall to maintain thorn woodland in neighboring situations, Smith (1972) suggested the existence of heavy metals in the rock substrata; but there is not yet any further evidence to sustain this hypothesis, though undoubtedly the responsible factor is linked to a particular type of geological formation.

Another type of desert community that covers a wide extent of flat country in northern Venezuela appears on heavy soil developed on old Quaternary terraces. This desert community scarcely covers more than 5 percent of the ground and is composed mainly of species of *Jatropha*, *Opuntia*, *Lemaireocereus*, and *Ipomoea*, together with some annual herbs. Though this community is rather common in several parts of the Caribbean arid area, a satisfactory explanation for its occurrence has not been given for it, either.

Some more easily understood types of Desert formation are the salt deserts near the coast and the sand deserts of dunes and beaches. Salt deserts are almost everywhere characterized by low, shrubby Chenopodiaceae, such as *Salicornia* and *Heterostachys*; while sand deserts show a dominance of geophytes together with some shrubby species of *Lycium*, *Castela*, *Opuntia*, and *Acacia*.

Floristic Composition and Diversity

The floristic inventory of the Caribbean arid vegetation has not yet been made. My list of plant families and genera has been compiled from several sources (Boldingh, 1914; Tamayo, 1941 and 1967; Dugand, 1941 and 1970; Pittier et al., 1947; Croizat, 1954; Marcuzzi, 1956; Stoffers, 1956; Cuatrecasas, 1958; Trujillo, 1966) as well as from direct field knowledge of this vegetation and flora.

Table 4-1 presents a list of 94 families and 470 genera which have been reported from this area. Both figures must be taken as rough approximations of the regional total flora, because this arid flora is still not well known and in many areas plant collections are lacking; there are also some overrepresentations in the tabulated figures, because many of the listed taxa collected in the arid region surely belong to various riparian forests and therefore are not strictly part of the arid Caribbean flora. The total number of species is still more imprecisely known; a figure of 1,000 will give an idea of the magnitude of the species diversity in this vegetation.

If the floristic richness and diversity in more restricted areas is taken into consideration, the following figures are obtained: a thorough

Table 4-1. *Families and Genera of Flowering Plants Reported from the Caribbean Dry Region*

Family	Genera
Acanthaceae	*Anisacanthus, Anthacanthus, Dicliptera, Elytraria, Justicia, Odontonema, Ruellia, Stenandrium*
Achatocarpaceae	*Achatocarpus*
Aizoaceae	*Mollugo, Sesuvium, Trianthema*
Amaranthaceae	*Achyranthes, Alternanthera, Amaranthus, Celosia, Cyathula, Froelichia, Gomphrena, Iresine, Pfaffia, Philoxerus*
Amaryllidaceae	*Agave, Crinum, Fourcroya, Hippeastrum, Hymenocallis, Hypoxis, Zephyranthes*
Anacardiaceae	*Astronium, Mauria, Metopium, Spondias*
Apocynaceae	*Aspidosperma, Echites, Forsteronia, Plumeria, Prestonia, Rauvolfia, Stemmadenia, Thevetia*
Araceae	*Philodendron*
Aristolochiaceae	*Aristolochia*

Asclepiadaceae	*Asclepias, Calotropis, Cynanchum, Gompho-carpus, Gonolobus, Ibatia, Marsdenia, Meta-stelma, Omphalophthalmum, Sarcostemma*
Bignoniaceae	*Amphilophium, Anemopaegma, Arrabidaea, Bignonia, Clytostoma, Crescentia, Distictis, Lundia, Memora, Pithecoctenium, Tabebuia, Tecoma, Xylophragma*
Bombacaceae	*Bombacopsis, Bombax, Cavanillesia, Pseudobombax*
Boraginaceae	*Cordia, Heliotropium, Rochefortia, Tourne-fortia*
Bromeliaceae	*Aechmea, Bromelia, Pitcairnia, Tillandsia, Vriesia*
Burseraceae	*Bursera, Protium*
Cactaceae	*Acanthocereus, Cephalocereus, Cereus, Hylocereus, Lemaireocereus, Mammillaria, Melocactus, Opuntia, Pereskia, Phyllocactus, Rhipsalis*
Canellaceae	*Canella*
Capparidaceae	*Belencita, Capparis, Cleome, Crataeva, Morisonia, Steriphoma, Stuebelia*
Caryophyllaceae	*Drymaria*
Celastraceae	*Hippocratea, Maytenus, Pristimera, Rhacoma, Schaefferia*
Chenopodiaceae	*Atriplex, Chenopodium, Heterostachys, Salicornia*
Cochlospermaceae	*Amoreuxia, Cochlospermum*
Combretaceae	*Bucida, Combretum*
Commelinaceae	*Callisia, Commelina, Tripogandra*
Compositae	*Acanthospermum, Ambrosia, Aster, Balti-mora, Bidens, Conyza, Egletes, Eleutheran-thera, Elvira, Eupatorium, Flaveria,*

	Gundlachia, Isocarpha, Lactuca, Lagascea, Lepidesmia, Lycoseris, Mikania, Oxycarpha, Parthenium, Pectis, Pollalesta, Porophyllum, Sclerocarpus, Simsia, Sonchus, Spilanthes, Synedrella, Tagetes, Trixis, Verbesina, Vernonia, Wedelia
Convolvulaceae	*Bonomia, Cuscuta, Evolvulus, Ipomoea, Jacquemontia, Merremia*
Cruciferae	*Greggia*
Cucurbitaceae	*Bryonia, Ceratosanthes, Corallocarpus, Doyerea, Luffa, Melothria, Momordica, Rytido-stylis*
Cyperaceae	*Bulbostylis, Cyperus, Eleocharis, Fimbristylis, Hemicarpha, Scleria*
Elaeocarpaceae	*Muntingia*
Erythroxylaceae	*Erythroxylon*
Euphorbiaceae	*Acalypha, Actinostemon, Adelia, Argithamnia, Bernardia, Chamaesyce, Cnidoscolus, Croton, Dalechampsia, Ditaxis, Euphorbia, Hippomane, Jatropha, Julocroton, Mabea, Manihot, Pedilanthus, Phyllanthus, Sebastiania, Tragia*
Flacourtiaceae	*Casearia, Hecatostemon, Laetia, Mayna*
Gentianaceae	*Enicostemma*
Gesneriaceae	*Kohleria, Rechsteineria*
Goodeniaceae	*Scaevola*
Gramineae	*Andropogon, Anthephora, Aristida, Bouteloua, Cenchrus, Chloris, Cynodon, Dactyloctenium, Digitaria, Echinochloa, Eleusine, Eragrostis, Eriochloa, Leptochloa, Leptothrium, Panicum, Pappophorum, Paspalum, Setaria, Sporobolus, Tragus, Trichloris*
Guttiferae	*Clusia*

Hernandaceae	*Gyrocarpus*
Hydrophyllaceae	*Hydrolea*
Krameriaceae	*Krameria*
Labiatae	*Eriope, Hyptis, Leonotis, Marsypianthes, Ocimum, Perilomia, Salvia*
Lecythidaceae	*Chytroma, Lecythis*
Leguminosae (Caesalpinoideae)	*Bauhinia, Brasilettia, Brownea, Caesalpinia, Cassia, Cercidium, Haematoxylon, Schnella*
Leguminosae (Mimosoideae)	*Acacia, Calliandra, Cathormium, Desmanthus, Inga, Leucaena, Mimosa, Piptadenia, Pithecellobium, Prosopis*
Leguminosae (Papilionoideae)	*Abrus, Aeschynomene, Benthamantha, Callistylon, Canavalia, Centrosema, Crotalaria, Dalbergia, Dalea, Desmodium, Diphysa, Erythrina, Galactia, Geoffraea, Gliricidia, Humboldtiella, Indigofera, Lonchocarpus, Machaerium, Margaritolobium, Myrospermum, Peltophorum, Phaseolus, Piscidia, Platymiscium, Pterocarpus, Rhynchosia, Sesbania, Sophora, Stizolobium, Stylosanthes, Tephrosia*
Lennoaceae	*Lennoa*
Liliaceae	*Smilax, Yucca*
Loasaceae	*Mentzelia*
Loganiaceae	*Spigelia*
Loranthaceae	*Oryctanthus, Phoradendron, Phthirusa, Struthanthus*
Lythraceae	*Ammannia, Cuphea, Pleurophora, Rotala*
Malpighiaceae	*Banisteria, Banisteriopsis, Brachypteris, Bunchosia, Byrsonima, Heteropteris, Hiraea, Malpighia, Mascagnia, Stigmatophyllum, Tetrapteris*

Malvaceae	*Abutilon, Bastardia, Cienfuegosia, Hibiscus, Malachra, Malvastrum, Pavonia, Sida, Thespesia, Urena, Wissadula*
Melastomaceae	*Miconia, Tibouchina*
Meliaceae	*Trichilia*
Menispermaceae	*Cissampelos*
Moraceae	*Brosimum, Chlorophora, Ficus, Helicostylis*
Myrtaceae	*Anamomis, Pimenta, Psidium*
Nyctaginaceae	*Allionia, Boerhavia, Mirabilis, Naea, Pisonia, Torrubia*
Ochnaceae	*Sauvagesia*
Oenotheraceae	*Jussiaea*
Olacaceae	*Schoepfia, Ximenia*
Oleaceae	*Forestiera, Linociera*
Opiliaceae	*Agonandra*
Orchidaceae	*Bifrenaria, Bletia, Brassavola, Brassia, Catasetum, Dichaea, Elleanthus, Epidendrum, Gongora, Habenaria, Ionopsis, Maxillaria, Oncidium, Pleurothallis, Polystachya, Schombergkia, Spiranthes, Vanilla*
Oxalidaceae	*Oxalis*
Palmae	*Bactris, Copernicia*
Papaveraceae	*Argemone*
Passifloraceae	*Passiflora*
Phytolaccaceae	*Petiveria, Rivinia, Seguieria*
Piperaceae	*Peperomia, Piper*
Plumbaginaceae	*Plumbago*
Polygalaceae	*Bredemeyera, Monnina, Polygala, Securidaca*
Polygonaceae	*Coccoloba, Ruprechtia, Triplaris*

Portulacaceae	*Portulaca*, *Talinum*
Ranunculaceae	*Clematis*
Rhamnaceae	*Colubrina*, *Condalia*, *Gouania*, *Krugiodendron*, *Zizyphus*
Rubiaceae	*Antirrhoea*, *Borreria*, *Cephalis*, *Chiococca*, *Coutarea*, *Diodia*, *Erithalis*, *Ernodea*, *Guettarda*, *Hamelia*, *Machaonia*, *Mitracarpus*, *Morinda*, *Psychotria*, *Randia*, *Rondeletia*, *Sickingia*, *Spermacoce*, *Strumpfia*
Rutaceae	*Amyris*, *Cusparia*, *Esenbeckia*, *Fagara*, *Helietta*, *Pilocarpus*
Sapindaceae	*Allophylus*, *Cardiospermum*, *Dodonaea*, *Paullinia*, *Serjania*, *Talisia*, *Thinouia*, *Urvillea*
Sapotaceae	*Bumelia*, *Dipholis*
Scrophulariaceae	*Capraria*, *Ilysanthes*, *Scoparia*, *Stemodia*
Simarubaceae	*Castela*, *Suriana*
Solanaceae	*Bassovia*, *Brachistus*, *Capsicum*, *Cestrum*, *Datura*, *Lycium*, *Nicotiana*, *Physalis*, *Solanum*
Sterculiaceae	*Ayenia*, *Buettneria*, *Guazuma*, *Helicteres*, *Melochia*, *Waltheria*
Theophrastaceae	*Jacquinia*
Tiliaceae	*Corchorus*, *Triumfetta*
Turneraceae	*Piriqueta*, *Turnera*
Ulmaceae	*Celtis*, *Phyllostylon*
Urticaceae	*Fleurya*
Verbenaceae	*Aegiphila*, *Bouchea*, *Citharexylon*, *Clerodendrum*, *Lantana*, *Lippia*, *Phyla*, *Priva*, *Stachytarpheta*, *Vitex*
Violaceae	*Rinorea*
Vitaceae	*Cissus*

| Zingiberaceae | *Costus* |
| Zygophyllaceae | *Bulnesia*, *Guaiacum*, *Kallstroemia*, *Tribulus* |

---

floristic survey of a dry forest community in the lower Magdalena valley in Colombia, with an annual rainfall of 720 mm (Dugand, 1970), gives a total of 55 families, 154 genera, and 187 species of flowering plants in a stand of less than 300 ha. For the three small islands of Curaçao, Aruba, and Bonaire, with a total area of 860 km^2, Boldingh (1914) gives a list of 79 families, 239 genera, and 391 species of flowering plants, excluding the mangroves as the only local formation not belonging to the dry types.

As we can see in table 4-1, the best-represented families in total number of genera are the Leguminosae (50), Compositae (33), Euphorbiaceae (20), and Rubiaceae (19). Other well-represented families are the Amaranthaceae, Malvaceae, Malpighiaceae, Cactaceae, Verbenaceae, Orchidaceae, and Asclepiadaceae; almost all of them are typical of warm, arid floras everywhere.

If we compare now the floristic richness of this Caribbean dry region to the flora of North and South American middle-latitude deserts, we obtain roughly equivalent figures. In fact, Johnson (1968) gives a total of 278 genera and 1,084 species for the Mojave and Colorado deserts of California; Shreve (1951) reports 416 genera for the whole Sonoran desert, while Morello (1958) gives a list of 160 genera from the floristically less known Monte desert in Argentina. We can see then, that, in spite of a smaller total area, the Caribbean dry flora is as rich as other American desert floras.

Johnson (1968) gives a list of monotypic or ditypic genera of the Mojave-Colorado deserts, considered according to the ideas of Stebbins and Major (1965) to be old relict taxa. That list includes 60 species belonging to 56 genera. Applying this same criterion to the flora of the Caribbean desert I have recognized only 14 relict endemic species—a number that, even if it represents a gross underestimate, is significantly smaller than the preceding one (see table 4-2).

Concerning the geographic distribution and centers of diversification of the Caribbean arid taxa, it is not possible to proceed here to a detailed analysis because of the fragmentary knowledge of plant distribution in tropical America. However, I have tried to give a preliminary analysis based on only a few best-known families.

Taking the Compositae for instance, one of the most diversified families within this vegetation, I took the data on its distribution from Aristeguieta (1964) and Willis (1966). The species of 19 genera occurring in the Caribbean dry lands could be considered as widely distributed weeds, whose areas also extend to arid climates. Six genera are very rich genera with a few species also occurring in arid vegetation: *Eupatorium*, *Vernonia*, *Mikania*, *Aster*, *Verbesina*, and *Simsia*; 2 genera (*Lepidesmia* and *Oxycarpha*) are monotypic taxa endemic to the Caribbean coasts; while the remaining 6 genera (*Pollalesta*, *Egletes*, *Baltimora*, *Gundlachia*, *Lycoseris*, and *Sclerocarpus*) are small-to-medium-sized taxa restricted to tropical America, with some species characteristic of arid plant formations. We see, then, that in this family an important proportion of the species that occur in the arid vegetation may be considered as weeds (19 out of 33 genera); one part (6/33) has originated from widely distributed and very rich genera, some of whose species have succeeded in colonizing arid habitats also; while the remaining part, about a quarter of the genera of Compositae occurring in arid vegetation, is formed of species belonging to genera of more restricted distribution and lesser adaptive radiation, whose presence in this arid flora may be indicative of the adaptation to arid conditions of an ancient Neotropical floristic stock— in some cases, as in the two monotypic endemics, probably through a rather long evolution in contact with similar environmental stress.

In all events, the Compositae, a very important family in the temperate and cold American deserts, neither shows a similar degree of differentiation in the arid Caribbean flora nor occupies a prominent role in these tropical plant communities.

Another family whose taxonomy and geographical distribution is rather well known, the Bromeliaceae (Smith, 1971), has five genera inhabiting the Caribbean arid lands; three of them, *Pitcairnia*, *Vriesia*, and *Aechmea*, are very rich genera (150 to 240 species) mainly growing in humid vegetation types but with a few species also entering dry plant formations. None of them is exclusive to the arid types. *Tillandsia*, a great and polymorphous genus of more than 350 species adapted to nearly all habitat types from the epiphytic types in the rain forests to the xeric terrestrial plants of extreme deserts, has 15

Table 4-2. *Relictual Endemic Species Occurring in the Caribbean Dry Region*

Family	Species
Asclepiadaceae	*Omphalophthalmum ruber* Karst.
Capparidaceae	*Belencita hagenii* Karst.
Capparidaceae	*Stuebelia nemorosa* (Jacq.) Dugand
Compositae	*Lepidesmia squarrosa* Klatt
Compositae	*Oxycarpha suaedaefolia* Blake
Cucurbitaceae	*Anguriopsis (Doyerea) margaritensis* Johnson
Leguminosae	*Callistylon arboreum* (Griseb.) Pittier
Leguminosae	*Humboldtiella arborea* (Griseb.) Hermann
Leguminosae	*Humboldtiella ferruginea* (H.B.K.) Harms.
Leguminosae	*Margaritolobium luteum* (Johnson) Harms.
Leguminosae	*Myrospermum frutescens* Jacq.
Lennoaceae	*Lennoa caerulea* (H.B.K.) Fourn.
Rhamnaceae	*Krugiodendron ferreum* (Vahl.) Urb.
Rubiaceae	*Strumpfia maritima* Jacq.

species recorded in the Caribbean arid lands; 14 of them are widely distributed species also occurring in dry formations. Only 1 species, *T. andreana*, growing on bare rock, seems strictly confined to dry plant formations. The fifth genus, *Bromelia*, a medium-sized genus of about 40 species, has 4 species growing in deciduous forests in the Caribbean that also extend their areas to the drier plant formations. As we can see by the distribution patterns of this old Neotropical family, the degree of speciation that has occurred in response to aridity in the Caribbean region seems to be minimal. This fact is in sharp contrast with the behavior of this family in other South American deserts, such as the Monte and the Chilean-Peruvian coastal deserts, where it has reached a good degree of diversification.

Let us take as a last example a typical family of arid lands, the Zygophyllaceae, recently studied by Lasser (1971) in Venezuela; it has four genera growing in the Caribbean arid region of which two,

Kallstroemia and *Tribulus*, are weedy genera of widely distributed species on bare soils and in dry habitats. The other two genera, *Bulnesia* and *Guaiacum*, are typical elements of arid and semiarid Neotropical plant formations. *Bulnesia* has its maximal diversification in semiarid and arid zones of temperate South America; while only one species, *B. arborea*, has reached the deciduous forests and thorn woodlands of northern South America, but without extending even to the nearby islands. But it is a dominant tree in many thorn woodland communities of northern South America. *Guaiacum* is a peri-Caribbean genus with several species from Florida to Venezuela, some of them exclusively restricted to arid coastal vegetation. In summary, this small family, whose species are frequently restricted to dry regions, does not show in the Caribbean arid flora the same degree of differentiation it has attained in southern South America, but it has nevertheless distinctly arid species, some originating from the south, such as *Bulnesia*, others from the north, such as *Guaiacum*.

Conclusions

As a conclusion, I wish to point out the most significant facts that follow from the preceding data. We have seen that in northern South America and in the nearby Caribbean islands a region of dry climates exists, which includes semiarid and arid climatic types, wherein five different plant formations occur. Considering the major environmental feature acting upon plant and animal life in this area, that is, the strong annual water deficit, these ecosystems seem subjected to water stress of comparable intensity and extension to that influencing living organisms in the extratropical South and North American deserts. If this water stress constitutes the directing selective force in the evolution of plant species and vegetation forms, the evolutionary framework would be comparable in tropical and extratropical American deserts. If, therefore, significant differences in speciation and vegetation features between these ecosystems could be detected, either they ought to be attributed to a different period of evolution under similar selective pressures, in which case the tropical and temperate American deserts would be of noncomparable geological age, or they could be attributed to the action on the evolution of these species of other environmental factors linked to the latitudinal difference between these deserts.

As many floristic and ecological features of these two types of ecosystems do not seem to be quite similar, even at a preliminary qualitative level of comparison, both previous hypotheses, that of differential age and that of divergent environmental selection, could probably be true. This supposes that the ancestral floristic stock feeding all dry American warm ecosystems was not so different as to explain the actual divergences on the basis of this sole historical factor.

The structure and physiognomy of plant formations occurring in the Caribbean area under a severe arid climate do not seem to correspond strictly to most semidesertic or desertic physiognomies of temperate North and South America. Several plant associations show undoubtedly a high degree of physiognomic convergence, also emphasized by a close floristic affinity, as is the case of the thorn scrub communities dominated in all these regions by species of *Prosopis*, *Cercidium*, *Cereus*, and *Opuntia*. But the most widespread plant associations in temperate American deserts, which are the scrub communities where a mixture of evergreen and deciduous shrubs prevail, like the *Larrea divaricata–Franseria dumosa* association of the Sonoran desert or the *Larrea cuneifolia* communities of the Monte desert; or the communities characterized by aphyllous or subaphyllous shrubs or low trees, such as the *Bulnesia retama–Cassia aphylla* communities of South America or the various *Fouquieria* associations in North America, do not have a similar physiognomic counterpart in tropical America.

As I have already noted in a previous paper (1972), even the degree of morphoecological adaptation in tropical American arid species is significantly smaller than that exhibited by the temperate American desert flora. Such plant features as succulence, spines, or aphylly are widely represented in the desert floras of North and South America, but they appear much more restricted quantitatively in the tropical American arid flora where, for instance, only one family of aphyllous plants occurs, the Cactaceae, in contrast to eleven families in the Monte region of Argentina.

Concerning floristic diversity, the dry Caribbean vegetation has a richness comparable to North American warm-desert floras and perhaps a richer flora than the warm deserts of temperate South America. The tropical arid flora is highly heterogeneous in origin and affinities, with the most significant contribution coming from neighboring less-dry formations, particularly the Tropical Deciduous Forest and the Dry Evergreen Woodland, with an important contribution from cosmopolitan or subcosmopolitan weeds, and a variety of floristic elements whose area of greater diversification occurs in northern or southern latitudes.

Among the elements of direct tropical descent reaching the dry formations from the contiguous less-arid types, the species of wide ecological spectrum predominate, whose ecological amplitude extends from subhumid or seasonally wet climates to semiarid and arid plant formations. On the other hand, few of them show a narrow ecological amplitude, appearing thus restricted only to arid plant communities; and in the majority of these cases the species thus restricted occur in particular types of habitats, like sand beaches, dunes, coral reefs, saline soils, and rock outcrops.

There exist in the Caribbean dry flora some species which are old relictual endemic taxa, in the sense considered by Stebbins and Major (1965), but they are neither as numerous as in North American deserts nor characteristic of "normal" habitats or typical communities; they are, rather, typical species of particular edaphic conditions or characteristics of the less-extreme types, such as the deciduous forests and dry evergreen woodlands.

In summary, then, the speciation of the autochthonous tropical taxa has been important in subhumid or semiarid plant formations as well as in restricted dry habitats, but the arid flora has received only a minor contribution from this source.

In spite of the actual occurrence of a chain of arid islands along the Andes connecting the dry areas of Venezuela and Peru, where neighboring patches occur no more than 200 to 300 km apart, southern floristic affinities are not conspicuous among the families analyzed. Further arguments are available to support this lack of connection between Caribbean and southern South American deserts on the basis of the distributions of all genera of Cactaceae (Sarmiento, 1973, unpublished). The representatives of this typical family that live in the Caribbean region show a closer phylogenetic affinity with the Mexican and West Indian cactus flora, a looser relationship with the Brazilian cactus flora, and a much more restricted affinity with the Peruvian and Argentinean cactus flora.

This slight affinity between tropical American and southern South American dry floras, in spite of more direct biogeographical and paleogeographical connections, is a rather difficult fact to explain, particularly if we consider that some species of disjunct area between North and South America, *Larrea divaricata*, for example, originated in South America and later expanded northward (Hunziker et al., 1972). These species have therefore crossed tropical America, but have not remained there.

In contrast to the loose affinity with southern South America, a stronger relationship with the North American arid flora is easily discernible. The most noteworthy cases are those of the genera *Agave*, *Fourcroya*, and *Yucca*, richly diversified in Mexico and southwestern United States, that reach their southern limits in the dry regions of northern South America. There are many other cases of North American genera, characteristic of dry regions, extending southward to Venezuela, Colombia, or less commonly to Ecuador and northern Brazil.

We can thus infer from the above information that the origin and age of the Caribbean arid vegetation certainly seems heterogeneous. Some elements evolved in tropical dry environments; many are almost cosmopolitan; others came from the north; and a few also came from the south. Several migratory waves along different routes probably occurred during a rather long evolutionary history under similar environmental conditions. Though the Central American connection does not actually offer a natural bridge for arid-adapted species, and there is no evidence of the former existence of this type of biogeographical bridge, the northern affinity of many Caribbean desert elements may be more easily explained by resorting to a dry bridge across the Caribbean islands, from Cuba and Hispaniola through the Lesser Antilles to Venezuela, instead of a more hypothetical Central American pass.

Axelrod's model (1950) of gradual evolution of the arid flora and vegetation in southwestern North America from a Madro-Tertiary geoflora, with the most arid forms and the maximal widespread of arid plant formations occurring only during the Quaternary, does not seem to fit well with the evidence provided by the analysis of the arid Caribbean flora and vegetation. On the contrary, the ideas of Stebbins and Major (1965) about the existence of small arid pockets along the western mountains from the late Mesozoic upward, together with a much more agitated evolutionary history from that time on to the Quaternary, are probably in better agreement with these data, which account for a heterogeneous and polychronic origin of these elements.

Acknowledgments

It is a great pleasure for me to acknowledge all the intellectual stimulus, material help, arduous criticism, and audacious ideas received through frequent and passionate discussions of these topics with my colleague, Maximina Monasterio.

Summary

Tropical American arid vegetation, particularly the formations occurring along the Caribbean coast of northern South America and the small nearby islands, is still not well known. However, within the framework of a comparative analysis of all American dry areas, this region provides not only the interest of knowing the features of plant cover in the driest region of tropical America, but also the knowledge that this possibly may clarify many obscure points of Neotropical biogeography, such as the evolutionary history of arid plant formations and the origin of their flora.

The major points of Caribbean dry ecosystems dealt with in this paper are (a) geographical distribution and climatic conditions, mainly the annual water deficiency and some differential features between low- and middle-latitude climates; (b) physiognomy, structure, and morphoecological traits of each of the five plant formations occurring in that area; and (c) floristic richness, origin, and affinities of floristic elements.

On this basis some relevant facts are discussed, such as the lack of correspondence between arid vegetation physiognomies in tropical

and temperate American dry regions; the comparable floristic diversification; and the varied origin of its taxa, where most elements evolved
on the spot from a tropical drought-adapted stock. Some others are
cosmopolitan taxa; many came from North America; and a few came
from the south. This brief analysis leads to the hypothesis that tropical
American desert flora is, at least in part, of considerable age and
shows a heterogeneous origin, probably brought about by several
migratory events. All these facts seem to support Stebbins and
Major's ideas about the complex evolution of American dry flora and
vegetation.

References

Aristeguieta, L. 1964. Compositae. In *Flora de Venezuela, X*, ed. T.
    Lasser. Caracas: Instituto Botánico.
Axelrod, D. I. 1950. Evolution of desert vegetation in western North
    America. *Publs Carnegie Instn* 590:1–323.
Beard, J. S. 1944. Climax vegetation in tropical America. *Ecology* 25:
    127–158.
———. 1955. The classification of tropical American vegetation-
    types. *Ecology* 36:89–100.
Boldingh, I. 1914. *The flora of Curaçao, Aruba and Bonaire*, vol. 2.
    Leiden: E. J. Brill.
Croizat, L. 1954. La faja xerófila del Estado Mérida. *Universitas Emeri-
    tensis* 1:100–106.
Cuatrecasas, J. 1958. Aspectos de la vegetación natural de Colom-
    bia. *Revta Acad. colomb. Cienc. exact. fís. nat.* 10:221–268.
Dugand, A. 1941. Estudios geobotánicos colombianos. *Revta Acad.
    colomb. Cienc. exact. fís. nat.* 4:135–141.
———. 1970. Observaciones botánicas y geobotánicas en la costa
    del Caribe. *Revta Acad. colomb. Cienc. exact. fís. nat.* 13:415–
    465.
Hunziker, J. H.; Palacios, R. A.; de Valesi, A. G.; and Poggio, L. 1973.
    Species disjunctions in *Larrea*: Evidence from morphology, cyto-
    genetics, phenolic compounds and seed albumins. *Ann. Mo. bot.
    Gdn.* 59:224–233.

Johnson, A. W. 1968. The evolution of desert vegetation in western North America. In *Desert Biology*, ed. G. W. Brown, vol.1, pp. 101–140. New York: Academic Press.

Koeppen, W. 1923. *Grundriss der Klimakunde*. Berlin and Leipzig: Walter de Gruyter & Co.

Lahey, J. F. 1958. *On the origin of the dry climate in northern South America and the southern Caribbean*. Ph.D. dissertation, University of Wisconsin.

Lasser, T. 1971. Zygophyllaceae. In *Flora de Venezuela, III*, ed. T. Lasser. Caracas: Instituto Botánico.

Loveless, A. R., and Asprey, C. F. 1957. The dry evergreen formations of Jamaica I. The limestone hills of the south coast. *J. Ecol.* 45:799–822.

Lowe, C.; Morello, J.; Goldstein, G.; Cross, J.; and Neuman, R. 1973. Análisis comparativo de la vegetación de los desiertos subtropicales de Norte y Sud América (Monte-Sonora). *Ecologia* 1:35–43.

McGinnies, W. G.; Goldman, B. J.; and Paylore, P. 1968. *Deserts of the world, an appraisal of research into their physical and biological environments*. Tucson: Univ. of Ariz. Press.

Marcuzzi, G. 1956. Contribución al estudio de la ecologia del medio xerófilo Venezolano. *Boln Fac. Cienc. for.* 3:8–42.

Meigs, P. 1953. World distribution of arid and semiarid homoclimates. In *Reviews of research on arid zone hydrology*, 1:203–209. Paris: Arid Zone Programme, Unesco.

Morello, J. 1958. La provincia fitogeográfica del Monte. *Op. lilloana* 2:1–155.

Pittier, H. 1926. *Manual de las plantas usuales de Venezuela*. 2d ed. Caracas: Fundación Eugenio Mendoza.

Pittier, H.; Lasser, T.; Schnee, L.; Luces de Febres, Z.; and Badillo, V. 1947. *Catálogo de la flora Venezolana*. Caracas: Litografía Vargas.

Porras, O.; Andressen, R.; and Pérez, L. E. 1966. *Estudio climatológico de las Islas de Margarita, Coche y Cubagua, Edo. Nueva Esparta*. Caracas: Ministerio de Agricultura y Cria.

Sarmiento, G. 1972. Ecological and floristic convergences between seasonal plant formations of tropical and subtropical South America. *J. Ecol.* 60:367–410.

————. 1973. The historical plant geography of South American dry vegetation. I. The distribution of the Cactaceae. Unpublished.

Shreve, F. 1951. Vegetation of the Sonoran desert. *Publs Carnegie Instn* 591:1–178.

Smith, L. B. 1971. Bromeliaceae. In *Flora de Venezuela, XII*, ed. T. Lasser. Caracas: Instituto Botánico.

Smith, R. F. 1972. La vegetación actual de la región Centro Occidental: Falcón, Lara, Portuguesa y Yaracuy de Venezuela. *Boln Inst. for lat.-am. Invest. Capacit.* 39–40:3–44.

Stebbins, G. L., and Major, J. 1965. Endemism and speciation in the California flora. *Ecol. Monogr.* 35:1–35.

Stehlé, H. 1945. Los tipos forestales de las islas del Caribe. *Caribb. Forester* 6:273–416.

Stoffers, A. L. 1956. The vegetation of the Netherlands Antilles. *Uitg. natuurw. Stud-Kring Suriname* 15:1–142.

Tamayo, F. 1941. Exploraciones botánicas en la Peninsula de Paraguaná, Estado Falcón. *Boln Soc. venez. Cienc. nat.* 47:1–90.

————. 1967. El espinar costanero. *Boln Soc. venez. Cienc. nat.* 111: 163–168.

Thornthwaite, C. W. 1948. An approach toward a rational classification of climate. *Geogr. Rev.* 38:155–194.

Trujillo, B. 1966. *Estudios botánicos en la región semiárida de la Cuenca del Turbio, Cejedes Superior.* Mimeographed.

Willis, J. C. 1966. *A dictionary of the flowering plants and ferns.* Cambridge: At the Univ. Press.

5. Adaptation of Australian Vertebrates to Desert Conditions   A. R. Main

Introduction

It is an axiom of modern biology that organisms survive in the places where they are found because they are adapted to the environmental conditions there. Current thinking has often associated the more subtle adaptations with physiological attributes, and the analysis of physiology has been widely applied to desert-dwelling animals in order to better understand their adaptation. Results of these inquiries frequently do not produce complete or satisfying explanations of why or how organisms survive where they do, and it is possible that explanations couched in terms of physiology alone are too simplistic. Clearly, while physiology cannot be ignored, other factors, including behavioral traits, need to be taken into account.

Accordingly, this paper sets out to interpret the adaptations of Australian vertebrates to desert conditions in the light of the physiological traits, the species ecology, and the geological and evolutionary history of the biota. To the extent that the components of the biota are integrated, its evolution can be conceived of as analogous to the evolution of a population; thus migrations and extinctions are analogous to genetic additions and deletions; and change in the ecological role of a component of the biota, the analogue of mutation.

The biota has changed and evolved mainly as a result of (a) changes in location and disposition of the land mass, (b) changes in the environment consequent on (a) above, and (c) extinctions and accessions. In the course of these changes strategies for survival will also change and evolve. It is the totality of these strategies which constitutes the adaptations of the biota.

Change in Location of Australia

The present continent of Australia appears to have broken away from East Antarctica in late Mesozoic times and to have moved to its present position adjacent to Asia in middle Tertiary (Miocene) times. In the course of these movements southern Australia changed its latitude from about 70°S in the Cretaceous to about 30°S at present (Brown, Campbell, and Crook, 1968; Heirtzler et al., 1968; Le Pichon, 1968; Vine, 1966, 1970; Veevers, 1967, 1971).

Changes in Environment

Prior to the fragmentation discussed above, the tectonic plate that is now Australia probably had a continental-type climate except when influenced by maritime air. As movement to the north proceeded, extensive areas were covered by epicontinental seas, and, later, extensive fresh-water lake systems developed in the central parts of the present continent. As Australia changed its latitude, the continental climate was influenced by the temperature of the surrounding oceans and particularly the temperature, strength, and origin of the ocean currents which bathed the shores. The ocean currents would in turn be driven by the global circulation, and the variations in the strength of the circulation and its cellular structure have affected not only the strength of the currents but also the climate of the continent. Frakes and Kemp (1972) suggested that for these reasons the Oligocene was colder and drier than the Eocene.

The present location of Australia across the global high-pressure belt, coupled with the fact that ocean currents driven by the west-wind storm systems pass south of the continent, has meant the inevitable drying of the central lake systems and the onset of desert conditions in the interior of the continent. In the absence of marine fossils or volcanicity the precise timing of the stages in the drying of the continent is not possible. Stirton, Tedford, and Miller (1961, p. 23; and see also Stirton, Woodburne, and Plane, 1967) used the morphological evolution shown by marsupial fossils to infer possible age in terms of Lyellian epochs of the sedimentary beds in which marsupial fossils have been found. Ride (1971) tabulated the fossil evidence as it relates to macropods.

Two other events associated with the changed position of the continent have occurred concurrently: (a) the development of weathering profiles, especially duricrust formation, on the land surface; and (b) changes in the composition of the flora.

Weathering profiles capped with duricrust are widespread throughout Australia, and Woolnough (1928) believed this duricrust to be synchronously developed over an enormous area. Since the Upper Cretaceous Winton formation was capped by duricrust, Woolnough believed the episode to be of Miocene age. The climate at this time of peneplanation and duricrust formation was thought to be marked by well-defined wet and dry seasons, so that the more soluble material was leached away in the wet season, and less soluble and particularly colloidal fraction of the weathering products was carried to the surface and precipitated during the dry season. Recent work in Queensland where basalts overlie deep weathering profiles indicates that deep weathering took place earlier than early Miocene (Exon, Langford-Smith, and McDougall, 1970). Other workers suggested that, as the climate becomes progressively drier, weathering processes follow a sequence from laterite formation through silcrete formation to aeolian processes and dune formation (Watkins, 1967).

Biologically the significant aspect of duricrust formation is, however, the removal from, or binding within, the weathering profile of soluble plant nutrients. Beadle (1962a, 1966) showed experimentally that the woodiness which is so characteristic of Australian plants is to some extent related to the low phosphorus status of the soil. Australian soils are well known for their low phosphorus status (Charley and Cowling, 1968; Wild, 1958). It has been argued that the low phosphorus is due to the low status of the parent rocks (Beadle, 1962b) or to the leaching which occurred during the process of laterization (Wild, 1958).

Changes in the floral composition are indicated by the fossil and pollen record. Early in the Tertiary, pollen of southern beech (*Nothofagus*), in common with other pollen present in these deposits, suggests that a vegetation with a floral composition similar to that of present-day western Tasmania was widespread in southern Australia

(see fig. 5-1), for example, at Kojonup (McWhae et al., 1958); Coolgardie (Balme and Churchill, 1959); Nornalup, Denmark, Pidinga,
and Cootabarlow, east of Lake Eyre (Cookson, 1953; Cookson and
Pike, 1953, 1954); and near Griffith in New South Wales (Packham,
1969, p. 504).

Later the Lake Eyre deposits show a change, and the pollen record
is dominated by myrtaceous and grass pollen (Balme, 1962). By Pliocene times the fossil record is restricted to eastern Australia and suggests a cool rain forest with *Dacrydium, Araucaria, Nothofagus*, and
Podocarpus, which was later replaced by wet sclerophyll forest with
Eucalyptus resinifera (Packham, 1969, p. 547). This record is consistent with a drying of the climate; however, in Tasmania comparable
changes in the floral composition—that is, from *Nothofagus* forest to
myrtaceous shrub or *Eucalyptus* woodland with a grass understory—
result from fire (Gilbert, 1958; Jackson, 1965, 1968*a*, 1968*b*), and it
seems highly likely that associated with the undeniable deterioration
of the climate there occurred an increased incidence of fire.

Many authors have recognized that the present Australian flora not
only is adapted to periodic fires but also includes many species which
are dependent on fire for their persistence (Gardner, 1944, 1957;
Mount, 1964, 1969; Cochrane, 1968). At present many wild or bush
fires are intentionally lit or are the result of man's carelessness, but
every year there are many fires which are caused by lightning strike
(Wallace, 1966).

Fires are important in the Australian arid, semiarid, and seasonally
arid environments because it is principally from the ash beds resulting
from intense fires, and not from the slow decay of plant material, that
nutrients are returned to the soil. It is thought that the oily nature of the
common Australian shrubs and trees and their fire dependence
reflect an evolutionary adaptation to fire. There is no doubt that in the
past, in the absence of man, many intense fires were lit when lightning
strike ignited ample and highly inflammable fuel.

Apart from returning nutrients to the soil, fire appears to be an
important ecological factor in habitats ranging from the well-watered
coastal woodlands dominated by *Eucalyptus* forests to the hummock
grassland (dominated by *Triodia*) of the arid interior (Burbidge, 1960;
Winkworth, 1967). Numerous postfire successions occur depending
on the season of the burn, the quantity of fuel, and the frequency of

burning. As an ecological factor, in arid Australia fire is as ubiquitous as drought.

Not all the changes in the biota have been due to fire and the increasing aridity. Numerous elements of the flora must have invaded Australia and then colonized the arid sandy interior by way of littoral sand dunes (Gardner, 1944; Burbidge, 1960). This invasion of Australia could only have occurred after the collision of the Australian plate with Asia in Miocene times. Simultaneously these migrant plant species would have been accompanied by rodents and other vertebrates of Asian affinities which also invaded through similar channels (Simpson, 1961).

To summarize the foregoing, Australia arrived in its present position from much higher southern latitudes, and the change in latitude was associated with a change in climate which passed from being mild and uniform in early Tertiary through marked seasonality to severe and arid by the end of the Pleistocene. Associated with climatic change two things occurred: first, a removal of plant nutrients and the probable development of a "woody" flora, and, second, the concurrent appearance of fire as a significant ecological factor.

Extinctions and Accessions

The climatic changes led to numerous extinctions in the old vertebrate fauna, for example, the Diprotodontidae; to a marked development in macropod marsupials (Stirton, Tedford, and Woodburne, 1968; Woodburne, 1967), which are adapted to the low-nutrient-status fibrous plants; and to the radiation of those Asian invaders which could exploit the progressive development of an arid climate in central Australia. As a result of the events outlined above, the fauna of arid Australia consists of two elements:

1. An older one originating in a cool, high-latitude climate now adapted to or at least persisting under arid conditions. Ride (1964), in his review of fossil marsupials, placed *Wynyardia bassiana*, the oldest diprotodont marsupial known, as of Oligocene age. This was at a time when Australia still occupied a southern location far distant from Asia (Brown et al., 1968, p. 308), suggesting that marsupials are part of the old fauna not derived from Asia.

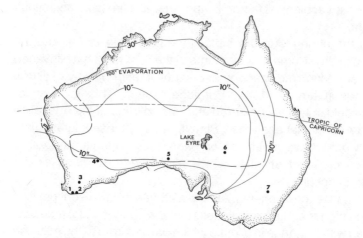

Fig. 5-1. Map of Australia showing approximate extent of arid zone as defined in Slatyer and Perry (1969). The northern boundary corresponds to the 30-inch (762 mm) isohyet, while the southern boundary corresponds with the 10-inch (242 mm) isohyet. The northern boundary of the 10-inch isohyet is also shown, as is the approximate boundary of the region experiencing evaporation of 100 inches (2,540 mm) or more per year. Localities from which fossil pollen recorded: 1. Nornalup; 2. Denmark; 3. Kojonup; 4. Coolgardie; 5. Pidinga; 6. Cootabarlow; 7. Lachlan River occurrence (near Griffith).

2. A younger element (not older than the time at which Australia collided with Asia) derived by evolution from migrants which established themselves on beaches. Rodents, agamid lizards, elapid snakes, and some bird groups fall in this category. These invasive episodes have continued up to the present.

The Australian Arid Environment

The extent of the Australian arid zone is shown in figure 5-1. This area is characterized by irregular rainfall, constant or seasonal shortage of water, high temperatures, and, for herbivores, recurrent seasonal inadequacy of diet. In many respects the arid area manifests in a more intense form the less intense seasonal or periodic droughts of the surrounding semiarid areas. Earlier it has been suggested that the present arid conditions are merely the terminal manifestation of climatic conditions which commenced to deteriorate in middle Tertiary times.

Desert Adaptation

A biota which has survived under increasingly arid conditions for such a long period of time might be expected to have evolved well-marked adaptations to high temperatures, shortage of water, and poor-quality diet. Yet one would not necessarily expect that all species would show similar or equal adaptive responses. The reason for this is that the incidence of intense drought is patchy; and, while some parts of arid Australia are always suffering drought, the areas suffering drought in the same season or between seasons are spatially discontinuous and often widely separated, so that it is conceivable that a mobile population could flee drought-stricken areas for more equable parts. Moreover, since fire as well as drought is ubiquitous in arid Australia, the benefits resulting from break of drought may be quite different according to whether an area has been recently burnt or not. Furthermore, these differences will be dissimilar depending on whether the biotic elements occur early or late in seral stages of the postfire succession. Indeed, many animal species which occupy the late seral or climax stages of the postfire succession could conceivably avoid much of the heat stress and resultant water shortage consequent upon evaporative cooling by behaviorally seeking out the cooler sites in the climax vegetation. All of the foregoing suggest that for the Australian arid biota we should not only look at the stressful factors (high temperature, water shortage, and quality of diet) but also determine the postfire seral stages occupied by the species.

Tolerance to the stressful factors is important because it affects the individual's ability to survive to reproduce. When individuals of a population reproduce successfully, the population persists. However, in its persistence a population will not maintain constant numerical abundance, because, for example, drought and fire will reduce numbers; and species favoring one seral stage of the postfire succession will, in any one locality, vary from being rare, then abundant, and finally again rare, as the preferred seral stage is passed through. For any species only detailed inquiry will show how population characteristics of reproductive capacity, age to maturity, longevity, and dispersal are related to the individual's ability to survive drought.

Individual Responses

Mammals

Among mammals, and marsupials especially, the macropods (kangaroos and wallabies) have received most study, but some work has been done on rodents, particularly *Notomys* (MacMillan and Lee, 1967, 1969, 1970). Both these groups are herbivores, and the macropods particularly show spectacular adaptations to the fibrous nature of their food plants and the attendant low-nutrition quality of the diet.

With the exception of the forest-dwelling *Hypsiprimnodon moschatus*, all kangaroos and wallabies so far investigated have a ruminantlike digestive system. This is an especially elaborated saccular development of the alimentary tract anterior to the true stomach. Within the sacculated "ruminal area" the fibrous ingesta are retained and fermented by a prolific bacterial flora and protozoan fauna. As a result of this activity, otherwise indigestible cellulose is broken down to material which can be metabolized by the kangaroo as an energy source (Moir, Somers, and Waring, 1956).

The bacteria of the gut need a nitrogen source in order to grow. In general, most natural diets in arid Australia are low in nitrogen; however, the bacteria of the kangaroo's gut are able to use urea as a nitrogen source (Brown, 1969) and so can supplement dietary nitrogen by recycling urea which would otherwise require water for its excretion in urine. The common arid-land kangaroo *Macropus robustus* (the euro) can remain in positive nitrogen balance on a diet which contains less than 1 g nitrogen per day for a euro of the average body weight of 12.8 kg (Brown and Main, 1967; Brown, 1968). However, it can supplement the dietary intake of nitrogen by recycling urea (Brown, 1969). In this connection the two common kangaroos (table 5-1) of arid Australia show contrasting solutions to the stresses imposed by heat, drought, and shortage of water, and in fact respond differently to seasonal stress (Main, 1971).

Body temperatures of marsupials are elevated at high ambient temperatures, but they are maintained below environmental tempera-

Table 5-1. *Comparison of Adaptations of Two Arid-Land Species of Kangaroos*

Characteristic	Red Kangaroo (*Megaleia rufa*)	Euro (*Macropus robustus*)
Coat	short, close; highly reflective	long, shaggy; not reflective
Preferred shelter	sparse shrubs	caves and rock piles
Diet	better fodder; higher nitrogen content	poorer fodder; lower nitrogen content
Temperature regulation	reflective coat; evaporative cooling	cool of caves or rock piles; evaporative cooling
Water shortage	acute shortage not demonstrated	unimportant except when shelter inadequate
Urea recycling	not pronounced; urea always high in urine	pronounced when fermentable energy as starch available, e.g., as seed heads of grasses
Electrolyte	higher in diet	lower in diet
Population characteristics	flock; locally nomadic	solitary; sedentary
Breeding	continuous	continuous

Source: Data from Dawson and Brown (1970), Storr (1968), and Main (1971).

tures. Temperature regulation under hot conditions appears to be costly in terms of water (Bartholomew, 1956; Dawson, Denny, and Hulbert, 1969; Dawson and Bennet, 1971).

MacMillan and Lee (1969) studied two species of desert-dwelling *Notomys* and interpreted their findings in terms of the contrasting habitats occupied, so that the salt-flat-dwelling *Notomys cervinus* has a kidney adapted to concentrating electrolytes while the sandhill-dwelling *N. alexis* was adapted to concentrating urea.

Birds

Because of their mobility, birds in an arid environment present a different set of problems to those presented by both mammals and lizards. They can and do fly long distances to watering places and, when water is available, may use it for evaporative cooling. Many of the adaptations are likely to be related to conservation of water so that the frequency of drinking is reduced. Fisher, Lindgren, and Dawson (1972) studied the drinking patterns of many species, including the zebra finch and budgerigar which have been shown under laboratory conditions to consume little or no water (Cade and Dybas, 1962; Cade, Tobin, and Gold, 1965; Calder, 1964; Greenwald, Stone, and Cade, 1967; Oksche et al., 1963).

It is likely that the ability of the budgerigar and the zebra finch to withstand water deprivation in the laboratory reflects their ability to survive in the field with minimum water intake and so exploit food resources which are distant from the available water. Johnson and Mugaas (1970) showed that both these species possess kidneys which are modified in a way which assists in water conservation.

Reptiles

Four responses of individuals to hot arid environments are readily measured: preferred temperatures, heat tolerance, rates of pulmonary and cutaneous water loss, and tolerance to dehydration.

Under field conditions nocturnal lizards, for example, geckos, have to tolerate the temperatures experienced in their daytime shelters. On the other hand, diurnal species, for example, agamids, such as *Amphibolurus*, have body temperatures higher than ambient temperatures during the cooler parts of the year and body temperatures cooler than ambient during the hotter season. The body temperatures recorded for field-caught animals indicate a specific constancy (Licht et al., 1966*b*) which comes about by a series of behavioral responses which range from body posture to avoidance reactions (Bradshaw and Main, 1968). A large series of data on body temperatures in the field has been presented by Pianka (1970, 1971*a*, and 1971*b*) and Pianka and Pianka (1970).

With the exception of *Diporophora bilineata*, with a mean body temperature in the field of 44.3°C (Bradshaw and Main, 1968), no species recorded a mean temperature above 39°C. However, arid-land species spend more of their time in avoidance reactions than do species from semiarid situations (Bradshaw and Main, 1968). Further information on preferred body temperature can be obtained by placing lizards in a temperature gradient and allowing them to choose a body temperature.

Data from neither the field-caught animals nor those selecting temperature in a gradient indicate any marked preference for exceptionally high temperatures on the part of most lizard species. However, in a situation where choice of temperature was not possible, it is conceivable that species from arid environments could tolerate higher temperatures for a longer period than species from less-arid situations. Bradshaw and Main (1968) compared *Amphibolurus ornatus*, a species from semiarid situations, with *A. inermis*, a species from arid areas, after acclimating them to 40°C and then exposing them to 46°C. Their mean survival times were 64 ± 5.6 and 62 ± 6.58 minutes respectively. There was no statistically significant difference in the survival time of each species. These results suggest that the major adaptation of *Amphibolurus* species to hot arid environments is likely to be the development of a pattern of behavioral avoidance of heat stress.

Not all lizards in hot arid situations show the pattern of *Amphibolurus* sp. and *Diporophora bilineata*, which when acclimated to 40°C can withstand an exposure of six hours to a body temperature of 46°C without apparent ill effect and survive for thirty minutes at 49°C (Bradshaw and Main, 1968). The nocturnal geckos, which must tolerate the temperature of their daytime refuge, show another pattern illustrated by *Heteronota binoei*, a species sheltering beneath litter; *Rhynchoedura ornata*, a species which frequently shelters in cracks and holes (deserted spider burrows) in bare open ground; and *Gehyra variegata*, a species sheltering beneath bark. Data for these three species are given in table 5-2. These data suggest that adaptation to high temperatures in Australian geckos is considerably modified by behavioral and habitat preferences and is not directly related to increased aridity in a geographical sense.

Bradshaw (1970) determined the respiratory and cutaneous components of the evaporative water loss in specimens of *Amphibolurus ornatus*, *A. inermis*, and *A. caudicinctus* (matched for body weight) held in the dry air at 35°C after being held under conditions which allowed them to attain their preferred body temperature by behavioral regulation. Bradshaw showed that evaporative water loss was greatest in *A. ornatus* and least in *A. inermis*. He also showed that losses by both pathways were reduced in the desert species. All differences were statistically significant. However, while the cutaneous component was greater than the respiratory in *A. ornatus*, it was less than the respiratory in *A. inermis*. Bradshaw also compared CO_2 production of uniformly acclimated *A. ornatus* with that of *A. inermis* and showed that CO_2 production and respiratory rate of *A. inermis* were significantly lower than *A. ornatus*. Bradshaw concluded that the greater water economy of desert-living *Amphibolurus* was achieved both by reduction in metabolic rate and change to a more impervious integument.

By means of a detailed field population study of *A. ornatus*, Bradshaw (1971) was able to show that individuals of the same cohort grew at different rates so that some animals matured in one, two, or three years. These have been referred to as fast- or slow-growing animals. Bradshaw, using marked animals of known growth history, showed that during summer drought there was a difference between fast- and slow-growing animals with respect to distribution of fluids and electrolytes. Slow-growing animals showed no difference when compared with fully hydrated animals except that electrolytes in plasma and skeletal muscle were elevated. Fast-growing animals, however, showed weight losses and changes in fluid volume. Weight losses greater than 20 percent of hydrated weight encroached upon the extracellular fluid volume; but the decrease in volume was restricted to the interstitial fluid, leaving the circulating fluid volume intact, that is, the blood volume and plasma volume remained constant. Earlier, Bradshaw and Shoemaker (1967) showed that the diet of *A. ornatus* consisted of sodium-rich ants and that during summer the lizards lacked sufficient water to excrete the electrolytes without using body water. Instead, the sodium ions were retained at an elevated level in the extracellular fluid which increased in volume by an isosmotic shift of fluid from the intracellular compartment. This sodium retention operates to protect fluid volumes when water is scarce and so enhances survival. Electrolytes were excreted following the occasional summer thunderstorm.

In his population study Bradshaw (1971) showed that only fast-growing animals died as a result of summer drought. Bradshaw (1970) extended his study of water and weight loss in field populations to other species including *A. inermis* and *A. caudicinctus*. As a result of this study he concluded that only males of *A. inermis* lost weight and that, in all species studied, fluid volumes were protected by the retention of sodium ions during periods when water was short. Bradshaw also showed that sodium retention occurred in *A. ornatus* in midsummer but only occurred in *A. inermis* and *A. caudicinctus* after long and intense drought.

Both *A. inermis* and *A. caudicinctus* complete their life cycle in a year (Bradshaw, 1973, personal communication; Storr, 1967). They are thus fast growing in the classification used to describe the life history of *A. ornatus*; but, either by a change in metabolic rate and integument or by some other means, they have avoided the deleterious effects associated with the rapid development of *A. ornatus*.

Population Response

The capacity and speed with which a species can occupy an empty but suitable habitat are related to its capacity to increase. Cole (1954) pointed out that time to maturity, litter size, and whether reproduction is a single episode or repeated throughout the female's life bear on the rate at which a population can increase; but he believed that reproduction early in life was most important for the population. No systematic recording of life-history data appears to have been undertaken in Australia, but such information is critical for understanding how populations persist in fluctuating environments. Whether the fluctuations are due to recurrent drought or to fire or seral stages of postfire succession is not too important, because following any of these events a population nucleus will have the opportunity to expand quickly into an empty but suitable habitat. Moreover, its chances of persisting are enhanced if it can very quickly occupy all favorable habitats at the maximum density because the random spatial distribution of the next drought or fire sequence will determine the sites of the next *refugium*.

The foregoing would suggest that modification of the life history, particularly early maturity, might be as important as physiological adaptation under Australian arid conditions. However, young or small animals are at a disadvantage because of the effects of metabolic body size compared with larger, older mature animals, and hence there is an advantage in late maturity and greater longevity, so that the risks of death which are related to metabolic body size are spread more favorably than they would be in a species in which each generation lived for only one year. Undoubtedly, natural selection will have produced adaptations of life history so that the foregoing apparent conflicts are resolved.

Several workers—MacArthur and Wilson (1967) and Pianka (1970) —have considered the response of populations to selection in terms of whether high fecundity and rapid development or individual fitness and competitive superiority have been favored. These two types of selection were referred to by MacArthur and Wilson (1967) as r-selection and K-selection. Pianka (1970) has tabulated the correlates of each type of selection.

King and Anderson (1971) pointed out that, if a cyclically changing environment varies over few generations, r-type selection factors will be dominant; on the other hand, in a changing environment which has a period of fluctuation many generations long, K-type selection will be dominant. In this connection we might consider quick maturity and large clutch size as manifesting the response to r-type selection; and slower maturity, smaller clutch size, well-marked display, and other devices for marking territory as representing responses to K-type selection.

Mertz (1971) showed that the response of a population to selection will be different depending on whether the population is increasing or declining. In the latter case selection favors the long-lived individual which continues to breed and is thus able to exploit any environmental amelioration even if it occurs late in life. This type of selection tends to produce long-lived populations.

Earlier it was suggested that Australia has been subjected to a pro-longed climate and fire-induced deterioration of the environment which might be expected to produce a response akin to that envisaged by Mertz (1971) and unlike the advantageous rapid development and early reproduction mentioned by Cole (1954). Selection for longevity is a special case in which competitive superiority is principally expressed in terms of a long reproductive life. Murphy (1968) showed this was as a consequence of uncertainty in survival of the prereproductive stages.

With the foregoing outline, it is possible to consider the little information known about the life histories of species from arid Australia in terms of whether they reflect selection during the past for capacity to increase, competitive efficiency related to carrying capacity, or longevity.

Mammals

Macropods. The fossil record suggests that in both Tertiary and post-Pleistocene times the macropods have increased their dominance of the fauna despite the general deterioration of the climate (Stirton et al., 1968). It has already been suggested that the ruminantlike digestion preadapted these species to the desert conditions. The highly developed ruminantlike digestion of macropods can be viewed as a device for delaying the death from starvation caused by a nutritionally inadequate diet. It is thus a device for maximizing physiological longevity once adulthood is achieved.

Among the marsupials there is considerable diversity in their life histories, but there appear to be tendencies toward longevity with respect to populations in arid situations as indicated in the two cases below:

1. In the typical mainland swampy situations the quokka (*Setonix brachyurus*) matures early and breeds continuously and is apparently not long lived. On the other hand, a population of this species on the relatively arid Rottnest Island is older than the mainland form when it first breeds. Breeding is seasonal, and so Rottnest animals tend to produce fewer offspring per unit time than the mainland form (Shield, 1965). Moreover, individuals from the island population tend to live seven to eight years, with a few females present and

still reproducing in their tenth year. The pollen record on Rottnest indicates that the environment has declined from a woodland to a coastal heath and scrubland over the past 7,000 years (Storr, Green, and Churchill, 1959). Despite the difference in detail, the modification in the life history of the quokka on the semiarid Rottnest Island achieves the same end as the red kangaroo and euro discussed below.

2. Typical arid-land species, such as the red kangaroo, *Megaleia rufa*, and the euro, *Macropus robustus*, have no defined season of breeding, and females are always carrying young except under very severe drought. Both these species tend to be long lived (Kirkpatrick, 1965), and females may still be able to bear young when approaching twenty years of age.

The breeding of the red kangaroo and euro suggests that adaptation of life history has centered around the metabolic advantages of large body size in a long-lived animal which is virtually capable of continuous production of offspring, some of which must by chance be weaned into a seasonal environment which permits growth to maturity.

Numerous workers have shown that macropod marsupials have lowered metabolic rates with which are associated reduced requirements for water, energy, and protein and a slower rate of growth. The first three of these are of advantage during times of drought; and, should the last contribute to longevity, it will also be advantageous, insofar as offspring have the potential to be distributed into favorable environments whenever they occur.

Rodents. The Australian rodents appear to have a typical rodent-type reproductive pattern with a high capacity to increase. They appear to be able to survive through drought because of their small size and capacity to persist as small populations in minor, favorable habitats.

Birds

Most bird species which have been studied physiologically belong to taxonomic groups which also occur outside Australia, for example, finches, pigeons, caprimulgids, and parrots. The information on which a comparative study of modifications of the life histories of the Australian forms with their old-world relatives could be based has not been assembled. However, several observations—for example, Cade et al. (1965), that Australian and African estrildine finches are markedly different in physiology, and Dawson and Fisher (1969), that the spotted nightjar (*Eurostopodus guttatus*), like all caprimulgids, has a depressed metabolism—are suggestive that the life histories of some Australian species (finches) might be highly modified, while others show only slight modification from their old-world relatives (caprimulgids); and these may, in a sense, be thought of as being pre-adapted to survival in arid Australia.

Keast (1959), in a review of the life-history adaptations of Australian birds to aridity, showed that the principal adaptation is opportunistic breeding after the break of drought when the environment can provide the necessities for successful rearing of young. Longevity of individuals in unknown, but the breeding pattern is consistent with selection which has favored longevity.

Fisher et al. (1972) observed that honeyeaters (Meliphagidae), which are widespread and common throughout arid Australia, are surprisingly dependent on water. The growth of these birds to maturity and their metabolism are not known, but these authors speculated that the dependence may be due in large part to the water loss attendant on the activity associated with the high degree of aggressive behavior exhibited by all species of honeyeaters.

The following speculation would be consistent with the observations of Fisher et al. (1972): Most honeyeaters frequent late and climax stages when the vegetation is at its maximum diversity with numerous sources of nectar and insects. Such a habitat preference would suggest that K-type selection would have operated in the past, and the advantages of obtaining and maintaining an adequate territory by aggressive display may outweigh any disadvantages of individual high water needs which were consequent upon the aggressive display.

Reptiles

Table 5-2 has been compiled from the information available on lizard physiology and biology. The information is not equally complete for all species tested; however, it does suggest that Australian desert

Table 5-2. *Physiological, Ecological, and Life-History Information for Selected Species of Australian Lizards*

Species	Mean Preferred Temperature	Mean Survival		Water Loss (mg/g/hr)	Seral Stage
		Minutes	Temperature (°C)		
Amphibolurus inermis	36.4[a]	102.8 2.0	46.0 48.0	1.05 at 35°C	burrows in early seres
A. caudicinctus	37.7[a]	92.8 45.0	46.0 47.0	1.80 at 35°C	rock piles in climax hummock grass land
A. scutulatus	38.2[a]	40.8 28.0	46.0 47.0	?	shady climax woodland
Diporophora bilineata	44.3[b]	360.0 29.5	46.0 49.0	?	fire disclimax
Moloch horridus	36.7[a]	?	?	?	late seres and climax

| Age to Maturity (yrs) | Reproduction | | Longevity (yrs) | Reference |
	No. of Clutches per Year	Eggs per Clutch (means)		
0.75	possibly 2	3.43	1	Licht et al., 1966a, 1966b; Pianka, 1971a; Bradshaw, personal communication
0.75	possibly 2	?	1	Licht et al., 1966a, 1966b; Storr, 1967; Bradshaw, 1970
?	possibly only 1	6.5	?	Licht et al., 1966a, 1966b; Pianka, 1971c
?	?	?	?	Bradshaw and Main, 1968
3–4	usually 1	6–7	6–20	Sporn, 1955, 1958, 1965; Licht et al., 1966b; Pianka and Pianka, 1970

| Species | Mean Preferred Temperature | Mean Survival | | Water Loss (mg/g/hr) | Seral Stag |
		Minutes	Temperature (°C)		
Gehyra variegata	35.3[a]	72.8 2.0	43.5 46.0	2.07 at 25°C 3.37 at 30°C 3.80 at 35°C	climax anc postclimax woodland
Heteronota binoei	30.0[ac]	162.0 0.0	40.5 43.5	0.27 at 30°C	climax wit litter
Rhynchoe-dura ornata	34.0[a]	55.3	46.0	?	holes in b soil in clin woodland

[a]In gradient.   [b]In field.   [c]May be too high—see Licht et al., 1966*b*.

species exhibit a wide range of tolerances to elevated temperatures. It is surprising, for example, that *Heteronota binoei* survives at all in the desert. Geckos, depending on the species, may have clutches of a single egg, but no species have clutches larger than two eggs; however, they may have one or two clutches each breeding season. *Heteronota binoei* and *Gehyra variegata* have respectively one and two clutches. *Heteronota binoei*, with an apparent preference for low temperatures and an inability to tolerate high temperatures, has adapted to the desert by its extremely low rate of water loss, behavioral attachment to sheltered climate situations, and, relative to *G. variegata*, early maturity and large clutch size (two eggs vs. one).

On the other hand, *G. variegata* is better adapted to high temperatures and, even though it is relatively poor at conserving water, is able to survive in the deteriorating and more exposed situations of the late climax and postclimax. Moreover, these physiological adapta-

| Age to Maturity (yrs) | Reproduction | | Longevity (yrs) | Reference |
	No. of Clutches per Year	Eggs per Clutch (means)		
2; breed in 3rd	2	1	mean 4.4	Bustard and Hughes, 1966; Licht et al., 1966a, 1966b; Bustard, 1968a, 1969; Bradshaw, personal communication
1.6 or 2.5	usually 1	2	mean 1.9	Bustard and Hughes, 1966; Licht et al., 1966a; Bustard, 1968b
?	?	?	?	Licht et al., 1966a, 1966b

tions are associated with a long adult life and thus enhance the possibility of favorable recruitment in any season where conditions are ameliorated so that eggs and young have an enhanced survival.

Among the agamids the information is not nearly as complete. *Amphibolurus inermis* and *A. caudicinctus* are early maturing, short-lived species relying on a high rate of reproduction to maintain the population and are thus the analogue of *H. binoei. Moloch horridus* and *A. scutulatus*, on the other hand, appear to be the analogue of *G. variegata*; and it is unfortunate that information on age to maturity and longevity of *A. scutulatus* is not available. One can only speculate on age to maturity and longevity of *Diporophora*, but it seems likely that recruitment would only be successful in years when summer cyclonic rain ameliorated environmental conditions; and one might guess that it is a long-lived animal.

It is interesting that the fast-maturing species either have a cool refuge in which the small young can establish themselves (*H. binoei* in climax) or a cool season in which they can grow to almost adult size (*A. inermis*, *A. caudicinctus*). In addition, these species have another adaptation in producing twin broods in each breeding season. Should there be a drought, the young from the first brood will almost certainly be lost. However, should the young be born into a season in which thunderstorms are common, they would be able to thrive under almost ideal conditions. Since the offspring from the second clutch are born late in the summer or early autumn, they are almost certain to survive regardless of the preceding summer conditions.

Discussion

The foregoing suggests that early in Tertiary times Australia underwent a change in position from higher (southern) to lower (tropical) latitudes. Stemming from this there has been a prolonged and disastrous change in climate toward increasing aridity. This has been accompanied by the increased incidence of fire as an ecological factor.

Much of the original biota has become extinct as the result of these changes, but there have been some additions from Asia. Both the old and new elements of the biota that have survived to the present have done so because they have been able to accommodate their individual physiology and population biology to the stresses imposed by climatic deterioration (drought) and fire.

The foregoing has been achieved by a series of complementary strategies as follows:

1. Physiological strategies
    a. Behavioral avoidance of stressful environmental factors
    b. Heat tolerance
    c. Ability to conserve water, including ability to handle electrolytes
    d. Ability to survive on diets of low nutritional value (herbivores)
2. Reproductive strategies
    a. High reproductive capacity, so enabling a population nucleus surviving after drought to rapidly repopulate the former range and to occupy all areas which could possibly form *refugia* in future droughts

    b. Increased competitive advantage by means of small well-tended broods of young and well-developed displays for holding territories

    c. Increased longevity, so that adults gain advantage from metabolic body size while young are produced over a span of years, so ensuring that at least some are born into a seasonal environment in which they can survive and become recruits to the adult population

It is thus apparent that vertebrates inhabiting arid parts of Australia display a diversity of individual adaptations to single components of the arid environment, and it is difficult to interpret the significance of experimental laboratory findings achieved as the result of simple single-factor experiments. For example, under experimental conditions, kangaroos and wallabies, if exposed to high ambient temperatures, use quantities of water in evaporative cooling (Bartholomew, 1956; Dawson et al., 1969; Dawson and Bennet, 1971).

Yet these arid-land species are capable of surviving intense drought conditions when the environment provides the appropriate shelter conditions. These may be postfire seral stages as needed by the hare wallaby, *Lagorchestes conspicillatus* (Burbidge and Main, 1971), or rock piles needed by the euro, *Macropus robustus*. Given that the euro and hare wallaby have shelter of the appropriate quality, both species are apparently well adapted to grow and reproduce on the low-quality forage which is available where they live. Moreover, both species are capable of reproduction at all seasons so that, while their reproductive potential is limited—because of having only one young at a time—they do maximize their reproductive potential by continuous breeding and by distributing the freshly weaned young at all seasons, which is particularly important in a seasonally unpredictable environment.

In general, while it is true that some species show a highly developed degree of adaptation to arid conditions, it is difficult to find a case which is unrelated to seral successional stages. A pronounced example of this is afforded by the lizard *Diporophora bilineata* and the gecko *Diplodactylus michaelseni*, which can withstand higher field body temperatures than any other Australian species but which appear to be abundant only in excessively exposed fire disclimax situations.

Most of the vertebrates which survive in the desert appear to do so not solely because of well-developed individual adaptation (tolerance) to the hot dry conditions of arid Australia, but because of habitat preference and population attributes which permit the species to cope, first, with the ecological consequences of fire and, second, with drought. In a sense, adaptation to fire has preadapted the vertebrates to drought.

Desert species have had to choose whether the ability of a population to grow is equivalent to ability to persist. Two circumstances can be envisaged in which ability to grow is equivalent to persistence: when rapid repopulation of an area after drought will ensure that all potential future *refugia* are occupied and when rapid population growth excludes other species from a resource.

In the desert where drought conditions are the norm, however, persistence is achieved by females replacing themselves with other females in their lifetime. This requires that juveniles must withstand or avoid desert conditions until they reach reproductive age. Seasonal amelioration of conditions in desert environments is notoriously unpredictable, and it seems that many Australian desert animals persist as populations because of long reproductive lives during which some young will be produced and grow to maturity.

In considering the individual and the population aspects of survival we should envisage the space occupied by an animal as providing scope for minimizing the environmental stresses of heat, water shortage, and poor-quality diet. An animal will choose to live in places where the stresses are least; when these are not available, it will select sites or opt for physiological responses which allow it to prolong the time to death. Urea recycling by macropods should be viewed in this light. When environmental amelioration occurs, it is taken as an opportunity to replenish the population by recruiting young.

Acknowledgments

Financial assistance is acknowledged from the University of Western Australia Research Grants Committee, the Australian Research Grants Committee, and Commonwealth Scientific and Industrial Research Organization. Professor H. Waring, Dr. S. D. Bradshaw, and Dr. J. C. Taylor kindly read and criticized the manuscript.

Summary

It is suggested that the Australian deserts developed as a consequence of the movement in Tertiary times of the continental plate from higher latitudes to its present position. An increasing incidence of wild fire is associated with the development of dry conditions. The vertebrate fauna has adapted to the development of deserts and incidence of fire at two levels: (a) the individual or physiological, emphasizing such strategies as behavioral avoidance of stressful conditions, conservation of water, tolerance of high temperatures, and, with macropod herbivores, ability to survive on low-quality forage and through the supplementation of nitrogen by the recycling of urea; and (b) the population, emphasizing reproductive strategies and longevity, so that young are produced over a long period of time thus enhancing the possibility of successful recruitment.

It is further suggested that survival of individuals and persistence of the population are only possible when the environment, especially the postfire plant succession, provides the space and scope for the implementation of the strategies which have evolved.

References

Balme, B. E. 1962. Palynological report no. 98: Lake Eyre no. 20 Bore, South Australia. In *Investigation of Lake Eyre*, ed. R. K. Johns and N. H. Ludbrook. *Rep. Invest. Dep. Mines S. Aust.* No. 24, pts. 1 and 2, pp. 89–102.

Balme, B. E., and Churchill, D. M. 1959. Tertiary sediments at Coolgardie, Western Australia. *J. Proc. R. Soc. West. Aust.* 42:37–43.

Bartholomew, G. A. 1956. Temperature regulation in the macropod marsupial *Setonix brachyurus. Physiol. Zoöl.* 29:26–40.

Beadle, N. C. W. 1962a. Soil phosphate and the delimitation of plant communities in Eastern Australia, II. *Ecology* 43:281–288.

———. 1962b. An alternative hypothesis to account for the generally low phosphate content of Australian soils. *Aust. J. agric. Res.* 13: 434–442.

———. 1966. Soil phosphate and its role in molding segments of the Australian flora and vegetation, with special reference to xeromorphy and sclerophylly. *Ecology* 47:992–1007.

Bradshaw, S. D. 1970. Seasonal changes in the water and electro-lyte metabolism of *Amphibolurus* lizards in the field. *Comp. Biochem. Physiol.* 36:689–718.

———. 1971. Growth and mortality in a field population of *Amphibolurus* lizards exposed to seasonal cold and aridity. *J. Zool., Lond.* 165:1–25.

Bradshaw, S. D., and Main, A. R. 1968. Behavioral attitudes and regulation of temperature in *Amphibolurus* lizards. *J. Zool., Lond.* 154: 193–221.

Bradshaw, S. D., and Shoemaker, V. H. 1967. Aspects of water and electrolyte changes in a field population of *Amphibolurus* lizards. *Comp. Biochem. Physiol.* 20:855–865.

Brown, D. A.; Campbell, K. S. W.; and Crook, K. A. W. 1968. *The geological evolution of Australia and New Zealand*. Oxford: Pergamon Press.

Brown, G. D. 1968. The nitrogen and energy requirements of the euro (*Macropus robustus*) and other species of macropod marsupials. *Proc. ecol. Soc. Aust.* 3:106–112.

———. 1969. Studies on marsupial nutrition. VI. The utilization of dietary urea by the euro or hill kangaroo, *Macropus robustus* (Gould). *Aust. J. Zool.* 17:187–194.

Brown, G. D., and Main, A. R. 1967. Studies on marsupial nutrition. V. The nitrogen requirements of the euro, *Macropus robustus*. *Aust. J. Zool.* 15:7–27.

Burbidge, A. A., and Main, A. R. 1971. Report on a visit of inspection to Barrow Island, November, 1969. *Rep. Fish. Fauna West. Aust.* 8:1–26.

Burbidge, N. T. 1960. The phytogeography of the Australian region. *Aust. J. Bot.* 8:75–211.

Bustard, H. R. 1968a. The ecology of the Australian gecko *Gehyra variegata* in northern New South Wales. *J. Zool., Lond.* 154:113–138.

———. 1968b. The ecology of the Australian gecko *Heteronota binoei* in northern New South Wales. *J. Zool., Lond.* 156:483–497.

———. 1969. The population ecology of the gekkonid lizard *Gehyra variegata* (Dumeril and Bibron) in exploited forests in northern New South Wales. *J. Anim. Ecol.* 38:35–51.

Bustard, H. R., and Hughes, R. D. 1966. Gekkonid lizards: Average ages derived from tail-loss data. *Science, N.Y.* 153:1670–1671.

Cade, T. J., and Dybas, J. A. 1962. Water economy of the budgerygah. *Auk* 79:345–364.

Cade, T. J.; Tobin, C. A.; and Gold, A. 1965. Water economy and metabolism of two estrildine finches. *Physiol. Zoöl.* 38:9–33.

Calder, W. A. 1964. Gaseous metabolism and water relations of the zebra finch *Taenopygia castanotis*. *Physiol. Zoöl.* 37:400–413.

Charley, J. L., and Cowling, S. W. 1968. Changes in soil nutrient status resulting from overgrazing in plant communities in semi-arid areas. *Proc. ecol. Soc. Aust.* 3:28–38.

Cochrane, G. R. 1968. Fire ecology in southeastern Australian sclerophyll forests. *Proc. Ann. Tall Timbers Fire Ecol. Conf.* 8:15–40.

Cole, La M. C. 1954. Population consequences of life history phenomena. *Q. Rev. Biol.* 29:103–137.

Cookson, I. C. 1953. The identification of the sporomorph *Phyllocladites* with *Dacrydium* and its distribution in southern Tertiary deposits. *Aust. J. Bot.* 1:64–70.

Cookson, I. C., and Pike, K. M. 1953. The Tertiary occurrence and distribution of *Podocarpus* (section *Dacrycarpus*) in Australia and Tasmania. *Aust. J. Bot.* 1:71–82.

———. 1954. The fossil occurrence of *Phyllocladus* and two other podocarpaceous types in Australia. *Aust. J. Bot.* 2:60–68.

Dawson, T. J., and Brown, G. D. 1970. A comparison of the insulative and reflective properties of the fur of desert kangaroos. *Comp. Biochem. Physiol.* 37:23–38.

Dawson, T. J.; Denny, M. J. S.; and Hulbert, A. J. 1969. Thermal balance of the macropod marsupial *Macropus eugenii* Desmarest. *Comp. Biochem. Physiol.* 31:645–653.

Dawson, W. R., and Bennet, A. F. 1971. Thermoregulation in the marsupial *Lagorchestes conspicillatus*. *J. Physiol., Paris* 63:239–241.

Dawson, W. R., and Fisher, C. D. 1969. Responses to temperature by the spotted nightjar (*Eurostopodus guttatus*). *Condor* 71:49–53.

Exon, N. R.; Langford-Smith, T.; and McDougall, I. 1970. The age and geomorphic correlations of deep-weathering profiles, silcrete, and basalt in the Roma-Amby Region Queensland. *J. geol. Soc. Aust.* 17:21–31.

Fisher, C. D.; Lindgren, E.; and Dawson, W. R. 1972. Drinking patterns and behaviour of Australian desert birds in relation to their ecology and abundance. *Condor* 74:111–136.

Frakes, L. A., and Kemp, E. M. 1972. Influence of continental positions on early Tertiary climates. *Nature, Lond.* 240:97–100.

Gardner, C. A. 1944. Presidential address: The vegetation of Western Australia. *J. Proc. R. Soc. West. Aust.* 28:xi–lxxxvii.

———. 1957. The fire factor in relation to the vegetation of Western Australia. *West. Aust. Nat.* 5:166–173.

Gilbert, J. M. 1958. Forest succession in the Florentine Valley, Tasmania. *Pap. Proc. R. Soc. Tasm.* 93:129–151.

Greenwald, L.; Stone, W. B.; and Cade, T. J. 1967. Physiological adjustments of the budgerygah (*Melopsettacus undulatus*) to dehydrating conditions. *Comp. Biochem. Physiol.* 22:91–100.

Heirtzler, J. R.; Dickson, G. O.; Herron, E. M.; Pitman, W. C.; and Le Pichon, X. 1968. Marine magnetic anomalies, geomagnetic field reversals, and motions of the ocean floor and continents. *J. geophys. Res.* 73:2119–2136.

Jackson, W. D. 1965. Vegetation. In *Atlas of Tasmania*, ed. J. L. Davis, pp. 50–55. Hobart, Tasm.: Mercury Press.

———. 1968*a*. Fire and the Tasmanian flora. In *Tasmanian year book no. 2*, ed. R. Lakin and W. E. Kellend. Hobart, Tasm.: Commonwealth Bureau of Census and Statistics, Hobart Branch.

———. 1968*b*. Fire, air, water and earth: An elemental ecology of Tasmania. *Proc. ecol. Soc. Aust.* 3:9–16.

Johnson, O. W., and Mugaas, J. N. 1970. Quantitative and organizational features of the avian renal medulla. *Condor* 72:288–292.

Keast, A. 1959. Australian birds: Their zoogeography and adaptation to an arid continent. In *Biogeography and ecology in Australia*, ed. A. Keast, R. L. Crocker, and C. S. Christian, pp. 89–114. The Hague: Dr. W. Junk.

King, C. E., and Anderson, W. W. 1971. Age specific selection, II. The interaction between r & K during population growth. *Am. Nat.* 105:137–156.

Kirkpatrick, T. H. 1965. Studies of Macropodidae in Queensland. 2. Age estimation in the grey kangaroo, the eastern wallaroo and the red-necked wallaby, with notes on dental abnormalities. *Qd J. agric. Anim. Sci.* 22:301–317.

Le Pichon, X. 1968. Sea-floor spreading and continental drift. *J. geophys. Res.* 73:3661–3697.

Licht, P.; Dawson, W. R.; and Shoemaker, V. H. 1966a. Heat resistance of some Australian lizards. *Copeia* 1966:162–169.

Licht, P.; Dawson, W. R.; Shoemaker, V. H.; and Main, A. R. 1966b. Observations on the thermal relations of Western Australian lizards. *Copeia* 1966:97–110.

MacArthur, R. H., and Wilson, E. O. 1967. *The theory of island biogeography.* Monographs in Population Biology, 1. Princeton: Princeton Univ. Press.

MacMillan, R. E., and Lee, A. K. 1967. Australian desert mice: Independence of exogenous water. *Science, N.Y.* 158:383–385.

———. 1969. Water metabolism of Australian hopping mice. *Comp. Biochem. Physiol.* 28:493–514.

———. 1970. Energy metabolism and pulmocutaneous water loss of Australian hopping mice. *Comp. Biochem. Physiol.* 35:355–369.

McWhae, J. R. H.; Playford, P. E.; Lindner, A. W.; Glenister, B. F.; and Balme, B. E. 1958. The stratigraphy of Western Australia. *J. geol. Soc. Aust.* 4:1–161.

Main, A. R. 1971. Measures of well-being in populations of herbivorous macropod marsupials. In *Dynamics of populations*, ed. P. J. den Boer and G. R. Gradwell, pp. 159–173. Wageningen: PUDOC.

Mertz, D. B. 1971. Life history phenomena in increasing and decreasing population. In *Statistical ecology, volume II: Sampling and modeling biological populations and population dynamics*, ed. G. P. Patil, E. C. Pielou, and W. E. Waters, pp. 361–399. University Park: Pa. St. Univ. Press.

Moir, R. J.; Somers, M.; and Waring, H. 1956. Studies in marsupial nutrition: Ruminant-like digestion of the herbivorous marsupial *Setonix brachyurus* (Quoy and Gaimard). *Aust. J. biol. Sci.* 9:293–304.

Mount, A. B. 1964. The interdependence of eucalypts and forest fires in southern Australia. *Aust. For.* 28:166–172.

———. 1969. Eucalypt ecology as related to fire. *Proc. Ann. Tall Timbers Fire Ecol. Conf.* 9:75–108.

Murphy, G. I. 1968. Pattern in life history and the environment. *Am. Nat.* 102:391–404.

Oksche, A.; Farner, D. C.; Serventy, D. L.; Wolff, F.; and Nicholls,

C. A. 1963. The hypothalamo-hypophysial neurosecretory system of the zebra finch, *Taeniopygia castanotis*. *Z. Zellforsch. mikrosk. Anat.* 58:846–914.

Packham, G. H., ed. 1969. The geology of New South Wales. *J. geol. Soc. Aust.* 16:1–654.

Pianka, E. R. 1969. Sympatry of desert lizards (*Ctenotus*) in Western Australia. *Ecology* 50:1012–1030.

———. 1970. On r and K selection. *Am. Nat.* 104:592–597.

———. 1971a. Comparative ecology of two lizards. *Copeia* 1971:129–138.

———. 1971b. Ecology of the agamid lizard *Amphibolurus isolepis* in Western Australia. *Copeia* 1971:527–536.

———. 1971c. Notes on the biology of *Amphibolurus cristatus* and *Amphibolurus scutulatus*. *West. Aust. Nat.* 12:36–41.

Pianka, E. R., and Pianka, H. D. 1970. The ecology of *Moloch horridus* (Lacertilia: Agamidae) in Western Australia. *Copeia* 1970:90–103.

Ride, W. D. L. 1964. A review of Australian fossil marsupials. *J. Proc. R. Soc. West. Aust.* 47:97–131.

———. 1971. On the fossil evidence of the evolution of the Macropodidae. *Aust. Zool.* 16:6–16.

Shield, J. W. 1965. A breeding season difference in two populations of the Australian macropod marsupial *Setonix brachyurus*. *J. Mammal.* 45:616–625.

Simpson, G. G. 1961. Historical zoogeography of Australian mammals. *Evolution* 15:431–446.

Slatyer, R. O., and Perry, R. A., eds. 1969. *Arid lands of Australia*. Canberra: Aust. Nat. Univ. Press.

Sporn, C. C. 1955. The breeding of the mountain devil in captivity. *West. Aust. Nat.* 5:1–5.

———. 1958. Further observations on the mountain devil in captivity. *West. Aust. Nat.* 6:136–137.

———. 1965. Additional observations on the life history of the mountain devil (*Moloch horridus*) in captivity. *West. Aust. Nat.* 9:157–159.

Stirton, R. A.; Tedford, R. D.; and Miller, A. H. 1961. Cenozoic stratigraphy and vertebrate palaeontology of the Tirari Desert, South Australia. *Rep. S. Aust. Mus.* 14:19–61.

Stirton, R. A.; Tedford, R. H.; and Woodburne, M. O. 1968. Australian Tertiary deposits containing terrestrial mammals. *Univ. Calif. Publs geol. Sci.* 77:1–30.

Stirton, R. A.; Woodburne, M. O.; and Plane, M. D. 1967. A phylogeny of Diprotodontidae and its significance in correlation. *Bull. Bur. Miner. Resour. Geol. Geophys. Aust.* 85:149–160.

Storr, G. M. 1967. Geographic races of the agamid lizard *Amphibolurus caudicinctus*. *J. Proc. R. Soc. West. Aust.* 50:49–56.

―――. 1968. Diet of kangaroos (*Megaleia rufa* and *Macropus robustus*) and merino sheep near Port Hedland, Western Australia. *J. Proc. R. Soc. West. Aust.* 51:25–32.

Storr, G. M.; Green, J. W.; and Churchill, D. M. 1959. The vegetation of Rottnest Island. *J. Proc. R. Soc. West. Aust.* 42:70–71.

Veevers, J. J. 1967. The Phanerozoic geological history of northwest Australia. *J. geol. Soc. Aust.* 14:253–271.

―――. 1971. Phanerozoic history of Western Australia related to continental drift. *J. geol. Soc. Aust.* 18:87–96.

Vine, F. J. 1966. Spreading of the ocean floor: New evidence. *Science, N.Y.* 154:1405–1415.

―――. 1970. Ocean floor spreading. *Rep. Aust. Acad. Sci.* 12:7–24.

Wallace, W. R. 1966. Fire in the Jarrah forest environment. *J. Proc. R. Soc. West. Aust.* 49:33–44.

Watkins, J. R. 1967. The relationship between climate and the development of landforms in the Cainozoic rocks of Queensland. *J. geol. Soc. Aust.* 14:153–168.

Wild, A. 1958. The phosphate content of Australian soils. *Aust. J. agric. Res.* 9:193–204.

Winkworth, R. E. 1967. The composition of several arid spinifex grasslands of central Australia in relation to rainfall, soil water relations, and nutrients. *Aust. J. Bot.* 15:107–130.

Woodburne, M. O. 1967. Three new diprotodontids from the Tertiary of the Northern Territory. *Bull. Bur. Miner. Resour. Geol. Geophys. Aust.* 85:53–104.

Woolnough, W. G. 1928. The chemical criteria of peneplanation. *J. Proc. R. Soc. N.S.W.* 61:17–53.

6. Species and Guild Similarity of North American Desert Mammal Faunas: A Functional Analysis of Communities   James A. MacMahon

Introduction

A major thrust of current ecological and evolutionary research is the analysis of patterns of species diversity or density in similar or vastly dissimilar community types. Such studies are believed to bear on questions concerned with the nature of communities and their stability (e.g., MacArthur, 1972), the concept of ecological equivalents or ecospecies (Odum, 1969 and 1971; Emlen, 1973), and, of course, the nature of the "niche" (Whittaker, Levin, and Root, 1973).

An approach emerging from this plethora is that of functional analysis of community components: attempts to compare the functionally similar community members, regardless of their taxonomic affinities. Root (1967, p. 335) coined the term *guild* to define "a group of species that exploit the same class of environmental resources in a similar way. This term groups together species without regard to taxonomic position, that overlap significantly in their niche requirements."

Guild is clearly differentiated from *niche* and *ecotope*, recently redefined and defined respectively (Whittaker et al., 1973, p. 335) as "applying 'niche' to the role of the species within the community, 'habitat' to its distributional response to intercommunity environmental factors, and 'ecotope' to its full range of adaptations to external factors of both niche and habitat." *Guild* groups parts of species' niches permitting intercommunity comparisons.

Without referring to the semantic problems, Baker (1971) used such a "functional" approach when he compared nutritional strategies of North American grassland myomorph rodents. Wiens (1973) developed a similar theme in his recent analysis of grassland bird communities, as did Wilson (1973) with an analysis of bat faunas, and Brown (1973) with rodents of sand-dune habitats.

This paper is an attempt to compare species and functional analyses of the small mammal component of North American deserts and to use these analyses to discuss some aspects of the broader ecological and evolutionary questions of "similarity" and function of communities.

Sites and Techniques

Sites

The data base for this study is simply the species lists for a number of desert or semidesert grassland localities in the western United States. The list for a locality represents those species that occur on a piece of landscape of 100 ha in extent. This size unit allows the inclusion of spatial heterogeneity.

The localities used, the data source, and the abbreviations to be used subsequently are Jornada Bajada (*j*), a Chihuahuan desert shrub community near Las Cruces, New Mexico, operated by New Mexico State University as part of the US/IBP Desert Biome studies; Jornada Playa (*jp*), a desert grassland and mesquite area a few meters from *j* operated under the same program; Portal, Arizona (*cc*), a semidesert scrub area studied extensively by Chew and Chew (1965, 1970); Santa Rita Experimental Range (*sr*), south of Tucson, Arizona, an altered desert grassland studied by University of Arizona personnel for the US/IBP Desert Biome; Tucson Silverbell Bajada (*t*), a typical Sonoran desert (*Larrea-Cereus-Cercidium*) locality northwest of Tucson, Arizona, operated as *sr*; Big Bend National Park (*bb*), a Chihuahuan desert shrub community near the park headquarters typified by Denyes (1956) and K. L. Dixon (1974, personal communication); Deep Canyon, California (*dc*), studied by Ryan (1968) and Joshua Tree National Monument (*jt*) studied by Miller and Stebbins (1964)—both *Larrea*-dominated areas in a transition from a Sonoran desert subdivision (Coloradan) but including many Mojave desert elements; Rock Valley (*rv*), northwest of Las Vegas, Nevada, on the

Fig. 6-1. The location of sites discussed (abbreviations explained in text) and outline of mammal provinces adapted from Hagmeier and Stults (1964): I. Artemisian; II. Mojavian; III. Navajonian; IV. Sonoran; V. Yaquinian; VI. Mapimi.

Atomic Energy Commission's Nevada Test Site, a Mojave desert shrub site operated as part of the US/IBP Desert Biome by personnel of the Environmental Biology Division of the Laboratory of Nuclear Medicine and Radiation Biology of the University of California, Los Angeles; and Curlew Valley (c), a Great Basin desert, sagebrush site of the US/IBP Desert Biome operated by Utah State University. The positions of the sites are summarized in figure 6-1. All sites have been visited and observed by me.

Analyses

Similarity was calculated using a modified form of Jaccard analysis (community coefficients) (Oosting, 1956; see also MacMahon and Trigg, 1972):

$$\frac{2w}{a+b} \times 100$$

where w is the number of species common to both faunas being compared, a is the number of species in the smaller fauna, and b is the number of species in the larger fauna.

Species similarity merely uses different taxa as units for calculations. Functional similarity uses functional units (guilds) based mainly on food habits and adult size of nonflying mammals, jack rabbit in size or smaller. The twelve desert guilds recognized, with examples of species from a single locality, include five granivores (two possible dormant-season divisions) (*Dipodomys spectabilis*, *D. merriami*, *Perognathus penicillatus*, *P. baileyi*, *P. amplus*); a "carnivorous" mouse (*Onychomys torridus*); a large and small browser (*Lepus californicus*, *Sylvilagus auduboni*); two micro-omnivores (*Peromyscus eremicus*, *P. maniculatus*); a "pack rat" (*Neotoma albigula*); and a diurnal medium-sized omnivore (*Citellus tereticaudus*). When grassland guilds are mentioned, two grazers are added to the above. Data for all pair-wise comparisons of sites are summarized in figure 6-2.

The list of mammals for all sites includes forty-seven species in fifteen genera. An additional fifteen or so species occur near the sites but were not collected on the prescribed areas.

SPECIES

FUNCTIONS	j	jp	cc	sr	t	bb	dc	jt	rv	c
j		67	67	41	41	56	33	30	20	14
jp	67		63	52	39	46	32	23	19	19
cc	79	63		60	56	54	32	29	19	14
sr	63	60	52		63	38	29	27	15	25
t	85	79	56	55		48	33	30	20	04
bb	80	62	85	69	85		40	44	48	08
dc	85	67	67	63	71	80		63	60	09
jt	86	69	69	65	73	89	73		73	08
rv	85	67	67	55	85	80	85	86		14
c	60	67	56	55	50	72	60	53	71	

Fig. 6-2. Similarity analysis (%) matrix derived from Jaccard analysis (see text): species comparisons above the diagonal, guild comparisons below the diagonal.

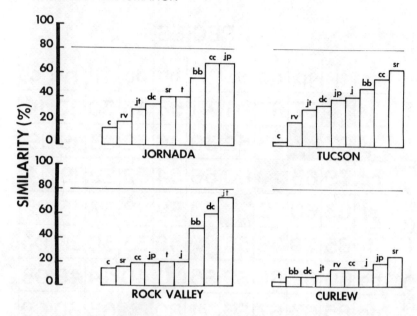

Fig. 6-3. Comparison of the similarity (%) of lists of nonflying mammal species on all sites to those of four "typical" North American desert sites: Jornada (j), Chihuahuan desert; Tucson (t), Sonoran desert; Rock Valley (rv), Mojave Desert; Curlew (c), Great Basin (cold desert).

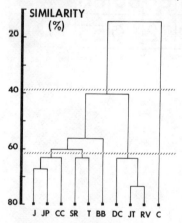

Fig. 6-4. Dendrogram showing relationships between species composition (maximum percent similarity, using Jaccard analysis) at North American sites (see text for abbreviations). The levels of significance for provinces (about 62%) and for superprovinces (about 39%), as defined by Hagmeier and Stults (1964), are marked.

Results and Discussion

Species Density

Figure 6-3 depicts the comparison of the species composition of all sites with that of each of the four "typical" desert sites of the US/IBP Desert Biome. These sites represent each of the four North American deserts: three "hot" deserts—Chihuahuan (*j*), Sonoran (*t*), Mojave (*rv*); one "cold" desert—Great Basin (*c*). Sites *jp*, *sr*, and *cc* are considered to have strong desert grassland affinities; all were "good" grasslands in historical times (Gardner, 1951; Lowe, 1964; Lowe et al., 1970). It is clear that none of the comparisons indicates high similarity (operationally defined as 80%). Comparison of figure 6-3 and figure 6-1 indicates that what similarity exists seems to be due to geographic proximity.

Maximum Species Similarity

The maximum species similarity of all sites is used to develop a dendrogram of relationships of sites similar to those of Hagmeier and Stults (1964) (fig. 6-4). The five groupings derived (using a 60% similarity level as was used by Hagmeier and Stults)—that is, *j*, *jp*, *cc*; *sr*, *t*; *bb*; *dc*, *jt*, *rv*; and *c*—do not follow closely the mammal provinces erected by Hagmeier and Stults (1964) and are redrawn here (fig. 6-1).

There is agreement between my data and those of Hagmeier and Stults at the superprovince level (about 38% similarity) which sets *c* (Artemisian) apart from all hot desert sites. The Artemisian province is equivalent to the Great Basin desert or cold desert in extent. The failure of this study and that of Hagmeier and Stults to agree may be due to the animal size limit used herein (smaller than jack rabbits) or the confined areal sample (100 ha vs. larger areas).

The groups defined here (at the province level) do seem to have some faunal meaning: there is a distinct Big Bend (*bb*) assemblage, a Sonoran desert group (*sr*, *t*), a Chihuahuan desert group (*j*, *jp*, *cc*), a Mojave desert group (*dc*, *jt*, *rv*), and a Great Basin desert group (*c*).

Some groups can be explained utilizing the evolution-biogeography discussion of Findley (1969) which differentiated eastern and western

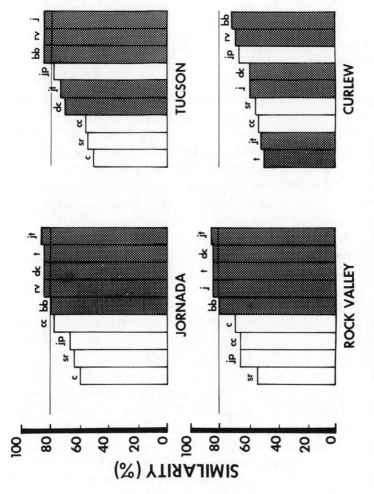

Fig. 6-5. Comparison of similarity in guilds of nonflying mammals at all sites. Presentation as in figure 6-3, except that the bars of all "hot" desert localities are shaded.

desert components meeting in southeastern Arizona—this coincides with the Sonoran desert versus the Chihuahuan desert above, with the cc site being intermediate. Figure 6-2 supports the intermediacy of cc. Findley postulated that the Deming Plain was a barrier to desert mammal movement in pluvial times, that it limited gene flow, and that it permitted speciation to the west and east. There is sharp differentiation of these two components from the Mojave and Great Basin, which is expected on the basis of their different geological histories, and also from the Big Bend, which might be more representative of the major portion of the Chihuahuan desert mammal fauna in Mexico.

Functional Diversity

When the forty-seven species of mammals are placed in guilds, rather than being treated as taxa, there is a high degree of similarity among all hot desert sites, but significant differences persist between the cold desert (c) and hot deserts (fig. 6-5).

A further indication of the biological soundness of guilds follows from a comparison of each desert with its geographically closest desert or destroyed grassland (jp, cc, or sr). This comparison generally indicates no increase in similarity whether using taxa or guilds (table 6-1). If some distinctly grassland guilds (grazers) are added

Table 6-1. *Similarity between Desert Grasslands and Closest Deserts*

Sites Compared	Similarity Index By Species	By Guilds
j-jp	67	67
j-cc	67	79
t-cc	56	56
t-sr	63	55

Note: Figures represent percent similarity coefficient (Jaccard analysis) of each desert (altered) grassland with its geographically closest desert on the basis of both species composition of fauna and functional groups (guilds).

(table 6-2), similarity of grasslands to each other rises from low levels to those considered significant. Nonsimilarity was then a problem of not including enough specifically grassland guilds.

Comparison of levels of similarity is significant and explains the operational definition of adequate similarity. Using mean similarity values (mean ± standard error) calculated from data in figure 6-2, functional (guild) similarity among hot deserts is 81.87 ± 1.43 percent; grassland to other grasslands, 87.33 ± 0.33 percent; hot deserts to grasslands, 66.32 ± 1.88 percent; cold desert to hot desert, 61.0 ± 3.69 percent; and cold desert to grassland, 59.33 ± 3.85 percent. Three functional categories are clear: hot desert, cold desert, and grassland. Cross comparison of the means of functional categories versus nonfunctional ones demonstrates significant (.001 level) differences. The Behrens-Fisher modification of the *t* test and Cochran's approximation of *t'* were used because of the heterogeneity of sample variances (Snedecor and Cochran, 1967).

Table 6-2. *Similarity among Altered Grassland Sites*

| Sites Compared | Similarity Indexes | | |
	By Species	By Guild	By Guilds (including two grassland guilds)
jp-cc	63	63	87
cc-sr	60	52	88
sr-jp	52	60	87

Note: Figures represent percent similarity coefficient (Jaccard analysis) between destroyed or altered grassland sites, using species composition and functional units (guilds) and adding two specifically grassland guilds.

General Discussion

Causes of Species Mixes

An implication of the data presented here is that any one of a number of species of small mammals may be functionally similar in a particular example of desert scrub community. It is well known that various species of mice in the genera *Microdipidops*, *Peromyscus*, and *Perognathus* may have overlapping ranges and be desert adapted, but seem to replace each other in specific habitats of various soil-texture characteristics (sandy to rocky) (Hardy, 1945; Ryan, 1968; Ghiselin, 1970). Soil surface strength, and vegetation height and density, explain relative densities of some desert small mammals (Rosenzweig and Winakur, 1969; Rosenzweig, 1973). Interspecific behavior is part of the partitioning of habitats by a number of desert rodents: *Neotoma* (Cameron, 1971); *Dipodomys* (Christopher, 1973); and a seven-species "community" (MacMillan, 1964). Brown (1973) and Brown and Lieberman (1973) attribute species diversity of sand-dune rodents to a mix of ecological, biogeographic, and evolutionary factors, a position similar to that I take for a broader range of desert conditions.

Many other factors are also involved in defining various axes of specific niches (*sensu* Whittaker et al., 1973) of desert mammals; these specific niche differences do not exclude the functional overlap of other niche components. These examples and others merely show how finely genetically plastic organisms can subdivide the environment. These differences need not change the basic role of the species.

If the important shrub community function is seed removal, it does not matter to the community what species does it. The removal could be by any one of several rodents differing in soil-texture preferences or perhaps by a bird or even ants. Ecological equivalents may or may not be genetically close. All niche axes of a species or population are not equally important to the functioning of the community.

Importance of Functional Approach

Since all three hot deserts have similar functional diversity but vary in basic ways (e.g., more rain, more vegetational synusia, biseasonal rainfall pattern in the Sonoran desert as compared to the Mojave), functional diversity does not correlate well with those factors thought to relate animal species diversity to community structure—that is, vertical and horizontal foliage complexity (Pianka, 1967; MacArthur, 1972).

The crux of the problem is that most measures of species diversity include some measure of abundance (Auclair and Goff, 1971). The analysis herein counts only presence or absence—a level of abstraction more general than species diversity measures.

This greater generality is justified, I believe, because it strikes closer to the problem of comparing community types which are basically similar (e.g., hot deserts) but include units each having undergone specific development in time and space (e.g., Mojave vs. Sonoran vs. Chihuahuan hot deserts). The analysis seeks to elucidate various levels of the least-common denominators of community function.

The guild level of abstraction has potential applied value. The generalization of forty-seven mammal species into only twelve functional groups may permit the development of suitable predictive computer models of "North American Hot Deserts." The process of accounting for the vagaries of every species makes the task of modeling cumbersome and may prohibit rapid expansion of this ecological tool. The abstraction of a large number of species into biologically determined "black boxes" is an acceptable compromise.

There is clear evidence that temporal variations in the density of mammal species may so affect calculations of species diversity that their use and interpretation in a community context are difficult (M'Closkey, 1972). While such variations are intrinsically interesting biologically, they should not prevent us from seeking more generally applicable, albeit less detailed, predictors of community organization.

Guilds, Niches, Species, Stability

The guilds chosen here were selected on the basis of subjective familiarity with desert mammals. These guilds may not be requisites for the community as a whole to operate. If one were able to perceive the *requisite* guilds of a community, several things seem reasonable. First, to be stable (able to withstand perturbations without changing basic structure and function) a community cannot lose a requisite guild. Species performing functions provide a stable milieu if, first, they are themselves resilient to a wide range of perturbations (i.e., no matter what happens they survive) or, second, each requisite function can be performed by any one of a number of species, despite niche differences among these species—that is, the community contains a high degree of functional redundancy, preventing or reducing changes in community characteristics. The importance of this was alluded to by Whittaker and Woodwell (1972), who cited the case of oaks replacing chestnuts after they were wiped out by the blight in the North American eastern deciduous forests, and on theoretical grounds by Conrad (1972).

Any stable community is some characteristic mix of resilient and redundant species. Species diversity per se may not then correlate well with stability. Stability might come with a number of species diversities as long as the requisite guilds are represented. Tropics and deserts might represent extremes of a large number of redundant species as opposed to fewer more resilient species, both mixes conferring some level of stability as witnessed by the historical persistence of these community types.

None of this implies that all species are requisite to a community, or that coevolution is the only path for community evolution. Many species may be "tolerated" by communities just because the community is well enough buffered that minor species have no noticeable effect. As long as they pass their genetic make-up on to a new generation, species are successful; they need not do anything for the community.

Acknowledgments

These studies were made possible by a National Science Foundation grant (GB 32139) and are part of the contributions of the US/IBP Desert Biome Program. I am indebted to the following people for data collected by them or under their supervision: W. Whitford, E. L. Cockrum, K. L. Dixon, F. B. Turner, B. Maza, R. Anderson, and D. Balph. F. B. Turner and N. R. French kindly commented on a version of the manuscript.

Summary

The nonflying, small mammal faunas of western United States deserts were compared (coefficient of community) on the basis of their species and guild (functional) composition. Guilds were derived from information on animal size and food habits.

It is concluded that the similarity among sites with respect to guilds, though the species may differ, is a result of a complex of evolutionary events and particular contemporary community characteristics of the specific sites. Functional similarity, based on functional groups (guilds), is rather constant among hot deserts and different between hot deserts and either cold desert or desert grassland.

The functional analysis describes only a part of the niche of an organism, but perhaps an important part. Such abstractions and generalizations of the details of the community's complexities permit mathematical modeling to progress more rapidly and allow address to the general question of community "principles."

Guilds required by communities to maintain community integrity against perturbations may be better correlates to community stability than the various measures of species diversity currently popular.

References

Auclair, A. N., and Goff, F. G. 1971. Diversity relations of upland forests in the western Great Lakes area. *Am. Nat.* 105:499–528.

Baker, R. H. 1971. Nutritional strategies of myomorph rodents in North American grasslands. *J. Mammal.* 52:800–805.

Brown, J. H. 1973. Species diversity of seed-eating desert rodents in sand dune habitats. *Ecology* 54:775–787.

Brown, J. H., and Lieberman, G. A. 1973. Resource utilization and co-existence of seed-eating desert rodents in sand dune habitats. *Ecology* 54:788–797.

Cameron, G. N. 1971. Niche overlap and competition in woodrats. *J. Mammal.* 52:288–296.

Chew, R. M., and Chew, A. E. 1965. The primary productivity of a desert shrub (*Larrea tridentata*) community. *Ecol. Monogr.* 35:355–375.

——. 1970. Energy relationships of the mammals of a desert shrub *Larrea tridentata* community. *Ecol. Monogr.* 40:1–21.

Christopher, E. A. 1973. Sympatric relationships of the kangaroo rats, *Dipodomys merriami* and *Dipodomys agilis*. *J. Mammal.* 54:317–326.

Conrad, M. 1972. Stability of foodwebs and its relation to species diversity. *J. theoret. Biol.* 32:325–335.

Denyes, H. A. 1956. Natural terrestrial communities of Brewster County, Texas, with special reference to the distribution of mammals. *Am. Midl. Nat.* 55:289–320.

Emlen, J. M. 1973. *Ecology: An evolutionary approach*. Reading, Mass: Addison-Wesley.

Findley, J. S. 1969. Biogeography of southwestern boreal and desert mammals. *Univ. Kans. Publs Mus. nat. Hist.* 51:113–128.

Gardner, J. L. 1951. Vegetation of the creosotebush area of the Rio Grande Valley in New Mexico. *Ecol. Monogr.* 21:379–403.

Ghiselin, J. 1970. Edaphic control of habitat selection by kangaroo mice (*Microdipodops*) in three Nevadan populations. *Oecologia* 4:248–261.

Hagmeier, E. M., and Stults, C. D. 1964. A numerical analysis of the distributional patterns of North American mammals. *Syst. Zool.* 13:125–155.

Hardy, R. 1945. The influence of types of soil upon the local distribution of some small mammals in southwestern Utah. *Ecol. Monogr.* 15:71–108.

Lowe, C. H. 1964. Arizona landscapes and habitats. In *The vertebrates of Arizona*, ed. C. H. Lowe, pp. 1–132. Tucson: Univ. of Ariz. Press.

Lowe, C. H.; Wright, J. W.; Cole, C. J.; and Bezy, R. L. 1970. Natural hybridization between the teiid lizards *Cnemidophorus sonorae* (parthenogenetic) and *Cnemidophorus tigris* (bisexual). *Syst. Zool.* 19:114–127.

MacArthur, R. H. 1972. *Geographical ecology*. New York: Harper & Row.

M'Closkey, R. T. 1972. Temporal changes in populations and species diversity in a California rodent community. *J. Mammal.* 53:657–676.

MacMahon, J. A., and Trigg, J. R. 1972. Seasonal changes in an old-field spider community with comments on techniques for evaluating zoosociological importance. *Am. Midl. Nat.* 87:122–132.

MacMillan, R. E. 1964. Population ecology, water relations and social behavior of a southern California semidesert rodent fauna. *Univ. Calif. Publs Zool.* 71:1–66.

Miller, A. H., and Stebbins, R. C. 1964. *The lives of desert animals in Joshua Tree National Monument*. Berkeley and Los Angeles: Univ. of Calif. Press.

Odum, E. P. 1969. The strategy of ecosystem development. *Science, N.Y.* 164:262–270.

———. 1971. *Fundamentals of ecology*. 3d ed. Philadelphia: W. B. Saunders Co.

Oosting, H. J. 1956. *The study of plant communities*. 2d ed. San Francisco: Freeman Co.

Pianka, E. R. 1967. On lizard species diversity: North American flatland deserts. *Ecology* 48:333–351.

Root, R. B. 1967. The niche exploitation pattern of the blue-gray gnatcatcher. *Ecol. Monogr.* 37:317–350.

Rosenzweig, M. L. 1973. Habitat selection experiments with a pair of co-existing heteromyid rodent species. *Ecology* 54:111–117.

Rosenzweig, M. L., and Winakur, J. 1969. Population ecology of desert rodent communities: Habitats and environmental complexity. *Ecology* 50:558–572.

Ryan, R. M. 1968. *Mammals of Deep Canyon*. Palm Springs, Calif.: Desert Museum.

Snedecor, G. W., and Cochran, W. G. 1967. *Statistical methods*. 6th ed. Ames: Iowa State Univ. Press.

Whittaker, R. H., and Woodwell, G. M. 1972. Evolution of natural communities. In *Ecosystem structure and function*, ed. J. Wiens, pp. 137–156. Corvallis: Oregon State Univ. Press.

Whittaker, R. H.; Levin, S. A.; and Root, R. B. 1973. Niche, habitat, and ecotope. *Am. Nat.* 107:321–338.

Wiens, J. A. 1973. Pattern and process in grassland bird communities. *Ecol. Monogr.* 43:237–270.

Wilson, D. E. 1973. Bat faunas: A trophic comparison. *Syst. Zool.* 22:14–29.

7. The Evolution of Amphibian Life Histories in the Desert   Bobbi S. Low

Introduction

Among desert animals amphibians are especially intriguing because at first glance they seem so obviously unsuited to arid environments. Most amphibians require the presence of free water at some stage in the life cycle. Their skin is moist and water permeable, and their eggs are not protected from water loss by any sort of tough shell. Perhaps the low number of amphibian species that live in arid regions reflects this.

Some idea of the variation in life-history patterns which succeed in arid regions is necessary before examining the environmental parameters which shape those life histories. Consider three different strategies. Members of the genus *Scaphiopus* found in the southwestern United States frequent short-grass plains and alkali flats in arid and semiarid regions and are absent from high mountain elevations and extreme deserts (Stebbins, 1951). Species of the genus breed in temporary ponds and roadside ditches, often on the first night after heavy rains. *Scaphiopus bombifrons* lays 10 to 250 eggs in a number of small clusters; *Scaphiopus hammondi* lays 300 to 500 eggs with a mean of 24 eggs per cluster; and *S. couchi* lays 350 to 500 eggs in a number of small clusters. All three species burrow during dry periods of the year.

The genus *Bufo* is widespread, and a large number of species live in arid and semiarid regions. *Bufo alvarius* lives in arid regions but, unlike *Scaphiopus*, appears to be dependent on permanent water. Stebbins (1951) notes that, while summer rains seem to start seasonal activity, such rains are not always responsible for this activity.

The mating call has been lost. Also, unlike *Scaphiopus*, this toad lays between 7,500 and 8,000 eggs in one place at one time.

Eleutherodactylus latrans occurs in arid and semiarid regions and does not require permanent water. It frequents rocky areas and canyons and may be found in crevices, caves, and even chinks in stone walls. In Texas, Stebbins (1951) reported that *E. latrans* becomes active during rainy periods from February to May. About fifty eggs are laid on land in seeps, damp places, or caves; and, unlike either *Scaphiopus* or *Bufo*, the male may guard the eggs.

These three life histories diverge in degree of dependence on water, speed of breeding response, degree of iteroparity, and amount and kind of reproductive effort per offspring or parental investment (Trivers, 1972). All three strategies may have evolved, and at least are successful, in arid regions.

Two forces will shape desert life histories. The first is the relatively high likelihood of mortality as a result of physical extremes. Considerable work has been done on the mechanics of survival in amphibians which live in arid situations (reviewed by Mayhew, 1968). Most of the research concentrated on physiological parameters like dehydration tolerance, ability to rehydrate from damp soil, and speed of rehydration (Bentley, Lee, and Main, 1958; Main and Bentley, 1964; Warburg, 1965; Dole, 1967); water retention and cocoon formation (Ruibal, 1962a, 1962b; Lee and Mercer, 1967); and the temperature tolerances of adults and tadpoles (Volpe, 1953; Brattstrom, 1962, 1963; Heatwole, Blasina de Austin, and Herrero, 1968). Bentley (1966), reviewing adaptations of desert amphibians, gave the following list of characteristics important to desert species:

1. No definite breeding season
2. Use of temporary water for reproduction
3. Initiation of breeding behavior by rainfall
4. Loud voices in males, with marked attraction of both males and females, and the quick building of large choruses
5. Rapid egg and larval development
6. Ability of tadpoles to consume both animal and vegetable matter
7. Tadpole cannibalism
8. Production of growth inhibitors by tadpoles

  9. High heat tolerance by tadpoles
10. Metatarsal spade for burrowing
11. Dehydration tolerance
12. Nocturnal activity

Bentley's list consists mostly of physiological or anatomical characteristics associated with survival in the narrow sense. Only two or three items involve special aspects of life cycles, and some characteristics as stated are not exclusive to desert forms. Most investigators have emphasized the problems of survival for desert amphibians for the obvious reason that the animal and its environment seem so ill-matched, and most investigators have emphasized morphological and physiological attributes because they are easier to measure. A notable exception to this principally anatomical or physiological approach is the work of Main and his colleagues (Main, 1957, 1962, 1965, 1968; Main, Lee, and Littlejohn, 1958; Main, Littlejohn, and Lee, 1959; Lee, 1967; Martin, 1967; Littlejohn, 1967, 1971) who have discussed life-history adaptations of Australian desert anurans. Main (1968) has summarized some general life-history phenomena that he considered important to arid-land amphibians including high fecundity, short larval life, and burrowing. However, as he implied, the picture is not simple. A surprising variety of successful life-history strategies exists in arid and semiarid amphibian species, far greater than one would predict from attempts (Bentley, 1966; Mayhew, 1968) to summarize desert adaptations in amphibians. If survival were the critical focus of selection, one might predict fewer successful strategies and more uniformity in the kinds of life histories successful in arid-land amphibians.

But succeeding in the desert, as elsewhere, is a matter of balancing risk of mortality against optimization of reproductive effort so that realized reproduction is optimized. As soon as survival from generation to generation occurs, selection is then working on differences in reproduction among the survivors, an important point emphasized by Williams (1966a) in arguing that adaptations should most often be viewed as the outcomes of better-versus-worse alternatives rather than as necessities in any given circumstance. The focus I wish to develop here is on the critical parameters shaping the evolution of life-history strategies and the better-versus-worse alternatives in each of a number of situations. Adaptations of desert amphibians have scarcely been examined in this light.

Life-History Components and Environmental Parameters

Wilbur, Tinkle, and Collins (1974), in an excellent paper on the evolution of life-history strategies, list eight components of life histories: juvenile and adult mortality schedules, age at first reproduction, reproductive life span, fecundity, fertility, fecundity-age regression, degree of parental care, and reproductive effort. The last two are included in Trivers's (1972) concept of parental investment. For very few, if any, anurans are all these parameters documented.

I will concentrate here on problems of parental investment, facultative versus nonfacultative responses, cryptic versus clumping responses, and shifts in life-history stages. How does natural selection act on these traits in different environments? What environmental parameters are actually significant?

Classifying ranges of environmental variation may seem at first like a job for geographers; but, even when one acknowledges that deserts may be hot or cold, seasonal or nonseasonal, that they may possess temporary or permanent waters, different vegetation, and different soils, I think a few parameters can be shown to have overriding importance. These are (a) the *range* of the variation in the environmental attributes I have just described—temperature, humidity, day length, and so on; (b) the *predictability* of these attributes; and (c) their *distribution*—the patchiness or grain of the environment (Levins, 1968). Wilbur et al. (1974) consider trophic level and successional position also as life-history determinants in addition to environmental uncertainty. At the intraordinal level, these effects may be more difficult to sort out; and, at present, data are really lacking for anurans.

Range

Obviously the overall range of variation in environmental parameters is important in shaping patterns of behavior or life history. The same life-history strategy will not be equally successful in an environment where, for instance, temperature fluctuates only 5° daily, as in an environment in which fluctuations may be as much as 20° to 30°C. The range of fluctuations may strongly affect selection on physiological adaptations and differences in survival. Ranges of variation, particularly in temperature and water availability, are extreme in the environments of desert amphibians; but such effects have been dealt with more fully than the others I wish to discuss, and so I will concentrate on other factors.

Predictability

It is probably true that deserts are less predictable than either tropical or temperate mesic situations; Bentley's (1966) list of adaptations reflected this characteristic. The terms *uncertainty* and *predictability* are generally used for physical effects—seasonality and catastrophic events, for instance—but may include both spatial and biotic components. In fact, both patchiness and the distribution of predation mortality modify uncertainty.

Two aspects of predictability must be distinguished, for they affect the relative success of different life-history strategies quite differently. Areas may vary in reliability with regard to when or where certain events occur, such as adequate rainfall for successful breeding. Further, the suitability of such events may vary—a rain or a warm spell, whenever and wherever it may occur, may or may not be suitable for breeding. In a northern temperate environment the succession of the seasons is predictable. For a summer-breeding animal some summers will be better than others for breeding; this is reflected, for example, in Lack's (1947, 1948) results on clutch-size variation in English songbirds from year to year (see also Klomp, 1970; Hussell, 1972). Most summers, however, will be at least minimally suitable, and relatively few temperate—mesic-area organisms appear to have evolved to skip breeding in poor years. On the other hand, in most deserts rain is less predictable not only in regard to when and where it occurs, but also in regard to its effectiveness. Perhaps this latter aspect of environmental predictability has not been sufficiently emphasized in terms of its role in shaping life histories.

It is probably sufficient to distinguish four classes of environments with regard to predictability.

1. Predictable and relatively unchanging environments, such as caves and to a lesser extent tropical rain forests.
2. Predictably fluctuating or cyclic environments, areas with diurnal and seasonal periodicities, like temperate mesic areas.
3. Acyclic environments, unpredictable with reference to the timing and frequency of important events like rain, but predictable in terms of their effectiveness. If an event occurs, either it always is effective or the organism can judge the effectiveness.
4. Noncyclic environments that give few clues as to effectiveness of events: for example, rainfall erratic in spacing, timing, and amount. Areas like the central Australian desert present this situation for most frogs.

Optimal life-history strategies will differ in these environments, and desert amphibians must deal not only with extremes of temperature and aridity that seem contrary to their best interests but also with high degrees of unpredictability in those same environmental features and with localized and infrequent periods suitable for breeding.

Environmental uncertainty may have significant effects on shifts in life histories and on phenotypic similarities between life-history stages. If the duration of habitat suitable for adults is uncertain, or frequently less than one generation, the evolution of very different larval stages, not dependent on duration of the adult habitat, will be favored. The very fact that anurans show complex metamorphosis, with very different larval and adult stages, suggests this has been a factor in anuran evolution. Wilbur and Collins (1973) have discussed ecological aspects of amphibian metamorphosis and the role of un-certainty in the evolution of metamorphosis. An effect of complex metamorphosis is to increase independence of variation in the likeli-hoods of success in different life stages. Selective forces in the vari-ous habitats occupied by the different life stages are more likely to change independently of one another. As I will show later, this situa-tion has profound effects on life-cycle patterns.

Predictable seasonality will favor individuals which breed seasonal-ly during the most favorable period. Those who breed early in the good season will produce offspring with some advantage in size and feeding ability, and perhaps food availability, over the offspring of later breeders. Females which give birth or lay eggs early may, furthermore, increase their fitness and reduce their risk of feeding and improving their condition during the good season (Tinkle, 1969). Fisher (1958) has shown that theoretical equilibrium will be reached when the numbers of individuals breeding per day are normally distributed, if congenital earliness of breeding and nutritional level are also normally distributed. Predation (see below) on either eggs or breeding adults may cause amphibian breeding choruses to become clumped in space (Hamilton, 1971) and time. The timing, then, of the breeding peaks will depend on the balance between the time required after conditions become favorable for animals to attain breeding condition and the pressure to breed early. Both seasonal temperature and seasonal rainfall differences may limit breeding, and most amphibians in North American mesic areas and seasonally dry tropics (Inger and Greenberg, 1956; Schmidt and Inger, 1959) appear to breed seasonally.

In predictable unchanging environments, two strategies may be effective, depending on the presence or absence of predation. If no predation existed, individuals in "uncrowded" habitats would be selected to mature early and breed whenever they mature, maximizing egg numbers and minimizing parental investment per offspring, while individuals in habitats of high interspecific competition would be selected for the production of highly competitive offspring. That is, neither climatic change nor predation would influence selection, and MacArthur and Wilson's (1967) suggestion of r- and K- trends may hold. The result would be that adults would be found in breeding condition throughout the year. In a study by Inger and Greenberg (1963), reproductive data were taken monthly from male and female *Rana erythraea* in Sarawak. Rain and temperature were favorable for breeding throughout the year. From sperm and egg counts and assessment of secondary sex characters, they determined that varying proportions of both sexes were in breeding condition throughout the year. The proportion of breeding bore no obvious relation to climatic factors. Inger and Greenberg suggested that this situation represented the "characteristic behavior of most stock from which modern species of frogs arose." If predation exists in nonseasonal environments, year-round breeding with cryptic behavior may be successful; but if predation is erratic or predictably fluctuating (rather than constant), a "selfish herd" strategy may be favored.

Situations in which important events are unpredictable lead to other strategies. Life where the environment is unpredictable not only as to when or where events will occur but also as to whether or not they will be effective is comparable to playing roulette on a wheel weighted in an unknown fashion. Two strategies will be at a selective advantage:

1. Placing a large number of small bets will be favored, rather than placing a small number of large bets, or placing the entire bet on one spin of the wheel. In other words, in such an unpredictable situation, one expects iteroparous individuals who will lay a few eggs each time there is a rain. A corollary to this prediction is that, when juvenile mortality is unpredictable, longer adult life as well as iteroparity will be favored (cf. Murphy, 1968).

2a. Any strategy will be favored which will help an individual to judge the effectiveness of an event (i.e., to discover the weighting of the wheel). The central Australian species of Cyclorana— in fact, most of the Australian deep-burrowing frogs—may represent such a case. During dry periods, Cyclorana platycephalus, for instance, burrows three to four feet deep in clay soils. Light rains have no effect on dormant frogs even when rain occurs right in the area, since much of it runs off and does not percolate through to the level where the frogs are burrowed. Any rain reaching the frogs, we may suppose, is likely to be sufficient for tadpoles to mature and metamorphose. Thus, whatever functions (sensu Williams, 1966a) burrowing may serve in Cyclorana, one effect is that selective advantage accrues to those burrowing deeply because reproductive effort is not expended on unsuitable events.

2b. Any behavior which makes events less random, enhancing positive effects or reducing the effects of catastrophic events, will be favored. For example, parents may be favored who lay their eggs in some manner that tends to reduce the impact of flooding on their offspring, such as by laying their eggs out of the water and in rocky crevices or up on leaves or in burrows. A number of leptodactylid frogs do this (table 7-1). Obviously, such a strategy would only be favored when it had the effect of making mortality nonrandom. In deserts, where humidity is low and evapotranspiration high, it would not appear to be a particularly effective strategy; in fact only one *Eleutherodactylus* (Stebbins, 1951) and one species of *Pseudophryne* (Main, 1965) living in arid regions appear to follow strategies of hiding their eggs (table 7-1).

When the timing of events is unpredictable, but their effectiveness is not, individuals who only respond to suitable events will obviously be favored. This situation probably never exists a priori but only because organisms living in environments unpredictable both as to timing and effectiveness will evolve to respond only to suitable events, as in the burrowing *Cyclorana*. Thus, environments in the No. 4 category above will slowly be transformed into No. 3 environments by changes in the organisms inhabiting them. This emphasizes the importance of describing environments in terms of the organisms.

In the evolution of life cycles in uncertain environments, one kind of evidence of "learning the weighting of the wheel" is the capability of quickly exploiting unpredictable breeding periods—for example, ability to start a reproductive investment quickly after a desert rainfall. Another is the ability to terminate inexpensively an investment that has become futile, such as the care of offspring begun during a rainfall that turns out to be inadequate. These are adaptations over and above iteroparity as such, which is a simpler strategy.

Uncertainty and Parental Care. The effect of uncertainty on degree and distribution of parental investment varies with the type of unpredictability. Some kinds of uncertainty, such as prey availability, apparently can be ameliorated by increased parental investment. Types of uncertainty arising from biotic factors, rather than physical factors, comprise most of this category. Thus, vertebrate predators as a rule should show lengthened juvenile life and high degree of parental care because the biggest and best-taught offspring are at an advantage.

Uncertainties which are catastrophic or otherwise not density dependent appear to favor minimization or delay of parental investment such that the cost of loss at any point before the termination of parental care is minimized. The limited distribution of parental investment in desert amphibians supports this suggestion (fig. 7-3), and it appears to be true not only for anurans, in which parental care varies but is

Table 7-1. Habitat, Clutch, and Egg Sizes of Various Anurans

Species	Habitat[a]	Adult Size (mm)	Site of Deposition[c]	Number of Eggs[d]
Ascaphidae				
Ascaphus truei	1D	30–40	2b	28–50/
Leiopelma hochstetteri	8D		4f	6–18/
Pelobatidae				
Scaphiopus bombifrons	1B	35	2a	10–250/ 10–50
S. couchi	1B	80	2a	350–500/ 6–24
S. hammondi	1B	38	2a	300–500/24
Bufonidae				
Bufo alvarius	1B	180	1,2a	7,500– 8,000/
B. boreas	1F	95	2a	16,500/
B. cognatus	1G	85	1,2a,2b	20,000/
B. punctatus	1G	55	2a	30–5,000/ 1–few

[a]Habitat: 1 = North America   A = Temporary ponds
2 = Central America   B = Permanent water, xeric areas
3 = South America   C = Permanent water, mesic areas
4 = Europe   D = Permanent streams
5 = Asia   E = Caves
6 = Africa   F = Mesic   F+ = Cloud or tropical rain forest
7 = Australia   G = Grasslands, savannahs, or subhumid corridor
8 = New Zealand
[b]Size of adult female.

Egg Size (mm)	Time to Hatch (hours)	Time to Metamorphose (days)	Time to Mature (years)	Reference
4.0–5.0	720	365+		Noble and Putnam, 1931; Slater, 1934; Stebbins, 1951
	30 days[e]			
	<48	36–40		Stebbins, 1951
1.4–1.6	9–72	18–28		Ortenburger and Ortenburger, 1926; Stebbins, 1951; Gates, 1957
1.0–1.62	38–120	51		Little and Keller, 1937; Stebbins, 1951; Sloan, 1964
1.4		30		Stebbins, 1951; Mayhew, 1968
1.5–1.7	48			Stebbins, 1951
1.2	53	30–45		Stebbins, 1951
1.0–1.3	72	40–60		Stebbins, 1951

[c]Deposition site:	1 = Temporary ponds		3b = Burrows, not requiring rain to hatch	
	2a = Permanent ponds		4a = Terrestrial (seeps, etc.)	
	2b = Permanent streams		4b = On leaves above water	
	3a = Burrows, requiring rain to hatch		4c = On submerged leaves	
			5 = With parent: brood pouch, on back, etc.	

[d]When eggs are laid in several clusters, figures represent total number laid/number per cluster.
[e]Larval development completed in egg.
[f]Tending behavior.   [h]Tadpoles burrow to water.
[g](W): winter (S): summer.   [i]Female digs tunnel to water.

Species	Habitat[a]	Adult Size (mm)	Site of Deposition[c]	Number of Eggs[d]
B. woodhousei	1G	130	1	25,600/
B. compactilis	1G	70	1	
B. microscaphus	1B	65	1	several thousand
B. regularis	6G	65[b]	1	23,000
B. rangeri	6G	105[b]	1,2a	
B. carens	6G	74–92[b]		10,000
B. angusticeps	6	65		650–850
B. gariepensis	6	55		100+
B. vertebralis	6	30		
Ansonia muellari	5D	31[b]		150
Phrynomeridae				
Phrynomerus bifasciatus bifasciatus	6G	65	1,2a	400–1,500
Microhylidae				
Gastrophryne carolinensis	1,2	20	2a	850
G. mazatlanensis	3	20	4	175–200
Breviceps adspersus adspersus	6	38	4a,3b	20–46

[a]Habitat:
1 = North America   A = Temporary ponds
2 = Central America   B = Permanent water, xeric areas
3 = South America   C = Permanent water, mesic areas
4 = Europe   D = Permanent streams
5 = Asia   E = Caves
6 = Africa   F = Mesic   F+ = Cloud or tropical rain forest
7 = Australia   G = Grasslands, savannahs, or subhumid corridor
8 = New Zealand
[b]Size of adult female.

Egg Size (mm)	Time to Hatch (hours)	Time to Metamorphose (days)	Time to Mature (years)	Reference
1.0–1.5	48–96	34–60		Mayhew, 1968; Blair, 1972
1.4	48			Stebbins, 1951
1.75–1.9				Stebbins, 1951
1.0	24–48	72–143		Power, 1927; Wager, 1965; Stewart, 1967
1.3	96	35–42		Stewart, 1967
1.6	72–96			Stewart, 1967
2.0				Wager, 1965
2.2	48			Wager, 1965
<1.0				Wager, 1965
2.15				Inger, 1954
.3–1.5	96	30		Stewart, 1967
	48	20–70	2	Stebbins, 1951
.2–1.4				Stebbins, 1951
1.5	28–42 days[e]			Wager, 1965

[c]Deposition site: 1 = Temporary ponds   3b = Burrows, not requiring rain to hatch
2a = Permanent ponds   4a = Terrestrial (seeps, etc.)
2b = Permanent streams   4b = On leaves above water
3a = Burrows, requiring   4c = On submerged leaves
rain to hatch   5 = With parent: brood pouch, on back, etc.

[d]When eggs are laid in several clusters, figures represent total number laid/number per cluster.
[e]Larval development completed in egg.
[f]Tending behavior.
[g](W): winter (S): summer.

[h]Tadpoles burrow to water.
[i]Female digs tunnel to water.

Species	Habitat[a]	Adult Size (mm)	Site of Deposition[c]	Number of Eggs[d]
B. a. pentheri	6	38	4a,3b	20
Hypopachus variolosus	2G	29–53[b]	1	30–50

Ranidae

Pyxicephalus adspersus	6G	115	1	3,000–4,000
P. delandii	6G	65	1	2,000–3,000
P. natalensis	6G	51	1	hundreds / 1–6
Ptychadena anchietae	6G	48–58[b]	1	200–300
P. oxyrhynchus	6G	57[b]	1	300–400
P. porosissima	6G	44[b]	1	?/1
Hildebrandtia ornata	6G	63.5	2	?/1
Rana fasciata fuellborni	6C	44.5[b]	4a	64/1–12
R. f. fasciata	6G	51	1,2	?/1
R. angolensis	6	76		thousands
R. fuscigula	6	127	2	1,000–15,000
R. wageri	6	51[b]	4c	120–1,000/ 12–100

[a]Habitat: 1 = North America   A = Temporary ponds
            2 = Central America   B = Permanent water, xeric areas
            3 = South America   C = Permanent water, mesic areas
            4 = Europe   D = Permanent streams
            5 = Asia   E = Caves
            6 = Africa   F = Mesic   F+ = Cloud or tropical rain forest
            7 = Australia   G = Grasslands, savannahs, or subhumid corridor
            8 = New Zealand
[b]Size of adult female.

Egg Size (mm)	Time to Hatch (hours)	Time to Metamorphose (days)	Time to Mature (years)	Reference
5.0	28–42 days[e]			Wager, 1965
	24			Wager, 1965
2.0	48	49		Stewart, 1967
1.5	72	35		Wager, 1965
1.2	96			Wager, 1965; Stewart, 1967
1.0	30			Wager, 1965; Stewart, 1967
1.3	48	42–56		Wager, 1965
1.0	48			Wager, 1965
1.4				Wager, 1965
2.0–3.0		730		Stewart, 1967
1.65		28–35		Wager, 1965
1.5	168			Wager, 1965
1.5	168–240	1,095		Wager, 1965
2.8	192–216			Wager, 1965

[c]Deposition site:

1	= Temporary ponds	3b	= Burrows, not requiring rain to hatch
2a	= Permanent ponds	4a	= Terrestrial (seeps, etc.)
2b	= Permanent streams	4b	= On leaves above water
3a	= Burrows, requiring rain to hatch	4c	= On submerged leaves
		5	= With parent: brood pouch, on back, etc.

[d]When eggs are laid in several clusters, figures represent total number laid/number per cluster.
[e]Larval development completed in egg.
[f]Tending behavior.
[g](W): winter (S): summer.

[h]Tadpoles burrow to water.
[i]Female digs tunnel to water.

Species	Habitat[a]	Adult Size (mm)	Site of Deposition[c]	Number of Eggs[d]
R. grayi	6B	45	3a	few hundred/1–few
R. catesbiana	1	205	2	10,000–25,000
R. pipiens	1	90	2	1,200–6,500
R. temporaria	1		1	1,500–4,000
R. tarahumarae	1,2	115	1,2	2,200
R. aurora aurora	1C	102	2a	750–1,300
R. a. cascadae	1C	95	2a	425
R. a. dratoni	1C	95	2a	2,000–4,000
R. boylei	1B,C	70	2a, b	900–1,000
R. clamitans	1B,C	102	2a, b	1,000–5,000
R. pretiosa pretiosa	1F	90	2	1,100–1,500
R. p. lutiventris	1F	90	2	2,400
R. sylvatica	1F	60	2a,1	2,000–3,000
Phrynobatrachus natalensis	6G	28–30[b]	1	200–400/25–50
P. ukingensis	6G	16[b]	1	
Anhydrophryne rattrayi	6	20[b]	3b	11–19
Natalobatrachus bonegergi	6F	38	4b	75–100
Arthroleptis stenodactylus	6	29–44[b]	3b	100/33

[a]Habitat: 1 = North America   A = Temporary ponds
2 = Central America   B = Permanent water, xeric areas
3 = South America   C = Permanent water, mesic areas
4 = Europe   D = Permanent streams
5 = Asia   E = Caves
6 = Africa   F = Mesic   F+ = Cloud or tropical rain forest
7 = Australia   G = Grasslands, savannahs, or subhumid corridor
8 = New Zealand
[b]Size of adult female.

Egg Size (mm)	Time to Hatch (hours)	Time to Metamorphose (days)	Time to Mature (years)	Reference
1.5	5–10	90–120		Wager, 1965
1.3	4–5	120–365	2–3	Stebbins, 1951
1.7	312–480	60–90	1–3	Stebbins, 1951
	336–504	90–180	3–5	Stebbins, 1951
2–2.2				Stebbins, 1951
3.04	192–480		3–4	Stebbins, 1951
2.25	192–480			Stebbins, 1951
2.1	192–480			Stebbins, 1951
2.2		90–120		Stebbins, 1951
1.5	72–144	90–360		Stebbins, 1951
2–2.8	96		2+	Stebbins, 1951
1.97				Stebbins, 1951
1.7–1.9	336–504	90		Stebbins, 1951
1.0(W)[g] 0.7(S)	48	28		Wager, 1965; Stewart, 1967
0.9		35		Stewart, 1967
2.6	28 days[e]			Wager, 1965
2.0	144–240	270		Wager, 1965; Stewart, 1967
2.0		e		Stewart, 1967

[c]Deposition site: 1 = Temporary ponds   3b = Burrows, not requiring rain to hatch
                      2a = Permanent ponds   4a = Terrestrial (seeps, etc.)
                      2b = Permanent streams 4b = On leaves above water
                      3a = Burrows, requiring 4c = On submerged leaves
                           rain to hatch   5 = With parent: brood pouch, on back, etc.

[d]When eggs are laid in several clusters, figures represent total number laid/number per cluster.
[e]Larval development completed in egg.
[f]Tending behavior.   [h]Tadpoles burrow to water.
[g](W): winter (S): summer.   [i]Female digs tunnel to water.

Species	Habitat[a]	Adult Size (mm)	Site of Deposition[c]	Number of Eggs[d]
A. wageri	6	25	3b	11–30
Arthroleptella lightfooti	6F	20	3b	40/5–8
A. wahlbergi	6F	28	3b	11–30
Cacosternum n. nanum	6G	20	4c	8–25/5–8
Chiromantis xerampelina	6F	60–87[b]	4b	150
Hylambates maculatus	6B	54–70[b]	4c	few hundred/1
Kassina wealii	6	40	1,4c	500/1
K. senegalensis	6	35–43[b]	1	400/1–few
Hemisus marmoratum	6F	38[b]	3fh	200
H. guttatum	6F	64[b]	3fi	2,000
Leptopelis natalensis	6F	64	4a	200
Afrixalus spinifrons	6F	22	4c	?/10–50
A. fornasinii	6F	30–40[b]	4b	40
Hyperolius punticulatus	6	32–43[b]	1	?/19
H. pictus	6	22.8–38[b]	4b	?/60–90
H. tuberilinguis	6	36–39[b]	4b	350–400

[a]Habitat:
1 = North America   A = Temporary ponds
2 = Central America   B = Permanent water, xeric areas
3 = South America   C = Permanent water, mesic areas
4 = Europe   D = Permanent streams
5 = Asia   E = Caves
6 = Africa   F = Mesic   F+ = Cloud or tropical rain forest
7 = Australia   G = Grasslands, savannahs, or subhumid corridor
8 = New Zealand
[b]Size of adult female.

Egg Size (mm)	Time to Hatch (hours)	Time to Metamorphose (days)	Time to Mature (years)	Reference
2.5	28 days[e]			Wager, 1965
4.5	10 days[e]			Stewart, 1967
2.5	e			Wager, 1965
0.9	48	5		Wager, 1965
1.8	120–144			Wager, 1965
1.5	144	300		Wager, 1965; Stewart, 1967
2.4	144	60		Wager, 1965; Stewart, 1967
1.5	144	90		Stewart, 1967
2.0	240			Wager, 1965
2.5				Wager, 1965
3.0				Wager, 1965
1.2	168	42		Wager, 1965
1.6–2.0				Wager, 1965; Stewart, 1967
2.5				Stewart, 1967
2.0	432	56		Stewart, 1967
1.3–1.5	96–120	60		Stewart, 1967

[c]Deposition site: 1 = Temporary ponds   3b = Burrows, not requiring rain to hatch
2a = Permanent ponds   4a = Terrestrial (seeps, etc.)
2b = Permanent streams   4b = On leaves above water
3a = Burrows, requiring   4c = On submerged leaves
rain to hatch   5 = With parent: brood pouch, on back, etc.
[d]When eggs are laid in several clusters, figures represent total number laid/number per cluster.
[e]Larval development completed in egg.
[f]Tending behavior.
[g](W): winter (S): summer.
[h]Tadpoles burrow to water.
[i]Female digs tunnel to water.

Species	Habitat[a]	Adult Size (mm)	Site of Deposition[c]	Number of Eggs[d]
H. pusillus	6	17–21[b]	1,2a	500/1–76
H. nasutus nasutus	6	20.6–23.8[b]	2	200/2–20
H. marmoratus nyassae	6	29–31[b]	2a	370
H. horstocki	6		2a	?/10–30
H. semidiscus	6	35	2a, 4c	200/30
H. verrucosus	6	29	2a	400/4–20

Leptodactylidae

Eleutherodactylus rugosus	2G		1	several thousand
Limnodynastes tasmaniensis	7F	39.4[b]		1,100
L. dorsalis dumerili	7F	61.5[b]		3,900
Leichriodus fletcheri	7	46.5[b]		300
Adelotus brevus	7	33.5[b]		270
Philoria frosti	7	49.2[b]		95
Helioporus albopunctatus	7	73.3[b]		480
H. eyrei	7F	54.0[b]		265–270
H. psammophilis	7	42–52		160

[a]Habitat:
  1 = North America   A = Temporary ponds
  2 = Central America   B = Permanent water, xeric areas
  3 = South America   C = Permanent water, mesic areas
  4 = Europe   D = Permanent streams
  5 = Asia   E = Caves
  6 = Africa   F = Mesic   F+ = Cloud or tropical rain forest
  7 = Australia   G = Grasslands, savannahs, or subhumid corridor
  8 = New Zealand
[b]Size of adult female.

Egg Size (mm)	Time to Hatch (hours)	Time to Metamorphose (days)	Time to Mature (years)	Reference
1.4–1.5	120	42		Stewart, 1967
0.8–2.2	120			Wager, 1965; Stewart, 1967
2.0	192			Wager, 1965; Stewart, 1967
1.0				Wager, 1965
1.0	108	60		Wager, 1965
1.3				Wager, 1965
4.0	24			Wager, 1965
1.47				Martin, 1967
1.7				Martin, 1967
1.7				Martin, 1967
1.5				Martin, 1967
3.9				Martin, 1967
2.75				Main, 1965; Lee, 1967
2.50–3.28				Main, 1965; Lee, 1967; Martin, 1967
3.75				Lee, 1967

cDeposition site:
1	= Temporary ponds	3b	= Burrows, not requiring rain to hatch
2a	= Permanent ponds	4a	= Terrestrial (seeps, etc.)
2b	= Permanent streams	4b	= On leaves above water
3a	= Burrows, requiring rain to hatch	4c	= On submerged leaves
		5	= With parent: brood pouch, on back, etc.

dWhen eggs are laid in several clusters, figures represent total number laid/number per cluster.
eLarval development completed in egg.
fTending behavior.
g(W): winter (S): summer.

hTadpoles burrow to water.
iFemale digs tunnel to water.

Species	Habitat[a]	Adult Size (mm)	Site of Deposition[c]	Number of Eggs[d]
H. barycragus	7	68–80		430
H. inornatus	7	55–65		180
Crinea rosea	7F	24.8[b]		26–32
C. leai	7F	21.1[b]		52–96
C. georgiana	7F	21.1[b]		70
C. insignifera	7F	19–21[b]		

Hylidae

Species	Habitat[a]	Adult Size (mm)	Site of Deposition[c]	Number of Eggs[d]
Hyla arenicolor	1	37	1	several hundred/1
H. regilla	1	55	1	500–1,250/ 20–25
H. versicolor	1		2	1,000–2,000
H. verrucigera	2		1	200
H. lancasteri	2F+	41.1[b]	2b,4b	20–23
H. myotympanum	2F+	51.6[b]		120
H. thorectes	2F+	70[b]	2	10
H. ebracata	2F+	36.5[b]	4b	24–76
H. rufelita	2F	60[b]	2	75–80
H. loquax	2F	45[b]	2	250
H. crepitans	2G	52.6[b]	1	
H. pseudopuma	2F	44.2[b]	4b	?/10
H. tica	2F+	38.9		

[a]Habitat: 1 = North America   A = Temporary ponds
2 = Central America   B = Permanent water, xeric areas
3 = South America   C = Permanent water, mesic areas
4 = Europe   D = Permanent streams
5 = Asia   E = Caves
6 = Africa   F = Mesic   F+ = Cloud or tropical rain forest
7 = Australia   G = Grasslands, savannahs, or subhumid corridor
8 = New Zealand
[b]Size of adult female.

Egg Size (mm)	Time to Hatch (hours)	Time to Metamorphose (days)	Time to Mature (years)	Reference
2.60				Lee, 1967
3.75				Lee, 1967
2.35	60+ days[e]			Main, 1957
1.66–2.03	149–174 days[e]		2	Main, 1957
0.97–1.3	130+ days[e]		1	Main, 1957
				Main, 1957
2.1		40–70		Stebbins, 1951
4.3	168–336		2	Stebbins, 1951
	96–120	45–65	1–3	Stebbins, 1951
2.0		89		Trueb and Duellman, 1970
5.0				Duellman, 1970
2.25				Duellman, 1970
1.22				Duellman, 1970
1.2–1.4				Duellman, 1970; Villa, 1972
1.8				Villa, 1972
				Villa, 1972
1.8				Villa, 1972
1.71	24	65–69		Villa, 1972
2.0				Villa, 1972

[c]Deposition site: 1 = Temporary ponds   3b = Burrows, not requiring rain to hatch
2a = Permanent ponds   4a = Terrestrial (seeps, etc.)
2b = Permanent streams   4b = On leaves above water
3a = Burrows, requiring   4c = On submerged leaves
rain to hatch   5 = With parent: brood pouch, on back, etc.

[d]When eggs are laid in several clusters, figures represent total number laid/number per cluster.
[e]Larval development completed in egg.
[f]Tending behavior.
[h]Tadpoles burrow to water.
[g](W): winter (S): summer.
[i]Female digs tunnel to water.

Species	Habitat[a]	Adult Size (mm)	Site of Deposition[c]	Number of Eggs[d]
Agalychnis colli-dryas	2	71	4*b*	40–110/ 11–78
A. annae	2	82.9[b]	4*b*	47–162
A. calcarifer	2	65.0[b]	4*b*	16
Smilisca cyanosticta	2F+	70[b]	2	1,147
S. baudinii	2G	76–90	1	2,620–3,32●
S. phaeola	2G	80		1,870–2,01●
Pachymedusa dacnicolor	2G	103.6[b]	4*b*	100–350
Hemiphractus panimensis	2F	58.7[b]	5[f]	12–14
Gastrotheca ceratophryne	2F	74.2	5[f]	9
Centrolenellidae				
Centrolenella fleischmanni	2F	19.2	4*b*	17–28

[a]Habitat: 1 = North America   A = Temporary ponds
             2 = Central America B = Permanent water, xeric areas
             3 = South America   C = Permanent water, mesic areas
             4 = Europe   D = Permanent streams
             5 = Asia   E = Caves
             6 = Africa   F = Mesic   F+ = Cloud or tropical rain forest
             7 = Australia   G = Grasslands, savannahs, or subhumid corridor
             8 = New Zealand
[b]Size of adult female.

generally low, but also for groups with high parental care, such as mammals. For example, marsupials have flourished in uncertain desert environments in central Australia where indigenous and introduced eutherians have not, even though the eutherian species prevail in areas of more predictable climate. In uncertain areas a premium

Egg Size (mm)	Time to Hatch (hours)	Time to Metamorphose (days)	Time to Mature (years)	Reference
2.3–5.0	96–240	50–80		Duellman, 1970; Villa, 1972
3.41				Villa, 1972
3.5				Villa, 1972
.22				Duellman, 1970
.3				Trueb and Duellman, 1970
				Duellman, 1970
				Duellman, 1970
5.0				Duellman, 1970
2.0				Duellman, 1970
.5	24	9		Villa, 1972

c Deposition site:
- 1 = Temporary ponds
- 2a = Permanent ponds
- 2b = Permanent streams
- 3a = Burrows, requiring rain to hatch
- 3b = Burrows, not requiring rain to hatch
- 4a = Terrestrial (seeps, etc.)
- 4b = On leaves above water
- 4c = On submerged leaves
- 5 = With parent: brood pouch, on back, etc.

d When eggs are laid in several clusters, figures represent total number laid/number per cluster.
e Larval development completed in egg.
f Tending behavior.
g (W): winter (S): summer.
h Tadpoles burrow to water.
i Female digs tunnel to water.

is set on strategies which will make breeding response facultative and reduce the cost of loss of offspring at any point. Facultative, rather than seasonal, delayed implantation (Sharman, Calaby, and Poole, 1966) and anoestrus condition during drought (Newsome, 1964, 1965, 1966) are examples. Also, I think, is the shape of the parental investment curve for marsupials, which is depressed to a remarkable degree in the initial stages (my unpublished data). This whole constellation of attributes provides facultativeness of response, capabilities for quick initiation of new investments, and less expense of termination at any point. While the classical arguments about marsupial proliferation in Australia have claimed that introduced eutherians "outcompete" marsupials (Frith and Calaby, 1969), they are probably able to do so only because they evolved their reproductive behavior in other kinds of environments. Most Australian environments may have consistently favored marsupialism over any step-by-step transitions toward placentalism. It may be worthwhile to reexamine the question in the light of a new framework.

Distribution

A third important environmental aspect is patchiness or graininess. Wet tropical areas, seemingly ideal from an amphibian's point of view, are basically rather fine grained environments. For instance, ponds, fields, and forest areas may interdigitate so that a single frog spends some time in each and may spend time in more than one pond. From an amphibian's point of view, most deserts are comparatively coarse grained. This does not mean that all the environmental patches are physically large (as may be implied in Levins's [1968] discussion) but that the suitable patches, of whatever size, are likely to be separated by large unsuitable or uninhabitable areas. Thus an individual is likely to spend its entire life in the same patch. For amphibians, widely separated permanent water holes in desert environments are islands and subject to the same selective pressures (MacArthur and Wilson, 1967).

Degrees of patchiness will have two major sorts of effects, on divergence rates and life-history strategies. In a coarse-grained or island model, as in the desert I have described, rates of speciation and extinction will both be higher than in a fine-grained environment. Thus, in some uncertain environments, if they are continually minimally inhabitable and also coarse grained, speciation and extinction rates, contrary to Slobodkin and Sanders's (1969) prediction, may be higher than in predictable environments, if those predictable areas are fine grained. This point, not considered by Slobodkin and Sanders, was raised by Lewontin (1969). Environmental uncertainty will affect populations in the coarse-grained situation much more than those in the fine-grained areas to the extent that there are differences in population sizes and isolation of populations. Slobodkin and Sanders considered only predictability, but predictability and patchiness, and their interaction, will influence the rate of speciation.

In very coarse grained models, because isolation is much more complete than in the fine-grained situation, immigration and emigration may be virtually nonexistent. The number of species in any suitable grain at any time will depend on infrequent past immigrations and will be lower than in the fine-grained model. Selection will be strong on several parameters, to be discussed below, but may be relaxed on characters, such as premating isolating mechanisms. Selection on these characters will be strongest in the fine-grained model where the number of sympatric species is higher. The desert coarse-grained situation is a model for the occurrence of character release (MacArthur and Wilson, 1967; Grant, 1972): populations founded by few individuals and on which selection on interspecific discrimination is relaxed. Thus, in the isolated desert populations described, one might predict that the variations in call characters (in males) and in call discrimination (in females) would be greater.

The distribution of suitable resources and the duration of this distribution will affect strategies of dispersal and competition. While density-dependent effects will operate here, the "r" and "K" parameters of Pianka (1970) and others are not sufficient indicators—a point made by Wilbur et al. (1974) for other groups of organisms.

Consider a pond suitable for breeding: it may be effectively isolated from other suitable areas, or other good ponds may be close or easy to reach. Dispersal ability will evolve to the degree that the cost-benefit ratio is favorable between the relative goodness of another pond and the risk incurred in getting there. Goodness relative to the home pond may be measured by a number of criteria: physical parameters, amount of competition from other species, and other conspecifics (Wilbur et al., 1974), amount of predation, and so on. The cost of reaching another pond and the probability of success in doing so may be correlated with distance, but other classic "barriers" (mountains, very dry areas) are also relevant. Both distance and barriers of low humidity and little free water are likely to be greater in arid regions than in tropical and temperate mesic areas.

If ponds are not totally isolated from each other and are relatively unchanging in "value," migration strategies will be more favored in finer-grained areas because the cost of migration is lower. If ponds are not isolated from each other, and their relative values fluctuate, the evolution of emigration strategies will depend in part on the persistence of ponds relative to the generation length of the frog. If ponds are temporary, and others are likely to be available, migration will be advantageous. The longer ponds last, the closer the situation approaches the "permanent pond" situation, where migration will be favored only in periods of high local population density. Some invertebrate groups, such as migratory locusts and crickets (Alexander, 1968), show phenotypic flexibility supporting this generalization; they increase the proportion of long-winged migratory offspring as the habitat deteriorates and in periods of high population density. Frog morphology does not alter in a comparable way, but dispersal behavior may show flexibility. I know of no pertinent data or studies, however.

In good patches like permanent waters, isolated from others, emigration will be disfavored. Increased parental investment will be favored only when it increases predictability in ways relevant to offspring success. Examination of table 7-1 shows that species with parental care and species laying large-yolked eggs occur in tropical and temperate areas but not generally in unpredictable areas. Since some of these species lay foamy masses not permeable to water, the aridity of desert areas alone is not sufficient to explain this distribution of strategies.

Two arid-region species do show parental investment in the form of larger or protected eggs. As previously described, *Eleutherodactylus latrans* females lay about fifty large eggs of 6–7.5 mm diameter on land or in caves (table 7-1; Stebbins, 1951); the males may guard the eggs. Since this frog lives largely in caves and rocky crevices, the microenvironment is far more stable and predictable than the zoogeography would suggest. The Australian *Pseudophryne occidentalis* lives by permanent waters with muddy rather than sandy soils. Eggs are laid in mud burrows near the edge of the water (Main, 1965). In both cases it appears that the nature of mortality is such that increased parental investment is successful. This may be related to the relatively higher physical stability of the microhabitat when compared to desert environments in general. The proportion of mortality due to catastrophes which parental care is ineffective to combat is relatively lower.

Mortality

Mortality may arise from a number of factors: foot shortages, predators (including parasites and diseases), and climate. An important consideration in what life-history strategy will prevail is whether the mortality is random (unpredictable) or nonrandom (predictable). Any cause of mortality could be either random or nonrandom in its effects, but mortality from biotic causes is probably less often random than mortality from physical factors and may be more effectively countered by strategies of parental investment.

Catastrophic mortality, which is essentially random rather than selective (even though it may be density dependent), will be more frequent in the coarse-grained desert environments I have described than in the tropics. An example would be heavy sudden floods which frequently occur after heavy rains in areas like central Australia and the southwestern United States. This kind of flood may wash eggs, tadpoles, and adults to flood-out areas which then dry up. The result may be devastating sporadic mortality for populations living in the path of such floods. Further, in terms of the animals themselves, environments may be predictable for certain stages in the life history and unpredictable for others. In animals like amphibians with complex metamorphosis, this difference can be particularly significant.

If any stage encounters significant uncertainty, one of two strategies should evolve: physical avoidance, such as hiding or development of protection in that stage, or a shift in life history to spend minimal time in the vulnerable stage (table 7-2). If survivorship is high for adults but uncertain and sometimes very low for tadpoles, one predicts strategies of: (a) long adult life, iteroparity, and reduced investment per clutch; (b) long egg periods and short tadpole periods; or (c) increased parental investment through hiding or tending behavior. Evolution of behavior like that of *Rinoderma darwini* may result from such pressure. The males appear to guard the eggs; when development reaches early tadpole stage, the males snap up the larvae, carrying them in the vocal sac until metamorphosis. Perhaps the extreme case is represented by the African *Nectophrynoides*, in which birth is viviparous.

In temporary waters in desert environments much uncertainty will be concentrated on aquatic stages, and two principal strategies should be evident in desert amphibians: increased iteroparity, longer adult life, and lower reproductive effort per clutch; and shifts in time spent in different stages, reducing time spent in the vulnerable stages. Short, variable lengths in egg and juvenile stages (table 7-1) will result.

Even in climatically more predictable areas, uncertainty of mortality may be concentrated on one stage. In some temperate urodele forms, Salthe (1969) suggested that success at metamorphosis correlated with size—that larger offspring were more successful. This in turn selected for lengthened time spent in aquatic stages.

Some generalizations are apparent from table 7-2. The important differences appear to be between uncertainty in juvenile stages and adult stages. All conditions of uncertain adult survival will lead to concentration of reproductive effort in one or a few clutches (semelparity or reduction of iteroparity). Uncertainty of survivorship in adult stages when combined with high predictability in juvenile stages may lead to the extreme conditions of neoteny and paedogenesis. Uncertainty in either or both juvenile stages leads to increased iteroparity and reduced reproductive effort per clutch.

Predation

Because predation is usually nonrandom, its effects on prey life histories will frequently differ from the effects of climate and other sources of mortality. An important point frequently overlooked is that, because predation and competition arise from biotic components of the system, they are not simply subsets of uncertainty. Their effects are more thoroughly related to density-dependent parameters. Some strategies will be effective which would not be advantageous in situations rendered uncertain solely by physical factors. Consider predation: strategies frequently effective in reducing predation-caused un-

Table 7-2. *Relative Uncertainty in Different Life-History Stages and Strategies of Selective Advantage*

Likelihood of Survival			Strategy
Egg	Tadpole	Adult	
high	high	low	semelparity or reduced iteroparity; large numbers of small eggs; no parental care
low	high	low	semelparity or reduced iteroparity; neoteny; quick hatching
high	low	low	semelparity or reduced iteroparity; large numbers of small eggs; no parental care; quick metamorphosis
high	low	high	iteroparity; large eggs, fewer eggs; avoidance of aquatic tadpole stage; parental care of tadpoles
low	high	high	iteroparity; tending, hiding of eggs; fewer eggs; viviparity
low	low	high	iteroparity; parental care, tending strategies; viviparity

certainty are those of spatial (Hamilton, 1971) and temporal clumping, increased parental investment (Trivers, 1972), and allelochemical effects. These strategies would be far less effective in increasing predictability of an environment rendered uncertain by physical factors.

Predation pressure may lead to hiding or tending eggs and consequent lowering of clutch size. Whether this is true or whether responses of increased fecundity (Porter, 1972; Szarski, 1972) prevail will depend on the nature of the predation. In the unusual case of a predator whose effect is limited, such as one which could eat no more than x eggs per nest, parents would gain by increased fecundity, making $(x+2)$ rather than $(x+1)$ eggs. However, m, the genotypic rate of increase, will be higher for these more fecund genotypes even in the absence of predation. Further, an increase in numbers of eggs laid implies either smaller eggs (in which case the predator may be able to eat $[x+2]$ eggs) or an increase in the size of the parent. In most cases, high fecundity carries a greater risk under increased predation—for example, by laying more eggs which are then lost or, in species like altricial birds with parental care, by incurring greater risk attempting to feed more offspring if they are not protected. In these cases, lowered fecundity and increased parental investment in caring for fewer eggs will be favored.

The strategies of hiding or protection and life-history shifts, which may follow from increased uncertainty in any stage, are also favored in the special case of uncertainty induced by predation. Predation concentrated on certain stages in the life cycle—on eggs, tadpoles, newly metamorphosed animals, or breeding adults—may lead to (a) quick hatching, tending, or hiding of eggs, as in *Scaphiopus* or *Helioporus* (table 7-1); (b) quick metamorphosis or tending of tadpoles, as in *Rhinoderma*; (c) cryptic behavior by newly metamorphosed animals (many species) or lengthened egg or tadpole stages with consequent greater size (and possibly reduced predation vulnerability) on metamorphosis, as in *Rana catesbiana* (table 7-1); or (d) cryptic behavior by adults or very clumped patterns of breeding behavior.

Length of the breeding season may also be strongly affected by the presence of predation. In fact, I think that the general shape of breeding-curve activities of many vertebrates may be related to predation. Fisher (1958) has shown that, if there is an optimal breeding time, a symmetrical curve will result. While restriction of resource availability, such as food or breeding resources, limits the seasonality of breeding and produces some clumping, such seasonal differences seem not to be sharp enough to explain the extreme temporal clumping of breeding and birth in many species. Temporary ponds of very short duration in arid regions are commonly assumed to show clumping for climatic reasons, but this is not certain; at any rate, the addition of predation to such a system should follow the same pattern as in any seasonal situation. In seasonal conditions a breeding-activity or birth curve may approach a normal curve, perhaps with a slight right-hand skew because earlier birth will give a size and food advantage to offspring and a risk advantage to parents. When predation on breeding adults or new young exists, however, two other pressures may cause both an increased right-hand skew and a sharper peak:

1. The advantage to those individuals which have offspring early before a generalized predator develops a specific search image.
2. The advantage to those individuals which breed and give birth or lay eggs when everyone else does—when, in other words, the predator food market is flooded. This constitutes a temporal "selfish herd" effect (Hamilton, 1971). Thus, if seasonality of resource availability exists so that thoroughly cryptic breeding is not of advantage, the curve of breeding or birth activity will tend under predation pressure to shift from a fairly normal distribution to a kurtotic curve with an abrupt beginning shoulder and a gentler trailing edge.

Despite their importance, predation effects on life histories have largely been ignored. This may be, in part, because the physical factors are so extreme that it seems sufficient to examine their effects on amphibian physiology and survival. Another reason predation effects may be slighted is that one ordinarily sees the end product of organisms which evolved with predation pressure, and the present-day descendents represent the most successful of the antipredation strategies. As a simple example, consider the large variety of substances found in the skin of most amphibians (Michl and Kaiser, 1963). A great variety exists, including such disparate compounds as urea, the bufadienolides, indoles, histamine derivatives, and polypeptides like caerulein (Michl and Kaiser, 1963; Erspamer, Vitali, and Roseghini, 1964; Anastasi, Erspamer, and Endean, 1968; Cei, Erspamer, and Roseghini, 1972; Low, 1972). The production of some of these compounds is energetically expensive; others are costly in terms of water economy (Cragg, Balinsky, and Baldwin, 1961; Balinsky, Cragg, and Baldwin, 1961). Why, then, do so many amphibians produce a wide variety of such costly compounds? Despite wide chemical variety most of these compounds share one striking attribute: they are either distasteful or have unpleasant physiological effects. Most irritate the mucous membranes. Bufadienolides and other cardiac glycosides have digitalislike effects on such predators as snakes as well as on mammals (Licht and Low, 1968). Caerulein differs in only two amino acids from gastrin and has similar effects (Anastasi et al., 1968), including the induction of vomiting.

Although I know of no good study of predation mortality in any desert amphibian, and demography data on amphibians are generally sparse (Turner, 1962), predation has been reported in every life-history stage (Surface, 1913; Barbour, 1934; Brockelman, 1969; Littlejohn, 1971; Szarski, 1972). It is obvious that there is selective advantage to tasting vile or being poisonous, and scattered studies show that successful predators on amphibians show adaptations of increased tolerance (Licht and Low, 1968) or avoidance of the poisonous parts (Miller, 1909; Wright, 1966; Schaaf and Garton, 1970).

Predation concentrated on adults will lead to the success of individuals which show cryptic behavior and color patterns as well as those which concentrate unpleasant compounds in their skins. Particularly poisonous or distasteful individuals with bright or striking color patterns may also be favored (Fisher, 1958). Two apparently opposite breeding strategies may succeed, depending on other factors discussed below. These are cryptic breeding behavior and temporally and spatially clumped breeding behavior.

Several strategies may evolve as a response to predation on eggs: eggs with foam coating, as in a number of Limnodynastes species (Martin, 1967, Littlejohn, 1971); eggs containing poisonous substances, as in Bufo (Licht, 1967, 1968); eggs hatching quickly, as in Scaphiopus (Stebbins, 1951; Bragg, 1965, summarizing earlier papers); and a clumping of egg laying or hiding or tending of eggs, as is done by a number of New World tropical species (table 7-1). If adults become poisonous and effectively invulnerable, they concomitantly become good protectors of the eggs.

The strategies of hiding or tending eggs involve a greater parental investment per offspring and result in a decrease in the total number of eggs laid (figs. 7-1 and 7-2). That a general correlation exists between strategies of parental care and numbers of eggs has been recognized for some time; but no pattern has been recognized, and explanations by herpetologists have verged on the teleological, such as those of Porter (1972).

Figures 7-1, 7-2, and 7-3 show the relationships of egg sige, female size, litter size, and predictability of habitat. Indeed, as the size of egg relative to the female increases, the clutch size decreases (table 7-1, fig. 7-1). This is as expected and correlates with results from other groups (Williams, 1966a, 1966b; Salthe, 1969; Tinkle, 1969). When habitat or egg-laying locality is shown on a graph plotting the ratio of egg size to female size against litter size (fig. 7-3), it is apparent that most of those species showing some increase in parental care, such as laying eggs in burrows or leaves or tending the eggs or tadpoles, lay fewer, larger eggs; these species without exception live in habitats of relatively high environmental predictability—tropical rain forests, caves, and so on (table 7-1). No species laying eggs in temporary ponds show such behavior. The species in areas of high predictability possess a variety of strategies of high parental investment per off-spring. As mentioned above, *Rhinoderma darwini* males carry the eggs in the vocal sac (Porter, 1972, and others). *Leiopelma hochstetteri* eggs are laid terrestrially and tended by one of the parents.

Females of several species of *Helioporus* lay eggs in a burrow excavated by the male, and the eggs await flooding to hatch (Main, 1965; Martin, 1967). Eggs of *Pipa pipa* are essentially tended by the female, on whose back they develop. Barbour (1934) and Porter (1972) reviewed a number of cases of parental tending and hiding strategies.

In situations (such as physical uncertainty or unpredictable predation) where increased parental investment per offspring is ineffective in decreasing the mortality of an individual's offspring, the minimum investment per offspring will be favored. In these cases, individuals which win are those which lay eggs in the peak laying period and in the middle of a good area being used by others. Any approaching predator should encounter someone else's eggs first. This strategy should be common in deserts and indeed appears to be (table 7-1). The costs of playing this temporal and spatial variety of "selfish herd" game (Hamilton, 1971) are that some aspects of intraspecific competition are maximized and predators may evolve to exploit the conspicuous "herd."

Three strategies would appear to be of selective advantage if predation is concentrated on the tadpole stage. One is the laying of larger or larger-yolked eggs producing larger and less-vulnerable tad-

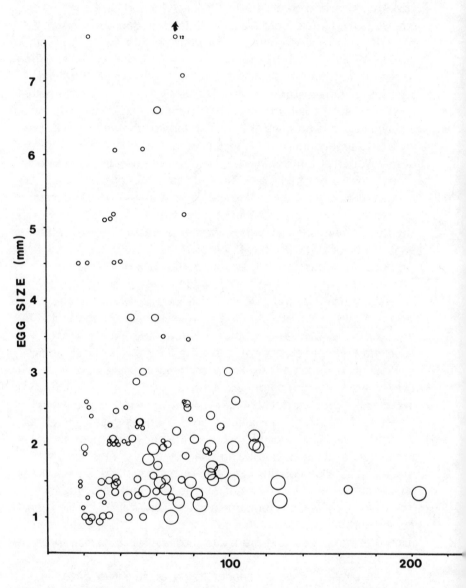

Fig. 7-1. Relationship of egg size to size of adult female for species from table 7-1. Size of circle indicates size of clutch:

o = ⟨ 500   ◯ = 1,000–10,000   ◯ = ⟩ 10,000
◯ = 500–1,000

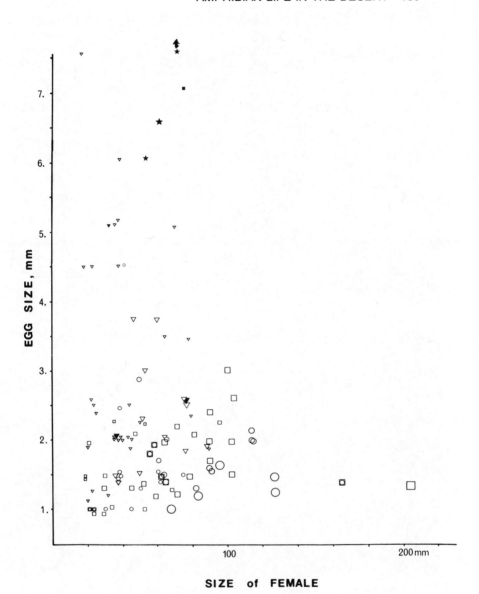

Fig. 7-2. Relationship of egg size to size of adult female. As in figure 7-1,
clutch size is shown by size of symbol. Solid symbols indicate tending behavior by a parent. Habitat of eggs:

○ = temporary water   △ = laid in burrows, wrapped in leaves, etc.
□ = permanent water   ★ = carried in brood pouch, on back, in
                                        vocal sac

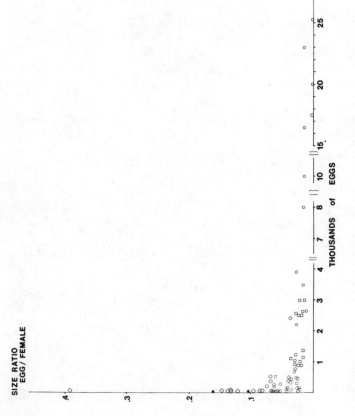

Fig. 7-3. Relationship of egg habitat to clutch size and relative size of eggs.
Habitat of eggs:

○ = temporary water   ◇ = laid on leaves, in burrows
□ = permanent water   ★ = carried by parent

poles at hatching. If seasonal or environmental conditions permit, producing offspring which spend a longer time as eggs may be successful. This would frequently involve strategies of hiding or tending eggs. In some species the entire development is completed while secreted so that on emergence offspring, in fact, are adults (*Leiopelma, Rhinoderma*). A third strategy is that of facultatively quick metamorphosis (*Bufo, Scaphiopus*), a strategy one might also expect to be favored in temporary ponds in desert situations. However, this advantage is balanced by intraspecific competition with its contingent selective advantage on size. So, in fact, what one would predict whenever genetic cost is not too great (Williams, 1966a) are facultative lengths of egg and larval periods and facultative hatching and metamorphosis. Thus, under strong predation and in more uncertain environments, one predicts an increase in facultativeness in these parameters. While this is predictable for both factors, studies of predation have not been able to separate out effects (De Benedictis, 1970). In amphibians of desert temporary ponds, either length of egg and larval periods are short or there is a large variation in reported lengths. Lengths of time to hatch in *S. couchi*, for example, range from 9 hours (Ortenburger and Ortenburger, 1926) to 48–72 hours (Gates, 1957); and in *S. hammondi*, from 38 (Little and Keller, 1937) to 120 hours (Sloan, 1964). The sizes at which these species metamorphose are highly variable (Bragg, 1965), suggesting that the strongest pressures of uncertainty center on the tadpole stages.

If predation is concentrated on juveniles, there will be an advantage to cryptic behavior by newly metamorphosed individuals. If predation is nonrandom and size related (as appears likely at metamorphosis when major predators are fish and other frogs), larger-sized individuals will be favored. If laying larger-yolked eggs results in larger offspring and increased offspring survivorship, this strategy will win. Certainly spending a longer time in the egg and tadpole stages, if predation is not heavily concentrated on these stages, will be favored. In species like *Rana catesbiana*, length of larval life is facultative and, for late eggs, is greater than a year. This appears to be involved with time required to reach a large enough size to be relatively invulnerable as a juvenile. While lengthened juvenile life cannot be selected for directly, conditions like predation on newly metamorphosed individuals are precisely those rendering lengthened periods in the tadpole stage advantageous.

Suggestions for Future Research

We can see that the interplay of these conditions is complex, and it is not necessarily a simple undertaking to predict strategies favored in each situation. Presently observed situations reflect the summation of a number of possibly conflicting selective advantages. Further, even though some biotic factors may be partially predictable from physical factors (e.g., in seasonal situations it is predictable that not only will food and breeding suitability be greater at some periods than others, but also at those same times predation will increase), others are not, and there is no single simple pattern.

The questions raised here are difficult to answer without further data, which are skimpy for anurans. Studies like Tinkle's (1969) on lizards or Inger and Greenberg's (1963) would afford comparative data for examination in the theoretical approach put forward here. For the most part, work on life histories in anurans has been zoogeographic and anecdotal. We haven't asked the right questions. Needed now are comparative studies similar to Tinkle's, between similar species in different habitats, and, in wide-ranging species, conspecific comparisons between habitats. We need to have:

1. Demographic data including length of life, time to maturity, age-specific fecundity, and degree of iteroparity, including number of eggs per clutch and number of clutches per year.
2. Ratio of egg size to female size.
3. Behavior: territoriality, tending behavior. (For example, Porter [1972] reported that *Rhinoderma darwini* males tend eggs that may not be their own. Such genetic altruism seems unlikely and needs further examination.)
4. Within wide-ranging species, comparative studies including, in addition to the above, work on mating-call parameters of males and discrimination of females.

Only when we begin to ask the above kinds of questions will we be able to develop an overall theoretical framework within which to view amphibian life histories. Many of the predictions and speculations discussed here seem obvious or trivial, but perhaps such attempts are necessary first steps toward a conceptual treatment of amphibian life histories.

Summary

Despite their normal requirement for an aqueous environment during the larval stage, a considerable number of amphibian species have adapted successfully to the desert environment. Possible methods of adaptation are considered, and their occurrence is reviewed. A number depend on modification of life histories, and attention is concentrated on these. Success depends on balancing the risk of mortality against the cost of reproductive effort.

Since desert environments are often less predictable than others, life-history strategies must take this uncertainty into account. This implies repeated small but prompt reproductive efforts and long adult life; behavior which enhances positive effects of random events, or reduces their negative effects, will be favored. Reduction in parental investment is generally advantageous in conditions of uncertainty in the physical environment.

From the amphibian point of view, the desert environment is patchy —coarse-grained—with high rates of speciation and extinction. Migration is favored where ponds are temporary and disfavored where they are permanent.

Mortality in the deserts is much more random than in mesic environments where it is dominated by predation. Reduction in the duration of vulnerable stages will then be advantageous. Responses to predation, however, have helped to shape amphibian life histories in the desert, as well as leading to production of noxious substances in many species. Differences in egg size and number per clutch may depend on likelihood of predation as against other hazards.

The importance of increased information about amphibian demography, and aspects of behavior related to it, is emphasized.

References

Alexander, R. D. 1968. Life cycle origins, speciations, and related phenomena in crickets. *Q. Rev. Biol.* 43:1–41.

Anastasi, A.; Erspamer, V.; and Endean, R. 1968. Isolation and amino acid sequence of caerulein, the active decapeptide of the skin of *Hyla caerulea*. *Archs Biochem. Biophys.* 125:57–68.

Balinsky, J. B.; Cragg, M. M.; and Baldwin, E. 1961. The adaptation of amphibian waste nitrogen excretion to dehydration. *Comp. Biochem. Physiol.* 3:236–244.

Barbour, T. 1934. *Reptiles and amphibians: Their habits and adaptations*. Boston and New York: Houghton Mifflin.

Bentley, P. J. 1966. Adaptations of Amphibia to desert environments. *Science, N.Y.* 152:619–623.

Bentley, P. J.; Lee, A. K.; and Main, A. R. 1958. Comparison of dehydration and hydration in two genera of frogs (*Helioporus* and *Neobatrachus*) that live in areas of varying aridity. *J. exp. Biol.* 35: 677–684.

Blair, W. F., ed. 1972. *Evolution in the genus "Bufo."* Austin: Univ. of Texas Press.

Bragg, A. N. 1965. *Gnomes of the night: The spadefoot toads*. Philadelphia: Univ. of Pa. Press.

Brattstrom, B. H. 1962. Thermal control of aggregation behaviour in tadpoles. *Herpetologica* 18:38–46.

———. 1963. A preliminary review of the thermal requirements of amphibians. *Ecology* 24:238–255.

Brockelman, W. Y. 1969. An analysis of density effects and predation in *Bufo americanus* tadpoles. *Ecology* 50:632–644.

Cei, J. M.; Erspamer, V.; and Roseghini, M. 1972. Biogenic amines. In *Evolution in the genus "Bufo,"* ed. W. F. Blair. Austin: Univ. of Texas Press.

Cragg, M. M.; Balinsky, J. B.; and Baldwin, E. 1961. A comparative study of the nitrogen secretion in some Amphibia and Reptilia. *Comp. Biochem. Physiol.* 3:227–236.

De Benedictis, P. A. 1970. "Interspecific competition between tadpoles of *Rana pipiens* and *Rana sylvatica*: An experimental field study." Ph.D. dissertation, University of Michigan.

Dole, J. W. 1967. The role of substrate moisture and dew in the water economy of leopard frogs, *Rana pipiens*. *Copeia* 1967:141–150.

Duellman, W. E. 1970. The hylid frogs of Middle America. *Monogr. Univ. Kans. Mus. nat. Hist.* 1:1–753.

Erspamer, V.; Vitali, T.; and Roseghini, M. 1964. The identification of new histamine derivatives in the skin of *Leptodactylus*. *Archs Biochem. Biophys.* 105:620–629.

Fisher, R. A. 1958. *The genetical theory of natural selection*. 2d rev. ed. New York: Dover.

Frith, H. J., and Calaby, J. H. 1969. *Kangaroos*. Melbourne: F. W. Cheshire.

Gates, G. O. 1957. A study of the herpetofauna in the vicinity of Wickenburg, Maricopa County, Arizona. *Trans. Kans. Acad. Sci.* 60:403–418.

Grant, P. R. 1972. Convergent and divergent character displacement. *J. Linn. Soc. (Biol.)* 4:39–68.

Hamilton, W. D. 1971. Geometry for the selfish herd. *J. theoret. Biol.* 31:295–311.

Heatwole, H.; Blasina de Austin, S.; and Herrero, R. 1968. Heat tolerances of tadpoles of two species of tropical anurans. *Comp. Biochem. Physiol.* 27:807–815.

Hussell, D. J. T. 1972. Factors affecting clutch-size in Arctic passerines. *Ecol. Monogr.* 42:317–364.

Inger, R. F. 1954. Systematics and zoogeography of Philippine Amphibia. *Fieldiana, Zool.* 33:185–531.

Inger, R. F., and Greenberg, B. 1956. Morphology and seasonal development of sex characters in two sympatric African toads. *J. Morph.* 99:549–574.

————. 1963. The annual reproductive pattern of the frog *Rana erythraea* in Sarawak. *Physiol. Zoöl.* 36:21–33.

Klomp, H. 1970. The determination of clutch-size in birds. *Ardea* 58: 1–124.

Lack, D. 1947. The significance of clutch-size. Pts. I and II. *Ibis* 89: 302–352.

————. 1948. The significance of clutch-size. Pt. III. *Ibis* 90:24–45.

Lee, A. K. 1967. Studies in Australian Amphibia. II. Taxonomy, ecology, and evolution of the genus *Helioporus* Gray (Anura: Leptodactylidae). *Aust. J. Zool.* 15:367–439.

Lee, A. K., and Mercer, E. H. 1967. Cocoon surrounding desert-dwelling frogs. *Science, N.Y.* 157:87–88.

Levins, R. 1968. *Evolution in changing environments*. Monographs in Population Biology, 2. Princeton: Princeton Univ. Press.

Lewontin, R. C. 1969. Comments on Slobodkin and Sanders "Contribution of environmental predictability to species diversity." *Brookhaven Symp. Biol.* 22:93.

Licht, L. E. 1967. Death following possible ingestion of toad eggs. *Toxicon* 5:141–142.

———. 1968. Unpalatability and toxicity of toad eggs. *Herpetologica* 24:93–98.

Licht, L. E., and Low, B. S. 1968. Cardiac response of snakes after ingestion of toad parotoid venom. *Copeia* 1968:547–551.

Little, E. L., and Keller, J. G. 1937. Amphibians and reptiles of the Jornada Experimental Range, New Mexico. *Copeia* 1937:216–222.

Littlejohn, M. J. 1967. Patterns of zoogeography and speciation by southeastern Australian Amphibia. In *Australian inland waters and their fauna*, ed. A. H. Weatherley, pp. 150–174. Canberra: Aust. Nat. Univ. Press.

———. 1971. Amphibians of Victoria. *Victorian Year Book* 85:1–11.

Low, B.S. 1972. Evidence from parotoid gland secretions. In *Evolution in the genus "Bufo,"* ed. W. F. Blair. Austin: Univ. of Texas Press.

MacArthur, R. H., and Wilson, E. O. 1967. *The theory of island biogeography*. Monographs in Population Biology, 1. Princeton: Princeton Univ. Press.

Main, A. R. 1957. Studies in Australian Amphibia. I. The genus *Crinia tschudi* in south-western Australia and some species from southeastern Australia. *Aust. J. Zool.* 5:30–55.

———. 1962. Comparisons of breeding biology and isolating mechanisms in Western Australian frogs. In *The evolution of living organisms*, ed. G. W. Leeper. Melbourne: Melbourne Univ. Press.

———. 1965. *Frogs of southern Western Australia*. Perth: West Australian Nat. Club.

———. 1968. Ecology, systematics, and evolution of Australian frogs. *Adv. ecol. Res.* 5:37–87.

Main, A. R., and Bentley, P. J. 1964. Water relations of Australian burrowing frogs and tree frogs. *Ecology* 45:379–382.

Main, A. R.; Lee, A. K.; and Littlejohn, M. J. 1958. Evolution in three genera of Australian frogs. *Evolution* 12:224–233.

Main, A. R.; Littlejohn, M. J.; and Lee, A. K. 1959. Ecology of Australian frogs. In *Biogeography and ecology in Australia*, ed. A. Keast, R. L. Crocker, and C. S. Christian. The Hague: Dr. W. Junk.

Martin, A. A. 1967. Australian anuran life histories: Some evolutionary and ecological aspects. In *Australian inland waters and their fauna*, ed. A. H. Weatherley, pp. 175–191. Canberra: Aust. Nat. Univ. Press.

Mayhew, W. W. 1968. Biology of desert amphibians and reptiles. In *Desert biology*, ed. G. W. Brown, vol. 1, pp. 195–356. New York and London: Academic Press.

Michl, H., and Kaiser, E. 1963. Chemie and Biochemie de Amphibiengifte. *Toxicon* 1963:175–228.

Miller, N. 1909. The American toad. *Am. Nat.* 43:641–688.

Murphy, G. I. 1968. Pattern in life history and the environment. *Am. Nat.* 102:391–404.

Newsome, A. E. 1964. Anoestrus in the red kangaroo, *Megaleia rufa*. *Aust. J. Zool.* 12:9–17.

————. 1965. The influence of food on breeding in the red kangaroo in central Australia. *CSIRO Wildl. Res.* 11:187–196.

————. 1966. Reproduction in natural populations of the red kangaroo *Megaleia rufa* in central Australia. *Aust. J. Zool.* 13:735–759.

Noble, C. K., and Putnam, P. G. 1931. Observations on the life history of *Ascaphus truei* Stejneger. *Copeia* 1931:97–101.

Ortenburger, A. I., and Ortenburger, R. D. 1926. Field observations on some amphibians and reptiles of Pima County, Ariz. *Proc. Okla. Acad. Sci.* 6:101–121.

Pianka, E. R. 1970. On r and K selection. *Am. Nat.* 104:592–597.

Porter, K. R. 1972. *Herpetology*. Philadelphia: W. B. Saunders Co.

Power, J. A. 1927. Notes on the habits and life histories of South African Anura with descriptions of the tadpoles. *Trans. R. Soc. S. Afr.* 14:237–247.

Ruibal, R. 1962a. The adaptive value of bladder water in the toad, *Bufo cognatus*. *Physiol. Zoöl.* 35:218–223.

————. 1962b. Osmoregulation in amphibians from heterosaline habitats. *Physiol. Zoöl.* 35:133–147.

Salthe, S. N. 1969. Reproductive modes and the number and size of ova in the urodeles. *Am. Midl. Nat.* 81:467–490.

Schaaf, R. T., and Garton, J. S. 1970. Racoon predation on the American toad, *Bufo americanus*. *Herpetologica* 26:334–335.

Schmidt, K. P., and Inger, R. F. 1959. Amphibia. *Explor. Parc natn. Upemba Miss. G. F. de Witt* 56.

Sharman, G. B.; Calaby, J. H.; and Poole, W. E. 1966. Patterns of reproduction in female diprotodont marsupials. *Symp. zool. Soc. Lond.* 15:205–232.

Slater, J. R. 1934. Notes on northwestern amphibians. *Copeia* 1934: 140–141.

Sloan, A. J. 1964. Amphibians of San Diego County. *Occ. Pap. S Diego Soc. nat. Hist.* 13:1–42.

Slobodkin, L. D., and Sanders, H. L. 1969. On the contribution of environmental predictability to species diversity. *Brookhaven Symp. Biol.* 22:82–96.

Stebbins, R. C. 1951. *Amphibians of western North America*. Berkeley and Los Angeles: Univ. of Calif. Press.

Stewart, M. M. 1967. *Amphibians of Malawi*. Albany: State Univ. of N.Y. Press.

Surface, H. A. 1913. The Amphibia of Pennsylvania. *Bi-m. zool. Bull. Pa Dep. Agric.* May–July 1913:67–151.

Szarski, H. 1972. Integument and soft parts. In *Evolution in the genus "Bufo,"* ed. W. F. Blair. Austin: Univ. of Texas Press.

Tinkle, D. W. 1969. The concept of reproductive effort and its relation to the evolution of life histories of lizards. *Am. Nat.* 103:501–514.

Trivers, R. L. 1972. Parental investment and sexual selection. In *Sexual selection and the descent of man*, ed. B. Campbell, pp. 136–179. Chicago: Aldine.

Trueb, L., and Duellman, W. E. 1970. The systematic status and life history of *Hyla verrucigera* Werner. *Copeia* 1970:601–610.

Turner, F. B. 1962. The demography of frogs and toads. *Q. Rev. Biol.* 37:303–314.

Villa, J. 1972. *Anfibios de Nicaragua*. Managua: Instituto Geográfico Nacional, Banco Central de Nicaragua.

Volpe, E. P. 1953. Embryonic temperature adaptations and relationships in toads. *Physiol. Zoöl.* 26:344–354.

Wager, V. A. 1965. *The frogs of South Africa*. Capetown: Purnell & Sons.

Warburg, M. R. 1965. Studies on the water economy of some Australian frogs. *Aust. J. Zool.* 13:317–330.

Wilbur, H. M., and Collins, J. P. 1973. Ecological aspects of amphibian metamorphosis. *Science, N.Y.* 182:1305.

Wilbur, H. M.; Tinkle, D. W.; and Collins, J. P. 1974. Environmental certainty, trophic level, and successional position in life history evolution. *Am. Nat.* 108:805–818.

Williams, G. C. 1966*a*. *Adaptation and natural selection: A critique of some current evolutionary thought*. Princeton: Princeton Univ. Press.

―――. 1966*b*. Natural selection, the costs of reproduction, and a refinement of Lack's principle. *Am. Nat.* 100:687–692.

Wright, J. W. 1966. Predation on the Colorado River toad, *Bufo alvarius*. *Herpetologica* 22:127–128.

8. Adaptation of Anurans to Equivalent Desert Scrub of North and South America

W. Frank Blair

Introduction

The occurrence of desertic environments at approximately the same latitudes in western North America and in South America provides an excellent opportunity to investigate comparatively the structure and function of ecosystems that have evolved under relatively similar environments. A multidisciplinary investigation of these ecosystems to determine just how similar they are in structure and function is presently in progress under the Origin and Structure of Ecosystems Program of the U.S. participation in the International Biological Program. The specific systems under study are the Argentine desert scrub, or Monte, as defined by Morello (1958) and the Sonoran desert of southwestern North America.

In this paper I will discuss the origins and nature of one component of the vertebrate fauna of these two xeric areas, the anuran amphibians. Pertinent questions are (a) How do the two areas compare in the degree of desert adaptedness of the fauna? (b) How do the two areas compare with respect to the size of the desert fauna? (c) What are the geographical origins of the various components of the fauna? and (d) What are the mechanisms of desert adaptation?

The comparison of the two desert faunas must take into account a number of major factors that have influenced their evolution. The most important among these would seem to be:

1. The nature of the physical environment of physiography and climate
2. The degree of similarity of the vegetation in general ecological aspect and in plant species composition

3. The size of each desert area
4. Possible sources of desert-invading species and the nature of adjacent biogeographic areas
5. The past history of the area through Tertiary and Pleistocene times
6. The evolutionary-genetic capabilities of available stocks for desert colonization

The Physical Environment

As defined by Morello (1958), the Monte extends through approximately 20° of latitude from 24°35'S in the state of Salta to 44°20'S in the state of Chubut and through approximately 7° of longitude from 69°50'W in Neuquen to 62°54'W on the Atlantic coast. The Sonoran desert occupies an area lying approximately between lat. 27° and 34°N and between long. 110° and 116°W (Shelford, 1963, fig. 15-1). Both areas are characterized by lowlands and mountains. The present discussion will deal principally with the lowland fauna.

Rainfall in both of the areas is usually less than 200 mm annually (Morello, 1958; Barbour and Díaz, 1972). Thus, availability of water is the most important factor determining the nature of the vegetation and the most important control limiting the invasion of these areas by terrestrial vertebrates.

The Vegetation

A more precise discussion of the vegetation of the Monte will be found elsewhere in this volume (Solbrig, 1975), so I will point out only that the general aspect is very similar in the two areas. The genera *Larrea*, *Prosopis*, and *Acacia* are among the most important components of the lowland vegetation and are principally responsible for this similarity of aspect. Various other genera are shared by the two areas. Some notably desert-adapted genera are found in one area but not in the other (Morello, 1958; Raven, 1963; Axelrod, 1970).

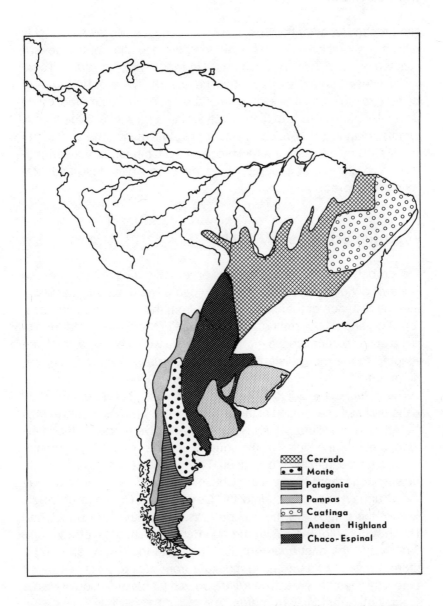

Fig. 8-1. Approximate distribution of xeric and subxeric areas in eastern and southern South America (adapted from Cabrera, 1953; Veloso, 1966; Sick, 1969).

Size of Area

The present areas of the Sonoran desert and the Monte are roughly similar in size. However, in considering the evolution of the desert-adapted fauna of the two continents, it is important to consider all contiguous desert areas. In this context the desertic areas of North America far exceed those that exist east of the Andes in South America. In South America there is only the Patagonian area with a cold desertic climate and the cold Andean Puna. In North America the addition of the Great Basin desert, the Mojave, and the Chihuahuan desert provides a much greater geographical expanse in which desert adaptations are favored.

Potential Sources of Stocks

The probability of any particular taxon of animal contributing to the fauna of either desert area obviously can be expected to decrease with the distance of that taxon's range from the desert area in question. This should be true not only because of the mere matter of distance but also because the more distant taxa would be expected to be adapted to the more distant and, hence, usually more different environments.

The nature of the adjacent ecological areas is, therefore, important to the process of evolution of the desert faunas. The Monte lies east of the Andean cordillera, which is a highly effective barrier to the interchange of lowland biota. To the south is the cold, desertic Patagonia, smaller in area than the Monte itself. To the east the Monte grades into the semixeric thorn forest of the Chaco, which extends into Paraguay and Uruguay and merges into the Cerrado and Caatinga of Brazil. East of the Chaco are the pampa grasslands between roughly lat. 31° and 38°S (fig. 8-1). With the huge area of Chaco, Cerrado, and Caatinga to the east and northeast, and with the Chaco showing a strong gradient of decreasing moisture from east to west, we might expect this eastern area to be a likely source for the evolution of Monte species of terrestrial vertebrates.

The geographical relationship of the Sonoran desert to possible source areas for invading species is very different from that of the Monte. Mountains are to the west, but beyond that little similarity exists. For one thing, the Sonoran desert is part of a huge expanse of desertic areas that stretches over 3,000 km from the southern part of the Chihuahuan desert in Mexico to the northern tip of the Great Basin desert in Oregon. To the east of these deserts in the United States, beyond the Rocky Mountain chain, are the huge central grasslands extending from the Gulf of Mexico into southern Canada. A similarity to the South American situation is seen, however, in the presence of a thorny vegetation type (the Mesquital), comparable to the Chaco, on the Gulf of Mexico lowlands of Tamaulipas and southern Texas. As in Argentina, a gradient of decreasing moisture exists westward from this Mesquital through the Chihuahuan desert and into the Sonoran desert. By contrast with the Monte, the Sonoran desert seems much more exposed to invasion by taxa which have adapted toward warm-xeric conditions in other contiguous areas.

Past Regional History

The present character of the two desert faunas obviously relates to the past histories of the two regions. For how long has there been selection for a xeric-adapted fauna in each area? What have been the effects on these faunas of secular climatic changes in the Tertiary and Pleistocene? These questions are difficult to answer with any great precision.

According to Axelrod (1948, p. 138, and other papers), "the present desert vegetation of the western United States, as typified by the floras of the Great Basin, Mohave and Sonoran deserts" is no older than middle Pliocene. Prior to the Oligocene, a Neotropical-Tertiary geoflora extended from southeastern Alaska and possibly Nova Scotia south into Patagonia (Axelrod, 1960) and began shrinking poleward as the continent became cooler and drier from the Oligocene onward. With respect to the Monte, Kusnezov (1951), as quoted by Morello (1958), believed that the Monte has existed without major change since "Eocene-Oligocene" times.

Arguments have been presented that there was a Gondwanaland

dry flora prior to the breakup of that land mass in the Cretaceous, which is represented today by xeric relics in southern deserts (Axelrod, 1970). It seems then that selection for xeric adaptation has been going on in the southern continent and, from paleobotanical evidence, in North America as well (Axelrod, 1970, p. 310) for more than 100 million years. However, major climatic changes have occurred in the geographic areas now known as the Monte and the Sonoran desert. The present desert floras of these two areas are combinations of the old relicts and of types that have evolved as the continents dried and warmed from the Oligocene onward (Axelrod, 1970).

One of the unanswered questions is where the desert-adapted biotas were at times of full glaciations in the Pleistocene. Martin and Mehringer (1965, p. 439) have addressed this question with respect to North American deserts and have concluded that "Sonoran desert plants may have been hard pressed." The question is yet unanswered. The desert plants presumably retreated southward, but the degree of compression of their ranges is unknown. Doubt also exists whether the Monte biota could have remained where it now is at peaks of glaciation in the Southern Hemisphere (Simpson Vuilleumier, 1971).

The Anurans

The number of species of frogs is not greatly different for the two deserts, and, as might be expected, both faunas are relatively small. As we define the two faunas on the basis of present knowledge, the Sonoran desert fauna includes eleven species representing four families and four genera, while that of the Monte includes fourteen species representing three families and seven genera (table 8-1). (Definition of the Monte fauna is less certain and more arbitrary than that of the Sonoran because of scarcity of data. The listings of Monte and Chacoan species used here are based largely on data from Freiberg [1942], Cei [1955a, 1955b, 1959b, 1962], Reig and Cei [1963], and Barrio [1964a, 1964b, 1965a, 1965b, 1968] and on my own observations. Species recorded from Patquia in the province of La Rioja and from Alto Pencoso on the San Luis–Mendoza border [Cei, 1955a, 1955b] are included in the Monte fauna as here considered.)

Table 8-1. *Anuran Faunas: Monte of Argentina and Sonoran Desert of North America*

Sonoran	Monte
Pelobatidae	Ceratophrynidae
Scaphiopus couchi	*Ceratophrys ornata*
S. hammondi	*C. pierotti*
	Lepidobatrachus llanensis
	L. asper
Bufonidae	Bufonidae
Bufo woodhousei	*Bufo arenarum*
B. cognatus	
B. mazatlanensis	
B. retiformis	
B. punctatus	
B. alvarius	
B. microscaphus	
Hylidae	
Pternohyla fodiens	
Ranidae	Leptodactylidae
Rana sp.	*Odontophrynus occidentalis*
(*pipiens* gp.)	*O. americanus*
	Leptodactylus ocellatus
	L. bufonius
	L. prognathus
	L. mystaceus
	Pleurodema cinerea
	P. nebulosa
	Physalaemus biligonigerus

The composition of the two faunas is phylogenetically quite dissimilar. The Sonoran is dominated by members of the genus *Bufo* with seven species. The Monte fauna is dominated by leptodactylids with nine species distributed among four genera of that family.

Ecological similarities are evident between the two pelobatids (*Scaphiopus couchi* and *S. hammondi*) of the Sonoran fauna and the four ceratophrynids (*Ceratophrys ornata*, *C. pierotti*, *Lepidobatrachus asper*, and *L. llanensis*) of the Monte. The Sonoran has a single fossorial hylid (*Pternohyla fodiens*); I have found no evidence of a Monte hylid. However, a remarkably xeric-adapted hylid, *Phyllomedusa sauvagei*, extends at least into the dry Chaco (Shoemaker, Balding, and Ruibal, 1972); and, because of these adaptations, it would not be surprising to find it in the Monte. The canyons of the desert mountains of the Sonoran and Monte have a single species of *Hyla* of roughly the same size and similar habits. In Argentina it is *H. pulchella*; in the United States it is *H. arenicolor*. These are not included in our faunal listing for the two areas. The Sonoran has a ranid (*Rana* sp. [*pipiens* gp.]); the family has penetrated only the northern half of South America (with a single species) from old-world origins and via North America, so has had no opportunity to contribute to the Monte fauna.

The origins of the Monte anuran fauna seem relatively simple. This fauna is principally a depauperate Chacoan fauna (table 8-2). At least thirty-seven species of anurans are included in the Chacoan fauna. Every species in the Monte fauna also occurs in the Chaco. Nine of the fourteen Monte species have ranges that lie mostly within the combined Chaco-Monte. The Monte fauna thus represents that component of a biota which has had a long history of adaptation to xeric or subxeric conditions and is able to occupy the western, xeric end of a moisture gradient that extends from the Atlantic coast west to the base of the Andes. Two of the Monte species (*Leptodactylus mystaceus* and *L. ocellatus*) are wide-ranging tropical species that reach both the Monte and the Chaco from the north or east. We are treating *Odontophrynus occidentalis* as a sub-Andean species (Barrio, 1964a), but the genus has the Chaco-Monte distribution; and since this species reaches the Atlantic coast in Buenos Aires province, there is no certainty that it evolved in the Monte. *Pleurodema nebulosa* of the Monte is listed by Cei (1955b, p. 293) as "a characteristic cordilleran form"; and, as mapped by Barrio (1964b), its range barely enters the Chaco, although other members of the same species group occur in the dry Chaco. *Pleurodema cinerea* is treated

Table 8-2. *Comparison of Chaco and Monte Anuran Faunas*

Monte	Chaco
	Hypopachus mulleri
Ceratophrys ornata	*Ceratophrys ornata*
C. pierotti	*C. pierotti*
Lepidobatrachus llanensis	*Lepidobatrachus llanensis*
	L. laevis
L. asper	*L. asper*
Pleurodema nebulosa	*Pleurodema nebulosa*
	P. quayapae
	P. tucumana
P. cinerea	*P. cinerea*
Physalaemus biligonigerus	*Physalaemus biligonigerus*
	P. albonotatus
Leptodactylus ocellatus	*Leptodactylus ocellatus*
	L. chaquensis
L. bufonius	*L. bufonius*
L. prognathus	*L. prognathus*
L. mystaceus	*L. mystaceus*
	L. sibilator
	L. gracilis
	L. mystacinus
Odontophrynus occidentalis	*Odontophrynus occidentalis*
O. americanus	*O. americanus*
Bufo arenarum	*Bufo arenarum*
	B. paracnemis
	B. major
	B. fernandezae
	B. pygmaeus
	Melanophryniscus stelzneri
	Pseudis paradoxus
	Lysapsus limellus

Monte	Chaco
	Phyllomedusa sauvagei
	P. hypochondrialis
	Hyla pulchella
	H. trachythorax
	H. venulosa
	H. phrynoderma
	H. nasica

Note: All species listed for Monte occur also in Chaco.

by Gallardo (1966) as a member of his fauna "Subandina." The genus ranges north to Venezuela.

The Sonoran anurans seemingly have somewhat more diverse geographical origins than those of the Monte, and they have been more thoroughly studied. Most of the ranges can be interpreted as ones that have undergone varying degrees of expansion northward following full glacial displacement into Mexico (Blair, 1958, 1965). Several of these (*Scaphiopus couchi*, *S. hammondi*, and *Bufo punctatus*) have a main part of their range in the Chihuahuan desert (table 8-3). *Bufo cognatus* ranges far northward through the central grasslands to Canada. Three species extend into the Sonoran from the lowlands of western Mexico. One of these is the fossorial hylid *Pternohyla fodiens*. Another, *B. retiformis*, is one of a three-member species group that ranges from the Tamaulipan Mesquital westward through the Chihuahuan desert into the Sonoran. The third, *B. mazatlanensis*, is a member of a species group that is absent from the Chihuahuan desert but is represented in the Tamaulipan thorn scrub. *Bufo woodhousei* has an almost transcontinental range. *Rana* sp. is an undescribed member of the *pipiens* group.

Two desert-endemic species occur in the Sonoran. One is *Bufo alvarius*, which appears to be an old relict species without any close living relative. *B. microscaphus* occurs in disjunct populations in the Chihuahuan, Sonoran, and southern Great Basin deserts. These populations are clearly relicts from a Pleistocene moist phase extension of the eastern mesic-adapted *B. americanus* westward into the present desert areas (A. P. Blair, 1955; W. F. Blair, 1957).

Table 8-3. *Comparison of Anuran Faunas of Sonoran Desert with Those of Chihuahuan Desert and Tamaulipan Mesquital*

Sonoran Desert	Chihuahuan-Tamaulipan
	Rhinophrynus dorsalis
	Hypopachus cuneus
	Gastrophryne olivacea
Scaphiopus hammondi	*Scaphiopus hammondi*
	S. bombifrons
S. couchi	*S. couchi*
	S. holbrooki
	Leptodactylus labialis
	Hylactophryne augusti
	Syrrhopus marnocki
	S. campi
	Bufo speciosus
Bufo cognatus	*B. cognatus*
B. punctatus	*B. punctatus*
	B. debilis
	B. valliceps
B. woodhousei	*B. woodhousei*
B. retiformis	
B. mazatlanensis	
B. alvarius	
B. microscaphus	
	Hyla cinerea
	H. baudini
	Pseudacris clarki
	P. streckeri
	Acris crepitans
Pternohyla fodiens	
Rana sp. (*pipiens* gp.)	*Rana* sp. (*pipiens* gp.)
	R. catesbeiana

Desert Adaptedness

If taxonomic diversity is taken as a criterion, the Monte fauna presents an impressive picture of desert adaptation. The genera *Odontophrynus* and *Lepidobatrachus* are both xeric adapted and are endemic to the xeric and subxeric region encompassed in this discussion. Three of the four leptodactylid genera which occur in the Monte (*Leptodactylus*, *Pleurodema*, and *Physalaemus*) are characterized by the laying of eggs in foam nests, either on the surface of the water or in excavations on land. This specialization may have a number of advantages, but one of the important ones would be protection from desiccation (Heyer, 1969).

In North America the only genus that can be considered a desert-adapted genus is *Scaphiopus*. This genus has two distinct subgeneric lines which, based on the fossil record, apparently diverged in the Oligocene (Kluge, 1966). Each subgenus is represented by a species in the Sonoran desert. Origin of the genus through adaptation of forest-living ancestors to grassland in the early Tertiary has been suggested by Zweifel (1956). *Pternohyla* is a fossorial hylid that apparently evolved in the Pacific lowlands of Mexico "in response to the increased aridity during the Pleistocene" (Trueb, 1970, p. 698). The diversity of *Bufo* species (*B. mazatlanensis*, *B. cognatus*, *B. punctatus*, and *B. retiformis*) that represent subxeric- and xeric-adapted species groups and the old relict *B. alvarius* implies a long history of *Bufo* evolution in arid and semiarid southwestern North America. Nevertheless, the total anuran diversity of xeric-adapted taxa compares poorly with that in South America.

The greater taxonomic diversity of desert-adapted South American anurans may be attributed to the Gondwanaland origin (Reig, 1960; Casamiquela, 1961; Blair, 1973) of the anurans and the long history of anuran radiation on the southern continent. The taxonomic diversity of anurans in South America vastly exceeds that in North America, which has an attenuated anuran fauna that is a mix of old-world emigrants (Ranidae, possibly Microhylidae) and invaders from South America (Bufonidae, Hylidae, and Leptodactylidae). The drastic effects of Pleistocene glaciations on North American environments may also account for the relatively thin anuran fauna of this continent.

Mechanisms of Desert Adaptation

Limited availability of water to maintain tissue water in adults and un-predictability of rains to permit reproduction and completion of the lar-val stage are paramount problems of desert anurans. Enough is known about the ecology, behavior, and physiology of the anurans of the two deserts to indicate the principal kinds of mechanisms that have evolved in the two areas.

With respect to the first of these two problems, two major and quite different solutions are evident in both desert faunas. One is to avoid the major issue by becoming restricted to the vicinity of permanent water in the desert environment. The other is to become highly fos-sorial, to evolve mechanisms of extracting water from the soil, and to become capable of long periods of inactivity underground. In the Sonoran desert three of the eleven species fit the first category. The *Rana* species is largely restricted to the vicinity of water throughout its range to the east and is a member of the *R. pipiens* complex, which is essentially a littoral-adapted group. Ruibal (1962*b*) studied a desert population of these frogs in California and regards their winter breed-ing as an adaptation to avoid the desert's summer heat. The relict endemic *Bufo alvarius* is smooth skinned and semiaquatic (Steb-bins, 1951; my data). The relict populations of *B. microscaphus* oc-cur where there is permanent water as drainage from the mountains or as a result of irrigation. Man's activities in impounding water for ir-rigation must have been of major assistance to these species in invad-ing a desert region without having to cope with the major water prob-lems of desert life. *Bufo microscaphus*, for example, exists in areas that have been irrigated for thousands of years by prehistoric cultures and more recently by European man (Blair, 1955). One species in the Monte fauna is there by this same adaptive strategy. *Leptodactylus ocellatus* offers a striking parallel to the *Rana* species. Its existence in the provinces of Mendoza and San Juan is attributed to extensive ag-ricultural irrigation (Cei, 1955*a*). That a second species, *B. arenarum*, fits this category is suggested by Ruibal's (1962*a*, p. 134) statement that "this toad is found near permanent water and is very common around human habitations throughout Argentina." However, Cei (1959*a*) has shown experimentally that *B. arenarum* from the Monte

(Mendoza) survives desiccation more successfully than *B. arenarum* from the Chaco (Córdoba), which implies exposure and adaptation to more rigorously desertic conditions for the former.

Most of the anurans of both desert faunas utilize the strategy of subterranean life to avoid the moisture-sapping environment of the desert surface. In the Sonoran fauna the two species of *Scaphiopus* have received considerable study. One of these, *S. couchi*, appears to have the greatest capacity for desert existence. Mayhew (1962, p. 158) found this species in southern California at a place where as many as three years might pass without sufficient summer rainfall to "stimulate them to emerge, much less successfully reproduce."

Mayhew (1965) listed a series of presumed adaptations of this species to desert environment:

1. Selection of burial sites beneath dense vegetation where reduced insolation reaching the soil means lower soil temperatures and reduced evaporation from the soil
2. Retention by buried individuals of a cover of dried, dead skin, thus reducing water loss through the skin
3. Rapid development of larvae—ten days from fertilization through metamorphosis (reported also by Wasserman, 1957)

Physiological adaptations of *S. couchi* (McClanahan, 1964, 1967, 1972) include:

1. Storage of urea in body fluids to the extent that plasma osmotic concentration may double during hibernation
2. Muscles showing high tolerance to hypertonic urea solutions
3. Rate of production of urea a function of soil water potential
4. Fat utilization during hibernation
5. Ability to tolerate water loss of 40–50 percent of standard weight
6. Ability to store up to 30 percent of standard body weight as dilute urine to replace water lost from body fluids

The larvae of *S. couchi* are more tolerant of high temperatures than anurans from less-desertic environments, and tadpoles have been observed in nature at water temperatures of 39° to 40°C (Brown, 1969).

Scaphiopus hammondi, as studied by Ruibal, Tevis, and Roig (1969) in southeastern Arizona, shows a pattern of desert adaptation generally comparable to that of *S. couchi* but with some difference in details. These spadefoots burrow underground in September to depths of up to 91 cm and remain there until summer rains come some nine months later. The burrows are in open areas, not beneath dense vegetation as reported for *S. couchi* by Mayhew (1965). *S. hammondi* can effectively absorb soil water through the skin and has greater ability to absorb soil moisture "than that demonstrated for any other amphibian" (Ruibal et al., 1969, p. 571). During the rainy season of July–August, the *S. hammondi* burrows to depths of about 4 cm.

Larval adaptations of *S. hammondi* include rapid development and tolerance of high temperatures (Brown, 1967*a*, 1967*b*), paralleling the adaptations of *S. couchi*.

The adaptations of *Bufo* for life in the Sonoran desert are less well known than those of *Scaphiopus*. Four of the nonsemiaquatic species escape the rigors of the desert surface by going underground. *Bufo cognatus* and *B. woodhousei* have enlarged metatarsal tubercles or digging spades, as in *Scaphiopus*. In southeastern Arizona, *B. cognatus* was found buried at the same sites as *S. hammondi* but in lesser numbers (Ruibal et al., 1969). McClanahan (1964) found the muscles of *B. cognatus* comparable to those of *S. couchi* in tolerance to hypertonic urea solutions, a condition which he regarded as a fossorial-desert adaptation. *Bufo punctatus* has a flattened body and takes refuge under rocks. It has been reported from mammal (*Cynomys*) burrows (Stebbins, 1951). *Bufo punctatus* has the ability to take up water rapidly from slightly moist surfaces through specialization of the skin in the ventral pelvic region ("sitting spot"), which makes up about 10 percent of the surface area of the toad (McClanahan and Baldwin, 1969). *Bufo retiformis* belongs to the arid-adapted *debilis* group of small but very thick-skinned toads (Blair, 1970).

The Sonoran desert species of *Bufo* have not evolved the accelerated larval development that is characteristic of *Scaphiopus*. Zweifel (1968) determined developmental rates for three species of *Scaphiopus*, three species of *Bufo, Hyla arenicolor*, and *Rana* sp. (*pipiens* gp.) in southeastern Arizona. The eight species fell into three groups: most rapid, *Scaphiopus*; intermediate, *Bufo* and *Hyla*; slowest, *Rana*. In my laboratory (table 8-4) *B. punctatus* from central Arizona showed no acceleration of development over the same species from the extreme eastern part of the range in central Texas. *Bufo cognatus* closely paralleled *B. punctatus* in duration of the lar-

Table 8-4. *Duration of Larval Stage of Four of the Sonoran Desert Species of* Bufo

Species	Locality of Origin	Days from Fertilization to Metamorphosis		Lab Stock No.
		First	50%	
B. punctatus	Wimberley, Texas	27	32	B64–173
B. punctatus	Mesa, Arizona	27	36	B64–325
B. cognatus	Douglas, Arizona	28	35	B64–234
B. mazatlanensis	Mazatlan, Sinaloa × Ixtlan, Nayarit	20	26	B63–87
B. alvarius	Tucson × Mesa, Arizona	36	53	B65–271
B. alvarius	Mesa, Arizona	29	33	B64–361

Note: Observations in a laboratory maintained at 24°–27° C.

val stage; *B. mazatlanensis* had a somewhat shorter larval life than these others; and *B. alvarius* spent a slightly longer period as tadpoles, but this could be accounted for by the fact that these are much larger toads. Overall, the impression is that these *Bufo* species have not shortened the larval stage as a desert adaptation. Tevis (1966) found that *B. punctatus* that were spawned in spring in Deep Canyon, California, required approximately two months for metamorphosis.

Developing eggs of *B. punctatus* and *B. cognatus* from Mesa, Ari-

zona, were tested for temperature tolerances by Ballinger and Mc-
Kinney (1966). Both of these desert species were limited by lower
maxima than was *B. valliceps*, a nondesert toad, from Austin, Texas.

The fossorial anurans of the Monte are much less well known than
those of the Sonoran desert. The ceratophrynids appear to be rather
similar to *Scaphiopus* in their desert adaptations. Both species of
Lepidobatrachus are reported to live buried (*viven enterrados*) and
emerge after rains (Reig and Cei, 1963). *Lepidobatrachus llanensis*
forms a cocoon made of many compacted dead cells of the stratum
corneum when exposed to dry conditions (McClanahan, Shoemaker,
and Ruibal, 1973). These anurans apparently live an aquatic existence as long as the temporary rain pools exist, in which respect they
differ from *Scaphiopus* species, which typically breed quickly and
leave the water. The skin of *L. asper* is described (Reig and Cei,
1963) as thin in summer (when they are aquatic) and thicker and more
granular in periods of drought. *Ceratophrys* reportedly uses the burrows of the viscacha (*Lagidium*), a large rodent (Cei, 1955*b*). However, *C. ornata* does bury itself in the soil, and one was known to
stay underground between four and five months and shed its skin
after emerging (Marcos Freiberg, 1973, personal communication).
Ceratophrys pierotti remains near the temporary pools in which it
breeds for a considerable time after breeding (my observations).
Odontophrynus at Buenos Aires makes shallow depressions and may
sit in these with only the head showing (Marcos Freiberg, 1973,
personal communication). *Leptodactylus bufonius* lives in dens or
natural cavities or in viscacha burrows (Cei, 1949, 1955*b*). *Pleurodema nebulosa* is a fossorial species with metatarsal spade that
spends a major portion of its lifetime living on land in burrows (Ruibal, 1962*a*; Gallardo, 1965). *Bufo arenarum* "winters buried up to
a meter in depth" (Gallardo, 1965, p. 67).

Phyllomedusa sauvagei of the dry Chaco, and possibly the Monte,
has achieved a high level of xeric adaptation by excreting uric acid
and by controlling water loss through the skin (Shoemaker et al.,
1972). Rates of water loss in this arboreal, nonfossorial hylid are comparable to those of desert lizards rather than to those of other anurans
(Shoemaker et al., 1972).

Ruibal (1962*a*) studied the osmoregulation of six of the Chaco-
Monte species and found that *P. nebulosa* is capable of producing

urine that is hypotonic to the lymph and to the external medium, thus enabling it to store bladder water as a reserve against dehydration. The others, including *P. cinerea*, *L. asper*, and *B. arenarum* of what we are calling the Monte fauna, produced urine that was essentially isotonic to the lymph and the external medium.

Reproductive Adaptations

One of the major hazards of desert existence for an anuran population is the unpredictability of rainfall to provide breeding pools. Two alternative routes are available. One is to be an opportunistic breeder, spending long periods of time underground but responding quickly when suitable rainfall occurs. The alternative is to breed only in permanent water, with the time of breeding presumably set by such cues as temperature or possibly photoperiod. Both strategies are found among the Sonoran desert anurans.

The two *Scaphiopus* species are the epitome of the first of these adaptive routes. *Bufo cognatus*, *B. retiformis*, and *Pternohyla fodiens* are also opportunistic breeders (Lowe, 1964; my data). Two species, *B. punctatus* and *B. woodhousei*, are opportunistic breeders or not, depending on the population. Both are opportunistic in Texas. In the Great Basin desert of southwestern Utah, these two species along with all other local anurans (*B. microscaphus*, *S. intermontanus*, *Hyla arenicolor*, and *Rana* sp. [*pipiens* gp.]) breed without rainfall (Blair, 1955; my data). Peak breeding choruses of *B. punctatus* and *B. alvarius* were found in a stock pond near Scottsdale, Arizona, in the absence of any recent rain (Blair and Pettus, 1954).

The Monte anurans, with the presumed exception of *Leptodactylus ocellatus*, appear to be opportunistic breeders (Cei, 1955a, 1955b; Reig and Cei, 1963; Gallardo, 1965; Barrio, 1964b, 1965a, 1965b). The apparent lesser development of the strategy of permanent water breeders could result from lesser knowledge of the behavior of the Monte anurans. However, the available evidence points to a real difference between the Monte and Sonoran desert faunas in degree of adoption of the habit of breeding in permanent water. *Leptodactylus ocellatus* of the Monte is ecologically equivalent to *R.* sp. (*pipiens* gp.) of the Sonoran desert; both are littoral adapted over a wide geographic range and have been able to penetrate their respective deserts by virtue of this adaptation where permanent water exists. There is no evidence that permanent water breeders are evolving from opportunistic breeders as in *B. punctatus*, *B. woodhousei*, and other North American desert species.

Foam Nests

One mechanism for desert adaptation, the foam nest, has been available for the evolution of the Monte fauna but not for the Sonoran desert fauna. Evolution of the foam-nesting habit has been discussed by various authors, especially Lutz (1947, 1948), Heyer (1969), and Martin (1967, 1970). The presumably more primitive pattern of floating the foam nest on the surface of the water is found among the Monte anurans in the genera *Physalaemus* and *Pleurodema* and in *Leptodactylus ocellatus*. The three other species of *Leptodactylus* in the Monte fauna lay their eggs in foam nests in burrows near water. These have aquatic larvae which are typically flooded out of the nests when pool levels rise with later rainfall. Heyer (1969) discussed advantages of the burrow nests over floating foam nests, among which the most important as adaptations to desert conditions are greater freedom from desiccation, and getting a head start on other breeders in the pool and thus being able to metamorphose earlier than others. Shoemaker and McClanahan (1973) investigated nitrogen excretion in the larvae of *L. bufonius* and found these larvae highly urotelic as an apparent adaptation to confinement in the foam-filled burrow versus the usual ammonotelism of anuran larvae.

Leptodactylids do reach the North American Mesquital (table 8-3), and one burrow-nesting species (*L. labialis*) reaches the southern tip of Texas. The other two genera both have direct, terrestrial development and hence would be unlikely candidates for desert adaptation. *Leptodactylus labialis* with a nesting pattern similar to that of *L. bufonius* would seem to be potential material for desert adaptation.

Cannibalism

An intriguing similarity between the two desert faunas is seen in the occurrence of cannibalism in both areas and in groups (ceratophrynids in South America, *Scaphiopus* in North America) that in other respects show rather similar patterns of desert adaptation.

In *S. bombifrons* and the closely related *S. hammondi*, some larvae have a beaked upper jaw and a corresponding notch in the lower as an apparent adaptation for carnivory (Bragg, 1946, 1950, 1956, 1961, 1964; Turner, 1952; Orton, 1954; Bragg and Bragg, 1959). The larvae of this type have been observed to be cannibalistic in *S. bombifrons* and suspected of being so in *S. hammondi* (Bragg, 1964). Cannibalism could be an important mechanism for concentrating food resources in a part of the population where these are limited and where there is a constant race against drying up of the breeding pool in the desert environment.

The ceratophrynids are much more cannibalistic than *Scaphiopus*. Both larvae and adults are carnivorous and cannibalistic (Cei, 1955*b*; Reig and Cei, 1963; my data). The head of the adult ceratophrynid is relatively large, with wide gape and with enlarged grabbing and holding teeth. Adult *Ceratophrys pierotti* are extremely voracious cannibals; one of these can quickly ingest another individual of its own body size.

Summary

The Monte of Argentina and the Sonoran desert of North America are compared with respect to their anuran faunas. Both deserts are roughly of similar size, but in North America there is a much greater extent of arid lands than in South America, with the Sonoran desert only a part of this expanse. Both deserts are at the dry end of moisture gradients that extend from thorn forest in the east to desert on the west.

Paleobotanical evidence suggests that xeric adaptation may have been occurring in South America prior to the breakup of Gondwanaland in the Cretaceous, while the North American deserts seem no older than middle Pliocene. Both desert systems must have been pressured and shifted during Pleistocene glacial maxima.

The anuran faunas of the two areas are similar in size, eleven species in the Sonoran desert, fourteen in the Monte. All anurans of the Monte occur also in the Chaco, and the fauna of the Monte is simply a depauperate Chacoan fauna. The origins of the Sonoran desert fauna are more diverse than this.

The Monte has the greatest taxonomic diversity, with seven genera versus four for the Sonoran desert. Two of the Monte genera (*Odontophrynus* and *Lepidobatrachus*) are truly desert and subxeric genera, but only one North American genus (*Scaphiopus*) fits this category. The presence of seven species of *Bufo* in the Sonoran desert implies a long history of desert adaptation by this genus in North America.

Mechanisms of desert adaptation are similar in the two areas. In each a littoral-adapted type (*Leptodactylus ocellatus* in the south, *Rana* sp. [*pipiens* gp.] in the north) has invaded the desert area by staying with permanent water. Additionally, the relict North American *B. alvarius* and *B. microscaphus* have followed the same strategy. Several of the North American species have abandoned opportunistic breeding in favor of breeding in permanent water, but no comparable trend is evident for the South American frogs. The most desert-adapted species in the North American desert is *Scaphiopus couchi*, which follows a pattern of highly fossorial life, opportunistic breeding with accelerated larval development, and physiological adaptations of adults to minimal water.

The ceratophrynids of the South American desert show parallel adaptations to those of *Scaphiopus*. In addition to other similarities, both groups employ some degree of cannibalism as an apparent adaptation to desert life.

References

Axelrod, D. I. 1948. Climate and evolution in western North America during middle Pliocene time. *Evolution* 2:127–144.

———. 1960. The evolution of flowering plants. In *Evolution after Darwin: Vol. 1 The evolution of life*, ed. S. Tax, pp. 227–305. Chicago: Univ. of Chicago Press.

———. 1970. Mesozoic paleogeography and early angiosperm history. *Bot. Rev.* 36:277–319.

Ballinger, R. E., and McKinney, C. O. 1966. Developmental temperature tolerance of certain anuran species. *J. exp. Zool.* 161:21–28.

Barbour, M. G., and Díaz, D. V. 1972. *Larrea* plant communities on bajada and moisture gradients in the United States and Argentina. *U.S./Intern. biol. Progn.: Origin and Structure of Ecosystems Tech. Rep.* 72–6:1–27.

Barrio, A. 1964*a*. Caracteres eto-ecológicos diferenciales entre *Odontophrynus americanus* (Dumeril et Bibron) y *O. occidentalis* (Berg) (Anura, Leptodactylidae). *Physis, B. Aires* 24:385–390.

———. 1964*b*. Especies crípticas del género *Pleurodema* que conviven en una misma área, identificados por el canto nupcial (Anura, Leptodactylidae). *Physis, B. Aires* 24:471–489.

———. 1965*a*. El género *Physalaemus* (Anura, Leptodactylidae) en la Argentina. *Physis, B. Aires* 25:421–448.

———. 1965*b*. Afinidades del canto nupcial de las especies cavicolas de género *Leptodactylus* (Anura, Leptodactylidae). *Physis, B. Aires* 25:401–410.

———. 1968. Revisión del género *Lepidobatrachus* Budgett (Anura, Ceratophrynidae). *Physis, B. Aires* 28:95–106.

Blair, A. P. 1955. Distribution, variation, and hybridization in a relict toad (*Bufo microscaphus*) in southwestern Utah. *Am. Mus. Novit.* 1722:1–38.

Blair, W. F. 1957. Structure of the call and relationships of *Bufo microscaphus* Cope. *Coepia* 1957:208–212.

———. 1958. Distributional patterns of vertebrates in the southern United States in relation to past and present environments. In *Zoogeography*, ed. C. L. Hubbs. *Publs Am. Ass. Advmt Sci*. 51:433–468.

———. 1965. Amphibian speciation. In *The Quaternary of the United States*, ed. H. E. Wright, Jr., and D. G. Frey, pp. 543–556. Princeton: Princeton Univ. Press.

———. 1970. Nichos ecológicos y la evolución paralela y convergente de los anfibios del Chaco y del Mesquital Norteamericano. *Acta zool. lilloana* 27:261–267.

———. 1973. Major problems in anuran evolution. In *Evolutionary biology of the anurans: Contemporary research on major problems*, ed. J. L. Vial, pp. 1–8. Columbia: Univ. of Mo. Press.

Blair, W. F., and Pettus, D. 1954. The mating call and its significance in the Colorado River toad (*Bufo alvarius* Girard). *Tex. J. Sci.* 6:72–77.

Bragg, A. N. 1946. Aggregation with cannibalism in tadpoles of

Scaphiopus bombifrons with some general remarks on the probable evolutionary significance of such phenomena. *Herpetologica* 3:89–98.

———. 1950. Observations on *Scaphiopus*, 1949 (Salientia: Scaphiopodidae). *Wasmann J. Biol.* 8:221–228.

———. 1956. Dimorphism and cannibalism in tadpoles of *Scaphiopus bombifrons* (Amphibia, Salientia). *SWest. Nat.* 1:105–108.

———. 1961. A theory of the origin of spade-footed toads deduced principally by a study of their habits. *Anim. Behav.* 9:178–186.

———. 1964. Further study of predation and cannibalism in spadefoot tadpoles. *Herpetologica* 20:17–24.

Bragg, A. N., and Bragg, W. N. 1959. Variations in the mouth parts in tadpoles of *Scaphiopus* (Spea) *bombifrons* Cope (Amphibia: Salientia). *SWest. Nat.* 3:55–69.

Brown, H. A. 1967*a*. High temperature tolerance of the eggs of a desert anuran, *Scaphiopus hammondi*. *Copeia* 1967:365–370.

———. 1967*b*. Embryonic temperature adaptations and genetic compatibility in two allopatric populations of the spadefoot toad, *Scaphiopus hammondi*. *Evolution* 21:742–761.

———. 1969. The heat resistance of some anuran tadpoles (Hylidae and Pelobatidae). *Copeia* 1969:138–147.

Cabrera, A. L. 1953. Esquema fitogeográfico de la República Argentina. *Revta Mus. La Plata (Nueva Serie), Bot.* 8:87–168.

Casamiquela, R. M. 1961. Un pipoideo fósil de Patagonia. *Revta Mus. La Plata Sec. Paleont. (Nueva Serie)* 4:71–123.

Cei, J. M. 1949. Costumbres nupciales y reproducción de un batracio caracteristico chaqueño (*Leptodactylus bufonius*). *Acta zool. lilloana* 8:105–110.

———. 1955*a*. Notas batracológicas y biogeográficas Argentinas, I–IV. *An. Dep. Invest. cient., Univ. nac. Cuyo.* 2(2):1–11.

———. 1955*b*. Chacoan batrachians in central Argentina. *Copeia* 1955:291–293.

———. 1959*a*. Ecological and physiological observations on polymorphic populations of the toad *Bufo arenarum* Hensel, from Argentina. *Evolution* 13:532–536.

———. 1959*b*. Hallazgos hepetológicos y ampliación de la distribución geográfica de las especies Argentinas. *Actas Trab. Primer Congr. Sudamericano Zool.* 1:209–210.

———. 1962. Mapa preliminar de la distribución continental de las "sibling species" del grupo *ocellatus* (género *Leptodactylus*). *Revta Soc. argent. Biol.* 38:258–265.

Freiberg, M. A. 1942. Enumeración sistemática y distribución geográfica de los batracios Argentinos. *Physis, B. Aires* 19:219–240.

Gallardo, J. M. 1965. Consideraciones zoogeográficas y ecológicas sobre los anfibios de la provincia de La Pampa Argentina. *Revta Mus. argent. Cienc. nat. Bernardino Rivadavia Inst. nac. Invest. Cienc. nat. Ecol.* 1:56–78.

———. 1966. Zoogeografía de los anfibios chaqueños. *Physis, B. Aires* 26:67–81.

Heyer, W. R. 1969. The adaptive ecology of the species groups of the genus *Leptodactylus* (Amphibia, Leptodactylidae). *Evolution* 23:421–428.

Kluge, A. G. 1966. A new pelobatine frog from the lower Miocene of South Dakota with a discussion of the evolution of the *Scaphiopus-Spea* complex. *Contr. Sci.* 113:1–26.

Kusnezov, N. 1951. *La edad geológica del régimen árido en la Argentina ségun los datos biológicos.* Geográfica una et varia, *Publnes esp. Inst. Estud. geogr., Tucumán* 2:133–146.

Lowe, C. H., ed. 1964. *The vertebrates of Arizona.* Tucson: Univ. of Ariz. Press.

Lutz, B. 1947. Trends toward non-aquatic and direct development in frogs. *Copeia* 1947:242–252.

———. 1948. Ontogenetic evolution in frogs. *Evolution* 2:29–39.

McClanahan, L. J. 1964. Osmotic tolerance of the muscles of two desert-inhabiting toads, *Bufo cognatus* and *Scaphiopus couchi*. *Comp. Biochem. Physiol.* 12:501–508.

———. 1967. Adaptations of the spadefoot toad, *Scaphiopus couchi*, to desert environments. *Comp. Biochem. Physiol.* 20:73–99.

———. 1972. Changes in body fluids of burrowed spadefoot toads as a function of soil potential. *Copeia* 1972:209–216.

McClanahan, L. J., and Baldwin, R. 1969. Rate of water uptake through the integument of the desert toad, *Bufo punctatus*. *Comp. Biochem. Physiol.* 29:381–389.

McClanahan, L. J.; Shoemaker, V. H.; and Ruibal, R. 1973. Evaporative water loss in a cocoon-forming South American anuran. Abstract of paper given at 53d Annual Meeting of American Society of Ichthyologists and Herpetologists, at San José, Costa Rica.

Martin, A. A. 1967. Australian anuran life histories: Some evolutionary and ecological aspects. In *Australian inland waters and their fauna*, ed. A. H. Weatherley, pp. 175–191. Canberra: Aust. Nat. Univ. Press.

———. 1970. Parallel evolution in the adaptive ecology of Leptodactylid frogs in South America and Australia. *Evolution* 24:643–644.

Martin, P. S., and Mehringer, P. J., Jr. 1965. Pleistocene pollen analysis and biogeography of the southwest. In *The Quaternary of the United States*, ed. H. W. Wright, Jr., and D. G. Frey, pp. 433–451. Princeton: Princeton Univ. Press.

Mayhew, W. W. 1962. *Scaphiopus couchi* in California's Colorado Desert. *Herpetologica* 18:153–161.

———. 1965. Adaptations of the amphibian, *Scaphiopus couchi*, to desert conditions. *Am. Midl. Nat.* 74:95–109.

Morello, J. 1958. La provincia fitogeográfica del Monte. *Op. lilloana* 2:1–155.

Orton, G. L. 1954. Dimorphism in larval mouthparts in spadefoot toads of the *Scaphiopus hammondi* group. *Copeia* 1954:97–100.

Raven, P. H. 1963. Amphitropical relationships in the floras of North and South America. *Q. Rev. Biol.* 38:141–177.

Reig, O. A. 1960. Lineamentos generales de la historia zoogeográfica de los anuros. *Actas Trab. Primer Congr. Sudamericano Zool.* 1:271–278.

Reig, O. A., and Cei, J. M. 1963. Elucidación morfológico-estadística de las entidades del género *Lepidobatrachus* Budgett (Anura, Ceratophrynidae) con consideraciones sobre la extensión del distrito chaqueño del dominio zoogeográfico subtropical. *Physis, B. Aires* 24:181–204.

Ruibal, R. 1962a. Osmoregulation in amphibians from heterosaline habitats. *Physiol. Zoöl.* 35:133–147.

———. 1962b. The ecology and genetics of a desert population of *Rana pipiens*. *Copeia* 1962:189–195.

Ruibal, R.; Tevis, L., Jr.; and Roig, V. 1969. The terrestrial ecology of the spadefoot toad *Scaphiopus hammondi*. *Copeia* 1969:571–584.

Shelford, V. E. 1963. *The ecology of North America*. Urbana: Univ. of Ill. Press.

Shoemaker, V. H., and McClanahan, L. J. 1973. Nitrogen excretion in the larvae of the land-nesting frog (*Leptodactylus bufonius*). *Comp. Biochem. Physiol.* 44A:1149–1156.

Shoemaker, V. H.; Balding, D.; and Ruibal, R. 1972. Uricotelism and low evaporative water loss in a South American frog. *Science, N.Y.* 175:1018–1020.

Sick, W. D. 1969. Geographical substance. In *Biogeography and ecology in South America*, ed. E. J. Fittkau, J. Illies, H. Klinge, G. H. Schwabe, and H. Sioli, 2: 449–474. The Hague: Dr. W. Junk.

Simpson Vuilleumier, B. 1971. Pleistocene changes in the fauna and flora of South America. *Science, N.Y.* 173:771–780.

Solbrig, O. T. 1975. The origin and floristic affinities of the South American temperate desert and semidesert regions. In *Evolution of desert biota*, ed. D. W. Goodall. Austin: Univ. of Texas Press.

Stebbins, R. C. 1951. *Amphibians of western North America*. Berkeley and Los Angeles: Univ. of Calif. Press.

Tevis, L., Jr. 1966. Unsuccessful breeding by desert toads (*Bufo punctatus*) at the limit of their ecological tolerance. *Ecology* 47: 766–775.

Trueb, L. 1970. Evolutionary relationships of casque-headed tree frogs with coossified skulls (family Hylidae). *Univ. Kans. Publs Mus. nat. Hist.* 18:547–716.

Turner, F. B. 1952. The mouth parts of tadpoles of the spadefoot toad, *Scaphiopus hammondi*. *Copeia* 1952:172–175.

Veloso, H. P. 1966. *Atlas florestal do Brasil*. Rio de Janeiro— Guanabara: Ministerio da Agricultura.

Wasserman, A. O. 1957. Factors affecting interbreeding in sympatric species of spadefoots (*Scaphiopus*). *Evolution* 11:320–338.

Zweifel, R. G. 1956. Two pelobatid frogs from the Tertiary of North America and their relationships to fossil and recent forms. *Am. Mus. Novit.* 1762:1–45.

———. 1968. Reproductive biology of anurans of the arid southwest, with emphasis on adaptation of embryos to temperature. *Bull. Am. Mus. nat. Hist.* 140:1–64.

Notes on the Contributors

John S. Beard was born in England and educated at Oxford University. After graduation he spent nine years in the Colonial Forest Service in the Caribbean and during this period studied for the degree of D.Phil. at Oxford. Soon after the end of the Second World War, he went to South Africa, where he was engaged in research on crop improvement in the wattle industry. In 1961 he was appointed director of King's Park in Perth, Western Australia, where, among other things, he was responsible for establishing a botanical garden. Ten years later he became director of the National Herbarium of New South Wales, a post from which he has recently retired. He is now devoting much of his time to the preparation of a series of detailed maps of Australian vegetation.

Dr. Beard has published books and papers on the vegetation of tropical America and has wide interests in plant ecology, biogeography, and systematics.

W. Frank Blair is professor of zoology at the University of Texas at Austin. He was born in Dayton, Texas. His first degree was taken at the University of Tulsa; he was awarded the M.S. at the University of Florida, and the Ph.D. at the University of Michigan. After eight years as a research associate there, he moved to a faculty position at the University of Texas, where he has been ever since.

At the inception of the International Biological Program, Dr. Blair became director of the Origin and Structure of Ecosystems section and, soon afterward, national chairman for the whole program. He also served as vice-president of the IBP on the international scale.

He has been involved in a wide range of personal research on vertebrate ecology and has worked extensively in Latin America as well as the United States. He is senior author of *Vertebrates of the United States* and has edited *Evolution in the Genus "Bufo"*—a subject on which much of his most recent research has concentrated.

David W. Goodall is Senior Principal Research Scientist at CSIRO Division of Land Resources Management, Canberra, Australia. Born and brought up in England, he studied at the University of London where he was awarded the Ph.D. degree, and, after a period of research in what is now Ghana, took up residence in Australia, of which country he is a citizen. He was awarded the D.Sc. degree of Melbourne University in 1953. He came to the United States in 1967 and the following year was invited to become director of the Desert Biome section of the International Biological Program then getting under way. This position he continued to hold until the end of 1973. During most of this period, and until the end of 1974, he held a position as professor of systems ecology at Utah State University.

His main research interests were initially in plant physiology, particularly in its application to agriculture and horticulture; but later he shifted his interest to plant ecology, especially statistical aspects of the subject, and to systems ecology.

Bobbi S. Low is associate professor of resource ecology at the University of Michigan. She was born in Kentucky and took her first degree at the University of Louisville and her doctorate at the University of Texas at Austin. After postdoctoral work at the University of British Columbia, she spent three years as a Research Fellow at Alice Springs, Australia, and returned to the United States in 1972.

Her main research interests have been in evolutionary ecology and in ecology of vertebrates in arid areas, both in the United States and in Australia.

James A. MacMahon is professor of biology at Utah State University and assistant director of the Desert Biome section of the International Biological Program. He was born in Dayton, Ohio, and took his first degree at Michigan State University and his doctorate at Notre Dame University, Indiana, in 1963. He then was appointed to a professorial position at the University of Dayton, and in 1971 he moved to Utah.

Though much of his research has been devoted to reptiles and Amphibia, he has also been concerned with plants, mammals, and invertebrates. In all these groups of organisms, he has mainly been interested in their ecology, particularly at community level, in relation to the arid-land environment.

A. R. Main is professor of zoology at the University of Western Australia. He was born in Perth; after military service during the Second World War, he returned there to take a first degree and then a doctorate at the University of Western Australia. He is a Fellow of the Australian Academy of Science.

He has done extensive research on the ecology of mammals and Amphibia in the Australian deserts, and he and his students have published numerous papers on the subject.

Guillermo Sarmiento is associate professor in the Faculty of Science, Universidad de Los Andes, Mérida, Venezuela. He was born in Mendoza, Argentina, and was educated at the University of Buenos Aires, where he was awarded a doctorate in 1965. He was appointed assistant professor and moved to Venezuela two years later.

His main research interests have been in tropical plant ecology, particularly as applied to savannah and to the vegetation of arid lands.

Otto T. Solbrig was born in Buenos Aires, Argentina. He took his first degree at the Universidad de La Plata and his Ph.D. at the University of California, Berkeley, in 1959. He worked at the Gray Herbarium, Harvard University, for seven years (during which period he became a U.S. citizen); after a period as professor of botany at the University of Michigan, he returned to Harvard University in 1969 as professor of biology and chairman of the Sub-Department of Organismic and Evolutionary Biology. Within the U.S. contribution to the International Biological Program, he served as director of the Desert Scrub subprogram of the Origin and Structure of Ecosystems section.

He has wide field experience in various parts of Latin America as well as in the United States. His main research interests have been in plant biosystematics, biogeography, and population biology.

Index